Viral Therapy of Cancer

Viral Therapy of Cancer

Editors

Kevin J. Harrington
Institute of Cancer Research, London, UK

Richard G. Vile
The Mayo Clinic, Rochester, MN, USA

Hardev S. Pandha
University of Surrey, Guildford, UK

John Wiley & Sons, Ltd

Telephone (+44) 1243 779777

Email (for orders and customer service enquiries): cs-books@wiley.co.uk
Visit our Home Page on www.wileyeurope.com or www.wiley.com

Other Wiley Editorial Offices

John Wiley & Sons Inc., 111 River Street, Hoboken, NJ 07030, USA

Jossey-Bass, 989 Market Street, San Francisco, CA 94103-1741, USA

Wiley-VCH Verlag GmbH, Boschstr. 12, D-69469 Weinheim, Germany

John Wiley & Sons Australia Ltd, 42 McDougall Street, Milton, Queensland 4064, Australia

John Wiley & Sons (Asia) Pte Ltd, 2 Clementi Loop #02-01, Jin Xing Distripark, Singapore 129809

John Wiley & Sons Canada Ltd, 6045 Freemont Blvd, Mississauga, Ontario, L5R 4J3

Wiley also publishes its books in a variety of electronic formats. Some content that appears in print may not be available in electronic books.

Library of Congress Cataloging-in-Publication Data

Viral therapy of cancer / edited by Kevin J. Harrington, Richard G. Vile.
and Hardev S. Pandha.
p. ; cm.
Includes bibliographical references.
ISBN 978-0-470-01922-1 (cloth : alk. paper)
1. Viruses–Therapeutic use. 2. Cancer–Treatment. I. Harrington, Kevin J., 1958- II. Vile, Richard G. III. Pandha, Hardev S.
[DNLM: 1. Oncolytic Virotherapy. 2. Gene Therapy. 3. Neoplasms–therapy. 4. Oncolytic Viruses. QZ 266 V8125 2007]
RC271.V567V52 2007
616.99'406–dc22

2007050277

British Library Cataloguing in Publication Data

A catalogue record for this book is available from the British Library

ISBN-978-0-470-01922-1 (H/B)

Typeset in 10/12 pt Times by Thomson Digital, Noida, India
Printed and bound in Great Britain by Antony Rowe Ltd., Chippenham, Wiltshire

Contents

Foreword **xiii**

Preface **xv**

Contributors **xvii**

1 Adenoviruses **1**
Kate Relph, Kevin J. Harrington, Alan Melcher and Hardev S. Pandha

1.1 Introduction 1
1.2 Viral structure and life cycle 1
1.3 Adenoviral vectors 5
1.4 Targeting adenoviral vectors 6
1.5 Clinical applications of adenoviral gene therapy 7
1.6 Adenoviral vectors for immunotherapy 7
1.7 Adenoviral vectors for suicide gene therapy 10
1.8 Adenoviral vectors for gene replacement therapy 11
1.9 Oncolytic adenoviral therapy 12
1.10 Adverse outcomes of adenoviral gene therapy 13
1.11 Summary 13
References 14

2 Application of HSV-1 vectors to the treatment of cancer **19**
Paola Grandi, Kiflai Bein, Costas G. Hadjipanayis, Darren Wolfe, Xandra O. Breakefield and Joseph C. Glorioso

2.1 Introduction 19
2.2 Basic biology of HSV 19
2.3 Replication competent or oncolytic vectors 24
2.4 Replication defective vectors 28
2.5 Amplicons 30
2.6 Impediments to the efficacy of HSV vectors for cancer gene therapy 32

2.7 Strategies to enhance the efficacy and specificity of HSV vectors for cancer gene therapy 36
2.8 Summary and conclusions 42
Acknowledgements 42
References 42

3 Adeno-associated virus **55**
Selvarangan Ponnazhagan

3.1 Introduction 55
3.2 Biology and life cycle of AAV 55
3.3 AAV serotypes 57
3.4 Production of recombinant AAV 57
3.5 Gene therapy for cancer treatment 57
3.6 Anti-oncogenic properties of AAV 58
3.7 Molecular chemotherapy studies with rAAV 59
3.8 AAV-mediated sustained transgene expression as a potential cancer gene therapy strategy 59
3.9 rAAV vectors have advantages in stimulating T helper 1/cytotoxic T lymphocyte responses 60
3.10 rAAV vectors can be used to initiate immune responses 61
3.11 Altering AAV tropism for tumour-specific delivery 62
3.12 Clinical trials involving rAAV 62
3.13 Conclusion 63
Acknowledgements 63
References 63

4 Retroviruses **69**
Simon Chowdhury and Yasuhiro Ikeda

4.1 Introduction 69
4.2 Structure of retroviral particles 69
4.3 Retroviral genome 69
4.4 Retroviral life cycle 70
4.5 Retroviral vectors 71
4.6 Safety of retroviral vectors: insertional mutagenesis 72
4.7 Gene therapy of X-linked SCID 72
4.8 Retroviral cancer gene therapy 75
4.9 Immunomodulatory approaches 78
4.10 Conclusions 79
References 80

5 Lentiviral vectors for cancer gene therapy **83**
Antonia Follenzi and Elisa Vigna

5.1 Development of lentiviral vectors (LV) 83
5.2 Targeting of transgene expression 85
5.3 Host immune responses to LV and their transgene 86
5.4 Transgenesis 87
5.5 Haematopoietic stem cell gene transfer 87

5.6 Cancer treatment by LV 89
5.7 Approved clinical trials using LV 91
5.8 Conclusions 91
References 91

6 Poxviruses as immunomodulatory cancer therapeutics **95**
Kevin J. Harrington, Hardev S. Pandha and Richard G. Vile

6.1 Introduction 95
6.2 General features of poxvirus structure and biology 95
6.3 Clinically applicable poxviruses 97
6.4 Poxviruses as potential cancer therapeutics 99
6.5 Clinical experience with poxviruses 102
6.6 Conclusions 110
References 110

7 Oncolytic herpes simplex viruses **115**
Guy R. Simpson and Robert S. Coffin

7.1 Introduction 115
7.2 Herpes simplex virology 115
7.3 Properties of HSV relevant to oncolytic virus therapy 117
7.4 Mutations giving tumour-selective replication 118
7.5 Oncolytic HSV expressing fusogenic membrane glycoproteins (FMG) 125
7.6 Prodrug activation therapy and oncolytic HSV 126
7.7 Combination of oncolytic HSV with immunomodulatory gene expression 127
7.8 Combination of conventional therapies with oncolytic HSV 128
7.9 Summary 129
Acknowledgement 130
References 130

8 Selective tumour cell cytotoxicity by reoviridae – preclinical evidence and clinical trial results **139**
Laura Vidal, Matt Coffey and Johann de Bono

8.1 Introduction 139
8.2 Reovirus structure 139
8.3 Reovirus replication 140
8.4 Reovirus and human infection 141
8.5 Oncolytic activitiy 142
8.6 Mechanism of reovirus-induced cytotoxicity 145
8.7 Preclinical experience 145
8.8 Immunogenicity 146
8.9 Clinical experience 146
8.10 Conclusions 147
References 148

9 Oncolytic vaccinia **151**
M. Firdos Ziauddin and David L. Bartlett

9.1 Introduction 151

9.2 Biology of vaccinia virus 151
9.3 Tumour selectivity and antitumour effect 153
9.4 Improving antitumour effects through bystander effects 160
9.5 Immune response to vaccinia and vaccinia immune evasion strategies 161
9.6 Virus-driven antitumour immune response 163
9.7 Imaging 164
9.8 Current and potential clinical applications 165
References 166

10 Newcastle Disease virus: a promising vector for viral therapy of cancer 171
Volker Schirrmacher and Philippe Fournier

10.1 Introduction 171
10.2 Structure, taxonomy, pathogenicity and oncolytic properties of NDV 171
10.3 Human application and safety 172
10.4 Tumour-selective replication of NDV 174
10.5 Virally based cancer immunotherapy and danger signals 174
10.6 NDV: a danger signal inducing vector 175
10.7 The human cancer vaccine ATV-NDV 176
10.8 Pre-existing antitumour memory T cells from cancer patients and their activation by antitumour vaccination with ATV-NDV 177
10.9 Clinical trials of antitumour vaccination with ATV-NDV 177
10.10 NDV-specific recombinant bispecific antibodies to augment antitumour immune responses 179
10.11 NDV-binding bispecific fusion proteins to improve cancer specific virus targeting 180
10.12 Recombinant NDV as a new vector for vaccination and gene therapy 180
10.13 Conclusion 181
References 182

11 Vesicular stomatitis virus 187
John Bell, Kelly Parato and Harold Atkins

11.1 Introduction 187
11.2 VSV: genomic organization and life cycle 187
11.3 Host range and pathogenesis of VSV infection 188
11.4 Control of VSV infection by the innate type I interferon response 189
11.5 Cancer cells are insensitive to type I interferon 190
11.6 VSV preferentially replicates in and lyses tumour cells *in vitro* 190
11.7 VSV attenuation: enhanced tumour selectivity and therapeutic index 192
11.8 Engineered/recombinant VSV 192
11.9 VSV effectively eradicates tumours *in vivo* 193
11.10 VSV and the host immune response 194
11.11 Host immunity vs. therapeutic efficacy 195
11.12 VSV is a potent vaccine 195
11.13 Innate sensing of VSV and the antitumour response 196
11.14 So what is a good oncolytic virus? 197
11.15 Future challenges for VSV 198
References 199

12 Measles as an oncolytic virus **205**
Adele Fielding

12.1 Introduction 205
12.2 Measles virus and the consequences of natural infection 205
12.3 MV vaccine 206
12.4 MV genetics and engineering 206
12.5 MV receptors 207
12.6 Animal models for the study of MV pathogenesis and oncolysis 207
12.7 Oncolytic activity of MV 208
12.8 Mechanism of specificity 208
12.9 Targeting MV entry 209
12.10 Enhancing the oncolytic activity of MV 210
12.11 Interactions with the immune system 210
12.12 Potential specific toxicities of clinical use of replicating attenuated MV 211
12.13 Clinical trials 211
12.14 Conclusions 212
References 212

13 Alphaviruses **217**
Ryuya Yamanaka

13.1 Introduction 217
13.2 RNA viruses as gene expression vectors 218
13.3 The biology of alphaviruses 218
13.4 Heterologous gene expression using alphavirus vectors 220
13.5 Cancer gene therapy strategies using alphavirus vectors 221
13.6 Alphavirus vector development for gene therapy application 223
13.7 Conclusions 224
References 225

14 Tumour-suppressor gene therapy **229**
Bingliang Fang and Jack A. Roth

14.1 Tumour-suppressor genes 229
14.2 Use of tumour-suppressing genes for cancer therapy 231
14.3 Clinical trials of *p53* gene replacement 232
14.4 Tumour-suppressor gene therapy in multimodality anticancer treatment 233
14.5 Future prospects 235
Acknowledgements 235
References 236

15 RNA interference and dominant negative approaches **241**
Charlotte Moss and Nick Lemoine

15.1 Introduction 241
15.2 Oligonucleotide agents 241
15.3 Mechanism of RNAi 242
15.4 RNAi and antisense compared 243

15.5 siRNA design 244
15.6 Off-target effects 244
15.7 Induction of innate immunity 246
15.8 Methods of delivery 247
15.9 Antisense 251
15.10 Dominant negative approaches 252
15.11 Research applications of siRNA 252
15.12 Therapeutic applications of siRNA 252
References 253

16 Gene-directed enzyme prodrug therapy **255**
Silke Schepelmann, Douglas Hedley, Lesley M. Ogilvie and Caroline J. Springer

16.1 Introduction 255
16.2 Enzyme-prodrug systems for GDEPT 255
16.3 Gene delivery vectors for GDEPT 262
16.4 Conclusions 268
References 269

17 Immunomodulatory gene therapy **277**
Denise Boulanger and Andrew Bateman

17.1 Introduction 277
17.2 Immunotherapy strategies using viral vectors 277
17.3 Viruses used as viral vectors in cancer immunotherapy 280
17.4 Clinical trials against specific TAA 283
17.5 Conclusions and future prospects 289
References 290

18 Antiangiogenic gene delivery **295**
Anita T. Tandle and Steven K. Libutti

18.1 Angiogenesis: role in tumour development and metastasis 295
18.2 Targeting tumour vasculature as an approach for cancer treatment 297
18.3 Viral vectors to deliver antiangiogenic gene products 299
18.4 Viral targeting 303
18.5 Concluding remarks 306
References 306

19 Radiosensitization in viral gene therapy **313**
Jula Veerapong, Kai A. Bickenbach and Ralph R. Weichselbaum

19.1 Introduction 313
19.2 Adenovirus 313
19.3 Adeno-associated viruses 314
19.4 Herpes simplex viruses 314
19.5 Enhancing the effect of radiation by delivering tumour suppressor genes 316
19.6 Virus-directed enzyme prodrug therapy 316

19.7 Conclusions 322
References 324

20 Radioisotope delivery **327**
Inge D.L. Peerlinck and Georges Vassaux

20.1 Introduction 327
20.2 History of iodine therapy 327
20.3 Genetic therapy 330
20.4 Conclusion 338
References 338

21 Radioprotective gene therapy: current status and future goals **341**
Joel S. Greenberger and Michael W. Epperly

21.1 Introduction 341
21.2 Organ-specific radiation protection: oral cavity/oropharynx 342
21.3 MnSOD-PL treatment reduces pulmonary irradiation damage 354
21.4 MnSOD-PL gene therapy down-modulates marrow cell migration to the lungs 357
21.5 MnSOD-PL systemic administration for radiation protection from TBI 358
21.6 Summary and future directions 359
References 360

22 Chemoprotective gene delivery **377**
Michael Milsom, Axel Schambach, David Williams and Christopher Baum

22.1 Introduction 377
22.2 The promise of chemoselection strategies 377
22.3 The limitations of chemoselection strategies 381
22.4 Which expression level of chemoprotective genes is appropriate? 384
22.5 Vector design to achieve optimal expression levels 385
22.6 Exploring side effects of continued transgene expression and insufficient chemoprotection 387
22.7 The future: inducible expression of drug resistance genes 388
Acknowledgements 389
References 389

Index **393**

Foreword

Cancer continues to represent a major global challenge despite advances made in the last 10 years that have seen improvements in survival rates for many of the common solid tumours. A number of cytotoxics, novel targeted agents, innovations in radiation oncology and new surgical techniques have been developed and all have played their part in the steady progress that has been made. However, some of the most important advances have come about due to better multidisciplinary working and successful multinational collaborations in clinical trials. Further work is required to optimize the standard anticancer modalities (surgery, radiotherapy, conventional chemotherapy and targeted agents) but even with the best efforts these are likely to yield little more than incremental gains in treatment outcomes.

The most significant change in oncology in the last 20 years has been our understanding of the molecular and genetic basis of cancer. In the early 1990s, this knowledge led to the development of an entirely new modality of treatment with a rationale based on fundamental molecular observations involving oncogenesis, immunology and intracellular signaling pathways. This new therapy was born out of the new biology, termed *gene therapy* and presented the biomedical community with the possibility of a quantum change in therapeutics. Suddenly there was the theoretical possibility of treating the root cause of a variety of diseases: not just cancer, but cardiovascular disorders, neurodegenerative conditions, inborn errors of metabolism and infectious diseases have all been the targets of this new therapeutic strategy.

Gene therapy represents the ultimate multidisciplinary activity. However, it should be regarded as a non-subject because it is more a series of scientific interdependencies coming together to achieve a particular therapeutic objective. *Viral Therapy of Cancer* illustrates this point very well with almost the entire gamut of bioscience and clinical expertise represented by the contributors. The book focuses on cancer and the use of viruses, both as vectors and as therapeutic agents, the latter strategy having grown out of the early days of gene therapy when viral vectors seemed to be the only possible way forward. The development of viral therapy demonstrates an important truth about gene therapy programmes: namely, that the field of gene therapy is not a strategy that should be judged simply by the triumphs or failures of clinical trials. It is a scientific activity of considerable consequence that spins out important scientific knowledge while at the same time making us question our current standard clinical trial methodologies which are not fit for all purposes, e.g. 'proof of principle' studies.

This book has been edited by three experts in the field of cancer gene therapy with experience of both laboratory and clinical research. The text bridges the gap between bench and bedside and will appeal to both basic scientists and clinicians with an interest in viral and gene therapy. The book is very comprehensive and deals with the biology, selectivity and clinical applications of the viruses that have been used as cancer therapeutics.

The multidisciplinary nature of gene therapy means that it is sometimes difficult for those involved; virologist, molecular biologist, clinician, nurse, pharmacist, safety officer, to get accessible information about those areas of the activity in which that they are not expert. This book provides the reader with an excellent and comprehensive account of all aspects of the use of viruses as cancer therapy.

Martin Gore PhD FRCP

Professor of Cancer Medicine
Royal Marsden Hospital and Institute of Cancer Research
Chairman, Gene Therapy Advisory Committee,
Department of Health (UK)

Preface

Treatment modalities for cancer have expanded well beyond the traditional approaches of surgery, radiotherapy and chemotherapy. There has been an enormous surge of interest in the use of biological therapies, facilitated by a seismic shift in our understanding of the molecular basis of cancer. Although the first gene therapy trial using a retroviral vector was undertaken more than fifteen years ago, gene transfer therapy for cancer still awaits its first great breakthrough in terms of prolonging life. Having fairly recently confirmed the role of certain viruses in tumorigenesis, there appears to be a natural justice that we should now try and harness viruses for cancer therapy. Until recently, we would never have contemplated the use of replication-competent viruses for the treatment of cancer and, in fact, much of the early work in the field was deliberately restricted to the evolution of non-replicating viral vectors capable of efficient gene transfer. However, in 2008 the landscape has changed immeasurably and we are looking at the use of a wide range of replication-competent viruses as potential anti-cancer agents. These agents include those, that occur in nature and others that have been specifically engineered to have specific cytotoxicity against cancer cells, either as single agents or in combination with other anti-cancer modalities. The range of potential agents presents a variety of tropisms and individual strengths and weaknesses. Progress in this field has been astonishing in the last decade and as a result we felt that a comprehensive textbook coherently presenting the advances with the individual viruses was timely.

We have attempted to present a text which will appeal to the clinician, clinician-scientist and basic scientist as well as to allied health professionals. The chapters review the mechanistic and clinical background to a range of viral therapies and are designed to proceed from basic science at the bench to the patient's bedside to give an up-to-date and realistic evaluation of a therapy's potential utility for the cancer patient. We anticipate intense clinical activity in this arena in the next few years with a very real prospect that virotherapy may establish a role in the standard treatment of both common and rare cancers.

We thank Dr Kate Relph for her enormous contribution in the editing of this book.

This volume would not have been possible without the support of our families and, so, we wish to dedicate it: to Sindy, Simran and Savneet; to Memy, Oriana and Sebastian; to Katie and Lila Rose.

Kevin J. Harrington,
Richard G.Vile and Hardev S. Pandha

Contributors

Harold Atkins
Ottawa Regional Cancer Centre Research Laboratories
503 Smyth Road
Ottawa, Ontario K1H IC4
Canada

David L. Bartlett
University of Pittsburgh Physicians Faculty
CNPAV 459 Pittsburgh
UPMC Cancer Pavilion
5150 Center Avenue
Pittsburgh, PA 15260, USA

Andrew Bateman
Division of Cancer Sciences
School of Medicine
Southampton General Hospital
Southampton SO16 64D, UK

Christopher Baum
Cincinnati Children's' Hospital
3333 Burnett Avenue
Cincinnati, OH 45229-3039, USA

Kiflai Bein
Department of Molecular Genetics and Biochemistry
University of Pittsburgh
BSTWR E1246
Pittsburgh, PA 15219, USA

John Bell
Ottawa Regional Cancer Centre Research Laboratories
503 Smyth Road
Ottawa, Ontario K1H IC4, Canada

Kai A. Bickenbach
University of Chicago
Duchessois Center for Advanced Medicine
5841 S Maryland Avenue
Chicago, IL 60637, USA

Denise Boulanger
The Somers Cancer Research Building
Southampton General Hospital
Mail Point 824, Tremona Road
Southampton SO16 64D, UK

Xandra O. Breakefield
Departments of Neurology and Radiology
Program in Neuroscience
Harvard Medical School
Boston, MA 02114, USA

Simon Chowdhury
Department of Medical Oncology
St George's Hospital
Blackshaw Road
London SW17 0QT, UK

Matt Coffey
Oncolytics Biotech Inc
210, 1167 Kensington Crescent NW
Calgary, AB T2N 1X7, Canada

Robert Coffin
BioVex Ltd
70 Milton Park
Abingdon OX14 4RX, UK

Johann de Bono
Centre for Cancer Therapeutics
Institute for Cancer Research
Royal Marsden Hospital
Downs Road
Sutton SM2 5PT, UK

Michael W. Epperly
Department of Radiation Oncology
University of Pittsburgh Cancer Institute
200 Lothrop Street
Pittsburgh, PA 15213, USA

Bingliang Fang
Department of Thoracic and Cardiovascular Surgery, Unit 445
The University of Texas M. D. Anderson Cancer Center
1515 Holcombe Boulevard
Houston, TX 77030, USA

Adele Fielding
Royal Free Hospital
Pond Street
London NW3 2QG, UK

Andrea Follenzi
Albert Einstein College of Medicine
Ullman Building
1300 Morris Park Avenue
Bronx, NY 10461, USA

Philippe Fournier
German Cancer Research Center
Division of Cellular Immunology
Im Neuenheimer Feld 280
69120 Heidelberg, Germany

Kevin J. Harrington
Targeted Therapy Laboroatory
Cancer Research UK Centre for Cell and Molecular Biology
Institute of Cancer Research
237 Fulham Road
London SW3 6JB, UK

Joseph C. Glorioso
Department of Molecular Genetics and Biochemistry
University of Pittsburgh
BSTWR E1246
Pittsburgh, PA 15260, USA

Paola Grandi
Department of Neurosurgery
University of Pittsburgh School of Medicine
Pittsburgh, PA 15261, USA

Joel S. Greenberger
Department of Radiation Oncology
University of Pittsburgh Cancer Institute
200 Lothrop Street
Pittsburgh, PA 15213, USA

Costas G. Hadjipanayis
Department of Neurosurgery
University of Pittsburgh School of Medicine
Pittsburgh, PA 15261, USA

Douglas Hedley
Cancer Research UK Centre for Cancer Therapeutics
The Institute of Cancer Research
15 Cotswold Road
Sutton SM2 5NG, UK

Yasuhiro Ikeda
Molecular Medicine Program
Guggenheim building 18-11c
Mayo Clinic College of Medicine
200 1st Street
Rochester, MN 55905, USA

Nick Lemoine
Cancer Research UK Clinical Centre
Barts and The London Queen Mary's School of Medicine and Dentistry
John Vane Science Centre
Charterhouse Square
London EC1M 6BQ, UK

Steven K. Libutti
National Cancer Institute
Suite 3036A
6116 Executive Road
MSC 8322
Bethesda, MD 20892-8322, USA

Michael Milsom
Division of Experimental Hematology
Cincinnati Children's Hospital Medical Center
Cincinnati, OH 45229-3039, USA

Charlotte Moss
Cancer Research UK Clinical Centre
Barts and The London Queen Mary's School of Medicine and Dentistry
John Vane Science Centre
Charterhouse Square
London EC1M 6BQ, UK

Lesley M. Ogilvie
Cancer Research UK Centre for Cancer Therapeutics
The Institute of Cancer Research
15 Cotswold Road
Sutton SM2 5NG, UK

Caroline J. Springer
The Institute of Cancer Research
123 Old Brompton Road
London SW7 3RP, UK

Hardev S. Pandha
Oncology Department
Postgraduate Medical School
University of Surrey
Guildford GU2 7WG, UK

Kelly Parato
Ottawa Regional Cancer Centre Research Labororatories
503 Smyth Road
Ottawa, Ontario K1H IC4, Canada

Inge D.L. Peerlinck
Centre for Molecular Oncology
Institute of Cancer and the CR-UK Clinical Centre
Barts and The London Queen Mary's School of Medicine and Dentistry
John Vane Science Centre
Charterhouse Square
London EC1M 6BQ, UK

Selvarangan Ponnazhagan
The University of Alabama in Birmingham
Lyons-Harrison Research Building
1530 3rd Avenue S
Birmingham, AL 35294-0007, USA

Kate Relph
Oncology Department
Postgraduate Medical School
University of Surrey
Guildford GU2 7WG, UK

Jack A. Roth
Department of Thoracic and Cardiovascular Surgery
Unit 445
The University of Texas M. D. Anderson Cancer Center
1515 Holcombe Boulevard
Houston, TX 77030, USA

Axel Schambach
Department of Experimental Hematology
Hannover Medical School
30625 Hannover, Germany

Silke Schepelmann
The Institute of Cancer Research
237 Fulham Road
London SW3 6JB, UK

Volker Schirrmacher
German Cancer Research Center
Division of Cellular Immunology
Im Neuenheimer Feld 280
69120 Heidelberg, Germany

Guy Simpson
Dept of Oncology
Postgraduate Medical School
University of Surrey
Daphne Jackson Road, Manor Park
Guildford GU2 5XH, UK

Caroline J. Springer
Institute of Cancer Research
123 Old Brompton Road
London SW7 3RP, UK

Anita Tandle
Advanced Technology Center
NCI Room 109G
8717 Grovemont Circle
Gaithersburg, MD 20892-4605, USA

Georges Vassaux
Centre for Molecular Oncology
Institute of Cancer and the CR-UK Clinical Centre
Barts and The London Queen Mary's School of Medicine and Dentistry
John Vane Science Centre
Charterhouse Square
London EC1M 6BQ, UK

Jula Veerapong
University of Chicago Duchessois Center for Advanced Medicine
5841 S Maryland Avenue
Chicago, IL 60637, USA

Laura Vidal
Centre for Cancer Therapeutics
Institute for Cancer Research, Royal Marsden Hospital
Downs Road
Sutton SM2 5PT, UK

Elisa Vigna
Institute for Cancer Research and Treatment
University of Torino
Strada Provinciale
10060 Candiolo
Torino, Italy

Richard G. Vile
Molecular Medicine Program
Guggenheim 1836
Mayo Clinic
200 1st Street
Rochester, MN 55902, USA

Ralph R. Weichselbaum
University of Chicago Duchessois Center for Advanced Medicine
5841 S Maryland Avenue
Chicago, IL 60637, USA

David Williams
Division of Experimental Hematology
Cincinnati Children's Hospital Medical Center
Cincinnati, OH 45229-3039, USA

Darren Wolfe
Diamyd Inc
100 Technology Drive
Pittsburgh, PA 15261, USA

Ryuya Yamanaka
Research Center of Innovative Cancer Therapy
Kurume University School of Medicine
Asahimachi 67
Kurume
Fukuoka 830-0011, Japan

M. Firdos Ziauddin
Division of Surgical Oncology
University of Pittsburgh Medical Center
Pittsburgh, PA 15260, USA

1

Adenoviruses

Kate Relph, Kevin J. Harrington, Alan Melcher and **Hardev S. Pandha**

1.1 Introduction

Adenoviral vectors are the most popular vehicles for gene transfer currently being used in worldwide clinical trials for cancer. Over the past decade our knowledge of the adenoviral lifecycle together with the discovery of novel tumour antigens has permitted the targeting of adenoviral vectors to specific tumours. Targeting adenoviral vectors to tumours is crucial for their use in clinical applications in order to allow for systemic administration and the use of reduced vector doses. In addition, novel approaches to tumour killing have also been explored which will have greater potency and selectivity than currently available treatments such as chemotherapy or radiation. This chapter discusses the basic concepts behind the use of adenoviral vectors for cancer gene therapy, their potential for clinical application and where possible reviews ongoing and completed clinical trials.

1.2 Viral structure and life cycle

Adenoviruses are a frequent cause of upper respiratory tract infections and have also been associated with gastroenteritis and pneumonia in young children. They were first isolated in 1953 by scientists trying to establish cell lines from adenoidal tissue of children removed during tonsillectomy, and since then more than 50 different serotypes have been identified (Table 1.1) (Hilleman and Werner, 1954). The adenoviruses have been classified into six subgroups based on sequence homology and their ability to agglutinate red blood cells (Shenk, 1996). Most adenoviral vectors are derived from Ad2 or Ad5 which have been well studied and noted for their safety: over 50 per cent of the population show antibodies to adenovirus serotype 5 suggesting that it is particularly safe.

Adenovirus is a non-enveloped, icosahedral virus of about 60–90 nm in diameter with a linear double stranded genome of about 30–40 kb (Figure 1.1) (Stewart *et al.*, 1993). The capsid consists of three major proteins, hexon (II), penton base (III), and a knobbed fibre (IV) along with a number of other minor proteins, VI, VII, IX, IIIa and IVa2. The virus genome has inverted terminal repeats (ITRs) and is associated with several proteins including a terminal protein (TP), which is attached to the 5′ end (Rekosh *et al.*, 1977), a highly basic protein VII and a small peptide termed *mu* (Anderson *et al.*, 1989). A further protein, V, links the DNA to the capsid via protein VI (Matthews and Russell, 1995).

The adenovirus life cycle essentially consists of the following steps. Virus entry into the cell is a two-stage process involving an initial interaction of the fibre protein with a range of cellular receptors, which include the major histocompatibility complex (MHC) class1 molecule and the coxsackie and adenovirus receptor CAR (Bergelson *et al.*, 1997). The CAR is a plasma membrane protein of 46 kDa belonging to the immunoglobulin family

Viral Therapy of Cancer Edited by Kevin J. Harrington, Richard G. Vile and Hardev S. Pandha

Table 1.1 Adenoviral serotypes

Group	Serotypes
A	12, 18, 31
B	3, 7, 11, 14, 16, 21, 34, 35, 50
C	1, 2, 5, 6
D	8–10, 13, 15, 17, 19, 20, 22–30, 32, 33, 36–39, 42–49, 51
E	4
F	40, 41

(Tomko *et al.*, 1997). Some cell types, such as those of haematopoietic origin, do not express CAR on their cell surface and appear to be refractory to adenoviral infection (Mentel *et al.*, 1997) suggesting that receptor recognition is one of the key factors in determining cell tropism. After initial interaction between the fibre knob and CAR the penton base protein then binds to the $\alpha_v\beta_3$ integrin family of cell surface heterodimers allowing internalization via receptor mediated endocytosis (Wickham *et al.*, 1993). Penetration into the cell involves phagocytosis into phagocytic vesicles, after which the toxic activity of the pentons ruptures the phagocytic vacuoles and releases the vesicles into the cytoplasm. Release of the virus into the cytoplasm is accompanied by a stepwise dismantling of the capsid by proteolysis of protein VI (Greber *et al.*, 1996). The partially dismantled viral particle is then delivered to the nucleus via microtubulin-assisted transport where the core-protein coated viral genome enters in through the nuclear pores.

Transcription of the adenoviral genome occurs in both early and late phases which occur before and after viral DNA replication respectively. A complex series of splicing events produces four early 'cassettes' of gene transcription termed E1, E2, E3 and E4 (Figure 1.2). The E1 proteins are divided into E1A and E1B. E1A is the first gene to

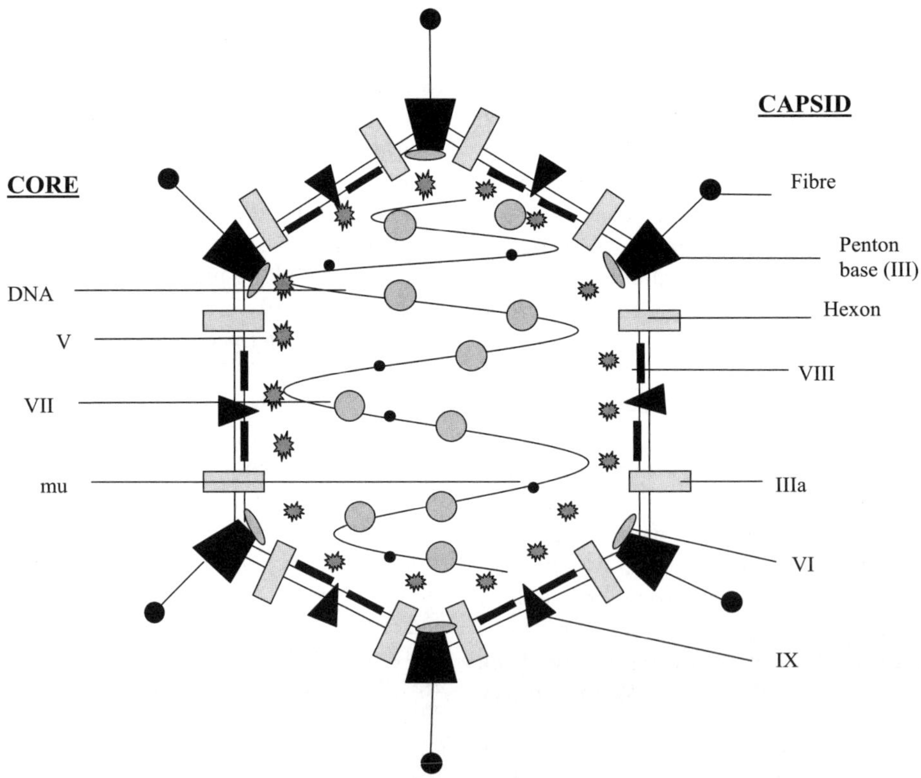

Figure 1.1 Structure of adenoviral capsid

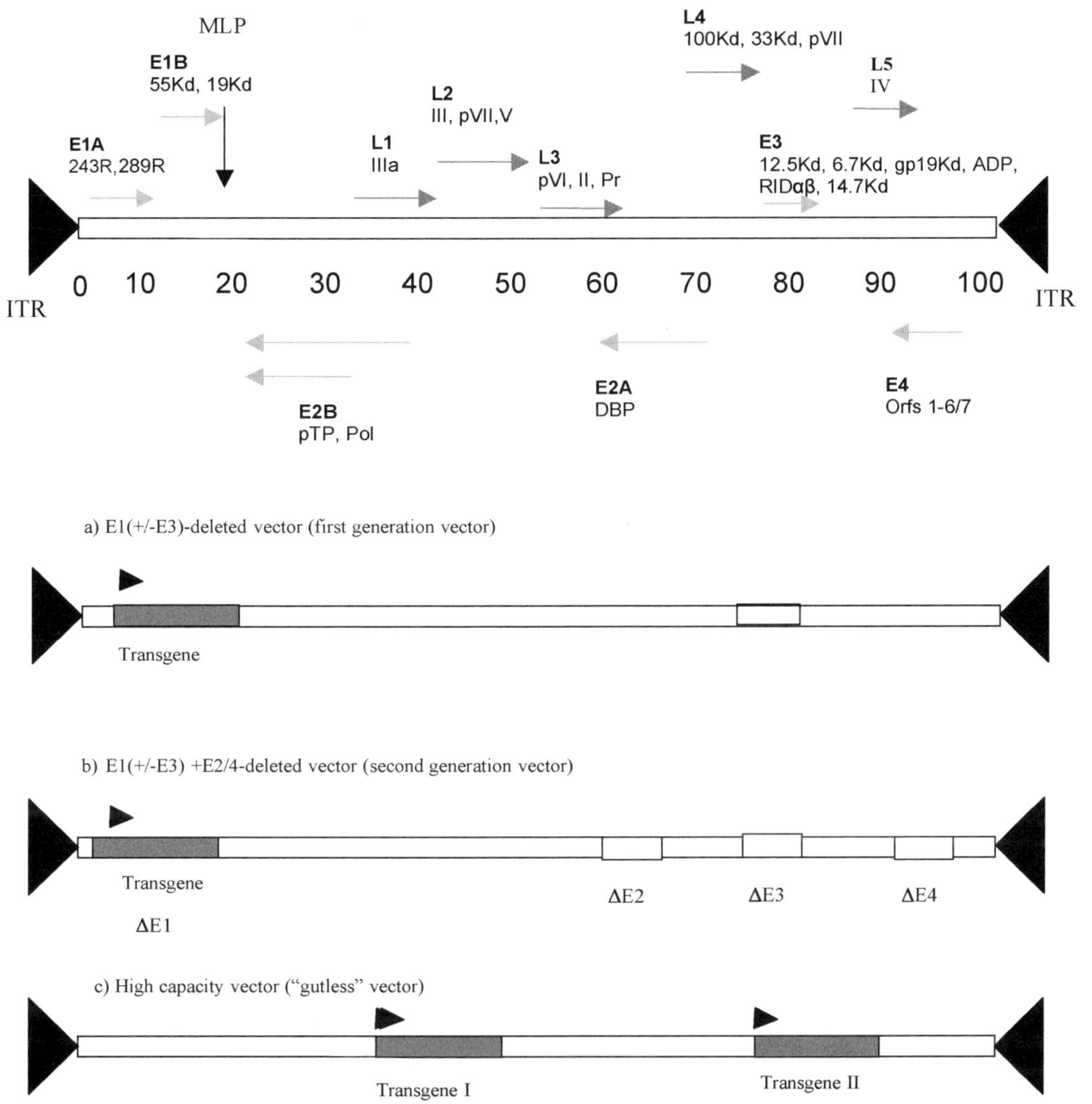

Figure 1.2 Schematic of adenoviral genome and adenoviral vectors. E1A must be removed to prevent recombinant virus from replicating. ITR, inverted terminal repeats

be expressed (Frisch and Mymryk, 2002) It encodes a transactivator for the transcription of the other early genes E1B, E2A, E2B, E3 and E4 but is primarily involved in many pathways to modulate cellular metabolism and make it more susceptible to viral replication (Table 1.2). E1A proteins interfere with cell division and regulation via direct and indirect action on a number of cellular proteins. For example E1A binds to the RB protein preventing it from binding to the transcription factor E2F. As a result E2F is transcriptionally active and can thus stimulate DNA synthesis. Also E1A maintains the stability of p53 via a variety of proteins and pathways including Mdm4, UBC9 and Sug1 (Table 1.2). E1A can directly bind and inhibit components involved in cell cycle control such as the cyclin dependent kinase inhibitor p21 (Chattopadhyay *et al.*, 2001).

Table 1.2 Some properties of E1A proteins

Property	Reference
Bind to p21 and related CDK inhibitors thereby stimulating cell division and growth	Chattopadhyay *et al.*, 2001
Bind to cyclins A and E-CDK complexes, which regulate passage to cell DNA synthesis	Faha *et al.*, 1993
Bind to the p300/CBP family of transactivators, which play a key role in regulating the transcription of many components of the cell cycle	Chakravati *et al.*, 1999
Binds to Rb and releases E2F- vital for synthesis of S-phase components as well as activation of E2 gene.	Brehm *et al.*, 1998
Interacts with multiprotein complex Sur-2, thereby stimulating the transcription of virus genes	Stevens *et al.*, 2002
Binds to the TATA-box binding protein to regulate transcription	Mazzarelli *et al.*, 1997
Induction of apoptosis via release of E2F which leads to increase in p53 and p19arf levels.	Hale and Braithwaite, 1999
Stabilises p53 via interaction with Sug1 a subunit of the proteasome complex that is required for proteolysis of p53	Grand *et al.*, 1999
Targets Mdm4 to stabilize tumour suppressor p53	Li *et al.*, 2004
Activates transcription of p73 and Noxa to induce apoptosis.	Flinterman *et al.*, 2005
Activates apoptosis by sensitizing cells to ionizing radiation, DNA damage, TNF and Fas ligand. Mediated by inhibiting the IκB kinases, which are critical for release of NFκB to nucleus and requires binding of E1A to P300/CBP	Shisler *et al.*, 1996
Binds to UBC9, a protein involved in the SUMO enzymatic pathway. Binding to E1A may interfere with SUMO modification of cellular proteins such as p53 and pRb	Desterro *et al.* 1999, Ledl *et al.* 2005

NFκB, nuclear factor κB.

It can also interact with a number of host factors involved in mediating chromatin structure including p400 (Fuchs *et al.*, 2001) and the histone acetyl transferases p300, pCAF and TRRAP/GCN5 (Lang and Hearing, 2003). Other early gene products are also involved in making the cell more refractory to viral replication. The E1B 19K protein is analogous to the Bcl-2 gene product and is concerned with increasing cell survival and ablating members of the Bax family which induce apoptosis (Han *et al.*, 1996). A second 55 kDa protein product of the E1B gene has been shown to interact with p53 reducing its transcription. The E1b protein has also been shown to block host mRNA transport to the cytoplasm (Pilder *et al.*, 1986). The E2 gene encodes proteins required for viral DNA replication, i.e. DNA polymerase, DNA-binding protein and the precursor of the terminal protein (de Jong *et al.*, 2003). Despite replicating in the nucleus the adenovirus need its own enzymatic machinery because of its complex chromosomal structure. The genome lacks telomeres and so the integrity of the ends of the DNA is maintained by a viral preterminal protein which is covalently linked to the 5′ end and acts as a primer for the viral DNA polymerase. The E3 genes encode a variety of transcripts involved in subverting the host defence mechanism (Wold and Chinnadurai, 2000). The E3-gp19K protein acts to prevent presentation of viral antigens by MHC class I pathway and therefore blocks cell lysis by cytotoxic T cells (Bennett *et al.*, 1999). One E3 protein is termed the adenovirus death protein (ADP) as it facilitates late cytolysis of the infected cell and thereby releases progeny virus more efficiently (Tollefson *et al.*, 1996a). The E4 proteins mainly facilitate virus mRNA metabolism and promote virus DNA replication and shut off of host protein synthesis (Halbert *et al.*, 1985).

Replication of the viral genome starts about 5–6 h after infection and is dependant on the inverted terminal repeats (ITRs) which act as the origins of replication. Adenovirus DNA replication has been studied extensively both *in vivo* (t.s. mutants in infected cells) and *in vitro* (nuclear extracts). At least three virus-encoded proteins are known to be involved in DNA replication: TP acts as a primer for initiation of synthesis. Ad DBP – a DNA-binding protein and Ad DNA Pol – 140 kDa DNA-dependent polymerase. The onset of DNA replication signals the pattern of transcription changes from early to late genes and only newly replicated DNA is used for late gene transcription. Late phase transcription is driven primarily through the major late promoter with five transcripts resulting from a complex series of splicing events. These transcripts are mainly used for the production of viral structural proteins. Encapsidation of the virus depends on the presence of a packaging signal near the 5′ end of the genome consisting of an AT-rich sequence. Intranuclear virion assembly starts about 8 h after infection and leads to the production of 10^4 to 10^5 progeny particles per cell, which can be released after final proteolytic maturation by cell lysis 30–40 h post-infection, completing the viral life cycle (Shenk, 1996).

1.3 Adenoviral vectors

Adenoviral vectors are attractive reagents for gene therapy because of their ability to transduce genes into a broad range of cells, and to infect both dividing and non-dividing cells (McConnell and Imperiale, 2004). Adenoviral vectors can accommodate large segments of DNA (up to 7.5 kb) and the viral genome rarely undergoes rearrangement meaning that inserted genes are maintained without change during virus replication. In addition, adenoviruses replicate episomally and do not insert their genome into that of the host cell ensuring less disruption of vital cellular genes and processes and reduced risk of insertional mutagenesis. This can, however, be a limitation in that transient expression of the therapeutic gene may be inadequate to treat chronic conditions such as cystic fibrosis. However, for situations in which short-term activity of the gene is needed, such as expression of suicide genes selectively in tumour cells, these viruses are suitable vectors. The adenoviral genes can be separated into two groups; the *cis*-genes, such as those responsible for the packaging signal, which must be carried by the virus itself, and the *trans*-genes which can generally be complemented and therefore replaced with ‘foreign’ DNA. The first generation of adenoviral vectors were used for the delivery of genes in monogenic disorders (Figure 1.2a). In these vectors the E1 region was removed to inhibit viral replication and make way for the therapeutic gene. Many of the first generation vectors also contained a deletion in the E3 region in order to allow for even greater transgenes to be incorporated. The E3 genes are dispensable for virus growth *in vitro* but some data suggests that E3 genes in vectors may be beneficial *in vivo* due to their ability to dampen the immune response (Bruder *et al.*, 1997). However, despite the removal of these regions of the viral genome there was still low-level transcription of viral genes, which led to a host cellular immune response and a reduction in the period of gene expression due to cell-mediated destruction of the transduced cells (Kay *et al.*, 1995; Yang *et al.*, 1995). In addition these types of vectors allowed the generation of E1 containing replication competent adenovirus (RCA) due to homologous recombination in 293 cells which further enhanced the adverse effects (Lochmuller *et al.*, 1994). In order to address these problems homologies between the vectors and the complementing cell lines have been reduced. Second generation adenoviral vectors have further deletions in E2a, E2b or E4 and have reduced immunogenicity and RCA generation (Figure 1.2b). Despite these improvements the complementing cell lines are difficult to engineer, can be difficult to grow and can lead to poor viral titers (Lusky *et al.*, 1998). As a result a third generation of adenoviral or gutless vectors have been created (Parks *et al.*, 1996) (Figure 1.2c). These have all of the viral genes deleted (except for the packaging signal) and replaced with the therapeutic gene of interest. They are therefore free from problems associated with immunogenicity and demonstrate long-term transgene expression. They are generated with a helper virus, which contains all of the genes necessary for viral replication but which contains a deletion in the packaging signal to ensure that it is not incorporated into the final vector. These vectors are still undergoing development in order to improve their purity and large-scale manufacture (Wu and Attai, 2000).

1.4 Targeting adenoviral vectors

Despite the fact that adenoviral vectors have many advantages over other gene transfer vehicles there are some problems associated with their use. The broad tropism of adenoviral vectors as well as being an advantage also represents an important limitation for their use in therapeutic applications. Animal studies have shown that adenoviral vectors do not remain confined to one compartment and are able to disseminate to distal sites with toxic effects that are most notable in the liver (Wang *et al.*, 2003; Yee *et al.*, 1996). This also restricts the systemic administration of the vectors due to the potential for toxicity in normal tissues (Brand *et al.*, 1997). In addition, important target tissues are often refractory to adenoviral infection leading to administration of increased doses of vector in an attempt to improve gene transfer. This in turn often leads to increased toxicity and enhanced humoral and cellular immune responses. Clearly there is a requirement for targeted adenoviral vectors in clinical applications in order to allow for systemic administration and the use of reduced vector doses, which will in turn reduce inflammatory, and immune responses (Mizuguchi and Hayakawa, 2004). Two main approaches have been taken in order to target expression of the therapeutic gene to the required tissue/tumour: (1) transductional targeting and (2) transcriptional targeting.

1.4.1 Transductional targeting of adenoviral vectors

The identification of the route by which human cells uptake adenovirus was an important step towards retargeting adenoviral vectors to different cell types, also known as transductional targeting. The adenovirus fibre knob anchors onto the surface of the target cell by means of the CAR and interaction of the capsid penton protein with integrins $\alpha_v\beta_3$ and $\alpha_v\beta_5$ on the surface of target cells allows internalization (Bergelson *et al.*, 1997; Wickham *et al.*, 1993). Most immortalized tumour cell lines express CAR and are therefore easily transduced by adenoviral vectors. However, certain studies have demonstrated that 50 per cent of primary epithelial cancers do not express CAR (Kasono *et al.*, 1999; Vanderkwaak *et al.*, 1999). This may account for some of the limited success with past clinical trials using adenoviral vectors. Transductional targeting may improve transfer of genes to particular cancer types, such as glioma, and in addition retargeting adenoviral vectors will permit the treatment of haematological malignancies because haematopoietic stem cells are known to lack CAR (Huang *et al.*, 1996).

There are many reports of retargeting of adenoviral vectors to tumour cells via the use of antibodies directed towards specific antigens on the surface of a particular tumour type (Barnett *et al.*, 2002). One group used a neutralizing anti-fibre antibody conjugated to an antibody directed against the epithelial cell adhesion molecule (EGP-2), which is highly expressed on the surface of a range of adenocarcinomas from the stomach, oesophagus, breast, ovary, colon and lung and its expression is limited in normal tissue. In this study the adenovirus specifically infected cancer cell lines expressing EGP-2 whilst gene transfer was dramatically reduced in EGP-2-negative cell lines. A recent study combines genetic ablation of native adenoviral tropism with redirection of viral binding to melanoma cells via a bispecific adaptor molecule (Nettelbeck *et al.*, 2004). This molecule consists of a bacterially expressed single chain diabody, scDb MelAd that binds to both the adenoviral fibre knob and to the high molecular weight melanoma associated antigen (HMWMAA), which is widely expressed on the surface of melanoma cells. This retargeting strategy mediated up to a 54-fold increase in adenoviral gene transfer to CAR-negative melanoma cells compared to a vector with native tropism.

Further targeting has been achieved by altering the structure of the fibre knob itself by inserting an arginine-glycine-aspartate (RGD) tripepetide

(Buskens *et al.*, 2003). Four oesophageal carcinoma cell lines and ten fresh surgical resection specimens were cultured and infected with either native adenovirus or retargeted adenovirus expressing the luciferase gene or green fluorescent protein to analyse gene transfer efficiencies. In both the cell lines and the primary cells more efficient gene transfer was seen with the retargeted virus. This phenomenon was less pronounced in normal cells.

1.4.2 Transcriptional targeting of adenoviral vectors

The targeting of gene expression to specific cell types/tissues can be achieved through the use of tumour or tissue specific promoters. This approach has been adopted in a range of studies targeting gene expression to tumours (Rots *et al.*, 2003; Haviv and Curiel, 2001). A recent study identified the cyclooxygenase-2 (cox-2) gene as a potential new target for melanoma gene therapy (Nettelbeck *et al.*, 2003). An adenoviral vector was constructed in which the cox-2 promoter drove the expression of a luciferase reporter gene. Melanoma cell lines, primary melanoma cells and normal melanocytes were infected with this novel vector. The results demonstrated activity of the cox-2 promoter in the melanoma cell lines and primary melanoma cells but not in non-malignant primary epidermal melanocytes. Several approaches have also considered the use of two different tumour specific promoters within the same vector in order to achieve a further degree of specificity. The second promoter is normally one that is a more general promoter which shows activity in a broad range of tumours such as the telomerase reverse transcriptase promoter.

In suicide gene therapy for cancer (discussed later) targeting is paramount to prevent unwanted toxicity. For example, the product of the thymidine kinase gene itself, without addition of the prodrug ganciclovir, has been shown to cause liver toxicity when under the control of the cytomegalovirus promoter (Yamamoto *et al.*, 2001). Several groups have therefore engineered adenoviral vectors to contain tissue/tumour specific regulatory elements in order to avoid these problems and target toxicity specifically to the transduced cells. One study used the prostate specific antigen promoter to target expression of HSV-TK to benign prostatic hyperplasia (Park *et al.*, 2003). This approach induced highly selective and definite ablation of epithelial cells in benign canine prostate.

Both transcriptional and transductional targeting have improved the efficacy of adenoviral vectors significantly. Some groups are now investigating the possibilities of combining these two approaches to further improve the specificity of adenoviral vectors. For example a combination of the tissue-specific SLP1 promoter and the ovarian cancer associated targeting adaptor protein, sCARfC6.5, which contains the CAR ectodomain and a single-chain antibody specific for c-erbB-2, increased the efficacy and specificity of adenoviral gene therapy for ovarian carcinoma (Barker *et al.*, 2003.

1.5 Clinical applications of adenoviral gene therapy

Advances in adenoviral vector technology have meant that there are now 140 clinical trials worldwide currently being conducted on various cancers using adenoviral vectors (*Journal of Gene Medicine* www.wiley.co.uk/wileychi/genemed). Table 1.3 gives details of seventeen completed gene therapy trials for cancer using adenoviral vectors. All of theses were phase I studies to test toxicity. Table 1.4 indicates some of the ongoing clinical phase II trials. Several approaches have been used to destroy the target tumour cells:

1.6 Adenoviral vectors for immunotherapy

T lymphocytes play a crucial role in the host's immune response to cancer. Although there is ample evidence for the presence of tumour-associated antigens on a variety of tumours, they are often unable to elicit an adequate antitumour response. Our increasing knowledge of the cellular interactions required to induce a specific antitumour response has led to the development of cancer vaccines which prime the host response and induce or enhance T-cell reactivity against tumour antigens. Gene-based strategies for

Table 1.3 A selection of completed phase I clinical trials using adenoviral vectors for the treatment of cancer

Investigator	Country	Cancer	Gene	No. of patients	Reference
Stewart	Canada	Breast, melanoma	IL-2	23	Stewart *et al.*, 1999
Tursz	France	Non-small cell lung carcinoma	Il-2	21	Griscelli *et al.*, 2003
Tursz	France	Non small cell lung carcinoma	Beta-gal	21	Griscelli *et al.*, 2003
Eck	USA	CNS	HSV-TK	N/C	N/C
Reid	USA	Anaplastic thyroid cancer	p53	N/C	N/C
Roth	USA	Non-small cell lung carcinoma	p53	N/C	N/C
Belani	USA	Hepatocellular carcinoma	p53	N/C	N/C
Belldegrun	USA	Prostate	p53	N/C	N/C
Hasenburg	Germany	Ovarian	HSV-tk	10	Hasenburg *et al.*, 2002
Kauczor	Germany	Non-small cell lung carcinoma	p53	6	Kauczor *et al.*, 1999
Fujiwara	Japan	Non-small cell lung carcinoma	p53		Fujiwara *et al.*, 1999
Boulay	Switzerland	Non small cell lung carcinoma	p53	N/C	N/C
–	Switzerland	Metastases from solid tumours	IFNγ	N/C	N/C
Albertini	UK	Melanoma	IFNγ	N/C	N/C
Lafollette	UK	Head and neck carcinoma	E1b del.	N/C	N/C
Lafollette	UK	Ovarian	E1b del.	16	Vasey *et al.*, 2002
Stewart	UK	Gastrointestinal cancer	p53	N/C	N/C

N/C = not stated.
Source: *Journal of Gene Medicine* website (http:www.wiley.co.uk/wileychi/genmed)

Table 1.4 A selection of ongoing phase I and II clinical trials with adenoviral vectors for the treatment of cancer

Principal Investigator	Country	Cancer targeted	Gene	Action of gene	Combination	Year approved	Route of administration
DeWeese	USA	Prostate	CV7606	Prostate-specific oncolysis	Radiotherapy	2001	Intraprostatic
Small	USA	Prostate	CV787	Prostate-specific oncolysis	Docetaxel	2001	Intravenous
Yoo	USA	Squamous head and neck carcinoma	p53	Tumour suppressor	Chemo	2001	Intratumoral
Cristofanilli	USA	Breast	p53	Tumour suppressor	Docetaxal + Doxorubicin	2001	Intratumoral
Schuler	Germany	Non-small cell lung carcinoma	p53	Tumour suppressor	Chemo	2001 Schiller *et al.*	N/C
N/C	Germany	Ovarian and tubal cancer	N/C	N/C	N/C	–	N/C
Gutierrez	Mexico	Cervical	N/C	N/C	N/C	–	N/C
Senzer	USA	Pancreatic	TNF	Cytokine	Chemo and radiotherapy	2002	Intratumoral
Senzer	USA	Oesophagus	TNF	Cytokine	Chemotherapy	2002	Intratumoral
Ross	USA	Non-small cell lung carcinoma	GM-CSF	Cytokine	Chemotherapy	2003	Intradermal
Kim	USA	Melanoma	MDA-7	Tumour suppressor	–	2003	Intratumoral
Hodi	USA	Melanoma	GM-CSF	Cytokine	–	2003	Intratumoral and subcutaneous
Libutti	USA	Colorectal	TNF	Cytokine	Chemo- and radiotherapy	2003	Intratumoral
Davies	USA	Non-small cell lung carcinoma, bronchioalveolar carcinoma	GM-CSF	Cytokine	–	2003	Intradermal
Deisseroth I	USA	Breast	MUC1CD154	Antigen	–	2004	Subcutaneous
Deisseroth I	USA	Prostate	MUC1CD154	Antigen	–	2004	Subcutaneous
ReidI	USA	Colorectal with liver metastasis	IFN-β	Cytokine	–	2004	Intravenous
Dinney I	USA	Bladder	IFN-α-2b	Cytokine	–	2005	Intravesical
Fisher I	USA	Pancreatic Adenocarcinoma	HSV-TK	Marker	Chemoradiation	2005	Intratumoral
Kim I	USA	Prostate	NIS	–	-	2005	Intratumoral

CV7606 and CV787, promoter and enhancer of PSA. N/C, not specified. MDA-7, Melanoma differentiation associated protein 7; IFN, interferon.
Source: *Journal of Gene Medicine* website (http:www.wiley.co.uk/wileychi/genmed)

immunotherapy of cancer include: *ex vivo* transduction of cytokine genes into tumour cells, direct transfer of cytokine genes into tumour cells or the transfer of tumour antigens or cytokine genes into dendritic cells.

Several clinical trials, both completed and ongoing, have involved the use of adenoviral vectors to transfer genes directly into the tumour (Tables 1.3 and 1.4). Stewart *et al.* (1999) conducted a phase I trial in which an E1, E3-deleted adenovirus encoding interleukin-2 (AdCAIL-2) was directly injected into subcutaneous deposits of melanoma or breast cancer. Twenty-three patients were injected at seven dose levels (10^7–10^{10} plaque-forming units, p.f.u). The side effects noted were minor and included local inflammation at the site of injection in 60 per cent of patients. Post-injection biopsies demonstrated tumour necrosis and lymphocytic infiltration with the predominant tumour-infiltrating cells being CD3- and CD8-positive. Vector derived sequences were detected in 14 of 18 biopsies examined 7 days after injection and vector derived interleukin-2 (IL-2) mRNA was detected in 80 per cent of 7-day biopsies from tumours injected with 10^8 p.f.u. of AdCAIL-2 or higher. IL-2 was detected by enzyme-linked immunosorbent assay in the tumour biopsies at 48 h but no protein was detected after 7 days. No vector sequences were detected before or after injection indicating the absence of replication competent virus. This trial concluded that this adenoviral vector was safe for delivery into humans and demonstrated successful transgene expression even in the face of preexisting immunity to adenovirus.

A second approach involved transducing autologous tumour cells *ex vivo* with granulocyte–macrophage colony-stimulating factor (GM-CSF). One phase I study carried out by Soiffer *et al.* (2003) tested the biologic activity of vaccination with irradiated, autologous melanoma cells engineered to secrete GM-CSF by adenoviral mediated gene transfer. Excised metastases were processed to single cells and transduced with adenoviral vector expressing GM-CSF, irradiated and then cryopreserved. For each autologous vaccine the average GM-CSF secretion was 745 ng/10^6 cells/24 h. Toxicity was restricted to grade 1 or 2 local skin reactions Vaccination elicited dense dendritic cell, macrophage, granulocyte, and lymphocyte infiltrates at injection sites in 19 of 26 assessable patients. Immunization stimulated the development of delayed-type hypersensitivity reactions to irradiated, dissociated, autologous, non-transduced tumour cells in 17 of 25 patients. Metastatic lesions that were resected after vaccination showed brisk or focal T-lymphocyte and plasma cell infiltrates with tumour necrosis in 10 of 16 patients. One complete, one partial, and one mixed response were noted. Ten patients (29 per cent) are alive, with a minimum follow-up of 36 months; four of these patients have no evidence of disease. It was concluded that vaccination with irradiated, autologous melanoma cells engineered to secrete GM-CSF by adenoviral-mediated gene transfer augments antitumour immunity in patients with metastatic melanoma.

1.7 Adenoviral vectors for suicide gene therapy

Conventional chemotherapeutic approaches to the treatment of cancer are non-selective and therefore cause toxicity in normal tissue as well as malignant tissue. Suicide gene therapy aims to achieve a high degree of selectivity through the use of gene-directed enzyme prodrug therapy (GDEPT) or GPAT (genetic prodrug activation therapy) (Niculescu-Duvaz *et al.*, 1998; Springer and Niculescu-Duvaz, 2000). This therapy involves a two-step treatment for solid tumours. First, a gene encoding a foreign enzyme is delivered to the tumour for expression. An inactive prodrug is then administered which becomes activated into a cytotoxic drug on encountering the foreign enzyme. As expression of the activating enzyme will not occur in every cell it is beneficial for the cytotoxic drug to exhibit a bystander effect, whereby it leaks out of the tumour cells to surrounding tumour cells not expressing the enzyme.

Studies using animal models have shown that adenoviral delivery of the herpes simplex virus thymidine kinase (HSV-tk) gene, which activates the prodrug ganciclovir, was one of the most successful approaches in treating experimental brain tumours (Chen *et al.*, 1994; Lanuti *et al.*, 1999).

There are several clinical trials which have tested the efficacy of suicide gene therapy in patients (Table 1.3). A recent phase I trial studied the adenoviral delivery of the HSV-tk gene together with administration of ganciclovir into 13 patients with advanced recurrent malignant brain tumours (Trask *et al.*, 2000). The study's main objective was to determine the safety of the treatment. Patients were injected intratumorally with a replication defective adenoviral vector expressing HSV-tk from the Rous sarcoma promoter (Adv.RSVtk). Vector concentrations used were either 2×10^9, 2×10^{10}, 2×10^{11} or 2×10^{12} virus particles per injection, followed by ganciclovir treatment. Patients tolerated doses of 2×10^{11} vector particles and below but patients treated with 2×10^{12} vector particles exhibited central nervous system toxicity with confusion, hyponatremia and seizures. One patient was still alive 29.2 months after the treatment. Two patients survived for greater than 25 months before succumbing to tumour progression. However, 10 patients died within 10 months of treatment, 9 from tumour progression and 1 with sepsis and endocarditis. A study carried out by Shalev *et al* (2000) found no toxicity after direct and repeated injection into the prostate of a replication defective adenovirus containing HSV-tk followed by ganciclovir. However, unlike the previous study the total amount of virus administered was 1×10^{10} IU in either one injection or as repeated injections with less virus.

1.8 Adenoviral vectors for gene replacement therapy

The role of p53 as a central mediator of the damage and cellular stress responses in the cell is well established (Fridman and Lowe, 2003). One of the most important functions of p53 is its ability to activate apoptosis on encountering DNA damage. Therefore disruption of this vital gene promotes tumour progression and desensitizes the tumour to both chemo- and radiotherapy (El-Deiry, 2003). The p53 gene is mutated in most human cancers and therefore represents an ideal target for gene replacement therapy. Preclinical studies have demonstrated that transient expression of a single potent tumour suppressor gene such as p53 is sufficient to mediate a therapeutic effect. Indeed, the transfer of a functional copy of the p53 gene into tumour cells is one of the most common strategies currently being evaluated in clinical trials using adenoviral vectors and is the predominant target in current phase III trials.

In a phase I trial conducted by Roth *et al.* (1998) administration of an adenoviral p53 vector (Adp53) to 21 patients with advanced non-small cell lung cancer resulted in little toxicity. The patients were given up to six intratumoral injections at monthly intervals which were well tolerated. Expression of the p53 gene was observed together with potentially useful clinical responses. Another phase I trial was conducted with an adenoviral vector expressing p53 (INGN201) in combination with cisplatin for the treatment of non-small cell lung cancer (Nemunaitis *et al.*, 2000b). Twenty-four patients (median age 64 years) received a total of 83 intratumoral injections with Adp53. The maximum dose administered was 1×10^{11} p.f.u. per dose. Transient fever related to Adp53 injection developed in eight patients. Seventeen patients achieved a best clinical response of stable disease, two patients achieved a partial response, four patients had progressive disease and one patient was not assessable. A phase II study evaluated the effect of INGN201 plus radiation on non-small cell lung carcinoma patients (Swisher *et al.*, 2003). Nineteen patients with non metastatic non-small cell lung cancer were treated with radiation therapy to 60 Gy over 6 weeks together with three intratumoral injections of Ad-p53 (INGN201). The most common adverse side effects were grade 1 or 2 fevers (79 per cent) and chills (53 per cent). Computed tomography and bronchoscopic findings at the primary injected tumour revealed complete response (1 of 19, 16 per cent), partial response (11 of 19, 58 per cent), stable disease (3 of 19, 16 per cent), progressive diseases (2of 19, 11 per cent) and not evaluable (2 of 19, 11 per cent). It seems that tumour cells expressing a functional p53 are more sensitive to chemotherapy and radiation than those lacking the gene. This heightened sensitivity is likely due to the ability of cells containing a functional p53 to undergo apoptosis more readily (Lowe 1997). The synergy between chemotherapy and radiation and gene therapy is also likely due to the fact that chemotherapy enhances expression of transgenes from adenoviral vectors with a wide range of promoters whilst radiation has been shown to improve transduction and duration of transgene expression (Stevens *et al.*, 1996). In addition to non-small cell lung cancer, clinical studies using INGN201 have been initiated in seven other tumour types (Table 1.4; Merritt *et al.*, 2001). Over 500 patients have been evaluated in these studies with seven different routes of administration. INGN201 has been well tolerated in all phase I and II studies completed by 2001. The majority of these patients received multiple intratumoral injections up to a dose of 2×10^{12} viral particles per injection which is the dose now being used in phase III studies. No toxic deaths were observed and the only adverse effects observed were fever in 60 per cent of patients and pain at the site of injection. Phase III trials are now underway using INGN201 with chemotherapy for the treatment of head and neck cancers and INGN201 together with chemoradiation therapy for non-small cell lung cancer with the primary goals being tumour free survival or at least tumour control with an impact on overall survival.

1.9 Oncolytic adenoviral therapy

Replication-selective oncolytic viruses (virotherapy) represent a novel and unique approach to the treatment of cancer (Wildner, 2003). Lytic viruses have evolved to infect cells, replicate, induce cell death, release viral particles and spread to surrounding tissue. Selective replication of the viruses within tumour tissue could increase the therapeutic index of these agents dramatically. In addition the fact that oncolytic viruses do not always induce cell death via classical apoptotic pathways makes the likelihood of cross-resistance with standard regimens such as chemo- or radiotherapy much less likely. Over the past decade advances in molecular biology have engineered these viruses to enhance their safety and antitumour potency.

Adenoviruses mediate cell death via several mechanisms. Viral proteins expressed late in the course of the viral lifecycle are directly cytotoxic. These include the E3 11.6 kDa adenovirus death protein (Tollefson *et al.*, 1996b) and E4ORF4 (Branton and Roopchand, 2001). Deletion of these gene products results in a significant delay in cell death. Expression of the E1A protein early in the adenovirus lifecycle makes the cells more refractory to killing via tumour necrosis factor (TNF). This effect is inhibited by the E3 proteins 10.4, 14.5 and 14.7. Deletion of these three E3 proteins leads to an increased TNF expression *in vivo* and enhanced cell sensitivity to TNF (Sparer *et al.*, 1996).

There are currently two main approaches to achieving tumour selective adenoviral replication. The first is via the use of tumour specific promoters, which are used to drive the expression of the E1A gene in tumour cells alone. E1A functions to stimulate S phase and to stimulate both viral and cellular genes that are critical for efficient viral replication (Whyte *et al.*, 1988). This approach has been studied in a phase I clinical trial which used the PSA promoter to drive the expression of the E1A gene in patients with locally-recurrent prostate carcinoma. This virus was termed CN706 (Calydon Pharmaceuticals, CA, USA) and was injected directly into the tumour. Similar approaches have been used by other groups to achieve selective replication in other tumour types including using the promoters from alphafetoprotein, carcinoembryonic antigen and MUC-1 (Hallenbeck *et al.*, 1999; Kurihara *et al.*, 2000)

One of the first clinical trials demonstrating antitumour efficacy in a specific cancer used a replication-conditional adenovirus. This virus, dl1520 also known as ONYX-015, is defective in the early regulator protein E1B which binds to and inactivates p53 to promote its own activation (Barker *et al.*, 2003). In normal cells p53 inactivates adenoviral replication but the exact mechanism by which it does this is still not clear. This mutated virus can infect and replicate in cells defective in p53 as well as cells with loss of p14ARF function (a protein that can mediate apoptosis by activation of p53). However, it cannot replicate in normal cells carrying wild-type p53 and an intact p53 pathway (Vollmer *et al.*, 1999; Lowe 1997). To date phase I and II trials have been conducted with virus alone or in combination with chemotherapy. dl1520 has been well tolerated at the highest practical doses that could be administered (2×10^{12}–2×10^{13}

particles) by intratumoral, intraperitoneal, intra-arterial and intravenous routes. Flu-like symptoms were the most common toxicities and were increased in patients receiving intravascular treatment (Ganly *et al.*, 2000). Two phase II trials enrolled a total of 40 patients with head and neck cancer (Nemunaitis *et al.*, 2000a). Despite a fairly aggressive injection regimen of six to eight daily needle passes for 5 consecutive days no objective responses were documented. Similarly no objective responses were noted in phase I or I/II trials in patients with pancreatic, colorectal or ovarian carcinomas (Mulvihill *et al.*, 2001). As a result combinations with chemotherapy were explored. Evidence for a potentially-synergistic interaction between oncolytic adenoviral therapy and chemotherapy has been obtained in multiple trials. Encouraging clinical data has been achieved in patients with recurrent head and neck cancer treated with intratumoral ONXY-015 in combination with cisplatin and 5-fluorouracil (Khuri *et al.*, 2000). Out of 30 patients treated an objective response (at least 50 per cent reduction in tumour size) was observed in 19 patients and a complete response was seen in 8 patients. Tumours as large as 10 cm regressed completely and none of the tumours that responded had progressed after a mean follow up of 5 months. Another phase I/II study used ONYX-015 in combination with fluorouracil to treat unresectable primary and secondary liver tumors. However only limited clinical response was seen (Habib *et al.*, 2000). One reason why the ONYX-015 vector has limited efficacy in some studies could be the lack of the CAR on the surface of target tumour cells preventing intratumoral spread (Douglas *et al.*, 2001).

1.10 Adverse outcomes of adenoviral gene therapy

To date clinical studies using adenoviral vectors have been associated with little toxicity and few serious side effects. However, in September 1999 an 18-year-old patient at the University of Pennsylvania with ornithine–cytosine transferase deficiency died as a direct consequence of adenoviral gene therapy (Lehrman, 1999). A first-generation replication defective adenovirus expressing the ornithine–cytosine transferase enzyme was administered through the hepatic artery. However, the viral titre used was very high (1×10^{14} virus particles per kg) and this led to systemic activation of the innate immune response mediated by antigen-presenting cells and macrophages leading to the release of cytokines. The patient developed a fever within 2 h, which was quickly followed by signs of liver dysfunction. Ammonia accumulated in the blood, followed by multi-organ failure and adult respiratory distress syndrome. This incident was the first death in 10 years of gene therapy clinical trials involving more than 3500 patients but still led many people to question the safety of gene therapy and prompted further considerations of treatment strategies. One of the very important aspects of adenoviral production, which is often underestimated, is virus quantitation. This is even more important in the case of products with an end use in clinical applications. There are currently several methods by which adenovirus can be quantitated. These include the measure of total viral particles, the measure of infectious units or titre and the measure of replication competent adenovirus by plaque assay to give p.f.u.. It is important that in addition to p.f.u. virus is measured in viral load which is 10 to 100 times higher. For example, 1×10^{11} p.f.u./kg is equivalent to 1×10^{14} viral particles/kg. As a consequence NIH and FDA are providing new guidelines and regulations. Vector manufacturing and clinical gene therapy protocols will have to meet new standards to improve the quality and safety of clinical trials (FDA/CBER March 6, 2000 letter: www.fda.gov/cber/ltr/gt030600.htm).

1.11 Summary

To date adenoviral vectors remain the gene transfer vehicles of choice. They are easy to manipulate, infect a broad range of human cells and are highly efficient in gene transfer compared to other vectors. Phase I clinical trials have demonstrated little toxicity and shown the approach to be generally safe. However, clinical efficacy has only so far been shown with replication competent adenoviruses. We therefore await the results of current phase III trials. Despite major advances in the field in the last 10 years there is still room for improvement. In addition to improving selective targeting and reducing immune stimulation the large scale production of adenoviral vectors for clinical trials is an area that also requires further research (Nadeau and Kamen, 2003). Areas include optimizing production conditions (suspension cultures seem to be preferred to adherent cultures), use of serum-free medium and accurate quantitation of viral particles will all improve the quality of the vectors used. If these problems are addressed and accurate and efficient clinical trials are conducted then adenoviral gene therapy for cancer represents a promising alternative to current treatment regimens.

References

Anderson CW, Young ME, Flint SJ (1989). Characterization of the adenovirus 2 virion protein, mu. *Virology* **172**, 506–12.

Barker DD, Berk AJ (1987). Adenovirus proteins from both E1B reading frames are required for transformation of rodent cells by viral infection and DNA transfection. *Virology* **156**, 107–21.

Barker SD, Dmitriev IP, Nettelbeck DM, *et al.* (2003). Combined transcriptional and transductional targeting improves the specificity and efficacy of adenoviral gene delivery to ovarian carcinoma. *Gene Ther* **10**, 1198–204.

Barnett BG, Crews CJ, Douglas JT (2002). Targeted adenoviral vectors. *Biochim Biophys Acta* **1575**, 1–14.

Bennett EM, Bennink JR, Yewdell JW, Brodsky FM (1999). Cutting edge: adenovirus E19 has two mechanisms for affecting class I MHC expression. *J Immunol.* **162**, 5049–52.

Bergelson JM, Cunningham JA, Droguett G, *et al.* (1997). Isolation of a common receptor for Coxsackie B viruses and adenoviruses 2 and 5. *Science* **275**, 1320–3.

Brand K, Arnold W, Bartels T, *et al.* (1997). Liver-associated toxicity of the HSV-tk/GCV approach and adenoviral vectors. *Cancer Gene Ther* **4**, 9–16.

Branton PE, Roopchand PE (2001). The role of adenovirus E4orf4 protein in viral replication and cell killing. *Oncogene.* **26**, 20(54):7855–7865.

Brehm A, Miska EA, McCance DJ, Reid JL, Bannister AJ, Kouzarides T (1998). Retinoblastoma protein recruits histone deacetylase to repress transcription. *Nature* **391**, 597–601.

Bruder JT, Jie T, McVey DL, Kovesdi I (1997). Expression of gp19K increases the persistence of transgene expression from an adenovirus vector in the mouse lung and liver. *J Virol* **71**, 7623–8.

Buskens CJ, Marsman WA, Wesseling JG, *et al.* (2003). A genetically retargeted adenoviral vector enhances viral transduction in esophageal carcinoma cell lines and primary cultured esophageal resection specimens. *Ann Surg* **238**, 815–24; discussion 825–826.

Chakravarti D, Ogryzko V, Kao HY, *et al.* (1999). A viral mechanism for inhibition of p300 and PCAF acetyltransferase activity. *Cell* **96**, 393–403.

Chattopadhyay D, Ghosh MK, Mal A. Harter ML (2001). Inactivation of p21 by E1A leads to the induction of apoptosis in DNA-damaged cells. *J Virol* **75**, 9844–56.

Chen SH, Shine HD, Goodman JC, Grossman RG, Woo SL (1994). Gene therapy for brain tumors: regression of experimental gliomas by adenovirus-mediated gene transfer *in vivo. Proc Natl Acad Sci U S A* **91**, 3054–3057.

de Jong RN, van der Vliet PC, Brenkman AB (2003). Adenovirus DNA replication: protein priming, jumping back and the role of the DNA binding protein DBP. *Curr Top Microbiol Immunol* **272**, 187–211.

Desterro JM, Rodriguez MS, Kemp GD (1999). Hay RT Identification of the enzyme required for activation of the small ubiquitin-like protein SUMO-1. *J Biol Chem* **274**, 10618–24.

Douglas JT, Kim M, Sumerel LA, Carey DE, Curiel DT (2001). Efficient oncolysis by a replicating adenovirus (ad) *in vivo* is critically dependent on tumor expression of primary ad receptors. *Cancer Res* **61**, 813–7.

El-Deiry WS (2003). The role of p53 in chemosensitivity and radiosensitivity. *Oncogene* **22**, 7486–7495.

Faha B, Harlow E, Lees E (1993). The adenovirus E1A-associated kinase consists of cyclin E-p33cdk2 and cyclin A-p33cdk2. *J Virol* **67**, 2456–65.

Flinterman M, Guelen L, Ezzati-Nik S, *et al.* (2005). E1A activates transcription of p73 and Noxa to induce apoptosis. *J Biol Chem* **280**, 5945–59. Epub 2004 Nov 29.

Fridman JS, Lowe SW (2003). Control of apoptosis by p53. *Oncogene* **22**, 9030–9040.

Frisch SM, Mymryk JS (2002). Adenovirus-5 E1A: paradox and paradigm. *Nat Rev Mol Cell Biol* **3**, 441–452.

Fuchs M, Gerber J, Drapkin R, *et al.* (2001). The p400 complex is an essential E1A transformation target. *Cell* **106**, 297–307.

Fujiwara T, Kataoka M, Kawamata O, Tanaka N (1999). [A phase I trial of adenoviral p53 gene therapy for non-small cell lung cancer]. *Nippon Geka Gakkai Zasshi* **100**, 749–755.

Ganly I, Kirn D, Eckhardt G, *et al.* (2000). A phase I study of Onyx-015, an E1B attenuated adenovirus, administered intratumorally to patients with recurrent head and neck cancer. *Clin Cancer Res* **6**, 798–806.

Grand RJ, Turnell AS, Mason GG, *et al.* (1999). Adenovirus early region 1A protein binds to mammalian SUG1-a regulatory component of the proteasome. *Oncogene* **18**, 449–58.

Greber UF, Webster P, Weber J, Helenius A (1996). The role of the adenovirus protease on virus entry into cells. *EMBO J* **15**, 1766–1777.

Griscelli F, Opolon P, Saulnier P, *et al.* (2003). Recombinant adenovirus shedding after intratumoral gene transfer in lung cancer patients. *Gene Ther* **10**, 386–395.

Habib NA, Sarraf CE, Mitry RR, *et al.* (2001). E1B-deleted adenovirus (dl1520) gene therapy for patients with primary and secondary liver tumors. *Hum Gene Ther* **12**, 219–226.

Halbert DN, Cutt JR, Shenk T (1985). Adenovirus early region 4 encodes functions required for efficient DNA replication, late gene expression, and host cell shutoff. *J Virol* **56**, 250–7.

Hale TK, Braithwaite AW (1999). The adenovirus oncoprotein E1a stimulates binding of transcription factor ETF to transcriptionally activate the p53 gene. *J Biol Chem* **274**, 23777–23786.

Hallenbeck PL, Chang YN, Hay C, *et al.* (1999). A novel tumor-specific replication-restricted adenoviral vector for gene therapy of hepatocellular carcinoma. *Hum Gene Ther* **10**, 1721–1733.

Han J, Sabbatini P, Perez D, Rao L, Modha D, White E (1996). The E1B 19K protein blocks apoptosis by interacting with and inhibiting the p53-inducible and death-promoting Bax protein. *Genes Dev* **10**, 461–477.

Hasenburg A, Fischer DC, Tong XW, *et al.* (2002). Adenovirus-mediated thymidine kinase gene therapy for recurrent ovarian cancer: expression of coxsackie-adenovirus receptor and integrins alphavbeta3 and alphavbeta5. *J Soc Gynecol Investig* **9**, 174–180.

Haviv YS, Curiel DT (2001). Conditional gene targeting for cancer gene therapy. *Adv Drug Deliv Rev* **53**, 135–54.

Hay RT, Freeman A, Leith I, Monaghan A, Webster A (1995). Molecular interactions during adenovirus DNA replication. *Curr Top Microbiol Immunol.* **199**, 31–48.

Hilleman MR, Werner JH (1954). Recovery of new agent from patients with acute respiratory illness. *Proc Soc Exp Biol Med* **85**, 183–188.

Huang S, Kamata T, Takada Y, Ruggeri ZM, Nemerow GR (1996). Adenovirus interaction with distinct integrins mediates separate events in cell entry and gene delivery to hematopoietic cells. *J Virol* **70**, 4502–4508.

Kasono K, Blackwell JL, Douglas JT, *et al.* (1999). Selective gene delivery to head and neck cancer cells via an integrin targeted adenoviral vector. *Clin Cancer Res* **5**, 2571–2579.

Kauczor HU, Schuler M, Heussel CP, *et al.* (1999). CT-guided intratumoral gene therapy in non-small-cell lung cancer. *Eur Radiol* **9**, 292–6.

Kay MA, Holterman AX, Meuse L, *et al.* (1995). Long-term hepatic adenovirus-mediated gene expression in mice following CTLA4Ig administration. *Nat Genet* **11**, 191–197.

Khuri FR, Nemunaitis J, Ganly I, *et al.* (2000). A controlled trial of intratumoral ONYX-015, a selectively-replicating adenovirus, in combination with cisplatin and 5-fluorouracil in patients with recurrent head and neck cancer. *Nat Med* **6**, 879–885.

Kurihara T, Brough DE, Kovesdi I and Kufe DW (2000). Selectivity of a replication-competent adenovirus for human breast carcinoma cells expressing the MUC1 antigen. *J Clin Invest* **106**, 763–771.

Lang SE, Hearing P (2003). The adenovirus E1A oncoprotein recruits the cellular TRRAP/GCN5 histone acetyltransferase complex. *Oncogene* **22**, 2836–41.

Lanuti M, Gao GP, Force SD, *et al.* (1999). Evaluation of an E1E4-deleted adenovirus expressing the herpes simplex thymidine kinase suicide gene in cancer gene therapy. *Hum Gene Ther* **10**, 463–475.

Ledl A, Schmidt D, Muller S (2005). Viral oncoproteins E1A and E7 and cellular LxCxE proteins repress SUMO modification of the retinoblastoma tumor suppressor. *Oncogene* **24**, 3810–3818.

Lehrman S (1999). Virus treatment questioned after gene therapy death. *Nature* **401**, 517–518.

Li Z, Day CP, Yang JY, *et al.* (2004). Adenoviral E1A targets Mdm4 to stabilize tumor suppressor p53. *Cancer Res* **64**, 9080–5.

Lochmuller H, Jani A, Huard J, *et al.* (1994).Emergence of early region 1-containing replication-competent adenovirus in stocks of replication-defective adenovirus recombinants (delta E1 + delta E3) during multiple passages in 293 cells. *Hum Gene Ther* **5**, 1485–1491.

Lowe SW (1997). Progress of the smart bomb cancer virus. *Nat Med* **3**, 606–608.

Lusky M, Christ M, Rittner K, *et al.* (1998).*In vitro* and *in vivo* biology of recombinant adenovirus vectors with E1, E1/E2A, or E1/E4 deleted. *J Virol* **72**, 2022–2032.

Martuza RL (2000). Conditionally replicating herpes vectors for cancer therapy. *J Clin Invest* **105**, 841–846.

Matthews DA, Russell WC (1995). Adenovirus protein-protein interactions: molecular parameters governing the binding of protein VI to hexon and the activation of the adenovirus 23K protease. *J Gen Virol* **76**, 1959–1969.

Mazzarelli JM, Mengus G, Davidson I, Ricciardi RP (1997). The transactivation domain of adenovirus E1A interacts with the C terminus of human TAF(II)135. *J Virol.* **71**(10):7978–7983.

McConnell MJ, Imperiale MJ (2004). Biology of adenovirus and its use as a vector for gene therapy. *Hum Gene Ther* **15**, 1022–1033.

Mentel R, Dopping G, Wegner U, Seidel W, Liebermann H, Dohner L (1997). Adenovirus-receptor interaction with human lymphocytes. *J Med Virol* **51**, 252–257.

Merritt JA, Roth JA, Logothetis CJ (2001). Clinical evaluation of adenoviral-mediated p53 gene transfer: review of INGN 201 studies. *Semin Oncol* **28**, 105–114.

Mizuguchi H, Hayakawa T (2004). Targeted adenovirus vectors. *Hum Gene Ther* **15**, 1034–1044.

Mulvihill S, Warren R, Venook A, *et al.* (2001). Safety and feasibility of injection with an E1B-55 kDa gene-deleted, replication-selective adenovirus (ONYX-015) into primary carcinomas of the pancreas: a phase I trial. *Gene Ther* **8**(4), 308–315.

Nadeau I, Kamen A (2003). Production of adenovirus vector for gene therapy. *Biotechnol Adv* **20**, 475–489.

Nemunaitis J, Ganly I, Khuri F, *et al.* (2000a). Selective replication and oncolysis in p53 mutant tumors with ONYX-015, an E1B-55kD gene-deleted adenovirus, in patients with advanced head and neck cancer: a phase II trial. *Cancer Res.* **60**, 6359–6366.

Nemunaitis J, Swisher SG, Timmons T, *et al.* (2000b). Adenovirus-mediated p53 gene transfer in sequence with cisplatin to tumors of patients with non-small-cell lung cancer. *J Clin Oncol* **18**, 609-622.

Nettelbeck DM, Rivera AA, Davydova J, Dieckmann D, Yamamoto M, Curiel DT (2003). Cyclooxygenase-2 promoter for tumour-specific targeting of adenoviral vectors to melanoma. *Melanoma Res* **13**, 287–292.

Nettelbeck DM, Rivera AA, Kupsch J, *et al.* (2004). Retargeting of adenoviral infection to melanoma: combining genetic ablation of native tropism with a recombinant bispecific single-chain diabody (scDb) adapter that binds to fiber knob and HMWMAA. *Int J Cancer* **108**, 136–145.

Niculescu-Duvaz I, Spooner R, Marais R, Springer CJ (1998). Gene-directed enzyme prodrug therapy. *Bioconjug Chem* **9**, 4–22.

Park HS, Cheon J, Cho HY, *et al.* (2003). *In vivo* characterization of a prostate-specific antigen promoter-based suicide gene therapy for the treatment of benign prostatic hyperplasia. *Gene Ther* **10**, 1129–1134.

Parks RJ, Chen L, Anton M, Sankar U, Rudnicki MA, Graham FL (1996). A helper-dependent adenovirus vector system: removal of helper virus by Cre-mediated excision of the viral packaging signal. *Proc Natl Acad Sci U S A* **93**, 13565–13570.

Pilder S, Moore M, Logan J, Shenk T (1986). The adenovirus E1B-55K transforming polypeptide modulates transport or cytoplasmic stabilization of viral and host cell mRNAs. *Mol Cell Biol* **6**, 470–476.

Reich N, Pine R, Levy D, Darnell JE Jr (1988). Transcription of interferon-stimulated genes is induced by adenovirus particles but is suppressed by E1A gene products. *J Virol* **62**, 114–119.

Rekosh DM, Russell WC, Bellet AJ, Robinson AJ (1977). Identification of a protein linked to the ends of adenovirus DNA. *Cell* **11**, 283–295.

Roth JA, Swisher SG, Merritt JA, *et al.* (1998). Gene therapy for non-small cell lung cancer: a preliminary report of a phase I trial of adenoviral p53 gene replacement. *Semin Oncol* **25**, 33–37.

Rots MG, Curiel DT, Gerritsen WR, Haisma HJ (2003). Targeted cancer gene therapy: the flexibility of adenoviral gene therapy vectors. *J Control Release* **87**, 159–165.

Shalev M, Kadmon D, Teh BS, *et al.* (2000).Suicide gene therapy toxicity after multiple and repeat injections in patients with localized prostate cancer. *J Urol* **163**, 1747–1750.

Shenk T (1996). *Adenoviridae: The Viruses and their Replication.* Lippincott-Raven, Philadelphia, pp 2111–2148.

Shisler J, Duerksen-Hughes P, Hermiston TM, Wold WS, Gooding LR (1996). Induction of susceptibility to tumor necrosis factor by E1A is dependent on binding to either p300 or p105-Rb and induction of DNA synthesis. *J Virol* **70**, 68–77.

Soiffer R, Hodi FS, Haluska F, *et al.* (2003).Vaccination with irradiated, autologous melanoma cells engineered to secrete granulocyte–macrophage colony-stimulating factor by adenoviral-mediated gene transfer augments antitumor immunity in patients with metastatic melanoma. *J Clin Oncol* **21**, 3343–3350.

Sparer TE, Tripp RA, Dillehay DL, Hermiston TW, Wold WS, Gooding LR (1996). The role of human adenovirus early region 3 proteins (gp19K, 10.4K, 14.5K, and 14.7K) in a murine pneumonia model. *J Virol* **70**, 2431–2439.

Springer CJ, Niculescu-Duvaz I (2000). Prodrug-activating systems in suicide gene therapy. *J Clin Invest* **105**, 1161–1167.

Stevens CW, Zeng M, Cerniglia GJ (1996). Ionizing radiation greatly improves gene transfer efficiency in mammalian cells. *Hum Gene Ther* **7**, 1727-1734.

Stevens JL, Cantin GT, Wang G, Shevchenko A, Berk AJ (2002). Transcription control by E1A and MAP kinase pathway via Sur2 mediator subunit. *Science* **296**, 755–758. Epub 2002 Apr 4.

Stewart AK, Lassam NJ, Quirt IC, *et al.* (1999). Adenovector-mediated gene delivery of interleukin-2 in metastatic breast cancer and melanoma: results of a phase I clinical trial. *Gene Ther* **6**, 350–363.

Stewart PL, Fuller SD, Burnett RM (1993). Difference imaging of adenovirus: bridging the resolution gap between X-ray crystallography and electron microscopy. *EMBO J* **12**, 2589–2599.

Swisher SG, Roth JA, Komaki R, *et al.* (2003). Induction of p53-regulated genes and tumor regression in lung cancer patients after intratumoral delivery of adenoviral p53 (INGN 201) and radiation therapy. *Clin Cancer Res* **9**, 93–101.

Tollefson AE, Ryerse JS, Scaria A, Hermiston TW, Wold WS (1996a). The E3–11.6-kDa adenovirus death protein (ADP) is required for efficient cell death: characterization of cells infected with adp mutants. *Virology* **220**, 152–162.

Tollefson AE, Scaria A, Hermiston TW, Ryerse JS, Wold LJ, Wold WS (1996b). The adenovirus death protein (E3–11.6K) is required at very late stages of infection for efficient cell lysis and release of adenovirus from infected cells. *J Virol* **70**, 2296–2306.

Tomko RP, Xu R, Philipson L (1997). HCAR and MCAR: the human and mouse cellular receptors for subgroup C adenoviruses and group B coxsackieviruses. *Proc Natl Acad Sci U S A* **94**, 3352–3356.

Trask TW, Trask RP, Aguilar-Cordova E, *et al.* (2000). Phase I study of adenoviral delivery of the HSV-tk gene and ganciclovir administration in patients with current malignant brain tumors. *Mol Ther* **1**, 195-203.

Vanderkwaak TJ, Wang M, Gomez-Navarro J, *et al.* (1999). An advanced generation of adenoviral vectors selectively enhances gene transfer for ovarian cancer gene therapy approaches. *Gynecol Oncol* **74**, 227–234.

Vasey PA, Shulman LN, Campos S, *et al.* (2002).Phase I trial of intraperitoneal injection of the E1B-55-kd-gene-deleted adenovirus ONYX-015 (dl1520) given on days 1 through 5 every 3 weeks in patients with recurrent/refractory epithelial ovarian cancer. *J Clin Oncol* **20**, 1562–1569.

Vollmer CM, Ribas A, Butterfield LH, *et al.* (1999). p53 selective and nonselective replication of an E1B-deleted adenovirus in hepatocellular carcinoma. *Cancer Res* **59**, 4369–4374.

Wang Y, Hu JK, Krol A, Li YP, Li CY, Yuan F (2003). Systemic dissemination of viral vectors during intratumoral injection. *Mol Cancer Ther* **2**, 1233–1242.

Webster A, Russell S, Talbot P, Russell WC, Kemp GD (1989). Characterization of the adenovirus proteinase: substrate specificity. *J Gen Virol* **70**, 3225–3234.

Whyte P, Ruley HE, Harlow E (1988). Two regions of the adenovirus early region 1A proteins are required for transformation. *J Virol* **62**, 257–265.

Wickham TJ, Mathias P, Cheresh DA, Nemerow GR (1993). Integrins alpha v beta 3 and alpha v beta 5 promote adenovirus internalization but not virus attachment. *Cell* **73**, 309–319.

Wildner O (2003). Comparison of replication selective oncolytic viruses for the treatment of human cancers. *Curr Opin Mol Ther* **5**(4), 351–361

Wold WM, Chinnadurai G (2000). *Adenovirus Proteins that Regulate Apoptosis.* Oxford University Press, Oxford, pp 200–232.

Wu N, Ataai MM (2000). Production of viral vectors for gene therapy applications. *Curr Opin Biotechnol* **11**, 205–208.

Yamamoto M, Alemany R, Adachi Y, Grizzle WE, Curiel DT (2001). Characterization of the cyclooxygenase-2 promoter in an adenoviral vector and its application for the mitigation of toxicity in suicide gene therapy of gastrointestinal cancers. *Mol Ther* **3**, 385–394.

Yang Y, Li Q, Ertl HC, Wilson JM (1995). Cellular and humoral immune responses to viral antigens create barriers to lung-directed gene therapy with recombinant adenoviruses. *J Virol* **69**, 2004–2015.

Yang Y, Wilson JM (1995). Clearance of adenovirus-infected hepatocytes by MHC class I-restricted CD4+ CTLs *in vivo*. *J Immunol* **155**, 2564–2570.

Yee D, McGuire SE, Brunner N, *et al.* (1996). Adenovirus-mediated gene transfer of herpes simplex virus thymidine kinase in an ascites model of human breast cancer. *Hum Gene Ther* **7**, 1251–1257.

2

Application of HSV-1 Vectors to the Treatment of Cancer

Paola Grandi, Kiflai Bein, Costas G. Hadjipanayis, Darren Wolfe, Xandra O. Breakefield and **Joseph C. Glorioso**

2.1 Introduction

Cancer remains one of the most important problems in human health. Advances in understanding the molecular bases of cancer and methods for early detection have greatly enhanced opportunities for therapeutic intervention. While some human tumours are now effectively treated by anti-cancer drugs, radiation and/or surgical methods, many tumour types remain unresponsive. Gene therapy should contribute to improved outcomes. Viruses deliver genes efficiently and considerable efforts have gone into the development of safe and effective viral vectors potentially useful as 'anti-cancer drugs'. Among these vectors, herpes simplex virus (HSV) has a number of biological features that support its utility for cancer treatment. Although significant hurdles remain, encouraging results from early phase clinical trials using HSV vectors suggest that advances in vector design and methods of delivery will likely provide effective therapies for certain tumour types, especially when applied in combination with currently available treatment modalities. In this review the salient features of HSV biology related to vector engineering and strategies for their use in anti-cancer therapy are described.

2.2 Basic biology of HSV

2.2.1 Introduction

HSV is a neurotropic virus that naturally occurs in humans. Nevertheless, the virus has a very broad host range and a large number of diverse animal species are susceptible to infection. This feature has allowed the development of animal models of human disease that can be tested for treatment using HSV vectors. There are two serotypes of HSV (type 1 and type 2) which have similar genome structures but differ in their prevalence for particular disease types. HSV-1 is most often associated with the common cold sore and herpes keratitis but can cause life-threatening encephalitis if infection spreads to the brain. HSV-2 has most often been associated with genital infections that can be spread to the newborn causing serious neonatal and disseminated disease. HSV-1 has been most extensively engineered as a gene transfer vector and much is known about its gene functions and molecular biology. For use in gene transfer, its most important features are its ability to infect cells with high efficiency and to deliver a large transgene 'payload'. Indeed, almost the entire genome can be replaced by non-HSV DNA and packaged into infectious particles. Of

Viral Therapy of Cancer Edited by Kevin J. Harrington, Richard G. Vile and Hardev S. Pandha

particular interest for vector engineering is the ability of HSV to persist in neurons as an episomal element which in natural infections appears to remain for the life of the host in a state of latency. HSV vectors also have a similar capability in animal models. During latent infection, there is little detectable expression of immediate early (IE), early (E) or late (L) viral proteins. Expression is limited to a set of non-translated RNA species, known as latency-associated transcripts (LATs) (Croen *et al.*, 1987; Rock *et al.*, 1987; Spivack and Fraser, 1987; Stevens *et al.*, 1987). A portion of the promoter regulating expression of LATs, LAP2, has been used for constructing HSV-1 vectors that allow long-term transgene expression in neurons (Goins *et al.*, 1999). This LAP2 element is capable of driving expression of therapeutic transgenes in both the central and peripheral nervous systems (Puskovic *et al.*, 2004; Chattopadhyay *et al.*, 2005). Although the wild type virus can reactivate from the latent state to cause recurrent disease and provide a mechanism for transmission to others by direct contact with a viral lesion, vectors are engineered to remove viral functions that allow virus growth in neurons thus blocking the potential for reactivation from latency. As a consequence, HSV vectors lose the capability to be transmitted to other hosts. For many gene therapy applications that involve nervous system disease, the vector can serve as a platform for therapeutic gene expression at the site where therapy is needed and transgene expression can be short or long term depending on the vector promoter employed (Goins *et al.*, 1999). For cancer applications, this biology is less important since the goal is to destroy tumour cells and thus targeting virus infection to the tumour becomes the paramount task. This review will describe some basic aspects of HSV biology as it relates to vector design, the types of vectors currently in use for cancer studies, and approaches to gene therapy. Although the majority of experience and greatest expectations for success utilize replication competent lytic vectors especially for treatment of brain tumours, other vector types may prove important and their testing in patients is anticipated.

2.2.2 Virus structure

HSV-1 is an enveloped double stranded DNA virus. The mature virus particle is 120–300 nm in size and it is composed of at least 34 virally encoded proteins (Homa and Brown, 1997; Mettenleiter, 2002). The structural components of the virus are: the (i) envelope, (ii) tegument, (iii) capsid and (iv) viral DNA genome (Figure 2.1A). The envelope contains a host-cell derived trilaminar lipid layer in which are embedded 10–12 glycoproteins (Spear *et al.*, 1993a and b; Steven and Spear, 1997; Kasamatsu and Nakanishi, 1998; Mettenleiter, 2002). These glycoproteins are responsible for host cell recognition and entry. Of these envelope glycoproteins, gB, gD, gH and gL are strictly required for viral infection *in vitro*, while gC, gE, gG, gI, gJ and gM are dispensable (Spear, 1993a and b; Steven and Spear, 1997). The tegument is a matrix of viral proteins that play an important role at different stages of the life cycle. The tegument contains proteins such as VP16 (virus protein 16), VP22, and virus host shut-off (vhs) function, which collectively are important for viral gene expression (Mackem and Roizman, 1982; Batterson and Roizman 1983; Campbell *et al.*, 1984), degradation of host cell mRNA (Read and Frenkel, 1983; Kwong and Frenkel 1989; Kwong *et al.*, 1988), viral particle assembly and inhibition of innate immune responses that repress virus gene expression (Smiley *et al.*, 2004). The icosahedral capsid is

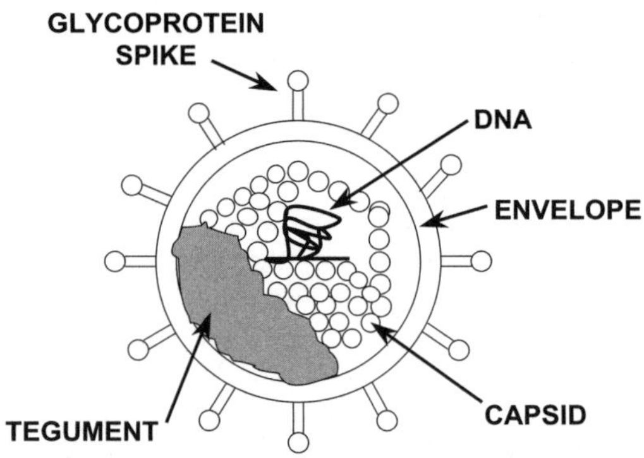

Figure 2.1 Schematic representation of HSV particle and genome. (A) Depiction of an HSV particle showing the major structural components. (B) Organization of the HSV genome. The unique long (UL) and short (US) genomic segments encode essential and accessory HSV gene products

composed of multiple structural proteins that encapsidate the viral genome (Homa and Brown, 1997; Newcomb *et al.*, 1999). The HSV-1 genome consists of a 152 kb linear double stranded DNA arranged as long and short unique segments (UL and US) each flanked by repeat sequences (ab, b′a′, ac, c′a′) (Figure 1B and Burton *et al.*, 2002).

HSV gene nomenclature is based upon the position of the gene within the long and short segments. Genes within the long segment are designated as UL1 to UL56 and genes in the short segment are designated as US1 to US12. The majority of virus genes are contiguous without introns thus facilitating their manipulation. Genes within the repeat regions are diploid in the genome and include two IE functions ICP0 (infected cell protein 0) and ICP4, a late (L) gene γ34.5 and the latency transcript gene. These genes are important to vector engineering and will be discussed below. Deletion of genes within the genomic region between the long and short segments has been used for the development of several replication defective vectors. It should be noted that the joint repeat region can be removed without preventing vector growth in cell culture since copies of these same genes are still represented at the genome ends. Joint removal prevents genome isomerization which normally occurs by recombination events between the repeats.

The HSV-1 genes can also be classified as essential or accessory to the virus life cycle *in vivo*, an important determinant of vector design. For example, the existence of two essential genes, ICP4 and glycoprotein D (US6), at the right hand end of the linear DNA provides an opportunity to readily manipulate this genomic region. Removal of an essential virus gene will prevent virus replication under any circumstances, requiring complementation of the missing function in cell lines engineered for this purpose. In contrast, the removal of accessory functions leads to virus attenuation and the ability to replicate efficiently is limited to certain cell types usually based on whether the cell is dividing or quiescent.

2.2.3 Viral infection

Initial virus binding to cell surface glycosaminoglycans (GAGs), primarily heparan sulfate (WuDunn and Spear, 1989; Fuller and Lee, 1992; Shieh *et al.*, 1992; Spear *et al.*, 1992; Gruenheid *et al.*, 1993; Herold *et al.*, 1994; Trybala *et al.*, 2002) (HS), is mediated by exposed domains of glycoproteins C (Tal-Singer *et al.*, 1995) and B (Herold *et al.*, 1995; Li *et al.*, 1995) (Figure 2.2). Together this binding represents approximately 85 per cent of the primary attachment activity with gC

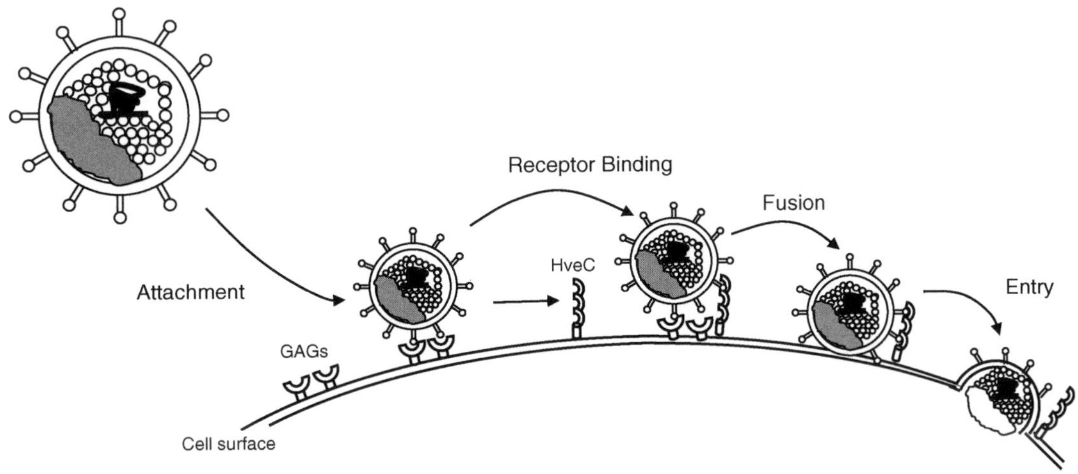

Figure 2.2 HSV cell attachment and entry. Binding of viral envelope glycoproteins to cell surface glycosaminoglycans is followed by interaction with one of several specific cell surface receptors (e.g. HveA). The molecular interactions lead to viral envelope fusion with the cell surface and HSV entry into the cell

contributing the majority of this function (Herold *et al.*, 1991; Laquerre *et al.*, 1998). Deletion of gC, and the HS binding domain of gB, in a single mutant virus (Laquerre *et al.*, 1998) impairs binding (i.e. slower kinetics) to an extent similar to the reduced binding to HS deficient cells by wild-type virus (Gruenheid *et al.*, 1993). However, virus adsorption is reduced but not eliminated in the absence of HS binding, indicating that other receptors are involved. Removal of the HS binding domain of gB does not interfere with the ability of gB to participate in virus penetration. Since GAGs are ubiquitous, it is thought that the initial binding of virus to these cell surface structures facilitates the virus envelope engagement with a specific entry receptor. Of interest is the fact that GAGs also are located in the extracellular matrix (ECM) *in vivo* and can impede extracellular virus spread in natural infections. Thus HSV spreads primarily in a cell-to-cell manner. The interaction of HSV with the ECM is an important consideration for intratumoral distribution of vector, an issue relevant to vector efficacy (discussed below). Infection of peripheral nerve terminals in the skin results in the HSV virus particle (VP26) transport along microtubules by a dynein-based molecular motor mechanism to peripheral nerve ganglion, the site of viral latency (Douglas *et al.*, 2004). This mechanism of non-replicative spread along a nerve tract provides a great advantage to the use of HSV vectors for peripheral nerve gene therapy.

Initial binding of HSV to cell surface HS is followed by gD-mediated binding to one of several principal and highly specific receptors, the essential event needed to activate the fusion mechanism for virus entry (Figure 2.2). gD-specific cognate receptors utilized for both virus attachment and penetration have been identified. The first herpesvirus entry mediator (HVEM or HveA) was identified by screening a cDNA expression library in HSV resistant CHO cells for clones that enabled virus infection. HveA was subsequently determined to be a member of the tumour necrosis factor-α (TNFα)/nerve growth factor (NGF) receptor family (Montgomery *et al.*, 1996). Domains of gD that potentially contribute to HveA binding have been identified in virus infection inhibition studies using monoclonal antibodies that recognize residues 11–19 and 222–252 (Nicola *et al.*, 1998). Consistent with these data, incubation of soluble gD with the host cell prior to infection can block infection by wild type virus, but mutants with single amino acid substitutions in the external domain at residues 25 (L25P) or 27 (Q27P, Rid1) are deficient for this blocking activity (Dean *et al.*, 1994), suggesting that these residues are crucial for binding to HveA. Deletion of residues 6–24 as well as a series of point mutations at the amino terminus of gD prevents binding and entry through HveA. In addition, mutations elsewhere in gD have been observed to impair HveA binding and entry. These results are in agreement with crystal structure data pinpointing the binding site for HveA and demonstrate that HveA binding to the N-terminus of gD results in a conformational change that may be responsible for inducing virus envelope fusion with the cell surface (Carfi *et al.*, 2001). Together, these studies suggest that HveA is recognized by a folded part of gD that can be affected by mutations spanning a substantial portion of the molecule. The consequence of the sequential attachment steps is fusion of the virus envelope with the cell surface membrane and subsequent virus entry.

A number of other gD entry mediators have been identified, including HveB (nectin-2) (Warner *et al.*, 1998), 3-*O*-sulfated HS (Shukla *et al.*, 1999), and HveC (nectin-1) (Geraghty *et al.*, 1998). HveC is a member of the immunoglobulin superfamily and has no structural relation to HveA (Geraghty *et al.*, 1998; Krummenacher *et al.*, 1998). Two potentially related but separate HveC binding sites appear to co-exist within gD. Certain substitutions within the amino terminus of gD that affect binding and entry through HveA also eliminate binding/entry via HveC. Additionally, residues 216–234 have been implicated in the HveC: gD interaction, in as much as monoclonal antibodies specific for an epitope in this region interfere with gD recognition of HveC but not HveA (Krummenacher *et al.*, 1998). Recent work has discovered the first mutants of gD that inactivate binding to HveC while preserving the recognition of HveA. Together, these specific gD mutants that eliminate entry through either HveC or HveA may prove useful in engineering gD for targeting tumours by attachment of novel ligands that recognize tumour-specific cell- surface markers (discussed below).

A role for gD in virus penetration is supported by evidence that attached virus can be neutralized by anti-gD antibody and virus mutants deleted for gD attach to cells but do not penetrate (Fuller and Spear, 1985; Highlander *et al.*, 1987; Ligas and Johnson, 1988). However, the specific binding afforded by gD could trigger penetration without direct involvement of this glycoprotein in the fusion process, a possibility supported by the finding that a soluble form of gD can induce virus entry (Cocchi *et al.*, 2004). Similarly a soluble form of the virus entry receptors can mediate virus entry supporting the notion that the interaction of gD with its receptor provides a signal to the fusion apparatus consisting of the remaining essential HSV glycoproteins, gH/gL and gB. Mutants deleted for either or all of these glycoproteins are blocked in virus penetration but are not defective in attachment. Both gB and gD have been shown to be capable of inducing syncytia if expressed on the cell surface at low pH, supporting a possible role for both molecules in fusion (Butcher *et al.*, 1990). Fuller and Lee (1992) proposed that entry involves a cascade of events in which gD initiates the fusion event where a fusion bridge is most likely formed by the action of gB followed by extension of the bridge and virus release requiring the activities of gH/gL. Recent evidence suggests that entry can be carried out in endosomes in addition to the cell surface (Nicola *et al.* 2003) but the glycoprotein requirements are the same. The key problem in designing retargeted vectors is to mutate gD to prevent recognition of the natural receptors while at the same time engineering into gD a novel ligand which upon binding to a non-HSV receptor preserves the appropriate entry mechanism.

2.2.4 Gene regulation and particle assembly

Following entry into the nucleus, the viral genes are expressed in a tightly regulated, interdependent temporal sequence (Honess and Roizman, 1974, 1975), reviewed by Roizman and Sears (1996). The first genes expressed without *de novo* viral protein synthesis are the IE genes ICP0, ICP4, ICP22, ICP27 and ICP47. The tegument protein VP16 is critical for rapid and high level expression of these genes. Upon envelope–plasmalemma fusion, VP16 enters the cytoplasm of the cell with the nucleocapsid, and it is transported to the host cell nucleus with the viral DNA genome. In its role as an activator of IE transcription, VP16 forms a multi-component complex with at least two cellular proteins, Oct-1 and HCF, which is targeted to specific upstream TAATGARAT enhancer recognition sequences in IE promoters (Wu *et al.*, 1994; Rajcani *et al.*, 2004). The highly acidic C-terminal domain of VP16 promotes transcription through recruitment of host RNA polymerase II and associated transcription initiation components. Recombinant viruses which are defective in this VP16 activity, either through mutations that impair complex formation or deletions within the activation domain, show highly reduced levels of IE transcription and significantly impaired replication (Greaves and O'Hare, 1990; Smiley and Duncan, 1997; Mossman and Smiley, 1999; Xiao *et al.*, 1994). While VP16 acts to boost the onset of lytic cycle gene expression, it is not essential for virus replication. The N-terminal sequences of VP16 are, however, absolutely required for assembly of infectious virus. VP16-null mutants can be propagated on a VP16 complementing cell line since the endogenously synthesized protein can be incorporated into the virions. In non-complementing cells there is a relatively moderate effect on gene expression and overall DNA replication, and although capsids are formed, the virions are defective for complete particle formation. As shown in Figure 2.3, IE gene expression is required for expression of early (E) genes, which primarily encode enzymes involved in DNA replication. Following viral DNA synthesis, expression of late (L) genes occurs; L genes mainly encode structural proteins for viral particle assembly (Roizman *et al.*, 2005).

The E gene products (including viral DNA polymerase, single-stranded DNA binding protein, origin binding protein, DNA primase helicase) form complexes with the parental viral DNA,

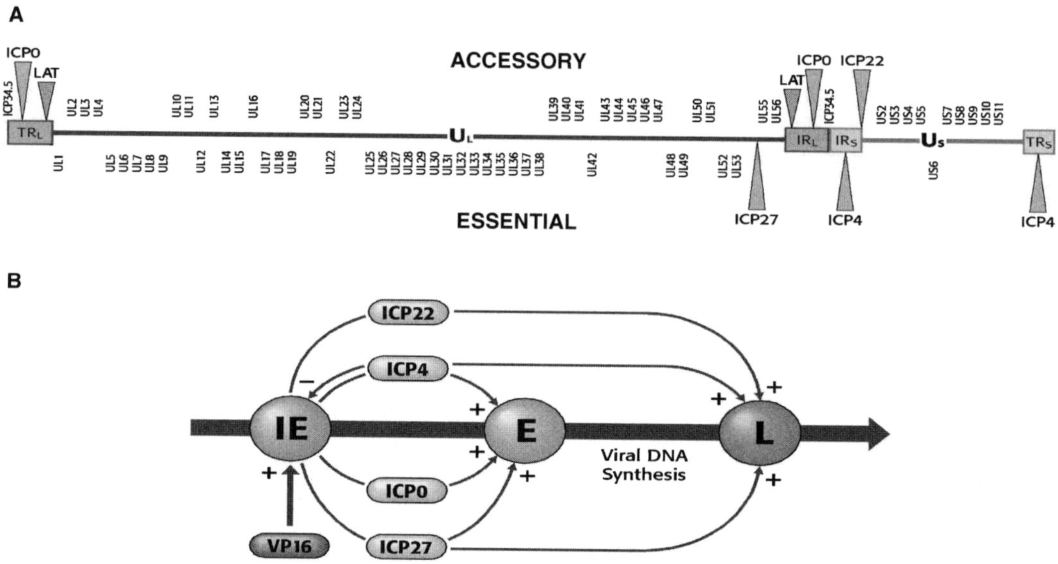

Figure 2.3 The cascade of HSV gene expression during lytic infection. Expression of HSV immediate early genes is critical for expression of early and late genes and viral assembly

and carry out replication at one or several of the three HSV origins of DNA synthesis (Mocarski and Roizman, 1982). Concomitant with initiation of DNA synthesis, L genes are expressed. Progeny DNA concatemers are cleaved into unit-length monomers and incorporated into capsids to form the nucleocapsids. Two *cis*-acting viral DNA elements, pac1 and pac2, direct precise cleavage and packaging of the HSV-1 genome into the capsid (Deiss *et al.*, 1986; Boehmer and Lehman, 1997). The nucleocapsids then mature into viral particles as they bud through the nuclear membrane with the concurrent acquisition of immature envelopes containing the viral glycoproteins. Envelopment proceeds through the Golgi and involves a membrane exchange and further glycoprotein processing prior to release from the Golgi. Just as the functional characterization of HSV genes has been instrumental to the development of replication competent and replication defective vectors, the identification of packaging signals has been crucial to the design of plasmid vectors (amplicons) that can be incorporated in mature virus particles and have played an important role in the design of vectors that are devoid of contaminating active virus (described below).

2.3 Replication competent or oncolytic vectors

2.3.1 Introduction

Oncolytic vectors are mutant viruses that replicate in tumour cells preferentially over normal cells (Figure 2.4A). Their selectivity can be substantial and is often related to whether the tumour cells are actively dividing and on the particular genetic and cellular changes that have led to tumour development. Oncolytic viruses have been used in early phase clinical trials for a variety of tumours including glioblastoma multiforme (GBM) (Ganly *et al.*, 2000; Khuri *et al.*, 2000; Markert *et al.*, 2000a; Rampling *et al.*, 2000; Papanastassiou *et al.*, 2002; Detta *et al.*, 2003; Liu *et al.*, 2003). Recent trials using genetically engineered viral strains, such as adenovirus (ONYX-015 and CV706) and HSV-1 (G207 and 1716), have been encouraging, showing these viruses to be relatively non-toxic for normal cells but remaining lytic in tumour cells. The targeting of tumour cells has not been absolute but it may be possible to improve tumour specificity by retargeting infection to tumour cell surface receptors that are more abundant in or

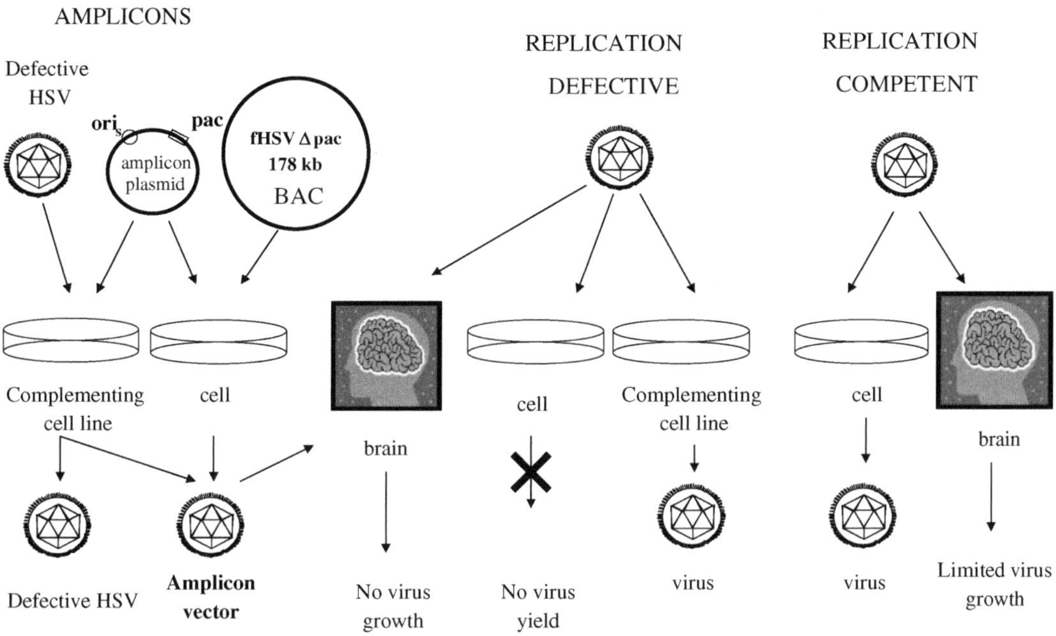

Figure 2.4 Generation of three major types of HSV vectors on the basis of the genes targeted for deletion. (A) Replication competent vectors are generated by deletion of accessory genes to replicate in tumor cells preferentially over normal cells. (B) Replication-defective vectors are generated by deletion of essential gene(s) to block virus growth. (c) Amplicon vectors are generated using plasmids bearing HSV origin of DNA replication (ori) and packaging signal (pac) to allow cloning of large transgenes

specific to the tumour cell membrane (described below).

The majority of studies using oncolytic HSV vectors involve treatment of patients with recurrent malignant gliomas. Malignant gliomas are the most common primary brain tumours and are almost universally fatal despite aggressive therapies including surgery, radiotherapy, and chemotherapy (Leibel and Sheline, 1987; Davis *et al.*, 1989; Nazzaro and Neuwelt, 1990; Quigley and Maroon, 1991; Warnick *et al.*, 1994; Prados *et al.*, 1996; Lacroix *et al.*, 2001; Selker *et al.*, 2002; Nelson *et al.*, 2003). Patients with GBM have a median survival of 12–18 months from initial diagnosis and 6–9 months after recurrence. Patients with anaplastic astrocytomas (AA) live longer with a median survival of 36–40 months after initial diagnosis and 12–18 months after recurrence. Standard management of these tumours includes biopsy and/or tumour resection, followed by external beam radiotherapy, with treatment doses of approximately 6000 cGy. The combination of radiotherapy and chemotherapy (temozolomide), chemoradiation, has significantly improved progression-free and overall survival in GBM patients (Stupp *et al.*, 2004). Partial responses to chemotherapy are seen in approximately 30 per cent of tumours, but no significant change in mortality has been shown. Infiltration of normal brain tissue is a hallmark of recurrent GBM and a key factor in preventing successful tumour resection. The majority of glioblastomas recur after treatment within 2 cm of their original tumour margin (Hochberg and Pruitt, 1980).

2.3.2 Vector design and application

The first oncolytic HSV-1 vector specifically designed for cancer treatment was dlsptk. The vector dlsptk was a thymidine kinase (tk) deletion mutant that displayed the ability to prolong survival in athymic mice implanted with xenografted human intracranial malignancies (Martuza *et al.*, 1991). This thymidine kinase-negative virus replicates in mitotic tumour cells, which upregulate their endogenous thymidine kinase enzyme levels. Despite demonstrating tumour-specific replication in mouse models, there were serious concerns about its safety, because this strain of virus is insensitive to the most potent anti-herpetic agents such as ganciclovir (GCV) and aciclovir (ACV) and high titers of dlsptk caused undesirable levels of neurotoxicity in mice. Hence, further development of HSV-1 vectors involved a search for genes that blocked virus replication in normal neurons but retained their replicative capacity in dividing cells.

The first of these vectors is hrR3, an HSV-1 mutant with an in-frame insertion of the *lacZ* gene into the UL39 (ICP6) gene encoding the large subunit of ribonucleotide reductase (RR). The viral RR product was expressed as an ICP6-*lacZ* fusion protein that retained no RR activity (Goldstein and Weller, 1988). Ribonucleotide reductase is a key enzyme for viral DNA synthesis in non-dividing cells but not in dividing cells where elevated levels of the host cell RR provide the deoxynucleotide pools needed as substrates for viral DNA synthesis. Intratumoral inoculation of hrR3 in rats bearing malignant gliosarcomas improved animal survival (Boviatsis *et al.*, 1994a). This improvement was further enhanced by GCV administration in which the natural virus TK product created an active form of the drug that functioned as a chain terminator during DNA synthesis (Boviatsis *et al.*, 1994b). Moreover, the RR-minus phenotype conferred hypersensitivity to GCV and ACV (Mineta *et al.*, 1994). In tumour cells mutated for the p16 or retinoblastoma gene in which the transcription factor E2F is activated, there is a resultant increase in RR activity (Elledge *et al.*, 1992) that predictably favored replication of HSV mutants lacking RR (Chiocca *et al.*, 2002).

Most viruses that produce double-stranded RNA, including HSV-1, have acquired genes that complement or counter the RAS, interferon (IFN), and double-stranded RNA-dependent protein kinase (PKR) pathways. In response to viral infection, PKR shuts off protein synthesis by phosphorylating eIF2a and thus blocks the production of viral gene products. PKR activity is regulated by protein phosphatase-1 which is activated by the HSV protein γ34.5 (He *et al.*, 1997). Thus, γ34.5 indirectly dephosphorylates eIF2α to allow viral protein synthesis to continue. An HSV-1 strain that fails to express a functional form of γ34.5 can therefore only replicate in cells with inactive PKR. *ras* activation by the epidermal growth-factor receptor (EGFR), v-Erb2, or platelet-derived growth-factor receptor (PDGFR) signaling inhibits protein kinase R (PKR) activity. Many glioblastoma cells have mutant forms of EGFR, for example, as part of the tumour phenotype and thus may contribute to effective replication of γ34.5 mutant HSV vectors. In addition to *ras*, glioma cells express heat-shock proteins (e.g. HSP90), which can also block activation of PKR (Todo *et al.*, 2000). Some tumour cells are also defective in IFN signaling, a second cellular strategy for activation of PKR. Tumour cells with defects in this signaling pathway allow a higher degree of viral replication than normal cells. PKR also activates the transcription factor nuclear factor (NF)-κB by inducing degradation of IκB that in turn allows nuclear factor-κB (NF-κB) to enter the nucleus and activate the transcription of pro-inflammatory genes that induce an immune response against viruses (Taddeo *et al.*, 2004). Tumour cells with defects in this signaling pathway allow a higher degree of viral replication than normal cells (Elledge *et al.*, 1992). Thus, inactivation of PKR favours HSV-1 mutant replication by allowing protein synthesis. Moreover, γ34.5 mutants cannot replicate in normal brain tissue (Bolovan *et al.*, 1994) and hence are non-pathogenic for highly sensitive primates such as *Aotus* monkeys following intracranial inoculation (Todo *et al.*, 2000). HSV vectors deleted for γ34.5 function have been highly touted for utility as oncolytic vectors.

The HSV vector 1716 is deleted for both copies of the gene encoding γ34.5 (MacLean *et al.*, 1991) and kills only dividing cells. HSV1716 replicates in and lyses glioblastoma cells but fails to replicate in normal post-mitotic brain cells (McKie *et al.*, 1996). When delivered intracerebrally, HSV-1716 has an LD50 in mice of approximately 7×10^6 plaque forming units (p.f.u.) compared to 1×10^1

p.f.u. for wild-type HSV demonstrating the significant reduction in toxicity to normal brain (MacLean *et al.*, 1991). Several preclinical studies have demonstrated that HSV-1716 can be effective in eliminating tumours in animal models (Randazzo *et al.*, 1995; Kesari *et al.*, 1995; Kucharczuk *et al.*, 1997). These encouraging results led to clinical trials in nine patients with relapsed malignant glioma (Rampling *et al.*, 2000). Following intratumoral inoculation of doses up to 10^5 p.f.u. there was no toxicity or evidence of encephalitis. Subsequent studies in patients with high-grade glioma showed that, following multiple intratumoral injections, HSV1716 replicates in tumours without causing toxicity in both HSV sero-positive and seronegative patients (Papanastassiou *et al.*, 2002). A pilot study with HSV1716 has also been conducted in patients with metastatic melanoma. The study suggested that injection of HSV1716 into subcutaneous nodules is non-toxic and may be beneficial to patients with metastatic melanoma (MacKie *et al.*, 2001).

R3616, a replication competent mutant of HSV-1, lacks 1000 bp from the coding domain of each copy of the $\gamma34.5$ genes (Chou *et al.*, 1990). R3616 is capable of replicating in glioma xenografts in nude mice and remains avirulent with no evidence of encephalitis. Enhanced tumour killing has been demonstrated when combined with ionizing radiation. Analysis of cells cultured from human ovarian tumours resistant to chemotherapy demonstrate effective oncolysis with R3616 suggesting that this vector may be beneficial in the treatment of chemotherapy-resistant tumours (Coukos *et al.*, 2000). GADD34, the mammalian homologue of the $\gamma34.5$ gene, is up-regulated in response to DNA damage caused by chemotherapeutic agents such as mitomycin C and thus treatment of tumour cells with this agent potentiates the replicative potential of $\gamma34.5$-deleted viruses (Bennett *et al.*, 2004).

More recent vectors have combined several mutations to increase safety and tumour specificity. G207 (MGH-1) is deleted in both copies of the $\gamma34.5$ with disruption of the UL39 gene by insertion of a *lacZ* gene (Mineta *et al.*, 1995; Kramm *et al.*, 1997). UL39 was mutated to further limit virus replication to proliferating cells and thus normal neurons and other quiescent cells would be protected. G207 has been demonstrated to selectively destroy glioblastoma cells while sparing primary rat astrocytes or neurons in culture (Chou *et al.*, 1990; Chou and Roizman, 1992; Kramm *et al.*, 1997). In addition, G207 was found to be non-virulent in both rodents and non-human primates (Herrlinger *et al.* 1998; Hunter *et al.*, 1999; Todo *et al.*, 2000; Varghese *et al.*, 2001; Schellingerhout *et al.*, 2000). G207 has been tested in a phase I trial in 21 patients with recurrent GBM (Markert *et al.*, 2000b). Vector administration was carried out by direct stereotactic injection into the tumour. Doses of up to 3×10^9 infectious units were well tolerated yet a maximum tolerated dose was not achieved. This trial showed that inoculation of oncolytic HSV, proved relatively safe in human brain without evidence of encephalitis. Analysis of tumour explants revealed the presence of replicating virus at 4–9 days after injection, and the amount of recovered virus exceeded the input dose in at least some patient samples.

As indicated in the NIH clinical trials database, G207 has also completed a dose ranging study in 65 glioma, astrocytoma, and glioblastoma patients with up to 1×10^{10} p.f.u. delivered (http://www.clinicaltrials.gov/ct/show/NCT00028158? order=2). The results of this study have not been published.

NV1020, originally developed as a vaccine against HSV-1 and HSV-2, is deleted for one copy of $\gamma34.5$ and the UL24 promoter (Wong *et al.*, 2001a, b). The U_L/U_S junction of NV1020 has a 5.2 kb fragment of HSV-2 DNA inserted and an exogenous copy of the thymidine kinase (tk) gene under the control of an IE promoter with a 700 bp deletion in the tk locus. These deletions severely attenuate virulence, although the function of only UL24 is disrupted. *In vivo* analysis with this vector has demonstrated safety as well as efficacy in a number of experimental models of bladder carcinoma (Cozzi *et al.*, 2001) squamous cell carcinoma of head and neck (Wong *et al.*, 2001a), and non-small cell lung cancer (Ebright *et al.*, 2002). NV1020 is currently recruiting patients in a phase I/II dose escalation clinical trial for treatment of liver metastases from colorectal adenocarcinoma (http://www.clinicaltrials.gov/ct/show/NCT00149396?order=1). Here the vector is delivered by intravascular injection and includes concomitant chemotherapy. A total of 27 patients may be treated with safety and biodistribution as primary endpoints and tumour response, antibody characterization, and disease progression as secondary endpoints.

HSV vectors with reduced toxicity can also be produced by deleting other genes. For example, removal of the non-essential gene $\gamma34.5$ and the gene encoding the transcriptional transactivation domain of the tegument viral protein VP16 reduces toxicity particularly when used in combination with other IE gene deletions greatly (Palmer *et al.*, 2000; Lilley *et al.*, 2001; Scarpini *et al.*, 2001). ICP47, unlike the other four IE gene products, is not cytotoxic. Deletion of ICP47 compromises the ability of the virus to block presentation of HSV antigens in infected cells [(York *et al.*, 1994; Fruh *et al.*, 1995), thus enhancing the ability of the immune system to curtail virus propagation and limiting more extended propagation in tumour cells (Todo *et al.*, 2001; Taneja *et al.*, 2001; Wong *et al.*, 2001b). However, it may be important to remove ICP47 from non-replicative vectors in order to enhance the development of specific immune responses directed against either HSV proteins or tumour antigens. Deletion of UL56 in combination with UL24 (NV1020) or ICP47 (NV1042, NV1034) resulted in vectors of reduced neurotoxicity yet they still retain the ability to destroy glial tumour cells (Wong *et al.*, 2001a and b; Meignier *et al.*, 1988). JS1/γ34.5-/ICP47-/GM-CSF (OncoVEX) was derived from a clinical isolate of HSV and is deleted in both copies of the $\gamma34.5$ gene plus ICP47 and it also expresses granulocyte–macrophage colony-stimulating factor (GM-CSF). The deletion of ICP47 changes the temporal expression of US11 to an IE gene as opposed to its normal expression as a late gene, enhancing viral replication in tumour cells. It is proposed that this change coupled with the insertion of GM-CSF should maximally stimulate the immune system following the release of tumour antigens generated by lytic viral replication. *In vivo* studies in mice have demonstrated the oncolytic properties of OncoVEX in addition to an enhanced tumour specific immune response following viral intratumoral injection (Liu *et al.*, 2003). OncoVEX is currently recruiting 60 patients in a phase II clinical trial for malignant melanoma patients (http://www.clinicaltrials.gov/ct/show/NCT00289016? order=1). Patients receive one dose of 1×10^6 p.f.u. followed by multiple doses of 1×10^8 p.f.u. with up to 24 doses delivered intratumorally; based on data from a previously completed 30 patient phase I Clinical trial (Hu *et al.*, 2006).

An alternate method of identifying vectors with the desired characteristics is simply to serially passage a base virus followed by screening isolates for the desired phenotype. One such serial passage isolate, HF10, has been used in a dose escalating phase I clinical trial in Japan in patients with breast cancer (Nakao *et al.*, 2004). A total of six patients received up to three doses of 5×10^5 p.f.u. Tumours were removed 14 days after vector delivery and found to be positive for viral antigens in breast tumour cells. The authors report that death was seen in 30–100 per cent of tumour cells overall.

2.4 Replication defective vectors

2.4.1 Introduction

Replication defective vectors can be created by deletion of any essential viral gene (Figure 2.4B). Depending on the role of the deleted function in virus replication, the virus will be blocked at that particular stage in virus growth. For example, deletion of the gene encoding the major capsid protein VP5 will result in expression of all virus genes but no particle assembly will occur. In contrast, ablation of an essential virus glycoprotein results in particle formation and virus release but the newly formed particles will not be capable of completing the process of virus entry. Current HSV non-replicative vectors contain multiple gene deletions. The effect of the various gene deletions on vector efficacy remains to be determined. However, besides minimizing vector-related toxicities, an added advantage of multiple deletions of HSV-1 genes is that insertion of multiple or large exogenous DNA sequences is enabled. Vectors that express up to five independent expression units have been produced (Krisky *et al.*, 1998a). Transgenes are inserted into HSV-1 vectors by homologous recombination (Krisky *et al.*, 1998a, b).

Replication-defective HSV-1 vectors are mainly used to deliver transgenes that curtail tumour cell growth. These include anti-angiogenic factors, tumour suppressor genes, pro-drug-activating genes (e.g. HSV-1 thymidine kinase) and immunostimulatory genes (McCormick, 2001). A potential problem related to the use of non-replicative vectors is their limited initial distribution which will impact their efficacy. Suboptimal levels of infection of tumour masses by these vectors are unavoidable. One possible solution to this technical hurdle is expression of transgenes (whether secreted or cytoplasmic) that exhibit bystander cell killing effects, either alone or in combination with chemo/radiotherapeutic approaches. An example of a gene product with bystander effects is provided by the HSV-1 thymidine kinase (described below). Another approach may be optimizing delivery of replication-defective viruses. In the brain, this may be achieved by convection-enhanced delivery (described below).

2.4.2 Vector design and application

Because of the temporal regulation of virus gene expression, the IE genes have received considerable attention in the engineering of replication defective vectors. Removal of essential IE genes leads to blockage of subsequent stages in the cascade. ICP4 and ICP27 are essential IE proteins because their expression is critical for virus replication in a permissive tissue culture environment. Thus, deletion of either ICP4 and/or ICP27 results in a virus that cannot replicate and deletion of both products results in a vector that fails to express early and late gene products (DeLuca *et al.*, 1985; McCarthy *et al.*, 1989; Wu *et al.*, 1996; Samaniego *et al.*, 1995, 1997, 1998). Virus culture in ICP4- and ICP27-expressing cells provides functional complementation for the deleted genes and allows temporal expression of viral genes and production of replication-defective HSV vectors. It should be pointed out that loss of IE genes in replication defective HSV-1 vectors requires that transgenes such as tk be placed under the transcriptional control of herpesvirus-related IE gene promoters in order to assure their expression.

In addition to controlling the cascade of virus gene expression, several IE genes have particular functions which greatly influence the intracellular milieu following expression. These include interference with innate immune responses and altering the cell cycle, induction of the DNA damage pathway, relieving repression of vector gene expression (ICP0), and interference with antigen presentation (ICP47). With the exception of ICP47, the IE gene products are reported to be toxic to cells in a dose-dependent manner (Johnson *et al.*, 1994; Wu *et al.*, 1996; Samaniego *et al.*, 1998). The removal of these genes affects vector host cell interactions that greatly influence the outcome of gene transfer and expression. Various IE deletion mutant combinations have been tested for cytotoxicity following infection of different cell types and for their role in transgene expression. A mutant lacking all five IE genes has been shown to be non-cytotoxic (Samaniego *et al.*, 1998) while mutants expressing only ICP0 have been found to cause cell cycle arrest and apoptosis (DeLuca *et al.*, 1985). Use of ICP4-null vectors revealed that ICP4 is important in negatively regulating other IE proteins, in particular modulating expression of the non-essential highly cytotoxic gene products ICP0 and ICP22. Deletion of either ICP0 or ICP22 genes reduces viral toxicity in the background of mutants defective for ICP4 and ICP27 (Samaniego *et al.*, 1998). With the exception of ICP0 expression vectors, mutants that express only one of the other IE genes do not arrest cell division or kill cells (Krisky and Glorioso, unpublished data). For example, in the absence of ICP0, an ICP4 expression vector shows highly reduced levels of ICP4 expression and fails to express early genes.

The deletion of viral genes that encode toxic gene products is important for ensuring vector safety, yet it has become apparent that replication defective vectors must retain at least minimum expression of ICP0 since in its absence, the expression of transgenes from the vector backbone are highly repressed. The exact mechanism for vector gene silencing is currently unknown although available evidence suggests that both virus activation of the DNA repair pathway and the induction of innate immune responses can influence transgene expression. For cancer gene therapy applications, the retention of ICP0 expression has several beneficial effects including high level transgene expression, tumour cell division arrest and the induction of tumour cell apoptosis (Niranjan *et al.*, 2003). Fortunately, the virus lytic genes are silenced in neurons without the loss of gene expression in tumour cells.

HSV-1 vector construction has not been limited to manipulation of genes and *cis*-acting elements that are critical during the lytic cycle. During latent neuronal infection the viral latency locus directs lifelong transcription. HSV-1 latency has been exploited to construct HSV-1 vectors that mediate long-term transgene expression (Goins *et al.*, 1994; Puskovic *et al.*, 2004). In the vector QLGD (ICP4 and ICP27 deleted, ICP22 and ICP47 converted to early genes), the gene encoding rat glial cell-line derived neurotrophic factor (GDNF) was expressed long-term under transcriptional control of the latency active promoter 2 (LAP2) at the UL41 locus. Expression of GDNF protected rodent dopaminergic neurons from toxic chemical insults up to six months after vector inoculation (Puskovic *et al.*, 2004). These findings raise the possibility of applying HSV-1 mediated prolonged transgene expression as a means to inhibit cancer recurrence. This might include, for example, the use of anti-angiogenic factors or genes that limit tumour cell spread. The use of non-replicative vectors in this manner can thus potentially be used in combination with replication competent vectors. In other words, the lytic ability of the oncolytic vectors to shrink tumour size early in treatment could be combined with the potential for non-replicative vectors to induce and sustain an environment that reduces the likelihood for tumour recurrence.

Delivery of suicide genes has been extensively tested in HSV-1-mediated anti-cancer gene therapy and HSV-1 thymidine kinase (TK) has been used most frequently (Spencer *et al.*, 2000; described below). In addition to TK, several other gene products have been tested in anti-cancer therapies. p53 is a tumour suppressor protein able to induce cell growth arrest. Rosenfeld and collaborators successfully restored the p53 function in medulloblastoma cell lines through the delivery of this wild-type protein, using an HSV vector (Rosenfeld *et al.*, 1995). Several studies identified TIMP-2 (tissue inhibitor of metalloproteinase-2) as a key factor in tumour angiogenesis. TIMP-2 suppression of angiogenesis occurs via inhibition of MT1- MMP (membrane type-matrix metalloproteinase)-dependent induction of vascular endothelial growth factor (VEGF) activity. The over-expression of this protein leads to down-regulation of VEGF expression. A defective HSV vector expressing TIMP-2 that demonstrated promising anti-tumoral potential for treatment of U87 gliomas has been reported (Hoshi *et al.*, 2000). Other therapeutic genes that have been incorporated into HSV-1 vectors include *IL4* (Lawler *et al.* 2006), *IL2* and *IL12* (Parker *et al.*, 2000; Carew *et al.*, 2001; Toda *et al.*, 2001; Zager *et al.*, 2001), *GM-CSF* (Herrlinger *et al.*, 2000; Toda *et al.*, 2000); *TNFα* (Moriuchi *et al.*, 1998), and soluble B7-1 (Todo *et al.*, 2001). We have constructed and characterized a replication defective HSV vector that expresses multiple transgenes and combined treatment with drug and radiotherapy (described in detail below).

2.5 Amplicons

2.5.1 Introduction

Standard HSV amplicon vectors consist of plasmids bearing an HSV origin of DNA replication (*ori*) and packaging signal (*pac*) which allows the amplicon DNA to be replicated and packaged as a concatenate into HSV virions in the presence of HSV helper functions (Spaete and Frenkel, 1982; Geller and Breakefield, 1988; Geller *et al.*, 1990; Lim *et al.*, 1996) (Figure 2.4C). These vectors can be packaged concomitantly with replication competent HSV helper virus or produced free of helper virus by co-transfection with an HSV genome that is deleted for *pac* signals and also either mutated/deleted for essential HSV gene(s) or too large to be incorporated into the virion, which has been cloned into a set of overlapping cosmids or a large capacity F-plasmid/BAC (Saeki *et al.*, 2003; Oehmig *et al.*, 2004). The advantages of amplicon vectors are: essentially no toxicity or antigenicity, as they express no virus proteins; a very large transgene capacity (up to 150 kb; Wade-Martins *et al.*, 2003); relatively high titres [$\sim 10^8$ transducing units (t.u.)/ml], and retention for up to months in non-dividing cells. Disadvantages include difficulty in producing large quantities of clinical grade vector and lack of retention in dividing tumour cells.

2.5.2 Vector design and application

Standard amplicon vectors deliver DNA to the cell nucleus as a circular 150 kb concatenate containing multiple copies of the amplicon, e.g. 15 copies of a 10 kb amplicon plasmid. This extra-chromosomal DNA is retained in non-dividing cells for up to several months, but is lost with cell division. Two strategies have been explored to increase retention of amplicon sequences in dividing cells. Incorporation of the EBNA-1 gene and the DNA origin of replication, *oriP*, of Epstein–Barr virus allows the amplicon DNA to replicate in tandem with the host cell genome through a number of cell divisions (Wang and Vos, 1996). Elements of the Tc1-like Sleeping Beauty transposon system incorporated into the amplicon can achieve random integration of transgene sequences into the cell genome with delivery of the transposase *in trans* (Bowers *et al.*, 2006). Site-specific integration of amplicon-encoded transgenes into the genomic target site for adeno-associated virus (AAV), AAVS1 (Kotin *et al.*, 1992) can be achieved by flanking the transgene cassette within the amplicon with AAV inverted terminal repeat sequences (ITRs) and including an AAV *rep* gene in the backbone (Heister *et al.*, 2002; Wang Y *et al.*, 2002). The AAV p5 promoter of the *rep* gene can also serve to mediate replicative amplification of transgene sequences in the presence of HSV-1 replication (Glauser *et al.*, 2006). Thus, through incorporation of other viral elements in amplicon vectors it is possible to achieve amplification of the transgene and integration into the host cell genome so as to boost transgene expression and provide inheritance to daughter cells, respectively.

HSV amplicon vectors have been tested in a variety of preclinical models in culture and in experimental tumours. A number of different therapeutic agents delivered by amplicon vectors have proven to be effective in killing tumour cells (Shah and Breakefield, 2006; Epstein *et al.*, 2005). These include toxic proteins, anti-angiogenic factors, and small inhibitory RNAs (siRNAs) used to directly infect tumour cells or the surrounding tissue. Toxic proteins include prodrug activating enzymes which convert a non-toxic prodrug to an active chemotherapeutic agent, including TK, which converts ganciclovir to a toxic nucleotide (Wang S, *et al.*, 2002), cytochrome P450B1 which converts 4-ipomeanol to alkylating metabolites (Rainov *et al.*, 1998), and a combination of TK/ganciclovir and cytosine deaminase, the latter of which converts 5-fluorocytosine to the chemotherapeutic drug 5-fluorouracil and which together exert a synergistic action (Aghi *et al.*, 2000). Apoptosis promoting proteins encoded in amplicon vectors have included FasL and FADD (Ho *et al.*, 2006) and a secreted form of TRAIL which extends the therapeutic zone to adjacent, non-infected tumour cells (bystander effect; Shah *et al.*, 2003). Other therapeutic proteins tested in the amplicon venue include: the HSV ICP0 which causes necrosis of tumour cells, while sparing normal cells (Cuchet *et al.*, 2005); replacement of the entire CDKN2 locus encoding p16 and p14, which is commonly deleted in glioblastomas (Inoue *et al.*, 2004); the measles virus fusogenic membrane glycoprotein (Hoffman *et al.*, 2006); and p53 which is mutated in a number of cancers (Barzilai *et al.*, 2006). Other strategies include modification of the surrounds of the tumour cells, such as delivery of TIMP-2, which can block breakdown of the extracellular matrix thus restricting invasion of tumour cells into normal tissue. Growth of hepatoma and adenocarcinoma tumours was also curtailed by blocking neovascularization through expression of a dominant-negative soluble vascular endothelial receptor, sFlk-1 (Pin *et al.*, 2004; Reinblatt *et al.*, 2005). Glioblastoma growth was inhibited by delivery of a ds hairpin RNA directed against the epidermal growth factor receptor (EGFR), which is frequently hyperactive in tumour cells with consequent apoptosis (Saydam *et al.*, 2005). Given the limited amplicon vector titer that can be injected into an experimental tumour model – typically 1×10^5 in 5 μl, it is surprising how effective these vectors can be, suggesting high infectivity of cells *in vivo*. In cases where transgene expression may be toxic to normal cells it is possible to promote expression of transgenes selectively in tumour cells using promoter/enhancer elements which are upregulated under conditions of hypoxia found in tumours (Reinblatt *et al.*, 2004) or during cell division (Ho *et al.*, 2004).

Many studies have explored the use of HSV amplicon vectors in cancer vaccination paradigms. The high infectivity of HSV virions combined with no impairment of antigen-presenting functions and prolonged transgene expression achieved in dendritic cells with amplicon vectors makes them a strong candidate for vaccination (for review see Ribas *et al.*, 2005; Santos *et al.*, 2006). Amplicon vectors can be used to express tumour antigens or immune enhancing cytokines with delivery to dendritic cells *ex vivo* or tumour cells *ex vivo* or *in vivo*. Dendritic cells infected with amplicon vectors expressing antigenic proteins can induce a mixed lymphocyte reaction and priming of nave T cells (Nunez *et al.*, 2004). Immunization of mice with dendritic cells infected *ex vivo* with an amplicon vector expressing the prostate-specific antigen protected mice from a challenge with prostate carcinoma cells (Willis *et al.*, 2001). A wide spectrum of enhancing agents has been explored by infecting tumour cells with amplicons expressing: B7.1 and CD40L or LIGHT for B-cell chronic lymphocytic leukemia (Tolba *et al.*, 2001, 2002); CD70 and IL2 for lymphoplastic leukaemia (Zibert *et al.*, 2005); B7.1 and RANTES (regulated on activation, normal, T-cell expressed and secreted) for lymphoma (Kutubuddin *et al.*, 1999); secondary lymphoid tissue chemokine (SLC) and CD40L for lymphoma and adenocarcinoma (Tolba *et al.*, 2002); RANTES, B7.1 and GM-CSF for colorectal cancer (D'Angelica *et al.*, 1999; Delman *et al.*, 2002); interleukin-12 (IL-12) for hepatoma (Jarnagin *et al.*, 2000), IL-2 for squamous cell carcinoma (Carew *et al.*, 2001) and GM-CSF for melanoma and glioblastoma (Toda *et al.*, 2000; Herrlinger *et al.*, 2000). The high potency of this approach suggests that cancer vaccination may be the first clinical use of amplicon vectors.

Amplicon vectors can also be combined with recombinant HSV vectors to improve therapeutic effect. For example, recombinant HSV vectors can act as helper virus for amplicon vectors thus allowing them to be propagated in tandem. In this scenario replication conditional virus can cause lysis of tumour cells while amplicon vectors can express therapeutic transgenes. Co-propagation has been demonstrated in culture and can be enhanced by placing an essential virus gene in the amplicon vector in combination with a recombinant virus deleted for the same gene (Pechan *et al.*, 1999) and expression of transgenes from the amplicon vector can be increased by placing them under an HSV early viral gene promoter (Zhang *et al.*, 2006). Although co-propagation of amplicon and recombinant virus vectors has not been documented *in vivo*, several studies indicate that the use of these two vectors in combination enhances the therapeutic effect. Co-delivery of the γ34.5 gene under a neural progenitor promoter active in brain tumour cells via an amplicon vector was found to increase replication of a γ34.5 deleted HSV virus (G207) in culture and to increase therapeutic efficacy in a xenograft non-small cell lung cancer model (Kanai *et al.*, 2006). Several studies also suggest augmentation of immune response to tumours by combining amplicon vectors expressing cytokines and oncolytic virus. In syngeneic colorectal carcinoma and hepatocellular carcinoma models, co-injection of G207 and an amplicon vector expressing IL-2 gave marked anti-tumour efficacy via an immune mediated mechanism. In fact just the presence of a helper virus can stimulate responses to tumour antigens via HSV amplicon-mediated expression of CD70 and IL-2 in a lymphoblastic leukemia model (Zibert *et al.*, 2005).

2.6 Impediments to the efficacy of HSV vectors for cancer gene therapy

2.6.1 Introduction

Thus far only the oncolytic vector type has been used for cancer treatment of patients. Despite the fact that some patients showed significant tumour regression in phase I trials, in most cases the majority of tumours were unresponsive. While the exact reasons underlying incomplete oncolytic virus-mediated tumour killing are unknown, six factors are likely to contribute: (i) vector distribution is limited to the site of inoculation, (ii) poor virus growth in some tumours suggesting the need to explore other mutant backgrounds, (iii) the pre-existence or rapid establishment of an anti-viral state on infection that limits virus production, spread or transgene expression, (iv) resistance to apoptotic mechanisms that involve bystander killing of uninfected tumour cells, (v) inadequate tumour specificity and (vi) the use of single modality treatments.

2.6.2 Extracellular matrix and vector distribution

Gliomas, in particular, are highly invasive and display a wide diversity of histological features. They are derived from glial support cells in the brain and the vast majority of gliomas are thought to be of astrocytic origin (Reardon and Wen, 2006). Even low-grade gliomas eventually infiltrate the entire brain, a feature that complicates the development of successful therapies. Molecular mechanisms of brain tumour invasion involve modification of receptor-mediated adhesive properties of tumour cells (Platten *et al.*, 2001; Lefranc *et al.*, 2005), degradation and remodeling of extracellular matrix (ECM) by tumour-secreted metalloproteinases (Lefranc *et al.*, 2005,), and creation of an intercellular space for invasion of tumour cells (Chintala and Rao, 1996; Chintala *et al.*, 1996). Spatial and temporal regulation of ECM proteolytic degradation is maintained to ensure successful invasion of tumour cells through the ECM. *In vitro* and *in vivo* glioma models show that deposition of ECM components occurs at the confrontation zone between tumour cells and normal brain tissue. The brain is largely free of a well-defined ECM, except where mesodermal-derived endothelial cells invade the CNS to establish a tumour vasculature (Chintala and Rao, 1996). The parenchyma of the CNS, however, appears to be filled with a relatively amorphous matrix that contains mainly hyaluronic acid, collagen and other fibrous proteins (Goldbrunner *et al.*, 1999). It is this ECM that provides the principal barrier to effective vector distribution following intratumoral inoculation, a problem that might be overcome through use of matrix metalloproteinases to transiently breakdown the ECM and allow effective vector distribution within the tumour mass. Modification of the ECM can lead to a homogeneous spreading of the vector throughout the entire tumour mass and an increase in oncolytic activity (discussed below).

2.6.3 Innate immunity and virus growth

Vertebrates have developed a complex system of non-specific and specific defence mechanisms against invading pathogens such as viruses. However, viruses depend on living cells for replication and hence they have evolved elaborate mechanisms to evade host defence mechanisms. Some of the qualities that make HSV-1 attractive for vector development, e.g., efficient transduction efficiency and target cell killing, are dependent on evasion of innate immune mechanisms. Thus deletion of viral genes that counter innate immune response can be an impediment to vector efficacy. To inhibit host cell defence mechanisms that weaken the therapeutic potential of a vector, it may be important in human trials to supply the vector with a transgene that overcomes cellular defences. The removal of vector functions that block innate immune responses require special attention in vector design.

Interferon

The interferons (IFNs) are inducible secretory proteins with antiviral activity. Many viruses have therefore evolved strategies to evade the effects of IFNs by blocking IFNs production or its antiviral actions. IFNα and IFNβ are the common type I responses of cells infected by viruses. Immune cells secrete type II or IFNγ to activate the innate and adaptive immune response and aid in up-regulating the antiviral state in cells to limit virus replication (Katze *et al.*, 2002). This defence pathway begins with induction of IFNγ by stimulated macrophages and microglia. In the case of viral infection, the most common inducer of type I IFN is the accumulation of complementary viral

RNAs capable of annealing to form double-stranded RNA (dsRNA) (Mossmann *et al.*, 2002). dsRNA activates PKR and the IFN regulatory factor 3 (IRF3) that is translocated from the cytoplasm into the nucleus and activates the expression of IFNα and IFNβ (Akira *et al.*, 2006) The interaction of IFN with its receptors leads to activation of signaling pathways initiated by the phosphorylation of signal transducers and activators of transcription (STAT) proteins. The STAT proteins are phosphorylated by Janus (e.g., Jak1) and tyrosine (e.g. Tyk2) kinases associated with IFN receptors. STAT1 and STAT2 heterodimerize and bind to IRF 9 and form the IFN-stimulated growth factor 3 (ISGF3) complex. The ISGF3 complex is translocated into the nucleus where it binds to the IFN stimulated response elements (ISRE) inducing transcription of Type I IFN responsive genes.

Similarly, upon the binding of IFNγ, the receptor chains dimerize triggering autophosphorylation of Jak2 and transphosphorylation of Jak1 that allows binding of the STAT1 homodimer and subsequent phosphorylation. STAT1 translocates into the nucleus where it binds to DNA at GAS (gamma-activated sequences) elements and promotes transcription. Several viral proteins have specific anti-IFN functions: γ34.5 and US11 interfere with the activation of eIF2α (He *et al.*, 1997; Leib *et al.*, 2000; Poppers *et al.*, 2000); the viral host shut off (vhs) function encoded by UL41 interferes with induction of the Jak/STAT pathway (Chee and Roizman, 2004) and ICP0 can block IRF3 activation leading to a down-regulation of type I IFN response genes (Lin *et al.*, 2004). However, different tumour cells may vary in their ability to produce or respond to type I IFN, leading to the conclusion that tumour typing may become important when selection of the vector mutant background for tumour killing and intratumoral growth. One of the effects of IFNs binding to their receptors is the activation of PKR transcription. Mutant viruses that can not inhibit the PKR pathway (such as the γ34.5 deleted vectors), can not therefore replicate in these cells except if Ras, an inhibitor of PKR activation, is over-expressed. However, most of the glioma cell lines tested in our laboratory constitutively express the activated form of PKR, with subsequent block of HSV-γ34.5 negative vector replication.

Role of IDO in the inhibition of HSV growth in tumour cells

IDO, indoleamine-2,3-dioxygenase, catalyses the oxidative degradation of tryptophan (Taylor and Feng, 1991; Schrocksnadel *et al.*, 2006). The oxidative cleavage of the 2,3 double bond in the indole ring is the first, rate-limiting step in tryptophan catabolism and results in the production of kynurenine and quinolinate. It has been well established that IFNγ induces IDO activity and thus represents a potent anti-viral effector mechanism for the control of HSV replication. Kynurenine is also released from tumours and taken up by local effector T cells resulting in T-cell energy. In contrast, IDO-mediated tryptophan depletion is not involved in the anti-viral effects mediated by IFNα/β (Adams *et al.*, 2004). During viral infection, IDO inhibits the replication of HSV by depleting tryptophan. We have observed for example, that the addition of supra-physiological levels of exogenous tryptophan restores virus replication.

Bin1 is involved in the down-regulation of the IDO enzyme-encoding (*Indo*) gene (Ge *et al.*, 1999). Bin1 also interacts with the Myc box region at the N-terminus of the Myc oncoprotein and appears to function as a negative regulator of cell proliferation and malignancy through the inhibition of the oncogenic activity of *c-Myc* or mutant p53 (Muller and Prendergast, 2005). In cells where c-Myc is down-regulated, Bin1 promotes cell cycle exit and differentiation (Muller *et al.*, 2005). In contrast, if Myc is activated and cell cycle exit is blocked, then cytokine depletion will cause programmed cell death. Although Bin1 is normally ubiquitous, it is reduced or undetectable in many tumour types where its reintroduction will block tumour cell proliferation (Ge *et al.*, 1999). For example, loss or attenuation of Bin1 occurs in advanced breast cancer, prostate cancer, melanoma, astrocytoma, neuroblastoma, and colon cancer leading to the over-expression of IDO.

At least 10 different Bin1 splice isoforms exist in mammalian cells of which two are ubiquitously expressed, and the remainder is restricted to specific terminally differentiated tissues including neurons and skeletal muscle cells (Muller *et al.*, 2005). Deleting the *Bin1* gene from cells resulted in super induction of IDO gene expression by IFNγ (Muller and Prendergast, 2005). *In vitro* transformation of Bin1-null and Bin1- expressing primary mouse embryo keratinocytes with *c-Myc* and mutant *Ras* oncogenes produced cell lines with similar *in vitro* growth properties. However, when these cells were grafted subcutaneously into syngeneic animals, the Bin1-null cells formed large tumours, whereas the Bin1-expressing cells formed only indolent nodules (Muller *et al.*, 2005). The deregulation of IDO, which accompanies Bin1 loss in these cells, promotes tumorigenicity by enabling immune escape from T cells (Muller and Prendergast, 2005). The restoration of Bin1 function as a vector transgene or the use of one methyl-tryptophan (1MT), an inhibitor of IOD function, may therefore down-regulate IDO synthesis with the consequent improvement in oncolytic virus growth primarily by restoring intracellular availability of tryptophan.

NF-κB activation

NF-κB regulates oncogenesis, tumour progression and inhibits apoptosis through induction of anti-apoptotic proteins and/or suppression of pro-apoptotic genes (Nakanishi and Toi, 2005). Constitutive NF-κB activation, observed in many malignant tumours, protects the cells from apoptotic stimuli, including anti-cancer treatments (Baldwin, 2001). By inhibiting the anti-apoptotic NF-κB pathway with a non-degradable inhibitor of NF-κB, designated IκBaM, apoptotic pathways predominate resulting in increased cell death (Van Antwerp *et al.*, 1996; Moriuchi *et al.*, 2005). Thus, IκBaM transduction can potentially enhance HSV oncolytic efficacy. In most cell types, NF-κB is sequestered in the cytoplasm in complexes with inhibitory proteins called IκBa, IκBb, and IκBe. In response to diverse stimuli, including inflammatory cytokines, mitogens, bacterial LPS, and certain viral products (Barton and Medzhitov, 2003), active NF-κB is released and translocated to the nucleus as a result of proteolytic degradation of the IκB proteins. Phosphorylation of IκBs on Ser-32 and Ser-36 targets these molecules for degradation by the ubiquitin-26S proteasome pathway. Mutations in the phosphorylation sites inhibit IκBa degradation and thus have a dominant-negative (DN) effect on NF-κB activation. Malignant brain tumour models treated with a replication-defective HSV-1 vector expressing the IκBa phosphorylation mutant IκBaM results in tumour cell destruction. No neurotoxicity was associated with this treatment, proving the safety of this strategy. By converting the anti-apoptotic state of the glioma cells to a pro-apoptotic state, vector-infected cells should exhibit an enhanced susceptibility to apoptotic mechanisms. Also, NF-κB activation increases cytokine and chemokine synthesis that in turn attracts more inflammatory cells and increases innate immune responses. The expression of an IκBaM should assist virus replication and enhance the effects of anti-tumour mechanisms.

Vector efficacy and resistance to apoptotic mechanisms

Programmed cell death or apoptosis plays an important role in the maintenance of normal tissue growth, while a dysregulation of apoptotic machinery causes cancer cell survival and proliferation. The restoration of apoptotic pathways in tumour cells is hence a therapeutic strategy to retard tumour growth (Fesik *et al.*, 2005). Earlier studies using this rationale demonstrate that tumour necrosis factor (TNF) and FAS ligand induce death receptor-mediated apoptosis in tumour cells (Barnhart *et al.*, 2003). However, side effects including hypotension and septic shock have discouraged the use of these ligands for treatment of brain tumours. The TNFα-related apoptosis-inducing ligand (TRAIL) is an alternative apoptosis inducing therapeutic ligand for malignant gliomas (Kaufmann and Vaux, 2003). TRAIL receptors (TRAILR-1 and TRAILR-2) are expressed in glioma tumour cells (Rieger *et al.*, 1998). Moreover, TRAIL does not induce apoptosis in non-tumorigenic cells and a safer, optimized soluble recombinant TRAIL with minimal *in-vivo* side effects has been developed (Kelley *et al.*, 2001). LY294002 as well as TRAIL can be useful drugs as a combined therapy with HSV vectors. A secreted form of the tumour apoptosis-inducing protein encoded in an HSV amplicon vector has been shown to result in tumour regression following a series of intratumoral injections of the vector. The protein kinase encoded in US3 and UL14 chaperone have anti-apoptotic activity (Nishiyama and Murata, 2002; Yamauchi *et al.*, 2003). ICP0 also has E3 ubiquitin ligase activity linked to proteasome-dependent degradation of cellular proteins (Hagglund *et al.*, 2002) and which can induce apoptosis (e.g. p53).

2.7 Strategies to enhance the efficacy and specificity of HSV vectors for cancer gene therapy

2.7.1 Introduction

Several strategies are being pursued to improve the efficacy of anti-cancer HSV vectors. These include novel methods of delivery to improve vector distribution, tumour selectivity through virus retargeting and exploiting properties critical to tumour cells to promote vector efficacy. In addition, the development of multi-modal approaches that exploit gene therapy in combination with more standard treatments should also improve therapeutic outcome.

2.7.2 Improving vector distribution

ECM

The MMPs are associated with degradation of the ECM, including the basement membrane, which is a specialized matrix composed of type IV collagen, vitronectin, laminin, entactin, proteoglycans, and glycosaminoglycans. This basement membrane serves as a barrier between tissue compartments. It was initially believed that the MMPs, via breakdown of the physical barrier, were primarily involved in tumour invasion, entry and exit of tumour cells from the circulation, and local migration at metastatic sites (Kondraganti *et al.*, 2000). However, there is growing evidence that the MMPs have an expanded role, as they are important for the creation and maintenance of a microenvironment that facilitates growth and angiogenesis of tumours at primary and metastatic sites (Chakraborti *et al.*, 2003). An important concept is that cells do not indiscriminately release proteases. Proteases, such as MMPs, are secreted and anchored to the cell membrane, thereby targeting their catalytic activity to specific substrates within the pericellular space (Nelson *et al.*, 2000).

Among the MMPs, MMP-9 specifically targets type IV collagen, a major component of the basement membrane (Choe *et al.*, 2002). The co-injection of oncolytic virus with MMP9 into the brain tumour may induce the spread of the virion particles to sites distant from the site of injection into the brain tumours. MMP9 action does not affect virus infectivity. It is unlikely that MMP9 will increase tumour cell infiltration for two main reasons: (i) the area affected by collagenase also will be reached by the oncolytic viruses that will infect and kill the cells, and (ii) the presence of high concentrations of pure collagenase will not establish the correct milieu to induce cell migration. During invasion, the tumour cells must change the expression of cytoskeletal proteins, cell adhesion molecules, and matrix-degrading proteases. Tumour cell migration is also influenced by migrating signals, such as ECM components, either by chemo-and/or haptotactic mechanisms.

Convection-enhanced delivery (CED)

Delivery of HSV-1 vectors in all clinical trials has been by multiple manual stereotactic intratumoral or peritumoral injections after surgical resection (Harrow *et al.*, 2004; Markert *et al.*, 2000a, b). Viral particles accumulate adjacent to the needle tract, and limited dispersal of particles occurs by diffusion. The binding of viral particles to the heparan sulfate proteoglycans found abundantly in the extracellular matrix and glycocalyx in the brain may contribute to limited dispersal (WuDunn and Spear, 1989). Convection-enhanced delivery (CED) is an approach developed to overcome the obstacles associated with current CNS agent delivery (Bobo *et al.*, 1994; Morrison *et al.*, 1994) and is increasingly used to distribute therapeutic agents for treatment of malignant gliomas.

Currently, multiple clinical trials involve CED for the treatment of recurrent GBM (Weingart *et al.*, 2002; Kunwar *et al.*, 2003; Voges *et al.*, 2003 Weaver and Laske, 2003). In CED, a small hydrostatic pressure differential, imposed by a syringe pump to distribute infusate directly to small or large regions of the CNS, is used in a safe, reliable, targeted, and homogeneous manner (Croteau *et al.*, 2005). CED relies on bulk flow that is driven by a small gradient to distribute molecules within the interstitial spaces of the CNS. Convection is not limited by the infusate's molecular weight, concentration, or diffusivity (Bobo *et al.*, 1994; Morrison *et al.*, 1994; Strasser *et al.*, 1995). Convection-enhanced delivery of HSV-1 contributes to optimal vector delivery to brain. Cell infection and distribution of HSV-1 appears relatively homogeneous in the brain after CED with a moderate dose of virus (3×10^7 p.f.u.). Spread of viral particles occurs within the gray matter and along white matter tracts in the brain. In addition, minimal viral reflux occurs along the catheter needle tract maximizing the viral dose available for cell infection.

2.7.3 Limiting local inflammation

There are a number of local and intracellular responses to virus infection which can limit virus replication and spread within the tumour. The development of innate immune responses within the tumour cells (IFNα/β) and the attraction of inflammatory cells that produce IFNγ can inhibit virus growth. In addition, local inflammatory cells attracted to the tumour can release other mediators of innate immunity. Using a syngeneic model, Wakimoto and collaborators (2004) showed that: (i) systemic pre-treatment of animal models of brain tumours with the immunosuppressive drug cyclophosphamide (CPA) promotes survival of oncolytic virus within tumours, and (ii) in the absence of CPA, a rapid and significant loss of oncolytic virus was observed in infected tumours. Concomitant with improved virus growth, these investigators observed the influx of fewer mononuclear cells into the tumour mass, which produce IFNγ and encourage the induction of innate immunity. The use of anti-inflammatory drugs or vectors armed with genes that reduce cellular innate immune responses should improve the growth of oncolytic vectors. For non-replicating vectors, the infiltration of inflammatory cells in response to the expression of inflammatory cytokines (e.g. TNFα) may provide potent anti-cancer cytolytic activity.

2.7.4 Virus re-targeting

The basic approach to engineer tumour selective HSV-1 vectors has focused on deletion of HSV-1 non-essential genes that are necessary for replication in normal quiescent cells, but not in tumour cells. Studies conducted with ribonucleotide reductase and γ34.5 deletion mutants are encouraging. In addition, an important aspect of vector design involves the manipulation of the earliest events in the virus replication cycle beginning with attachment and entry into host cells (Figure 2.2). For most vectors, the manipulation of these early events would be important for targeting the appropriate cell for infection, especially for cancer applications. Although considerable effort has been expended to understand the details of HSV infection, there are still gaps in our knowledge and effective and efficient targeting of HSV has not been achieved. This is largely because the process of virus attachment and entry requires the activities of multiple HSV glycoproteins that appear to act in a coordinated and sequential fashion (Spear, 1993a, b; Mettenleiter *et al.*, 1994). To accomplish targeted delivery, vectors carrying engineered ligands must be developed that seek out specific cell types based on recognition of cell-specific surface receptors while at the same time are unable to carry out infection through any of the normal virus receptors. Successful development of this technology depends on: (i) the identification of cell-specific surface receptor(s), and (ii) the ability to modify viral surface proteins for recognition of novel receptors without compromising infectivity. Since the mechanism of viral entry impacts on re-engineering virus infection, a sufficient understanding of the natural process of virus entry into the cell is essential. As an alternative strategy, a bi-specific adaptor approach may be used in which a soluble form of the natural receptor such as nectin-1a is fused to a novel ligand for binding to the cell surface while at the other end containing a novel ligand for binding to the gD of the virion in a manner to trigger virus entry. For example we have reported efficient targeting of an HSV-1 vector to gD-receptor deficient, EGF-receptor-expressing cells using an anti-EGFR scFv linked to the V-domain of nectin-1 (Nakano *et al.*, 2005).

Various attempts to alter the host range of different viruses have been reported, including retroviral (Kasahara *et al.*, 1994; Cosset *et al.*, 1995; Cosset and Russell, 1996; Nilson *et al.*, 1996; Valsesia-Wittmann *et al.*, 1996; Morling *et al.*, 1997; Fielding *et al.*, 1998; Chadwick *et al.*, 1999), AAV (Girod *et al.*, 1999) and adenoviral vectors (Wickham *et al.*, 1995, 1996a, 1996b, 1997; Rogers *et al.*, 1997; Grill *et al.*, 2001; Mizuguchi *et al.*, 2001; Belousova *et al.*, 2003). These approaches include the incorporation of a specific binding ligand or single-chain antibody (scFv) into a viral surface protein to target the vector to cells expressing the cognate receptor for that ligand and the inclusion of an RGD motif in a natural viral capsid protein (Girod *et al.*, 1999; Martin *et al.*, 1999) that resulted in the ability to infect normally refractory cells. Recent studies have introduced novel binding peptides into the region of the adenovirus fiber knob to achieve retargeting (Mizuguchi *et al.*, 2001; Belousova *et al.*, 2003). In a similar manner, HSV-1 was retargeted by the introduction of ligands into gC (Argnani *et al.*, 2004; Grandi *et al.*, 2004), but these viruses were still capable of binding to HveA/C-bearing cells since the natural tropism, via an interaction with gD, was not ablated. HSV-1 was targeted to IL-13Rα2-expressing cells by a virus in which the IL-13 ligand was fused to gC and into the amino terminal HveA binding site of gD (Zhou *et al.*, 2002). However, the IL-13:gD chimeric molecule still retained the ability to bind to and enter cells expressing HveC/nectin-1. gD has been also successfully engineered to recognize HER2/neu, the human epidermal growth factor receptor2/neuregulin that is over-expressed in highly malignant mammary and ovarian tumours (Menotti *et al.* 2006). This vector can infect cells lacking the HSV receptors. However in the presence of cells carrying the HveC receptor, the virus remains capable of infecting through this normal receptor.

Other target receptors of interest include EGFRvIII that is highly expressed on glioma cells (Wikstrand *et al.*, 1995). The mutant receptor EGFRvIII has a deletion in its extracellular domain that results in the formation of a new, tumour-specific antigen found in glioblastomas, breast carcinomas, and other tumours. The tumour specific peptide MR1 (mutant receptor 1) (Humphrey *et al.*, 1990; Wikstrand *et al.*, 1995) has high-affinity binding to the EGFRvIII and does not show any detectable binding to the wild-type receptor.

With the aim of retargeting HSV vectors to tumour cells expressing EGFRvIII, the HS binding domain of gC has been replaced with the tumour-specific peptide coding sequence, MR1. We showed an increase of infectability to the cell bearing the EGFRvIII (U87delta EGFR) compared to the parental cell lines (U87) that do not express the modified receptor.

These reports demonstrate the potential utility of this approach for specific targeting of HSV vectors. However, gD remains intact and capable of mediating virus attachment and entry via the natural receptors (e.g. HveC). Ultimately, the achievement of tumour-specific retargeting will require the engineering of an active gD binding/entry molecule in which the natural receptor binding activity of gD has been ablated. In addition, the virus avidity for tumour cells can be increased by substituting the tumour binding ligand for the HS binding domain of gC.

2.7.5 Promoter-mediated tumour cell selectivity of HSV-1 vectors

The specificity for virus replication and therapeutic gene expression in tumour cells can also be enhanced by cellular properties that are unique to the tumour microenvironment. In the oncolytic vector Myb34.5, both copies of the HSV-1 ribonucleotide reductase and γ34.5 genes were deleted. One copy of γ34.5, placed under the control of B-myb promoter, was reintroduced into the vector backbone to target lytic replication to tumour cells (Chung *et al.*, 1999). B-Myb is a transcription factor up-regulated in the G1/S phase of the cell cycle, and placing the γ34.5 gene under this promoter enhances its replicative capacity in dividing tumour cells by 1000–10 000-fold as compared to normal, non-dividing cells.

To further restrict growth of HSV-1 vectors to specific tissues, the HSV-1 vectors G92A and d12. CALP were constructed by inserting albumin or calponin promoter-driven ICP4 into the thymidine kinase gene of the HSV-1 mutant d120, respectively. In d120, both copies of the ICP4 gene had been deleted. By placing the essential gene encoding ICP4 under albumin promoter, G92A efficiently replicated in hepatoma cell lines expressing albumin and in subcutaneous xenografts of human hepatoma cells, but not in albumin non-expressing tumour cell lines or non-hepatoma subcutaneous tumours. Similarly, in d12CALP, cell-specific replication was limited to leiomyosarcoma in which calponin expression was augmented (Miyatake, 2002). Although deletion of the thymidine kinase gene would not be desirable in an ideal vector, G92A and d12CALP demonstrated the potential of promoter-driven tissue-directed tumour selectivity of HSV-1 vectors. Using this approach, DF3/MUC1 promoter/enhancer sequence was used to regulate expression of γ34.5 to drive replication of an HSV-1 mutant (DF3γ34.5) preferentially in DF3/MUC1-positive breast cancer cells (Kasuya *et al.*, 2004).

2.7.6 Consideration of novel mutant combinations

Thus far the number of mutant backgrounds for design of HSV vectors has generally been quite extensive but more mutant viruses require testing for the oncolytic vectors. In particular, the removal of virus genes that overcome cellular defence mechanisms in normal cells may not be required in tumour cells. Much attention has been paid to the use of vectors that are defective for the γ34.5 gene (described above). However, many tumour types are not defective for the PKR pathway and indeed PKR is often constitutively active. In these cases γ34.5 is required for virus growth. Several approaches have been used to address this problem including retention of the γ34.5 gene but under transcriptional control of a tumour specific promoter (Chung *et al.*, 1999). Alternatively, a virus mutant engineered to express the US11 gene as an IE product provides a second approach to overcoming the PKR-mediated block in viral protein synthesis (Mulvey *et al.*, 2003). A second alternative is simply to leave the γ34.5 gene intact and remove other accessory viral functions that are not required for virus replication in tumours. These include UL56 (Takakuwa *et al.*, 2003; Kimata *et al.*, 2003; Teshigahara *et al.*, 2004), US3 (Fink *et al.*, 1992; Nishiyama 2004), ICP0 (Krisky *et al.*, 1998a; Mullerad *et al.*, 2005) and ICP22 (Krisky *et al.*, 1998a). ICP0 deletion mutants, for example, have been shown to replicate in some tumour types and are effective at tumour destruction in animal models (Hummel *et al.*, 2005). Future studies may determine that particular vector mutant backgrounds are suited to particular tumours to be most effective and as this field develops, it is likely that more attention will be paid to this approach.

2.7.7 Combined therapeutic approaches to enhance tumour killing

Combination of genes and drugs

Anti-cancer drugs have been tested for use with the oncolytic HSV vector G207. The HSV-1 vector G207 in combination with cisplatin (Chahlavi *et al.*, 1999) and vincristine (Nakano *et al.*, 2001) produced synergistic therapeutic effects in experimental models of human squamous cell carcinoma of head and neck origin and in alveolar rhabdomyosarcoma, respectively. A similar synergistic effect was also shown in the treatment of cancer with mitomycin C and G207 (Toyoizumi *et al.*, 1999; Bennett *et al.*, 2004). However, the most extensively studied combined therapeutic approach has been the use of the viral tk and GCV (Moolten *et al.*, 1986 and 1994; Ezzeddine *et al.*, 1991; Culver *et al.*, 1992; Barba *et al.*, 1993; Caruso *et al.*, 1993; Eck *et al.*, 1996; Ram *et al.*, 1993; Trask *et al.*, 2000).

The HSV-1 thymidine kinase phosphorylates deoxypyrimidines with broader substrate specificity than the corresponding cellular enzyme. The non-toxic pro-drug ganciclovir (GCV) is a substrate for HSV tk, but not cellular tk. Following phosphorylation by HSV TK, phosphorylated

GCV becomes incorporated into replicating DNA and causes premature strand termination in proliferating cells. Thus, phosphorylated GCV is selectively cytotoxic to actively dividing cells such as cancer cells. In addition, the action of HSV TK on ganciclovir generates a bystander cytotoxic effect whereby the toxic product can diffuse into neighboring uninfected cancer cells (Carroll *et al.*, 1997; Marconi *et al.*, 2000). Thus, a low level of transduction of tumour cells with HSV-1 TK would not be a serious limitation, as surrounding replicating cells will be killed by the diffusion of activated ganciclovir. However, the HSV-TK/GCV system needs further improvement because the GCV is not effective against non-dividing cells and dividing cells distal to the application site.

Other pro-drugs and enzymes include CD (cytosine deaminase) (Nakamura *et al.*, 2001), 5-FC (5-fluorocytosine), PNP (*Escherichia coli* purine nucleoside phosphorylase) (Bharara *et al.* 2005), and cytochrome P450 2B1 (Chase *et al.*, 1998). The activity of the latter can be enhanced by additional inclusion of human intestinal carboxyesterase which can convert the topoisomerase I inhibitor, irinotecan (CPT11) into an anti-cancer metabolite which is synergistic with the active metabolites of cyclophosphamide. Neither of these byproducts blocks replication of the oncolytic vector in tumour cells (Tyminski *et al.*, 2005). Incorporation of genes that confer anti-angiogenic activities into virus genome augment the anti-tumour effect of oncolytic vectors by decreasing tumour angiogenesis (Mullen *et al.*, 2004).

Another strategy involved the use of anti-cancer drugs to enhance vector replication in tumours. Treatment with fluorodeoxyuridine (FUDR), an inhibitor of cellular thymidylate synthase that is responsible for conversion of dUMP (deoxyuracil monophosphate) and CDP (cytosine diphosphate) to dTTP, stimulates mammalian ribonucleotide reductase activity due to loss of feedback inhibition by dTTP. Increased ribonucleotide reductase activity therefore enhances replication of a UL39 (RR) mutant (Petrowsky *et al.*, 2001). cDNAs with more specific anti-cancer, rather than anti-viral, action might therefore provide better choices for addition to the oncolytic vector genome. For example, the rat cytochrome P450 2B1 gene product metabolizes the pro-drug cyclophosphamide into the active anti-cancer and immunosuppressive metabolite, phosphoramide mustard. However, cyclophosphamide metabolites, although causing tumour cytotoxicity, do not eliminate or inhibit viral replicative ability (Aghi *et al.*, 1999). It is likely that these differential effects might be due to the immunosuppressive properties of the metabolites, which can block the anti-viral immune responses. These metabolites also alkylate DNA every 100–150 kb, thereby causing a greater number of cytotoxic DNA cross-links in the mammalian cell genome than in the much smaller viral genome. Similarly, anti-cancer effects are enhanced when the pro-drug 5-FC is combined with a UL39 mutant virus that expresses yeast cytosine deaminase (Nakamura *et al.*, 2001). This enzyme converts 5-FC to the active anti-cancer agent, 5-fluorouracil (5-FU). This enhancement is possibly due to the fact that one of 5-FU's metabolites inhibits cellular thymidylate synthase, thereby decreasing dTTP levels. This, in turn, will remove feedback inhibition on mammalian ribonucleotide reductase levels, thus augmenting replication of the UL39 mutant virus. The *E. coli* gene, purine nucleoside phosphorylase (PNP) in combination with 6-methylpurine-2′-deoxyriboside (MeP-dR) has recently been shown to be superior to the tk-GCV in killing of uninfected neighboring cells even those which are non-dividing (Martiniello-Wilks *et al.*, 1998). This gene combination has been reported to be very effective in brain tumour models using adenoviral vectors (Martiniello-Wilks *et al.*, 1998) for delivery although this gene-drug combination has not been tested using HSV vectors.

Work in our laboratory has explored the use of a highly defective HSV vector (NC2), capable of expressing four transgene products, for treatment of animal models of glioblastoma. The multi-gene vector expresses 4 transgene products, HSV-ICP0, HSV-TK, connexin-43 (Cx-43) and TNFα. In initial animal studies, an early generation vector expressing HSV-TK as an IE gene driven by the ICP4 promoter was used to study suicide gene therapy (SGT) in rodents. Although overall animal survival time increased by 50 per cent in the presence of GCV pro-drug (Moriuchi *et al.*, 2000), the long-term benefits were minimal and it was evident that the bystander effect needed to be expanded. We accomplished this by engineering an 'armed' vector expressing Cx-43, a gap junction protein weekly expressed in gliomas (Shinoura *et al.*, 1996). HSV vector-mediated expression of Cx-43 resulted in enhanced SGT, with a 33 per cent survival rate at 70 days (Marconi *et al.*, 2000). Further, we engineered other 'armed' HSV vectors expressing the cytokine TNFα, which has previously been shown to enhance radiotherapy as well as being a potent activator of the immune system, and dramatically enhanced SGT in TNFα-sensitive brain tumours (Moriuchi *et al.*, 1998). The combined use of SGT/TNFα therapy with gamma knife radiosurgery (GKR) resulted in 89 per cent animal survival out to over 75d, with 75 per cent of those animals being tumour-free (Niranjan *et al.*, 2000). Together these studies suggest that combination therapy would yield the greatest efficacy in patient studies. Final pre-clinical studies comparing the NC2 vector to previous vectors, in combination with GCV and GKR, in an immunocompetent rat model of GBM demonstrated that multi-modal treatment represents the most effective therapy in support of the Phase-I clinical trial (Niranjan *et al.*, 2003).

Combined use of HSV-1 and ionizing radiation (IR) therapy

Increased efficacy with the use of HSV-1 vectors in combination with IR has been shown in different malignancies (Advani *et al.*, 1998; Niranjan *et al.*, 2000; Blank *et al.*, 2002; Chung *et al.*, 2002; Markert, 2004; Adusumilli *et al.*, 2005; Hadjipanayis and DeLuca, 2005; Kim *et al.*, 2005; Jarnagin *et al.*, 2006). Both replication-defective and oncolytic HSV-1 vectors have been used in combination with radiation. Our group has used replication-defective constructs in combination with gamma-knife stereotactic radiosurgery or standard whole-brain irradiation in a mouse glioma model (Niranjan *et al.*, 2000; Hadjipanayis and DeLuca, 2005). Radiation treatment of patients with malignant gliomas includes a fractionated plan for a total dose of 60 Gy to the targeted area with a 3 cm margin around the targeted tumour. This IR therapy is the primary adjuvant treatment modality and represents the standard of care, resulting in modestly increased patient survival (Walker *et al.*, 1980).

The replication-defective HSV-1 construct, d106, which solely expresses the IE protein, ICP0, has been shown to enhance the radiosensitivity of human glioblastoma cells (Hadjipanayis and DeLuca, 2005). Degradation of the catalytic subunit of DNA-dependent protein kinase (DNA-PK_{CS}) by ICP0 has been implicated in the inhibition of DNA double-strand break (DSB) repair after IR treatment of malignant glioma cells. It has been firmly established that DNA-PK_{CS} plays an important role in DNA end joining, especially DNA DSB after IR. Prior reports have shown that inhibition or deficiency in DNA-PK_{CS} leads to decreased DNA DSB repair and increased radiosensitivity both *in vitro* and *in vivo* (Kurimasa *et al.*, 1999; Lees-Miller *et al.*, 1995; Veuger *et al.*, 2003). The degradation of DNA-PK_{CS} by ICP0 occurs by the ubiquitin-dependent proteosome degradation pathway (Lees-Miller *et al.*, 1996, Parkinson *et al.*, 1999). The decrease in survival of these cells is in part due to the induction of apoptosis by ICP0 and IR treatment.

Several studies have investigated the effect of combining IR therapy with HSV-1 oncolytic vectors for the treatment of a variety of cancers. These include: human malignant gliomas using R3616 (Goldstein and Weller, 1988a, b), hepatomas using R7020 (Chung *et al.*, 2002), cervical cancer and squamous cell carcinoma of the head and neck using G207 (Blank *et al.*, 2002; Kim *et al.*, 2005), cholangiocarcinoma using NV-1023 (Jarnagin *et al.*, 2006) and lung cancer cell lines using NV-1066 (Adusumilli *et al.*, 2005). Advani *et al.* (1998) have shown the conditionally-replicative HSV-1 mutant, R3616, has greater oncolytic effects and increased replication when exposed to IR (Advani *et al.*, 1998). In their study, human U87-MG xenografts in mice resulted in a significantly greater reduction in tumour volume or total regression when tumours were inoculated with the R3616 mutant and irradiated. Increased spread of the virus was seen with in-situ hybridization with DNA probes to the virus. Other studies have confirmed the enhanced tumoricidal effect of HSV when combined with IR (Bradley *et al.*, 1999; Markert *et al.*, 2000a).

2.8 Summary and conclusions

HSV vectors appear very promising for treatment of cancer. Knowledge of the biology of HSV-1, the molecular biology of tumour cells and the novel interactions of these genetically altered cells with HSV provides a unique opportunity to engineer HSV-1 as a highly selective and potent anti-cancer vector potentially tailored to individual patient tumours. The use of viruses with improved initial intratumoral distribution and 'armed' with genes that both overcome innate immune responses and promote an intracellular pro-apoptotic state should improve the potency of HSV cancer gene therapy vectors without compromising safety. Brain tumours may be particularly good targets for HSV oncolytic vectors because of the vector neurotropism and their remarkable ability to be engineered for replication within these tumours without toxicity for normal brain. They are the only vector system that has provided early suggestions of anti-tumour efficacy having these unique attributes. Nevertheless, it is likely that novel HSV vectors can be engineered for attacking other tumour types with similar capabilities. In addition, the non-replicative HSV vectors and amplicon-based systems should also add to the arsenal of potential treatment vehicles. These vectors might be used in combination with the replicative vectors to modify the local environment in a manner to discourage tumour metastasis for example, or under circumstances where vector gene expression is intended for the induction of anti-cancer immunity. Finally, we are convinced that multi-modal therapies will be essential to successful treatment of solid tumours even where there is infiltration into normal tissue. These considerable advancements in the design and application of HSV vectors should provide a leap forward in enhancing anti-tumour potency, instilling confidence that gene therapy will become part of the standard of care for cancer patients in the near future.

Acknowledgements

This work was supported by the following grants: NIH/NCI CA69246, CA86355, CA119298-01; NIH/HINDS NS40923-04, NIH/NHLBI HL66949-065U54, NIH/NIAMS AR050733-03 NIH/NIDDK DK119298-01.

References

Adams O, Besken K, Oberdorfer C, MacKenzie CR, Takikawa O, Daubener W (2004). Role of indoleamine-2,3-dioxygenase in alpha/beta and gamma interferon-mediated antiviral effects against herpes simplex virus infections. *J Virol* **78**, 2632–2636.

Adusumilli PS, Stiles BM, Chan MK, *et al.* (2005). Radiation therapy potentiates effective oncolytic viral therapy in the treatment of lung cancer. *Ann Thorac Surg* **80**, 409–416.

Advani SJ, Sibley GS, Song PY, *et al.* (1998). Enhancement of replication of genetically engineered herpes simplex viruses by ionizing radiation: a new paradigm for destruction of therapeutically intractable tumors. *Gene Ther* **5**, 160–165.

Aghi M, Chou TC, Suling K, Breakefield XO, Chiocca EA (1999). Multimodal cancer treatment mediated by a replicating oncolytic virus that delivers the oxazaphosphorine/rat cytochrome P450 2B1 and ganciclovir/herpes simplex virus thymidine kinase gene therapies. *Cancer Res* **59**, 3861–3865.

Aghi M, Hochberg F, Breakefield XO (2000). Prodrug activation enzymes in cancer gene therapy. *J Gene Med* **2**, 148–164.

Akira S, Uematsu S, Takeuch, O (2006). Pathogen recognition and innate immunity. *Cell* **124**, 783–801.

Argnani R, Boccafogli L, Marconi PC, Manservigi R (2004). Specific targeted binding of herpes simplex virus type 1 to hepatocytes via the human hepatitis B virus preS1 peptide. *Gene Ther* **11**, 1087–1098.

Baldwin AS (2001). Control of oncogenesis and cancer therapy resistance by the transcription factor NF-kappaB. *J Clin Invest* **107**, 241–246.

Barba D, Hardin J, Ray J, Gage FH (1993). Thymidine kinase-mediated killing of rat brain tumors. *J Neurosurg* **79**, 729–735.

Barnhart BC, Lee JC, Alappat EC, Peter ME (2003). The death effector domain protein family. *Oncogene* **22**, 8634–8644.

Barton GM, Medzhitov R (2003). Linking Toll-like receptors to IFN-alpha/beta expression. *Nat Immunol* **4**, 432–433.

Barzilai A, Zivony-Elbom I, Sarid R, Noah E, Frenkel N (2006). The herpes simplex virus type 1 vhs-UL41 gene secures viral replication by temporarily evading apoptotic cellular response to infection: Vhs-UL41 activity might require interactions with elements of cellular mRNA degradation machinery. *J Virol* **80**, 505–513.

Batterson W, Roizman B (1983). Characterization of the herpes simplex virion-associated factor responsible for the induction of alpha genes. *J Virol* **46**, 371–377.

Belousova N, Korokhov N, Krendelshchikova V, *et al.* (2003). Genetically targeted adenovirus vector directed to CD40-expressing cells. *J Virol* **77**, 11367–11377.

Bennett JJ, Adusumilli P, Petrowsky H, *et al.* (2004). Up-regulation of GADD34 mediates the synergistic anticancer activity of mitomycin C and a gamma134.5 deleted oncolytic herpes virus (G207). *FASEB J* **18**, 1001–1003.

Bharara S, Sorscher EJ, Gillespie GY, *et al.* (2005). Antibiotic-mediated chemoprotection enhances adaptation of *E. coli* PNP for herpes simplex virus-based glioma therapy. *Hum Gene Ther* **16**, 339–347.

Blank SV, Rubin SC, Coukos G, Amin KM, Albelda SM, Molnar-Kimber KL (2002). Replication-selective herpes simplex virus type 1 mutant therapy of cervical cancer is enhanced by low-dose radiation. *Hum Gene Ther* **13**, 627–639.

Bobo RH, Laske DW, Akbasak A, Morrison PF, Dedrick RL, Oldfield EH (1994). Convection-enhanced delivery of macromolecules in the brain. *Proc Natl Acad Sci U S A* **91**, 2076–2080.

Boehmer PE, Lehman IR (1997). Herpes simplex virus DNA replication. *Annu Rev Biochem* **66**, 347–384.

Bolovan CA, Sawtell NM, Thompson RL (1994). ICP34.5 mutants of herpes simplex virus type 1 strain 17syn+ are attenuated for neurovirulence in mice and for replication in confluent primary mouse embryo cell cultures. *J Virol* **68**, 48–55.

Boviatsis EJ, Chase M, Wei MX, *et al.* (1994a). Gene transfer into experimental brain tumors mediated by adenovirus, herpes simplex virus, and retrovirus vectors. *Hum Gene Ther* **5**, 183–191.

Boviatsis EJ, Park JS, Sena-Esteves M, *et al.* (1994b). Long-term survival of rats harboring brain neoplasms treated with ganciclovir and a herpes simplex virus vector that retains an intact thymidine kinase gene. *Cancer Res* **54**, 5745–5751.

Bowers WJ, Mastrangelo MA, Howard DF, Southerland HA, Maguire-Zeiss KA, Federoff HJ (2006). Neuronal precursor-restricted transduction via *in utero* CNS gene delivery of a novel bipartite HSV amplicon/transposase hybrid vector. *Mol Ther* **13**, 580–588.

Bradley JD, Kataoka Y, Advani S, *et al.* (1999). Ionizing radiation improves survival in mice bearing intracranial high-grade gliomas injected with genetically modified herpes simplex virus. *Clin Cancer Res* **5**, 1517–1522.

Burton EA, Fink DJ, Glorioso JC (2002). Gene delivery using herpes simplex virus vectors. *DNA Cell Biol* **21**, 915–936.

Butcher M, Raviprakash K, Ghosh HP (1990). Acid pH-induced fusion of cells by herpes simplex virus glycoproteins gB and gD. *J Biol Chem* **265**, 5862–5868.

Campbell ME, Palfreyman JW, Preston CM (1984). Identification of herpes simplex virus DNA sequences which encode a *trans*-acting polypeptide responsible for stimulation of immediate early transcription. *J Mol Biol* **180**, 1–19.

Carew JF, Kooby DA, Halterman MW, Kim SH, Federoff HJ, Fong Y (2001). A novel approach to cancer therapy using an oncolytic herpes virus to package amplicons containing cytokine genes. *Mol Ther* **4**, 250–256.

Carfi A, Willis SH, Whitbeck JC, *et al.* (2001). Herpes simplex virus glycoprotein D bound to the human receptor HveA. *Mol Cell* **8**, 169–179.

Carroll NM, Chase M, Chiocca EA, Tanabe KK (1997). The effect of ganciclovir on herpes simplex virus-mediated oncolysis. *J Surg Res* **69**, 413–417.

Caruso M, Panis Y, Gagandeep S, Houssin D, Salzmann JL, Klatzmann D (1993). Regression of established macroscopic liver metastases after *in situ* transduction of a suicide gene. *Proc Natl Acad Sci U S A* **90**, 7024–7028.

Chadwick MP, Morling FJ, Cosset FL, Russell SJ (1999). Modification of retroviral tropism by display of IGF-I. *J Mol Biol* **285**, 485–494.

Chahlavi A, Todo T, Martuza RL, Rabkin SD (1999). Replication-competent herpes simplex virus vector G207 and cisplatin combination therapy for head and neck squamous cell carcinoma. *Neoplasia* **1**, 162–169.

Chakraborti S, Mandal M, Das S, Mandal A, Chakraborti T (2003). Regulation of matrix metalloproteinases: an overview. *Mol Cell Biochem* **253**, 269–285.

Chase M, Chung RY and Chiocca EA (1998). An oncolytic viral mutant that delivers the CYP2B1 transgene and augments cyclophosphamide chemotherapy. *Nat Biotechnol* **16**, 444–448.

Chattopadhyay M, Wolfe D, Mata M, Huang S, Glorioso JC, Fink DJ (2005). Long-term neuroprotection achieved with latency-associated promoter-driven herpes simplex virus gene transfer to the peripheral nervous system. *Mol Ther* **12**, 307–313.

Chee AV, Roizman B (2004). Herpes simplex virus 1 gene products occlude the interferon signaling pathway at multiple sites. *J Virol* **78**, 4185–4196.

Chintala SK, Rao JK (1996). Invasion of human glioma: role of extracellular matrix proteins. *Front Biosci* **1**, d324–339.

Chintala SK, Sawaya R, Gokaslan ZL, Fuller G, Rao JS (1996). Immunohistochemical localization of extracellular matrix proteins in human glioma, both *in vivo* and *in vitro*. *Cancer Lett* **101**, 107–114.

Chiocca EA (2002). Oncolytic viruses. *Nat Rev Cancer* **2**, 938–950.

Choe G, Park JK, Jouben-Steele L, *et al.* (2002). Active matrix metalloproteinase 9 expression is associated with primary glioblastoma subtype. *Clin Cancer Res* **8**, 2894–2901.

Chou J, Roizman B (1992). The gamma 1(34.5) gene of herpes simplex virus 1 precludes neuroblastoma cells from triggering total shutoff of protein synthesis characteristic of programed cell death in neuronal cells. *Proc Natl Acad Sci U S A* **89**, 3266–3270.

Chou J, Kern ER, Whitley RJ, Roizman B (1990). Mapping of herpes simplex virus-1 neurovirulence to gamma 134.5, a gene nonessential for growth in culture. *Science* **250**, 1262–1266.

Chung RY, Saeki Y, Chiocca EA (1999). B-myb promoter retargeting of herpes simplex virus gamma34.5 gene-mediated virulence toward tumor and cycling cells. *J Virol* **73**, 7556–7564.

Chung SM, Advani SJ, Bradley JD, *et al.* (2002). The use of a genetically engineered herpes simplex virus (R7020) with ionizing radiation for experimental hepatoma. *Gene Ther* **9**, 75–80.

Cocchi F, Fusco D, Menotti L, *et al.* (2004). The soluble ectodomain of herpes simplex virus gD contains a membrane-proximal pro-fusion domain and suffices to mediate virus entry. *Proc Natl Acad Sci U S A* **101**, 7445–7450.

Cosset FL, Russell SJ (1996). Targeting retrovirus entry. *Gene Ther* **3**, 946–956.

Cosset FL, Morling FJ, Takeuchi Y, Weiss RA, Collins MK, Russell SJ (1995). Retroviral retargeting by envelopes expressing an N-terminal binding domain. *J Virol* **69**, 6314–6322.

Coukos G, Makrigiannakis A, Kang EH, Rubin SC, Albelda SM, Molnar-Kimber KL (2000). Oncolytic herpes simplex virus-1 lacking ICP34.5 induces p53-independent death and is efficacious against chemotherapy-resistant ovarian cancer. *Clin Cancer Res* **6**, 3342–3353.

Cozzi PJ, Malhotra S, McAuliffe P, *et al.* (2001). Intravesical oncolytic viral therapy using attenuated, replication-competent herpes simplex viruses G207 and Nv1020 is effective in the treatment of bladder cancer in an orthotopic syngeneic model. *FASEB J* **15**, 1306–1308.

Croen KD, Ostrove JM, Dragovic LJ, Smialek JE, Straus SE (1987). Latent herpes simplex virus in human trigeminal ganglia. Detection of an immediate early gene 'anti-sense' transcript by in situ hybridization. *N Engl J Med* **317**, 1427–1432.

Croteau D, Walbridge S, Morrison PF, *et al.* (2005). Real-time *in vivo* imaging of the convective distribution of a low-molecular-weight tracer. *J Neurosurg* **102**, 90–97.

Cuchet D, Ferrera R, Lomonte P, Epstein AL (2005). Characterization of antiproliferative and cytotoxic properties of the HSV-1 immediate-early ICPo protein. *J Gene Med* **7**, 1187–1199.

Culver KW, Ram Z, Wallbridge S, Ishii H, Oldfield EH, Blaese RM (1992). *In vivo* gene transfer with retroviral vector-producer cells for treatment of experimental brain tumors. *Science* **256**, 1550– 1552.

D'Angelica M, Tung C, Allen P, *et al.* (1999). Herpes simplex virus (HSV)-mediated ICAM-1 gene transfer abrogates tumorigenicity and induces anti-tumor immunity. *Mol Med* **5**, 606–616.

Davis LW (1989). Malignant glioma – a nemesis which requires clinical and basic investigation in radiation oncology. *Int J Radiat Oncol Biol Phys* **16**, 1355–1365.

Dean HJ, Terhune SS, Shieh MT, Susmarski N, Spear PG (1994). Single amino acid substitutions in gD of herpes simplex virus 1 confer resistance to gD-mediated interference and cause cell-type-dependent alterations in infectivity. *Virology* **199**, 67–80.

Deiss LP, Chou J and Frenkel N (1986). Functional domains within the a sequence involved in the cleavage-packaging of herpes simplex virus DNA. *J Virol* **59**, 605–618.

Delman KA, Zager JS, Bennett JJ, *et al.* (2002). Efficacy of multiagent herpes simplex virus amplicon-mediated immunotherapy as adjuvant treatment for experimental hepatic cancer. *Ann Surg* **236**, 337–342.

DeLuca NA, McCarthy AM, Schaffer PA (1985). Isolation and characterization of deletion mutants of herpes simplex virus type 1 in the gene encoding immediate-early regulatory protein ICP4. *J Virol* **56**, 558–570.

Detta A, Harland J, Hanif I, Brown SM, Cruickshank G (2003). Proliferative activity and *in vitro* replication of HSV1716 in human metastatic brain tumours. *J Gene Med* **5**, 681–689.

Douglas MW, Diefenbach RJ, Homa FL, *et al.* (2004). Herpes simplex virus type 1 capsid protein VP26 interacts with dynein light chains RP3 and Tctex1 and plays a role in retrograde cellular transport. *J Biol Chem* **279**, 28522–28530.

Ebright MI, Zager JS, Malhotra S, *et al.* (2002). Replication-competent herpes virus NV1020 as direct treatment of pleural cancer in a rat model. *J Thorac Cardiovasc Surg* **124**, 123–129.

Eck SL, Alavi JB, Alavi A, *et al.* (1996). Treatment of advanced CNS malignancies with the recombinant adenovirus H5.010RSVTK: a phase I trial. *Hum Gene Ther* **7**, 1465–1482.

Elledge SJ, Zhou Z, Allen JB (1992). Ribonucleotide reductase: regulation, regulation, regulation. *Trends Biochem Sci* **17**, 119–123.

Epstein AL, Marconi P, Argnani R, Manservigi R (2005). HSV-1-derived recombinant and amplicon vectors for gene transfer and gene therapy. *Curr Gene Ther* **5**, 445–458.

Ezzeddine ZD, Martuza RL, Platika D, *et al.* (1991). Selective killing of glioma cells in culture and *in vivo* by retrovirus transfer of the herpes simplex virus thymidine kinase gene. *New Biol* **3**, 608–614.

Fesik SW (2005). Promoting apoptosis as a strategy for cancer drug discovery. *Nat Rev Cancer* **5**, 876–885.

Fielding AK, Maurice M, Morling FJ, Cosset FL, Russell SJ (1998). Inverse targeting of retroviral vectors: selective gene transfer in a mixed population of hematopoietic and nonhematopoietic cells. *Blood* **91**, 1802–1809.

Fink DJ, Sternberg LR, Weber PC, Mata M, Goins WF, Glorioso JC (1992). *In vivo* expression of beta-galactosidase in hippocampal neurons by HSV-mediated gene transfer. *Hum Gene Ther* **3**, 11–19.

Fruh K, Ahn K, Djaballah H, *et al.* (1995). A viral inhibitor of peptide transporters for antigen presentation. *Nature* **375**, 415–418.

Fuller AO, Lee WC (1992). Herpes simplex virus type 1 entry through a cascade of virus-cell interactions requires different roles of gD and gH in penetration. *J Virol* **66**, 5002–5012.

Fuller AO, Spear PG (1985). Specificities of monoclonal and polyclonal antibodies that inhibit adsorption of herpes simplex virus to cells and lack of inhibition by potent neutralizing antibodies. *J Virol* **55**, 475–482.

Ganly I, Kirn D, Eckhardt G, *et al.* (2000). A phase I study of Onyx-015, an E1B attenuated adenovirus, administered intratumorally to patients with recurrent head and neck cancer. *Clin Cancer Res* **6**, 798–806.

Ge K, DuHadaway J, Du W, Herlyn M, Rodeck U, Prendergast GC (1999). Mechanism for elimination of a tumor suppressor: aberrant splicing of a brain-specific exon causes loss of function of Bin1 in melanoma. *Proc Natl Acad Sci U S A* **96**, 9689–9694.

Geller AI, Breakefield XO (1988). A defective HSV-1 vector expresses *Escherichia coli* beta-galactosidase in cultured peripheral neurons. *Science* **241**, 1667–1669.

Geller AI, Keyomarsi K, Bryan J, Pardee AB (1990). An efficient deletion mutant packaging system for defective herpes simplex virus vectors: potential applications to human gene therapy and neuronal physiology. *Proc Natl Acad Sci U S A* **87**, 8950–8954.

Geraghty RJ, Krummenacher C, Cohen GH, Eisenberg RJ, Spear PG (1998). Entry of alphaherpesviruses mediated by poliovirus receptor-related protein 1 and poliovirus receptor. *Science* **280**, 1618–1620.

Girod A, Ried M, Wobus C, *et al.* (1999). Genetic capsid modifications allow efficient re-targeting of adeno-associated virus type 2. *Nat Med* **5**, 1052–1056.

Glauser DL, Ackermann M, Saydam O, Fraefel C (2006). Chimeric herpes simplex virus/adeno-associated virus amplicon vectors. *Curr Gene Ther* **6**, 315–324.

Goins WF, Sternberg LR, Croen KD, *et al.* (1994). A novel latency-active promoter is contained within the herpes simplex virus type 1 UL flanking repeats. *J Virol* **68**, 2239–2252.

Goins WF, Lee KA, Cavalcoli JD, *et al.* (1999). Herpes simplex virus type 1 vector-mediated expression of nerve growth factor protects dorsal root ganglion neurons from peroxide toxicity. *J Virol* **73**, 519–532.

Goldbrunner RH, Bernstein JJ, Tonn JC (1999). Cell-extracellular matrix interaction in glioma invasion. *Acta Neurochir* **141**, 295–305.

Goldstein DJ, Weller SK (1988a). Herpes simplex virus type 1-induced ribonucleotide reductase activity is dispensable for virus growth and DNA synthesis: isolation and characterization of an ICP6 lacZ insertion mutant. *J Virol* **62**, 196–205.

Goldstein DJ, Weller SK (1988b). Factor(s) present in herpes simplex virus type 1-infected cells can compensate for the loss of the large subunit of the viral ribonucleotide reductase: characterization of an ICP6 deletion mutant. *Virology* **166**, 41–51.

Grandi P, Wang S, Schuback D, *et al.* (2004). HSV-1 virions engineered for specific binding to cell surface receptors. *Mol Ther* **9**, 419–427.

Greaves RF and O'Hare P (1990). Structural requirements in the herpes simplex virus type 1 transactivator Vmw65 for interaction with the cellular octamer-binding protein and target TAATGARAT sequences. *J Virol* **64**, 2716–2724.

Grill J, Van Beusechem VW, Van Der Valk P, *et al.* (2001). Combined targeting of adenoviruses to integrins and epidermal growth factor receptors increases gene transfer into primary glioma cells and spheroids. *Clin Cancer Res* **7**, 641–650.

Gruenheid S, Gatzke L, Meadows H, Tufaro F (1993). Herpes simplex virus infection and propagation in a mouse L cell mutant lacking heparan sulfate proteoglycans. *J Virol* **67**, 93–100.

Hadjipanayis CG, DeLuca NA (2005). Inhibition of DNA repair by a herpes simplex virus vector enhances the radiosensitivity of human glioblastoma cells. *Cancer Res* **65**, 5310–5316.

Hagglund R, Van Sant C, Lopez P, Roizman B (2002). Herpes simplex virus 1-infected cell protein 0 contains two E3 ubiquitin ligase sites specific for different E2 ubiquitin-conjugating enzymes. *Proc Natl Acad Sci USA* **99**, 631–636.

Harrow S, Papanastassiou V, Harland J, *et al.* (2004). HSV1716 injection into the brain adjacent to tumour following surgical resection of high-grade glioma: safety data and long-term survival. *Gene Ther* **11**, 1648–1658.

He B, Gross M, Roizman B (1997). The gamma(1)34.5 protein of herpes simplex virus 1 complexes with protein phosphatase 1alpha to dephosphorylate the alpha subunit of the eukaryotic translation initiation factor 2 and preclude the shutoff of protein synthesis by double-stranded RNA-activated protein kinase. *Proc Natl Acad Sci U S A* **94**, 843–848.

Heister T, Heid I, Ackermann M, Fraefel C (2002). Herpes simplex virus type 1/adeno-associated virus hybrid vectors mediate site-specific integration at the adeno-associated virus preintegration site, AAVS1, on human chromosome 19. *J Virol* **76**, 7163–7173.

Herold BC, WuDunn D, Soltys N, Spear PG (1991). Glycoprotein C of herpes simplex virus type 1 plays a principal role in the adsorption of virus to cells and in infectivity. *J Virol* **65**, 1090–1098.

Herold BC, Visalli RJ, Susmarski N, Brandt CR, Spear PG (1994). Glycoprotein C-independent binding of herpes simplex virus to cells requires cell surface heparan sulphate and glycoprotein B. *J Gen Virol* **75**, 1211–1222.

Herold BC, Gerber SI, Polonsky T, Belval BJ, Shaklee PN, Holme K (1995). Identification of structural features of heparin required for inhibition of herpes simplex virus type 1 binding. *Virology* **206**, 1108–1116.

Herrlinger U, Kramm CM, Aboody-Guterman KS, *et al.* (1998). Pre-existing herpes simplex virus 1 (HSV-1) immunity decreases, but does not abolish, gene transfer to experimental brain tumors by a HSV-1 vector. *Gene Ther* **5**, 809–819.

Herrlinger U, Jacobs A, Quinones A, *et al.* (2000). Helper virus-free herpes simplex virus type 1 amplicon vectors for granulocyte-macrophage colony-stimulating factor-enhanced vaccination therapy for experimental glioma. *Hum Gene Ther* **11**, 1429–1438.

Highlander SL, Sutherland SL, Gage PJ, Johnson DC, Levine M, Glorioso JC (1987). Neutralizing monoclonal antibodies specific for herpes simplex virus glycoprotein D inhibit virus penetration. *J Virol* **61**, 3356–3364.

Ho IA, Hui KM, Lam PY (2004). Glioma-specific and cell cycle-regulated herpes simplex virus type 1 amplicon viral vector. *Hum Gene Ther* **15**, 495–508.

Ho IA, Hui KM, Lam PY (2006). Targeting proliferating tumor cells via the transcriptional control of therapeutic genes. *Cancer Gene Ther* **13**, 44–52.

Hochberg FH, Pruitt A (1980). Assumptions in the radiotherapy of glioblastoma. *Neurology* **30**, 907–911.

Hoffmann D, Bangen JM, Bayer W, Wildner O (2006). Synergy between expression of fusogenic membrane proteins, chemotherapy and facultative virotherapy in colorectal cancer. *Gene Ther* **13**, 1534–1544.

Homa FL, Brown JC (1997). Capsid assembly and DNA packaging in herpes simplex virus. *Rev Med Virol* **7**, 107–122.

Honess RW, Roizman B (1974). Regulation of herpesvirus macromolecular synthesis. I. Cascade regulation of the synthesis of three groups of viral proteins. *J Virol* **14**, 8–19.

Honess RW, Roizman B (1975). Proteins specified by herpes simplex virus. XIII. Glycosylation of viral polypeptides. *J Virol* **16**, 1308–1326.

Hoshi M, Harada A, Kawase T, Uyemura K, Yazaki T (2000). Antitumoral effects of defective herpes simplex virus-mediated transfer of tissue inhibitor of metalloproteinases-2 gene in malignant glioma U87 *in vitro*: consequences for anti-cancer gene therapy. *Cancer Gene Ther* **7**, 799–805.

Hu JC, Coffin RS, Davis CJ, *et al.* (2006). A phase I study of OncoVEXGM-CSF, a second-generation oncolytic herpes simplex virus expressing granulocyte macrophage colony-stimulating factor. *Clin Cancer Res* **12**(22), 6737–6747.

Hummel JL, Safroneeva E, Mossman KL (2005). The role of ICP0-Null HSV-1 and interferon signaling defects in the effective treatment of breast adenocarcinoma. *Mol Ther* **12**, 1101–1110.

Humphrey PA, Wong AJ, Vogelstein B, *et al.* (1990). Anti-synthetic peptide antibody reacting at the fusion junction of deletion-mutant epidermal growth factor receptors in human glioblastoma. *Proc Natl Acad Sci USA* **87**, 4207–4211.

Hunter WD, Martuza RL, Feigenbaum F, *et al.* (1999). Attenuated, replication-competent herpes simplex virus type 1 mutant G207: safety evaluation of intracerebral injection in nonhuman primates. *J Virol* **73**, 6319–6326.

Inoue R, Moghaddam KA, Ranasinghe M, Saeki Y, Chiocca EA, Wade-Martins R (2004). Infectious delivery of the 132 kb CDKN2A/CDKN2B genomic DNA region results in correctly spliced gene expression and growth suppression in glioma cells. *Gene Ther* **11**, 1195–1204.

Jarnagin WR, Delman K, Kooby D, *et al.* (2000). Neoadjuvant interleukin-12 immunogene therapy protects against cancer recurrence after liver resection in an animal model. *Ann Surg* **231**, 762–771.

Jarnagin WR, Zager JS, Hezel M, *et al.* (2006). Treatment of cholangiocarcinoma with oncolytic herpes simplex virus combined with external beam radiation therapy. *Cancer Gene Ther* **13**, 326–334.

Johnson PA, Wang MJ, Friedmann T (1994). Improved cell survival by the reduction of immediate-early gene expression in replication-defective mutants of herpes simplex virus type 1 but not by mutation of the virion host shutoff function. *J Virol* **68**, 6347–6362.

Kanai R, Eguchi K, Takahashi M, *et al.* (2006). Enhanced therapeutic efficacy of oncolytic herpes vector G207 against human non-small cell lung cancer-expression of an RNA-binding protein, Musashi1, as a marker for the tailored gene therapy. *J Gen Med* **8**, 1329–1340.

Kasahara N, Dozy AM, Kan YW (1994). Tissue-specific targeting of retroviral vectors through ligand-receptor interactions. *Science* **266**, 1373–1376.

Kasamatsu H, Nakanishi A (1998). How do animal DNA viruses get to the nucleus? *Annu Rev Microbiol* **52**, 627–686.

Kasuya H, Pawlik TM, Mullen JT, *et al.* (2004). Selectivity of an oncolytic herpes simplex virus for cells expressing the DF3/MUC1 antigen. *Cancer Res* **64**, 2561–2567.

Katze MG, He Y, Gale MJ (2002). Viruses and interferon: a fight for supremacy. *Nat Rev Immunol* **2**, 675–687.

Kaufmann SH, Vaux DL (2003). Alterations in the apoptotic machinery and their potential role in anticancer drug resistance. *Oncogene* **22**, 7414–7430.

Kelley SK, Harris LA, Xie D, *et al.* (2001). Preclinical studies to predict the disposition of Apo2L/tumor necrosis factor-related apoptosis-inducing ligand in humans: characterization of *in vivo* efficacy, pharmacokinetics, and safety. *J Pharmacol Exp Ther* **299**, 31–38.

Kesari S, Randazzo BP, Valyi-Nagy T, *et al.* (1995). Therapy of experimental human brain tumors using a neuroattenuated herpes simplex virus mutant. *Lab Invest* **73**, 636–648.

Khuri FR, Nemunaitis J, Ganly I, *et al.* (2000). a controlled trial of intratumoral ONYX-015, a selectively-replicating adenovirus, in combination with cisplatin and 5-fluorouracil in patients with recurrent head and neck cancer. *Nat Med* **6**, 879–885.

Kim SH, Wong RJ, Kooby DA, *et al.* (2005). Combination of mutated herpes simplex virus type 1 (G207 virus) with radiation for the treatment of squamous cell carcinoma of the head and neck. *Eur J Cancer* **41**, 313–322.

Kimata H, Takakuwa H, Goshima F, *et al.* (2003). Effective treatment of disseminated peritoneal colon cancer with new replication-competent herpes simplex viruses. *Hepatogastroenterology* **50**, 961–966.

Kondraganti S, Mohanam S, Chintala SK, *et al.* (2000). Selective suppression of matrix metalloproteinase-9 in human glioblastoma cells by antisense gene transfer impairs glioblastoma cell invasion. *Cancer Res* **60**, 6851–6855.

Kotin RM, Linden RM, Berns KI (1992). Characterization of a preferred site on human chromosome 19q for integration of adeno-associated virus DNA by non-homologous recombination. *EMBO J* **11**, 5071- 5078.

Kramm CM, Chase M, Herrlinger U, *et al.* (1997). Therapeutic efficiency and safety of a second-generation replication-conditional HSV1 vector for brain tumor gene therapy. *Hum Gene Ther* **8**, 2057–2068.

Krisky DM, Marconi PC, Oligino TJ, *et al.* (1998a). Development of herpes simplex virus replication-defective multigene vectors for combination gene therapy applications. *Gene Ther* **5**, 1517–1530.

Krisky DM, Wolfe D, Goins WF, *et al.* (1998b). Deletion of multiple immediate-early genes from herpes simplex virus reduces cytotoxicity and permits long-term gene expression in neurons. *Gene Ther* **5**, 1593–1603.

Krummenacher C, Nicola AV, Whitbeck JC, *et al.* (1998). Herpes simplex virus glycoprotein D can bind to poliovirus receptor-related protein 1 or herpesvirus entry mediator, two structurally unrelated mediators of virus entry. *J Virol* **72**, 7064–7074.

Kucharczuk JC, Randazzo B, Chang MY, *et al.* (1997). Use of a 'replication-restricted' herpes virus to treat experimental human malignant mesothelioma. *Cancer Res* **57**, 466–471.

Kunwar S (2003). Convection enhanced delivery of IL13-PE38QQR for treatment of recurrent malignant glioma: presentation of interim findings from ongoing phase 1 studies. *Acta Neurochir* Suppl **88**, 105–111.

Kurimasa A, Kumano S, Boubnov NV, *et al.* (1999). Requirement for the kinase activity of human DNA-dependent protein kinase catalytic subunit in DNA strand break rejoining. *Mol Cell Biol* **19**, 3877–3884.

Kutubuddin M, Federoff HJ, Challita-Eid PM, *et al.* (1999). Eradication of pre-established lymphoma using herpes simplex virus amplicon vectors. *Blood* **93**, 643–654.

Kwong AD and Frenkel N (1989). The herpes simplex virus virion host shutoff function. *J Virol* **63**, 4834–4839.

Kwong AD, Kruper JA, Frenkel N (1988). Herpes simplex virus virion host shutoff function. *J Virol* **62**, 912–921.

Lacroix M, Abi-Said D, Fourney DR, *et al.* (2001). A multivariate analysis of 416 patients with glioblastoma multiforme: prognosis, extent of resection, and survival. *J Neurosurg* **95**, 190–198.

Laquerre S, Argnani R, Anderson DB, Zucchini S, Manservigi R, Glorioso JC (1998). Heparan sulfate proteoglycan binding by herpes simplex virus type 1 glycoproteins B and C, which differ in their contributions to virus attachment, penetration, and cell-to-cell spread. *J Virol* **72**, 6119–6130.

Lawler SE, Peruzzi PP, Chiocca EA (2006). Genetic strategies for brain tumor therapy. *Cancer Gene Ther* **13**, 225–233.

Lees-Miller SP, Godbout R, Chan DW, *et al.* (1995). Absence of p350 subunit of DNA-activated protein kinase from a radiosensitive human cell line. *Science* **267**, 1183–1185.

Lees-Miller SP, Long MC, Kilvert MA, Lam V, Rice SA, Spencer CA (1996). Attenuation of DNA-dependent protein kinase activity and its catalytic subunit by the herpes simplex virus type 1 transactivator ICP0. *J Virol* **70**, 7471–7477.

Lefranc F, Brotchi J and Kiss R (2005). Possible future issues in the treatment of glioblastomas: special emphasis on cell migration and the resistance of migrating glioblastoma cells to apoptosis. *J Clin Oncol* **23**, 2411–2422.

Leib DA, Machalek MA, Williams BR, Silverman RH, Virgin HW (2000). Specific phenotypic restoration of an attenuated virus by knockout of a host resistance gene. *Proc Natl Acad Sci U S A* **97**, 5684–5686.

Leibel SA, Sheline GE (1987). Radiation therapy for neoplasms of the brain. *J Neurosurg* **66**, 1–22.

Li Y, van Drunen Littel-van den Hurk S, Babiuk LL, Liang X (1995). haracterization of cell-binding properties of bovine herpesvirus 1 glycoproteins B, C, and D: identification of a dual cell-binding function of gB. *J Virol* **69**, 4758–4768.

Ligas M, Johnson D (1988). A herpes simplex virus mutant in which glycoprotein D sequences are replaced by β-galactosidase sequences binds to but is unable to penetrate into cells. *J Virol* **62**, 1486–1494.

Lilley CE, Groutsi F, Han Z, *et al.* (2001). Multiple immediate-early gene-deficient herpes simplex virus vectors allowing efficient gene delivery to neurons in culture and widespread gene delivery to the central nervous system *in vivo*. *J Virol* **75**, 4343–4356.

Lim F, Hartley D, Starr P, *et al.* (1996). Generation of high-titer defective HSV-1 vectors using an IE 2 deletion mutant and quantitative study of expression in cultured cortical cells. *Biotechniques* **20**, 460–469.

Lin R, Noyce RS, Collins SE, Everett RD, Mossman KL (2004). The herpes simplex virus ICP0 RING finger domain inhibits IRF3- and IRF7-mediated activation of interferon-stimulated genes. *J Virol* **78**, 1675–1684.

Liu BL, Robinson M, Han ZQ, *et al.* (2003). ICP34.5 deleted herpes simplex virus with enhanced oncolytic, immune stimulating, and anti-tumour properties. *Gene Ther* **10**, 867–874.

Mackem S, Roizman B (1982). Structural features of the herpes simplex virus alpha gene 4, 0, and 27 promoter-regulatory sequences which confer alpha regulation on chimeric thymidine kinase genes. *J Virol* **44**, 939–949.

MacKie RM, Stewart B, Brown SM (2001). Intralesional injection of herpes simplex virus 1716 in metastatic melanoma. *Lancet Oncol* **357**, 525–526.

MacLean AR, ul-Fareed M, Robertson L, Harland J, Brown SM (1991). Herpes simplex virus type 1 deletion variants 1714 and 1716 pinpoint neurovirulence-related sequences in Glasgow strain 17+ between immediate early gene 1 and the 'a' sequence. *J Gen Virol* **72**, 631–639.

Marconi P, Tamura M, Moriuchi S, *et al.* (2000). Connexin 43-enhanced suicide gene therapy using herpesviral vectors. *Mol Ther* **1**, 71–81.

Markert JM (2004). Biologic warfare for a good cause: HSV-1 anti-tumor therapy. *Clin Neurosurg* **51**, 73–80.

Markert JM, Gillespie GY, Weichselbaum RR, Roizman B, Whitley RJ (2000a). Genetically engineered HSV in the treatment of glioma: a review. *Rev Med Virol* **10**, 17–30.

Markert JM, Medlock MD, Rabkin SD, *et al.* (2000b). Conditionally replicating herpes simplex virus mutant, G207 for the treatment of malignant glioma: results of a phase I trial. *Gene Ther* **7**, 867–874.

Martin A, Benichou D, Chao SF, Cohen LM, Lemon SM (1999). Maturation of the hepatitis A virus capsid protein VP1 is not dependent on processing by the 3Cpro proteinase. *J Virol* **73**, 6220–6227.

Martiniello-Wilks R, Garcia-Aragon J, Daja MM, *et al.* (1998). *In vivo* gene therapy for prostate cancer: preclinical evaluation of two different enzyme-directed prodrug therapy systems delivered by identical adenovirus vectors. *Hum Gene Ther* **9**, 1617–1626.

Martuza RL, Malick A, Markert JM, Ruffner KL, Coen DM (1991). Experimental therapy of human glioma by means of a genetically engineered virus mutant. *Science* **252**, 854–856.

McCarthy AM, McMahan L, Schaffer PA (1989). Herpes simplex virus type 1 ICP27 deletion mutants exhibit altered patterns of transcription and are DNA deficient. *J Virol* **63**, 18–27.

McCormick F (2001). Cancer gene therapy: fringe or cutting edge? *Nat Rev Cancer* **1**, 130–141.

McKie EA, MacLean AR, Lewis AD, *et al.* (1996). Selective *in vitro* replication of herpes simplex virus type 1 (HSV-1) ICP34.5 null mutants in primary human CNS tumours – evaluation of a potentially effective clinical therapy. *Br J Cancer* **74**, 745–752.

Meignier B, Longnecker R, Mavromara-Nazos P, Sears AE, Roizman B (1988). Virulence of and establishment of latency by genetically engineered deletion mutants of herpes simplex virus 1. *Virology* **162**, 251–254.

Menotti L, Cerretani A, Campadelli-Fiume G (2006). A herpes simplex virus recombinant that exhibits a single-chain antibody to HER2/neu enters cells through the mammary tumor receptor, independently of the gD receptors. *J Virol* **80**, 5531–5539.

Mettenleiter TC (1994). Initiation and spread of A-herpesvirus infections. *Trends Microbiol* **2**, 2–3.

Mettenleiter TC (2002). Herpesvirus assembly and egress. *J Virol* **76**, 1537–1547.

Mineta T, Rabkin SD, Martuza RL (1994). Treatment of malignant gliomas using ganciclovir-hypersensitive, ribonucleotide reductase-deficient herpes simplex viral mutant. *Cancer Res* **54**, 3963–3966.

Mineta T, Rabkin SD, Yazaki T, Hunter WD, Martuza RL (1995). Attenuated multi-mutated herpes simplex virus-1 for the treatment of malignant gliomas. *Nat Med* **1**, 938–943.

Miyatake S (2002). Gene therapy using tissue-specific replication competent HSV. *Hum Cell* **15**, 130–137.

Mizuguchi H, Kay MA, Hayakawa T (2001). Approaches for generating recombinant adenovirus vectors. *Adv Drug Deliv Rev* **52**, 165–176.

Mocarski ES, Roizman B (1982). Structure and role of the herpes simplex virus DNA termini in inversion, circularization and generation of virion DNA. *Cell* **31**, 89–97.

Montgomery RI, Warner MS, Lum BJ, Spear PG (1996). Herpes simplex virus-1 entry into cells mediated by a novel member of the TNF/NGF receptor family. *Cell* **87**, 427–436.

Moolten FL (1986). Tumor chemosensitivity conferred by inserted herpes thymidine kinase genes: paradigm for a prospective cancer control strategy. *Cancer Res* **46**, 5276–5281.

Moolten FL (1994). Drug sensitivity ('suicide') genes for selective cancer chemotherapy. *Cancer Gene Ther* **1**, 279–287.

Moriuchi S, Oligino T, Krisky D, *et al.* (1998). Enhanced tumor cell killing in the presence of ganciclovir by herpes simplex virus type 1 vector-directed coexpression of human tumor necrosis factor-alpha and herpes simplex virus thymidine kinase. *Cancer Res* **58**, 5731–5737.

Moriuchi S, Krisky DM, Marconi PC, *et al.* (2000). HSV vector cytotoxicity is inversely correlated with effective TK/GCV suicide gene therapy of rat gliosarcoma. *Gene Ther* **7**, 1483–1490.

Moriuchi S, Glorioso JC, Maruno M, *et al.* (2005). Combination gene therapy for glioblastoma involving herpes simplex virus vector-mediated codelivery of mutant IkappaBalpha and HSV thymidine kinase. *Cancer Gene Ther* **12**, 487–496.

Morling FJ, Peng KW, Cosset FL, Russell SJ (1997). Masking of retroviral envelope functions by oligomerizing polypeptide adaptors. *Virology* **234**, 51–61.

Morrison PF, Laske DW, Bobo H, Oldfield EH, Dedrick RL (1994). High-flow microinfusion: tissue penetration and pharmacodynamics. *Am J Physiol* **266**, R292–305.

Mossman KL, Smiley JR (1999). Truncation of the C-terminal acidic transcriptional activation domain of herpes simplex virus VP16 renders expression of the immediate-early genes almost entirely dependent on ICP0. *J Virol* **73**, 9726–9733.

Mossman KL (2002). Activation and inhibition of virus and interferon: the herpesvirus story. *Viral Immunol* **15**, 3–15.

Mullen JT, Donahue JM, Chandrasekhar S, *et al.* (2004). Oncolysis by viral replication and inhibition of angiogenesis by a replication-conditional herpes simplex virus that expresses mouse endostatin. *Cancer* **101**, 869–877.

Muller AJ, Prendergast GC (2005). Marrying immunotherapy with chemotherapy: why say IDO? *Cancer Res* **65**, 8065–8068.

Muller AJ, DuHadaway JB, Donover PS, Sutanto-Ward E, Prendergast GC (2005). Inhibition of indoleamine 2,3-dioxygenase, an immunoregulatory target of the cancer suppression gene Bin1, potentiates cancer chemotherapy. *Nat Med* **11**, 312–319.

Mullerad M, Bochner BH, Adusumilli PS, *et al.* (2005). Herpes simplex virus based gene therapy enhances the efficacy of mitomycin C for the treatment of human bladder transitional cell carcinoma. *J Urol* **174**, 741–746.

Mulvey M, Poppers J, Sternberg D, Mohr I (2003). Regulation of eIF2alpha phosphorylation by different functions that act during discrete phases in the herpes simplex virus type 1 life cycle. *J Virol* **77**, 10917–10928.

Nakamura H, Mullen JT, Chandrasekhar S, Pawlik TM, Yoon SS, Tanabe KK (2001). Multimodality therapy with a replication-conditional herpes simplex virus 1 mutant that expresses yeast cytosine deaminase for intratumoral conversion of 5-fluorocytosine to 5-fluorouracil. *Cancer Res* **61**, 5447–5452.

Nakanishi C, Toi M (2005). Nuclear factor-kappaB inhibitors as sensitizers to anticancer drugs. *Nat Rev Cancer* **5**, 297–309.

Nakano K, Todo T, Chijiiwa K, Tanaka M (2001). Therapeutic efficacy of G207, a conditionally replicating herpes simplex virus type 1 mutant, for gallbladder carcinoma in immunocompetent hamsters. *Mol Ther* **3**, 431–437.

Nakano K, Asano R, Tsumoto K, *et al.* (2005). Herpes simplex virus targeting to the EGF receptor by a gD-specific soluble bridging molecule. *Mol Ther* **11**, 617–626.

Nakao A, Kimata H, Imai T, *et al.* (2004). Intratumoral injection of herpes simplex virus HF10 in recurrent breast cancer. *Ann Oncol* **15**, 988–989.

Nazzaro JM, Neuwelt EA (1990). The role of surgery in the management of supratentorial intermediate and high-grade astrocytomas in adults. *J Neurosurg* **73**, 331–344.

Nelson AR, Fingleton B, Rothenberg ML, Matrisian LM (2000). Matrix metalloproteinases: biologic activity and clinical implications. *J Clin Oncol* **18**, 1135–1149.

Nelson AL, Algon SA, Munasinghe J, *et al.* (2003). Magnetic resonance imaging of patched heterozygous and xenografted mouse brain tumors. *J Neurooncol* **62**, 259–267.

Newcomb WW, Homa FL, Thomsen DR, *et al.* (1999). Assembly of the herpes simplex virus procapsid from purified components and identification of small complexes containing the major capsid and scaffolding proteins. *J Virol* **73**, 4239–4250.

Nicola AV, Ponce de Leon M, Xu R, *et al.* (1998). Monoclonal antibodies to distinct sites on herpes simplex virus (HSV) glycoprotein D block HSV binding to HVEM. *J Virol* **72**, 3595–3601.

Nicola AV, McEvoy AM, Straus SE (2003). Roles for endocytosis and low pH in herpes simplex virus entry into HeLa and Chinese hamster ovary cells. *J Virol* **77**, 5324–5332.

Nilson BH, Morling FJ, Cosset FL, Russell SJ (1996). Targeting of retroviral vectors through protease-substrate interactions. *Gene Ther* **3**, 280–286.

Niranjan A, Moriuchi S, Lunsford LD, *et al.* (2000). Effective treatment of experimental glioblastoma by HSV vector-mediated TNF alpha and HSV-tk gene transfer in combination with radiosurgery and ganciclovir administration. *Mol Ther* **2**, 114–120.

Niranjan A, Wolfe D, Tamura M, *et al.* (2003). Treatment of rat gliosarcoma brain tumors by HSV-based multigene therapy combined with radiosurgery. *Mol Ther* **8**, 530–542.

Nishiyama Y (2004). Herpes simplex virus gene products: the accessories reflect her lifestyle well. *Rev Med Virol* **14**, 33–46.

Nishiyama Y, Murata T (2002). Anti-apoptotic protein kinase of herpes simplex virus. *Trends Microbiol* **10**, 105–107.

Nunez R, Fraefel C, Suter M, Nunez-Liman A, Liou HC, Ackerman M (2004). ssessment of a cellular vaccination approach consisting of crawling dendritic cells (CDCs) transduced with HSV-1-Deltapac vectors. *Immunol Res* **30**, 105–124.

Oehmig A, Fraefel C, Breakefield XO (2004). Update of herpesvirus amplicon vectors. *Mol Ther* **10**, 630–643.

Palmer JA, Branston RH, Lilley CE, *et al.* (2000). Development and optimization of herpes simplex virus vectors for multiple long-term gene delivery to the peripheral nervous system. *J Virol* **74**, 5604–5618.

Papanastassiou V, Rampling R, Fraser M, *et al.* (2002). The potential for efficacy of the modified (ICP 34.5(-)) herpes simplex virus HSV1716 following intratumoral injection into human malignant glioma: a proof of principle study. *Gene Ther* **9**, 398–406.

Parker JN, Gillespie GY, Love CE, Randall S, Whitley RJ, Markert JM (2000). Engineered herpes simplex virus expressing IL-12 in the treatment of experimental murine brain tumors. *Proc Natl Acad Sci U S A* **97**, 2208–2213.

Parkinson J, Lees-Miller SP, Everett RD (1999). Herpes simplex virus type 1 immediate-early protein vmw110 induces the proteasome-dependent degradation of the catalytic subunit of DNA-dependent protein kinase. *J Virol* **73**, 650–657.

Pechan PA, Herrlinger U, Aghi M, Jacobs A, Breakefield XO (1999). Combined HSV-1 recombinant and amplicon piggyback vectors: replication-competent and defective forms, and therapeutic efficacy for experimental gliomas. *J Gene Med* **1**, 176–185.

Petrowsky H, Roberts GD, Kooby DA, *et al.* (2001). Functional interaction between fluorodeoxyuridine-induced cellular alterations and replication of a ribonucleotide reductase-negative herpes simplex virus. *J Virol* **75**, 7050–7058.

Pin RH, Reinblatt M, Bowers WJ, Federoff HJ, Fong Y (2004). Herpes simplex virus amplicon delivery of a hypoxia-inducible angiogenic inhibitor blocks capillary formation in hepatocellular carcinoma. *J Gastrointest Surg* **8**, 812–822.

Platten M, Wick W and Weller M (2001). Malignant glioma biology: role for TGF-beta in growth, motility, angiogenesis, and immune escape. *Microsc Res Tech* **52**, 401–410.

Poppers J, Mulvey M, Khoo D, Mohr I (2000). Inhibition of PKR activation by the proline-rich RNA binding domain of the herpes simplex virus type 1 Us11 protein. *J Virol* **74**, 11215–11221.

Prados MD, Schold SC, Spence AM, *et al.* (1996). Phase II study of paclitaxel in patients with recurrent malignant glioma. *J Clin Oncol* **14**, 2316–2321.

Puskovic V, Wolfe D, Goss J, *et al.* (2004). Prolonged biologically active transgene expression driven by HSV LAP2 in brain *in vivo*. *Mol Ther* **10**, 67–75.

Quigley MR, Maroon JC (1991). The relationship between survival and the extent of the resection in patients with supratentorial malignant gliomas. *Neurosurgery* **29**, 385–388.

Rainov NG, Dobberstein KU, Sena-Esteves M, *et al.* (1998). New prodrug activation gene therapy for cancer using cytochrome P450 4B1 and 2-aminoanthracene/4-ipomeanol. *Hum Gene Ther* **9**, 1261–1273.

Rajcani J, Andrea V, Ingeborg R (2004). Peculiarities of herpes simplex virus (HSV) transcription: an overview. *Virus Genes* **28**, 293–310.

Ram Z, Culver KW, Walbridge S, Blaese RM, Oldfield EH (1993). *In situ* retroviral-mediated gene transfer for the treatment of brain tumors in rats. *Cancer Res* **53**, 83–88.

Rampling R, Cruickshank G, Papanastassiou V, *et al.* (2000). Toxicity evaluation of replication-competent herpes simplex virus (ICP 34.5 null mutant 1716) in patients with recurrent malignant glioma. *Gene Ther* **7**, 859–866.

Randazzo BP, Kesari S, Gesser RM, *et al.* (1995). Treatment of experimental intracranial murine melanoma with a neuroattenuated herpes simplex virus 1 mutant. *Virology* **211**, 94–101.

Read GS, Frenkel N (1983). Herpes simplex virus utants defective in the virion-associated shutoff of host polypeptide synthesis and exhibiting abnormal synthesis of alpha (immediate early) viral polypeptides. *J Virol* **46**, 498–512.

Reardon DA, Wen PY (2006). Therapeutic advances in the treatment of glioblastoma: rationale and potential role of targeted agents. *Oncologist* **11**, 152–164.

Reinblatt M, Pin RH, Federoff HJ, Fong Y (2004). Utilizing tumor hypoxia to enhance oncolytic viral therapy in colorectal metastases. *Ann Surg* **239**, 892–899.

Reinblatt M, Pin RH, Bowers WJ, Federoff HJ, Fong Y (2005). Herpes simplex virus amplicon delivery of a hypoxia-inducible soluble vascular endothelial growth factor receptor (sFlk-1) inhibits angiogenesis and tumor growth in pancreatic adenocarcinoma. *Ann Surg Oncol* **12**, 1025–1036.

Ribas A (2005). Genetically modified dendritic cells for cancer immunotherapy. *Curr Gene Ther* **5**, 619–628.

Rieger J, Naumann U, Glaser T, Ashkenazi A, Weller M (1998). APO2 ligand: a novel lethal weapon against malignant glioma? *FEBS Lett* **427**, 124–128.

Rock DL, Nesburn AB, Ghiasi H, *et al.* (1987). Detection of latency-related viral RNAs in trigeminal ganglia of rabbits latently infected with herpes simplex virus type 1. *J Virol* **61**, 3820–3826.

Rogers BE, Douglas JT, Ahlem C, Buchsbaum DJ, Frincke J, Curiel DT (1997). Use of a novel cross-linking method to modify adenovirus tropism. *Gene Ther* **4**, 1387–1392.

Roizman B and Sears AE (1996). Herpes simplex viruses and their replication. In: Fields B, Knipe DM and Howley PM (eds), *Fundamental Virology*. Lippincott-Raven, Philadelphia, PA. pp. 1043–1107.

Roizman B, Gu H, Mandel G (2005). The first 30 minutes in the life of a virus: unREST in the nucleus. *Cell Cycle* **4**, 1019–1021.

Rosenfeld MR, Meneses P, Dalmau J, Drobnjak M, Cordon-Cardo C and Kaplitt MG (1995). Gene transfer of wild-type p53 results in restoration of tumor-suppressor function in a medulloblastoma cell line. *Neurology* **45**, 1533–1539.

Saeki Y, Breakefield XO, Chiocca EA (2003). Improved HSV-1 amplicon packaging system using ICP27-deleted, oversized HSV-1 BAC DNA. In: Machida CA (ed.), *Viral Vectors for Gene Therapy, Methods and Protocols*. Humana Press, Totowa, NJ. pp. 51–60.

Samaniego LA, Webb AL, DeLuca NA (1995). Functional interactions between herpes simplex virus immediate-early proteins during infection: gene expression as a consequence of ICP27 and different domains of ICP4. *J Virol* **69**, 5705–5715.

Samaniego LA, Wu N, DeLuca NA (1997). The herpes simplex virus immediate-early protein ICP0 affects transcription from the viral genome and infected-cell survival in the absence of ICP4 and ICP27. *J Virol* **71**, 4614–4625.

Samaniego LA, Neiderhiser L, DeLuca NA (1998). Persistence and expression of the herpes simplex virus genome in the absence of immediate-early proteins. *J Virol* **72**, 3307–3320.

Santos K, Duke CM, Dewhurst S (2006). Amplicons as vaccine vectors. *Curr Gene Ther* **6**, 382–392.

Saydam O, Glauser DL, Heid I, *et al.* (2005). Herpes simplex virus 1 amplicon vector-mediated siRNA targeting epidermal growth factor receptor inhibits growth of human glioma cells *in vivo*. *Mol Ther* **12**, 803–812.

Scarpini CG, May J, Lachmann RH, Preston CM, Dunnett SB, Torres EM, Efstathiou S (2001). Latency associated promoter transgene expression in the central nervous system after stereotaxic delivery of replication-defective HSV-1-based vectors. *Gene Ther* **8**, 1057–1071.

Schellingerhout D, Rainov NG, Breakefield XO, Weissleder R (2000). Quantitation of HSV mass distribution in a rodent brain tumor model. *Gene Ther* **7**, 1648–1655.

Schrocksnadel K, Wirleitner B, Winkler C, Fuchs D (2006). Monitoring tryptophan metabolism in chronic immune activation. *Clin Chim Acta* **364**, 82–90.

Selker RG, Shapiro WR, Burger P, *et al.* (2002). and the Brain Tumor Cooperative Group. The Brain Tumor Cooperative Group NIH Trial 87-**01**, a randomized comparison of surgery, external radiotherapy, and carmustine versus surgery, interstitial radiotherapy boost, external radiation therapy, and carmustine. *Neurosurgery* **51**, 343–355.

Shah K, Tang Y, Breakefield X, Weissleder R (2003). Real-time imaging of TRAIL-induced apoptosis of glioma tumors *in vivo*. *Oncogene* **22**, 6865–6872.

Shah K, Breakefield XO (2006). HSV amplicon vectors for cancer therapy. *Curr Gene Ther* **6**, 361–370.

Shieh MT, WuDunn D, Montgomery RI, Esko JD, Spear PG (1992). Cell surface receptors for herpes simplex virus are heparan sulfate proteoglycans. *J Cell Biol* **116**, 1273–1281.

Shinoura N, Chen L, Wani MA, *et al.* (1996). Protein and messenger RNA expression of connexin43 in astrocytomas: implications in brain tumor gene therapy. *J Neurosurg* **84**, 839–845.

Shukla D, Liu J, Blaiklock P, *et al.* (1999). A novel role for 3-*O*-sulfated heparan sulfate in herpes simplex virus 1 entry. *Cell* **99**, 13–22.

Smiley JR, Duncan J (1997). Truncation of the C-terminal acidic transcriptional activation domain of herpes simplex virus VP16 produces a phenotype similar to that of the in1814 linker insertion mutation. *J Virol* **71**, 6191–6193.

Smiley JR (2004). Herpes simplex virus virion host shutoff protein: immune evasion mediated by a viral RNase? *J Virol* **78**, 1063–1068.

Spaete RR, Frenkel N (1982). The herpes simplex virus amplicon: a new eucaryotic defective-virus cloning-amplifying vector. *Cell* **30**, 295–304.

Spear PG (1993a). Membrane fusion induced by herpes simplex virus. In: Bentz, J. (ed.), *Viral Fusion Mechanisms*. CRC Press, Boca Raton, FL, pp. 201–232.

Spear PG (1993b). Entry of alphaherpesviruses into cells. *Semin Virol* **4**, 167–180.

Spear PG, Shieh MT, Herold BC, WuDunn D, Koshy TI (1992). Heparan sulfate glycosaminoglycans as primary cell surface receptors for herpes simplex virus. *Adv Exp Med Biol* **313**, 341–353.

Spencer DM (2000). Developments in suicide genes for preclinical and clinical applications. *Curr Opin Mol Ther* **2**, 433–440.

Spivack JG, Fraser NW (1987). Detection of herpes simplex virus type 1 transcripts during latent infection in mice. *J Virol* **61**, 3841–3847.

Steven AC, Spear PG (1997). Herpesvirus capsid assembly and envelopment. In: Chiu W, Burnett R and Garcea R (eds), *Structural Biology of Viruses*. Oxford University Press, New York, NY.

Stevens JG, Wagner EK, Devi-Rao GB, Cook ML, Feldman LT (1987). RNA complementary to a herpesvirus alpha gene mRNA is prominent in latently infected neurons. *Science* **235**, 1056–1059.

Strasser JF, Fung LK, Eller S, Grossman SA, Saltzman WM (1995). Distribution of 1,3-bis(2-chloroethyl)-1-

nitrosourea and tracers in the rabbit brain after interstitial delivery by biodegradable polymer implants. *J Pharmacol Exp Ther* **275**, 1647–1655.

Stupp R, Mason WP, Van Den Bent MJ, *et al.* (2004). Concomitant and adjuvant temozolomide (TMZ) and radiotherapy (RT) for newly diagnosed glioblastoma multiforme (GBM). Conclusive results of a randomized phase III trial by the EORTC Brain) hybrid amplicon vector improves the stability of transgene expression in human cells by site-specific integration. *J Virol* **76**, 7150–7162.

Warner MS, Geraghty RJ, Martinez WM, *et al.* (1998). A cell surface protein with herpesvirus entry activity (HveB) confers susceptibility to infection by mutants of herpes simplex virus type 1, herpes simplex virus type 2, and pseudorabies virus. *Virology* **246**, 179–189.

Warnick RE, Prados MD, Mack EE, *et al.* (1994). A phase II study of intravenous carboplatin for the treatment of recurrent gliomas. *J Neurooncol* **19**, 69–74.

Weaver M, Laske DW (2003). Transferrin receptor ligand-targeted toxin conjugate (Tf-CRM107) for therapy of malignant gliomas. *J Neurooncol* **65**, 3–13.

Weingart J, Strauss LC, Grossman SA, Markett J (2002). Phase I/II study: intratumoral infusion of IL13-PE38QQR cytotoxin for recurrent supratentorial malignant glioma. *Neuro-Oncology* **4**, 379.

Wickham TJ, Carrion ME, Kovesdi I (1995). Targeting of adenovirus penton base to new receptors through replacement of its RGD motif with other receptor-specific peptide motifs. *Gene Ther* **2**, 750–756.

Wickham TJ, Roelvink PW, Brough DE, Kovesdi I (1996a). Adenovirus targeted to heparan-containing receptors increases its gene delivery efficiency to multiple cell types. *Nat Biotechnol* **14**, 1570–1573.

Wickham TJ, Segal DM, Roelvink PW, *et al.* (1996b). Targeted adenovirus gene transfer to endothelial and smooth muscle cells by using bispecific antibodies. *J Virol* **70**, 6831–6838.

Wickham TJ, Tzeng E, Shears LLN, *et al.* (1997). Increased *in vitro* and *in vivo* gene transfer by adenovirus vectors containing chimeric fiber proteins. *J Virol* **71**, 8221–8229.

Wikstrand CJ, Hale LP, Batra SK, *et al.* (1995). Monoclonal antibodies against EGFRvIII are tumor specific and react with breast and lung carcinomas and malignant gliomas. *Cancer Res* **55**, 3140–3148.

Willis RA, Bowers WJ, Turner MJ, *et al.* (2001). Dendritic cells transduced with HSV-1 amplicons expressing prostate-specific antigen generate antitumor immunity in mice. *Hum Gene Ther* **12**, 1867–1879.

Wong RJ, Kim SH, Joe JK, Shah JP, Johnson PA, Fong Y (2001a). Effective treatment of head and neck squamous cell carcinoma by an oncolytic herpes simplex virus. *J Am Coll Surg* **193**, 12–21.

Wong RJ, Patel SG, Kim S, *et al.* (2001b). Cytokine gene transfer enhances herpes oncolytic therapy in murine squamous cell carcinoma. *Hum Gene Ther* **12**, 253–265.

Wu N, Watkins SC, Schaffer PA, DeLuca NA (1996). Prolonged gene expression and cell survival after infection by a herpes simplex virus mutant defective in the immediate-early genes encoding ICP4, ICP27, and ICP22. *J Virol* **70**, 6358–6369.

Wu TJ, Monokian G, Mark DF, Wobbe CR (1994). Transcriptional activation by herpes simplex virus type 1 VP16 *in vitro* and its inhibition by oligopeptides. *Mol Cell Biol* **14**, 3484–3493.

Wudunn D, Spear P (1989). Initial interaction of herpes simplex virus with cells is binding to heparan sulfate. *J Virol* **63**, 52–58.

Xiao H, Pearson A, Coulombe B, *et al.* (1994). Binding of basal transcription factor TFIIH to the acidic activation domains of VP16 and p53. *Mol Cell Biol* **14**, 7013–7024.

Yamauchi Y, Daikoku T, Goshima F, Nishiyama Y (2003). Herpes simplex virus UL14 protein blocks apoptosis. *Microbiol Immunol* **47**, 685–689.

York IA, Roop C, Andrews DW, Riddell SR, Graham FL, Johnson DC (1994). A cytosolic herpes simplex virus protein inhibits antigen presentation to CD8+ T lymphocytes. *Cell* **77**, 525–535.

Zager JS, Delman KA, Malhotra S, *et al.* (2001). Combination vascular delivery of herpes simplex oncolytic viruses and amplicon mediated cytokine gene transfer is effective therapy for experimental liver cancer. *Mol Med* **7**, 561–568.

Zhang L, Daikoku T, Ohtake K, *et al.* (2006). Establishment of a novel foreign gene delivery system combining an HSV amplicon with an attenuated replication-competent virus, HSV-1 HF10. *J Virol Methods* **137**, 177–183.

Zhou G, Ye GJ, Debinski W, Roizman B (2002). Engineered herpes simplex virus 1 is dependent on IL13Ralpha 2 receptor for cell entry and independent of glycoprotein D receptor interaction. *Proc Natl Acad Sci U S A* **99**, 15124–15129.

Zibert A, Thomassen A, Muller L, *et al.* (2005). Herpes simplex virus type-1 amplicon vectors for vaccine generation in acute lymphoblastic leukemia. *Gene Ther* **12**, 1707–1717.

3

Adeno-associated Virus

Selvarangan Ponnazhagan

3.1 Introduction

Adeno-associated virus (AAV) is a non-pathogenic human parvovirus originally identified three decades ago (Atchison *et al.*, 1965). Several unique features of AAV including non-pathogenicity, replication incompetency and low-immunogenicity have prompted great interest towards adapting AAV-based recombinant vectors for the correction of monogenic diseases. However, a better understanding, both in properties of AAV vectors and the molecular pathology of human cancers, has expanded the utility of AAV in cancer gene therapy. This chapter will provide a brief overview on biology of AAV, its utility as gene therapy vector and the potential of recombinant AAV in cancer gene therapy.

3.2 Biology and life cycle of AAV

AAV is a helper-dependent parvovirus. For a replicative life cycle, AAV requires the presence of a helper virus, hence, is also known as dependovirus. The helper functions are provided by adenovirus, herpesvirus or vaccinia virus (Hoggan *et al.*, 1966; Buller *et al.*, 1981; Schlehofer *et al.*, 1984). In the absence of a helper virus, AAV integrates into host genome and establishes a latent cycle. When a latently infected cell encounters infection by any of the helper viruses, the integrated AAV genome rescues itself and undergoes a productive lytic cycle. AAV contains a genome of approximately 4.7 kb of single-stranded DNA (Srivastava *et al.*, 1983). Both positive and negative strands of the viral DNA are equally packaged in icosahedral capsids (Hermonat *et al.*, 1984). The genome of AAV encodes two proteins namely Rep, which is a non-structural protein involved in the rescue and replication of the virus, and Cap that forms capsid shell within which the replicated genome is packaged. Three different promoters have been identified in the wild type AAV genome. Based on their relative position, they are named as p5, p19 and p40 (Laughlin *et al.*, 1979; Lusby and Berns, 1982; Green and Roeder, 1980a, b) (Figure 3.1). Tanscripts from the p5 and p19 promoters produce four different species of Rep proteins by alternate splicing and transcript from p40 produces three different capsid proteins. Rep68 and Rep78 are produced from the p5 promoter as spliced and unspliced forms respectively while Rep52 and Rep40 are produced from the p19 promoter, similarly (Mendelson *et al.*, 1986; Trempe *et al.*, 1992).

The *Cap* gene encodes three different proteins namely VP-1, VP-2 and VP-3 using different initiation codons (Srivastava *et al.*, 1983). Among the three capsid proteins, VP-3 is the predominant form and represents approximately 90 per cent of the assembled capsid. In addition to

Viral Therapy of Cancer Edited by Kevin J. Harrington, Richard G. Vile and Hardev S. Pandha

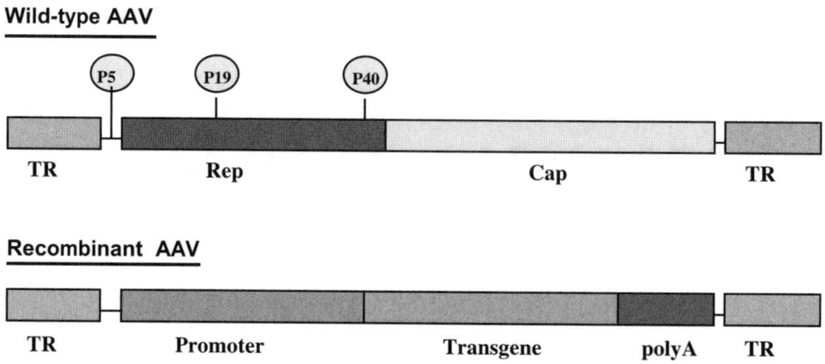

Figure 3.1 Genomic organization of wild-type and recombinant AAV. The wild-type AAV consists of two major open reading frames (ORF). The left ORF expresses four replication proteins (Rep) from two promoters, situated at map units 5 (P5) and 19 (P19) and the right ORF expresses three capsid proteins (Cap) from a single promoter, situated at map unit 40 (P40). The genome is flanked by two terminal repeat sequences (TR). In a typical recombinant AAV, the *Rep* and *Cap* genes are replaced by a cassette containing a transgene, promoter and polyadenylation signal (polyA)

the *Rep* and *Cap* genes, the AAV genome also contains two terminal repeat sequences (TRs) on either end of the genome that are approximately 140 bases in length each. The TRs are sole elements required for rescue, replication, packaging, and integration of AAV (Srivastava *et al.*, 1983; Berns, 1990; Muzyczka, 1992).

Recombinant (r) AAV on the other hand is devoid of many features of the wild type virus. rAAV does not encode any wild type viral genes and hence, is less immunogenic compared to other commonly used viral vectors (Samulski *et al.*, 1989; Jooss *et al.*, 1998). Only the ITRs of the wild-type virus are retained in rAAV. Since rAAV lack the *rep* gene, they are totally replication defective even in the presence of a helper virus. The genome of rAAV does not integrate efficiently in host chromosome but persists in non-dividing target cells as circular concatamers (Yang *et al.*, 1999).

Development of novel molecular strategies and identification of newer serotypes increases the potential of AAV vectors for gene therapy. Conventional recombinant AAV vectors deliver a single-stranded (*ss*) DNA genome, which must be converted by host-cell-mediated DNA synthesis to double-stranded (ds) DNA for active expression. At high multiplicities of infection, hybridization of complementary DNA strands from separate virions may also generate active ds templates for gene expression (Im and Muzyczka, 1990; Owens *et al.*, 1993). This requirement for the formation of duplex DNA has proven to be an important limiting factor for AAV transduction. The requirement for complementary-strand synthesis or recruitment by base pairing with a co-infecting complementary genome has been shown to severely reduce its efficiency in many cell types (Ferrari *et al.*, 1996; Fisher *et al.*, 1996; Nakai *et al.*, 2000). More recently, studies have indicated that this limitation could be by-passed by packaging both strands of rAAV genome as an inverted repeat forming a ds DNA structure (McCarty *et al.*, 2001, 2003; Wang *et al.*, 2003). This unique structure allows an intramolecular base-pairing reaction to generate a ds DNA template for gene expression in the absence of DNA synthesis in the target cell. Whereas transduction by AAV-2 in mouse liver was only 1–5 per cent at a dose where vector DNA was detected in most of the hepatocytes, when the AAV genome was packaged as a dsDNA, a single injection of the ds AAV resulted in 25–50 per cent transduction of liver cells (Nakai *et al.*, 2000). Using CEA as transgene, we have recently demonstrated that the advantage of ds AAV during initial stages of transduction is mainly through rapid conversion to concatemeric form, and less disintegration of dsDNA compared to *ss* DNA (Ren *et al.*, 2005).

Recently, a number of new serotypes of AAV have been identified by screening human and non-human primate tissues for the presence of rescuable AAV genomes. These efforts have resulted in over 40 genomic variants (Gao *et al.*, 2002, 2003, 2004). It appears that there are several major clades of related variants. Within these clades, many other individual variants appear to have been generated by recombination within the capsid coding sequence between different parental serotypes (Gao *et al.*, 2004). Based on the heterogeneity in amino-acid composition of the capsid protein, some of the known serotypes use different cellular receptors for internalization (Ding *et al.*, 2005). Further, pre-existing immunity in humans to a few isolates obtained from non-human primates (AAV7 and AAV8 in particular) has been relatively low. Such repertoire of AAV serotypes with differing transduction mechanisms and pre-exiting immunity in humans increases the potential for clinical utility.

3.3 AAV serotypes

So far, at least 10 AAV serotype capsids have been tested in preclinical studies with varying infectivity in different target cells (Gao *et al.*, 2005). Heparin sulfate proteoglycan has been identified as the cellular attachment receptor for AAV2 and $\alpha V\beta 5$ integrin and fibroblast growth factor receptor-1 (FGFR-1) have been reported as coreceptors (Summerford and SAmulski, 1998; Summerford *et al.*, 1999; Qing *et al.*, 1999). Subsequent studies have identified receptors for some of the other serotypes. Sialic acid with different linkage forms has been attributed to serve as binding protein for serotypes 1, 4, 5 and 6 (WAlters *et al.*, 2001; Seller *et al.*, 2006; Wu *et al.*, 2006). Difference in cellular binding characteristics of different serotypes is also exemplified in serological properties, which provide greater opportunity for repeat injections if required in gene therapy applications. Many of the earlier preclinical studies with rAAV have utilized AAV2-based vectors, which demonstrated sustained transgene expression in a variety of target tissues such as muscle, neuronal cells and eye without any vector-related toxicity and host immune responses. However, clinical application of this serotype vector has been hampered by low efficiency gene transfer and prevalent pre-existing immunity in human population (Manno *et al.*, 2006). Encouraging studies from preclinical animal models with alternate AAV serotypes have provided high-efficiency gene transfer to several target tissues *in vivo* thereby increasing the potential of rAAV in human gene therapy.

3.4 Production of recombinant AAV

Cloning of the AAV genome into a plasmid vector facilitated a wide range of molecular manipulations that led to the understanding of several key events in AAV biology (Samulski *et al.*, 1982). The crucial role of TRs in AAV life cycle had been shown initially in experiments using rAAV plasmids containing heterologous gene sequences flanked by AAV-TRs. Transfection of the rAAV plasmids into human cells resulted in successful rescue, replication and packaging of infectious, mature virions by transcomplementing AAV *Rep* and *Cap* genes from a non-rescuable plasmid together with genes encoding adenoviral proteins necessary for helper functions (Samulski *et al.*, 1989). Several modifications in rAAV production and purification steps involved generation of packaging cell lines (Clark *et al.*, 1995; Tamayose *et al.*, 1996; Gao *et al.*, 1998; Inoue and Russell, 1998), cloning of helper plasmids containing necessary adenoviral genes to eliminate any wild type adenovirus in AAV preparations (Xiao *et al.*, 1998; Grimm *et al.*, 1998; Matsushita *et al.*, 1998; Collaco *et al.*, 1999), large-scale cultures in bioreactors and purification using affinity columns and HPLC. These advancements have resulted in high-titre rAAV yields necessary for *in vivo* studies including human clinical trials (High, 1991).

3.5 Gene therapy for cancer treatment

Gene therapy offers a potentially useful approach for the treatment of cancers since a variety of genes controlling molecular processes can be introduced by gene transfer, which can in principle arrest tumour growth, angiogenesis, invasion and/ or metastasis. Currently available cancer gene therapy methods can be broadly divided into those that exert immediate cytotoxicity on tumour cells and those that regulate events that lead to either correction of underlying defects in tumour cells at a molecular level or enhancing the ability of the host immune system to recognize tumour cells for T-cell-mediated killing. A majority of the cytotoxic gene therapy involves delivery of genes encoding enzymes such as thymidine kinase and cytosine deaminase and following it with administration of non-toxic prodrugs, which are eventually converted to cytotoxic intermediates in the cells that express the transgene.

Genetic correction of molecular defects in tumour cells has also been attempted. Identification of genes that contribute to oncogenic transformation of cells presents an opportunity to use these gene products for treatment and as potential prevention targets. Genes that are implicated in carcinogenesis include dominant oncogenes such as members of *ras* family and tumour suppressor genes including p53 (Bishop, 1991; Weinberg, 1991). While inactivation of dominant oncogene products at the transcriptional level with anti-sense RNA may block their production, proper expression of tumour suppressor genes through gene transfer appears to be required to suppress the growth of tumour cells or lead to apoptosis and necrosis.

Different gene therapy approaches are also being employed to enhance the host immunity against tumour cells. One strategy has been to vaccinate 'the host' with tumour cells, which have been modified *ex vivo* by transfer of genes encoding cytokines, tumour-associated antigens or portions of the major histocompatability complex. A variety of such molecules have shown promising results in controlling tumour growth in animal models. These include interferon (IFN)-α, IFN-γ, tumour necrosis factor-α (TNF-α), macrophage colony-stimulating factor (MCSF), granulocyte–macrophage colony-stimulating factor (GM-CSF), interleukin (IL)-1, IL-3, IL-5, IL-6, IL-7, IL-10 and IL-12 (Tepper and Mule, 1994; Pardoll, 1995; Jaffee, 1999; Tagawa, 2000). The *in vitro* growth characteristics of many tumour cells are not affected by cytokine gene transfer thus confirming that suppression of tumorigenicity *in vivo* is caused by interaction of host immune defense system in addition to the expression of these molecules (Tting *et al.*, 1997). Another approach has been to immunize targeting tumour-specific or tumour-associated antigens. While initial approaches of genetic immunization targeted muscle cells for expression, processing and presentation of the antigen, subsequent studies have effectively employed antigen-presenting cells, particularly the dendritic cells, for transduction with tumour-associated antigen genes (Tting *et al.*, 1997; Boczkowski *et al.*, 1996; Nair *et al.*, 1998; De Veerman *et al.*, 1999; Brossart *et al.*, 1997; Tillman *et al.*, 1999). Further, transfer of costimulatory molecules such as B7.1 and B7.2 have also shown significant enhancements of host immunity (van Gool *et al.*, 1995). There is considerable interest in the use of gene transfer to enhance tumour homing or tumour cell-killing by adoptive transfer of *ex vivo* expanded tumour infiltrating lymphocytes [TILs, (Hwu *et al.*, 1993)].

3.6 Anti-oncogenic properties of AAV

Interestingly, wild type AAV has been known to provide anti-tumour effects in certain cancers. Infection of herpes simplex virus-transformed hamster tumour cells with AAV delayed the appearance of palpable tumours and increased the survival time of the animals (Cukor *et al.*, 1975). Since then, several reports have confirmed the inhibition by AAV of viral oncogenesis by a variety of DNA viruses, including bovine papillomavirus-1 (Hermonat *et al.*, 1998), human papillomavirus (HPV)-16 (Hermonat, 1994a; Horer *et al.*, 1995; Hermonat *et al.*, 1997), and Epstein–Barr virus (Wu *et al.*, 1999). Evidence from several reports also suggested that AAV infection might protect against human cervical cancer, in part, by interfering with HPV-induced tumorigenesis (Zhan *et al.*, 1999) although studies of Stickler *et al.* (1999) reported a lack of correlation of between AAV infection and cervical tumourigenesis in a Jamaican population.

Understanding of molecular mechanisms directing the anti-tumour properties of AAV identified a role for Rep78 in the inhibition of oncogenic transformation, specifically the downregulation of human c-*fos* and c-*myc* proto-oncogene promoters by Rep78 (Hermonat, 1994b). Inhibition of HPV-16 P97 promoter activity (Zhan *et al.*, 1999) may partially account for the tumour inhibitory property of Rep78 in cervical cancer cells. Saunders et al reported that while Rep68 and Rep78 inhibited the growth of primary, immortalized, and transformed cells, Rep52 and Rep40 did not (Sauden *et al.*, 2000). Further Rep68 induced cell cycle arrest in G_1 and G_2 with elevated cyclin dependent kinase inhibitor p21 and reduced cyclin E-, A- and B1-associated kinase activities. Rep78 was also found to arrest the cell cycle, preventing S-phase progression by binding to the hypophosphorylated retinoblastoma protein (Schmidt *et al.*, 2000; Batchu *et al.*, 2002). The regulatory differences between Rep78 and Rep68 have been mapped to the C-terminal zinc finger domain of Rep78. Despite the significance of Rep78 and Rep68 in tumour suppression, potential utilization of Rep as a therapeutic molecule is limited by its toxicity (Schmidt *et al.*, 2000). Thus, further advancements in highly tumour-specific delivery and/or expression of *Rep* gene is required before Rep can be used as a therapeutic molecule.

3.7 Molecular chemotherapy studies with rAAV

Delivery of a gene-encoded toxin into cancer cells to achieve tumour cell killing usually performed by activation of a prodrug. This approach has focused mainly on delivery of the herpes simplex virus thymidine kinase (HSV-TK) gene. Although a majority of both preclinical and clinical gene therapy studies using molecular chemotherapy approaches have been conducted with recombinant adenoviral vectors, AAV-mediated *in vivo* studies have also indicated therapeutic benefits for tumour regression. Selective killing of alpha fetoprotein (AFP)-positive hepatocellular carcinoma cells by AAV-mediated gene transfer of HSV-TK gene was reported in a mouse model using an albumin promoter and an AFP enhancer (Su *et al.*, 1996). Further work by the same group also reported therapeutic efficacy and a bystander effect of AAV-mediated intratumoral delivery of the HSV-TK gene followed with treatment using GCV (Sue *et al.*, 1997). Interestingly, it has been reported that enhancement of tumour cell killing is possible with a rAAV containing the HSV-TK gene along with IL-2 gene compared to transduction of vector containing only the HSV-TK gene (Su *et al.*, 2000). Thus, it is possible to enhance antitumour effects by delivering two different therapeutic genes in the same vector. Although there is a size constraint in the packaging of foreign genes in rAAV, most of the therapeutic genes in the context of cancer therapy are well within the packaging limits of rAAV either alone or in tandem. Further, recent studies have demonstrated expression from large genes, packaged in two different rAAV by *trans*-splicing events (McCarty *et al.*, 2001, 2003; Wang *et al.*, 2003). Similar *in vivo* therapeutic effects of AAV-mediated delivery of the HSV-TK gene has also been reported in an experimental glioma model (Mizunno *et al.* 1998).

It has been reported that the efficiency of rAAV transduction of primary tumour material, derived from malignant melanoma and ovarian carcinoma, is significantly higher (>90 per cent) than that seen in established tumour cells of the same derivation in culture (Maass *et al.*, 1998). This observation suggests that it is possible to utilize rAAV in direct targeting of tumour cells for an effective killing by approaches such as molecular chemotherapy, cytokine gene transfer, and inactivation of proto-oncogene expression. In addition, studies by Su *et al.* (2000) using an AAV-TK-IL-2 vector reported disappearance of the rAAV genome following GCV treatment and regression of the transduced hepatocellular carcinoma.

3.8 AAV-mediated sustained transgene expression as a potential cancer gene therapy strategy

Although sustained transgene expression by rAAV is of significant advantage in phenotypic correction of monogenic diseases, this feature could potentially benefit cancer gene therapy in strategies targeting normal cells that can mediate anti-tumour effects. For example, it is now well established that tumour growth and metastasis are dependent upon recruitment of a functional blood supply by a process known as tumour angiogenesis and indeed, the 'angiogenic phenotype' correlates negatively with prognosis in many human solid tumours (Folkman, 1971; Folkman *et al.*, 1996). Anti-angiogenic therapies target different steps of the angiogenic process, ranging from inhibition of expression of angiogenic molecules, through over-expression of anti-angiogenic factors, to direct targeting of tumour endothelial cells using endogenous angiogenic inhibitors or artificially constructed targeting ligands (Feldman and Libutti, 2000). Potential advantages of anti-angiogenic gene therapy are sustained expression of the anti-angiogenic factors (Feldman and Libutti, 2000). Initial studies of anti-angiogenic gene therapy were conducted with adenoviral vectors. Expression of anti-angiogenic factors mediated by adenovirus-based vectors is limited by an effective host immune response and also secondary to the transient nature of transgene expression. AAV, on the other hand, possesses desirable features for anti-angiogenic gene therapy.

The advantages of rAAV over other vectors for anti-angiogenic gene therapy are many. First, AAV is a non-pathogenic vector with a very limited host immune response. Second, in terminally differentiated target cells, AAV establishes stable expression of transgenic factors. Third, most of the anti-angiogenic genes are within the capacity to be cloned in AAV either independently or in tandem. Provision of two different anti-angiogenic genes from the same vector has been shown to increase therapeutic benefit since different anti-angiogenic factors signal through different metabolic pathways (Ponnazhagan *et al.*, 2004a; Isayeva *et al.*, 2005). In the last few years, several interesting preclinical studies have indicated the potential of rAAV in anti-angiogenic cancer gene therapy. A vast majority of them tested the potential of anti-angiogenic factors endostatin and/or angiostatin (Ponnazhagan *et al.*, 2004a; Isayeva *et al.*, 2005; Lalani *et al.*, 2004; Xu *et al.*, 2003; Noro *et al.* 2003; Subramanian *et al.*, 2005, 2006). Results from these studies are highly encouraging with anti-tumour activity on the growth of different types of cancers. Whereas reports indicated the advantage of sustained expression of either angiostatin or endostatin from rAAV vectors by a single injection in xenograft and orthotopic models of melanoma (Lalani *et al.*, 2004), lymphoma (Xu *et al.*, 2003), pancreatic cancer (Noro *et al.* 2003), and ovarian cancer (Subramanian *et al.*, 2005, 2006), combination of angiostatin and endostatin from a single AAV significantly improved therapeutic effects over vectors encoding either of the factors (Ponnazhagan *et al.*, 2004a). In studies of intraperitoneal model of ovarian cancer, application of the vector encoding both the factors was highly effective when combined with chemotherapy (Lalani *et al.*, 2004). A rAAV encoding endostatin has been shown to provide similar combination effect with chemotherapy in a mouse model of ovarian cancer (Subramanian *et al.*, 2005). Angiostatin and endostatin gene therapy by rAAV has also been shown to increase therapeutic efficacy when combined with radiation therapy (Shi *et al.*, 2003).

In addition to angiostatin and endostatin, rAAV encoding vascular endothelial growth factor receptor (VEGFR)-1 (sFlt-1), tissue factor pathway inhibitor (TFPI)-2, VEGFR-2 and IFN-β have shown anti-angiogenic efficacy in preclinical animal models of cancer (Davidoff *et al.*, 2005; Mahendra *et al.*, 2005; Streck *et al.*, 2005; Yanamandra *et al.*, 2005).

These studies provide encouraging results for possible clinical translation of sustained anti-angiogenic gene therapy.

3.9 rAAV vectors have advantages in stimulating T helper 1/cytotoxic T lymphocyte responses

Other viral vectors that have been used to express tumour antigens, such as adenovirus, poxvirus, and vaccinia virus induce a vigorous stimulation of the innate immune system immediately after their administration, which is then followed by the induction of the adaptive immune response (Kwak *et al.*, 2003; Basak *et al.*, 2004; Liu *et al.*, 2004). The early inflammatory response is independent of the expression of transgene, but the antigen-specific response can be directed to both viral and transgenic proteins. As a consequence, transgene expression is only transient. rAAV, on the other hand, do not elicit strong vector-specific T-cell response compared to other vectors since they do not encode vector structural proteins (Ponnazhagan *et al.*, 2001a; Conlon and Flotte, 2004). Further, the onset of transgene expression does not occur immediately following vector delivery, a time crucial for innate effector mechanisms. Thus, rAAV have advantages over other vectors in eliciting strong, persistent anti-tumour cellular immune response when optimally induced. Studies have shown that expression of antigens in peripheral tissues must be relatively high to facilitate DC priming of nave $CD8^+$ T cells by cross presentation (Kurts and Miller, 1998). Generating cytotoxic T lymphocytes (CTL) to subdominant epitopes, in particular, requires a high antigen dose and persistent antigen cross-presentation. Furthermore, the *in vivo* duration of the lytic activity of CTL parallels the kinetics of antigen presentation (Nelson *et al.*, 2000). A high dose of antigen has recently been shown to promote the development of T helper 1 (Th1) responses, whereas a low dose favoured Th2 responses (Boonstra *et al.*, 2003). The low immunogenicity of rAAV and its ability to stably express a neo-antigen in the absence of other viral vector antigens lessens the premature elimination of transduced cells by vector-specific CTL (and other mechanisms). Inflammatory cytokines, such as RANTES and interferon-inducible protein 10 (IP-10), IL-8, membrane inducible protein (MIP)-1β and MIP-2, are not induced by rAAV (Zeiss *et al.*, 2002). In addition, the expression of the neo-antigen in the absence of other viral vector antigens should minimize antigenic competition for major histocompatibility complex (MHC) class I peptide loading.

3.10 rAAV vectors can be used to initiate immune responses

A few studies have recently demonstrated that AAV can serve as a vaccine vector (Xin *et al.*, 2001). A single injection of a rAAV expressing human immunodeficiency virus (HIV)-1 *env*, *tat* and *rev* genes induced HIV-1 MHC class I-restricted CTL activity and serum immunoglobulin G (IgG) and faecal secretory IgA antibodies in mice (Xin *et al.*, 2001). Although intranasal (i.n.) administration was most effective, mucosal immune responses could be effected with intramuscular (i.m.), intraperitoneal (i.p.), and subcutaneous (s.c.) administration. A single i.m. administration of rAAV expressing simian immunodeficiency virus (SIV) elicited SIV-specific T cells and antibodies in macaques. Furthermore, immunized animals were able to significantly restrict replication of a live, virulent SIV challenge (Johnson *et al.*, 2005). A single i.m. injection of mice with a rAAV vector expressing herpes simplex virus type 2 glycoprotein B led to the generation of both MHC-class-I-restricted CTL and antibody responses (Manning *et al.*, 1997). In this model, rAAV-mediated immunization was more potent than plasmid DNA or protein in generating antibody responses. Studies in mice with rAAV expressing the influenza virus haemagglutinin and rAAV expressing β-galactosidase (β-gal) model antigens have demonstrated the importance of cross presentation of the neo-antigen to DC with respect to the generation of CTL responses (Zhang *et al.*, 2000; Sarukhan *et al.*, 2001). That rAAV can be used to elicit ovalbumin (Ova)-specific immune responses has also been demonstrated in mouse models. Mice injected with rAAV expressing Ova developed potent Ova-specific CTL and antibodies. The CTL response after i.p. administration of rAAV-Ova protected mice against a subsequent tumour challenge with an Ova-transfected B16 melanoma cell line (Brockstedt *et al.*, 1999). A single i.m. injection of a rAAV vector containing a HPV-16 *E7* CTL epitope and heat shock protein in mice could efficiently eliminate E7-expressing tumour cells (Liu *et al.*, 2000a). Recent studies have shown that AAV expressing CEA in combination with plasmid adjuvants can elicit antitumour immune responses (Ponnazhagan *et al.*, 2004a).

Initial studies in a mouse model of i.m. injection suggested that rAAV delivered by this route failed to transduce dendritic cells, the most potent antigen-presenting cells (Jooss *et al.*, 1998). Reports by Brockstedt *et al.* (1999) however, indicated generation of antibody-mediated and T cell-mediated immunity against rAAV-encoded ovalbumin delivered intramuscularly and intraperitoneally. Further studies by Zhang *et al.* (2000) reported that while mature murine DCs are refractory to AAV transduction, immature DCs are still transducible and that the transduction yields are lower in the absence of adenovirus coinfection.

Although these characteristics may limit one's ability to test rAAV in an *ex vivo* immunotherapy strategy in a murine system by genetic transfer of a potent tumour antigen gene into DCs, recent studies indicate the possibility of transducing DC prior to differentiation. The potential of such a strategy has been recently reported utilizing human DCs *in vitro*. In these studies, transfer of the IL-4 gene into human peripheral blood monocytes and culturing of these cells with GM-CSF resulted in differentiation to potent DCs (Liu *et al.*, 2000b). We have recently determined that transduction of a rAAV encoding the firefly luciferase in monocytes, following differentiation with IL-4 and GM-CSF, resulted in a robust increase in transgene expression in differentiated DCs (Ponnazhagan *et al.*, 2001a). Using fluorescent *in situ* hybridization analysis, we were also able to identify the transgene in potent DCs 10 days after transduction (Ponnazhagan *et al.*, 2001a). Similar to our earlier findings in human bone marrow-derived $CD34^+$ cells (Ponnazhagan *et al.*, 1997), we also observed differences in AAV transduction of DCs obtained from different individuals (Ponnazhagan *et al.*, 2001b). By adapting multiple transductions during DC culture and combining it with the use of serotype 6 capsids, packaging AAV DNA as a self-complementary structure and activation of rAAV transduced DC using agonist antibody and CpG oligodinucleotide, we reported *in vivo* immunostimulatory activity of rAAV transduced mouse DC (Aldrich *et al.*, 2006). DC transduced with rAAV expressing *E7* has also been shown to generate CTL capable of lysing E7-expressing cervical cancer cells (Chiriva-Internati *et al.*, 2002).

3.11 Altering AAV tropism for tumour-specific delivery

Although rAAV vectors transduce both dividing and non-dividing cells transcending species barrier, it is increasingly clear that there is wide variation in transduction efficiency among different cell types.

Recent studies have also indicated that cell-specific targets can be exploited as alternate entry pathways for AAV infection. Initial studies with targeted-AAV involved genetic and conjugate modifications of vector tropism. While genetic modifications of the capsid involve addition of DNA sequences representing targeting ligands (Yang *et al.*, 1998; Girod *et al.*, 1999; Rabinowitz *et al.*, 1999; Bartlett *et al.*, 2000; Wu *et al.*, 2000), conjugate modifications are based on developing stable, small molecular bridges to attach vector to a targeting moiety. Successful retargeting of AAV using antibody and biotin conjugates has resulted in significant increase in gene transfer (Ponnazhagan *et al.*, 2002; Arnold *et al.*, 2006). By mutational analysis, studies have also identified potential regions in the AAV capsid that may be used in genetic modifications (Girod *et al.*, 1999; Rabinowitz *et al.*, 1999; Bartlett *et al.*, 2000; Wu *et al.*, 2000).

Hajitou *et al.* (2006) recently established a system containing cis-elements from AAV and single-stranded bacteriophage to target integrin-positive cells and showed that such a chimeric vector resulted in superior tumour transduction over phage and that incorporation of terminal repeats of AAV also increased the longevity of transgene structure in cells. Further, they demonstrated that temporal dynamics and spatial heterogeneity of gene expression mediated by targeted AAVP can be monitored by positron emission tomography. Thus, development of rAAV capable of targeted transduction of cancer cells *in vivo* will further increase the utility of this vector system in cancer gene therapy.

3.12 Clinical trials involving rAAV

In the last few years, AAV vectors have been tested in phase I and phase II clinical trials with varying outcome (Carter, 2006; Warrington and Herzog, 2006). Most of these trials involved monogenic diseases including cystic fibrosis, haemophilia, α1-antitrypsin deficiency, Canaan disease, muscular dystrophy, congenital blindness, Alzheimer's disease and Parkinson's disease (Carter, 2006; Warrington and Herzog, 2006). Most of them have used AAV-2 serotype based vectors. While none of these trials have shown adverse effects due to vector toxicity, clinical outcomes have been modest. Clinical trials in cystic fibrosis patients have demonstrated safety of the vector delivery to the nasal sinus and bronchial epithelium and a partial correction of electrophysiologic defect and improved pulmonary function (Wagner *et al.*, 1998, 1999; Moss *et al.*, 2004). Using skeletal muscle or liver as platforms for transgene expression, trials have been conducted on haemophilia B patients (Kay *et al.*, 2000; Manno *et al.*, 2003, 2006). Results of these trials indicated modest increase in factor IX production, which was only transient, lasting for 8 months. Two of the patients showed T cell responses to AAV capsids (Mingozzi *et al.*, 2007). Results of the clinical trials with AAV so far although have not indicated successful therapeutic outcomes, indicate great potential for this vector system through modifications such as use of alternate serotypes with low immunogenicity and better transduction capabilities, transient immunosuppression and repeat administration. These modifications should provide great advantage in the application of rAAV in cancer gene therapy.

3.13 Conclusion

It is becoming increasingly clear that rAAV is a potential addition to other vectors for cancer gene therapy. Although a majority of preclinical studies with rAAV have mostly centered on the correction of genetic defects, studies in the last few years indicate the potential of rAAV in cancer gene therapy. It is also becoming apparent that for genetic therapy of cancer to be successful, a wide spectrum of target molecules and cells may be effectively employed. Features of AAV such as stable transgene expression, high-efficiency transduction, absence of host immunity, and native tumour suppressor properties, suggest that these properties can be exploited in therapeutic and preventive cancer gene therapy strategies. Further advances in the basic biology of the vector should lead to the development of new generation, high-efficiency and cell-specific vectors, which in turn, may lead to the emergence of novel vector paradigms advancing future cancer gene therapy applications.

Acknowledgements

Financial support of the National Institutes of Health grants R01CA90850, R01CA98817 and the US Army Department of Defense grants PC020372 and PC050949 is gratefully acknowledged.

References

Aldrich WA Ren C, White AF, *et al.* (2000). Enhanced transduction of mouse bone marrow-derived dendritic cells by repetitive infection with self-complementary adeno-associated virus 6 combined with immunostimulatory ligands. *Gene Ther* **13**, 29–39.

Arnold GS, Sasser AK, Stachler MD, Bartlett JS (2006). Metabolic biotinylation provides a unique platform for the purification and targeting of multiple AAV vector serotypes. *Mol Ther* **14**, 97–106.

Atchison RW, Casto BC, Hammon WM (1965). Adenovirus-asociated defective virus particles. *Science* **194**, 754–756.

Bartlett JS, Kleinschmidt J, Boucher RC, Samulski RJ (2000). Targeted adeno-associated virus vector transduction of nonpermissive cells mediated by a bispecific F(ab′gamma)2 antibody. *Nat Biotechnol* **17**, 181–186.

Basak SK, Kiertscher SM, Harui A, Roth MD (2004). Modifying adenoviral vectors for use as gene-based cancer vaccines. *Viral Immunol* **17**, 182–196.

Batchu RB, Shammas MA, Wang JY, Freeman J, Rosen N, Munshi NC (2002). Adeno-associated virus protects the retinoblastoma family of proteins from adenoviral-induced functional inactivation. *Cancer Res* **62**, 2982–2985.

Berns KI (1990). Parvovirus replication. *Microbiol Rev* **54**, 316–329.

Bishop JM (1991). Molecular themes in oncogenesis. *Cell* **64**, 235–248.

Boczkowski D, Nair SK, Snyder D, Gilboa E (1996). Dendritic cells pulsed with RNA are potent antigen-presenting cells *in vitro* and *in vivo*. *J Exp Med* **184**, 465–472.

Boonstra A, Asselin-Paturel C, Gilliet M, Crain C, Trinchieri G, Liu YJ, O'Garra A (2003). Flexibility of mouse classical and plasmacytoid-derived dendritic cells in directing T helper type 1 and 2 cell development: dependency on antigen dose and differential toll-like receptor ligation. *J Exp Med* **197**, 101–109.

Brockstedt DG, Podsakoff GM, Fong L, Kurtzman G, Mueller-Ruchholtz W, Engleman EG (1999). Induction of immunity to antigens expressed by recombinant adeno-associated virus depends on the route of administration. *Clin Immunol* **92**, 67–75.

Brossart P, Goldrath AW, Butz EA, Martin S, Bevan MJ (1997). Virus-mediated delivery of antigenic epitopes into dendritic cells as a means to induce CTL. *J Immunol* **158**, 3270–3276.

Buller RM, Janik JE, Sebring ED, Rose JA (1981). Herpes simplex virus types 1 and 2 completely help adenovirus-associated virus replication. *J Virol* **40**, 241–247.

Carter BJ (2006). Adeno-associated virus vectors in clinical trials. *Hum Gene Ther* **16**, 541–550.

Chiriva-Internati M, Liu Y, Salati E, *et al.* (2002). Efficient generation of cytotoxic T lymphocytes against cervical cancer cells by adeno-associated virus/human papillomavirus type 16 E7 antigen gene transduction into dendritic cells. *Eur J Immunol* **32**, 30–38.

Clark KR, Voulgaropoulou F, Fraley DM, Johnson PR (1995). Cell lines for the production of recombinant adeno-associated virus. *Hum Gene Ther* **6**, 1329–1341.

Collaco RF, Cao X, Trempe JP (1999). A helper virus-free packaging system for recombinant adeno-associated virus vectors. *Gene* **238**, 397–405.

Conlon TJ, Flotte TR (2004). Recombinant adeno-associated virus vectors for gene therapy. *Exp Opin Biol Ther* **4**, 1093–10101.

Cukor G, Blacklow NR, Kibrick S, Swan IC (1975). Effect of adeno-associated virus on cancer expression by herpesvirus-transformed hamster cells. *J Natl Cancer Inst* **55**, 957–959.

Davidoff AM, Ng CY, Zhang Y, *et al.* (2005). Careful decoy receptor titering is required to inhibit tumor angiogenesis while avoiding adversely altering VEGF bioavailability. *Mol Ther* **11**, 300–310.

De Veerman M, Heirman C, Van Meirvenne S, *et al.* (1999). Retrovirally transduced bone marrow-derived dendritic cells require CD4+ T cell help to elicit protective and therapeutic antitumor immunity. *J Immunol* **162**, 144–151.

Ding W, Zhang L, Yan Z, Engelhardt JF (2005). Intracellular trafficking of adeno-associated viral vectors. *Gene Ther* **12**, 873–880.

Feldman AL, Libutti SK (2000). Progress in antiangiogenic gene therapy of cancer. *Cancer* **89**, 1181–1194.

Ferrari FK, Samulski T, Shenk T, Samulski RJ (1996). Second-strand synthesis is a rate limiting step for efficient transduction by recombinant adeno-associated virus vectors. *J Virol* **70**, 3227–3234.

Fisher KJ, Gao GP, Weitzman MD, DeMatteo R, Burda JF, Wilson JM (1996). Transduction with recombinant adeno-associated virus for gene therapy is limited by leading-strand synthesis. *J Virol* **70**, 520–532.

Folkman J (1971). Tumor angiogenesis: therapeutic implications. *N Engl J Med* **285**, 1182–1186.

Folkman J, Cole P, Zimmerman S (1996). Tumor behavior in isolated perfused organs: *in vitro* growth and metastases of biopsy material in rabbit tyroid and canine intestinal segment. *Ann Surg* **164**, 491–502.

Gao G, Alvira MR, Somanathan S, *et al.* (2003). Adeno-associated viruses undergo substantial evolution in primates during natural infections. *Proc Natl Acad Sci U S A.* **100**, 6081–6086.

Gao G, Vandenberghe LH, Alvira MR, Lu Y, Calcedo R, Zhou X, Wilson JM (2004). Clades of adeno-associated viruses are widely disseminated in human tissues. *J Virol* **78**, 6381–6388.

Gao G, Vandenberghe LH, Wilson JM (2005). New recombinant serotypes of AAV vectors. *Curr Gene Ther* **5**, 285–297.

Gao GP, Alvira MR, Wang L, Calcedo R, Johnston J, Wilson JM (2002). Novel adeno-associated viruses from rhesus monkeys as vectors for human gene therapy. *Proc Natl Acad Sci U S A* **99**, 11854–11859.

Gao GP, Qu G, Faust LZ, *et al.* (1998). High-titer adeno-associated viral vectors from a Rep/Cap cell line and hybrid shuttle virus. *Hum Gene Ther* **9**, 2353–2362.

Girod A, Ried M, Wobus C, *et al.* (1999). Genetic capsid modifications allow efficient re-targeting of adeno-associated virus type 2. *Nat Med* **5**, 1052–1056.

Green MR, Roeder RG (1980a). Definition of a novel promoter for the major adenovirus-associated virus mRNA. *Cell* **22**, 231–242.

Green MR, Roeder RG (1980). Transcripts of the adeno-associated virus genome: mapping of the major RNAs. *J Virol* **36**, 79–92.

Grimm D, Kern A, Rittner K, Kleinschmidt JA (1998). Novel tools for the production and purification of recombinant adeno-associated virus vectors. *Hum Gene Ther* **9**, 2745–2760.

Hajitou A, Trepel M, Lilley CE, *et al.* (2006). A hybrid vector for ligand-directed tumor targeting and molecular imaging. *Cell* **125**, 385–398.

Hermonat PL (1994a). Adeno-associated virus inhibits human papillomavirus type 16: a viral interaction implicated in cervical cancer. *Cancer Res* **54**, 2278–2281.

Hermonat PL (1994b). Down-regulation of the human c-fos and c-myc proto-oncogene promoters by adeno-associated virus Rep78. *Cancer Lett* **81**, 129–136.

Hermonat PL, Labow MA, Wright R, Berns KI, Muzyczka N (1984). Genetics of adeno-associated virus: isolation and preliminary characterization of adeno-associated virus type 2 mutants. *J Virol* **51**, 611–619.

Hermonat PL, Plott RT, Santin AD, Parham GP, Flick JT (1997). Adeno-associated virus Rep78 inhibits oncogenic transformation of primary human keratinocytes by a human papillomavirus type 16-ras chimeric. *Gynecol Oncol* **66**, 487–494.

Hermonat PL, Meyers C, Parham GP, Santin AD (1998). Inhibition/stimulation of bovine papillomavirus by adeno-associated virus is time as well as multiplicity dependent. *Virology* **247**, 240–250.

High KA. Clinical gene transfer studies for hemophilia B (2004). *Semin Thromb Hemost* **30**, 257–267.

Hoggan MD, Blacklow NR, Rowe WP (1966). Studies of small DNA viruses found in various adenovirus preparations: physical, biological and immunological characteristics. *Proc Natl Acad Sci U S A* **55**, 1467–1474.

Horer M, Weger S, Butz K, Hoppe-Seyler F, Geisen C, Kleinschmidt JA (1995). Mutational analysis of adeno-associated virus Rep protein-mediated inhibition of heterologous and homologous promoters. *J Virol* **69**, 5485–5496.

Hwu P, Yannelli J, Kriegler M, *et al.* (1993). Functional and molecular characterisation of tumor-infiltrating lymphocytes transduced with tumor necrosis factor-alpha cDNA for the gene therapy of cancers in humans. *J Immunol* **150**, 4104–4115.

Im DS, Muzyczka N (1990). The AAV origin binding protein Rep68 is an ATP-dependent site-specific endonuclease with DNA helicase activity. *Cell* **61**, 447–457.

Inoue N, Russell DW (1998). Packaging cells based on inducible gene amplification for the production of adeno-associated virus vectors. *J Virol* **72**, 7024–7031.

Isayeva T, Ren C, Ponnazhagan S (2005). Recombinant adeno-associated virus 2-mediated anti-angiogenic prevention in a mouse model of intraperitoneal ovarian cancer. *Clin Cancer Res* **11**, 1342–1347.

Jaffee EM (1999). Immunotherapy of cancer. *Ann N Y Acad Sci* **886**, 67–72.

Johnson PR, Schnepp BC, Connell MJ, *et al.* (2005). Novel adeno-associated virus vector vaccine restricts replication of simian immunodeficiency virus in macaques. *J Virol* **79**, 955–965.

Jooss K, Yang Y, Fisher KJ, Wilson JM (1998). Transduction of dendritic cells by DNA viral vectors directs the immune response to transgene products in muscle fibers. *J Virol* **5**, 4212–4223.

Kay MA, Manno CS, Ragni MV, *et al.* (2000). Evidence for gene transfer and expression of factor IX in haemophilia B patients treated with an AAV vector. *Nat Genet* **24**, 257–261.

Kurts CJ, Miller FAP, Subramaniam RM, Carbone FR, Heath WR (1998). Major histocompatibility complex class I-restricted cross presentation is biased toward high-dose antigens and those released during cellular destruction *J Exp Med* **188**, 409–414.

Kwak H, Horig H, Kaufman HL (2003). Poxviruses as vectors for cancer immunotherapy. *Curr Opin Drug Discov Devel* **6**, 161–168.

Lalani AS, Chang, B, Lin J, *et al.* (2004). Anti-tumor efficacy of human angiostatin using liver-mediated adeno-associated virus gene therapy. *Mol Ther* **9**, 55–66.

Laughlin CA, Westphal H, Carter BJ (1979). Spliced adenovirus-associated virus RNA. *Proc Natl Acad Sci U S A* **76**, 5567–5571.

Liu DW, Tsao YP, Kung JT, *et al.* (2000). Recombinant adeno-associated virus expressing human papillomavirus type 16 E7 peptide DNA fused with heat shock protein DNA as a potential vaccine for cervical cancer. *J Virol* **74**, 2888–2894.

Liu M, Acres B, Balloul JM, *et al.* (2004). Gene-based vaccines and immunotherapeutics. *Proc Natl Acad Sci U S A* **101** (Suppl 2):14567–14571.

Liu Y, Santin AD, Mane M, *et al.* (2000). Transduction and utility of the granulocyte-macrophage colony-stimulating factor gene into monocytes and dendritic cells by adeno-associated virus. *J Interferon Cytokine Res* **20**, 21–30.

Lusby E, and Berns, KI (1982). Mapping of the 5′ termini of two adeno-associated virus 2 RNAs in the left half of the genome. *J Virol* **41**, 518–526.

Maass G, Bogedain C, Scheer U, *et al.* (1998). Recombinant adeno-associated virus for the generation of autologous, gene-modified tumor vaccines: evidence for a high transduction efficiency into primary epithelial cancer cells. *Hum Gene Ther* **9**, 1049–1059.

Mahendra G, Mahasreshti P, Curiel DT, *et al.* (2005). Anti-angiogenic gene therapy through adeno-associated virus 2-mediated stable expression of soluble Flt-1 receptor *Cancer Gene Ther* **12**, 26–34.

Manning WC, Paliard X, Zhou S, *et al.* (1997). Genetic immunization with adeno-associated virus vectors expressing herpes simplex virus type 2 glycoproteins B and D. *J Virol* **71**, 7960–7962.

Manno CS, Chew AJ, Hutchison S, *et al.* (2003). AAV-mediated factor IX gene transfer to skeletal muscle in patients with severe hemophilia B. *Blood* **101**, 2963–2972.

Manno CS, Pierce GF, Arruda VR, *et al.* (2006). Successful transduction of liver in hemophilia by AAV-Factor IX and limitations imposed by the host immune response. *Nat Med* **12**, 342–347.

Matsushita T, Elliger S, Elliger C, *et al.* (1998). Adeno-associated virus vectors can be efficiently produced without helper virus. *Gene Ther* **5**, 938–945.

McCarty DM, Monahan PE, Samulski RJ (2001). Self-complementary recombinant adeno-associated virus (scAAV) vectors promote efficient transduction independently of DNA synthesis. *Gene Ther* **8**, 1248–1254.

McCarty DM, Fu H, Monahan PE, Toulson CE, Naik P, Samulski RJ (2003). Adeno-associated virus terminal repeat (TR) mutant generates self-complementary vectors to overcome the rate-limiting step to transduction *in vivo*. *Gene Ther* **10**, 2112–2118.

Mendelson E, Trempe JP, Carter BJ (1986). Identification of the trans-acting Rep proteins of adeno-associated virus by antibodies to a synthetic oligopeptide. *J Virol* **60**, 823–832.

Mingozzi F, Maus MV, Hui DJ, *et al.* (2007). CD8(+) T-cell responses to adeno-associated virus capsid in humans. *Nat Med* **13**, 419–422.

Mizuno M, Yoshida J, Colosi P, Kurtzman G (1998). Adeno-associated virus vector containing thymidine kinase gene causes complete regression of intracerebrally implanted human gliomas in mice, in conjunction with ganciclovir administration. *Jpn J Cancer Res* **89**, 76–80.

Moss RB, Rodman D, Spencer LT, *et al.* (2004). Repeated adeno-associated virus serotype 2 aerosol-mediated cystic fibrosis transmembrane regulator gene transfer to the lungs of patients with cystic fibrosis: a multicenter, double-blind, placebo-controlled trial. *Chest* **125**, 509–521.

Muzyczka N (1992). Use of adeno-associated virus as a general transduction vector for mammalian cells. *Curr Top Microbiol Immunol* **158**, 97–129.

Nair SK, Boczkowski D, Morse M, Cumming RI, Lyerly HK, Gilboa E (1998). Induction of primary carcinoembryonic antigen (CEA)-specific cytotoxic T lymphocytes *in vitro* using human dendritic cells transfected with RNA. *Nat Biotechnol* **16**, 364–369.

Nakai H, Storm TA, Kay MA (2000). Recruitment of single-stranded recombinant adeno-associated virus vector genomes and intermolecular recombination are responsible for stable transduction of liver *in vivo*. *J Virol* **74**, 9451–9463.

Nelson D, Bundell C, Robinson B (2000). *In vivo* cross-presentation of a soluble protein antigen: kinetics, distribution, and generation of effector CTL recognizing dominant and subdominant epitopes. *J Immunol* **165**, 6123–6132.

Noro T, Miyake K, Suzuki-Miyake N, *et al.* (2004). Adeno-associated viral vector-mediated expression of endostatin inhibits tumor growth and metastasis in an orthotopic pancreatic cancer model in hamsters. *Cancer Res* **64**, 7486–7690.

Owens RA, Weitzman MD, Kyostio SR, Carter BJ (1993). Identification of a DNA-binding domain in the amino terminus of adeno-associated virus Rep proteins. *J Virol* **63**, 997–1005.

Pardoll DM (1995). Paracrine cytokine adjuvants in cancer immunotherapy. *Annu Rev Immunol* **13**, 399–415.

Ponnazhagan S, Mukherjee P, Wang X-S, *et al.* (1997). Adeno-associated virus 2-mediated transduction of primary human bone marrow derived CD34+ hematopoietic progenitor cells: Donor variation and correlation of expression with cellular differentiation. *J Virol* **71**, 8262–8267.

Ponnazhagan S, Curiel DT, Shaw DR, Alvarez RD, Siegal GP (2001a). Adeno-associated virus for cancer gene therapy. *Cancer Res* **61**, 6313–6321.

Ponnazhagan S, Mahendra G, Curiel DT, Shaw DR (2001b). Adeno-associated virus type 2-mediated transduction of human monocyte-derived dendritic cells: implications for *ex vivo* immunotherapy. *J Virol* **75**, 9493–9501.

Ponnazhagan S, Mahendra G, Kumar S, Thompson JA, Casillas Jr M (2002). Conjugate-based targeting of recombinant adeno-associated virus-2 using avidin-linked ligands. *J Virol* **76**, 12900–12907.

Ponnazhagan S, Mahendra G, Lima J, *et al.* (2004a). Augmentation of antitumor activity of a recombinant adeno-associated virus carcinoembryonic antigen vaccine with plasmid adjuvant. *Hum Gene Ther* **15**, 856–864.

Ponnazhagan S, Mahendra G, Kumar S, *et al.* (2004b). Adeno-associated virus 2-mediated anti-angiogenic cancer gene therapy: long-term efficacy of a vector encoding angiostatin and endostatin over vectors encoding a single factor. *Cancer Res* **64**, 1781–1787.

Qing KY, Mah C, Hansen J, Zhou SZ, Dwark, VJ, Srivastava A (1999). Human fibroblast growth factor receptor 1 is a co-receptor for infection by adeno-associated virus 2. *Nat Med* **5**, 71–77.

Rabinowitz JE, Xiao W, Samulski RJ (1999). Insertional mutagenesis of AAV2 capsid and the production of recombinant virus. *Virology* **265**, 274–285.

Ren C, Kumar S, Shaw DR, Ponnazhagan S (2005). Genomic stability of self-complementary adeno-associated virus 2 during early stages of transduction in mouse muscle *in vivo*. *Hum Gene Ther* **16**, 1047–57.

Samulski RJ, Berns KI, Tan M, Muzyczka N (1982). Cloning of adeno-associated virus into pBR322: rescue of intact virus from the recombinant plasmid in human cells. *Proc Natl Acad Sci USA* **79**, 2077–2081.

Samulski RJ, Chang LS, Shenk T (1989). Helper-free stocks of recombinant adeno-associated viruses: normal integration does not require viral gene expression. *J Virol* **63**, 3822–3838.

Sarukhan A, Soudais C, Danos O, Jooss K (2001). Factors influencing cross-presentation of non-self antigens expressed from recombinant adeno-associated virus vectors. *J Gene Med* **3**, 260–270.

Sauden P, Vlach J, Beard P (2000). Inhibition of S-phase progression by adeno-associated virus Rep78 protein is mediated by hypophosphorylated pRb. *EMBO J* **19**, 4351–4361.

Schlehofer JR, Ehrbar M, zur Hausen H (1984). Vaccinia virus, herpes simplex virus, and carcinogens induce DNA amplification in a human cell line and support replication of a helpervirus dependent parvovirus. *Virology* **152**, 110–117.

Schmidt M, Afione S, Kotin RM (2000). Adeno-associated virus type 2 Rep78 induces apoptosis through caspase activation indepedently of p53. *J Virol* **74**, 9441–9450.

Seiler MP, Miller AD, Zabner J, Halbert CL (2006). Adeno-associated virus types 5 and 6 use distinct receptors for cell entry. *Hum Gene Ther* **17**, 10–19.

Shi W, Teschendorf C, Muzyczka N, Sieman, DW (2003). Gene therapy delivery of endostatin enhances the treatment efficacy of radiation. *Radiother Oncol* **66**, 1–9.

Srivastava A, Lusby EW, Berns KI (1983). Nucleotide sequence and organization of adeno-associated virus 2 genome. *J Virol* **45**, 555–564.

Streck CJ, Dickson PV, Ng CY, *et al.* (2005). Adeno-associated virus vector-mediated systemic delivery of IFN-beta combined with low-dose cyclophosphamide affects tumor regression in murine neuroblastoma models. *Clin Cancer Res* **11**, 6020–6029.

Strickler HD, Viscidi R, Escoffery C, *et al.* (1999). Adeno-associated virus and development of cervical neoplasia. *J Med Virol* **59**, 60–65.

Su H, Chang JC, Xu SM, Kan YW (1996). Selective killing of AFP-positive hepatocellular carcinoma cells by adeno-associated virus transfer of the herpes simplex virus thymidine kinase gene. *Hum Gene Ther* **7**, 463–470.

Su H, Lu R, Chang JC, Kan JW (1997). Tissue-specific expression of herpes simplex virus thymidine kinase gene delivered by adeno-associated virus inhibits the growth of human hepatocellular carcinoma in athymic mice. *Proc Natl Acad Sci USA* **94**, 13891–13896.

Su H, Lu R, Ding R, Kan YW (2000). Adeno-associated viral-mediated gene transfer to hepatoma: thymidine kinase/interleukin 2 is more effective in tumor killing in a non-ganciclovir (GCV)-treated thanin GCV-treated animals. *Mol Ther* **1**, 509–525.

Subramanian IV, Ghebre R, Ramakrishnan S (2005). Adeno-associated virus-mediated delivery of a mutant endostatin suppresses ovarian carcinoma growth in mice. *Gene Ther* **12**, 30–38.

Subramanian IV, Bui Nguyen TM, Truskinovsky AM, Tolar J, Blazar BR, Ramakrishnan S (2006). Adeno-associated virus-mediated delivery of a mutant endostatin in combination with carboplatin treatment inhibits orthotopic growth of ovarian cancer and improves long-term survival. *Cancer Res* **66**, 4319–4328.

Summerford C, Samulski RJ (1998). Membrane-associated heparan sulfate proteoglycan is a receptor for adeno-associated virus type 2 virions. *J Virol* **72**, 1438.

Summerford C, Bartlett JS, Samulski RJ (1999). $\alpha V\beta 5$ integrin: A co-receptor for adeno-associated virus 2 infection. *Nat Med* **5**, 78–83.

Tagawa M (2000). Cytokine therapy for cancer. *Curr Pharm Des* **6**, 681–699.

Tamayose K, Hirai Y, Shimada T (1996). A new strategy for large-scale preparation of high-titer recombinant adeno-associated virus vectors by using packaging cell lines and sulfonated cellulose column chromatography. *Hum Gene Ther* **7**, 507–513.

Tepper RI, Mule JJ (1994). Experimental and clinical studies of cytokine gene-modified tumor cells. *Hum Gene Ther* **5**, 153–164.

Tillman BW, deGruijl TD, Luykx-de Bakker SZ, *et al.* (1999). Maturation of dendritic cells accompanies high-efficiency gene transfer by a CD40-targeted adenoviral vector. *J Immunol* **162**, 6378–6383.

Tillman BW, Hayes TL, de Gruijl TD, Douglas JT, Curiel DT (2000). Adenoviral vectors targeted to CD40 enhance the efficacy of dendritic cell-based vaccination against human papillomavirus 16-induced tumor cells in a murine model. *Cancer Res* **60**, 5456–5463.

Trempe JP, Mendelson E, Carter BJ (1987). Characterization of adeno-associated virus rep proteins in human cells by antibodies raised against rep expressed in *Escherichia coli*. *Virology* **161**, 18–28.

Tting T, Storkus WJ, Lotze MT (1997). Gene-based strategies for the immunotherapy of cancer. *J Mol Med* **75**, 478–491.

van Gool SW, Barcy S, Devos S, *et al.* (1995). CD80 (B7-1) and CD86 (B7-2): potential targets for immunotherapy? *Res Immunol* **146**, 183–196.

Wagner JA, Reynolds T, Moran ML, *et al.* (1998). Efficient and persistent gene transfer of AAV-CFTR in maxillary sinus. *Lancet* **351**, 1702–1703.

Wagner JA, Messner AH, Moran ML, *et al.* (1999). Safety and biological efficacy of an adeno-associated virus vector-cystic fibrosis transmembrane regulator (AAV-CFTR) in the cystic fibrosis maxillary sinus. *Laryngoscope* **109**, 266–274.

Walters RW, Yi SM, Keshavjee S, *et al.* (2001). Binding of adeno-associated virus type 5 to 2,3-linked sialic acid is required for gene transfer. *J Biol Chem* **276**, 20610–20616.

Wang Z, Ma HI, Li J, Sun L, Zhang J, Xiao X (2003). Rapid and highly efficient transduction by double-stranded adeno-associated virus vectors *in vitro* and *in vivo*. *Gene Ther* **10**, 2105–2111.

Warrington KH, Herzog RW (2006). Treatment of human disease by adeno-associated viral gene transfer. *Hum Genet* **119**, 571–603.

Weinberg RA (1991). Tumor suppressor genes. *Science* **254**, 1138–1147.

Wu FY, Wu CY, Lin CH, Wu CH (1999). Suppression of tumorigenicity in cervical carcinoma HeLa cells by an episomal form of adeno-associated virus. *Int J Oncol* **15**, 101–106.

Wu P, Xiao W, Conlon T, *et al.* (2000). Mutational analysis of the adeno-associated virus type 2 (AAV2) capsid gene and construction of AAV2 vectors with altered tropism. *J Virol* **74**, 8635–8647.

Wu Z, Miller E, Agbandje-McKenna M, Samulski RJ (2006). Alpha2,3 and alpha2,6 N-linked sialic acids facilitate efficient binding and transduction by adeno-associated virus types 1 and 6. *Virology* **80**, 9093–9103.

Xiao X, Li J, Samulski RJ (1998). Production of high-titer recombinant adeno-associated virus in the absence of helper adenovirus. *J Virol* **72**, 2224–2232.

Xin KQ, Urabe M, Yang J, *et al.* (2001). A novel recombinant adeno-associated virus vaccine induces a long-term humoral immune response to human immunodeficiency virus. *Hum Gene Ther* **12**, 1047–1061.

Xu R, Sun X, Tse LY, *et al.* (2003). Long-term expression of angiostatin suppresses metastatic liver cancer in mice. *Hepatology* **37**, 1451–1460.

Yanamandra N, Kondraganti S, Gondi CS, *et al.* (2005). Recombinant adeno-associated virus (rAAV) expressing TFPI-2 inhibits invasion, angiogenesis and tumor growth in a human glioblastoma cell line. *Int J Cancer* **115**, 998–1005.

Yang J, Zhou W, Zhang Y, Zidon T, Ritchie T, Engelhardt JF (1999). Concatamerization of adeno-associated virus circular genomes occurs through intermolecular recombination. *J Virol* **73**, 9468–9477.

Yang Q, Mamounas M, Yu G, *et al.* (1998). Development of novel cell surface CD34-targeted recombinant adeno-associated virus vectors for gene therapy. *Hum Gene Ther* **9**, 1929–1937.

Zaiss AK, Liu Q, Bowen GP, Wong NC, Bartlett JS, Muruve DA (2002). Differential activation of innate immune responses by adenovirus and adeno-associated virus vectors. *J Virol* **76**, 4580–4590.

Zhan D, Santin AD, Liu Y, *et al.* (1999). Binding of the human papillomavirus type 16 p97 promoter by the adeno-associated virus Rep78 major regulatory protein correlates with inhibition. *J Biol Chem* **274**, 31619–31624.

Zhang Y, Chirmule N, Gao Gp, Wilson J (2000). CD40 ligand-dependent activation of cytotoxic T lymphocytes by adeno-associated virus vectors *in vivo*: role of immature dendritic cells. *J Virol* **74**, 8003–8010.

4

Retroviruses

Simon Chowdhury and **Yasuhiro Ikeda**

4.1 Introduction

The *Retroviridae* comprise a diverse family of enveloped animal viruses with single strand positive sense diploid RNA genomes. Whilst retroviruses have been identified in a wide range of vertebrate hosts (Herniou *et al.*, 1998) they all share certain similarities with regards to their RNA genome, viral structure, mode of entry and replication. The unique feature of these viruses is the ability to convert their viral RNA into double stranded DNA using reverse transcriptase. Reverse transcriptase is an RNA-dependent DNA polymerase enzyme first identified in 1970 in murine and avian RNA tumour viruses (Baltimore, 1970; Temin and Mizutani, 1970; Herniou *et al.*, 1998). Subsequently, these viruses were named 'retro'viruses as they replicate by converting their viral genomic RNA into double stranded DNA (dsDNA). The DNA copy of the viral genome then integrates into the host chromosomal DNA as a provirus, which is subsequently transcribed and translated by host cellular mechanisms to provide the necessary proteins for packaging full length RNA transcripts into progeny virions.

4.2 Structure of retroviral particles

Most retroviruses are 80–130 nm in diameter with a similar structural morphology. The virion core consists of two molecules of genomic RNA that are associated with the nucleocapsid (NC) protein (NC-RNA complex) and a tRNA molecule, which primes the start of reverse transcription. The NC-RNA complex is surrounded by the capsid (CA) protein containing copies of three retroviral enzymes: reverse transcriptase (RT), protease (PR) and integrase (IN), which are all required in the early stages of infection. The matrix (MA) protein surrounds the capsid proteins or 'core' and lies just below the envelope that is derived from the host cell membrane. The retroviral envelope is studded with envelope glycoproteins arranged in oligomeric complexes that contain three or four heterodimers composed of a transmembrane (TM) component associated with a surface unit (SU) glycoprotein (Hunter, 1997).

4.3 Retroviral genome

The full length RNA genome varies between 7–13 kb with two identical molecules carried in a single virion. Conceptually it is easier to discuss the retroviral genome organization in terms of the DNA that is integrated into cellular DNA, as this places the promoter, the RNA start site and the polyadenylation site in the same position as found in typical host cells.

The retrovirus genome is organized with protein coding regions flanked by regulatory domains known as long terminal repeats (LTRs). LTRs are identical sequences which can be divided into three

Viral Therapy of Cancer Edited by Kevin J. Harrington, Richard G. Vile and Hardev S. Pandha

elements: U3, derived from sequences unique to the 3′ end of viral RNA; R derived from repeat sequences identical to both ends of viral RNA; and U5, derived from sequences unique to the 5′ end of viral RNA. The LTRs are generated during reverse transcription and as a result, proviral DNA is longer than genomic viral RNA. The LTRs contain sequences important in the control of viral transcription and post-transcriptional modifications, such as promoters, multiple enhancers and polyadenylation signals (Vogt, 1997).

The essential viral genes *gag*, *pro*, *pol* and *env* occupy the body of the DNA (Vogt, 1997). They are present in all retroviruses and for 'simple' retroviruses, such as gammaretroviruses, they are the only genes, where as 'complex' retroviruses, such as lentiviruses and spumaviruses, express other accessory genes required to co-ordinate and regulate viral replication. The *gag* gene encodes the internal structure protein of the virus. Gag is proteolytically processed into the mature proteins MA (matrix), CA (capsid), NC (nucleocapsid) and sometimes other proteins whose function remains uncertain e.g. p12 for murine leukaemia virus (MLV). The *pro* gene encodes the viral protease (PR) that acts late in assembly of viral particles by cleaving the proteins encoded by *gag, pro, pol* and sometimes *env. pol* encodes the enzymes reverse transcriptase, which has both DNA polymerase and RNase H activity, and integrase, which mediates replication of the genome. *env* encodes the surface unit glycoprotein and transmembrane protein of the virus, which form a complex that determines viral envelope tropism.

4.4 Retroviral life cycle

A complete understanding of the retroviral life cycle is crucial in the development of retroviral vectors. The life cycle can be divided into two distinct phases. The first involving attachment, entry, reverse transcription and integration with the host DNA and the second involving synthesis of viral proteins and genomic RNA, assembly and budding of infectious virions from the host cell.

The first step is attachment of the retrovirus to the host cell. The viral envelope glycoprotein dictates the host range of the virus through its interactions with the receptors on the target cells. Attachment of the envelope SU binding domain to a specific receptor triggers conformational changes within the SU. This leads to exposure of the hydrophobic 'fusion peptide' within the TM domain that mediates fusion of the cellular and viral membranes and results in viral core entry into the cell cytoplasm (Hunter, 1997).

In the cytoplasm viral genomic RNA (as part of the nucleoprotein complex) is reverse transcribed by activated RT (Hunter, 1997). Once DNA synthesis is complete viral integrase cleaves the two terminal nucleotides from the 3′ end of the linear double stranded viral DNA, creating recessed 3′-OH groups. In simple retroviruses like MLV, entry of the viral nucleoprotein complex into the nucleus occurs during mitosis when the nuclear membrane is dissociated (Roe *et al.*, 1993). However, in lentiviruses, nuclear entry can also occur during interphase by active transport through the nuclear pore allowing infection of non-dividing cells (Weinberg *et al.*, 1991).

Binding of viral DNA with host DNA is initiated by integrase, which uses the 3′-OH groups at the end of the viral DNA (in the integrase-viral complex) to attack the phosphodiester bonds on the host chromosomal DNA. Once in the host chromosome, integration is completed by cellular enzymes, which remove mismatched bases, repair single strand gaps and ligate host/viral DNA (Lee and Craigie, 1994).

The provirus is transcribed by the host cell machinery as a cellular gene. Viral mRNA, like its cellular counterpart, is 5′ capped and 3′ polyadenylated ensuring its stable export to the cytoplasm. *env* genes are translated from mRNA that has been spliced by cellular spliceosomes, while unspliced mRNAs are used for either *gag* and *pol* translation or incorporated into new virions as genomic RNA.

Retroviral proteins can be translated as a large Gag-Pro-Pol fusion polyprotein precursor or as a single Gag polyprotein or as an Env polyprotein. In MLV a read-through mechanism occurs to encode *gag* and *pol* in the same reading frame and thus produce a Gag-Pro-Pol fusion protein (Yoshinaka *et al.*, 1985). This translational mechanism ensures that correct amounts of proteins are packaged into mature virions, typically 2000 copies of Gag and 20–50 copies of Pol.

Retroviral assembly involves interactions between viral proteins, genomic proteins and host cell tRNA. Genomic RNA encapsidation and dimerization, tRNA packaging, viral protein and RNA transport to an assembly site at the host cell plasma membrane and envelope acquisition during budding are events common to all retroviruses. In C type retroviruses, such as MLV and lentiviruses, Gag-Pro-Pol core polyproteins and genomic RNA migrate to the plasma membrane, directed by the Gag MA domain, where they assemble before budding. Retroviral RNA contains *cis*-acting signals that allow it to be encapsidated by Gag polyproteins during virus assembly. This process requires interactions between the packaging signal, Ψ, or the encapsidation signal (E) and the Gag polyprotein (Berkowitz *et al.*, 1996).

The final stages in the retroviral life cycle involve the maturation of viral particles. In most retroviruses the core particles have an 'immature' morphology irrespective of their assembly pathway. Gag and Gag-Pro-Pol polyprotein precursors are cleaved by viral proteases to initiate maturation (Vogt, 1997). The timing of this protease-mediated cleavage is crucial and usually occurs at late assembly or just after budding. Premature processing results in intracytoplasmic accumulation of mature viral protein forms, which may interfere with normal particle assembly. Cleavage is critical in the production of infectious viruses, as demonstrated by mutations in the protease-coding domain (Stewart *et al.*, 1990) or the use of protease inhibitors (Sommerfelt, 1999), otherwise non-infectious virus containing unprocessed Gag and Gag-Pro-Pol is produced. After cleavage, the viruses take on a mature form in which the core is detached from the membrane, as opposed to the immature morphology in which Gag precursor proteins are anchored by the MA protein to the inner face of the viral envelope.

4.5 Retroviral vectors

Vectors based on members of the retrovirus family have been amongst the most commonly used for cancer gene therapy. The majority of the retroviral vectors used in gene therapy models are derived from the MLV and were amongst the first viral vectors to be used in human gene therapy trials (Blaese *et al.*, 1995). MLV has a small, simple and well-characterized genome, which allows extensive vector manipulation. To produce retroviral vectors all of the protein-encoding sequences are removed from the virus and replaced by the transgene of interest. The essential *cis*-acting sequences such as the packing sequence have to be included in the vector construct. The viral sequences necessary for reverse transcription of the vector RNA and integration of the proviral DNA, such as the LTRs, the transfer RNA-primer binding site, and the polypurine tract (PPT) also have to be present for efficient gene transduction and integration (Vogt, 1997). Thus the retroviral vector is replication-deficient as it is incapable of making the proteins required for additional rounds of replication.

Viral proteins needed for infection are provided *in trans* in the packaging cell line. Retroviral packaging constructs are either transiently transfected into the packaging cells or a cell line is established that stably expresses the viral proteins. The packaging constructs are modified to reduce the chances of generating replication-competent virus through recombination in the packaging cells. To further decrease the risk of developing helper virus in the packaging cells a split genome packaging structure was developed. Here two packaging constructs, one containing gag and pol and the other containing env are used (Danos and Mulligan, 1988, Markowitz *et al.*, 1988). This not only increases the safety of retroviral vectors but also facilitates the pseudotyping of retroviral vectors with different envelope constructs.

Retroviral vectors possess several features that make them suitable for gene therapy. They are able to integrate into the target cell chromatin and thus have the potential for long-term gene expression. Although integration does not guarantee stable expression of the transduced gene, it is an effective way for the genetic information to be maintained in a self-renewing tissue and in the clonal outgrowth of a stem cell. The design of the retroviral vector means that target cells do not express viral proteins, which are responsible for most of the pathological and immunological consequences of viral infection. Thus gene transduction is usually well tolerated. Vector design allows up to 8 kb of exogenous DNA to be inserted in place of the viral genes (Kay *et al.*, 2001). Recent packaging cell lines produce titres of virus above 10^7 transducing particles per ml and are suitable to be scaled-up for manufacturing large amounts of vector free from replication competent viruses for clinical use.

A critical limitation to the use of C-type retroviral vectors is their inability to infect non-dividing cells (Vogt, 1997). As discussed previously, nuclear entry of simple retroviruses requires disruption of the nuclear membrane to allow the pre-integration complex to gain access to the chromatin (Roe *et al.*, 1993) and is thus dependent on target cell mitosis shortly after entry (Miller *et al.*, 1990). This may provide an element of 'inherent targeting' for cancer gene therapy as tumour cells may be the only replicating cells e.g. central nervous system tumours. Recently, a nuclear localization signal was engineered in the matrix protein of an avian C-type retrovirus, spleen necrosis virus (SNV), to enable an SNV vector to transduce non-proliferating cells (growth-arrested human T lymphocytes and quiescent primary monocyte-derived macrophages) (Parveen *et al.*, 2000). However, most work has concentrated on the use of lentiviruses which are able to penetrate an intact nuclear membrane and transduce non-dividing cells. This subclass of retroviruses is discussed in a later chapter.

Other limitations of retroviral vectors have become apparent with their use in human gene therapy trials (Thomas *et al.*, 2003). These include low and variable particle titres, lack of vector targeting to specific cell types and genomic loci and relatively inefficient, position-dependent transcription. This has lead to substantial progress in vector development based on a deeper understanding of the biology of retroviruses and target cells. However, concerns remain most notably the risk of insertional mutagenesis as discussed below.

4.6 Safety of retroviral vectors: insertional mutagenesis

Integration is a potentially mutagenic event with the possibility of activating or inactivating cellular genes, including oncogenes or tumour-suppressor genes. A classical example of retroviral insertional mutagenesis is the avian leukosis virus (ALV). The ALV provirus integrates adjacent to the a gene encoding a cellular transcription factor (c-myc) gene and transcription, initiating from a viral promoter, causes enhanced expression of c-myc, leading to neoplastic transformation in chickens (Hayward *et al.*, 1981). Thus, one of the key advantages of retroviral vectors, integration of the transgene into the host chromosome with sustained gene expression, is also a disadvantage with the potential for insertional mutagenesis.

The first example of transformation by a gene therapy vector was described in a murine model using a retroviral vector (Hacein-Bey-Abina *et al.*, 2002). However, more recently, this issue has come to prominence with the development of T-cell leukaemia in children treated with retroviral gene therapy for X-linked severe combined immune deficiency (X-SCID), which will be discussed in greater detail below.

4.7 Gene therapy of X-linked SCID

X-SCID is an inherited disorder characterized by an early block in T and natural killer (NK) lymphocyte differentiation. This results in the absence of both T and B cells leading to severe and recurrent infections that are usually fatal in the first years of life. The block is caused by mutations of the gene encoding the γc cytokine receptor subunit of interleukin 2, 4, 7, 9 and 15 receptors, which participate in the delivery of growth, survival and differentiation signals to early lymphoid precursors. Bone marrow transplantation (BMT) can be used to successfully treat X-SCID, but works best when there is a fully compatible donor. Unfortunately this is the case in under one-third of X-SCID children. In unmatched recipients BMT is associated with significant morbidity (e.g. graft failure, graft versus host disease and lymphoma) and mortality.

The lack of therapeutic options in X-SCID led to the development of a gene therapy programme and trial by Fischer and colleagues (Cavazzana-Calvo *et al.*, 2000). Bone marrow stem cells were obtained from a selected group of affected children, cultured with growth factors and transfected on three consecutive days with a Moloney derived retroviral vector carrying the γc gene. Ten out of the first 11 patients achieved effective and life-saving immune reconstitution (Hacein-Bey-Abina *et al.*, 2002). The patients recovered well and were able to lead a normal life with apparent immune reconstitution. Expression of the γc gene was detected in T and NK cells with T, B and NK cell counts and function comparable to age matched controls (Cavazzana-Calvo *et al.*, 2000). Thus, gene therapy was able to correct the disease phenotype and from a clinical perspective the patients had been cured by this pioneering treatment.

However, 30 months after treatment one of the patients developed a monoclonal γ-δ T-cell lymphoproliferative disorder (leukaemia like disorder) (2003). Subsequently, two further children from this trial also developed T-cell leukaemias. The leukaemias appear to be caused by insertional mutagenesis, i.e. retroviral activation of a cellular oncogene at the site of insertion. The gene LIM domain only 2 (*LMO2*) is thought to be the responsible oncogene in these cases. *LMO2* is located on chromosome 11 and is normally involved in the control of blood cell proliferation and differentiation. *LMO2* has been shown to be overexpressed in certain types of T-cell leukaemias and its expression was elevated in the cases of leukaemia from this trial. It is thought that the malignant T cells in these patients are derived from single transduced cells in which the retrovirus genome has inserted near, or in, the LMO2 oncogene activating overexpression of *LMO2* (Kohn *et al.*, 2003).

The complication of leukaemia has not occurred in any other clinical trial of retroviral gene therapy or in any large animal model that used retroviral vectors to modify haematopoietic stem cells. However, as mentioned previously, leukaemia has been linked to vector integration by retroviral vectors in mouse models (Hacein-Bey-Abina *et al.*, 2002; Modlich *et al.*, 2005). Several factors may have contributed to the development of leukaemia in patients involved in the X-SCID trial. These include the high level of engraftment and expansion of genetically modified cells, unique properties of the haematopoietic stem and progenitor cells in the bone marrow of X-SCID patients, the inherent immune deficiency of X-SCID patients and the transferred gene itself. The γc gene itself is ideal for gene replacement as it provides a stimulus to growth and survival but this may also have contributed to the malignant transformation. Thus, X-SCID retroviral gene therapy may be one that is at particular risk of insertional mutagenesis for the reasons outlined above.

The development of leukaemia in this trial was unexpected. It was believed that the risk of inducing cancer with replication-defective viruses, such as those used in gene therapy, was very low. This is because carcinogenesis requires multiple synergizing mutations in the same cell. Thus, although a retrovirus can undoubtedly cause activation of a single oncogene, a single round of infection would be unlikely to activate multiple oncogenes in the same cell. This view is supported by the fact that replication-defective retroviral vectors that are not carrying oncogenes do not usually cause cancer in animal models and have been harmless in a large number of gene-therapy trials (Berns, 2004).

The leukaemias seen in the X-SCID trial challenge this view. A possible explanation is that the γc gene itself acts as an oncogene, and thus, insertion of the transgene near LMO2 constitutes a ‘double hit’ in promoting tumour development (Berns, 2004). Although an examination of the T-cell clones from the patients in the X-SCID trial provided no data to support this theory, supporting data has come from another source (Dave *et al.*, 2004). Dave and colleagues screened the Mouse Retroviral Cancer Gene database, analysing the sequences of more than 3000 insertions from nearly 1000 retrovirally induced haematopoietic tumours. They found two insertions near LMO2 and two near the endogenous γc gene. One tumour contained two clonal insertions: one near the γc gene and the other near LMO2. They concluded that the γc gene can act as an oncogene when under control of a retroviral promoter and that the γc gene and *LMO2* can act as collaborating oncogenes (Dave *et al.*, 2004).

This data helps to explain why leukaemia developed in the X-SCID trial. The insertion of the γc gene near LMO2 represents a 'double hit', leaving the transduced cells potentially only one mutation away from overt tumour development. The vast expansion of clonal populations of cells in this trial, although critical to the success of the therapy, is likely to increase the likelihood that additional genetic faults will develop. These findings are good news for the field of retroviral gene therapy as most therapeutic genes do not have oncogenic potential. The risk of developing cancer in gene therapy protocols will depend on several factors including the oncogenic potential of the therapeutic gene, its preference for integration in loci that harbour proto-oncogenes or tumour-suppressor genes as well as patient and protocol related factors. The ability to prevent such adverse events will depend on the availability of improved vectors, the findings of additional studies and a thorough assessment of the potential oncogenic capacity of the transgene.

In the light of the scientific information from the three children who developed leukaemia in the X-SCID trial European recommendations on the biosafety of gene therapy were made (Gonin *et al.*, 2005). These included:

1. Low vector dose to minimize the number of vector genomes inserted per cell.
2. Increased understanding of the potency of vector elements (i.e. promoter and transgene) in terms of activation of undesired cell function.
3. Improved characterization of target cells before and after transduction to avoid reinfusion with pathogenic proviral integration.
4. Replacing onco-retroviruses with other non-integrative vector systems.
5. Improved animal models.
6. Improved dialogue between researchers and regulatory authorities.

These recommendations as well as the recognition of the oncogenic potential of the transgene should decrease the risk of insertional mutagenesis in future gene therapy trials. Further use of current gene-transfer methods for the treatment of X-SCID poses a complex dilemma in the consideration of the potential risks and benefits. Initial recommendations from the United States Food and Drug Administration (FDA) Biological Response Modifiers Advisory Committee (BRMAC) state that this form of therapy should not be the first line of treatment for X-SCID, but it can be considered in the absence of other options such as matched bone-marrow transplantation (Check, 2003).

As stated above, X-SCID retroviral gene therapy is at particular risk of insertional mutagenesis. It is important to recognize that the risks could be different for each disease, each therapeutic gene and every individual patient. Every subsequent gene therapy trial involving the use of retroviral (or other integrating) vectors should carefully assess the risk–benefit ratio. In particular the risks of gene therapy should be weighed carefully against the risks and efficacy of existing treatments. It is important to remember that many treatments, such as organ transplantation, which are no longer considered experimental, are associated with significant morbidity and mortality.

Understanding the risk of oncogenesis by vector integration requires further investigation into the mechanisms that underlie transformation. This will hopefully lead to the development of approaches to minimize the likelihood of leukaemia or tumour formation. One of the first priorities is to analyse the site-selection patterns of integration for different vectors. For gammaretroviruses it is established that integration is biased towards DNAse I hypersensitive chromatin (Vijaya *et al.*, 1986; Rohdewohld *et al.*, 1987). This also appears to be the case for human immunodeficiency virus-1 (HIV-1) (Schroder *et al.*, 2002). This bias results in preferential integration in gene-rich regions, particularly in or near actively transcribed genes. This preference is likely to be conserved in the replication defective vectors derived from these viruses. The potential sites and frequency of integration will probably differ between vector types as well as between target cells and further research is needed to determine likely sites that will aid decisions about future treatments.

It is possible to safeguard against potential insertional mutagenesis by inserting a suicide gene (see later for full explanation of suicide gene therapy) into the therapeutic retroviral vector for selective elimination of transduced cells. Thus, if the therapeutic gene causes malignant transformation by insertional mutagenesis then these cells can be selectively eliminated by utilizing the suicide gene. In a study from Uchiyama and colleagues, B-cell lines from two X-SCID patients were transduced with bicistronic retroviral vector carrying human γc chain cDNA and the suicide gene herpes simplex virus thymidine kinase gene (HSV-tk) (Uchiyama *et al.*, 2006). After confirmation of functional reconstitution of the γc chain, the cells were treated with ganciclovir (GCV). The γc chain positive cells were eliminated under low concentration without cytotoxicity on untransduced cells and had not reappeared at 5 months. Furthermore, the γc chain transduced cells were still sensitive to GCV after five months. This elegant experiment demonstrates the ability of suicide gene therapy to control gene expression although further *in vivo* studies are required to assess feasibility of this approach in clinical trials.

One unforeseen benefit from retroviral insertional mutagenesis has been the ability to identify oncogenes that are important for both human and rodent carcinogenesis (Nakamura, 2005). This method has the ability to reveal not only primary oncogenes but also cooperative genes that might be affected as second or third hits in multistep carcinogenesis. With the use of the retrovirus-mediated gene transfer systems, retroviral insertional mutagenesis may help to provide invaluable information to understand genetic interaction in complex mechanisms of carcinogenesis.

4.8 Retroviral cancer gene therapy

Cancer gene therapy is the transfer of nucleic acids into tumour or normal cells to eliminate or reduce tumour burden by direct cell-killing, immunomodulation, or correcting genetic errors to reverse the malignant state. Cancer remains the most common disease treated in current gene therapy trials with 67 per cent of all current trials (data obtained from www.wiley.co.uk/genetherapy/clinical). Retroviruses were previously the most commonly used vectors in gene therapy trials (Thomas *et al.*, 2003) but are now second behind adenoviral vectors (data obtained from www.wiley.co.uk/genetherapy/clinical). The number of completed and ongoing retroviral cancer gene therapy trials is exhaustive and we will concentrate on key trials that illustrate the potential as well as the limitations of retroviral vectors (Table 4.1).

4.8.1 Suicide genes

Suicide genes are enzyme encoding genes which, once expressed, allow the cancer cell to metabolize a harmless prodrug into a toxic metabolite. Examples include herpes simplex thymidine kinase (HSV-tk) which initiates the phosphorylation of

Table 4.1 Advantages and disadvantages of retroviral vectors.

Advantages
Relatively high transfection efficiency
Stable integration of the transgene leading to long-term expression
Absence of immunogenic viral proteins in the target cells
Only transduces dividing cells thus 'inherent targeting' for cancer gene therapy
Pre-existing host immunity unlikely
Most widely studied vector system *in vitro* and *in vivo*
Disadvantages
Insertional mutagenesis: leukaemic transformation in X-SCID trial
Low titre (infectious units/ml) $\sim 10^{6-7}$ i.u./ml
Only transduces dividing cells
Relatively small transgene insert size ($\sim$8 kb)
Poor *in vivo* delivery and lack of efficacy in clinical trials
Possibility of homologous recombination resulting in replication competent viruses

non-toxic drugs aciclovir and ganciclovir to their toxic triphosphate forms and cytosine deaminase which converts the 5-fluorocytosine into the toxic agent 5-fluorouracil. A bystander effect, where by neighbouring non-transduced cells are also killed, is commonly seen with suicide gene therapy and may be mediated by either a local or immune effect or a combination of the two (McCormick, 2001).

Replication-defective recombinant retroviruses were the first vectors to be used in human clinical trials of suicide gene therapy (Ram *et al.*, 1997). They had the perceived advantage of selectively transducing only dividing cells which would restrict gene expression to tumour cells and spare normal brain parenchyma. The immediate limitation of retroviral vectors was the inability to highly concentrate the vector and thus limit the volume for injection to a reasonable size. Also, the vectors only survive *in vivo* for a short period of time relative to the rate of cell division (needed for retroviral entry into the nucleus). To overcome these limitations, vector-producing cells derived from murine fibroblasts were constructed to release the retroviral vector containing the suicide gene (HSV-tk). The aim was that the producer cells would release retrovirus encoding HSV-tk over a sustained period of time (several days), before being rejected by the host immune system. The tumour cells should be the only cells replicating at this site and thus they should be the only cells susceptible to retroviral infection. Subsequent systemic treatment with the prodrug ganciclovir would allow the cells encoding the HSV-tk gene to initiate its metabolism to its toxic triphosphate form. This would potentially kill the retroviral producer cells, the transduced cells and hopefully bystander tumour cells.

Retroviral suicide gene therapy was successfully achieved *in vivo* in the rat model (Culver *et al.*, 1992). Rats with a cerebral glioma were given an intratumoral stereotactic injection of murine fibroblasts that were producing a retroviral vector in which the herpes simplex virus thymidine kinase (HSV-tk) gene had been inserted. After 5 days during which the HSV-tk retroviral vectors that were produced in situ transduced the neighbouring proliferating glioma cells, the rats were treated with ganciclovir. Gliomas in the ganciclovir- and vector-treated rats regressed completely both macroscopically and microscopically (Culver *et al.*, 1992). However, there was only limited success when this approach was used to treat 15 patients with progressive growth of recurrent malignant brain tumours (Ram *et al.*, 1997). Antitumour activity was seen in five small tumours (1.4 ± 0.5 ml), but there appeared to be limited gene transfer to tumours suggesting that this effect was due to 'bystander' mechanisms from the vector-producing cells. The injected producer cells remained stuck in close proximity to the injecting needle and tk cDNA transfer by the retroviral vector was limited to a few cells away. The response of only very small tumours showed the feasibility of this approach, but highlighted the need to improve delivery of the therapeutic gene.

Several enzyme-prodrug combinations are being evaluated but HSV-tk has been most widely evaluated and has progressed farthest into the clinic with many phase I and II trials currently in progress. The first phase III trial in cancer gene therapy used a retroviral vector and has recently reported (Rainov, 2000). This multicentre trial randomized 248 patients with newly diagnosed, previously untreated glioblastoma multiforme (GBM) to standard therapy (surgical resection and radiotherapy) or standard therapy plus adjuvant gene therapy (using retroviral vectors) introduced during surgery. The same HSV-tk vector producing cells, as previously described, were injected at surgery followed by 2 weeks of ganciclovir. Progression-free median survival in the gene therapy group was 180 days compared with 183 days in control subjects. Median survival was 365 versus 354 days, and 12 month survival rates were 50 per cent versus 55 per cent in the gene therapy and control groups respectively. None of these differences were statistically significant. Thus, the adjuvant gene therapy improved neither time to progression nor overall survival time, although the feasibility and biosafety of this retroviral gene therapy approach were further supported. The failure of this specific protocol is probably due to the limited delivery of HSV-tk genes to the tumour cells. In addition, the current mode of manual injection of vector-producing cells with a non-migratory phenotype limits the distribution of these cells, and subsequently the retroviral vectors that they produce, to the immediate vicinity of the needle track. Further evaluation of this retroviral mediated gene therapy strategy needs to incorporate refinements such as improved delivery of vectors to the tumour cells, non-invasive *in vivo* assessment of transduction rates and improved delivery of the prodrug across the blood–brain and blood–tumour barrier to the transduced tumour cells.

These studies show that retrovirus producer cells and the HSV-tk ganciclovir system can be used safely. There has been no evidence of systemic toxicity from the virus or producer cells. However, the efficacy of this therapy has been limited. The retroviral vector has limited distribution in brain tumours and so has shown antitumour activity predominantly in tumours $\leq$1.5 ml in volume. This size limitation would restrict clinical applications to tumours that have been debulked, but in the phase III study of adjuvant therapy no benefit was seen for retroviral gene therapy. The retrovirus HSV-tk system established a proof of principle for its use but subsequent work has concentrated on the use of recombinant adenoviral vectors to over come some of the inherent limitations of retroviral vectors, most notably low titre (see Table 4.1).

4.8.2 Tumour suppressor genes

An alternative approach for cancer gene therapy is the restoration of tumour suppressor function, the loss of which is associated with many human malignancies (Greenblatt *et al.*, 1994). Among the tumour suppressors being considered for gene replacement strategies, p53 has been the focus of many groups for several reasons. p53 plays a pivotal role in the fate of a cell following DNA damage. It determines if the damaged cell will undergo growth arrest in order to repair itself (Levine, 1997) or if the cell will undergo programmed cell death or apoptosis because the damage is too extensive (Lowe *et al.*, 1993). Thus, loss of p53 significantly contributes to tumour development, tumour progression, and chemotherapeutic resistance. With respect to gene replacement, p53 is a potent inducer of cancer cell apoptosis despite the presence of multiple genetic changes in cancer cells (McCormick, 2001). Cancer gene therapy approaches have focused on replacing or even over expressing wild-type p53 in the hope that aberrant cell cycle control can once again be tightly regulated.

The initial p53 gene therapy experiments conducted by Roth and colleagues used retroviruses to deliver the tumour suppressor gene into various cancer cell lines. Wild type p53 was introduced using a retroviral vector into a non-small cell lung cancer (NSCLC) cell line with suppression of the tumour *in vitro* (Cai *et al.*, 1993). Efficient transduction of the wild-type p53 gene was achieved into human lung cancer cell lines H358a (deleted p53) and H322a (mutant p53). p53 mRNA and protein were detected in these cell lines 6 months after transduction by Northern and Western blot analyses. The successful restoration of the wild type p53 gene suppressed growth in the two transduced cell lines. Mixing experiments showed that transduced cells could reduce the growth rate of non-transduced cells; this reduction may have been mediated by factors shed into the supernatant of the transduced cell cultures.

These and other experiments provided ‘proof of principle’ with regards to p53 gene replacement strategies. Transformation of a normal cell to a malignant cell is causally related to the acquisition of a series of genetic defects. Replacement of tumour suppressor genes relies on the principle that restoration of a single defect will be effective in inhibiting tumour growth. The above experiments show that restoration of p53 alone is sufficient to arrest growth of an *in vitro* human lung cancer cell line. It was initially thought that the effects of delivering tumour suppressor genes would be cell autonomous with negligible effects on surrounding uninfected cells. This would mean that virtually every tumour cell would need to be infected, an enormous technical hurdle especially for disseminated disease. However, as discussed above, it may be that transduced cells are able to affect neighbouring non-transduced cells causing a ‘bystander effect’.

Retroviral vectors carrying wild-type p53 were subsequently shown to be able to suppress human lung cancer growth *in vivo* in a nude mouse model

(Fujiwara *et al.*, 1994). Irradiated BALB/c nu/nu mice were inoculated intratracheally with H226Br cells (mutant p53) and treated beginning 3 days later with an intratracheal injection of retroviral vector expressing p53 for 3 days. Thirty days after tumour cell inoculation, 62–80 per cent of the control mice showed macroscopic tumours of the right main stem bronchus. The retroviral vector suppressed H226Br tumour formation in 62–100 per cent of mice, and the effect was abrogated by dilution of the retroviral supernatant with inactive vector (Fujiwara *et al.*, 1994).

The first clinical trials delivering p53 were conducted by Roth and colleagues who used a retroviral vector carrying wild-type p53 to treat patients with non-small cell lung cancer who had failed other treatments (Roth *et al.*, 1996). The virus was administered intratumorally and caused no toxic side effects up to 5 months later. Wild type p53 was detected in lung biopsies by *in situ* hybridization and polymerase chain reaction amplification and apoptosis was increased in post-treatment biopsies. Nine patients were treated in this study with three showing tumour growth stabilization and three showing minor tumour regression. The retrovirus p53 delivery system established a proof of principle for tumour suppressor gene replacement strategies but subsequent work has concentrated on the use of other vectors such as recombinant adenoviral vectors to overcome the inherent limitations of retroviral vectors as discussed previously.

4.9 Immunomodulatory approaches

The failure of normal immune surveillance mechanisms is an integral part of the process of tumour development. Several mechanisms that allow cancer cells to evade host immunity have been identified. These include lower expression of major histocompatiiblity complex (MHC) class I and class II proteins, decreased growth and differentiation of effector immune cells and defects in expression of co-stimulatory molecules (Rosenberg, 2004). The use of vectors expressing genes to activate the host immune system or to bypass some of these defects by introducing genes that can alter the local immune microenvironment is an attractive anticancer strategy. This is because they have the potential to provide an amplified response that is truly systemic in nature. The types of genes include interleukins, tumour necrosis factor, interferon-γ (IFN-γ), granulocyte–macrophage colony stimulating factor (GM-CSF); genes which have immunostimulatory activity (e.g. MHC antigens), or T-cell co-stimulatory molecules such as B7.1 and B7.2. An alternative approach is to use cell-based cancer vaccines in which the individual is vaccinated with autologous tumour cells expressing vector-mediated cytokines, immunostimulatory or co-stimulatory molecules. Another approach uses vector-mediated gene transfer to T-lymphocytes or dendritic cells to augment their anti-tumour effector activity.

There are a large number of completed and ongoing retroviral cancer gene therapy trials using immunomodulatory approaches. These will be covered in greater detail in later chapters and we will concentrate on key trials that illustrate the potential as well as the limitations of using retroviral vectors. The era of clinical gene transfer began in May 1989 with the introduction of an antibiotic drug resistance gene into tumour-infiltrating lymphocytes (TILs) of patients with melanoma (Rosenberg *et al.*, 1990). This approach was used to optimize the use of TILs and define their *in vivo* distribution and survival. Retroviral-mediated gene transduction was used to introduce the gene coding for resistance to neomycin into human TILs before their infusion into patients thus using the new gene as a marker for the infused cells. This study demonstrated the feasibility and safety of using retroviral gene transduction for human gene therapy and highlighted the possibility of introducing alternative genes into TILs to improve their antitumour potency.

One approach to try to augment antitumour T-cell activity is by gene transfer of tumour-antigen specific T-cell receptors. It is now possible to clone the cDNAs encoding antigen-specific T cell receptors. Nishimura and colleagues cloned the T-cell receptor DNAs from a melanoma target antigen (MART-1) reactive $CD8^+$ TIL cell line, and demonstrated that plasmid-mediated gene transfer could reconstitute the tumour-antigen specificity of the T-cell receptor in a human T-cell line (Cole *et al.*, 1995). They subsequently showed that retroviral gene transfer could be used to introduce the MART-1 specific T-cell receptor in activated primary human T cells and demonstrated MART-1 specific cytokine release and cellular cytotoxicity in selected subclones (Clay *et al.*, 1999). A phase I study using an allogeneic, cytotoxic T lymphocyte cell line that recognizes MART-1-positive tumor cells through retroviral transduction with a T-cell receptor encoding gene, has been conducted in patients with metastatic melanoma (Duval *et al.*, 2006). Fifteen patients received a total of 24 treatment cycles with a total of 266 intratumoral injections of the T lymphocytes. One patient obtained a partial response, encompassing both metastases used and not used for intratumoral injections. In addition, regression of metastases used for injection in two patients and of metastases not used for injection in one patient was seen. This study showed that intratumoral injection of retrovirally modified T cells is feasible, safe, and capable of inducing tumour regression.

Retroviral vectors require that T cells be activated and replicating prior to viral infection. This therefore limits this approach to *ex vivo* strategies. To overcome the requirement for activation and cell division prior to gene transfer and allow *in vivo* gene transfer, alternative viral vector systems with the ability to transduce resting cell populations are increasingly being used. The majority of work has concentrated on the use of adeno-associated viruses and lentiviruses that have been successfully used to transduce resting T cells (Thomas *et al.*, 2003).

An alternative approach is to try to increase the immunogenicity of a tumour by the introduction of certain genes directly into tumour cells (Rosenberg, 2004). The most frequently used approach uses the introduction of cytokine genes such as interleukin (IL)-2, IL-4, IL-7, IL-12, INF-γ, tumour necrosis factor-α or GM-CSF into tumour cells. These manipulated cells are then injected intratumorally, subcutaneously or by other routes and are capable of producing high local concentrations of the cytokine. The use of autologous tumour cells is cumbersome and prone to technical problems in clinical trials, and because efficient vectors for *in vivo* transfection are not currently available, alternative methods have been developed. These include the use of allogeneic tumour cells or other autologous or allogeneic cells such as fibroblasts that can be transfected with cytokines prior to reinjection. The common principle of all these studies is the creation of an immunostimulatory environment in close proximity to tumour-specific antigens.

An example of this approach using a retroviral vector to deliver IL-12 to fibroblasts has recently been published by Kang and colleagues (Kang *et al.*, 2001). This phase I dose-escalation trial of peritumoral injections of IL-12-transduced autologous fibroblasts was performed in patients with disseminated cancer for whom effective treatment did not exist. Transient but clear reductions of tumour sizes were observed at the injected sites in four of nine cases, and at non-injected distant sites in one melanoma patient. These data indicate that gene therapy by peritumoral injection of IL-12-producing autologous fibroblasts is feasible, and promising in patients with advanced cancer.

Immunomodulatory approaches account for the vast majority of gene therapy trials with >40 per cent of the total number of current trials (data obtained from www.wiley.co.uk/genetherapy/clinical). Whilst retroviral vectors have helped to establish 'proof of principle' for this approach they have been superseded by other vectors that are better adapted to this approach. Examples of these include: adenoviral vectors that have higher titre than retroviral vectors and lentiviral vectors that are able to transduce critical non-dividing cells (such as dendritic cells or T cells). These and other vectors will be the mainstay of future trials of immunomodulatory cancer gene therapy.

4.10 Conclusions

Retroviral vector systems have dominated cancer gene therapy research in the past and it is likely that they will continue to have an important, albeit smaller, role in the future. Our knowledge of the biology of retroviruses and their target cells is continuing to increase and will allow improved vector systems. Progress in the field of transductional and transcriptional targeting will substantially impact on the quality and safety of cancer gene therapy approaches. The X-SCID trial has highlighted the risks of insertional mutagenesis and future retroviral studies will aim to decrease these risks for all integrating vectors. Importantly, many aspects of vector design established using simple retroviral vectors will also be applicable to lentiviral vectors and non-retroviral vectors, which are emerging as important alternatives for many approaches in cancer gene therapy and will substantially widen the perspectives of the field.

References

(2003) French gene therapy group reports on the adverse event in a clinical trial of gene therapy for X-linked severe combined immune deficiency (X-SCID). Position statement from the European Society of Gene Therapy. *J Gene Med* **5**, 82–84.

Baltimore D. RNA-dependent DNA polymerase in virions of RNA tumour viruses. *Nature* (1970) **226**, 1209–1211.

Berkowitz R, Fisher J, and Goff SP (1996). RNA packaging. *Curr Top Microbiol Immunol* **214**, 177–218.

Berns A (2004). Good news for gene therapy. *N Engl J Med* **350**, 1679–1680.

Blaese RM, Culver KW, Miller AD, *et al.* (1995). T lymphocyte-directed gene therapy for ADA- SCID: initial trial results after 4 years. *Science* **270**, 475–480.

Cai DW, Mukhopadhyay T, Liu Y, Fujiwara T, ROTH JA (1993). Stable expression of the wild-type p53 gene in human lung cancer cells after retrovirus-mediated gene transfer. *Hum Gene Ther* **4**, 617–624.

Cavazzana-Calvo M, Hacein-Bey S, De Saint Basile G, *et al.* (2000). Gene therapy of human severe combined immunodeficiency (SCID)-X1 disease. *Science* **288**, 669–672.

Check E (2003). Cancer risk prompts US to curb gene therapy. *Nature* **422**, 7.

Clay TM, Custer MC, Sachs J, Hwu P, Rosenberg SA, Nishimura MI (1999). Efficient transfer of a tumor antigen-reactive TCR to human peripheral blood lymphocytes confers anti-tumor reactivity. *J Immunol* **163**, 507–513.

Cole DJ, Weil DP, Shilyansky J, *et al.* (1995). Characterization of the functional specificity of a cloned T-cell receptor heterodimer recognizing the MART-1 melanoma antigen. *Cancer Res* **55**, 748–752.

Culver KW, Ram Z, Wallbridge S, Ishii H, Oldfield EH, Blaese RM (1992). *In vivo* gene transfer with retroviral vector-producer cells for treatment of experimental brain tumors. *Science* **256**, 1550–1552.

Danos O, Mulligan RC (1988). Safe and efficient generation of recombinant retroviruses with amphotropic and ecotropic host ranges. *Proc Natl Acad Sci USA* **85**, 6460–6464.

Dave UP, Jenkins NA, Copeland NG (2004). Gene therapy insertional mutagenesis insights. *Science* **303**, 333.

Duval L, Schmidt H, Kaltoft K, *et al.* (2006). Adoptive transfer of allogeneic cytotoxic T lymphocytes equipped with a HLA-A2 restricted MART-1 T-cell receptor: a phase I trial in metastatic melanoma. *Clin Cancer Res* **12**, 1229–1236.

Fujiwara T, Cai DW, Georges RN, Mukhopadhyay T, Grimm EA, Roth JA (1994). Therapeutic effect of a retroviral wild-type p53 expression vector in an orthotopic lung cancer model. *J Natl Cancer Inst* **86**, 1458–1462.

Gonin P, Buchholz CJ, Pallardy M, Mezzina M (2005). Gene therapy bio-safety: scientific and regulatory issues. *Gene Ther* **12**(Suppl 1): S146–152.

Greenblatt MS, Bennett WP, Hollstein M, Harris CC (1994). Mutations in the p53 tumor suppressor gene: clues to cancer etiology and molecular pathogenesis. *Cancer Res* **54**, 4855–4878.

Hacein-Bey-Abina S, Le Deist F, Carlier F, *et al.* (2002). Sustained correction of X-linked severe combined immunodeficiency by ex vivo gene therapy. *N Engl J Med* **346**, 1185–1193.

Hayward WS, Neel BG, Astrin SM (1981). Activation of a cellular onc gene by promoter insertion in ALV-induced lymphoid leukosis. *Nature* **290**, 475–480.

Herniou E, Martin J, Miller K, Cook J, Wilkinson M, Tristem M (1998). Retroviral diversity and distribution in vertebrates. *J Virol* **72**, 5955–5966.

Hunter E (1997). Viral entry and receptors. In: Coffin JM, Varmus HE (eds.) *Retroviruses*, Chapter 3. Cold Spring Harbor Laboratory Press, Cold Spring Harbor, NY, pp. 71–121.

Kang WK, Park C, Yoon HL, *et al.* (2001). Interleukin 12 gene therapy of cancer by peritumoral injection of transduced autologous fibroblasts: outcome of a phase I study. *Hum Gene Ther* **12**, 671–684.

Kay MA, Glorioso JC, Naldini L (2001). Viral vectors for gene therapy: the art of turning infectious agents into vehicles of therapeutics. *Nat Med* **7**, 33–40.

Kohn DB, Sadelain M, Glorioso JC (2003). Occurrence of leukaemia following gene therapy of X-linked SCID. *Nat Rev Cancer* **3**, 477–488.

Lee MS, Craigie R (1994). Protection of retroviral DNA from autointegration: involvement of a cellular factor. *Proc Natl Acad Sci U S A* **91**, 9823–9827.

Levine AJ (1997). p53, the cellular gatekeeper for growth and division. *Cell* **88**, 323–331.

Lowe SW, Ruley HE, Jacks T, Housman DE (1993). p53-dependent apoptosis modulates the cytotoxicity of anticancer agents. *Cell* **74**, 957–967.

Markowitz D, Goff S, Bank A (1988). A safe packaging line for gene transfer: separating viral genes on two different plasmids. *J Virol* **62**, 1120–1124.

McCormick F (2001). Cancer gene therapy: fringe or cutting edge? *Nat Rev Cancer* **1**, 130–141.

Miller DG, Adam MA, Miller AD (1990). Gene transfer by retrovirus vectors occurs only in cells that are actively replicating at the time of infection. *Mol Cell Biol* **10**, 4239–4242.

Modlich U, Kustikova OS, Schmidt M, *et al.* (2005). Leukemias following retroviral transfer of multidrug resistance 1 (MDR1) are driven by combinatorial insertional mutagenesis. *Blood* **105**, 4235–4246.

Nakamura T (2005). Retroviral insertional mutagenesis identifies oncogene cooperation. *Cancer Sci* **96**, 7–12.

Parveen Z, Krupetsky A, Engelstadter M, Cichutek K, Pomerantz RJ, Dornburg R (2000). Spleen necrosis virus-derived C-type retroviral vectors for gene transfer to quiescent cells. *Nat Biotechnol* **18**, 623–629.

Rainov NG (2000). A phase III clinical evaluation of herpes simplex virus type 1 thymidine kinase and ganciclovir gene therapy as an adjuvant to surgical resection and radiation in adults with previously untreated glioblastoma multiforme. *Hum Gene Ther* **11**, 2389–2401.

Ram Z, Culver KW, Oshiro EM, *et al.* (1997). Therapy of malignant brain tumors by intratumoral implantation of retroviral vector-producing cells. *Nat Med* **3**, 1354–1361.

Roe T, Reynolds TC, Yu G, Brown PO (1993). Integration of murine leukemia virus DNA depends on mitosis. *EMBO J* **12**, 2099–2108.

Rohdewohld H, Weiher H, Reik W, Jaenisch R, Breindl M (1987). Retrovirus integration and chromatin structure: Moloney murine leukemia proviral integration sites map near DNase I-hypersensitive sites. *J Virol* **61**, 336–343.

Rosenberg SA (2004). Shedding light on immunotherapy for cancer. *N Engl J Med* **350**, 1461–1463.

Rosenberg SA, Aebersold P, Cornetta K, *et al.* (1990). Gene transfer into humans – immunotherapy of patients with advanced melanoma, using tumor-infiltrating lymphocytes modified by retroviral gene transduction. *N Engl J Med* **323**, 570–578.

Roth JA, Nguyen D, Lawrence DD, *et al.* (1996). Retrovirus-mediated wild-type p53 gene transfer to tumors of patients with lung cancer. *Nat Med* **2**, 985–991.

Schroder AR, Shinn P, Chen H, Berry C, Ecker JR, Bushman F (2002). HIV-1 integration in the human genome favors active genes and local hotspots. *Cell* **110**, 521–529.

Sommerfelt MA (1999). Retrovirus receptors. *J Gen Virol* **80** (Pt 12), 3049–3064.

Stewart L, Schatz G, Vogt VM (1990). Properties of avian retrovirus particles defective in viral protease. *J Virol* **64**, 5076–5092.

Temin HM, Mizutani S (1970). RNA-dependent DNA polymerase in virions of Rous sarcoma virus. *Nature* **226**, 1211–1213.

Thomas CE, Ehrhardt A, Kay MA (2003). Progress and problems with the use of viral vectors for gene therapy. *Nat Rev Genet* **4**, 346–358.

Uchiyama T, Kumaki S, Ishikawa Y, *et al.* (2006). Application of HSVtk suicide gene to X-SCID gene therapy: ganciclovir treatment offsets gene corrected X-SCID B cells. *Biochem Biophys Res Commun* **341**, 391–398.

Vijaya S, Steffen DL, Robinson HL (1986). Acceptor sites for retroviral integrations map near DNase I-hypersensitive sites in chromatin. *J Virol* **60**, 683–692.

Vogt VM (1997). Retroviral virions and genomes. In: Coffin JM, Varmus HE (eds), *Retroviruses*, Chapter 2. Cold Spring Harbor Laboratory Press, Cold Spring Harbor, NY, pp. 27–71.

Weinberg JB, Matthews TJ, Cullen BR, Malim MH (1991). Productive human immunodeficiency virus type 1 (HIV-1) infection of nonproliferating human monocytes. *J Exp Med* **174**, 1477–1482.

Yoshinaka Y, Katoh I, Copeland T D, Oroszlan S (1985). Murine leukemia virus protease is encoded by the gag-pol gene and is synthesized through suppression of an amber termination codon. *Proc Natl Acad Sci U S A* **82**, 1618–1622.

5

Lentiviral Vectors for Cancer Gene Therapy

Antonia Follenzi and **Elisa Vigna**

5.1 Development of lentiviral vectors (LV)

To date lentiviral vectors (LV) represent one of the best tools for highly efficient gene transfer, in a wide type of gene therapy applications. These vectors are derived from lentivirus, a family of retroviruses involved in the development of disease related to an acquired status of immunodeficiency and characterized by the intrinsic ability to integrate their genome into the chromatin of cells, independently from the replicating status of the target cell. Until now, vectors derived both from primate lentiviruses (human immunodeficiency virus type 1, type 2, and simian immunodeficiency virus – HIV-1, HIV-2 and SIV) (reviewed in Gilbert and Wong-Staal, 2001) and non primate lentiviruses (feline immunodeficiency virus and equine infection anaemia virus – FIV, EIAV) (reviewed in Poeschla, 2003) have been developed. The most largely improved and most characterized of the lentiviral vectors are those derived from HIV-1. These represent the type of LV employed in the vast majority of the reported applications. However the use of non-primate vectors has been proposed to be more acceptable for future clinical applications, considering that the parental viruses are not pathogenic in humans. Non-primate lentiviruses have a different genetic organization from HIV, in number, types and functions of viral accessory genes and because genome expression is less dependent on the viral encoded transactivator (for a review see Curran and Nolan, 2002). However the latest generation of HIV-1 derived vectors (see below) have overcame the need for the presence of the accessory genes and Tat, and thus the HIV-1 derived vectors are now comparable to the non-primate vectors.

The safety issue that remains to be considered is the possibility of generating a replication competent retrovirus (RCR). However considering the great molecular knowledge and the number of clinically validated drugs available, the employment of an HIV-1 vector may be preferable, as some types of adverse events could be handled. HIV-1-derived vectors are characterized by efficient gene transfer and stable long-term transgene expression *in vivo* (for a review see Vigna and Naldini, 2000). Due to the fact that the parental virus induces a highly pathogenic disease in humans, a great deal of effort has been invested to generate particles with a high level of biosafety. The first goal of vector design was to separate the cis-acting sequences, required for the transfer of the vector genome to target cells, the trans-acting sequences, encoding the viral proteins required for vector packaging. Thanks to this design, viral proteins are not transferred to target cells (Naldini

Viral Therapy of Cancer Edited by Kevin J. Harrington, Richard G. Vile and Hardev S. Pandha

et al., 1996). This strategy was coupled to the use of an envelope derived from another virus, i.e. the vesicular stomatitis virus (VSV-G) (Naldini *et al.*, 1996), to get a vector tropism different from the parental virus which would normally infect $CD4^+$ cells. In a second phase of vector design large regions of HIV-1 sequence were deleted, maintaining only the genes strictly necessary to keep gene transfer efficacy. In the packaging constructs all the accessory genes, strongly related to the pathogenicity of the HIV-1 (Zufferey *et al.*, 1997) and the Tat gene have been eliminated (Dull *et al.*, 1998). In the transfer construct both the LTR regions have been modified. The 5′ LTR is chimeric, and completely independent from Tat activity (Dull *et al.*, 1998). The 3′ LTR promoter sequences present into the U3 region have been deleted generating a vector that upon integration into target cells is transcriptionally inactive (self inactivating – SIN – vector) (Zufferey *et al.*, 1998). In particular, the use of a SIN-vector with an inactive LTR is particularly important for reasons related to the insertional mutagenesis effect. This adverse effect was observed in some children during a gene therapy clinical trial for the treatment of X-linked immunodeficiency through the use of an LTR driven Mo-MLV retroviral vector. These patients developed a lymphoid tumor (Hacein-Bey-Abina *et al.*, 2003), due to insertion of the vector near to a proto-oncogene (LMO-2 gene). The authors showed that the proliferative disease was due to a series of concomitant causes among which one was the presence of the intact LTR, which was able to enhance LMO-2 expression (Cavazzana-Calvo *et al.*, 2005). As a result of this adverse event aspects relating to the oncogenic potential of the integrating vectors were investigated. High throughput studies concerning hot spots of integration by different types of viruses (Mitchell *et al.*, 2004) have underlined that while MLV preferentially integrates into the 5′ regions of genes, HIV-1 does not, thus perturbing less gene expression profile of the target cells. Recently Montini *et al.* (2006), validated a preclinical genotoxicity assay based on the use of tumour-prone mice and provided evidence of very low genotoxicity upon LVs integration, whilst retroviral vectors used in the same setting triggered acceleration of tumour formation. In relation to this, the introduction of the lowest number of integrated vector copies is a factor that can strongly reduce the oncogenic risk. Thus, in response to this requirement, the incorporation in the LV backbone sequences, such as the c-PPT (Follenzi *et al.*, 2000) and the wPRE (Zufferey *et al.*, 1999), that allow superior performances of the vector is a key point to achieve efficacy without losing safety.

The use of a SIN-vector assigns transgene expression control exclusively to the internal promoter. This is relevant to the versatility of the transfer vector ensuring that it can express the transgene using internal promoters characterized by different desirable features. One of the problems that could be faced is the requirement of multiple gene transfer. Examples of this requirement are: combination of genetic correction, transduced cell marking, selection of the genetic modified target cell, and reconstitution of multisubunit components. To provide a solution to these needs, Amendola *et al.*(2005) developed a LV with an expression cassette containing a synthetic bidirectional promoter able to allow highly coordinate expression of two distinct mRNAs. This type of promoter can be designed both for ubiquitous or tissue specific expression.

Another important characteristic, considered a priority for gene therapy, is the regulation of gene expression. Fine regulation in the amount of the transgene product, which can be adjusted to the therapeutic level needed by each single patient in particular pathology, is an aspect that could be desirable. In addition a vector that allows termination of the therapy according to the development of the pathology, together with the possibility to re-start the treatment if necessary, is an important feature related also to safety considerations.

Several kinds of regulated LV have been developed (Kafri *et al.*, 2000; Vigna *et al.*, 2002, 2005; Haack *et al.*, 2004; Koponen *et al.*, 2003; Galimi *et al.*, 2005). Regulation of gene expression can not only be in terms of exogenous gene induction, but also in terms of induction of endogenous gene silencing. In the last years the development of RNA interference technology has emerged as a powerful tool to explore gene functions and develop new experimental therapies (Behlke, 2006). A lentiviral RNAi library has been established that enables genome-scale loss of function screening for 27 000 mouse/human genes (Root *et al.*, 2006). Different types of LV expressing short hairpin RNA (shRNA) have been developed (Scherer and Rossi, 2004). Particular attention has been given to the system that allows conditional silencing (reviewed in Szulc *et al.*, 2006), and the system that allows concomitant and coordinate expression of a cDNA for gene replacement together with an shRNA for silencing an endogenous gene (Samakoglu *et al.*, 2006).

5.2 Targeting of transgene expression

Tissue-specific expression is essential in order to bring LV closer to the clinic. Tissue restricted vectors are more efficient and safer than constitutively expressing vectors. Using tissue specific promoters allows a better production and a higher biological activity of the therapeutic protein, due to the presence of all the elements that normally regulate the process. Moreover, the integrated exogenous gene is less prone to silencing (Chang *et al.*, 2006) assuring high and stable transgene expression, even when a low vector copy number is present in the cell. The power of this approach has been shown by replacing constitutively active promoters within the vector with the enhancer/promoter of the albumin gene, which is expressed only in hepatocytes. In spite of widespread transduction, the expression of the GFP transgene was confined to the hepatocytes in the liver and virtually absent in the spleen and bone marrow of treated mice (Follenzi *et al.*, 2002).

In a mouse model of mucopolysaccaridosis type I (MPS I), where the alpha-L-iduronidase (IDUA) activity is absent, LV with the albumin gene promoter selectively expressed IDUA in hepatocytes and limited transgene-specific immune responses allowing stable and prolonged expression of the IDUA enzyme and a partial correction of the pathology (Di Domenico *et al.*, 2006).

In order to develop a safe and effective LV gene transfer cassette for gene therapy of Wiskott-Aldrich syndrome (WAS), Dupre and colleagues assessed the potential of the *WAS* promoter to regulate expression of a human *WAS* cDNA transgene in multiple haematopoietic cell lineages and induce correction of the cellular defects. They initially validated the efficacy of gene expression cassettes utilizing the native *WAS* promoter both *in vivo* in murine haematopoietic progenitor cells and in WAS T lymphocytes and recently *in vitro* in multiple lineages of patient cells including haematopoietic progenitor cells. For the first time, it was shown that WAS patient's $CD34^+$ cells can be efficiently transduced with LV and that physiological levels of transgene can be achieved with *WAS* promoters while preserving their *in vitro* differentiation potential, important requisites for future clinical use. (Dupre *et al.*, 2004, 2006; Charrier *et al.*, 2007).

Another interesting approach to targeting transgene expression by LVs was developed by Klatzman's group. *In situ* gene correction of T lymphoid progenitors was obtained in the thymus of ZAP-70-deficient severe combined immunodeficient (SCID) mice with direct injection of T-cell specific ZAP-70-expressing LVs. This allowed long-term T-cell reconstitution and transgene expression in peripheral T lymphocytes and resulted in the long-term differentiation of mature T-cell receptor-$\alpha\beta^+$ thymocytes, indicating that the vector had integrated into progenitor cells. Moreover, peripheral ZAP-70-expressing T cells demonstrated a partially diversified receptor repertoire and were responsive to alloantigens *in vitro* and *in vivo*. Thus, intrathymic injection of a LV containing an expression cassette under the transcriptional control of T-cell specific regulatory sequences derived from the CD4 gene could represent a simplified and potentially safer alternative to *ex vivo* gene-modified haematopoietic stem cell transplantation for gene therapy of T-cell immunodeficiencies (Adjali *et al.*, 2005).

In order to get tissue specificity, other strategies have also been explored. In particular, the use of different envelope glycoproteins to pseudotype the vector has been reported (reviewed in Bartosch and Cosset, 2004). In this context, engineered or naturally occurring proteins have been tested. The use of a wild type envelope can be exemplified by the application reported by Kobinger *et al.* (2001).

The authors aimed to develop a vector suitable for treatment of chronic lung disease such as cystic fibrosis and assembled vector particles using the envelope of the filovirus. Using this strategy, they achieved a high tropism of the LVs for the airway epithelia in contrast to cells infected at a very low level using vector particles pseudotyped with the classic VSV-G protein. Further modifications of the envelope include incorporation of antibody and fusogenic protein on the surface of the LV (Yang *et al.*, 2006). The antibody directs the cell targeting, while the fusion protein allows the cell entry of the vector. To date, the use of lineage-restricted promoters represents the strategy by which the most satisfactory results have been obtained until now.

5.3 Host immune responses to LV and their transgene

In the use of LV the major barrier to stable gene transfer is the innate antiviral responses and the development of transgene-specific immunity. In studies of gene therapy for inherited diseases, such as the haemophilias, a successful outcome has been precluded by the development of immune responses against the vector and transgene product. The *de novo* expression of a protein, which is lacking in the patient because of deletions or other mutations in its genes, may result in immune responses leading to the clearance of the transduced cells and to the formation of antibodies that inhibit the activity of secreted factors. The main problem is the direct expression of the transgene product within professional antigen-presenting cells (APC) of the immune system. Intracellular proteins are continuously sampled in most cell types and exposed to the cell surface as peptides bound to major histocompatibility complex (MHC) class I molecules (Gromme and Neefjes, 2002). Secreted proteins are taken up by APC, processed and presented in the context of MHC class II molecules. Depending on the gene delivery method, antigen presentation by professional or non-professional APC, such as tissue cells, and by the direct or indirect pathway, may be favoured (Brown and Lillicrap, 2002). Thus, the immunological consequences of gene transfer are expected to vary with the type of transgene, vector, and target cells, the average level of transgene expression and the genetic and immunological, background of the host.

Remarkably, the use of tissue specific promoters has helped to reduce immune responses after vector delivery. We showed that hepatocyte-specific LV expression limited transgene-specific immune responses both in the case of green fluorescent protein (GFP) and in the case of the secreted hF.IX, allowing stable long-term expression. Prolonged expression and mitigated immune response to secreted hF.IX was also demonstrated when hepatocyte-specific promoters were used in the context of adenoviral and AAV vectors (Fields *et al.*, 2000, 2001; Herzog *et al.*, 2002). However, the protection from immune clearance afforded by hepatocyte-selective LV expression was not complete. We observed both mouse strain-dependence, in the case of GFP, and the occurrence of few low-titre antibody responses against hF.IX among similarly treated syngenic mice. However, this approach developed may suffice when a low frequency of antigen-specific lymphocyte precursors is expected, such as in the case of normal mice expressing allogenic F.IX and, in the future, in haemophilia B patients at low risk of inhibitor formation (Follenzi *et al.*, 2004). Interestingly, Brown and colleagues (Brown *et al.*, 2006) showed a new gene transfer system that exploits the endogenous microRNA (miRNA) machinery for transgene regulation. By using LV-mediated delivery, *in vivo* gene transfer was possible, and, they provided some of the first *in situ* data of miRNA activity in an adult mammal. By using miRNA regulation to de-target transgene expression from haematopoietic lineages, it was possible to prevent immune-mediated vector clearance and enable stable gene transfer. Recently, the same authors addressed the role of innate host response following *in vivo* LV delivery in mice (Brown *et al.*, 2007). They found that interferon-$\alpha\beta$ (IFN-$\alpha\beta$) strongly inhibits transduction efficiency, specifically within the liver, and contributes to immune-mediated clearance of transduced cells and plays a major role in preventing stable gene transfer. These results indicate that the minimum requirement for stimulating this response is an infectious LV particle and because the response is transient, the developing of strategies to prevent it, such as IFN-$\alpha\beta$ antibodies or IFN-$\alpha\beta$ receptor antagonists, should serve to improve the effectiveness and stability of LV-mediated gene transfer for gene therapy purposes.

It is likely that more sophisticated vector engineering and/or a combined pharmacological or genetic manipulation of the immune system may be needed for gene transfer to fully escape immune recognition and successful gene therapy.

5.4 Transgenesis

Viral transgenesis can be obtained only through integrating vectors able to reach high level of exogenous gene expression without eliciting the phenomenon known as gene silencing, which frequently occurs during development. In theory, the LV meet all the requirements for transgenic animal generation. Indeed, Lois *et al.* (2002) and Pfeifer *et al.* (2002) first demonstrated the efficient use of the LV technology for this application. These authors could generate transgenic mice and rats by delivering the vector into the pre-implantation embryo at the zygote and morula stage. Even though the currently available technology allows a relatively easy production of transgenic mice, thus not requiring the viral transgenesis, this is not the case for the transgenic animal generation of other species, where new ways to face the problem are required. LV transgenesis have addressed this issue, by generation of large animals such as pigs and cats (Hofmann *et al.*, 2004, 2006). All this data opened the way to use LV for the large-scale preparations of human proteins and for the even more interesting application of generating a vast variety of animal disease models (Ewerling *et al.*, 2006), including the ones requiring gene silencing, by using LV expressing shRNA (see above).

5.5 Haematopoietic stem cell gene transfer

The therapeutic potential of haematopoietic stem cell (HSC) gene therapy can be fully exploited only by reaching efficient gene transfer into HSC without compromising their biological properties. Retroviruses, both oncoretroviral and lentiviral, have been the preferred vectors for gene transfer into HSC because they stably integrate into cells. Although HSC can be transduced by LV in short *ex vivo* culture, they display low permissivity to the vector, requiring cytokine stimulation to reach high-frequency transduction (Ailles *et al.*, 2002). Using stringent assays of competitive xenograft repopulation, Santoni *et al.* (2006) showed that early-acting cytokines synergistically enhanced human HSC gene transfer by LV without impairing engraftment and repopulation capacity. Using S-phase suicide assays, it was shown that trans- duction enhancement by cytokines was not dependent on cell cycle progression and that LV can transduce quiescent HSC. Pharmacological inhibition of the proteasome during transduction dramatically enhanced HSC gene transfer, allowing very high levels of vector integration in their progeny in vivo. Thus, LV are effectively restricted at a post-entry step by the activity of this proteolytic complex and cytokine stimulation down-regulated proteasome activity in haematopoietic progenitors, enhancing permissiveness to LV gene transfer and establishing improved conditions for HSC-based gene therapy.

In order to develop efficient LV transduction protocol without preconditioning, Worsham *et al.* (2006), demonstrated that efficient transduction of bone marrow HSC could be achieved by *in situ* delivery of a LV through intra-bone marrow injection in mice without preconditioning. Transgene-expressing mesenchymal stem cells were also observed to retain multiple differentiation potential in both injected and secondary bone marrow (BM) transplant recipient mice. This approach may potentially provide a new technology for disease treatment and represents an interesting new tool to study adult stem cell plasticity and the nature of unperturbed haematopoiesis.

LV-mediated gene transfer into haematopoietic cells has long had been considered an attractive option for the treatment of genetic diseases, selected malignant diseases, and acquired immune deficiency syndrome. An example is the recent success of genetically modified HSC by LV to target gene therapy to the nervous system and correction of metachromatic leukodystrophy (MLD) in the mouse model of the disease (Biffi *et al.*, 2004). The authors showed extensive reconstitution of central nervous system microglia and peripheral nervous system macrophages by the transgene-expressing progeny of long-term repopulating HSC. By transplanting HSC transduced with the *ARSA* gene, they fully reconstituted enzyme activity in the haematopoietic system of MLD mice and prevented the development of functional and neuropathological manifestations of MLD in the mouse model when treatment begins at presymptomatic stage. Recently, the same authors reported that the same approach can correct already established neurologic disease manifestations and neuronal damage when applied to symptomatic MLD mice (Biffi *et al.*, 2006). The curative effects observed in affected mice strongly support clinical testing of gene therapy by LV in the bone marrow of MLD patients.

Another interesting application of gene therapy is the genetic defects in the adenosine deaminase (ADA) gene responsible for about 15 to 20 per cent of SCID. Recently it was developed a successful human clinical protocol using autologous haematopoietic stem/progenitor cells transduced with gammaretroviral vectors encoding for ADA. Results of this trial showed that gene therapy with bone marrow (BM) $CD34^+$ cells resulted in correction of both the immune and metabolic defects of ADA-SCID children (Aiuti *et al.*, 2002). However, despite the excellent safety record of all ADA-SCID gene transfer protocols, the adverse events occurred in the SCID-X1 GT trials Hacein-Bey-Abina *et al.* (2003) have raised general concerns on the potential risks of gammaretroviral vectors.

In this respect, SIN LV have an advanced safety profile over non-SIN gammaretroviral vectors and may thus reduce the risk of insertional oncogenesis. Mortellaro *et al.* (2006) assessed the efficacy of *ex vivo* BM gene therapy by LV, compared with transplantation of $ADA^{+/+}$ BM cells. They showed that LV-mediated ADA gene transfer rescues $ADA^{-/-}$ mice from lethality and corrects both their metabolic and immunological defects, similar to BM transplants. In addition, LV have been shown to be superior to gammaretroviral vectors in infecting human candidate haematopoietic stem cells and maintaining sustained transgene expression, particularly in short-term transduction protocols that induce minimal cell manipulation.

The haemophilias occupy a special place among disorders amenable to gene therapy because their clinical manifestations are attributable to the lack of a single protein that circulates in minute amounts in the plasma. Factor VIII (FVIII) was expressed by LV in haematopoietic stem cells derived from human cord blood and transplanted in FVIII-deficient mice FVIII (Kootstra *et al.*, 2003). Activated partial thromboplastin time levels in transplanted mice seemed to improve over the next 2 months but eventually reached pretreatment levels, which coincided with the development of inhibitory antibodies against FVIII. In a separate study, xenotransplantation of lentivirally transduced $CD34^+$ into non-obese diabetic NOD/SCID mice did not result in circulating FVIII levels. However, NOD/SCID repopulating cell (SRC)-derived human monocytes isolated from BM of these mice secreted functional recombinant FVIII after culture ex vivo. In contrast, SRC-derived human lymphocytes did not secrete FVIII, indicating that FVIII production is restricted to specific haematopoietic cell types, particularly the myeloid lineage (Tiede *et al.*, 2003).

So far the use of HSC has not been really successful for haemophilia gene therapy by LV; most probably additional target cells are needed for this purpose.

Instead, the use of LV was pivotal for the cure of mouse models of human β-thalassaemia and sickle cell disease (Imren *et al.*, 2002, 2004; May *et al.*, 2000, 2002; Pawliuk *et al.*, 2001; Rivella *et al.*, 2003; Rivella and Sadelain, 2002). The globin LV used in most of these experiments contain regulatory elements of the human β-globin LCR that upregulate globin synthesis, as well as deletion of sequences within the second intervening sequence (IVS-2) that cause aberrant processing of β-globin mRNA. Recent evidence indicates a high transduction rate of human HSC with high vector-derived globin expression in their erythroid progeny. The success of recent reports in lentiviral gene transfer of the human βglobin gene into human HSC have led to begin a human clinical trial of human β-globin gene therapy for β-thalassaemia (Bank *et al.*, 2005).

The use of haematopoietic stem cells can be crucial in the gene therapy for HIV-1 by introducing into these cells genes that inhibit replication of HIV-1 using LV. An interesting study (Bahner *et al.*, 2007) investigates the delivery of a dominant-negative *rev* gene humanized by LV. This vector was used to transduce primary human $CD34^+$ haematopoietic progenitor cells and yielded high-level transduction without toxicity and conferred potent inhibition of HIV-1. These results are aligned with the recent success of the first clinical trial for HIV-1 using LV.

5.6 Cancer Treatment by LV

The gene therapy goal in the treatment of cancer is to provide a highly targeted approach to the disease, currently unachieved with conventional treatments. Cell transformation and cancer can be approached from different point of views. As cancer is a genetic disease, one approach could be to provide direct gene correction, especially in the case of mutation of a tumor suppressor gene. Another approach could be to suppress an over-expressed oncogene. For cancer gene therapy factors that must be taken in account are related to the multi-step nature of the malignant process and to the requirement of genetic correction in virtually 100 per cent of the transformed cells.

In those tumors characterized by the accumulation of genetic lesions due to mutations of DNA repair genes, complementation by wild type gene could be considered. A preclinical study for the treatment of Fanconi anaemia (FA) using RV has been reported (Galimi *et al.*, 2002). A LV was used to deliver a normal FA gene into HSC from FA knock-out mice *ex vivo*. The LV corrected cells were used to reconstitute bone marrow in the same FA knock-out mice. After an *in vivo* selection of transduced cells, the reconstituted mice became fully resistant to DNA-damaging agents. This study provides evidence for the successful approach of gene therapy in the treatment of FA. The use of the LV is a key point of the protocol because, as discussed above, this type of vector is peculiar for its ability to infect long-term repopulating stem cells.

Mutation of oncogenes is crucial to tumour development, making the inhibition of the mutated oncogene itself a rational therapeutic target. The use of small interfering RNAs is a relatively new technology that allows gene specific down-regulation of expression. Proof of concept of this approach has been provided by Taulli *et al.* (2006) who used cells derived from embryonal and alveolar rhabdomyosarcomas and specifically silenced the Met oncogene with an anti-Met shRNA inserted in an inducible LV. The authors showed inhibition of the transformed phenotype both *in vitro* and *in vivo*. A parallel but not identical approach, using the delivery of specific small RNA, is based on the activation of the double-strand RNA-dependent protein kinase (PKR). This enzyme is normally activated in the cells upon viral infection by the presence of double stranded RNA molecules and it acts as a potent growth inhibitor, inducing cell death through different pathways (for a review see Garcia *et al.*, 2006). Selective activation of PKR can be used to kill cancer cells. This is possible when cancer cells themselves are expressing mutated oncogenes carrying deletions or rearrangements. A synthetic RNA designed so that it will not target the corresponding wild type oncogene can specifically recognize these sequences. Upon the formation of the dsRNA (selectively occurring in cancer cells), PKR will exert its inhibitory activity. Similarly, LV have been used to deliver an anti-sense RNA unique to the exon junction that give rise to a truncated form of the EGF receptor present in certain types of glioblastoma. When such LV are injected into mouse brain tumors, a strong inhibition of tumour growth has been obtained (Shir and Levitzki, 2002).

Tumour angiogenesis is crucial for both the progression and metastasis of the primary tumour. Thus, anti-angiogenic treatment could represent an effective therapeutic strategy to achieve suppression of tumour growth both in early and late stage disease. For a variety of reasons, compared to tumour cells endothelial cells are a more attractive target. First, the inhibition of vessel formation; even targeting a limited number of endothelial cells, can affect a large number of cells in the tumour. Second, endothelial cells have a low mutation rate compared to tumour cells, thus it will be unlikely that they will acquire a phenotype resistant to therapeutics. Third, an efficient therapy to prevent tumour vascularization will act on a large number of tumours, independently of tumour cell phenotype. LV mediated gene transfer of endogenous antiangiogenic factors, such as angiostatin and endostatin, has achieved inhibition of endothelial growth *in vitro* (Shichinohe *et al.*, 2001) and *in vivo* (Balaggan *et al.*, 2006; Yin *et al.*, 2002). Effective inhibition of tumour growth by antiangiogenic factor gene transfer was obtained by Pfeifer *et al.* (2000), using LV expressing PEX, a non catalytic fragment of matrix metalloproteinase 2, and by Indraccolo *et al.* (2005) who delivered a LV expressing Interferon-alpha systemically. Although angiogenesis is an infrequent event in the adult life, targeting endothelial cells in tumour vasculature is more preferable, as it will allow fewer side effects. A strategy proposed by De Palma (De Palma *et al.*, 2003) is based on the use of a LV, engineered to deliver Tie-2 promoter-driven cell specific expression to HSC. As bone marrow contributes different cell types to the tumour stroma, the transplantation of genetically modified bone marrow progenitor cells represents a vehicle for the transport of gene therapy to tumours. With this approach they identify a subset of mononuclear cells (TEM cells) present selectively at the angiogenic site. By delivering a suicide gene, TEM cells were selectively eliminated achieving substantial inhibition of angiogenesis and consequently impairment of tumour growth, avoiding systemic toxicity.

Lentiviral vectors have also been developed for immunotherapeutic strategies. Tumour cells can be differentiated from their normal counterpart due to the fact that they express tumour associated antigens (TAA). The presence of these molecules, in principle, could render the tumour a target of the immune system, but an effective anti-tumour immune response is lacking. This is probably due to the low level and the low avidity of the TAA specific cytotoxic T lymphocytes and by immune evasion strategies exerted by the tumour, such as down regulation of the MHC and local production of inhibitory cytokines. In this context an effective anticancer vaccine should ideally not only activate tumour specific immunity directed against TAA but also should provide immuno-stimulatory signals to break the tolerance. Dendritic cells (DC) are considered to play a central role in the immune response. As immature precursors, they infiltrate the tissues and continuously capture antigens. Here they undergo maturation, up regulate co-stimulatory molecules and prime both naïve T cytotoxic and helper lymphocytes. After recovery of DC from the patient the cells could be loaded with antigen and/or co-stimulatory molecules, and then re-administered to the patient to elicit the immune-response against the tumour. Sources of immature DC include peripheral blood, where $CD14^+$ progenitors cells are present, and $CD34^+$ stem cells from the bone marrow. In both cases the immature cells are quiescent, thus the use of the LV is highly preferable. Moreover LV do not express viral protein in target cells. This feature significantly reduces the anti-vector immune response that could suffocate the immune response against the tumour and block the possibility of repeat immunization. In addition LV-mediated gene transfer does not affect DC maturation, plasticity, and antigen presentation functions while other types of viral vectors do (Drillien *et al.*, 2000; Morelli *et al.*, 2000). The use of viral vector to engineer DC is preferable compared to direct peptides or protein pulsing. This is due to the fact that presentation of endogenously expressed antigens is more efficient. In fact, He and co-authors provided the direct comparison of the two types of systems (He *et al.*, 2005) and demonstrated that vaccination with LV transduced DC achieved a more potent antitumour activity. Many studies have reported the use of LV to genetically modify DC (for a review see (Dullaers and Thielemans, 2006). Among them, work from Cui and colleagues (Cui *et al.*, 2003) outlined the necessity to have efficient DC re-localization to the secondary lymphoid organs, after cell re-infusion following *ex vivo* gene transfer, to ensure T-cell activation occurs. They proposed to transduce HSC, and before re-infusion, to sub-lethally irradiate the recipients. This procedure allowed an enrichment of the donor derived DC expressing the transgene that in turn elicited a substantial expansion and activation of specific T cells able to act against an aggressive established tumor. Of interest is also the method described by Kim *et al.* (Kim *et al.*, 2005) where the LVs were directly administered *in vivo* in three distinct murine tumor models. *In vivo* LV-transduced DCs effectively presented and stimulated the T cell-mediated immunity with a significant therapeutic anti-tumour activity.

5.7 Approved clinical trials using LV

Several clinical trials using LV indicate promise for gene transfer to human cells. A phase I open-label non-randomized clinical trial was carried out using a LV based on HIV (Levine *et al.*, 2006). After extensive preclinical safety tests (Manilla *et al.*, 2005) and approval by institutional review boards, the Food and Drug Administration (FDA), the Cellular, Tissue, and Gene Therapies Advisory Committee, and the National Institutes of Health Office of Biotechnology Activities, a phase I open label nonrandomized clinical trial, was initiated to investigate the safety and tolerability of autologous T cells modified with the LV VRX496. In this trial the safety of a conditionally replicating HIV-1-derived vector expressing an antisense gene against the HIV envelope was investigated. Five subjects with chronic HIV infection who had failed to respond to at least two antiviral regimens were enrolled. All patients received a single intravenous infusion of gene-modified autologous CD4 T cells. Viral loads were stable, and one subject exhibited a sustained decrease in viral load. CD4 counts remained steady or increased in four subjects, and sustained gene transfer was observed. Self-limiting mobilization of the vector was observed in four of five patients. So far, there has been no evidence of serious adverse events, such as immunogenicity or insertional mutagenesis after 21–36 months of observation. Immune function improved in four subjects. Two phase II trials testing the safety and tolerability of single and repeated doses of VRX496 are currently underway after the completion of the previous phase I and preliminary results from these trials should be available in 2007. Recent success in the long-term correction of mouse models of human β-thalassaemia and sickle cell anaemia by LV and evidence of high gene transfer and expression in transduced human haematopoietic cells have led to the first human phaseI/II clinical trial of lentivirus-mediated gene therapy of the β-haemoglobinopathies: sickle cell disease and β-thalassaemia (Imren *et al.*, 2004; Pawliuk *et al.*, 2001). A LentiGlobin vector containing a β-globin gene that produces modified haemoglobin that can be distinguished from normal haemoglobin has been used. The LentiGlobin vector is SIN and contains large elements of the β-globin locus control region as well as chromatin insulators and other features that should prevent untoward events. The study is ongoing in Paris with 10 patients enrolled in the protocol, five with sickle cell disease and five with β-thalassaemia (Bank *et al.*, 2005).

5.8 Conclusions

The goal of gene therapy is to transfer and stably express curative genes in the desired population of cells without adverse effects. Continued efforts to improve the vector design, additional studies in larger animal models and more clinical trials are warranted to explore the full potential of LV for gene therapy.

References

Adjali O, Marodon G, Steinberg M, *et al.* (2005). *In vivo* correction of ZAP-70 immunodeficiency by intrathymic gene transfer. *J Clin Invest* **115**, 2287–2295.

Ailles L, Schmidt M, Santoni de Sio FR, *et al.* (2002). Molecular evidence of lentiviral vector-mediated gene transfer into human self-renewing, multi-potent, long-term NOD/SCID repopulating haematopoietic cells. *Mol Ther* **6**, 615–626.

Aiuti A, Slavin S, Aker M, *et al.* (2002) Correction of ADA-SCID by stem cell gene therapy combined with nonmyeloablative conditioning. *Science* **296**, 2410–2413.

Amendola M, Venneri MA, Biffi A, *et al.* (2005). Coordinate dual-gene transgenesis by lentiviral vectors carrying synthetic bidirectional promoters. *Nat Biotechnol* **23**, 108–116.

Bahner I, Sumiyoshi T, Kagoda M, *et al.* (2007). Lentiviral vector transduction of a dominant-negative Rev gene into human CD34(+) haematopoietic progenitor cells potently inhibits human immunodeficiency virus-1 replication. *Mol Ther* **15**, 76–85.

Balaggan KS, Binley K, Esapa M (2006). *et al.* EIAV vector-mediated delivery of endostatin or angiostatin inhibits angiogenesis and vascular hyperpermeability in experimental CNV. *Gene Ther* **13**, 1153–1165.

Bank A, Dorazio R, Leboulch P (2005). A phase I/II clinical trial of beta-globin gene therapy for beta-thalassemia. *Ann N Y Acad Sci* **1054**, 308–316.

Bartosch B, Cosset FL (2004). Strategies for retargeted gene delivery using vectors derived from lentiviruses. *Curr Gene Ther* **4**, 427–443.

Behlke MA (2006). Progress towards *in vivo* use of siRNAs. *Mol Ther* **13**, 644–670.

Biffi A, De Palma M, Quattrini A, *et al.* (2004). Correction of metachromatic leukodystrophy in the mouse model by transplantation of genetically modified haematopoietic stem cells. *J Clin Invest* **113**, 1118–1129.

Biffi A, Capotondo A, Fasano S, *et al.* (2006). Gene therapy of metachromatic leukodystrophy reverses neurological damage and deficits in mice. *J Clin Invest* **116**, 3070–3082.

Brown BD, Lillicrap D (2002). Dangerous liaisons: the role of 'danger' signals in the immune response to gene therapy. *Blood* **100**, 1133–1140.

Brown BD, Venneri MA, Zingale A, *et al.* (2006). Endogenous microRNA regulation suppresses transgene expression in haematopoietic lineages and enables stable gene transfer. *Nat Med* **12**, 585–591.

Brown BD, Sitia G, Annoni A, *et al.* (2007). *In vivo* administration of lentiviral vectors triggers a type I interferon response that restricts hepatocyte gene transfer and promotes vector clearance. *Blood* **109**, 2797–2805.

Cavazzana-Calvo M, Lagresle C, Hacein-Bey-Abina S, Fischer A (2005). Gene therapy for severe combined immunodeficiency. *Annu Rev Med* **56**, 585–602.

Chang AH, Stephan MT, Sadelain M (2006). Stem cell-derived erythroid cells mediate long-term systemic protein delivery. *Nat Biotechnol* **24**, 1017–1021.

Charrier S, Dupre L, Scaramuzza S, *et al.* (2007). Lentiviral vectors targeting WASp expression to haematopoietic cells, efficiently transduce and correct cells from WAS patients. *Gene Ther* **14**, 415–428.

Cui Y, Kelleher, E., Straley, E., *et al.* (2003). Immunotherapy of established tumors using bone marrow transplantation with antigen gene – modified haematopoietic stem cells. *Nat Med* **9**, 952–958.

Curran MA, Nolan GP (2002). Nonprimate lentiviral vectors. *Curr Top Microbiol Immunol* **261**, 75–105.

De Palma M, Venneri MA, Roca C, Naldini L (2003). Targeting exogenous genes to tumor angiogenesis by transplantation of genetically modified haematopoietic stem cells. *Nat Med* **9**, 789–795.

Di Domenico C, Di Napoli D, Gonzalez YRE, *et al.* (2006). Limited transgene immune response and long-term expression of human alpha-L-iduronidase in young adult mice with mucopolysaccharidosis type I by liver-directed gene therapy. *Hum Gene Ther* **17**, 1112–1121.

Drillien R, Spehner D, Bohbot A, Hanau D (2000). Vaccinia virus-related events and phenotypic changes after infection of dendritic cells derived from human monocytes. *Virology* **268**, 471–481.

Dull T, Zufferey R, Kelly M, *et al.* (1998). A third-generation lentivirus vector with a conditional packaging system. *J Virol* **72**, 8463–8471.

Dullaers M, Thielemans K (2006). From pathogen to medicine: HIV-1-derived lentiviral vectors as vehicles for dendritic cell based cancer immunotherapy. *J Gene Med* **8**, 3–17.

Dupre L, Marangoni F, Scaramuzza S, *et al.* (2006). Efficacy of gene therapy for Wiskott–Aldrich syndrome using a WAS promoter/cDNA-containing lentiviral vector and nonlethal irradiation. *Hum Gene Ther* **17**, 303–313.

Dupre L, Trifari S, Follenzi A, *et al.* (2004). Lentiviral vector-mediated gene transfer in T cells from Wiskott–Aldrich syndrome patients leads to functional correction. *Mol Ther* **10**, 903–915.

Ewerling S, Hofmann A, Klose R, *et al.* (2006). Evaluation of laser-assisted lentiviral transgenesis in bovine. *Transgenic Res* **15**, 447–454.

Fields PA, Kowalczyk DW, Arruda VR, *et al.* (2000). Role of vector in activation of T cell subsets in immune responses against the secreted transgene product factor IX. *Mol Ther* **1**, 225–235.

Fields PA, Arruda VR, Armstrong E, *et al.* (2001). Risk and prevention of anti-factor IX formation in AAV-mediated gene transfer in the context of a large deletion of F9. *Mol Ther* **4**, 201–210.

Follenzi A, Ailles LE, Bakovic S, *et al.* (2000). Gene transfer by lentiviral vectors is limited by nuclear translocation and rescued by HIV-1 pol sequences. *Nat Genet* **25**, 217–222.

Follenzi A, Sabatino G, Lombardo A, *et al.* (2002). Efficient gene delivery and targeted expression to hepatocytes *in vivo* by improved lentiviral vectors. *Hum Gene Ther* **13**, 243–260.

Follenzi A, Battaglia M, Lombardo A, *et al.* (2004). Targeting lentiviral vector expression to hepatocytes limits transgene-specific immune response and establishes long-term expression of human antihaemophilic factor IX in mice. *Blood* **103**, 3700–3709.

Galimi F, Noll M, Kanazawa Y, *et al.* (2002). Gene therapy of Fanconi anaemia: preclinical efficacy using lentiviral vectors. *Blood* **100**, 2732–2736.

Galimi F, Saez E, Gall J, *et al.* (2005). Development of ecdysone-regulated lentiviral vectors. *Mol Ther* **11**, 142–148.

Garcia MA, Gil J, Ventoso I, *et al.* (2006). Impact of protein kinase PKR in cell biology: from antiviral to antiproliferative action. *Microbiol Mol Biol Rev* **70**, 1032–1060.

Gilbert JR, Wong-Staal F (2001). HIV-2 and SIV vector systems. *Somat Cell Mol Genet* **26**, 83–98.

Gromme M, Neefjes J (2002). Antigen degradation or presentation by MHC class I molecules via classical and non-classical pathways. *Mol Immunol* **39**, 181–202.

Haack K, Cockrell AS, Ma H, *et al.* (2004). Transactivator and structurally optimized inducible lentiviral vectors. *Mol Ther* **10**, 585–596.

Hacein-Bey-Abina S, von Kalle C, Schmidt M, *et al.* (2003). A serious adverse event after successful gene therapy for X-linked severe combined immunodeficiency. *N Engl J Med* **348**, 255–256.

He Y, Zhang J, Mi Z, *et al.* (2005). Immunization with lentiviral vector-transduced dendritic cells induces strong and long-lasting T cell responses and therapeutic immunity. *J Immunol* **174**, 3808–3817.

Herzog RW, Fields PA, Arruda VR, *et al.* (2002). Influence of vector dose on factor IX-specific T and B cell responses in muscle-directed gene therapy. *Hum Gene Ther* **13**, 1281–1291.

Hofmann A, Zakhartchenko V, Weppert M, *et al.* (2004). Generation of transgenic cattle by lentiviral gene transfer into oocytes. *Biol Reprod* **71**, 405–409.

Hofmann A, Kessler B, Ewerling S, *et al.* (2006). Epigenetic regulation of lentiviral transgene vectors in a large animal model. *Mol Ther* **13**, 59–66.

Imren S, Payen E, Westerman KA, *et al.* (2002). Permanent and panerythroid correction of murine beta thalassemia by multiple lentiviral integration in hematopoietic stem cells. *Proc Natl Acad Sci U S A* **99**, 14380–14385.

Imren S, Fabry ME, Westerman KA, *et al.* (2004). High-level beta-globin expression and preferred intragenic integration after lentiviral transduction of human cord blood stem cells. *J Clin Invest* **114**, 953–962.

Indraccolo S, Tisato V, Tosello V, *et al.* (2005). Interferon-alpha gene therapy by lentiviral vectors contrasts ovarian cancer growth through angiogenesis inhibition. *Hum Gene Ther* **16**, 957–970.

Kafri T, van Praag H, Gage FH, Verma IM (2000). Lentiviral vectors: regulated gene expression. *Mol Ther* **1**, 516–521.

Kim JH, Majumder N, Lin H, *et al.* (2005). Induction of therapeutic antitumour immunity by *in vivo* administration of a lentiviral vaccine. *Hum Gene Ther* **16**, 1255–1266.

Kobinger GP, Weiner DJ, Yu QC, Wilson JM (2001). Filovirus-pseudotyped lentiviral vector can efficiently and stably transduce airway epithelia *in vivo*. *Nat Biotechnol* **19**, 225–230.

Kootstra NA, Matsumura R, Verma IM (2003). Efficient production of human FVIII in hemophilic mice using lentiviral vectors. *Mol Ther* **7**, 623–631.

Koponen JK, Kankkonen H, Kannasto J, *et al.* (2003). Doxycycline-regulated lentiviral vector system with a novel reverse transactivator rtTA2S-M2 shows a tight control of gene expression *in vitro* and *in vivo*. *Gene Ther* **10**, 459–466.

Levine BL, Humeau LM, Boyer J, *et al.* (2006). Gene transfer in humans using a conditionally replicating lentiviral vector. *Proc Natl Acad Sci U S A* **103**, 17372–17377.

Lois C, Hong EJ, Pease S, *et al.* (2002). Germline transmission and tissue-specific expression of transgenes delivered by lentiviral vectors. *Science* **295**, 868–872.

Manilla P, Rebello T, Afable C, *et al.* (2005). Regulatory considerations for novel gene therapy products: a review of the process leading to the first clinical lentiviral vector. *Hum Gene Ther* **16**, 17–25.

May C, Rivella S, Callegari J, *et al.* (2000). Therapeutic haemoglobin synthesis in beta-thalassaemic mice expressing lentivirus-encoded human beta-globin. *Nature* **406**, 82–86.

May C, Rivella S, Chadburn A, Sadelain M (2002). Successful treatment of murine beta-thalassemia intermedia by transfer of the human beta-globin gene. *Blood* **99**, 1902–1908.

Mitchell RS, Beitzel BF, Schroder AR, *et al.* (2004). Retroviral DNA integration: ASLV, HIV, and MLV show distinct target site preferences. *PLoS Biol* **2**, E234.

Montini E, Cesana D, Schmidt M, *et al.* (2006). Haematopoietic stem cell gene transfer in a tumorprone mouse model uncovers low genotoxicity of lentiviral vector integration. *Nat Biotechnol* **24**, 687–696.

Morelli AE, Larregina AT, Ganster RW, *et al.* (2000). Recombinant adenovirus induces maturation of dendritic cells via an NF-kappaB-dependent pathway. *J Virol* **74**, 9617–9628.

Mortellaro A, Hernandez RJ, Guerrini MM, *et al.* (2006). *Ex vivo* gene therapy with lentiviral vectors rescues adenosine deaminase (ADA)-deficient mice and corrects their immune and metabolic defects. *Blood* **108**, 2979–2988.

Naldini L, Blomer U, Gallay P, *et al.* (1996). *In vivo* gene delivery and stable transduction of nondividing cells by a lentiviral vector. *Science* **272**, 263–267.

Pawliuk R, Westerman KA, Fabry ME, *et al.* (2001). Correction of sickle cell disease in transgenic mouse models by gene therapy. *Science* **294**, 2368–2371.

Pfeifer A, Kessler T, Silletti S, *et al.* (2000). Suppression of angiogenesis by lentiviral delivery of PEX, a noncatalytic fragment of matrix metalloproteinase 2. *Proc Natl Acad Sci U S A* **97**, 12227–12232.

Pfeifer A, Ikawa M, Dayn Y, Verma IM (2002). Transgenesis by lentiviral vectors: lack of gene silencing in mammalian embryonic stem cells and preimplantation embryos. *Proc Natl Acad Sci U S A* **99**, 2140–2145.

Poeschla EM (2003). Non-primate lentiviral vectors. *Curr Opin Mol Ther* **5**, 529–540.

Rivella S, May C, Chadburn A, *et al.* (2003). A novel murine model of Cooley anemia and its rescue by lentiviral-mediated human beta-globin gene transfer. *Blood* **101**, 2932–2939.

Rivella S, Sadelain M (2002). Therapeutic globin gene delivery using lentiviral vectors. *Curr Opin Mol Ther* **4**, 505–514.

Root DE, Hacohen N, Hahn WC, *et al.* (2006). Genome-scale loss-of-function screening with a lentiviral RNAi library. *Nat Methods* **3**, 715–719.

Samakoglu S, Lisowski L, Budak-Alpdogan T, *et al.* (2006). A genetic strategy to treat sickle cell anemia by coregulating globin transgene expression and RNA interference. *Nat Biotechnol* **24**, 89–94.

Santoni de Sio FR, Cascio P, Zingale A, *et al.* (2006). Proteasome activity restricts lentiviral gene transfer into hematopoietic stem cells and is down-regulated by cytokines that enhance transduction. *Blood* **107**, 4257–4265.

Scherer L, Rossi JJ (2004). Therapeutic applications of RNA interference: recent advances in siRNA design. *Adv Genet* **52**, 1–21.

Shichinohe T, Bochner BH, Mizutani K, *et al.* (2001). Development of lentiviral vectors for anti-angiogenic gene delivery. *Cancer Gene Ther* **8**, 879–889.

Shir A, Levitzki A (2002). Inhibition of glioma growth by tumor-specific activation of double-stranded RNA-dependent protein kinase PKR. *Nat Biotechnol* **20**, 895–900.

Szulc J, Wiznerowicz M, Sauvain MO, *et al.* (2006). A versatile tool for conditional gene expression and knockdown. *Nat Methods* **3**, 109–116.

Taulli R, Scuoppo C, Bersani F, *et al.* (2006). Validation of met as a therapeutic target in alveolar and embryonal rhabdomyosarcoma. *Cancer Res* **66**, 4742–4749.

Tiede A, Eder M, von Depka M, *et al.* (2003). Recombinant factor VIII expression in haematopoietic cells following lentiviral transduction. *Gene Ther* **10**, 1917–1925.

Vigna E, Naldini L (2000). Lentiviral vectors: excellent tools for experimental gene transfer and promising candidates for gene therapy. *J Gene Med* **2**, 308–316.

Vigna E, Cavalieri S, Ailles L, *et al.* (2002). Robust and efficient regulation of transgene expression *in vivo* by improved tetracycline-dependent lentiviral vectors. *Mol Ther* **5**, 252–261.

Vigna E, Amendola M, Benedicenti F, *et al.* (2005). Efficient Tet-dependent expression of human factor IX *in vivo* by a new self-regulating lentiviral vector. *Mol Ther* **11**, 763–775.

Worsham DN, Schuesler T, von Kalle C, Pan D (2006). *In vivo* gene transfer into adult stem cells in unconditioned mice by in situ delivery of a lentiviral vector. *Mol Ther* **14**, 514–524.

Yang L, Bailey L, Baltimore D, Wang P (2006). Targeting lentiviral vectors to specific cell types *in vivo*. *Proc Natl Acad Sci U S A* **103**, 11479–11484.

Yin G, Liu W, An P, *et al.* (2002). Endostatin gene transfer inhibits joint angiogenesis and pannus formation in inflammatory arthritis. *Mol Ther* **5**, 547–554.

Zufferey R, Nagy D, Mandel RJ, *et al.* (1997). Multiply attenuated lentiviral vector achieves efficient gene delivery *in vivo*. *Nat Biotechnol* **15**, 871–875.

Zufferey R, Dull T, Mandel RJ, *et al.* (1998). Self-inactivating lentivirus vector for safe and efficient *in vivo* gene delivery. *J Virol* **72**, 9873–9880.

Zufferey R, Donello JE, Trono D, Hope TJ (1999). Woodchuck hepatitis virus posttranscriptional regulatory element enhances expression of transgenes delivered by retroviral vectors. *J Virol* **73**, 2886–2892.

6

Poxviruses as Immunomodulatory Cancer Therapeutics

Kevin J. Harrington, Hardev S. Pandha and **Richard G. Vile**

6.1 Introduction

The poxviruses are a complex family of DNA viruses that can be divided into the subfamilies Chordopoxvirinae and Entomopoxvirinae on the basis of their ability to infect vertebrates and insects, respectively (Moss, 2007) (Table 6.1). This large group of viruses is responsible for important human and animal infections and many can cross the species divide from animals to humans to cause zoonoses. Poxvirus species from amongst the Orthopoxvirus, Parapoxvirus, Yatapoxvirus and Molluscipoxvirus genera have been shown to cause human illnesses (reviewed by Damon, 2007). The most notorious poxvirus was variola virus, which caused the human disease smallpox. Studies of this virus initially led Jenner to report that an infectious agent responsible for skin and mucosal lesions in cows (and in humans involved in their husbandry) could be used to 'vaccinate' against smallpox. Remarkably, variola was completely eradicated in 1977 at the end of a worldwide vaccination campaign in which many millions of people received attenuated poxviruses. The agent (vaccinia virus) that was used in the final stages of the vaccination programme has now found a new lease of life as a potential vector for viral therapy of cancer (and for vaccination against other infectious diseases, including malaria and human immunodeficiency virus).

In this chapter, we shall review the features of poxviruses that make them attractive as cancer therapeutics. Much of the discussion will focus on modified forms of vaccinia virus and on the avipoxviruses (particularly fowlpox and canarypox). The oncolytic poxviruses are reviewed in detail in a separate chapter and, so, this chapter will be largely devoted to the use of poxviruses as vehicles for gene delivery. Much of this discussion will deal with their potential role in immunomodulatory gene therapy.

6.2 General features of poxvirus structure and biology

Poxviruses are large, barrel-shaped agents that measure up to 360 nm in length and, as such, they are just discernable by light microscopy. By electron microscopy, they have a complex internal structure that consists of a core, which has a dumbbell shape, and two lateral bodies that are located between the concavities of the core and the outer membrane (Heuser, 2005) (Figure 6.1). They have a linear, covalently-closed, double-stranded DNA genome of 130–300 kilobase pairs (kbp) which is smallest

Viral Therapy of Cancer Edited by Kevin J. Harrington, Richard G. Vile and Hardev S. Pandha

Table 6.1 Classification of poxviruses

Subfamily	Genus	Species
Chordopoxvirinae	Orthopoxvirus	Variola
		Vaccinia
		Cowpox
		Monkeypox
	Parapoxvirus	Orf
		Auzduk disease
	Avipoxvirus	Canarypox
		Fowlpox
	Capripoxvirus	Goatpox
		Sheeppox
	Leporipoxvirus	Myxoma
		Rabbit fibroma
	Suipoxvirus	Swinepox
	Molluscipoxvirus	Molluscum contagiosum
	Yatapoxvirus	Tanapox
		Yaba monkey tumour
Entomovirinae	Alphaentomopoxvirus	Melontha melontha
	Betaentomopoxvirus	Amsacta moorei
	Gammaentomopoxvirus	Chrionimus luridus

in the parapoxviruses and largest in the avipoxviruses. The genome contains identical, oppositely oriented inverted terminal repeats (ITRs) that include A/T-rich incompletely base-paired hairpin loops (Garon *et al.*, 1978). Approximately 100 genes are conserved in all chordopoxviruses (Upton *et al.*, 2003) – the highly conserved genes that are involved in replication tend to be clustered in the central region of the genome while the more variable genes are localised in the end regions.

Poxviruses can exist in two different forms depending on the nature of the membranes that enclose them: mature virions (MV) are the basic infectious form of poxviruses; and extracellular enveloped virions (EV) consist of MV enclosed within an additional membrane layer. The virions contain a large number of polypeptides that can be subdivided into those associated with either the viral surface or the core. The latter group of almost 50 polypeptides includes enzymes that play a key

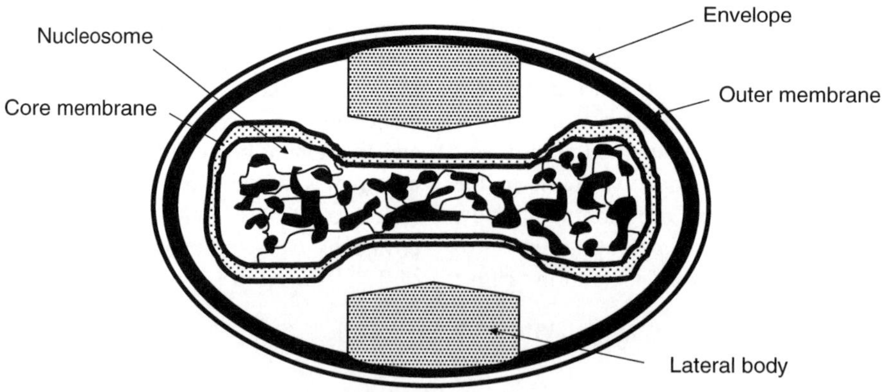

Figure 6.1 Schematic representation of the structure of poxviruses.

role in early mRNA biosynthesis within minutes of cellular entry.

The mechanism of cellular entry of poxviruses remains to be fully elucidated, but it appears that there may be pathways that involve both membrane fusion and endosomal entry. To further complicate matters, the MV and EV forms of poxviruses may access cells by using different mechanisms. MV contain proteins that are capable of binding to cellular glycosaminoglycans, such as chondroitin sulfate and heparan sulfate, but it is by no means clear that attachment to these proteins plays an important role in virus entry and they should not be regarded as viral receptors. For EV, it would seem that membrane fusion is most important and this requires initial unveiling of the additional outer membrane (the EV wrapper) (Senkevich *et al.*, 2004). In vaccinia, EV seems to play a dominant role in cell-to-cell transmission through a process that involves their adherence to cell surface projections formed by polymerization of cytoskeletal actin filaments. The relative importance of these various mechanisms of *in vitro* cellular entry to the situation *in vivo* remains to be determined.

The expression of genes from poxviruses represents a carefully controlled and highly orchestrated series of events. The mRNAs that encode the various viral proteins can be divided into early, intermediate and late groups (reviewed by Moss, 2007). Early mRNAs (capped, methylated and polyadenylated) are produced directly by proteins that are contained within infectious poxvirus particles. They begin to appear in the cell within minutes of cellular entry after the viral cores have been transported on microtubules to cytoplasmic sites of transcription (Baldick and Moss, 1993). These mRNA species are produced before the onset of DNA replication. The intermediate mRNAs are produced after the onset of DNA replication and encode late-stage transcription factors, RNA helicase NPH II and a DNA binding protein. The late mRNAs are produced after the intermediate mRNAs have peaked and are involved in the production of viral structural and functional proteins required for formation of virions (reviewed by Moss, 2007).

DNA replication takes place exclusively in the cytoplasm (Pennington and Follett, 1974) in discrete foci called factory areas. DNA replication commences approximately 1–2 h after infection with vaccinia and ultimately results in the formation of 10 000 copies of the genome per infected cell (Salzman, 1960). Of these, about 50 % will be packaged into virions.

Poxviruses exert profound effects on the cells that they infect, including shutdown of host protein synthesis, avoidance of immune recognition and attack, alteration of the cellular cytoskeleton and inhibition of apoptosis. Each of these effects is mediated by viral gene products with the aim of allowing the virus to achieve a maximally productive infection without interference from the host. The switch from host to viral protein synthesis seems to be mediated largely by inhibition of host mRNA synthesis and transport coupled with rapid degradation of host mRNA species. In regard to avoidance of host immune responses, poxviruses are able to deploy an enormous range of defences. These include an inhibitor of the classic and alternative pathways of complement activation, secreted proteins that bind type I and type II interferons, soluble 'decoy' receptors for interleukin-18 (IL-18), IL-1β and chemokine inhibitors (reviewed by Moss, 2007). Poxviruses exploit the cellular cytoskeleton in a number of different ways: on cellular entry they initially use microtubules to transport them to sites of active transcription of viral mRNAs; the viral A36 protein is involved in the use of microtubules to transport newly packaged viral particles from assembly sites to the plasma membrane for release of EV; efficient cell-to-cell spread is ensured by localization of EV on the tips of motile microvilli in a process that is dependent on the viral A33R, A34R and A36R genes. Poxviruses are able to prevent apoptosis in a number of ways, many of which are coordinated by the viral CrmA gene product. CrmA can function as an IL-1β converting enzyme (ICE, caspase 1) inhibitor, a caspase 8 inhibitor and can interfere with signalling through the Fas and tumour necrosis factor death receptors.

6.3 Clinically applicable poxviruses

6.3.1 Vaccinia virus

Vaccinia virus (VV) is the prototype member of the genus Orthopoxvirus. Its origin is unknown

Table 6.2 Smallpox vaccination: adverse event rates in 1968 (number per 10^6 vaccines) (Adapted from Lane JM *et al.* (1969) *N. Engl. J. Med.* **281** 1201–1208, courtesy of Massachusetts Medical Society)

	National Survey[a]		Ten-State Survey[b]	
	All primary vaccines	Vaccines ≥1 year old	All primary vaccines	Vaccines ≥1 year old
	Serious, but not life-threatening reactions:			
Inadvertent inoculation	25.4	27.1	529.2	532.0
Generalized vaccinia	23.4	17.7	241.5	222.8
Erythema multiforme	N/A	N/A	164.6	131.3
Total number of serious, but not life-threatening reactions:	48.8		935.3*	
	Life-threatening reactions:			
Postvaccinal encephalitis	2.9	2.4	12.3	8.6
Progressive vaccinia (vaccinia necrosum)	0.9	1.0	1.5	1.7
Eczema vaccinatum	10.4	10.6	38.5	41.5
Total number of life-threatening reactions:	14.2*		52.3*	
Deaths	1.1*	0.6	1.5	None reported

*Adverse event statistics cited are marked with an asterisk.

and no natural host for the virus has been identified. Vaccinia virus has a large and complex particle containing a single linear double-stranded DNA genome of 186 kbp with inverted terminal repeats and terminal hairpin loops. The DNA genome of a number of strains of vaccinia virus has been sequenced and encodes approximately 150–200 proteins. As with the other poxviruses, many of the essential genes are found within the more highly conserved central portion of the genome, while genes that are non-essential for replication and morphogenesis (in cell culture) are located closer to the ends. Vaccinia virus replicates entirely in the cytoplasm of infected cells and the virus particle contains the enzymes (RNA polymerase, polyA polymerase, capping enzyme and methylating enzymes) required for the synthesis of capped (methylated) polyadenylated mRNAs. Vaccinia virus was used very extensively in the latter stages of the vaccination campaign against smallpox and information from this experience provides an almost unparalleled database of safety information (albeit from the rather limited standpoint of cutaneous inoculation) (Table 6.2). Overall, life-threatening reactions to vaccinia virus inoculation were reported in 14.2 to 52.3 cases per million and death was attributable to vaccinia virus in between 1.1 and 1.5 cases per million (Lane *et al.*, 1969, 1970).

6.3.2 Modified vaccinia Ankara

Although millions of humans were vaccinated against smallpox without adverse effects using a conventional live vaccinia virus, there was a small, but quantifiable, risk of disseminated infection in the young, in immune compromised patients and in those with eczema or other diffuse skin conditions. During the eradication campaign, several attenuated strains of vaccinia virus were developed. One of these agents, modified vaccinia Ankara (MVA) has been shown to be extremely attenuated compared to the wild type virus. MVA was originally derived from the vaccinia strain Ankara by more than 500 serial passages in primary chicken embryo fibroblasts (Moss, 1996). It has six major genomic deletions compared to the parental vaccinia genome and these changes severely impair its ability to replicate in mammalian cells. Viral replication is blocked at a late stage during cellular infection but viral and recombinant protein synthesis is unimpaired even during this abortive infection (Sutter and Moss, 1992). In animal models, recombinant MVA vectors have proven to be non-virulent and their use in vaccination protocols has been seen to result in protective immunity against infectious diseases and cancer (Hirsch *et al.*, 1996; Wyatt *et al.*, 1996; Carroll *et al.*, 1997; Schneider *et al.*, 1998; Hanke *et al.*, 1999; McShane *et al.*, 2001). They have also been shown to be safe in the setting of immunosuppression. Macaques that had been immune suppressed by total body irradiation, anti-thymocyte globulin treatment or measles virus infection did not show clinical, pathological or haematological abnormalities following injection of up to 10^9 plaque-forming units (p.f.u.) of MVA. In addition, after injection no replication-competent MVA was isolated from the animals (Stittelaar *et al.*, 2001). During the final stages of the smallpox eradication programme, MVA was administered to over 120 000 people and no significant side-effects were recorded, even though this agent was specifically targeted to those at high-risk of adverse reactions (Moss, 1996).

MVA is now being seen as an increasingly attractive option for therapeutic vaccination strategies because of its potential to deliver antigens to the immune system without causing serious local or systemic complications.

6.3.3 Avipoxviruses

There are a large number of avipoxviruses which are named after the bird species in which they have been shown to cause disease. They are large viruses that measure approximately 330 nm in length and have a genome of about 300 kbp. From the point of view of gene therapy approaches, the two most important members of the genus are fowlpox and canarypox, which have both been developed for clinical trial usage. An important feature of these agents is their inability to set up productive infections in mammalian cells, although they are capable of serving the function of presenting antigen to the immune system.

6.3.4 Yatapoxviruses

The recognition of the fact that pre-existing vaccination-induced immunity to vaccinia viruses is likely to curtail the clinical usefulness of these vectors has led to attempts to derive vectors which will not previously have been seen by the human immune system. Unfortunately, the host range of poxviruses that do not cross-react with orthopoxviruses is rather limited and, although members of the avipoxvirus genus and entomopoxvirus subfamily will infect and express genes in human cells, they will not replicate in human cells (Li *et al.*, 1997; Hu *et al.*, 2001). One possible solution to this problem is to use members of the Yatapoxvirus genus (comprising Yaba monkey tumour virus, Tanapox and Yaba-like disease (YLD) virus), which have been shown to be responsible for zoonotic infections in caretakers handling infected monkeys. In such cases, replicating virus has been recovered from cutaneous lesions (Grace and Mirand, 1965). YLD virus was first recognized in 1965–1966 in monkey handlers in Oregon, California and Texas and was seen to cause a brief fever and self-limiting firm, necrotic maculopapular nodules. Hu *et al.* (2001) explored the potential of YLD in an ovarian cancer model. YLD virus did not cross-react with vaccinia virus antibodies, replicated efficiently in human cells and could be purified to high titre. Furthermore, the agent was genetically manipulable and was able to express genes regulated by a synthetic promoter designed for use in orthopoxviruses. Further development of this group of agents may represent a promising avenue for the generation of clinically useful poxvirus vectors.

6.4 Poxviruses as potential cancer therapeutics

Poxviruses have many attractive features as potential agents for cancer therapy. In general terms, they can be used either as oncolytic agents or as vehicles for gene delivery. Clearly, these two potential roles are not necessarily mutually exclusive. In particular, it is conceivable that oncolytic poxviruses may be able to deliver an immunostimulatory gene(s) at the same time as causing lytic cell death and this may promote the generation of a protective immune response against tumour tissue that may be augmented by the presence of viral constituents acting as immune adjuvants. The use of poxviruses as oncolytic agents is reviewed in detail elsewhere in this volume and, so, will not be discussed at length here. Nonetheless, because of the potential overlap between the use of poxviruses for oncolysis and gene delivery, some brief comments will be made.

6.4.1 Poxviruses as oncolytic agents

For the purposes of this discussion only vaccinia virus will be considered – because the other clinically applicable poxviruses can not replicate in mammalian cells. Vaccinia virus has many characteristics that are desirable in an oncolytic virus:

- it has a short, well-characterized life cycle and spreads very effectively and rapidly from cell to cell (at least *in vitro*);
- it is highly cytolytic for a broad range of tumour cell types;
- it is genetically stable;
- it can be genetically modified and has a large packaging capacity (>25 kbp) for exogenous genes (either to augment the immune response or to enhance cytotoxicity);
- it is an excellent expression vector and reliably yields high levels of protein;
- large-scale production of high titres of infectious virus is possible;
- it lacks a natural host and does not cause any known disease in humans;
- throughout its entire life cycle it remains confined to the cytoplasm and, therefore, does not pose a risk of insertional mutagenesis through integration within the host genome (in direct contrast to retroviruses and lentiviruses);
- it has been used very extensively as a smallpox vaccine in millions of people with a well-documented side effect profile (summarized in Table 6.2);
- in the rare event of the development of a local or systemic vaccinial infection, there are a number of drugs (vaccinia immunoglobulin (VIG), cidofovir) available that are highly effective treatments;
- vaccinia virus has previously been administered intravenously without serious adverse effects.

In the field of oncolytic viral vectors, vaccinia virus has some potential advantages over other vectors – the viral particle contains all the enzymes required to initiate viral mRNA synthesis and its replication in the cytoplasm ensures that there is no risk of integration into host genomic DNA. By careful choice of promoters, expression can be engineered to be early, late or both in the viral life cycle, and the levels of expression can similarly be modified from low to high. Vaccinia virus can be engineered such that it is conditionally replication-competent in cancer cells. The most common modification of the vaccinia virus genome is insertional inactivation of the thymidine kinase (TK) gene, which is non-essential for replication in cell culture but essential in non-dividing cells *in vivo*. TK deletion effectively limits the virus's ability to replicate to cells (such as tumour cells) that contain large intracellular nucleotide pools (Puhlmann *et al.*, 2000). Almost all vaccinia virus vectors used in cancer gene therapy trials have a deleted TK gene. A favourable by-product of this process is the fact that the genetic sequence that is inserted into the vaccinia virus genome can encode a therapeutic gene product. Because of the large genomic size of vaccinia virus, it can accommodate very large DNA fragments and offers the prospect of including complex, multiple inserts and their regulatory sequences.

The main disadvantage of vaccinia virus as a gene therapy vector is its immunogenicity and potential complications of its use in immunocompromised individuals or those with skin conditions such as eczema. However, genetic engineering has reduced some of the safety concerns. Vaccinia has been further modified in an attempt to reduce its virulence and increase its tumour specificity. For example, deletion of the vaccinia growth factor (VGF) gene reduces virulence (Buller *et al.*, 1988a). VGF secreted from infected cells predisposes adjacent cells to viral infection (Buller *et al.*, 1988b). In addition, VGF has homology to the cellular growth factors epidermal growth factor (EGF) and transforming growth factor-alpha (TGF-α). Secreted VGF binds to epidermal growth factor receptors and stimulates the Ras pathway. Therefore, VGF deletion effectively restricts viral replication to cancer cells with a constitutively active Ras pathway (de Magalhaes *et al.*, 2001; Andrade *et al.*, 2004). Engineered vaccinia virus with double deletions of TK and VGF can replicate as well as wild type or single deletants in dividing cells, but shows significantly restricted growth in resting cultures. Growth was restricted to tumour cells *in vivo* with evidence of tumour regression in colon cancer xenografts (McCart *et al.*, 2001). Furthermore, vaccinia virus with deletions in two host range/anti-apoptotic genes (SPI-1 and SPI-2) (Guo *et al.*, 2005) have been shown to replicate preferentially in transformed or p53-null cells and to exhibit good tumour specificity *in vivo*. For a review on development of replication-selective oncolytic poxviruses and their use in cancer therapy see Zeh and Bartlett (2002) and Thorne *et al.* (2005).

6.4.2 Poxviruses as anticancer immunogens

Although many cancers display tumour-associated antigens (TAA) that can be recognized by the humoral and cellular limbs of the immune system, the immune system rarely mounts a clinically meaningful antitumour response. Indeed, there are extensive data showing that tumours actively evade immune surveillance both by reducing their immunogenicity and by antagonizing the ability of the immune system to mount an effective response (Melcher *et al.*, 1999). If it were possible to enlist the services of the immune system, there would be a number of potential therapeutic benefits: (1) the inherent specificity of the immune response would limit normal tissue damage; (2) an immune response generated at one site should prime the immune system to react to disease deposits at other local and distant sites, yielding a potent bystander effect; (3) the signal amplification inherent in the immune system would allow a small immunogenic stimulus to trigger a large response; (4) once established, anti-tumour immunity would persist through the generation of memory cells, which would prevent disease recurrence.

There are a number of ways in which poxviruses have the potential to stimulate or reactivate an

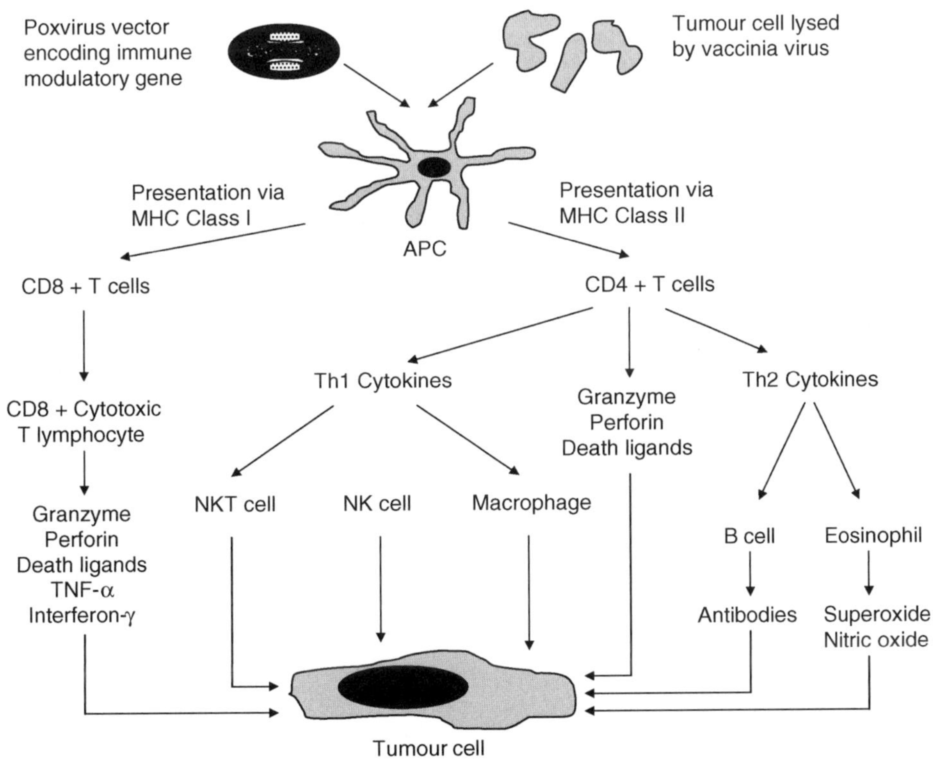

Figure 6.2 Potential immunostimulatory effects o poxviruses through gene expression or tumour cell oncolysis

anti-tumour response. The essential goals of the process are to generate specific $CD8^+$ cytotoxic T lymphocytes (CTL) and/or tumour-specific antibodies that recognize TAA and kill cells which express them. In addition, the adaptive immune system can interact with components of the innate immune system (such as macrophages, NK cells and NKT cells) and prime them to kill cancer cells through a variety of mechanisms (Figure 6.2). Therefore, immunomodulatory strategies can be designed in which poxvirus vectors deliver therapeutic genes directly to components of the immune system (e.g. antigen-presenting cells (APC)) or in which the viral vectors kill cancer cells through oncolysis in an immunostimulatory environment that promotes uptake of TAA by APC.

The means by which delivery of poxviruses could be used to generate or enhance the immune response against tumours can be summarized thus: (1) intratumoral delivery of viruses that express cytokine genes that increase trafficking of APC and T cells in to the tumour and enhance their ability to mount an immune response (Tepper and Mule, 1994); (2) intratumoral delivery of viruses encoding co-stimulatory molecules, such as B7.1 and B7.2, with the goal of enhancing the ability of CTL to recognize, engage and kill tumour cells (Dohring *et al.*, 1994); (3) tumour cell killing by oncolytic viruses with subsequent release of TAA, which can be processed by APC and presented to immune effector cells in an environment which is perceived as 'dangerous'; (4) delivery of viruses that encode TAA to aid the generation of tumour-specific CTL and antibody responses (this does not require intratumoral delivery) (Vile *et al.*, 1994; Hall *et al.*, 1998). Poxviruses are ideal agents to serve all of the above functions. They can readily be modified to carry therapeutic genes and to express them in large amounts in tumour and immune cells. In addition, they are potently oncolytic in their own right and excite strong immune responses.

6.5 Clinical experience with poxviruses

A large number of clinical studies have been conducted with poxvirus vectors – and many more are in the developmental stage. The vaccinia virus vectors that have been (and will be) used in clinical studies have the option of exploiting the twin potentials of oncolytic activity and gene delivery, although most studies have restricted their use to gene delivery to normal cells of the immune system. In contrast, by virtue of their inability to replicate in mammalian cells, avipoxvirus vectors have been used exclusively as a means of delivering immunogenic genes.

Despite these fundamental differences in replication-competence in human cells, there exists a strong rationale for the use of both vaccinia virus and fowlpox virus vectors during a course of vaccination. The use of sequential therapy, first with a vaccinia virus and then in later courses with a fowlpox virus has been shown to be a safe and effective means of generating a robust immune response to a protein that is encoded by the recombinant viruses (Li *et al.*, 1993; McShane *et al.*, 2001; Hodge *et al.*, 2003). In many examples described below, this so-called prime-boost strategy has been exploited in clinical trials.

6.5.1 Oncolytic poxviruses with immunomodulatory activity

VV-GM-CSF (JX-594)

A phase I dose escalation trial of vaccinia virus expressing human granulocyte–macrophage colony-stimulating factor (GM-CSF) was conducted in patients with refractory recurrent melanoma (Mastrangelo *et al.*, 1999). The recombinant vaccinia virus (rVV-GM-CSF) used in this study was constructed by inserting the genes for GM-CSF and for β-galactosidase (lacZ) into the TK locus. Therefore, this agent, called JX-594, has been engineered to be selectively replication-competent in tumour cells. All seven of the patients who were initially enrolled in the trial had previously received smallpox vaccination but, in view of safety concerns and to confirm immune competence, all were re-vaccinated (scarification at 2.5×10^5 p.f.u.). Thereafter, the patients received intralesional injections of rVV-GM-CSF into dermal or subcutaneous tumours twice a week for 6 weeks with escalating doses between 1×10^4 to 2×10^7 p.f.u. per lesion. Patients with stable or responding disease were maintained on treatment until complete tumour resolution or disease progression. In the initial study of seven patients there were: two non-responders; three mixed responses (regression of treated and untreated dermal metastases but progression of disease elsewhere); one partial response and one complete response. The treatment was tolerated extremely well and toxicity was infrequent, dose-dependant and mild (flu-like symptoms that resolved within 24 h). Changes in white blood cell counts were not demonstrated. Injection site inflammation was reported at vector doses equal to or greater than 10^7 p.f.u. per lesion. Reverse-transcription polymerase chain reaction (RT-PCR) confirmed intratumoral expression of GM-CSF, with mRNA detectable from 18 h after virus administration. The mRNA levels appeared to be dose dependant. However, there was no further increase in GM-CSF mRNA levels with repeat administrations of the vector and GM-CSF was not detected in the patients' serum. Patients developed an anti-vaccinia virus humoral immune response 14–21 days following re-vaccination and antibody titres reached a plateau at 4–6 weeks. Unsurprisingly, antibodies were also generated to β-galactosidase and they were detectable at 3–6 weeks post-vaccination. There was, however, no correlation between antibody titres and disease progression. Biopsies were taken from selected lesions and analyses confirmed that the injected lesions were infiltrated with $CD4^+$ and $CD8^+$ T cells which were positive for the T-cell activation markers CD3 and CD45RO. Tissue macrophages, eosinophils and activated B cells were also present in injected lesions. Importantly, regression of uninjected lesions was associated with T-cell infiltration, which supports the hypothesis that treatment was able to induce a systemic anti-tumour immune response. JX-594 is currently in phase I clinical trial via the intravenous route in patients with advanced cancers.

VV-IL-2

A conditionally replication-competent (TK-deleted) vaccinia virus expressing human IL-2 was assessed in clinical trial by Mukherjee *et al.* (2000). Six patients with treatment-refractory malignant pleural mesothelioma received repeated intratumoral injections of VV-IL-2 over a 12-week period. The initial viral dose was 10^5 p.f.u. but in later patients in the trial this was escalated to a ceiling dose of 10^7 p.f.u.. The treatment was tolerated extremely well and no significant toxicities were documented. The study agent was not excreted by any of the patients and was only detected in tumour tissue from one of the patients. VV-IL-2 mRNA expression was detectable for up to 3 weeks after intratumoral injection, but was maximal at 1–3 days. In regard to immune responses, a T-cell infiltrate was documented in 50 % of tumour biopsies and all patients generated an anti-vaccinia antibody response. There were no tumour responses.

VV-B7.1

A number of trials have investigated using vaccinia virus (or other poxvirus vectors) to express immune co-stimulatory molecules. For example, in melanoma local T-cell tolerance may result in down-regulation of co-stimulatory molecules such as B7.1. By using VV to express B7.1 it is hoped to overcome T-cell tolerance resulting in T-cell proliferation, activation and immune clearance of tumours. In a phase I trial of 12 patients, the first six received monthly intralesional injections of rVV-B7.1 at 4.26×10^7 p.f.u. and the next six 4.26×10^8 p.f.u. (Kaufman *et al.*, 2005). The treatment was well tolerated with symptoms limited to low-grade fever, myalgia and fatigue. Interestingly, two patients developed vitiligo (indicative of an anti-melanocyte immune response). Two patients had an objective partial response and one patient had a complete response. Six other patients had stable disease (including some cases of remote metastatic tumours). All patients had previously been vaccinated against smallpox and had low, but measurable, pre-existing anti-vaccinia antibody titres that were boosted following treatment. As expected, anti-vaccinia T-cell responses were increased following treatment but it was also possible to detect T cells that were responsive to melanoma-specific antigens (MART-1, gp100). Real-time RT-PCR analysis of gene expression demonstrated a correlation between increased IFN-γ and CD8 and tumour regression.

A VV vector expressing the T-cell co-stimulatory molecules B7.1, intracellular adhesion molecule-1 (ICAM-1), and leukocyte function associated molecule-3 (LFA-3) (together referred to as TRICOM) has also been used for intralesional injection in patients with melanoma (Kaufman *et al.*, 2006) in a phase I dose escalation trial (5.1×10^6, 5.1×10^7, 5.1×10^8 p.f.u.). As in previous studies, the treatment was well tolerated and a clinical response was seen in 31 % of patients with one complete response (for at least 22 months). Once again, evidence of melanocyte-specific immunity was seen with one patient developing grade 1 vitiligo.

6.5.2 Poxviruses as gene delivery vectors for immunomodulatory therapy

Vaccinia virus-based studies

VV-MUC1/IL-2 (TG1031, TG4010) Recombinant VV expressing polymorphic epithelial mucin (MUC-1) and IL-2 has been investigated as a potential treatment of breast (Scholl *et al.*, 2000) and prostate cancer (Pantuck *et al.*, 2004). MUC-1 is an excellent potential target for immunotherapy of cancer since aberrantly glycosylated forms of this protein are expressed by a range of tumours. Studies were conducted in patients with prostate and breast cancers with a vaccinia vector, VV-MUC1/IL-2 (TG1031) (Scholl *et al.*, 2000). The agent was tolerated extremely well and resulted in the generation of a T-cell proliferative response against MUC1 in 1 of 9 patients with advanced metastatic breast cancer. In a subsequent phase II study of TG1031 in 31 patients with breast cancer, two patients achieved a partial response, including one patient with hepatic metastasis (Scholl *et al.*, 2003). Thereafter, the same group developed a safer and potentially more effective agent, TG4010, based on the MVA strain which they tested in a phase I study (Rochlitz *et al.*, 2003). Like its precursor, TG4010 contained sequences for both MUC1 and IL2. A total of 13 patients were treated at various dose levels – six at 5×10^6 p.f.u., three at 5×10^7 p.f.u. and four at 1×10^8 p.f.u.. The virus was administered by repeated intramuscular injection once every 3 weeks. The agent was very well tolerated and the main adverse events were injection site pain and coryzal symptoms. Four of the 13 patients showed evidence of disease stabilization for between 6 and 9 months, whereas a single patient with lung cancer experienced significant reduction in metastatic lesions that was maintained for more than a year.

VV-CEA The use of VV expressing carcinoembryonic antigen (CEA) as a single vaccine (McAneny *et al.*, 1996; Conry *et al.*, 1999) or in prime boost studies in combination with another non-replicating poxvirus vector such as fowlpox (Marshall *et al.*, 1999, 2000) has been investigated in a number of studies. CEA is expressed in a wide range of tumour types including colorectal, gastric, pancreatic, breast and non-small cell lung cancers. From these studies it was determined that VV and other poxvirus vectors were well tolerated, resulted in the generation of anti-vaccinia antibodies that appeared to reduce the effectiveness of the vaccine on subsequent inoculation and, in some cases, resulted in increased specific CEA T-cell responses.

A recent phase I study in CEA-expressing tumours used VV to co-express CEA with immunostimulatory molecules in combination with a prime-boost protocol with an identical fowlpox vector, with or without recombinant GM-CSF (Marshall *et al.*, 2005). The aims were to determine if either a simple or prime-boost vaccination schedule was tolerated by the patients; if the use of T-cell co-stimulatory molecules (B7.1, ICAM-1, LFA-3 (TRICOM)) enhanced the T-cell response; if immunogenicity to a tumour antigen (CEA) was enhanced by using a CEA with a modification in the human leukocyte antigen (HLA)-A2 CEA CAP-1 epitope (CAP1-6D); and if recombinant GM-CSF enhanced the recruitment of dendritic cells to the vaccination site. Patients received vaccinations (intradermally or subcutaneously) with either fowlpox-CEA(6D)-TRICOM, vaccinia-CEA(6D)-TRICOM followed by fowlpox-CEA(6D)-TRICOM or vaccinia-CEA(6D)-TRICOM followed by fowlpox-CEA(6D)-TRICOM with GM-CSF (100 μg protein). The dose of fowlpox-CEA(6D)-TRICOM was escalated from 4×10^6 to 4×10^8 p.f.u. and vaccinia-CEA(6D)-TRICOM was escalated from 1.2×10^6 p.f.u. to 1.2×10^8 p.f.u.. Vaccinations were administered every 28 days (for four doses) and then once every 3 months. GM-CSF was given at the time of vaccination and for the next 3 days. Toxicity was restricted to skin reactions at the vaccine site, regional lymphadenopathy, fatigue and mild coryzal symptoms. All HLA-A2 positive patients developed a T-cell response to CEA (as demonstrated by enzyme-linked immunospot assay (ELISPOT) for interferon-γ). Antibody responses to CEA and GM-CSF were variable and no antibodies to the TRICOM proteins were detected. All patients showed increased antibody titres to VV and most showed increased titres to fowlpox. Forty per cent of patients had stable disease for at least 4 months with 24 % having stable disease for at least 6 months and one patient had a complete response. Nineteen per cent of patients had tumour marker responses as shown by decreasing or stable serum CEA levels.

Morse *et al.* (2005) conducted a phase I study to determine the safety, immunological and clinical efficacy of a dendritic cell vaccine modified to overexpress costimulatory molecules and TAA. Dendritic cells that had been modified *ex vivo* by a recombinant fowlpox vector encoding CEA and the TRICOM triad of costimulatory molecules were administered as one or two cycles of four tri-weekly subcutaneous/intradermal injections. Fourteen patients (11 colorectal cancer, 3 non-small cell lung cancer) were entered and 12 completed at least one cycle of immunization with no evidence of grade 3 or 4 toxicity directly attributable to the immunizations. One patient had a fall in CEA level from 46 to 6.8 that was accompanied by a minor regression of lymphadenopathy. Five other patients had stable disease for at least 3 months. There was an increase in the frequency of CEA-specific T cells as measured by ELISPOT in 10 patients. There was an association between the peak frequency of CEA-specific T cells (both $CD4^+$ and $CD8^+$) and either minor response or stable disease after one cycle of therapy. A second cycle was not associated with higher T-cell frequencies.

VV-PSA A number of clinical trials have investigated VV expressing prostate specific antigen (PSA) in patients following prostatectomy and/or radiotherapy (Sanda *et al.*, 1999; Eder *et al.*, 2000; Gulley *et al.*, 2002; Kaufman *et al.*, 2004). In the study by Sanda *et al.* (1999), patients with androgen-sensitive recurrence of prostate cancer after radical prostatectomy were treated with recombinant VV (rVV-PSA) at 2.65×10^7 (3 patients) or 2.65×10^8 PFU (3 patients) delivered intradermally. Vaccination commenced on day 7 after luteinizing hormone-releasing hormone (LHRH) analogue treatment (an anti-androgen) which was then suspended. The aim of the study was to monitor serum PSA levels (related to serum testosterone restoration) and anti-PSA antibody generation. The treatment was associated with very little toxicity which was limited to vaccine site erythema, fever and chills that occurred 24–48 h following administration. One patient's serum PSA remained undetectable for over 8 months following restoration of serum testosterone levels. The other patients' PSA levels began to rise within 0–2 months of restoration of serum testosterone. Serum anti-PSA antibodies were detected in only one patient. In a further study, there was little treatment-associated toxicity when the virus was administered either subcutaneously (higher doses) or by dermal scarification (lower doses) monthly for up to 3 months (Gulley *et al.*, 2002). Increased numbers of PSA-reactive T cells were observed but no antibodies were generated to PSA and no objective tumour responses were documented. In a study by Eder *et al.* (2000), 3 monthly doses were given to men with rising PSA levels following radical prostatectomy, radiation therapy or both. The dose escalation scheme was: 2.65×10^6 (6 patients), 2.65×10^7 (6 patients) and 2.65×10^8 p.f.u. (11 patients) and 10 patients also received 250 $\mu g/m^2$ GM-CSF. As with previous studies, the only toxicity that was seen was a mild cutaneous reaction. In men receiving $\geq 2.65 \times 10^7$ p.f.u., injection-site pustule formation or erythema occurred after the initial dose. GM-CSF-related toxicity (grade 3 fever and tachycardia) was reported in one patient and required a reduction in the dose of GM-CSF. In the remaining nine patients, only mild GM-CSF-related toxicity was seen. In 42 % of patients, the PSA levels were stable for at least 6 months and in nine patients they remained stable for between 11 and 25 months. In addition, 19 of the 33 patients showed a decrease in PSA levels at some stage during treatment and this did not correlate with anti-PSA antibody generation. A specific T-cell response to PSA was also demonstrated in a number of patients.

Arlen *et al.* (2006) have assessed the effect of concomitant treatment with docetaxel (plus dexamethasone) on the immune response to a vaccinia/fowlpox-based vaccination strategy in 28 patients with metastatic androgen-independent prostate. The vaccination regimen comprised a recombinant vaccinia virus expressing PSA (rVV-PSA) admixed with a recombinant vaccinia virus expressing the B7.1 costimulatory gene (rVV-B7.1) and sequential booster vaccinations with recombinant fowlpox virus containing the PSA gene (rF-PSA). GM-CSF was administered with each vaccination. Patients were randomized to receive either vaccine and weekly docetaxel or vaccine alone. Patients on the vaccine alone arm were allowed to cross over to receive single-agent docetaxel in the event of disease progression. The median increase in T-cell precursors reactive to PSA (measured by ELISPOT) was greater than three-fold in each arm after 3 months of therapy. Immune responses were also detected to other prostatic TAA in the period after vaccination. Eleven patients who developed progressive disease on the vaccine alone arm crossed over to receive docetaxel at the time of progression and achieved a median progression-free survival of 6.1 months on docetaxel (compared with 3.7 months for historical controls). These data suggest that vaccination may modulate the subsequent effect of conventional anticancer therapeutics.

DiPaola *et al.* (2006) conducted a phase I study in 10 patients with androgen-independent prostate cancer (with or without metastatic disease) to evaluate the safety and immunogenicity of a novel vaccinia/fowlpox vaccine that incorporated the PSA gene sequence and TRICOM. Patients initially received 2×10^8 p.f.u. of the recombinant vaccinia virus vaccine (PROSTVAC-V) followed by 1×10^9 p.f.u. of the booster recombinant fowlpox virus (PROSTVAC-F), each containing gene sequences for PSA and TRICOM. The most commonly reported adverse events were injection site reactions and fatigue. There were no grade 3 or 4 adverse events. Four patients maintained stable disease (with less than 25 % increase in PSA) throughout the 8-week study period. Anti-vaccinia antibody titres increased in all patients, but there was no evidence of generation of an anti-PSA antibody response.

Gulley *et al.* (2005) reported data from a randomized phase II trial designed to test the ability of a PSA-encoding poxvirus vaccine to induce a specific T-cell response when combined with radiotherapy in patients with localized prostate cancer. Thirty patients were randomized in a 2:1 ratio to vaccine plus radiotherapy or radiotherapy alone. Patients who received vaccination did so using a prime-boost approach in which recombinant vaccinia encoding PSA (rVV-PSA) plus recombinant vaccinia containing B7.1 (rVV-B7.1) were used initially and then followed by monthly booster vaccines with recombinant fowlpox encoding PSA. Vaccination was given in combination with local GM-CSF and low-dose systemic IL-2. Standard external beam radiotherapy was delivered between the fourth and the sixth vaccinations. Seventeen of the 19 patients in the combination arm were able to complete all eight vaccinations, 13 of whom had an increase in PSA-specific T cells of at least three-fold. These data compared with no detectable increase in PSA-specific T cells in those patients treated with radiotherapy alone. Equally importantly, there was evidence of generation of T cells that recognized prostatic TAA that were not encoded by the vaccine and this was taken as circumstantial evidence of immune-mediated tumour killing.

Kaufman *et al.* (2004) conducted a randomized phase II trial to evaluate the feasibility and tolerability of a prime/boost vaccine strategy using vaccinia virus and fowlpox virus expressing PSA in 64 patients with biochemical progression after local therapy for prostate cancer. Patients were randomly assigned to receive four vaccinations with fowlpox-PSA (rFP-PSA), three rFP-PSA vaccines followed by one vaccinia-PSA (rVV-PSA) vaccine, or one rVV-PSA vaccine followed by three rFP-PSA vaccines. The prime/boost schedule was well tolerated with few adverse events. A significant proportion of men remained free of PSA (45.3 %) and clinical (78.1 %) evidence of progression after 19 months follow-up. There was a trend for a better outcome in the treatment group that received a priming dose of rVV-PSA. No significant increases in anti-PSA antibody titres were detected and 46 % of patients demonstrated an increase in PSA-reactive T cells.

VV-Tyrosinase Tyrosinase is a melanocyte-specific enzyme that catalyses a key step in melanin biosynthesis. It has, therefore, been identified as a legitimate target for vaccination in an attempt to treat malignant melanoma (MM). Lindsay *et al.* (2006) reported data from two clinical trials that were conducted to evaluate the clinical efficacy and immunologic impact of vaccination against tyrosinase (in combination with systemic IL-2 administration) in patients with metastatic MM. Full-length tyrosinase was used as an immunogen in an attempt to generate diverse immune responses against a melanoma-specific antigen. A prime/boost vaccination strategy with recombinant vaccinia virus (rVV-Tyrosinase) and fowlpox virus (rFP-Tyrosinase) vectors encoding tyrosinase was initially studied in a randomized three-arm phase II trial, in which the vaccines were given alone or at the same time as low- or high-dose IL-2. In a further phase II trial, all patients received the same vaccines and high-dose IL-2 sequentially rather than concurrently. A total of 64 patients were treated on these trials. There were 8 partial responses (12.5 %), all of which occurred in patients who received the high-dose IL-2. Some of the patients achieved a mixed tumour response or overall regression that fell below the threshold for a partial response. Laboratory studies confirmed that immune responses against tyrosinase had been generated in 3 of 49 (6 %) patients tested serologically, in 3 of 23 (13 %) patients tested for T cell recognition of individual tyrosinase peptides and in 4 of 16 (25 %) patients tested for T-cell recognition of full-length tyrosinase protein with real-time reverse transcription-PCR techniques. Despite this promising evidence of modulation of the immune response to tyrosinase, it was felt that the clinical responses were not significantly different from what would have been expected from treatment with IL-2 alone.

VV-NY-ESO-1 NY-ESO-1 is a cancer/testis antigen expressed by a range of human malignancies. An analysis of the safety and immunogenicity of recombinant vaccinia (rVV-NY-ESO-1) and fowlpox viruses (rFP-NY-ESO-1) expressing NY-ESO-1 was conducted in 36 patients with a variety of tumour types (Jager *et al.*, 2006). Each of the agents was initially tested singly at two different dose levels and then in a prime-boost setting with rVV-NY-ESO-1 followed by rFP-NY-ESO-1. The vaccines were well tolerated either singly or in combination. Specific antibodies to NY-ESO-1 and/or specific $CD8^+$ and $CD4^+$ T-cell responses against a broad range of NY-ESO-1 epitopes were induced by a course of at least four vaccinations at monthly intervals in a high proportion of patients. $CD8^+$ T-cell clones derived from five vaccinated patients were shown to be able lyse NY-ESO-1-expressing melanoma target cells in ^{51}Cr release assays. The authors concluded that in a number of patients with melanoma, the natural course of the disease was favourably influenced by vaccination although there was no particular difference between either of the single vaccination schedules or the combination.

VV-HPV16/18 A number of studies (Baldwin *et al.*, 2003; Corona Gutierrez *et al.*, 2002, 2004; Davidson *et al.*, 2003, 2004) have investigated vaccinia viruses expressing human papillomavirus (HPV)-16 and -18 E6 and E7 proteins as a treatment for vulval and vaginal intraepithelial neoplasia either as a sole vaccination or in a prime boost setting with a non-poxvirus vector boost (Davidson *et al.*, 2004). The treatment was well tolerated and some clinical responses were observed. In addition, the authors reported the generation of HPV 16 E6 and E7 T-cell responses.

MVA-Based Studies

MVA-5T4 (TroVax) 5T4 is a TAA that is expressed on the surface of a wide variety of human adenocarcinomas, including colorectal cancers. In preclinical murine studies, a recombinant modified vaccinia Ankara virus (TroVax) expressing human 5T4 induced protection against challenge with a murine colon cancer cell line expressing human 5T4 (CT26-h5T4) (Harrop *et al.*, 2006a). The antitumour activity was durable and dependent on the presence of $CD4^+$ (but not $CD8^+$) T cells. This same agent was subsequently evaluated in a phase I/II study in 17 evaluable patients with colorectal cancer who received TroVax doses ranging from 5×10^7 up to 5×10^8 p.f.u. at 0, 4, and 8 weeks (Reinis, 2004; Harrop *et al.*, 2006b). The study agent was well tolerated in all patients and no serious adverse events were considered attributable to vaccination. Of the 17 evaluable patients, 16 showed 5T4-specific cellular responses and 14 had detectable antibody levels following vaccination. The vector was able to boost 5T4-specific immune responses even in the presence of MVA neutralizing antibodies. Disease stabilization from 3–18 months was observed in five patients, all of whom were shown to mount 5T4-specific immune responses. Analysis of time to progression and patient survival revealed a positive correlation between the generation of an antibody response against 5T4, but the same was not true of an anti-MVA antibody response.

MVA-tyrosinase A MVA expressing human tyrosinase (MVA-hTyr) has been assessed in a phase I adjuvant study in 20 patients with surgically treated stage II malignant melanoma (Meyer *et al.*, 2005). Patients were vaccinated three times at 4-weekly intervals with 5×10^8 p.f.u. of MVA-hTyr that was delivered by combined intradermal and subcutaneous injection. Safety measures demonstrated that the vaccination strategy was safe and well tolerated. The outcomes measures included ELISPOT assessment of responses to the viral vector, known HLA class I-restricted tyrosinase peptides and to dendritic cells transfected with tyrosinase mRNA. A marked response to the viral vector was seen, as indicated by an increase in the frequency of MVA-specific $CD4^+$ and $CD8^+$ T cells and an increase in anti-MVA antibody titres. Regrettably, no tyrosinase-specific T-cell or antibody responses were seen in response to vaccination with MVA-hTyr.

MVA-polyepitope versus melanoma A phase I clinical trial has been reported in which patients received vaccinations with recombinant plasmid DNA and/or MVA encoding seven melanoma tumour antigen CTL epitopes (the so-called Mel3 string) (Smith *et al.*, 2005). The epitopes encoded by the Mel3 string were: tyrosinase$_{1–9}$; melan-A$_{26–35}$ analogue; tyrosinase$_{369–377}$; MAGE-3$_{168–176}$; MAGE-3$_{271–279}$; MAGE-1$_{161–169}$; NY-ESO-1$_{155–167}$; and a murine H2-D^b restricted influenza virus nucleoprotein epitope. Patients were enrolled if they were HLA-A*0201-positive and had undergone surgical excision of melanoma. They received either a prime-boost DNA/MVA regimen or an MVA-only regimen. CTL responses were monitored by *ex vivo* tetramer analysis for the high-affinity melan-A$_{26–35}$ analogue epitope and were generated in two of six patients who received DNA/MVA and four of seven patients who received MVA only. The magnitude of the CTL responses was low (0.03–0.19 % of $CD8^+$ T cells) but ELISPOT analysis and *in vitro* proliferation assays confirmed that the CTLs that were generated possessed effector functions. Responses were seen irrespective of prior vaccination against smallpox.

Avipoxvirus-based studies

ALVAC-IL-12 A canarypox virus vector encoding the immunostimulatory cytokine IL-12 has been assessed in a phase I protocol in patients with metastatic melanoma deposits that were amenable to local injection (Triozzi *et al.*, 2005a). The study protocol was designed to evaluate the safety and activity of intratumoral administration of avipoxvirus vectors. The virus dose administered was varied depending on the volume of the tumour mass to be injected, such that nodules less than 2 cm in diameter received 1×10^6 $TCID_{50}$ and larger nodules received 2×10^6 $TCID_{50}$. This treatment strategy was tolerated extremely well and no dose-limiting toxicities were seen. Levels of IL-12 and IFN-γ were increased in the injected lesions of four patients and in the serum of three patients in comparison to the saline-injected controls. One patient experienced a complete regression of the injected lesion and adjacent uninjected in-transit disease. As expected, all patients developed an antibody response to the ALVAC vector.

ALVAC-B7.1/IL-12 ALVAC-B7.1 (encoding the immune co-stimulatory molecule B7.1) and ALVAC-IL12 have been used in combination (Triozzi *et al.*, 2005b). Fourteen patients with metastatic melanoma who had subcutaneous nodules received four intratumoral injections of one or both vectors over a ten day period. Nine patients received escalating doses of up to 2.5×10^9 p.f.u. of ALVAC-B7.1 and a further five patients received 2.5×10^9 p.f.u. ALVAC-B7.1 combined with ALVAC-IL-12 at a $TCID_{50}$ of 2×10^6. Treatment was well tolerated and toxicity was reported as mild to moderate, limited to injection site inflammation and fever, chills, myalgia and fatigue. Higher levels of B7.1 mRNA were observed in ALVAC-B7.1-injected tumours than in the saline-injected controls. However, the levels of intratumoral vascular endothelial growth factor (VEGF) and immunosuppressive IL-10 were also greater in the ALVAC-B7.1- and ALVAC-IL12-injected tumours compared with the controls. All patients developed antibody responses to ALVAC. No objective tumour responses were seen. These data demonstrated that intratumoral injections of ALVAC-B7.1 and ALVAC-IL12 were safe and tolerable and resulted in measurable biological effects. However, it is important to appreciate that the resulting response to the intratumoral delivery of ALVAC vectors involved the production of factors that may suppress their anti-tumour immunologic activity.

ALVAC-EpCAM/KSA Colorectal carcinoma cells express the TAA epithelial cellular adhesion molecule (Ep-CAM)/KSA. This surface marker has been exploited as a therapeutic target in passive immunotherapy protocols using anti-EpCAM monoclonal antibodies with promising results (Riethmuller *et al.*, 1998). These observations have resulted in the evaluation of Ep-CAM as a potential target for active specific immunotherapy by expression of the TAA from an avipox viral vector (ALVAC-KSA) (Ullenhag *et al.*, 2003). Twelve patients with colorectal cancer who had undergone radical surgery with no evidence of residual macroscopic disease (stages I, II, and III) were recruited. The first 6 patients were immunized with three 3-weekly injections of ALVAC-KSA ($1 \times 10^{7.1}$ $TCID_{50}$ per immunization) and the next six patients received the same schedule of ALVAC-KSA together with adjuvant GM-CSF (75 μg/day for 4 days). The vaccination schedule was well tolerated and the main side effect was mild local skin reaction. In the ALVAC-KSA group a weak T-cell response was generated in two of the six patients. In the ALVAC-KSA/GM-CSF group a marked IFN-γ response was induced in five of the six patients. Interestingly, the T-cell response seemed to appear relatively late and reached a peak 4–5 months after vaccination. No anti-EpCAM antibodies were detected. Most of the patients had pre-existing T-cell responses (IFN-γ) against the vector and this was boosted after vaccination. All of the patients developed high titres of anti-ALVAC antibodies.

ALVAC-p53 Since p53 is overexpressed in a large number of tumour types, it has been seen as a rational antigen target for immunotherapeutic approaches. To this end, Menon *et al.* (2003) performed a study in which a canarypox virus encoding wild type p53 was injected intravenously in 16 patients with metastatic colorectal cancer. Patients were treated in dose escalation cohorts that received three 3-weekly injections at $1 \times 10^{6.5}$, 1×10^7 and $1 \times 10^{7.5}$ TCID50, respectively. The only significant treatment-related toxicity was fever and this was grade 1 or 2 in 93 % of cases. Anti-ALVAC immune responses, both humoral and cellular, were induced in all patients. Pre-existing anti-p53 antibodies were detectable before vaccination in seven patients and were induced in a further three patients (one in each dose level cohort) after vaccination. Two patients developed anti-p53 cellular immune responses. Only one patient who received a further round of three vaccinations had stable disease and all the other patients developed progressive disease.

Fowlpox-gp100 Rosenberg *et al.* (2003) evaluated immunological responses and therapeutic efficacy of fowlpox vaccines encoding the melanoma TAA, gp100, in patients with metastatic melanoma. In three consecutive trials, patients received immunizations with recombinant fowlpox viruses encoding three different forms of gp100. These consisted of: (i) the native, full-length gp100 molecule; (ii) a modified gp100 molecule with two amino acids altered to increase binding to HLA-A*0201 molecules; and (iii) a minigene construct encoding a single, modified endoplasmic reticulum-targeted epitope (gp100: 209–217(210M)) targeted to the endoplasmic reticulum. Patients were also eligible to receive IL-2 in the event of tumour progression. None of the patients showed immunoreactivity to gp100 prior to fowlpox immunization. Only one of seven patients developed reactivity after receiving fowlpox encoding native gp100. Ten of 14 patients who received the fowlpox encoding the gp100 that was modified to alter binding to the HLA molecule exhibited a response against the native gp100 molecule. Finally, 12 of 16 patients generated immune responses to gp100 after immunization with the modified minigene construct. Despite these impressive immune responses, there was only one partial response in the group of 46 patients receiving virus without IL-2. Importantly, once patients had progressed, none who had the full-length or modified full-length forms of gp100 showed a response to IL-2, but 6 of 12 patients who had received the fowlpox encoding the minigene showed objective responses to IL-2. These data support further investigation of combined IL-2 and recombinant fowlpox virus vectors encoding modified epitopes from TAA.

6.6 Conclusions

The data reviewed in this chapter demonstrate that poxviruses have enormous flexibility as vectors for delivery of anticancer immunotherapeutics. In addition, in the case of vaccinia virus vectors, they offer the opportunity to combine immunomodulation/stimulation with direct viral oncolysis. As has been reviewed, there have been a large number of clinical trials of poxvirus vectors, although most of these have been in the setting of phase I toxicity studies in patients with disseminated, treatment-refractory disease. In addition, most of the studies have recruited patients who have received prior anticancer therapies including cytotoxic chemotherapy, which may have had an impact on their ability to mount a robust immune response to the vaccine. Despite these considerations, a large number of studies have reported tumour-specific T-cell and humoral responses. However, as yet, these immune effects have not translated into consistent antitumoral effects. The next wave of development of poxvirus-based vaccines should focus on bringing these agents to the clinic at an earlier stage in the patient's treatment schedule. In particular, an emphasis on treatment in the adjuvant setting when the disease burden is minimal seems most likely to yield therapeutic effects. In addition, combination (or sequential) scheduling with conventional anticancer therapeutics should be an area for intensive future investigation.

References

Andrade AA, Silva PN, Pereira AC, *et al.* (2004). The vaccinia virus-stimulated mitogen-activated protein kinase (MAPK) pathway is required for virus multiplication. *Biochem J* **381**, 437–446.

Arlen PM, Gulley JL, Parker C, *et al.* (2006). A randomized phase II study of concurrent docetaxel plus vaccine versus vaccine alone in metastatic androgen-independent prostate cancer. *Clin Cancer Res* **12**, 1260–1269.

Baldick CJ Jr, Moss B (1993). Characterisation and temporal regulation of mRNAs encoded by vaccinia virus intermediate stage genes. *J Virol* **67**, 3515–3527.

Baldwin PJ, van der Burg SH, Boswell CM, *et al.* (2003). Vaccinia-expressed human papillomavirus 16 and 18 E6 and E7 as a therapeutic vaccination for vulval and vaginal intraepithelial neoplasia. *Clin Cancer Res* **9**, 5205–5213.

Buller RM, Chakrabarti S, Cooper JA, Twardzik DR, Moss B (1988a). Deletion of the vaccinia virus growth factor gene reduces virus virulence. *J Virol* **62**, 866–874.

Buller RM, Chakrabarti S, Moss B (1988b). Cell proliferative response to vaccinia virus is mediated by VGF. *Virology* **164**, 182–192.

Carroll MW, Overwijk WW, Chamberlain RS, Rosenberg SA, Moss B, Restifo NP (1997). Highly attenuated modified vaccinia virus Ankara (MVA) as an effective recombinant vector: a murine tumor model. *Vaccine* **15**, 387–394.

Conry RM, Khazaeli MB, Saleh MN, *et al.* (1999). Phase I trial of a recombinant vaccinia virus encoding carcinoembryonic antigen in metastatic adenocarcinoma: comparison of intradermal versus subcutaneous administration. *Clin Cancer Res* **5**, 2330–2337.

Corona Gutierrez CM, Tinoco A, Lopez Contreras M, *et al.* (2002). Clinical protocol. A phase II study: efficacy of the gene therapy of the MVA E2 recombinant virus in the treatment of precancerous lesions (NIC I and NIC II) associated with infection of oncogenic human papillomavirus. *Hum Gene Ther* **13**, 1127–1140.

Corona Gutierrez CM, Tinoco A, Navarro T, *et al.* (2004). Therapeutic vaccination with MVA E2 can eliminate precancerous lesions (CIN 1, CIN 2, and CIN 3) associated with infection by oncogenic human papillomavirus. *Hum Gene Ther* **15**, 421–431.

Damon IK (2007). Poxviruses. In: Knipe DM, Howley PM (eds), *Fields Virology*, 5th edition. Lippincott, Williams & Wilkins, Philadelphia, PA, pp 2947–2975.

Davidson EJ, Boswell CM, Sehr P, *et al.* (2003). Immunological and clinical responses in women with vulval intraepithelial neoplasia vaccinated with a vaccinia virus encoding human papillomavirus 16/18 oncoproteins. *Cancer Res* **63**, 6032–6041.

Davidson EJ, Faulkner R, Sehr P, *et al.* (2004). Effect of TA-CIN (HPV 16 L2E6E7) booster immunisation in vulval intraepithelial neoplasia patients previously vaccinated with TA-HPV (vaccinia virus encoding HPV 16/18 E6E7). *Vaccine* **22**, 2722–2729.

de Magalhaes JC, Andrade AA, Silva PN, *et al.* (2001). A mitogenic signal triggered at an early stage of vaccinia virus infection: implication of MEK/ERK and protein kinase A in virus multiplication. *J Biol Chem* **276**, 38353–38360.

DiPaola R, Plante M, Kaufman H, *et al.* (2006). A phase I trial of Pox PSA vaccines (PROSTVAC(R)-VF) with B7-1, ICAM-1, and LFA-3 co-stimulatory molecules (TRICOM^{TM}) in patients with prostate cancer. *J Transl Med* **4**, 1.

Dohring C, Angman L, Spagnoli G, Lanzavecchia A (1994). T-helper- and accessory-cell-independent cytotoxic responses to human tumor cells transfected with a B7 retroviral vector. *Int J Cancer* **57**, 754–759.

Eder JP, Kantoff PW, Roper K, *et al.* (2000). A phase I trial of a recombinant vaccinia virus expressing prostate-specific antigen in advanced prostate cancer. *Clin Cancer Res* **6**, 1632–1638.

Garon CF, Barbosa E, Moss B (1978). Visualization of an inverted terminal repetition in vaccinia virus DNA. *Proc. Natl. Acad. Sci. U S A* **75**, 4863–4867.

Grace JT Jr, Mirand EA (1965). Yaba virus infection in humans. *Exp Med Surg* **23**, 213–216.

Gulley J, Chen AP, Dahut W, *et al.* (2002). Phase I study of a vaccine using recombinant vaccinia virus expressing PSA (rV-PSA) in patients with metastatic androgen-independent prostate cancer. *Prostate* **53**, 109–117.

Gulley JL, Arlen PM, Bastian A, *et al.* (2005). Combining a recombinant cancer vaccine with standard definitive radiotherapy in patients with localized prostate cancer. *Clin Cancer Res* **11**, 3353–3362.

Guo ZS, Naik A, O'Malley ME, *et al.* (2005). The enhanced tumor selectivity of an oncolytic vaccinia lacking the host range and antiapoptosis genes SPI-1 and SPI-2. *Cancer Res* **65**, 9991–9998.

Hall SJ, Sanford MA, Atkinson G, Chen SH (1998). Induction of potent antitumor natural killer cell activity by herpes simplex virus-thymidine kinase and ganciclovir therapy in an orthotopic mouse model of prostate cancer. *Cancer Res* **58**, 3221–5.

Hanke T, Samuel RV, Blanchard TJ, *et al.* (1999). Effective induction of simian immunodeficiency virus-specific cytotoxic T lymphocytes in macaques by using a multiepitope gene and DNA prime-modified vaccinia virus Ankara boost vaccination regimen. *J Virol* **73**, 7524–7532.

Harrop R, Ryan MG, Myers KA, Redchenko I, Kingsman S, Carroll MW (2006a). Active treatment of murine tumors with a highly attenuated vaccinia virus expressing the tumor associated antigen 5T4 (TroVax) is CD4+ T cell dependent and antibody mediated. *Cancer Immunol Immunother* **55**, 1081–1090.

Harrop R, Connolly N, Redchenko I, *et al.* (2006b). Vaccination of colorectal cancer patients with modified vaccinia Ankara delivering the tumor antigen 5T4 (TroVax) induces immune responses which correlate with disease control: a phase I/II trial. *Clin Cancer Res* **12**, 3416–3424.

Heuser J (2005). Deep-etch EM reveals that the early poxvirus envelope is a single membrane bilayer stabilized by a geodetic 'honeycomb' surface coat. *J Cell Biol* **69**, 269–283.

Hirsch VM, Fuerst TR, Sutter G, *et al.* (1996). Patterns of viral replication correlate with outcome in simian immunodeficiency virus (SIV)-infected macaques:

effect of prior immunization with a trivalent SIV vaccine in modified vaccinia virus Ankara. *J Virol* **70**, 3741–52.

Hodge JW, Poole DJ, Aarts WM, *et al.* (2003). Modified vaccinia virus Ankara recombinants are as potent as vaccinia recombinants in diversified prime and boost vaccine regimens to elicit therapeutic antitumor responses. *Cancer Res* **63**, 7942–7949.

Hu Y, Lee J, McCart JA, *et al.* (2001). Yaba-like disease virus: an alternative replicating poxvirus vector for cancer gene therapy. *J Virol* **75**, 10300–10308.

Jager E, Karbach J, Gnjatic S, *et al.* (2006). Recombinant vaccinia/fowlpox NY-ESO-1 vaccines induce both humoral and cellular NY-ESO-1-specific immune responses in cancer patients. *Proc Natl Acad Sci U S A* **103**, 14453–14458.

Kaufman HL, Wang W, Manola J, *et al.* (2004). Phase II randomized study of vaccine treatment of advanced prostate cancer (E7897): a trial of the Eastern Cooperative Oncology Group. *J Clin Oncol* **22**, 2122–2132.

Kaufman HL, Deraffele G, Mitcham J, *et al.* (2005). Targeting the local tumor microenvironment with vaccinia virus expressing B7.1 for the treatment of melanoma. *J Clin Invest* **115**, 1903–1912.

Kaufman HL, Cohen S, Cheung K, *et al.* (2006). Local delivery of vaccinia virus expressing multiple costimulatory molecules for the treatment of established tumors. *Hum Gene Ther* **17**, 239–244.

Lane JM, Ruben FL, Neff JM, Millar JD (1969). Complications of smallpox vaccinations, 1968: national surveillance in the United States. *N Engl J Med* **281**, 1201–1208.

Lane JM, Ruben FL, Neff JM, Millar JD (1970). Complications of smallpox vaccination, 1968: results of ten statewide surveys. *J Infect Dis* **122**, 303–309.

Li S, Rodrigues M, Rodriguez D, *et al.* (1993). Priming with recombinant influenza virus followed by administration of recombinant vaccinia virus induces CD8+ T-cell-mediated protective immunity against malaria. *Proc Natl Acad Sci U S A* **90**, 5214–5218.

Li Y, Hall RL, Moyer RW (1997). Transient, nonlethal expression of genes in vertebrate cells by recombinant entomopoxviruses. *J Virol* **71**, 9557–9562.

Lindsey KR, Gritz L, Sherry R, *et al.* (2006). Evaluation of prime/boost regimens using recombinant poxvirus/tyrosinase vaccines for the treatment of patients with metastatic melanoma. *Clin Cancer Res* **12**, 2526–2537.

Marshall JL, Hawkins MJ, Tsang,KY, *et al.* (1999). Phase I study in cancer patients of a replication-defective avipox recombinant vaccine that expresses human carcinoembryonic antigen. *J Clin Oncol* **17**, 332–337.

Marshall JL, Hoyer RJ, Toomey MA, *et al.* (2000). Phase I study in advanced cancer patients of a diversified prime-and-boost vaccination protocol using recombinant vaccinia virus and recombinant nonreplicating avipox virus to elicit anti-carcinoembryonic antigen immune responses. *J Clin Oncol* **18**, 3964–3973.

Marshall JL, Gulley JL, Arlen PM, *et al.* (2005). Phase I study of sequential vaccinations with fowlpox-CEA (6D)-TRICOM alone and sequentially with vaccinia-CEA(6D)-TRICOM, with and without granulocyte–macrophage colony-stimulating factor, in patients with carcinoembryonic antigen-expressing carcinomas. *J Clin Oncol* **23**, 720–731.

Mastrangelo MJ, Maguire HC Jr, Eisenlohr L, *et al.* (1999). Intratumoral recombinant GM-CSF-encoding virus as gene therapy in patients with cutaneous melanoma. *Cancer Gene Ther* **6**, 409–422.

McAneny D, Ryan CA, Beazley RM, Kaufman HL (1996). Results of a phase I trial of a recombinant vaccinia virus that expresses carcinoembryonic antigen in patients with advanced colorectal cancer. *Ann Surg Oncol* **3**, 495–500.

McCart JA, Ward JM, Lee J, *et al.* (2001). Systemic cancer therapy with a tumor-selective vaccinia virus mutant lacking thymidine kinase and vaccinia growth factor genes. *Cancer Res* **61**, 8751–8757.

McShane H, Brookes R, Gilbert SC, Hill AV (2001). Enhanced immunogenicity of CD4(+) T cell responses and protective efficacy of a DNA-modified vaccinia virus Ankara prime-boost vaccination regimen for murine tuberculosis. *Infect Immun* **69**, 681–686.

Melcher A, Gough M, Todryk S, Vile R (1999). Apoptosis or necrosis for tumor immunotherapy: what's in a name? *J Mol Med* **77**, 824–833.

Menon AG, Kuppen PJK, van den Burg SH, *et al.* (2003). Safety of intravenous administration of a canarypox virus encoding the human wild-type p53 gene in colorectal cancer patients. *Cancer Gene Ther* **10**, 509–517.

Meyer RG, Britten CM, Siepmann U, *et al.* (2005). A phase I vaccination study with tyrosinase in patients with stage II melanoma using recombinant modified vaccinia virus Ankara (MVA-hTyr). *Cancer Immunol Immunother* **54**, 453–467.

Morse MA, Clay TM, Hobeika AC, *et al.* (2005). Phase I study of immunization with dendritic cells modified with fowlpox encoding carcinoembryonic antigen and costimulatory molecules. *Clin Cancer Res* **11**, 3017–3024.

Moss B (1996). Genetically engineered poxviruses for recombinant gene expression, vaccination, and safety. *Proc Natl Acad Sci U S A* **93**, 11341–11348.

Moss B (2007). Poxviridae: the viruses and their replication. In: Knipe DM, Howley PM (eds), *Fields Virology*, 5th edition. Lippincott Williams & Wilkins, Philadelphia, PA, pp. 2905–2945.

Mukherjee S, Haenel T, Himbeck R, *et al.* (2000). Replication-restricted vaccinia as a cytokine gene therapy vector in cancer: persistent transgene expression despite antibody generation. *Cancer Gene Ther* **7**, 663–670.

Pantuck AJ, van Ophoven A, Gitlitz BJ, *et al.* (2004). Phase I trial of antigen-specific gene therapy using a recombinant vaccinia virus encoding MUC-1 and IL-2 in MUC-1-positive patients with advanced prostate cancer. *J Immunother* **27**, 240–253.

Pennington TH, Follett EA (1974). Vaccinia virus replication in enucleated BSC-1 cells: particle production and synthesis of viral DNA and proteins. *J Virol* **13**, 488–493.

Puhlmann M, Brown CK, Gnant M, *et al.* (2000). Vaccinia as a vector for tumor-directed gene therapy: biodistribution of a thymidine kinase-deleted mutant. *Cancer Gene Ther* **7**, 66–73.

Reinis M (2004). Technology evaluation: TroVax, Oxford BioMedica. *Curr Opin Mol Ther* **6**, 436–444.

Riethmuller G, Holz E, Schlimok G, *et al.* (1998). Monoclonal antibody therapy for resected Dukes' C colorectal cancer: seven-year outcome of a multicentre randomized trial. *J Clin Oncol* **16**, 1788–1794.

Rochlitz C, Figlin R, Squiban P, *et al.* (2003). Phase I immunotherapy with a modified vaccinia virus (MVA) expressing human MUC1 as antigen-specific immunotherapy in patients with MUC1-positive advanced cancer. *J Gene Med* **5**, 690–699.

Rosenberg SA, Yang JC, Schwartzentruber DJ, *et al.* (2003). Recombinant fowlpox viruses encoding the anchor-modified gp100 melanoma antigen can generate antitumor immune responses in patients with metastatic melanoma. *Clin Cancer Res* **9**, 2973–2980.

Salzman NP (1960). The rate of formation of vaccinia deoxyribonucleic acid and vaccinia virus. *Virology* **10**, 150–152.

Sanda MG, Smith DC, Charles LG, *et al.* (1999). Recombinant vaccinia-PSA (PROSTVAC) can induce a prostate-specific immune response in androgen-modulated human prostate cancer. *Urology* **53**, 260–266.

Schneider J, Gilbert SC, Blanchard TJ, *et al.* (1998). Enhanced immunogenicity for CD8+ T cell induction and complete protective efficacy of malaria DNA vaccination by boosting with modified vaccinia virus Ankara. *Nat Med* **4**, 397–402.

Scholl S, Squiban P, Bizouarne N, *et al.* (2003). Metastatic breast tumour regression following treatment by a gene-modified vaccinia virus expressing MUC1 and IL-2. J. *Biomed Biotechnol* **3**, 194–201.

Scholl SM, Balloul JM, Le Goc G, *et al.* (2000). Recombinant vaccinia virus encoding human MUC1 and IL2 as immunotherapy in patients with breast cancer. *J Immunother* **23**, 570–580.

Senkevich TG, Ward BM, Moss B (2004). Vaccinia virus A28L gene encodes an essential protein component of the virion membrane with intramolecular disulfide bonds formed by the viral cytoplasmic redox pathway. *J Virol* **78**, 2348–2356.

Smith CL, Dunbar PR, Mirza F, *et al.* (2005). Recombinant modified vaccinia Ankara primes functionally activated CTL specific for a melanoma tumor antigen epitope in melanoma patients with a high risk of disease recurrence. *Int J Cancer* **113**, 259–266.

Stittelaar KJ, Kuiken T, de Swart RL, *et al.* (2001). Safety of modified vaccinia virus Ankara (MVA) in immune-suppressed macaques. *Vaccine* **19**, 3700–3709.

Sutter G, Moss B (1992). Nonreplicating vaccinia vector efficiently expresses recombinant genes. *Proc Natl Acad Sci U S A* **89**, 10847–10851.

Tepper RI, Mule JJ (1994). Experimental and clinical studies of cytokine gene-modified tumor cells. *Hum Gene Ther* **5**, 153–64.

Thorne SH, Bartlett DL, Kirn DH (2005). The use of oncolytic vaccinia viruses in the treatment of cancer: a new role for an old ally. *Curr Gene Ther* **5**, 429–443.

Triozzi PL, Strong TV, Bucy RP, *et al.* (2005a). Intratumoral administration of a recombinant canarypox virus expressing interleukin 12 in patients with metastatic melanoma. *Hum Gene Ther* **16**, 91–100.

Triozzi PL, Allen KO, Carlisle RR, Craig M, LoBuglio AF, Conry RM. Phase I study of the intratumoural administration of recombinant canarypox viruses expressing B7.1 and interleukin 12 in patients with metastatic melanoma. *Clin Cancer Res* 2005b; **11**, 4168–4175.

Ullenhag GJ, Frodin JE, Mosolits S, *et al.* (2003). Immunization of colorectal carcinoma patients with a recombinant canarypox virus expressing the tumor antigen Ep-CAM/KSA (ALVAC-KSA) and granulocyte macrophage colony- stimulating factor induced a tumor specific cellular immune response. *Clin Cancer Res* **9**, 2447–2456.

Upton C, Slack S, Hunter AL, Ehlers A, Roper RL (2003). Poxvirus orthologous clusters: toward defining the minimum essential poxvirus genome. *J Virol* **77**, 7590–7600.

Vile RG, Nelson JA, Castleden S, Chong H, Hart IR (1994). Systemic gene therapy of murine melanoma using tissue specific expression of the HSVtk gene involves an immune component. *Cancer Res* **54**, 6228–6234.

Wyatt LS, Shors ST, Murphy BR, Moss B (1996). Development of a replication-deficient recombinant vaccinia virus vaccine effective against parainfluenza virus 3 infection in an animal model. *Vaccine* **14**, 1451–1458.

Zeh HJ, Bartlett DL (2002). Development of a replication-selective, oncolytic poxvirus for the treatment of human cancers. *Cancer Gene Ther* **9**, 915–921.

7

Oncolytic herpes simplex viruses

Guy R. Simpson and **Robert S. Coffin**

7.1 Introduction

A number of oncolytic viral vectors are under clinical development for cancer therapy. Herpes simplex virus type 1 (HSV-1), deleted for ICP34.5 to provide tumour selectivity, has demonstrated particular promise in this regard, showing tumour selective replication and necrosis in a wide variety of tumour types, without damaging normal tissues, in both clinical and pre-clinical models. Enhanced activity has been observed when a number of therapeutic genes have been inserted into various versions of oncolytic HSV. These include cytokines, fusogenic membrane glycoproteins and prodrug activating genes which provide multiple modality cancer therapies intended to increase systemic antitumour immune responses, enhance the effectiveness of local tumour control, or enhance the effectiveness of other therapies such as chemotherapy or radiation. In addition, oncolytic HSVs in general have been demonstrated to not only be compatible with traditional therapies with respect to the combined safety profile, but also in most cases improve or potentiate the effects of the other therapies such that either additive or synergistic effects are observed. This chapter examines the properties of HSV, which are relevant to oncolytic virus therapy; discusses the various strategies that have been used to genetically modify the virus and further enhance efficacy through the insertion of therapeutic genes, and summarizes the preclinical and clinical results with oncolytic HSV for cancer treatment so far.

7.2 Herpes simplex virology

HSV-1 virus infection is highly prevalent worldwide, serum studies having also shown that antibody prevalence increases with age (Nahmias *et al.*, 1990). The natural route of HSV-1 infection involves uptake of the virus by skin epithelial cells, whereupon the virus lytic cycle genes are expressed and the virus replicates (reviewed in Roizman and Knipe, 2001). Following lytic replication the virus infects sensory nerve terminals and is transported retrogradely to the nerve cell body. The virus can then either undergoes a further round of replication or the virus can enter latency. Sensory neurons are the natural site of herpes latency and as such the virus is thought of as neurotropic. However, *in vitro* and *in vivo* HSV-1 has been shown to infect a broad range of cell types as well as neurons. The lytic, replicative cycles of HSV-1 and HSV-2 are responsible for the symptoms of oral and genital herpes, respectively (Whitley, 2001). HSV-1 can also infect the central nervous system to cause encephalitis, but this is very rare, only occurring in approximately one to two cases per million per year (Kennedy, 1984).

HSV-1 is a double-stranded DNA virus with a genome of 152kb, encoding over 80 polypeptides (reviewed in Roizman and Knipe, 2001), see Figure 7.1. The HSV genome comprises two unique regions, short and long, each of which is flanked by inverted terminal repeats. The genes

Viral Therapy of Cancer Edited by Kevin J. Harrington, Richard G. Vile and Hardev S. Pandha

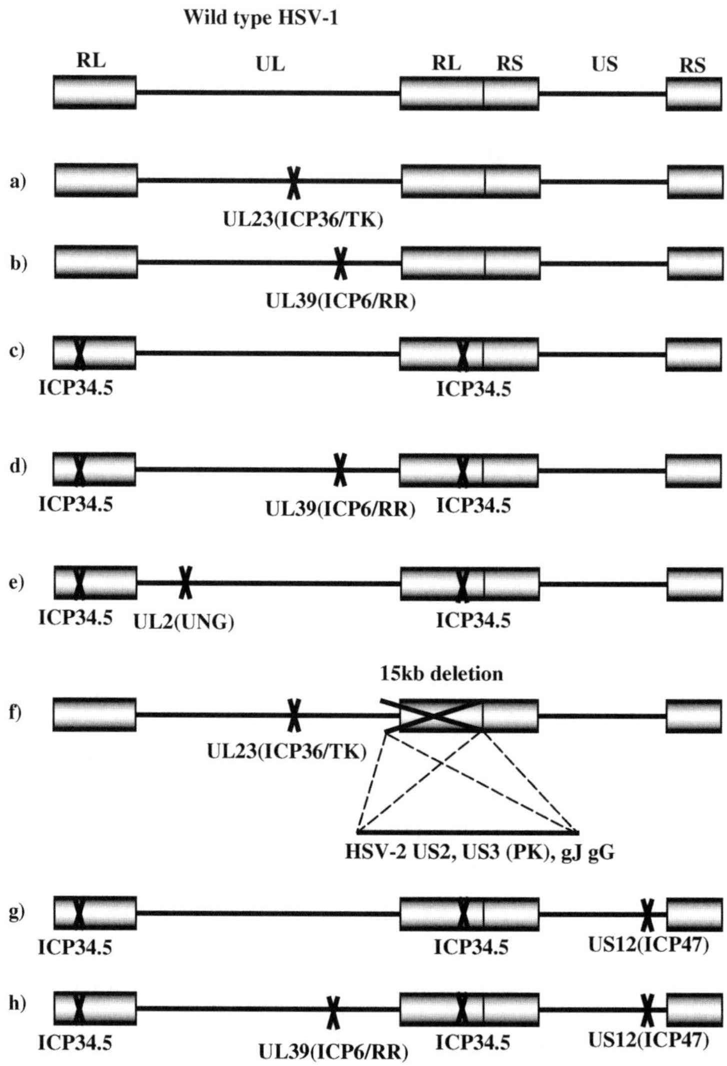

Figure 7.1 HSV-1 has a linear, double-stranded DNA genome of 152kb encoding more than 80 genes. The genome is composed of unique long (UL) and unique short (US) segments, which are flanked by inverted repeats (terminal repeat of the short segment, RS) (terminal repeat of the long segment, RL). HSV-1 mutants that have been tested for oncolytic use are shown. Nucleotide metabolism mutants are represented by (a) thymidine kinase (TK) (Kaplitt *et al.*, 1994) and (b) ribonucleotide reductase (RR) (Goldstein *et al.*, 1988). Neurovirulence mutants (ICP34.5-) are represented by (c); examples are 1716 (MacLean *et al.*, 1991) and R3616 (Chou *et al.*, 1990). A combination of mutations in the neurovirulence (ICP34.5-) and nucleotide metabolism genes is indicated in (d) and (e). (d) G207, an HSV-1 mutant with both copies of ICP34.5 deleted and the ICP6 gene (encoding ribonucleotide reductase) inactivated (Mineta 1995). (e) U3616, which was generated by deletion of the US2 ORF (Uracil-*N*-glycoslase) in a previously described ICP34.5- mutant (R3616) (Pyles *et al.*, 1997). Attenuation of HSV-1 has also been achieved by including sequences from HSV-2 (Meignier *et al.*,1988, 1990) (f). NV1020 has a 15 kb deletion that extends from the 3′ end of UL55 across the junction of the long and short components of the HSV-1 genome to the promoter region of ICP4 and replaced with sequences from HSV-2. Dramatically improved growth in tumour cells is achieved with ICP34.5 deleted viruses when ICP47 is also deleted so as to place the US11 gene under the control of the ICP47 promoter (g). Examples of such mutants are SUP (Mohr *et al.* 1996) and OncoVex (Liu *et al.*, 2003). ICP47 mutation has also been included in G207, to produce G47 Delta (h) (Todo *et al.*, 2001)

contained within the inverted repeats are present in two copies per viral genome. The unique segments can invert relative to each other, thus yielding four possible isomers. Approximately half of the 80 proteins encoded by HSV are essential for virus replication *in vitro*, the rest being necessary for full pathogenesis in the host.

7.3 Properties of HSV relevant to oncolytic virus therapy

Mutation of a number of genes allows tumour selective virus replication and blocks virulence

Because wild type HSV-1 can infect replicating or quiescent cells, certain genes can be deleted to render the virus non-virulent, which also results in selective replication in tumour cells (Martuza *et al.*, 1991; MacLean *et al.*, 1991; McKie *et al.*, 1996; Todo *et al.*, 2001; Rampling *et al.*, 2000). The functional results of these mutations are discussed below.

Ease of manipulation

Foreign DNA can be inserted into the HSV genome by homologous recombination techniques (Roizman and Jenkins, 1985). This property is not exclusive to HSV, but the lytic nature of HSV-1 aids fast plaque purification. An alternative method used by some groups is to employ the bacterial artificial chromosome (BAC), a single-copy plasmid that can stably retain a large size (300 kb) DNA as an insert (Shizuya *et al.*, 1992). BAC plasmids have been used to propagate the entire HSV-1 genome in *Escherichia coli*, allowing easy genetic manipulation (Saeki *et al.*, 2001; Horsburgh *et al.*, 1999).

Oncolytic HSV can infect a broad range of human tumour cell types

The following studies have shown that oncolytic HSV infects and replicates *in vitro* and *in vivo* in a wide range of human tumour cell types: glioma (Chambers *et al.*, 1995; Andreansky *et al.*, 1996; Andreansky *et al.*, 1997; Samoto *et al.*, 2002; Detta *et al.*, 2003), colon carcinoma (Carroll *et al.*, 1996), retinoblastoma (Nicolo and Chiocca, 1998), epithelial ovarian cancer (Coukos *et al.*, 1999), colorectal cancer (Kooby *et al.*, 1999; Reinblatt *et al.*, 2004), prostate cancer (Walker *et al.*, 1999; Cozzi *et al.*, 2002), non-small cell lung cancer (Toyoizumi *et al.*, 1999), gallbladder carcinoma (Nakano *et al.*, 2001), head and neck squamous cell carcinoma (Wong, 2001), oesophageal adenocarcinoma (Stiles *et al.*, 2003), breast cancer (Teshigahara *et al.*, 2004; Pin *et al.*, 2004; Liu and Rabkin, 2005), thyroid cancers (Yu *et al.*, 2004), hepatocellular carcinoma (Pin *et al.*, 2004) and rhabdomyosarcoma (Currier *et al.*, 2005).

HSV-1 is a highly lytic virus, resulting in tumour cell death

The inherent cytotoxicity of this virus, if harnessed and made to be selective by genetic manipulations, makes this virus an ideal candidate for developing viral oncolytic approach.

HSV does not integrate into the host genome

This avoids the possibility of activating proto-oncogenes in the host genome, as has recently been described with integrating vectors (Fischer *et al.*, 2002).

Ease of production

Oncolytic HSV can easily be grown in culture to the titres required for clinical use.

Large insert capacity

HSV allows the insertion of up to 30 kb of exogenous DNA (Longnecker *et al.*, 1988). This capacity allows the insertion of multiple therapeutic genes such as prodrug activating genes, immune-stimulatory genes and fusogenic membrane glycoprotein genes to increase tumour killing.

Viral replication control

Most versions of oncolytic HSV [(1716, MacLean *et al.*, 1991)(G207, Mineta *et al.*, 1995) (OncoVex, Liu *et al.*, 2003)] retain an intact thymidine kinase gene. As a result, many anti-herpes drugs, which require the thymidine kinase gene for activity,

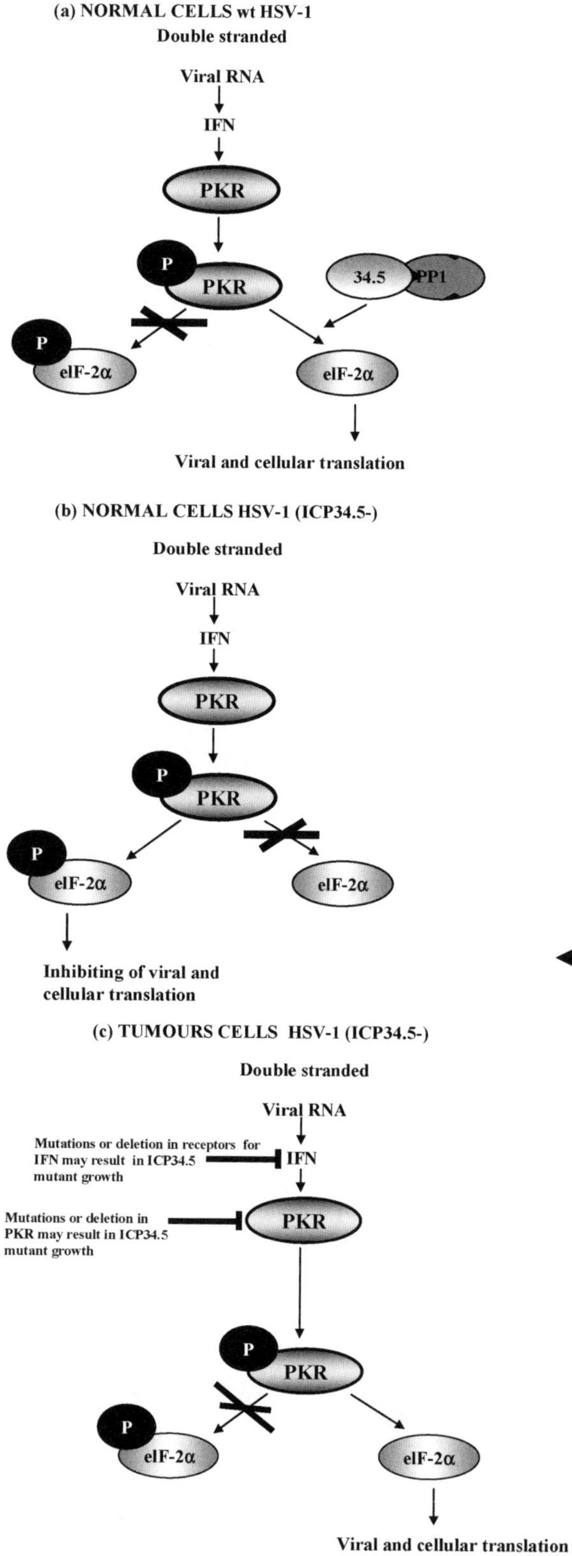

[e.g. ganciclovir (GCV), aciclovir (ACV), 1-(2′-deoxy-2′-fluoro-β-D-arabinofuranosyl)-5-iodouracyl (FIAU)] are available to prevent virus replication if this were thought to be necessary in a patient.

7.4 Mutations giving tumour-selective replication

Viruses with single mutations

Nucleotide metabolism gene mutants

HSV encodes a number of enzymes which are involved in nucleotide metabolism. These include thymidine kinase (TK) (Kaplitt *et al.*, 1994), ribonucleotide reductase (RR) (Goldstein *et al.*, 1988) and uracil N-glycosylase (UNG) (Pyles and Thompson, 1994). The cellular homologues of these genes have similar functions, but they are not normally expressed in post-mitotic cells. In contrast cancer cells have increased levels of these proteins and therefore HSV mutants carrying mutations in these genes preferentially grow in cancer tissue as compared to normal tissue. TK has a broad range of substrates and is capable of phosphorylating both purines and pyrimidines. Nucleoside analogues that are phosphorylated by TK serve as antiviral compounds (Fyfe *et al.*, 1978).

Figure 7.2 The host viral response in normal and tumour cells to infection with wild-type HSV and ICP34.5 mutant viruses. (a) Wild type HSV-1 can replicate in normal cells due to its ability to inhibit the cellular anti-viral response by the action of ICP34.5. After infection, the presence of the virus within the cell produces an IFN response which leads to an up regulation of PKR, the cell phosphorylates PKR, which in turn phosphorylates elF-2α, which would lead to the shutoff of protein translation. However, ICP34.5 prevents this by binding to protein phosphatase 1 (PP1) and redirecting it to stop the phosphorylation of elF-2α. (b) When ICP34.5 mutants infect normal cells, PKR is up regulated and phosphorylated leading to a cessation of translation. Therefore the ICP34.5 mutants cannot replicate (c). The general deregulation of the cellular IFN response pathways and increased levels of elF-2α within tumour cells limits the antiviral response allowing replication of ICP34.5 mutants

Table 7.1 Summary of oncolytic HSV-1 clinical trials

HSV-1 vectors	Tumour types	Phase/country	Viral dose	Route	No of patients	No of patients with response	Reference
1716 (ICP34.5-)	Glioma	I/UK	10^{3}– 3×10^{5} p.f.u.	Intratumoral	9	4 patients alive 14 months after treatment	Rampling *et al.*,2000
G207 (ICP6-, ICP34.5-)	Glioma	I/USA	10^{5}– 3×10^{9} p.f.u.	Intratumoral	21	8/21 reduction tumour volume	Markert *et al.*, 2000
1716 (ICP34.5-)	Melanoma	I/UK	10^{5} p.f.u.	Intratumoral	5	1/5 with flatenned nodule	MacKie *et al.*, 2001
1716 (ICP34.5-)	Glioma	I/UK	10^{5} p.f.u.	Intratumoral	12	Not stated	Paanatassiou *et al.*, 2002
NV1020 (ICP34.5-(1 copy), UL24, UL56, TK-)	CRC liver	I/USA	1.3×10^{6} – 1.3×10^{7} p.f.u.	Intratumoral	9	Decrease in expression CEA in 9/9 patients	Fong *et al.*, 2002
1716 (ICP34.5-)	Glioma	I/UK	10^{5} p.f.u.	Intratumoral	12	3/12 patients remain alive for 15, 18 and 22 months and clinically stable post surgery and 1716 injection.	Harrow *et al.*, 2004
OncoVEX^{GM-CSF} (ICP34.5-, ICP47- GM-CSF+)	Cutaneous and subcutaneous metastases	I/UK	10^{6}–10^{8} p.f.u./ml	Intratumoral	30	Flattening and necrosis of injected and uninjected adjacent lesions	Hu *et al.*, 2006

p.f.u., plaque-forming units; CEA, carcinoembryonic antigen; CRC, colorectal carcinoma.

A HSV-1 mutant (*dlsptk*) containing a deletion in the TK gene was tested against brain tumours as a proof of principle of this concept (Martuza *et al.*, 1991). Athymic mice were injected intracerebrally with human glioma cells and then treated with *dlsptk*. The treated animals showed increased survival over control animals (Martuza *et al.*, 1991), but neurotoxicity was seen at high virus titres. Critics pointed out that the absence of TK is a major drawback as it prevents virus replication being controlled in a clinical situation with the anti-herpes drugs such as aciclovir (ACV) and ganciclovir (GCV), if that were necessary.

The UL39 gene encodes the large subunit of the viral nucleotide reductase (ICP6). Ribonucleotide reductase catalyses the reduction of ribonucleotide diphosphates to the corresponding deoxynucleotide diphosphates. Deletion of this gene results in preferential replication in proliferating cancer cells with high levels of endogenous nucleotide reductase activity (Goldstein and Weller, 1988). Such mutants exhibit decreased neurovirulence (Cameron *et al.*, 1988; Yamada *et al.*, 1991) and nude mice injected with the human tumour cell line U-87MG and then treated with an ICP6$^-$ mutant showed increased survival over control animals (Mineta *et al.*, 1994; Boviatsis *et al.*, 1994).

Mutation to the neurovirulence gene (ICP34.5$^-$)

The existence of a neurovirulence locus in the long repeat region of the HSV genome is well documented (MacLean *et al.*, 1991; Taha *et al.*, 1989a, b). This phenotype has been specifically assigned to the *RL-1* gene (Chou *et al.*, 1990, Dolan *et al.*, 1992) and its encoded protein ICP34.5 (Chou *et al.*, 1990). Expression of this protein facilitates viral replication in non-dividing cells, such as adult neurons (Robertson *et al.*, 1992, Whitley *et al.*, 1993). ICP34.5$^-$ mutant viruses are also avirulent (Chou *et al.*, 1990; MacLean *et al.*, 1991; Taha *et al.*, 1989a, b). The LD_{50} of many strains of wild type HSV-1 is less than 300 plaque-forming units (p.f.u.) following intracranial delivery. In contrast upwards of 10^6-10^9 p.f.u. of ICP34.5 mutant viruses have been safely injected intracranially into mouse, rat, non-human primate, and human brains (Chou et al., 1990; MacLean *et al.*, 1991; Mineta *et al.*, 1995a; Hunter *et al.*, 1999; Markert *et al.*, 2000; Rampling *et al.*, 2000; Sundaresan *et al.*, 2000; Simpson *et al.*, 2006). An ICP34.5$^-$ mutant (1714) was identified in studies of spontaneous mutants of wild type HSV that had lost their neurovirulence properties. The 1714 virus had a number of deletions and mutations, including a 759bp deletion in ICP34.5. When this particular deletion was introduced into 17syn$^+$ backbone, creating virus strain 1716, a loss in neurovirulence was observed (MacLean *et al.*, 1991) demonstrating the *RL1* gene to be responsible.

A common cellular response to viral infection is activation by phosphorylation of the interferon inducible protein kinase R (PKR) (Clemens *et al.*, 1997). PKR is activated by double-stranded RNA (dsRNA) binding to two dsRNA motifs in its amino terminus to promote dimerization and subsequent *trans*-autophosphorylation of PKR (Wu and Kaufman, 1997). Activated PKR in turn phosphorylates the translation initiation factor (eIF-2α), resulting in inhibition of translation of viral transcripts (Clemens *et al.*, 1997). In addition PKR phosphorylation causes nuclear factor (NF)-κB activation, which possibly contributes to interferon dependent antiviral effects such as apoptosis (Kumar *et al.*, 1997).

Most viruses studied to date block phosphorylation of eIF-2α by encoded proteins or RNAs. Some virus proteins bind to dsRNA thus preventing PKR activation, for example E3L of vaccinia virus (Davies *et al.*, 1993; Carroll *et al.*, 1993; Yuwen et al., 1993), α3 protein of reovirus (Imani and Jacobs, 1988; Langland *et al.*, 1994; Beattie *et al.*, 1995) and NS1 protein of influenza (Lu *et al.*, 1995). The adenovirus VA1 RNA acts as a competitive inhibitor of dsRNA thus preventing PKR activation (Ghadge 1994). Another mechanism of inhibition of PKR is through direct binding of viral proteins or RNAs to PKR presumably displacing PKR from the ribosome, for example Epstein–Barr virus EBERS-1-2 RNAs (Sharp *et al.*, 1993) and human immunodeficiency virus Tar (Park *et al.*, 1994). The vaccinia K3L (Romano *et al.*, 1998) shares homology with eIF2a and acts as a pseudosubstrate preventing eIF2a phosphorylation. Finally, the polio virus synthesizes a protease that causes degradation of PKR (Black *et al.*, 1993).

Like the other viral genes discussed above, the function of HSV ICP34.5 is to disrupt the host antiviral defence mechanisms. However, ICP34.5 does not target PKR itself but instead forms a complex with protein phosphatase 1, which is then directed to dephosphorylate eIF-2α, promoting translation of viral transcripts and subsequently inhibiting the induction of apoptosis and promoting infection (Roizman and Markowitz, 1997, He *et al.*, 1997). The carboxyl-terminal domain of ICP34.5 is homologous to the corresponding domain of a conserved mammalian protein called growth arrest and DNA damage 34 protein (GADD34). GADD34 can substitute for the corresponding domain in ICP34.5 blocking the effects of the PKR/eIF-2α pathway (Brown *et al.*, 1997). It has also been shown that ICP34.5 binds to a cell cycle protein, proliferating cell nuclear antigen (PCNA) (Rampling *et al.*, 2000). In so doing an environment is provided in which the virus can replicate. In tumour cells where functional PCNA levels are high, ICP34.5 is not required for HSV replication whereas in neurons, in which PCNA levels are low, ICP34.5 is an absolute requirement for production of infectious progeny virus (Rampling *et al.*, 2000).

ICP34.5$^-$ mutants in which both copies of the gene are mutated are incapable of replicating in neurons, but can replicate in and destroy glioma cells *in vitro* and *in vivo* (Andreansky *et al.*, 1996). This suggests that deletion of the ICP34.5 gene somehow allows the virus to specifically target cancer cells while sparing normal tissue (Rampling *et al.*, 2000; Andreansky *et al.*, 1996). The precise mechanism for growth of ICP34.5 mutants in each tumour type is not fully understood, but it is known from knockout mouse studies that deletions and mutations in PKR and the interferon receptors allow ICP34.5 mutant growth (Leib *et al.*, 1999, 2000), and that these mutations and deletions have been found in a number of tumour types (Haus, 2000). Work from 2001 hypothesized that the higher levels of Ras activation found in transformed cells as compared to normal cells inhibits or reverses eIF-2α phosphorylation, thereby allowing viral protein synthesis and virus replication (Farassati *et al.*, 2001). Thus cancer cells with an activated Ras signalling pathway would be susceptible targets for ICP34.5$^-$ mutants, but normal or tumour cells without an activated Ras pathway would be resistant. However, more recently this hypothesis has been disproved as no correlation has been found between Ras status and virus susceptibility in various panels of tumour cells (Sarinella *et al.*, 2006; Mahller *et al.*, 2006). In pancreatic cell lines it has also been shown that PKR is fully functional and able to phosphorylate eIF2α, which suggests the permissiveness of a ICP34.5$^-$ mutant, seen in these cells, is not explained by a defective PKR activity (Sarinella *et al.*, 2006). To establish a mechanism for this PKR independent permissiveness, the pancreatic tumour cell lines were treated with specific inhibitors of signal transduction pathways. The results suggest that mis-regulation of PI 3-kinase within pancreatic tumour cell lines in this case may have led to deregulated protein translation independent from PKR activity, allowing efficient replication of 34.5-deleted HSV-1. Another broader range study examined 12 cell lines derived from a variety of human tumours, including prostate, pancreatic, breast, colorectal, hepatoma and fibrosarcoma and again showed no correlation between viral yields and the Ras genotype of these cells (Smith *et al.*, 2006). The early work leading to the Ras hypothesis was carried out in mouse cells artificially over-expressing the Ras oncogene, a phenotype that does not accurately model human tumour cells and probably explains the misleading results obtained (Farassati *et al.*, 2001; Sarinella *et al.*, 2006; Smith *et al.*, 2006). Smith *et al.*, (2006) went on to show a correlation between viral growth and activation of MEK which together suggests that the deregulation of different and/or multiple pathways in different tumour types allows tumour selective growth of ICP34.5 mutants. It is important to note that oncolytic HSV containing only a deletion in both copies of the ICP34.5 gene such as R3616 (Chou *et al.*, 1990) and 1716 (MacLean *et al.*, 1991) are now considered relatively primitive oncolytic viruses because they fail to replicate in a number tumour cell types (reviewed in Mohr, 2005), unlike more advanced multiply mutated viruses in development now. The reason for this is that ICP34.5 mutants cannot take advantage of the other HSV-1 genes including US11 (reviewed in Mohr 2005) that also act on the interferon/PKR/eIF-2α pathway. If the virus is engineered such that US11 is expressed as an immediate early (IE) rather than as a late gene, this increases tumour selective virus replication and, therefore, tumour killing *in vitro* and *in vivo* without restoring the abrogation of virulence of the virus provided by the mutation of ICP34.5 (Mohr *et al.*, 2001). As a result, work to determine the mechanism of selectivity of viruses only deleted for ICP34.5 (Farassati *et al.*, 2001; Sarinella *et al.*, 2006; Smith *et al.*, 2006) is not directly relevant to more advanced viruses where US11 is expressed as an IE gene. Such viruses include the SUP mutant (Mohr and Gluzman, 1996; Mulvey *et al.*, 1999; Cassady *et al.*, 1998a), G47 Delta (Todo *et al.*, 2001) and OncoVEX (Liu *et al.*, 2003).

Viruses with only ICP34.5^{-} mutated include 1716 (MacLean *et al.*, 1991) and R3616 (Chou *et al.*, 1990), both of which have been used successfully to treat various animal tumour models (McKie *et al.*, 1996, Todo *et al.*, 2001). However, while these mutants replicate selectively in tumour cells, replication in tumour cells is somewhat impaired as compared to wild type HSV, which limits the efficacy seen (McKie *et al.*, 1996, Todo *et al.*, 2001). The first phase I clinical trial with 1716 was carried out on nine patients with recurrent gliomas using low doses of virus (10^3–10^5p.f.u.) (Rampling *et al.*, 2000). No virus was detected by polymerase chain reaction from blood samples in five patients and none exhibited any virus-associated adverse events. Promisingly, four patients were alive 14 months after treatment. A subsequent clinical trial using 1716 was carried out in 12 patients with high grade glioma. Each patient was injected intratumorally with 10^5 p.f.u. of virus, following which the tumour was surgically removed after 4–9 days (Papanastassiou *et al.*, 2002). Two lines of evidence suggested that virus replication occurred in the tumour. First, infectious virus was recovered from two patients at higher levels than inoculated. Second, the gene product of UL42 was detected by immunohistochemical staining UL42 encodes a late viral protein not present in virions and thus would not be expected to be detected without virus DNA replication. (Papanastassiou *et al.*, 2002). As in the first clinical trial, no virus-associated adverse events were seen (Papanastassiou *et al.*, 2002). A third clinical trial in five melanoma patients who each received between one and four injections of 10^3 p.f.u. into a single melanoma nodule resulted in three out of five patients showing evidence of tumour necrosis and the presence of HSV antigens by immunohistochemical staining (MacKie *et al.*, 2001). In a more recent clinical study 12 patients with high-grade glioma were treated by surgery following which 1716 was injected at eight to 10 sites in the surrounding brain. Three patients remained alive and clinically stable 15, 18 and 22 months later (Harrow *et al.*, 2004).

Oncolytic HSV containing multiple mutations

Double ICP34.5^{-}/ICP6^{-} mutants

Single deletion oncolytic HSV mutants such as 1716 have shown safety as well as indications of efficacy in phase I clinical trials (Rampling *et al.*, 2000; MacKie *et al.*, 2001; Papanastassiou *et al.*, 2002; Harrow *et al.*, 2004). However, in a cautious approach to the then unknown issue of the safety of ICP34.5-deleted HSV when injected into the human brain, another group decided to initially maximize safety by the combined mutation of both ICP34.5- and ICP6-generating virus strain G207. G207 was constructed from laboratory strain F with both copies of ICP34.5 deleted and its ICP6 gene inactivated by the insertion of LacZ (Mineta *et al.*, 1995b). ICP6 is a large subunit of viral nucleotide reductase. Ribonucleotide reductase catalyses the reduction of ribonucleotide diphosphates to the corresponding deoxynucleotide diphosphates. Deletion of this gene compels the virus to replicate preferentially in proliferating cancer cells with high levels of endogenous nucleotide reductase activity (Goldstein and Weller, 1988), but also reduces virus replication even here. Initially G207 was used to target glioma tumours but studies *in vitro* and *in vivo* have shown that it is reasonably effective against a wide range of solid tumour cell types including breast, colon, head and neck, melanoma, pancreas and prostrate cancers (Toda *et al.*,

2002). In immune competent mice inoculation of G207 into a local tumour induced systemic anti-tumour immunity, leading to the regression of distant tumours and the resistance of the animals to further challenge by tumour cells (Toda *et al.*, 1999). G207 has been put through extensive pre-clinical toxicology studies using two animal models: BALB/c mice and New World owl monkeys *Aotus nancymai*. HSV-1 has a natural tropism for the central nervous system and on this basis extensive toxicity studies were conducted after its administration into the brain. In mice doses up to 1×10^7 p.f.u. were inoculated directly into the brain, liver, prostate, and by intravenous delivery resulting in no adverse affects (Sundaresan *et al.*, 2000; Varghese *et al.*, 2001). In *Aotus* monkeys a dose of 10^3 p.f.u. of the parental strain F causes encephalitis and serious morbidity within 5 days. In contrast a dose of 10^9 p.f.u. of G207 inoculated intracerebrally did not result in detectable pathology (Hunter *et al.*, 1999).

The first phase I clinical trial with G207 was carried out in 21 recurrent glioma patients (Markert *et al.*, 2000) who received doses ranging from 1×10^5 to 3×10^9 p.f.u. All of the patients tolerated G207 without any serious adverse events attributed to the virus. Eight patients exhibited reductions in tumour volume from 4 days to 1 month post inoculation. Two patients were still alive over 4 years after treatment, but one of the patients died from a cerebral infarction at 10 months post-treatment. However, an autopsy showed that this death was probably unrelated to treatment. None of the other patients who died displayed any evidence of virus-induced pathology at post-mortem. Therefore, a good safety profile was demonstrated.

A syncytial mutant of G207 (called Fu-10) has also been obtained, by subjecting the virus to random mutagenesis through inclusion of the thymidine analogue BrdUrd in the media during growth in Vero cells (Fu and Zhang, 2002). *In vitro* characterization of Fu-10 showed a fusogenic phenotypic on a variety of human tumour cell lines, which resulted in large plaque formation as well as higher cell death than the parental G207 (Fu *et al.*, 2002). Systemic administration of Fu-10 into mice with established lung metastatic breast cancer resulted in effective therapy (Fu *et al.*, 2002) although the site of the mutation(s) causing the effect was not identified.

HSV-1/HSV-2 intertypic recombinants

NV1020 (also known as R7020) was originally developed as a Herpes vaccine but was not further pursued for this purpose. Later studies also demonstrated that it had some promise as an oncolytic agent (Meignier *et al.*, 1988, 1990). In this virus the joint region of the long (L) and short (S) regions of the HSV-1 genome is deleted, including one copy of ICP34.5, UL34 and UL56 and replaced with a fragment of HSV-2 containing US2, US3 (PK), gJ and gG (Meignier *et al.*, 1988). This novel junction also contains an exogenous copy of the HSV-1 TK gene under the control of the ICP4 promoter and a 3.7 kb duplication of the UL5/6 sequence. NV1020 has a 700 base pair deletion in the endogenous thymidine kinase locus that overlaps the promoter upstream of the UL24 gene, preventing its expression. This mutant retains one intact copy of the ICP34.5 gene and thus the precise reason for its attenuation is unclear. A direct comparison between G207 and NV1020 in 22 different human tumour lines found that the cytotoxicity of NV1020 was greater than G207 at a lower virus dose (McAuliffe *et al.*, 2000). As the virus does express one copy of ICP34.5, neurovirulence might have been expected to be problematic. However, no toxicity has been shown in rodents or primates (Meignier *et al.*, 1988, 1990). NV1020 was tolerated in owl monkeys at doses 10 000 fold higher than the lethal dose of wild type HSV-1 (Meignier *et al.*, 1990) and has shown oncolytic properties in a wide range of non-CNS tumour models such as prostate, pancreas, and head and neck tumours (Wong *et al.*, 2001; Advani *et al.*,1998).

ICP34.5$^-$, US2 double mutants

Uracil-*N*-glycoslase (UNG), encoded by the US2 gene, is an enzyme involved in the cleavage of the N-glycosidic bond linking uracil to the deoxyribose sugar during viral DNA synthesis (Caradonna *et al.*, 1987). This gene product is not essential for replication in growing cells in culture, but is involved in oral pathogenesis and reactivation from latency (Pyles and Thompson, 1994). U3616 was generated by the insertion of the LacZ gene into the US2 ORF (Pyles *et al.*, 1997) in a previously described ICP34.5⁻ mutant (R3616). U3616 was tested in primary embryonic rat dorsal root ganglia neural cultures and showed no replication or toxicity. In contrast, the wild type strain F and R3616 both replicated, leading to the eventual destruction of the cultures (Pyles *et al.*, 1997). This suggests that the UNG mutation renders the virus less able to replicate in non-dividing cells compared to mutation of ICP34.5 alone. U3616 was tested in human tumour xenograft (DAOY, SK-M) established in SCID mice and showed similar efficacy to the ICP34.5 mutant R3616. However, this virus has not been progressed into clinical trials.

ICP34.5⁻, ICP47 double mutants

When an ICP34.5 deleted virus was serially passaged in tumour cells a novel mutant appeared, which exhibited dramatically improved growth properties in tumour cells (Mohr and Gluzman, 1996). This so called suppressor mutant virus (SUP) contained an additional mutation that overcomes the protein synthesis block by altering the expression profile of US11, which encodes a viral RNA binding protein, from a late gene to an immediate early gene (Mohr *et al.*, 1996; Mulvey *et al.*, 1999; Cassady *et al.*, 1998a). The altered regulation of US11 in the SUP mutant takes place because of a deletion in the US12 gene encoding ICP47, which places US11 under the control of the ICP47 immediate early promoter. Accumulation of US11 at early times during infection inhibits the activation of the cellular PKR kinase and allows protein synthesis to proceed in the absence of the ICP34.5 gene product (Mulvey *et al.*, 1999; Cassady *et al.*, 1998b). Inhibition of PKR is mediated through an RNA binding domain in the carboxy terminus of the US11 gene product (Poppers *et al.*, 2000). A physical complex between US11 and PKR has been observed in infected cells, and this protein-protein interaction may also play a role in inhibiting PKR activation (Cassady *et al.*, 1998a, b). This activity of US11 is comparable to the vaccinia virus E3L protein, which has been shown to bind viral RNA and form a physical complex with PKR, both of which are thought to inhibit PKR activation (Romano *et al.*, 1998). The neurovirulence of the SUP mutant was tested by intracerebral injection of immuno-competent mice and the virus showed that, like the ICP34.5 single mutant, it was severely attenuated (Mohr *et al.*, 2001). Thus, it appears that the multifunctional nature of ICP34.5 means that the attenuation that occurs following its deletion is not restored by replacing only one of these functions (i.e. PKR activity), but that this does dramatically increase tumour selective virus replication.

Wild type HSV-1 infection causes down regulation of major histocompatibility complex (MHC) class I expression on the surface of infected cells (Jennings *et al.*, 1985; Hill *et al.*, 1995). The binding of ICP47 to the transporter associated with antigen presentation (TAP) blocks peptide transport in the endoplasmic reticulum and loading of MHC class I molecules (York *et al.*, 1994; Hill *et al.*, 1995; Fruh *et al.*, 1995). Consequently human tumour cells infected with ICP47- mutants express high levels of MHC class I on their surface compared to wild type HSV infected cells (Todo *et al.*, 2001; Liu *et al.*, 2003). This would be expected to improve any antitumour immune response following intratumoral injection of the virus due to the presentation of tumour antigens at much higher levels on the surface of both tumour cells and HSV infected antigen presenting cells. Deletion of ICP47 has been included in G207 (Mineta *et al.*, 1995b) to give G47Delta. G47Delta has been demonstrated to give both enhanced antigen expression and enhanced anti-tumour activity due to increased expression of US11 (Todo *et al.*, 2001; Taneja *et al.*, 2001).

In order to develop oncolytic HSV with greater tumour selective replicative ability, clinical isolates were tested for their ability to replicate in and kill human tumour cell lines as compared to the previously used laboratory strains (Liu *et al.*, 2003). Both clinical isolates showed greater tumour cell killing than serially passaged laboratory strains, suggesting that they provided a better starting point for the development of an oncolytic virus. ICP34.5 (MacLean *et al.*, 1991; Chou *et al.*, 1990) and ICP47 were then deleted from one of these clinical virus strains resulting in tumour selectivity, the expression of US11 gene as an IE rather than a L gene to further increase tumour replication (Taneja *et al.*, 2001), and increased antigen presentation (Hill *et al.*, 1995; Todo *et al.*, 2001; Liu *et al.*, 2003). Both the use of the clinical isolate, and the increased expression of US11 were shown to increase tumour shrinkage in mouse tumour models (Liu *et al.*, 2003). Finally, the gene encoding granulocyte–macrophage colony-stimulating factor (GM-CSF) was inserted into this virus in place of ICP34.5, and this was demonstrated to increase the antitumour immune response generated such that un-injected as well as injected tumours could be cured in mouse models. This virus was called OncoVEX$^{GM\text{-}CSF}$ (Liu *et al.*, 2003). Expression of GM-CSF has previously been shown to induce myeloid precursor cells to proliferate and differentiate, is a recruiter and stimulator of dendritic cells and has shown promise in preclinical and clinical trials in cancer (Wong *et al.*, 2001; Toda *et al.*, 2000; Parker *et al.*, 2000; Bennett *et al.*, 2001; Andreansky *et al.*, 1998).

A phase I clinical trial has been completed using OncoVEX$^{GM\text{-}CSF}$ by intratumoral injection in patients with cutaneous or subcutaneous deposits of breast, head and neck and gastrointestinal cancer and malignant melanoma (Hu *et al.*, 2006). The virus was generally well tolerated with local inflammation, erythema and febrile responses being the main side effects seen, which were expected from previous studies with oncolytic viruses. Virus replication and GM-CSF expression were observed, as was considerable tumour necrosis, including in tumours adjacent to those which had been injected with the virus (Hu *et al.*, 2006). Some evidence of a more distant, potentially immune-mediated effect, was observed as in some cases distant tumours became inflamed (Hu *et al.*, 2006). Following these promising results, OncoVEX$^{GM\text{-}CSF}$ is currently in a number of phase II studies in individual tumour types.

7.5 Oncolytic HSV expressing fusogenic membrane glycoproteins (FMG)

A new approach to cancer therapy is the transduction of tumour cells with viral fusogenic membrane glycoproteins (FMG) (Bateman *et al.*, 2000). A number of different viruses kill their target cells by causing fusion between infected and non-infected cells, via the interaction between the viral envelope and its receptor. This fusion of uninfected cells with infected cells also provides a bystander effect.

The cytotoxic activity of three envelope genes from different groups of viruses were originally evaluated (Bateman *et al.*, 2000), these include the rhabdoviral VSV-G envelope gene, the combination of F and H genes from measles virus and a mutated version of the retroviral Gibbon Ape leukaemia virus (GALV *env* R-) (Bateman *et al.*, 2000). In this early *in vitro* study all three membrane glycoproteins showed a high level of cell killing, with GALV *env* R- performing the best. Truncation of the GALV protein by removing 16 amino acids in the transmembrane R-peptide, which normally serves to restrict fusion of the envelope until it is cleaved during viral infection renders it constitutively highly fusogenic and therefore cytotoxic to human tumour cells (Diaz *et al.*, 2000).

Early attempts to produce adenovirus viruses encoding GALV *env* R- failed as rapid cell fusion inhibited virus replication (Diaz *et al.*, 2000). Therefore retroviral and lenti-viral delivery was investigated both of which encountered problems in generating high titre virus stocks. Even so, these viruses were tested against human tumour xenografts in nude mice resulting in a considerable reduction in tumour size (Diaz *et al.*, 2000). Several immune stimulatory molecules associated with a stress response, such as heat shock proteins (Melcher *et al.*, 1998), the NK cell receptor (Groh *et al.*, 1998) and murine interferon-γ were detected in treated mice, suggesting that the immune system may partly contribute to the mode of action of GALV *env* R- even in immunocompromised animals such as nude mice.

The problems of expression were solved by encoding GALV *env* R- in HSV, initially inserted into strain G207 (see above) (Fu *et al.*, 2003) and later a second generation virus based on a clinical isolate of HSV and in which US11 expression levels were increased, OncoVEX^{GALV} (Simpson *et al.*, 2006). These showed no impairment of viral replication such that virus stocks could easily be produced (Simpson *et al.*, 2006). The expression of GALV *env* R- caused syncytia formation in tumour cells of a variety of origins including from colon, brain, pancreas, lung, liver, prostate and connective tissue cancers (Fu *et al.*, 2003; Simpson *et al.*, 2006). This fusogenic property also translated into a significantly increased tumour cell killing as compared to previous versions of oncolytic HSV vectors (Fu *et al.*, 2003; Simpson 2006).

In rodent tumour models, improved tumour shrinkage was seen with an approximately 5–10-fold reduction in the virus dose required to give equivalent tumour shrinkage or cure (Fu *et al.*, 2003, Simpson *et al.*, 2006). There was no evidence of toxicity in normal tissue associated with GALV expression, including when the virus was directly injected into the rat brain (Simpson *et al.*, 2006).

GALV expression has also been combined with expression of the highly potent prodrug activating Fcy::Fur enzyme, a fusion of yeast cytosine deaminase gene and yeast uracil phosphoribosyltransferase gene, to generate OncoVEX$^{GALV/CD}$ (Simpson *et al.*, 2006). In the presence of 5-fluorocytosine (5-FC), OncoVEX$^{GALV/CD}$ showed further improved tumour shrinkage *in vivo* (Simpson *et al.*, 2006) resulting from the conversion of 5-FC to 5-FU (the active drug) locally in the tumour by Fcy::Fur (Simpson *et al.*, 2006). Viruses expressing pro-drug activating enzymes are discussed in more detail below.

7.6 Prodrug activation therapy and oncolytic HSV

Current ionizing radiation and chemotherapy approaches to cancer therapy destroy dividing cells because tumour cells divide more rapidly than normal cells. These therapies have to be balanced between causing maximum damage to the cancer cells, while keeping toxicity in normal host cells within acceptable levels. Prodrug activation therapy (suicide gene therapy) strives to deliver genes to cancer cells, which convert non-toxic prodrugs into active chemotherapeutic agents. The net gain is that a systemically administered prodrug can be converted in to high local concentrations of an active anticancer drug in the tumour, sparing surrounding normal cells and systemic toxic effects.

The herpes simplex virus thymidine kinase gene (HSV-TK) is the most widely studied enzyme used for prodrug activation experiments. This system is based on the metabolism of certain purine nucleosides [ganciclovir (GCV), aciclovir (ACV) and 1-(2′-deoxy-2′-fluoro-β-D-arabinofuranosyl)-5-iodouracyl (FIAU)] by the HSV-TK gene, which have previously demonstrated clinical efficacy for the treatment of HSV-1 in humans. These prodrug metabolites are able to stall DNA synthesis by inhibiting DNA polymerase and by incorporation into DNA causing chain termination (Elion *et al.*, 1977; Elion *et al.*, 1980; Davidson *et al.*, 1981; Mar *et al.*, 1985), thus killing dividing cells (Elion, 1980). Most of these studies have been carried with thymidine kinase alone, i.e. not in the context of the HSV-1 genome, for example plasmid delivery (Kuriyama *et al.*, 1996; Pope *et al.*, 1997) or other viral delivery: retrovirus (Yoshida *et al.*, 1995, Yang *et al.*, 1996)/adenoviruses (Chen *et al.*, 1994, Matsukura *et al.*, 1999, Tanaka 1997, Wildner *et al.*, 1999; Ohwada *et al.*, 1998; Rosenfeld *et al.*, 1997; DiMaio *et al.*, 1994). These non-HSV genome TK/prodrug studies have shown effective tumour killing *in vitro* and *in vivo* (review in Yazawa *et al.*, 2002). Oncolytic HSV-1 vectors with intact TK should also activate these prodrugs such that infected cells are killed by both the activated prodrug active and oncolysis. This may of course be counterproductive as replication of the oncolytic HSV may also be inhibited. However, early reports in glioma models demonstrated that with a ribonucleotide reductase HSV mutant (hrR3) that administration of GCV did increase animal survival (Mineta *et al.*, 1994; Carroll *et al.*, 1997). In contrast no enhanced cell killing was seen with the same HSV mutant and prodrug on colon carcinoma (Yoon *et al.*, 1998; Carroll *et al.*, 1997), neuroblastoma (Todo *et al.*, 2000), glioma

(Aghi *et al.*, 1999) and liver metastases (Nakamura *et al.*, 2001; Pawlik *et al.*, 2002) *in vivo*. In conclusion most studies have shown very little or no synergy between oncolytic HSV vectors and TK mediated prodrug activation, due to the inhibition of virus replication, which occurs. A number of other suicide genes have been expressed from oncolytic HSV including yeast cytosine deaminase (Nakamura *et al.*, 2001; Simpson *et al.*, 2006), rat p450 CYP2B1 (Chase *et al.*, 1998, Aghi *et al.*, 1999; Ichikawa *et al.*, 2001; Pawlik *et al.*, 2002; Tyminski *et al.*, 2005) and human intestinal carboxylesterase (Tyminski *et al.*, 2005).

E. coli/yeast enzyme cytosine deaminase (CD) and 5-fluorocytosine (5-FC) are a commonly utilized enzyme prodrug combination. CD deaminates the anti-fungal compound 5-fluorocytosine (5-FC) into 5-fluorouracil (5-FU), a highly toxic compound, widely used as a cancer chemotherapeutic agent (Moolten, 1994). This enzyme has been found in prokaryotes and lower eukaryotes, but appears to be absent in higher eukaryotes (Austin and Huber, 1993, Kilstrup *et al.*, 1989). Consequently, mammalian cells are resistant to 5-FC but not 5-FU (Mullen *et al.*, 1992). Expression of yeast CD from an HSV-1 oncolytic virus mediates intra-tumoral conversion of 5-FC to 5-FU, which results in enhanced tumour cell killing compared with the backbone virus alone (Nakamura *et al.*, 2001; Simpson *et al.*, 2006). *In vitro* studies have shown increased killing in cell lines derived from lung cancer (A549, H460), pancreatic cancer (CAPAN-1, MIA PACA-2, BXPC-3) and colon cancer (HCT 116, HT-29, SW620) without inhibiting virus replication (Nakamura *et al.*, 2001; Simpson *et al.*, 2006). It has been suggested that the difference between the effects of GCV and 5-FC relate to differences in the mechanism of action between their respective active metabolites. GCV that has been phosphorylated acts as a termination nucleotide therefore blocking both viral and genomic DNA synthesis. The mechanism of 5-FU cytotoxicity is unclear, because it is converted to several metabolites which each have different biochemical actions (Nakamura *et al.*, 2001). However, much interest has been placed in a metabolite 5-fluorodeoxyuridylate that inhibits thymidylate synthase, which has been suggested to inhibit cellular DNA synthesis more than viral DNA synthesis (Nakamura *et al.*, 2001). OncoVEX$^{GALV/CD}$, described above, expressing both a highly potent version of CD (Fcy::Fur) and the fusogenic glycoprotein from gibbon ape leukaemia virus (GALV) gave the best tumour control of any oncolytic virus so far developed (Simpson *et al.*, 2006).

Expression of the rat p450 enzyme from hrR3 (rRp450), in the presence of cyclophosphamide, has also shown oncolysis and prodrug activation *in vitro* without inhibiting virus replication in both colon carcinoma and glioma cells (Chase *et al.*, 1998; Aghi *et al.*, 1999; Ichikawa *et al.*, 2001; Pawlik *et al.*, 2002; Tyminski *et al.*, 2005). *In vivo* studies on liver metastases and glioma models revealed a substantial decrease in the tumour burden in all animals treated with rRp450 (+cyclophosphamide) compared to controls (Chase *et al.*, 1998; Aghi *et al.*, 1999; Ichikawa *et al.*, 2001; Pawlik *et al.*, 2002). More recently p450 has been co-expressed from an ICP34.5- and ICP6- mutant with another prodrug activating gene, intestinal carboxylesterase (Tyminski *et al.*, 2005). This new oncolytic virus (MGH2) displays increased anti-tumour efficacy against human glioma cell lines (U251, T98G) when combined with the prodrugs, cyclophosphamide and CPT-11 (Tyminski *et al.*, 2005). *In vivo* studies using MGH2 in glioma models have indicated an additive benefit of each of the prodrugs (Tyminski *et al.*, 2005).

7.7 Combination of oncolytic HSV with immunomodulatory gene expression

Oncolytic HSV can have two modes of action, direct oncolysis and an indirect immune effect (review in Varghese and Rabkin, 2002). To enhance this indirect effect a number of groups have explored expressing immune stimulatory molecules, including IL-12 (Bennett *et al.*, 2001; Parker *et al.*, 2000; Wong *et al.*, 2004; Ino *et al.*, 2006; Varghese *et al.*, 2006), IL-18 (Ino *et al.*, 2006; Fukuhara *et al.*, 2005), GM-CSF (Wong *et al.*, 2001; Liu *et al.*, 2003) and soluble B7.1 (Toda *et al.*, 2001; Ino *et al.*, 2006; Fukuhara *et al.*, 2005) from various oncolytic HSV viruses. When

GM-CSF was inserted into the OncoVEX virus (see above) an enhanced anti-tumour effect was seen in immune competent mice where significant effects on both injected and un-injected tumours were seen (Liu *et al.*, 2003). Mice were also then also protected from re-challenge with tumour cells (Liu *et al.*, 2003). Interleukin (IL)-12 expression has demonstrated increased survival in an intracranial neuroblastoma model where significant infiltration of $CD4^+$ and $CD8^+$ T cells and macrophages was shown (Parker *et al.*, 2000). NV1020 has also been used to express IL-12. NV1020-IL-12 was tested on two pulmonary metastasis models (squamous cell carcinoma cancer, prostate cancer) with improved tumour control and survival demonstrated (Wong *et al.*, 2004; Varghese *et al.*, 2006). The efficacy of G47Delta expressing either soluble B7.1 or IL-12 or IL-18 has been tested in the Neuro2a murine tumour model with improved results (Ino *et al.*, 2006). It was also shown that the triple combination of G47Delta-IL-12, G47Delta-IL-18, G47Delta-B7.1 gave the greatest efficacy in a poorly immunogenic tumour model (Ino *et al.*, 2006). An alternate approach to combining viruses that express immuno-stimulatory molecules together, is to express more than one molecule from the same virus. B7.1 and IL-18 were expressed from G47Delta using an internal ribosomal entry site (IRES) sequence which resulted in significantly increased inhibition of the growth of the TRAMP-C2 tumour model as compared to the individual molecules alone (Fukuhara *et al.*, 2005).

7.8 Combination of conventional therapies with oncolytic HSV

Oncolytic viruses such as 1716, G207 and OncoVEX$^{GM\text{-}CSF}$ have shown promise in phase I clinical trials (Rampling *et al.*, 2000; MacKie *et al.*, 2001; Papanatassiou *et al.*, 2002; Harrow *et al.*, 2004; Markert *et al.*, 2000; Hu *et al.*, 2006). Conventional therapies such as chemotherapy or radiotherapy are the mainstay of cancer treatment, but in contrast to oncolytic viruses, they often have a narrow therapeutic index, coupled with limiting high dose toxicities, severely restricting their effectiveness. Combined treatments are often used due to the fact that single agents are more likely to promote the development of resistance. On this basis a number of oncolytic HSVs such as 1716, G207, NV1066, OncoVEX$^{GM\text{-}CSF}$ have been combined with chemotherapeutic agents (Toyoizumi *et al.*, 1999; Chahlavi *et al.*, 1999; Cinatl *et al.*, 2003; Bennett *et al.*, 2004; Mullerad *et al.*, 2005; Adusumilli *et al.*, 2006) or ionizing radiation (Advani *et al.*, 1998; Bradley *et al.*, 1999; Chung *et al.*, 2002).

Enhanced efficacy was seen when HSV-1 G207 or OncoVEX$^{GM\text{-}CSF}$ were combined with cisplatin *in vivo* in head and neck squamous cell carcinoma models (Chahlavi *et al.*, 1999, Han *et al.*, 2006). The G207/cisplatin study resulted in a 100 per cent cure rate in contrast to 42 per cent with G207 or 14 per cent with cisplatin (Chahlavi *et al.*, 1999). It has also been shown that cisplatin induced the expression of an active mammalian cellular homology to the ICP34.5 gene (GADD34), therefore enhancing the replication and cytotoxicity of the virus (Adusumilli *et al.*, 2006). This provides a cellular basis for combination therapy with cisplatin and ICP34.5 mutants to treat malignant pleural mesothelioma (and possibly squamous cell carcinoma) and achieve synergistic efficacy, while minimizing dosage and toxicity (Adusumilli *et al.*, 2006). G207 was also combined with vincristine, resulting in a complete regression of alveolar rhabdomyosarcoma tumours *in vivo* in five of eight animals. In contrast, none of the control animals given single agent therapy were cured (Cinatl *et al.*, 2003). Mitomycin C has been studied in combination with 1716, G207 and NV1066, in non-small cell lung cancer, gastric cancer and bladder cancer. 1716 demonstrated efficacy both *in vitro* and *in vivo* against non-small cell lung cancer cells, which was increased in combination with mitomycin C (Toyoizumi *et al.*, 1999). HSV therapy also synergistically enhanced the cytotoxicity of mitomycin C to two bladder cancer cell lines (Mullerad *et al.*, 2005). This synergistic effect allowed a considerable dose reduction of the two agents, minimizing toxicity (Mullerad *et al.*, 2005). Mitomycin C has demonstrated a similar pharmacological effect to cisplatin in inducing the expression of the cellular GADD34 gene, resulting in the enhanced replication of ICP34.5 deleted HSV in a gastric carcinomatosis model *in vitro* and *in vivo* (Bennett *et al.*, 2004). A detailed study has also demonstrated that OncoVEX combined with paclitaxel in breast cancer cell lines, is truly synergistic *in vitro* (Han *et al.*, 2006).

Ionizing radiation is another primary tool in cancer care. The combination of ICP34.5 mutants with ionizing radiation on both flank and intracranial glioma models has shown synergity resulting in a reduction of tumour volumes and longer survival time as compared to either single treatment alone (Advani *et al.*, 1998; Bradley *et al.*, 1999). The tumour response with G207 and radiation was more ambiguous, as in a study in a prostate cancer model no enhanced efficacy was seen (Jorgensen *et al.*, 2001), whereas in cervical cancer and colorectal cancer models, increased efficacy was seen (Blank *et al.*, 2002; Stanziale *et al.*, 2002; Kim *et al.*, 2005). The increased efficacy in the colorectal cancer model was shown to be dependent on the increased expression of cellular ribonucleotide reductase caused by the ionizing radiation, therefore compensating for the ICP6 gene which is deleted in G207 (Stanziale *et al.*, 2002).

Interestingly, therapeutic doses of both chemotherapy and ionizing radiation do not appear to inhibit the replication of oncolytic HSV, nor add to toxicity. This suggests that a combination of oncolytic HSV and conventional therapies may translate into enhanced efficacy in the clinic.

7.9 Summary

Herpes Simplex virus has a number of properties making it a promising candidate for development as an oncolytic agent for tumour therapy. It can easily be manipulated to allow both tumour selective virus growth and the insertion of therapeutic genes. HSV is a highly lytic virus, which can infect and rapidly kill a broad range of human tumour cell types. In addition, the replication of oncolytic HSVs that retain an intact thymidine kinase gene can also be controlled with standard anti-herpetic drugs if this were found to be necessary during clinical use.

Tumour selective virus growth can be achieved by making a number of mutations to the HSV genome. However, the most effective mutation providing this property is to the gene encoding ICP34.5. This has proved to be safe by intracranial injection in both animals and in human clinical trials. Clinical trials with ICP34.5-mutated HSV have also shown that these viruses are well tolerated by intratumoral injection in head and neck, melanoma and breast cancer patients (Hu *et al.*, 2006; MacKie *et al.*, 2001) and in colorectal cancer liver metastases patients by hepatic artery infusion (Fong *et al.*, 2002). Indications of clinical activity have also been observed (Rampling *et al.*, 2000; Markert *et al.*, 2000; MacKie *et al.*, 2001; Paanatassiou *et al.*, 2002; Fong *et al.*, 2002; Harrow *et al.*, 2004; Hu *et al.*, 2006).

Oncolytic viruses have two modes of action, direct oncolysis and the induction of an antitumour immune response. The antitumour immune response can be enhanced by the deletion of ICP47, which otherwise reduces the levels of MHC class I expression on the surface of the infected cells (Jennings *et al.*, 1985; Hill *et al.*, 1995; York *et al.*, 1994; Fruh *et al.*, 1995). Consequently tumour cells infected with ICP47-mutants are better targeted by the immune system (Todo *et al.*, 2001; Liu *et al.*, 2003). Deletion of the ICP47 gene also places the US11 gene under the control of the ICP47 immediate early promoter, which results in dramatically improved growth properties in tumours without reducing the level of virus attenuation achieved by the deletion of ICP34.5 (Mohr *et al.*, 1996, 2001). Hence, the most potent versions of oncolytic HSV have both ICP34.5 and ICP47 deleted so as to increase the expression of US11 in target cells.

Cancer is commonly treated with a combination of different therapies. In order to provide multimodality therapy with oncolytic HSV, a range of therapeutic genes have been inserted in order to enhance the therapeutic effect. These include fusogenic membrane glycoproteins, prodrug activating genes and immunomodulatory genes. Each of these have enhanced the therapeutic effect seen, and are thus promising for clinical use. One of these viruses, expressing GM-CSF, is currently in phase II clinical trials.

It is important that oncolytic viruses are compatible with traditional therapies if they are to be used together with current treatment approaches. Pre-clinical work has demonstrated that oncolytic HSV therapy in general enhances the effects of a range of chemotherapeutic drugs and radiation therapy, either additively or synergistically depending on the treatment used, and additional toxicity is not usually observed. In no case has it been demonstrated that oncolytic HSV reduces the effects of a pre-existing treatment, which is also an important consideration for clinical use.

In the last 6 years the results of seven phase I clinical trials with oncolytic HSV have been published, which show that such viruses are well tolerated and providing early indications of efficacy in controlling tumour growth. Later stage clinical trials are now underway in a number of tumour types with a number of the viruses discussed above. It is anticipated that at least some of these will begin to demonstrate that oncolytic HSV may be therapeutically beneficial for cancer treatment and that in the future these viruses may be used in routine clinical practice, particularly in tumour types where local control of the tumour would be expected to directly impact patient survival. In summary, the development of oncolytic HSV has reached an exciting stage where whether or not they can be clinically useful will soon be demonstrated.

Acknowledgement

We would like to thank Dr Matt Booth and Dr Georgina Platt for proof reading this manuscript.

References

Adusumilli PS, Chan MK, Chun YS, *et al.* (2006). Cisplatin-induced GADD34 upregulation potentiates oncolytic viral therapy in the treatment of malignant pleural mesothelioma. *Cancer Biol Ther* **5**, 48–53.

Advani SJ, Sibley GS, Song PY, *et al.* (1998). Enhancement of replication of genetically engineered herpes simplex viruses by ionizing radiation: a new paradigm for destruction of therapeutically intractable tumors. *Gene Ther* **5**, 160–165.

Aghi M, Chou TC, Suling K, Breakefield XO, Chiocca EA (1999). Multimodal cancer treatment mediated by a replicating oncolytic virus that delivers the oxazaphosphorine/rat cytochrome P450 2B1 and ganciclovir/herpes simplex virus thymidine kinase gene therapies. *Cancer Res* **59**, 3861–3865.

Andreansky S, Soroceanu L, Flotte ER, *et al.* (1997). Evaluation of genetically engineered herpes simplex viruses as oncolytic agents for human malignant brain tumors. *Cancer Res* **57**, 1502–1509.

Andreansky S, He B, van Cott J, *et al.* (1998). Treatment of intracranial gliomas in immunocompetent mice using herpes simplex viruses that express murine interleukins. *Gene Ther* **5**, 121–130.

Andreansky SS, He B, Gillespie GY, *et al.* (1996). The application of genetically engineered herpes simplex viruses to the treatment of experimental brain tumors. *Proc Natl Acad Sci U S A* **93**, 11313–11318.

Austin EA, Huber BE (1993). Localization of the codA gene on the *Escherichia coli* chromosome. *J Bacteriol* **175**, 3685–3686.

Bateman A, Bullough F, Murphy S, *et al.* (2000). Fusogenic membrane glycoproteins as a novel class of genes for the local and immune-mediated control of tumor growth. *Cancer Res* **60**, 1492–1497.

Beattie E, Denzler KL, Tartaglia J, Perkus ME, Paoletti E, Jacobs BL (1995). Reversal of the interferon-sensitive phenotype of a vaccinia virus lacking E3L by expression of the reovirus S4 gene. *J Virol* **69**, 499–505.

Bennett JJ, Malhotra S, Wong RJ, *et al.* (2001). Interleukin 12 secretion enhances antitumor efficacy of oncolytic herpes simplex viral therapy for colorectal cancer. *Ann Surg* **233**, 819–826.

Bennett JJ, Adusumilli P, Petrowsky H, *et al.* Up-regulation of GADD34 mediates the synergistic anticancer activity of mitomycin C and a gamma134.5 deleted oncolytic herpes virus (G207). *FASEB J* **200418**, 1001–1003.

Black TL, Barber GN, Katze MG (1993). Degradation of the interferon-induced 68 000-M(r) protein kinase by poliovirus requires RNA. *J Virol* **67**, 791–800.

Blank SV, Rubin SC, Coukos G, Amin KM, Albelda SM, Molnar-Kimber KL (2002). Replication-selective herpes simplex virus type 1 mutant therapy of cervical cancer is enhanced by low-dose radiation. *Hum Gene Ther* **13**, 627–639.

Boviatsis EJ, Park JS, Sena-Esteves M, *et al.* (1994). Long-term survival of rats harboring brain neoplasms treated with ganciclovir and a herpes simplex virus vector that retains an intact thymidine kinase gene. *Cancer Res* **54**, 5745–5751.

Bradley JD, Kataoka Y, Advani S, *et al.* (1999). Ionizing radiation improves survival in mice bearing intracranial high-grade gliomas injected with genetically modified herpes simplex virus. *Clin Cancer Res* **5**, 1517–1522.

Brown SM, Harland J, MacLean AR, Podlech J, Clements JB (1994). Cell type and cell state determine differential in vitro growth of non-neurovirulent ICP34.5-negative herpes simplex virus types 1 and 2. *J Gen Virol* **75** (Pt 9), 2367–2377.

Brown SM, MacLean AR, McKie EA, Harland J (1997). The herpes simplex virus virulence factor ICP34.5 and the cellular protein MyD116 complex with proliferating cell nuclear antigen through the 63-amino-acid domain conserved in ICP34.5, MyD116, and GADD34. *J Virol* **71**, 9442–9449.

Cameron JM, McDougall I, Marsden HS, Preston VG, Ryan DM, Subak-Sharpe JH (1988). Ribonucleotide reductase encoded by herpes simplex virus is a determinant of the pathogenicity of the virus in mice and a valid antiviral target. *J Gen Virol* **69** (10), 2607–2612.

Caradonna S, Worrad D, Lirette R (1987). Isolation of a herpes simplex virus cDNA encoding the DNA repair enzyme uracil-DNA glycosylase. *J Virol* **61**(10), 3040–3047.

Carroll K, Elroy-Stein O, Moss B, Jagus R (1993). Recombinant vaccinia virus K3L gene product prevents activation of double-stranded RNA-dependent, initiation factor 2 alpha-specific protein kinase. *J Biol Chem* **268**(17), 12837–12842.

Carroll NM, Chiocca EA, Takahashi K, Tanabe KK (1996). Enhancement of gene therapy specificity for diffuse colon carcinoma liver metastases with recombinant herpes simplex virus. *Ann Surg* **224**(3), 323–329.

Carroll NM, Chase M, Chiocca EA, Tanabe KK (1997). The effect of ganciclovir on herpes simplex virus-mediated oncolysis. *J Surg Res* **69**(2), 413–417.

Cassady KA, Gross M, Roizman B (1998a). The herpes simplex virus US11 protein effectively compensates for the gamma1(34.5) gene if present before activation of protein kinase R by precluding its phosphorylation and that of the alpha subunit of eukaryotic translation initiation factor 2. *J Virol* **72**(11), 8620–8626.

Cassady KA, Gross M, Roizman B (1998b). The second-site mutation in the herpes simplex virus recombinants lacking the gamma134.5 genes precludes shutoff of protein synthesis by blocking the phosphorylation of eIF-2alpha. *J Virol* **72**(9), 7005–7011.

Chahlavi A, Todo T, Martuza RL, Rabkin SD (1999). Replication-competent herpes simplex virus vector G207 and cisplatin combination therapy for head and neck squamous cell carcinoma. *Neoplasia* **1**(2), 162–169.

Chambers R, Gillespie GY, Soroceanu L, *et al.* (1995). Comparison of genetically engineered herpes simplex viruses for the treatment of brain tumors in a scid mouse model of human malignant glioma. *Proc Natl Acad Sci U S A* **92**(5), 1411–1415.

Chase M, Chung RY, Chiocca EA (1998). An oncolytic viral mutant that delivers the CYP2B1 transgene and augments cyclophosphamide chemotherapy. *Nat Biotechnol* **16**(5), 444–448.

Chen SH, Shine HD, Goodman JC, Grossman RG, Woo SL (1994). Gene therapy for brain tumors: regression of experimental gliomas by adenovirus-mediated gene transfer in vivo. *Proc Natl Acad Sci U S A* **91**(8), 3054–3057.

Chou J, Kern ER, Whitley RJ, Roizman B (1990). Mapping of herpes simplex virus-1 neurovirulence to gamma 134.5, a gene nonessential for growth in culture. *Science* **250**(4985), 1262–1266.

Chung SM, Advani SJ, Bradley JD, *et al.* (2002). The use of a genetically engineered herpes simplex virus (R7020) with ionizing radiation for experimental hepatoma. *Gene Ther* **9**(1), 75–80.

Cinatl J, Jr, Cinatl J, Michaelis M, *et al.* (2003). Potent oncolytic activity of multimutated herpes simplex virus G207 in combination with vincristine against human rhabdomyosarcoma. *Cancer Res* **63**(7), 1508–1514.

Clemens MJ (1997). PKR–a protein kinase regulated by double-stranded RNA. *Int J Biochem Cell Biol* **29**(7), 945–949.

Coukos G, Makrigiannakis A, Kang EH, *et al.* (1999). Use of carrier cells to deliver a replication-selective herpes simplex virus-1 mutant for the intraperitoneal therapy of epithelial ovarian cancer. *Clin Cancer Res* **5** (6), 1523–1537.

Cozzi PJ, Burke PB, Bhargav A, *et al.* (2002). Oncolytic viral gene therapy for prostate cancer using two attenuated, replication-competent, genetically engineered herpes simplex viruses. *Prostate* **53**(2), 95–100.

Currier MA, Adams LC, Mahller YY, Cripe TP (2005). Widespread intratumoral virus distribution with fractionated injection enables local control of large human rhabdomyosarcoma xenografts by oncolytic herpes simplex viruses. *Cancer Gene Ther* **12**(4), 407–416.

Davidson RL, Kaufman ER, Crumpacker CS, Schnipper LE (1981). Inhibition of herpes simplex virus transformed and nontransformed cells by acycloguanosine: mechanisms of uptake and toxicity. *Virology* **113**(1), 9–19.

Davies MV, Chang HW, Jacobs BL, Kaufman RJ (1993). The E3L and K3L vaccinia virus gene products stimulate translation through inhibition of the double-stranded RNA-dependent protein kinase by different mechanisms. *J Virol* **67**(3), 1688–1692.

Detta A, Harland J, Hanif I, Brown SM, Cruickshank G (2003). Proliferative activity and in vitro replication of HSV1716 in human metastatic brain tumours. *J Gene Med* **5**(8), 681–689.

Diaz RM, Bateman A, Emiliusen L, *et al.* (2000). A lentiviral vector expressing a fusogenic glycoprotein for cancer gene therapy. *Gene Ther* **7**(19), 1656–1663.

DiMaio JM, Clary BM, Via DF, Coveney E, Pappas TN, Lyerly HK (1994). Directed enzyme pro-drug gene therapy for pancreatic cancer in vivo. *Surgery* **116**(2), 205–213.

Dolan A, McKie E, MacLean AR, McGeoch DJ (1992). Status of the ICP34.5 gene in herpes simplex virus type 1 strain 17. *J Gen Virol* **73**(4), 971–973.

Elion GB, Furman PA, Fyfe JA, de Miranda P, Beauchamp L, Schaeffer HJ (1977). Selectivity of action of an antiherpetic agent, 9-(2-hydroxyethoxymethyl) guanine. *Proc Natl Acad Sci U S A* **74**(12), 5716–5720.

Elion GB (1980). The chemotherapeutic exploitation of virus-specified enzymes. *Adv Enzyme Regul* **18**, 53–66.

Farassati F, Yang AD, Lee PW (2001). Oncogenes in Ras signalling pathway dictate host-cell permissiveness to herpes simplex virus 1. *Nat Cell Biol* **3**(8), 745–750.

Fischer A, Hacein-Bey S, Cavazzana-Calvo M (2002). Gene therapy of severe combined immunodeficiencies. *Nat Rev Immunol* **2**(8), 615–621.

Fong Y, Jarnagin WR, Stanziale SF, *et al.* NV1020 phase I clinical trial, 38th Annual Meeting of the American Society of Clinical Oncology, 2002.

Fruh K, Ahn K, Djaballah H, *et al.* (1995). A viral inhibitor of peptide transporters for antigen presentation. *Nature* **375**(6530), 415–418.

Fu X, Zhang X (2002). Potent systemic antitumor activity from an oncolytic herpes simplex virus of syncytial phenotype. *Cancer Res* **62**(8), 2306–2312.

Fu X, Tao L, Jin A, Vile R, Brenner MK, Zhang X (2003). Expression of a fusogenic membrane glycoprotein by an oncolytic herpes simplex virus potentiates the viral antitumor effect. *Mol Ther* **7**(6), 748–754.

Fukuhara H, Ino Y, Kuroda T, Martuza RL, Todo T (2005). Triple gene-deleted oncolytic herpes simplex virus vector double-armed with interleukin 18 and soluble B7–1 constructed by bacterial artificial chromosome-mediated system. *Cancer Res* **65**(23), 10663–10668.

Fyfe JA, Keller PM, Furman PA, Miller RL, Elion GB (1978). Thymidine kinase from herpes simplex virus phosphorylates the new antiviral compound, 9-(2-hydroxyethoxymethyl)guanine. *J Biol Chem* **253**(24), 8721–8727.

Ghadge GD, Malhotra P, Furtado MR, Dhar R, Thimmapaya B (1994). *In vitro* analysis of virus-associated RNA I (VAI RNA), inhibition of the double-stranded RNA-activated protein kinase PKR by VAI RNA mutants correlates with the *in vivo* phenotype and the structural integrity of the central domain. *J Virol* **68**(7), 4137–4151.

Goldstein DJ, Weller SK (1988). Herpes simplex virus type 1-induced ribonucleotide reductase activity is dispensable for virus growth and DNA synthesis: isolation and characterization of an ICP6 lacZ insertion mutant. *J Virol* **62**(1), 196–205.

Groh V, Steinle A, Bauer S, Spies T (1998). Recognition of stress-induced MHC molecules by intestinal epithelial gammadelta T cells. *Science* **279**(5357), 1737–1740.

Harrow S, Papanastassiou V, Harland J, *et al.* (2004). HSV1716 injection into the brain adjacent to tumour following surgical resection of high-grade glioma: safety data and long-term survival. *Gene Ther* **11** (22), 1648–1658.

Haus O (2000). The genes of interferons and interferon-related factors: localization and relationships with chromosome aberrations in cancer. *Arch Immunol Ther Exp (Warsz)* **48**(2), 95–100.

He B, Gross M, Roizman B (1997). The gamma(1)34.5 protein of herpes simplex virus 1 complexes with protein phosphatase 1alpha to dephosphorylate the alpha subunit of the eukaryotic translation initiation factor 2 and preclude the shutoff of protein synthesis by double-stranded RNA-activated protein kinase. *Proc Natl Acad Sci U S A* **94**(3), 843–848.

Hill A, Jugovic P, York I, Russ G, *et al.* (1995). Herpes simplex virus turns off the TAP to evade host immunity. *Nature* **375**(6530), 411–415.

Horsburgh BC, Hubinette MM, Qiang D, MacDonald ML, Tufaro F (1999). Allele replacement: an application that permits rapid manipulation of herpes simplex virus type 1 genomes. *Gene Ther* **6**(5), 922–930.

Hu JCC, Coffin RS, Davis CJ, *et al.* (2006). A Phase I study with a second generation oncolytic herpes simplex virus expressing GM-CSF (OncoVEX^{GMCSF}), *Clinical Cancer Research* **15**, 6737–47.

Hunter WD, Martuza RL, Feigenbaum F, *et al.* (1999). Attenuated, replication-competent herpes simplex virus type 1 mutant G207: safety evaluation of intracerebral injection in nonhuman primates. *J Virol* **73**(8), 6319–6326.

Ichikawa T, Petros WP, Ludeman SM, *et al.* (2001). Intraneoplastic polymer-based delivery of cyclophosphamide for intratumoral bioconversion by a replicating oncolytic viral vector. *Cancer Res* **61**(3), 864–868.

Imani F, Jacobs BL (1988). Inhibitory activity for the interferon-induced protein kinase is associated with the reovirus serotype 1 sigma 3 protein. *Proc Natl Acad Sci U S A* **85**(21), 7887–7891.

Ino Y, Saeki Y, Fukuhara H, Todo T (2006). Triple combination of oncolytic herpes simplex virus-1 vectors armed with interleukin-12, interleukin-18, or soluble B7-1 results in enhanced antitumor efficacy. *Clin Cancer Res* **12**(2), 643–652.

Jennings SR, Rice PL, Kloszewski ED, Anderson RW, Thompson DL, Tevethia SS (1985). Effect of herpes simplex virus types 1 and 2 on surface expression of class I major histocompatibility complex antigens on infected cells. *J Virol* **56**(3), 757–766.

Jorgensen TJ, Katz S, Wittmack EK, *et al.* (2001). Ionizing radiation does not alter the antitumor activity of herpes simplex virus vector G207 in subcutaneous tumor models of human and murine prostate cancer. *Neoplasia* **3**(5), 451–456.

Kaplitt MG, Tjuvajev JG, Leib DA, *et al.* (1994). Mutant herpes simplex virus induced regression of tumors growing in immunocompetent rats. *J Neurooncol* **19** (2), 137–147.

Kennedy PG (1984). Herpes simplex virus and the nervous system. *Postgrad Med J* **60**(702), 253–259.

Kilstrup M, Meng LM, Neuhard J, Nygaard P (1989). Genetic evidence for a repressor of synthesis of cytosine deaminase and purine biosynthesis enzymes in *Escherichia coli*. *J Bacteriol* **171**(4), 2124–2127.

Kim SH, Wong RJ, Kooby DA, *et al.* (2005). Combination of mutated herpes simplex virus type 1 (G207 virus) with radiation for the treatment of squamous cell carcinoma of the head and neck. *Eur J Cancer* **41**(2), 313–322.

Kooby DA, Carew JF, Halterman MW, *et al.* (1999). Oncolytic viral therapy for human colorectal cancer and liver metastases using a multi-mutated herpes simplex virus type-1 (G207). *FASEB J* **13**(11), 1325–1334.

Kumar A, Yang YL, Flati V, *et al.* (1997). Deficient cytokine signaling in mouse embryo fibroblasts with a targeted deletion in the PKR gene: role of IRF-1 and NF-kappaB. *EMBO J* **16**(2), 406–416.

Kuriyama S, Nakatani T, Masui K, *et al.* (1996). Evaluation of prodrugs ability to induce effective ablation of cells transduced with viral thymidine kinase gene. *Anticancer Res* **16**(5A), 2623–2628.

Langland JO, Pettiford S, Jiang B, Jacobs BL (1994). Products of the porcine group C rotavirus NSP3 gene bind specifically to double-stranded RNA and inhibit activation of the interferon-induced protein kinase PKR. *J Virol* **68**(6), 3821–3829.

Leib DA, Harrison TE, Laslo KM, Machalek MA, Moorman NJ, Virgin HW (1999). Interferons regulate the phenotype of wild-type and mutant herpes simplex viruses *in vivo*. *J Exp Med* **189**(4), 663–672.

Leib DA, Machalek MA, Williams BR, Silverman RH, Virgin HW (2000). Specific phenotypic restoration of an attenuated virus by knockout of a host resistance gene. *Proc Natl Acad Sci U S A* **97**(11), 6097–6101.

Liu BL, Robinson M, Han ZQ, *et al.* (2003). ICP34.5 deleted herpes simplex virus with enhanced oncolytic, immune stimulating, and anti-tumour properties. *Gene Ther* **10**(4), 292–303.

Liu RB, Rabkin SD (2005). Oncolytic herpes simplex virus vectors for the treatment of human breast cancer. *Chin Med J (Engl)* **118**(4), 307–312.

Longnecker R, Roizman B, Meignier B (1988). Herpes simplex virus vectors: properties of a prototype vaccine strain a suitable for use as a vector. In: Longnecker R, Roizman B, Meignier B, Gluzman Y (eds), Cold Spring Harbor Laboratory Press, Cold Spring Harbor, NY, p. 68.

Lu Y, Wambach M, Katze MG, Krug RM (1995). Binding of the influenza virus NS1 protein to double-stranded RNA inhibits the activation of the protein kinase that phosphorylates the elF-2 translation initiation factor. *Virology* **214**(1), 222–228.

MacKie RM, Stewart B, Brown SM (2001). Intralesional injection of herpes simplex virus 1716 in metastatic melanoma. *Lancet* **357**(9255), 525–526.

MacLean AR, ul-Fareed M, Robertson L, Harland J, Brown SM (1991). Herpes simplex virus type 1 deletion variants 1714 and 1716 pinpoint neurovirulence-related sequences in Glasgow strain 17+ between immediate early gene 1 and the 'a' sequence. *J Gen Virol* 72 (Pt 3), 631–639.

Mahller YY, Rangwala F, Ratner N, Cripe TP (2006). Malignant peripheral nerve sheath tumors with high and low Ras-GTP are permissive for oncolytic herpes simplex virus mutants. *Pediatr Blood Cancer* **46**(7), 745–754.

Mar EC, Chiou JF, Cheng YC, Huang ES (1985). Inhibition of cellular DNA polymerase alpha and human cytomegalovirus-induced DNA polymerase by the triphosphates of 9-(2-hydroxyethoxymethyl) guanine and 9-(1,3-dihydroxy-2-propoxymethyl)guanine. *J Virol* **53**(3), 776–780.

Markert JM, Medlock MD, Rabkin SD, *et al.* (2000). Conditionally replicating herpes simplex virus mutant, G207 for the treatment of malignant glioma: results of a phase I trial. *Gene Ther* **7**(10), 867–874.

Martuza RL, Malick A, Markert JM, Ruffner KL, Coen DM (1991). Experimental therapy of human glioma by means of a genetically engineered virus mutant. *Science* **252**(5007), 854–856.

Matsukura N, Hoshino A, Igarashi T, *et al.* (1999). *In situ* gene transfer and suicide gene therapy of gastric cancer induced by *N*-ethyl-*N*′-nitro-*N*-nitrosoguanidine in dogs. *Jpn J Cancer Res* **90**(9), 1039–1049.

McAuliffe PF, Jarnagin WR, Johnson P, Delman KA, Federoff H, Fong Y (2000). Effective treatment of pancreatic tumors with two multimutated herpes simplex oncolytic viruses. *J Gastrointest Surg* **4**(6), 580–588.

McKie EA, MacLean AR, Lewis AD, *et al.* (1996). Selective *in vitro* replication of herpes simplex virus type 1 (HSV-1) ICP34.5 null mutants in primary human CNS tumours – evaluation of a potentially effective clinical therapy. *Br J Cancer* **74**(5), 745–752.

Meignier B, Longnecker R, Roizman B (1988). *In vivo* behavior of genetically engineered herpes simplex viruses R7017 and R7020: construction and evaluation in rodents. *J Infect Dis* **158**(3), 602–614.

Meignier B, Martin B, Whitley RJ, Roizman B (1990). *In vivo* behavior of genetically engineered herpes simplex viruses R7017 and R7020. II. Studies in immunocompetent and immunosuppressed owl monkeys (*Aotus trivirgatus*). *J Infect Dis* **162**(2), 313–321.

Melcher A, Todryk S, Hardwick N, Ford M, Jacobson M, Vile RG (1998). Tumor immunogenicity is determined by the mechanism of cell death via induction of heat shock protein expression. *Nat Med* **4**(5), 581–587.

Mineta T, Rabkin SD, Martuza RL (1994). Treatment of malignant gliomas using ganciclovir-hypersensitive, ribonucleotide reductase-deficient herpes simplex viral mutant. *Cancer Res* **54**(15), 3963–3966.

Mineta T, Rabkin SD, Yazaki T, Hunter WD, Martuza RL (1995a). Attenuated multi-mutated herpes simplex virus-1 for the treatment of malignant gliomas. *Nat Med* **1**(9), 938–943.

Mineta T, Tabuchi K, Martuza RL (1995b). [Gene therapy for brain tumors: experimental treatment of malignant brain tumors using recombinant herpes simplex virus type 1 (series 7)]. *No Shinkei Geka* **23**(4), 285–292.

Mohr I, Gluzman Y (1996). A herpesvirus genetic element which affects translation in the absence of the viral GADD34 function. *EMBO J* **15**(17), 4759–4766.

Mohr I, Sternberg D, Ward S, Leib D, Mulvey M, Gluzman Y (2001). A herpes simplex virus type 1 gamma34.5 second-site suppressor mutant that exhibits enhanced growth in cultured glioblastoma cells is severely attenuated in animals. *J Virol* **75**(11), 5189–5196.

Mohr I (2005). To replicate or not to replicate: achieving selective oncolytic virus replication in cancer cells through translational control. *Oncogene* **24**(52), 7697–7709.

Moolten FL (1994). Drug sensitivity (suicide) genes for selective cancer chemotherapy. *Cancer Gene Ther* **1** (4), 279–287.

Mullen CA, Kilstrup M, Blaese RM (1992). Transfer of the bacterial gene for cytosine deaminase to mammalian cells confers lethal sensitivity to 5-fluorocytosine: a negative selection system. *Proc Natl Acad Sci U S A* **89**(1), 33–37.

Mullerad M, Bochner BH, Adusumilli PS, *et al.* (2005). Herpes simplex virus based gene therapy enhances the efficacy of mitomycin C for the treatment of human bladder transitional cell carcinoma. *J Urol* **174**(2), 741–746.

Mulvey M, Poppers J, Ladd A, Mohr I (1999). A herpesvirus ribosome-associated, RNA-binding protein confers a growth advantage upon mutants deficient in a GADD34-related function. *J Virol* **73**(4), 3375–3385.

Nahmias AJ, Lee FK, Beckman-Nahmias S (1990). Sero-epidemiological and /sociological patterns of herpes simplex virus infection in the world. *Scand J Infect Dis Suppl* **69**, 19–36.

Nakamura H, Mullen JT, Chandrasekhar S, Pawlik TM, Yoon SS, Tanabe KK (2001). Multimodality therapy with a replication-conditional herpes simplex virus 1 mutant that expresses yeast cytosine deaminase for intratumoral conversion of 5-fluorocytosine to 5-fluorouracil. *Cancer Res* **61**(14), 5447–5452.

Nakano K, Todo T, Chijiiwa K, Tanaka M (2001). Therapeutic efficacy of G207, a conditionally replicating herpes simplex virus type 1 mutant, for gallbladder carcinoma in immunocompetent hamsters. *Mol Ther* **3**(4), 431–437.

Nicolo M, Chiocca EA (1998). Marker gene transfer and oncolysis of human Y79 retinoblastoma cells mediated by herpes simplex virus mutants. *Ophthalmic Res* **30**(1), 30–36.

Ohwada A, Hirschowitz EA, Crystal RG (1996). Regional delivery of an adenovirus vector containing the *Escherichia coli* cytosine deaminase gene to provide local activation of 5-fluorocytosine to suppress the growth of colon carcinoma metastatic to liver. *Hum Gene Ther* **7**(13), 1567–1576.

Papanastassiou V, Rampling R, Fraser M, *et al.* (2002). The potential for efficacy of the modified (ICP 34.5 (–)) herpes simplex virus HSV1716 following intratumoural injection into human malignant glioma: a proof of principle study. *Gene Ther* **9**(6), 398–406.

Park H, Davies MV, Langland JO, *et al.* (1994). TAR RNA-binding protein is an inhibitor of the interferon-induced protein kinase PKR. *Proc Natl Acad Sci USA* **91**(11), 4713–4717.

Parker JN, Gillespie GY, Love CE, Randall S, Whitley RJ, Markert JM (2000). Engineered herpes simplex virus expressing IL-12 in the treatment of experimental murine brain tumors. *Proc Natl Acad Sci USA* **97**(5), 2208–2213.

Pawlik TM, Nakamura H, Mullen JT, *et al.* (2002). Prodrug bioactivation and oncolysis of diffuse liver metastases by a herpes simplex virus 1 mutant that expresses the CYP2B1 transgene. *Cancer* **95**(5), 1171–1181.

Pin RH, Reinblatt M, Fong Y (2004). Employing tumor hypoxia to enhance oncolytic viral therapy in breast cancer. *Surgery* **136**(2), 199–204.

Pin RH, Reinblatt M, Fong Y (2004). Utilizing alphafetoprotein expression to enhance oncolytic viral therapy in hepatocellular carcinoma. *Ann Surg* **240**(4), 659–665.

Pope IM, Poston GJ, Kinsella AR (1997). The role of the bystander effect in suicide gene therapy. *Eur J Cancer* **33**(7), 1005–1016.

Poppers J, Mulvey M, Khoo D, Mohr I (2000). Inhibition of PKR activation by the proline-rich RNA binding domain of the herpes simplex virus type 1 Us11 protein. *J Virol* **74**(23), 11215–11221.

Pyles RB, Thompson RL (1994). Evidence that the herpes simplex virus type 1 uracil DNA glycosylase is required for efficient viral replication and latency in the murine nervous system. *J Virol* **68**(8), 4963–4972.

Pyles RB, Warnick RE, Chalk CL, Szanti BE, Parysek LM (1997). A novel multiply-mutated HSV-1 strain for the treatment of human brain tumors. *Hum Gene Ther* **8**(5), 533–544.

Rampling R, Cruickshank G, Papanastassiou V, *et al.* (2000). Toxicity evaluation of replication-competent herpes simplex virus (ICP 34.5 null mutant 1716) in patients with recurrent malignant glioma. *Gene Ther* **7** (10), 859–866.

Reinblatt M, Pin RH, Fong Y (2004). Carcinoembryonic antigen directed herpes viral oncolysis improves selectivity and activity in colorectal cancer. *Surgery* **136**(3), 579–584.

Robertson LM, MacLean AR, Brown SM (1992). Peripheral replication and latency reactivation kinetics of the non-neurovirulent herpes simplex virus type 1 variant 1716. *J Gen Virol* **73**(4), 967–970.

Roizman B, Jenkins FJ (1985). Genetic engineering of novel genomes of large DNA viruses. *Science* **229** (4719), 1208–1214.

Roizman B, Knipe DM (2001). Herpes simplex viruses and their replication. In: Knipe DM, Howley PM (eds), *Fields Virology*. Lippincott Williams & Wilkins, Philadelphia, pp. 2399–2459.

Roizman B, Markovitz N (1997). Herpes simplex virus virulence: the functions of the gamma (1)34.5 gene. *J Neurovirol* **3**(Suppl 1):S1–S2.

Romano PR, Zhang F, Tan SL, *et al.* (1998). Inhibition of double-stranded RNA-dependent protein kinase PKR by vaccinia virus E3: role of complex formation and the E3 N-terminal domain. *Mol Cell Biol* **18**(12), 7304–7316.

Rosenfeld ME, Vickers SM, Raben D, *et al.* (1997). Pancreatic carcinoma cell killing via adenoviral mediated delivery of the herpes simplex virus thymidine kinase gene. *Ann Surg* **225**(5), 609–618.

Saeki Y, Fraefel C, Ichikawa T, Breakefield XO, Chiocca EA (2001). Improved helper virus-free packaging system for HSV amplicon vectors using an ICP27-deleted, oversized HSV-1 DNA in a bacterial artificial chromosome. *Mol Ther* **3**(4), 591–601.

Samoto K, Ehtesham M, Perng GC, *et al.* (2002). A herpes simplex virus type 1 mutant with gamma 34.5 and LAT deletions effectively oncolyses human U87 glioblastomas in nude mice. *Neurosurgery* **50**(3), 599–605.

Sarinella F, Calistri A, Sette P, Palu G, Parolin C (2006). Oncolysis of pancreatic tumour cells by a gamma34.5-deleted HSV-1 does not rely upon Ras-activation, but on the PI 3-kinase pathway. *Gene Ther* **13**(14), 1080–1087.

Sharp TV, Schwemmle M, Jeffrey I, *et al.* (1993). Comparative analysis of the regulation of the interferon-inducible protein kinase PKR by Epstein–Barr virus RNAs EBER-1 and EBER-2 and adenovirus VAI RNA. *Nucleic Acids Res* **21**(19), 4483–4490.

Shizuya H, Birren B, Kim UJ, *et al.* (1992). Cloning and stable maintenance of 300-kilobase-pair fragments of human DNA in *Escherichia coli* using an F-factor-based vector. *Proc Natl Acad Sci U S A* **89**(18), 8794–8797.

Simpson GR, Han Z, Liu B, Wang Y, Campbell G, Coffin RS (2006). Combination of a fusogenic glycoprotein, prodrug activation, and oncolytic herpes simplex virus for enhanced local tumor control. *Cancer Res* **66**(9), 4835–4842.

Smith KD, Mezhir JJ, Bickenbach K, *et al.* (2006). Activated MEK suppresses activation of PKR and enables efficient replication and *in vivo* oncolysis by Deltagamma(1)34.5 mutants of herpes simplex virus 1. *J Virol* **80**(3), 1110–1120.

Stanziale SF, Petrowsky H, Joe JK, *et al.* (2002). Ionizing radiation potentiates the antitumor efficacy of oncolytic herpes simplex virus G207 by upregulating ribonucleotide reductase. *Surgery* **132**(2), 353–359.

Stiles BM, Bhargava A, Adusumilli PS, *et al.* (2003). The replication-competent oncolytic herpes simplex mutant virus NV1066 is effective in the treatment of esophageal cancer. *Surgery* **134**(2), 357–364.

Sundaresan P, Hunter WD, Martuza RL, Rabkin SD (2000). Attenuated, replication-competent herpes simplex virus type 1 mutant G207: safety evaluation in mice. *J Virol* **74**(8), 3832–3841.

Taha MY, Clements GB, Brown SM (1989a). The herpes simplex virus type 2 (HG52) variant JH2604 has a 1488 bp deletion which eliminates neurovirulence in mice. *J Gen Virol* **70**(11), 3073–3078.

Taha MY, Clements GB, Brown SM (1989b). A variant of herpes simplex virus type 2 strain HG52 with a 1.5 kb deletion in RL between 0 to 0.02 and 0.81 to 0.83 map units is non-neurovirulent for mice. *J Gen Virol* **70** (Pt 3):705–716.

Tanaka T (1997). Effect of adenoviral-mediated thymidine kinase transduction and ganciclovir therapy on tumor-associated endothelial cells. *Neurol Med Chir (Tokyo)* **37**(10), 730–737.

Taneja S, MacGregor J, Markus S, Ha S, Mohr I (2001). Enhanced antitumor efficacy of a herpes simplex virus mutant isolated by genetic selection in cancer cells. *Proc Natl Acad Sci U S A* **98**(15), 8804–8808.

Teshigahara O, Goshima F, Takao K, *et al.* (2004). Oncolytic viral therapy for breast cancer with herpes simplex virus type 1 mutant HF 10. *J Surg Oncol* **85** (1), 42–47.

Toda M, Rabkin SD, Kojima H, Martuza RL (1999). Herpes simplex virus as an in situ cancer vaccine for the induction of specific anti-tumor immunity. *Hum Gene Ther* **10**(3), 385–393.

Toda M, Martuza RL, Rabkin SD (2000). Tumor growth inhibition by intratumoral inoculation of defective herpes simplex virus vectors expressing granulocyte–macrophage colony-stimulating factor. *Mol Ther* **2**(4), 324–329.

Toda M, Iizuka Y, Kawase T, Uyemura K, Kawakami Y (2002). Immuno-viral therapy of brain tumors by combination of viral therapy with cancer vaccination using a replication-conditional HSV. *Cancer Gene Ther* **9**(4), 356–364.

Todo T, Rabkin SD, Martuza RL (2000). Evaluation of ganciclovir-mediated enhancement of the antitumoral effect in oncolytic, multimutated herpes simplex virus type 1 (G207) therapy of brain tumors. *Cancer Gene Ther* **7**(6), 939–946.

Todo T, Martuza RL, Rabkin SD, Johnson PA (2001). Oncolytic herpes simplex virus vector with enhanced MHC class I presentation and tumor cell killing. *Proc Natl Acad Sci U S A* **98**(11), 6396–6401.

Toyoizumi T, Mick R, Abbas AE, Kang EH, Kaiser LR, Molnar-Kimber KL (1999). Combined therapy with chemotherapeutic agents and herpes simplex virus type 1 ICP34.5 mutant (HSV-1716) in human non-small cell lung cancer. *Hum Gene Ther* **10**(18), 3013–3029.

Tyminski E, Leroy S, Terada K, *et al.* (2005). Brain tumor oncolysis with replication-conditional herpes simplex virus type 1 expressing the prodrug-activating genes, CYP2B1 and secreted human intestinal carboxylesterase, in combination with cyclophosphamide and irinotecan. *Cancer Res* **65**(15), 6850–6857.

Varghese S, Newsome JT, Rabkin SD, *et al.* (2001). Preclinical safety evaluation of G207, a replication-competent herpes simplex virus type 1, inoculated intraprostatically in mice and nonhuman primates. *Hum Gene Ther* **12**(8), 999–1010.

Varghese S, Rabkin SD (2002). Oncolytic herpes simplex virus vectors for cancer virotherapy. *Cancer Gene Ther* **9**(12), 967–978.

Varghese S, Rabkin SD, Liu R, Nielsen PG, Ipe T, Martuza RL (2006). Enhanced therapeutic efficacy of IL-12, but not GM-CSF, expressing oncolytic herpes simplex virus for transgenic mouse derived prostate cancers. *Cancer Gene Ther* **13**(3), 253–265.

Walker JR, McGeagh KG, Sundaresan P, Jorgensen TJ, Rabkin SD, Martuza RL (1999). Local and systemic therapy of human prostate adenocarcinoma with the conditionally replicating herpes simplex virus vector G207. *Hum.Gene Ther* **10**(13), 2237–2243.

Whitley RJ, Kern ER, Chatterjee S, Chou J, Roizman B (1993). Replication, establishment of latency, and induced reactivation of herpes simplex virus gamma 1 34.5 deletion mutants in rodent models. *J Clin Invest* **91**(6), 2837–2843.

Whitley RJ (2001). Herpes simplex viruses. In: Knipe DM, Howley PM (eds), *Fields Virology*. Lippincott Williams & Wilkins, Philadelphia, pp. 2461–2509.

Wildner O, Blaese RM, Morris JC (1999). Therapy of colon cancer with oncolytic adenovirus is enhanced by the addition of herpes simplex virus-thymidine kinase. *Cancer Res* **59**(2), 410–413.

Wong RJ, Kim SH, Joe JK, Shah JP, Johnson PA, Fong Y (2001). Effective treatment of head and neck squamous cell carcinoma by an oncolytic herpes simplex virus. *J Am Coll Surg* **193**(1), 12–21.

Wong RJ, Patel SG, Kim S, *et al.* (2001). Cytokine gene transfer enhances herpes oncolytic therapy in murine squamous cell carcinoma. *Hum Gene Ther* **12**(3), 253–265.

Wong RJ, Chan MK, Yu Z, *et al.* (2004). Angiogenesis inhibition by an oncolytic herpes virus expressing interleukin 12. *Clin Cancer Res* **10**(13), 4509–4516.

Wu S, Kaufman RJ (1997). A model for the double-stranded RNA (dsRNA)-dependent dimerization and activation of the dsRNA-activated protein kinase PKR. *J Biol Chem* **272**(2), 1291–1296.

Yamada Y, Kimura H, Morishima T, Daikoku T, Maeno K, Nishiyama Y (1991). The pathogenicity of ribonucleotide reductase-null mutants of herpes simplex virus type 1 in mice. *J Infect Dis* **164**(6), 1091–1097.

Yang L, Hwang R, Pandit L, Gordon EM, Anderson WF, Parekh D (1996). Gene therapy of metastatic pancreas cancer with intraperitoneal injections of concentrated retroviral herpes simplex thymidine kinase vector supernatant and ganciclovir. *Ann Surg* **224**(3), 405–414.

Yazawa K, Fisher WE, Brunicardi FC (2002). Current progress in suicide gene therapy for cancer. *World J Surg* **26**(7), 783–789.

Yoon SS, Carroll NM, Chiocca EA, Tanabe KK (1998). Cancer gene therapy using a replication-competent herpes simplex virus type 1 vector. *Ann.Surg* **228**(3), 366–374.

York IA, Roop C, Andrews DW, Riddell SR, Graham FL, Johnson DC (1994). A cytosolic herpes simplex virus protein inhibits antigen presentation to CD8+ T lymphocytes. *Cell* **77**(4), 525–535.

Yoshida K, Kawami H, Yamaguchi Y, *et al.* (1995). Retrovirally transmitted gene therapy for gastric carcinoma using herpes simplex virus thymidine kinase gene. *Cancer* **75**(6 Suppl), 1467–1471.

Yu Z, Eisenberg DP, Singh B, Shah JP, Fong Y, Wong RJ (2004). Treatment of aggressive thyroid cancer with an oncolytic herpes virus. *Int J Cancer* **112**(3), 525–532.

Yuwen H, Cox JH, Yewdell JW, Bennink JR, Moss B (1993). Nuclear localization of a double-stranded RNA-binding protein encoded by the vaccinia virus E3L gene. *Virology* **195**(2), 732–744.

8

Selective Tumour Cell Cytotoxicity by Reoviridae – Preclinical Evidence and Clinical Trial Results

Laura Vidal, Matt Coffey and **Johann de Bono**

8.1 Introduction

The Reoviridae are a family of viruses that include viruses that infect the gastrointestinal and respiratory systems. The name Reoviridae was first recognized in 1959 and is a derivation of respiratory enteric orphan viruses. 'Orphan viruses' are viruses that are not associated with any known disease. Although the Reoviridae family has been associated with a number of diseases, the original name remains in use. The family Reoviridae consists of six genera, three of which infect animals (including humans) named rotavirus, orbivirus and reovirus, with the remaining three infecting only plants and/or insects. Rotaviruses are an important cause of infantile diarrhoea and enteritis as well as being significant veterinary pathogens; Orbiviruses are predominately veterinary pathogens with the one notable exception being the virus that causes Colorado tick fever virus (CTFV), which is capable of infecting human subjects. Reovirus is, however, believed to be associated with minimal respiratory or enteric symptoms in humans and is also not known to cause any significant veterinary pathology (Tyler and Fields, 1996). Reovirus is ubiquitous in terms of geographic distribution. Isolates have been obtained throughout the world from untreated sewage, river and stagnant waters (Stanley, 1974; Adams *et al.*, 1982; Ridinger *et al.*, 1982). This wide distribution probably explains why exposure to reovirus is very common in the human population. However, reoviral infections are considered to be benign and usually asymptomatic in humans.

Reovirus has also been shown to selectively induce tumour cytotoxicity. This with the lack of pathogenicity has led to the evaluation of reovirus as a promising oncolytic therapy. Clinical trials are underway to evaluate the potential of this oncolytic virus as an anticancer agent.

8.2 Reovirus structure

The most unique feature of the Reoviridae family is the presence of a double-stranded RNA (dsRNA) genome. They all have similar morphology: non-enveloped, icosahedral capsid with a double shell of proteins.

Human reovirus contains 10 segments of dRNA in three distinct size classes: L – encodes proteins designated λ; M – encodes proteins designated μ; and S – encodes proteins designated σ. The reovirus virions measure approximately 85 nm in diameter and are composed of eight structural

Viral Therapy of Cancer Edited by Kevin J. Harrington, Richard G. Vile and Hardev S. Pandha

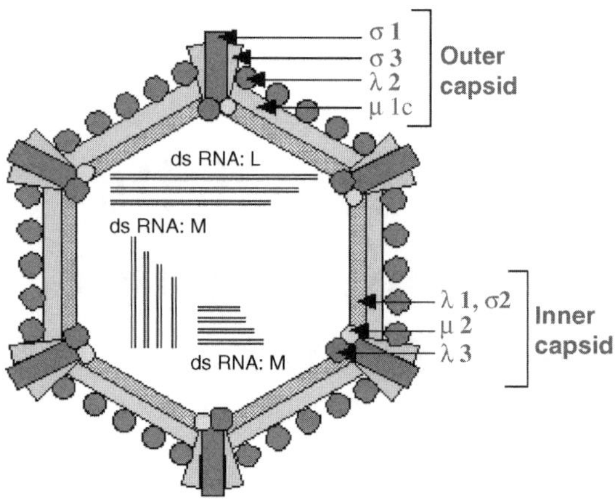

Figure 8.1 Reovirus structure. Picture donated by Dr Alan J. Cann, Department of Biology, University of Leicester (Microbiology @ Leicester)

proteins (Figure 8.1). Five of these ($\lambda 1$, $\lambda 2$, μ2, $\sigma 2$, $\lambda 3$) compose the inner capsid protein (ICP), known as the core, which remains intact after viral penetration. The core is an elaborate transcriptionally competent machine whose crystal structure has been described (Reinisch *et al.*, 2000). The $\lambda 1$ shell encloses the 10 dsRNA segments of the viral genome and the components of the viral transcriptase/replicase, $\lambda 3$ and μ2. The $\sigma 2$ protein functions as another main constituent of the core, while also anchoring the outer shell to the core and participating in core assembly (Coombs *et al.*, 1994). The reovirus ICP transcribes plus-strand copies from each of the ten genomic segments packaged within it, adds a methylated guanosine cap to the 5′ end of each transcript and exports the mature mRNA into the cytoplasm of the infected cells (Borsa *et al.*, 1981)). The fifth component of the core includes $\lambda 2$, which form pentameric turret-like structures, functioning in assembly of the outer capsid and in RNA capping (Dryden *et al.*, 1993)). The synthesized mRNAs are capped the instant they emerge from the active site of the polymerase. The guanosine cap is essential for RNA stability and for the ribosome to recognize the viral RNA. The function of the outer-shell proteins, μ1, $\sigma 2$ and $\sigma 3$ is to introduce the ICP into the cytoplasm. Approximately 600 heterodimeric complexes of the $\sigma 3$ and μ1 proteins compose the virion outer shell. The remaining $\sigma 1$ protein serves as the viral attachment protein (Furlong *et al.*, 1988). This $\sigma 1$ protein is a long, fibre-like molecule with head and tail morphology and several defined regions of flexibility within its tail. The tail is inserted in the turrets formed by the $\lambda 2$ pentons, whereas the head projects away from the virion surface (Fraser *et al.*, 1990).

Other dsRNA viruses possess structures similar to the $\lambda 1$ shell, and this appears to be a common feature of nearly all viruses in the Reoviridae family. The other seven structural proteins, particularly those of the outer shell appear to vary among dsRNA viruses, associated with different strategies for infection and spread. In particular, there are three serotypes of reovirus (type 1 Lang, type 2 Jones, type 3 Abney and type 3 Dearing), based on their antibody neutralization and haemagglutination–inhibitory activity. These strains also differ in targeting distinct cell types *in vivo* and their potential for inducing apoptosis in murine models. These differences are determined by the viral S1 gene that encodes the viral attachment protein $\sigma 1$ (Tyler *et al.*, 1995).

8.3 Reovirus replication

The reovirus infection cycle proceeds through different stages: attachment, endocytic uptake, outer shell uncoating and transmembrane penetration into the cytoplasm, where replication occurs (Nirbert *et al.*, 1996). Reovirus cell attachment is mediated by the outer capsid protein $\sigma 1$. Viral entry then occurs following receptor-mediated endocytosis, after reovirus binds to the cell surface sialic acid (Paul *et al.*, 1989) and junctional adhesion molecule-1 (JAM-1) (Barton *et al.*, 2001). JAM is a transmembrane protein overexpressed in many cell types that are targeted for reovirus infection (Liu *et al.*, 2000). Within the endosome, virions are converted to infectious subvirion particles (ISVP). This involves degradation of the outer capsid $\sigma 3$ protein by endosomal proteases, the breaking of $\mu 1$ protein into small fragments and conformational changes of the cell attachment protein $\sigma 1$. The infectivity of these ISVPs is similar to that of the virions (Nirbert *et al.*, 1995). ISVPs can originate intracellularly within the endosomes of infected cells or through the action of other proteolytic extracellular enzymes such as the ones found in the gastrointestinal tract. The ISVP need to be further processed in the endosomes to form the non-infectious but transcriptionally active core particle. These processes permit the ISVPs to fuse to the endosomal membrane, presumably through the action of the exposed μ1 protein and facilitate the core particle penetration to the host cell cytoplasm (Tosteson *et al.*, 1993). The $\sigma 3$ protein serves as a protective cap for μ1 in virions, and its previous removal by endosomal proteolysis facilitates membrane penetration by ISVP.

Transcription occurs within the core or inner capsid; the core particle is able to activate the viral RNA-dependent RNA polymerase. Then, the distinct (+)-sense capped primary transcripts corresponding to each of the ten genomic dsRNA (+) strands exit from the core into the cytoplasm through channels in the $\lambda 2$ core spike pentons (Tao *et al.*, 2002). Primary transcripts and protein products are then brought together to form RNA which undergoes secondary transcription and translation to result in the generation of more virions for release (Figure 8.2).

8.4 Reovirus and human infection

Reovirus has been isolated from the human respiratory and enteric tracts. This infection is usually asymptomatic, but can infrequently be associated with upper respiratory symptoms that resemble flu-like symptoms or mild gastrointestinal symptoms such as diarrhoea. Approximately 70–100 per cent of the adult population has anti-reovirus antibodies, reflecting previous exposure to reovirus probably during infancy, as shown by seroconversion data (less than 25 per cent in children <5 years old, to greater than 50 per cent in those 20–30 years old) (Jackson and

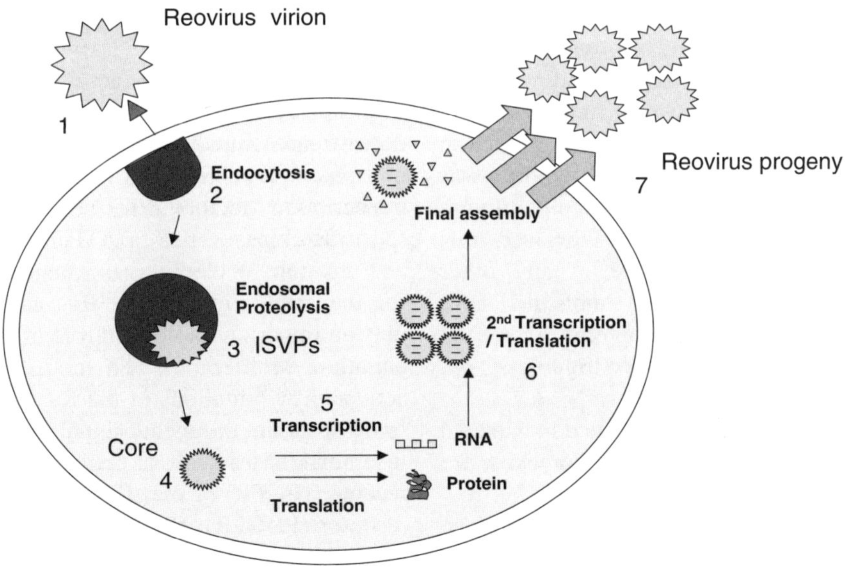

Figure 8.2 Reovirus cell cycle. The reovirus infection cycle proceeds through different stages: (1) cell attachment via $\sigma 1$ interactions; (2) receptor-mediated endocytosis; (3) conversion of virions to infectious subvirion particles (ISVP) within the endosome through the degradation of the outer capsid σ 3 protein and conformational changes of the cell attachment protein σ 1; (4) processing of ISVP in the endosomes to form the non-infectious but transcriptionally active core particle, penetrating to the host cell cytoplasm; (5) primary transcription occurs within reovirus cores through activation of the viral RNA-dependent RNA polymerase and release of capped mRNAs; (6) formation of RNA from primary transcripts and protein products are then brought together to form RNA which undergo secondary transcription and translation to (7) release of newly generated viral particles.

Muldoon, 1973, Stanley, 1974). There are several studies that support the lack of significant reovirus pathology in healthy adults. Trials in human volunteers demonstrated that reovirus infection resulted in minor upper respiratory symptoms (Rosen, 1963). In this study, 27 healthy males were administered intranasal inoculations of one of the three serotypes of reovirus and monitored for 23 days for signs of symptomatic illness. Of the nine men receiving reovirus type 1, only three showed signs or symptoms of illness, which included malaise, rhinorrhoea, cough, sneezing, pharyngitis and headache. These signs and symptoms typically had an onset within 24 to 48 h and lasted from 4 to 7 days. There was also one report of 'loose stools' in another volunteer. Reovirus type 2 inoculations were associated with similar signs, symptoms and durations in three of the nine volunteers with such exposure. Two of the nine volunteers receiving reovirus type 3 inoculations developed mild rhinitis. Overall, none of the subjects showed any other physical signs or symptoms, or any significant laboratory abnormality. Microbiological examinations of nose, throat, and faeces were all negative for viral shedding.

An aetiological link between human central nervous system disease and reovirus infection has been postulated after serotype 3 reovirus was isolated in post-mortem samples from a child with meningitis (Tyler *et al.*, 2004). However, this finding has not been confirmed in any other studies. Reovirus has been also implicated as a cause of upper respiratory symptoms, diarrhoea and exanthemas in children (Rodriguez *et al.*, 1977).

Reoviridae have also been implicated in neonatal extrahepatic biliary atresia (EHBA) (Morecki *et al.*, 1982). Preclinical *in vivo* studies indicate that reovirus infection induces the oily fur syndrome in newborn mice (Wilson *et al.*, 1994), which results in lipid-rich stools and characteristic oily hair due to biliary obstruction. Results from human studies trying to associate both events have, however, been unable to provide further evidence for this and no consensus has been reached. Large studies have reported a 62 per cent incidence of antibodies to reovirus serotype 3 in infants with extra hepatic biliary atresia, 52 per cent incidence of reovirus titers in patients with idiopathic neonatal hepatitis and <12 per cent in normal infants or those with other cholestatic disorders (Glaser *et al.*, 1984). However, it remains difficult to speculate about the causal contribution of reovirus to the aetiology of these diseases. Overall, the causal association of these isolated cases of hepatobiliary, neurological, respiratory or exanthematous diseases with reovirus infection remains uncertain and may reflect the wide distribution of reovirus in the environment.

8.5 Oncolytic activity

The first description of the oncolytic properties of these viruses reported that wild type reovirus replicated in certain transformed cell lines but not in normal cells (Hashiro *et al.*, 1977). Since then, several groups have confirmed that wild type reovirus is selectively replication competent, replicating preferentially in cells with an activated Ras pathway, either through Ras mutation or upregulated growth factor receptor signaling (Coffey *et al.*, 1998; Strong *et al.*, 1998). Ras is a small G protein, which when activated by guanosine triphosphate (GTP), leads to the activation of downstream signalling involving Raf kinases, phosphatidylinositol 3-kinase (PI3-kinase), the transcription factors Elk1 and nuclear factor (NF)-κB, kinases p38 and Janus kinase (JNK) and guanine nucleotide exchange factor (GEF) for the small protein Ral. Ras and its effectors play an important biological role in cellular differentiation, proliferation and motility. Ras can be activated by mutations of the Ras proto-oncogene or by upstream mitogenic signals, notably tyrosine receptor kinases such as epidermal growth factor receptor (EGFR) or platelet-derived growth factor receptor (PDGFR). Ras activating mutations promote angiogenesis, metastasis and loss of growth control and are present in 30–40 per cent of all human tumours (Bos, 1989), particularly in 80–90 per cent of pancreatic cancers, 40–50 per cent of colorectal cancers, 50 per cent of thyroid tumours, 30 per cent of myeloid leukaemias and 15–24 per cent of lung cancers.

Overexpressed or activated (e.g. by mutation) upstream mitogenic signals can result in elevated

Ras activity and may render cells permissive to reovirus replication. Examples of this include overexpression of EGFR in approximately 90 per cent of head and neck cancers, of PDGFR in 40 per cent of glioblastomas and of c-erbB2 in 25–30 per cent of breast cancers (Ben-Levy *et al.*, 1994; Hermanson *et al.*, 1996; Rogers *et al.*, 2005). Preclinical studies indicated that NIH-3T3 cells, naturally resistant to reovirus infection, became highly permissive to reovirus infection when transformed with v-erb oncogene, a truncated EGFR lacking the ligand binding extracellular domain but containing a constitutively active tyrosine kinase cytoplasmatic domain (Strong *et al.*, 1993). The activated EGFR triggers signalling of the Ras pathway and NIH-3T3 cells transformed by activated Ras become susceptible to reovirus infection. Further preclinical studies have identified the Ras/RalGEF/p38 as the major downstream molecular signalling pathway in determining host cell susceptibility to reovirus mediated oncolysis (Norman *et al.*, 2004).

The mechanisms behind the permissiveness of transformed cells to reovirus infection have not been fully elucidated. It has been reported, however, that this effect may be mediated, at least in part, through the failure of the activation of the double-stranded RNA-activated protein kinase (PKR) in Ras-activated cells that are exposed to reovirus infection (Meurs *et al.*, 1990). PKR is a serine/threonine kinase whose enzymatic activation requires dsRNA binding and consequent phosphorylation (Vorbuger *et al.*, 2004). PKR is implicated in transcriptional regulation, cell differentiation, signal transduction and tumour suppression although its main role is to contribute to the antiviral and antiproliferative response of interferon (IFN) following viral infections. PKR is expressed at low levels in non-IFN primed cells, but is significantly upregulated in response to IFN. When reovirus infects normal cells, PKR is activated and causes the arrest of viral protein translation, leading to the inhibition of cellular and viral protein synthesis. In contrast, in Ras activated cells, PKR remains inactivated because it is not phosphorylated and cannot abort viral translation, resulting in viral replication leading to oncolysis (Figure 8.3).

The innate immune response to viral infection results in a block in cellular translation. In the presence of dsRNA, whose concentration increases as a consequence of viral infection, the IFN-stimulated protein PKR, dimerizes and is activated via *trans*-autophosphorylation. Activated PKR stops translation initiation by phosphorylating serine 51 (S51) in the alpha-subunit of eIF-2α. This phosphorylation increases the affinity of eIF-2α for guanine nucleotide exchange factor eIF-2B and thus prevents the recycling of GDP for GTP. Since phosphorylated and GDP-bound eIF-2α can-

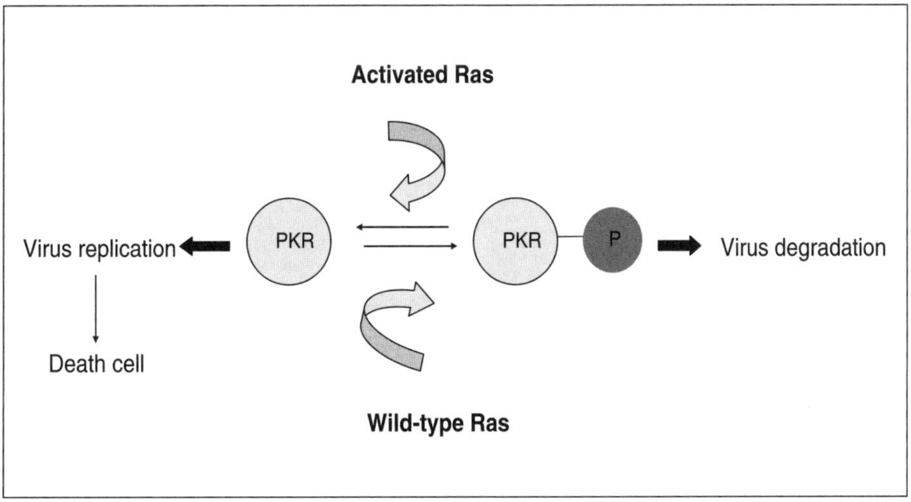

Figure 8.3 Ras signalling and reovirus infection

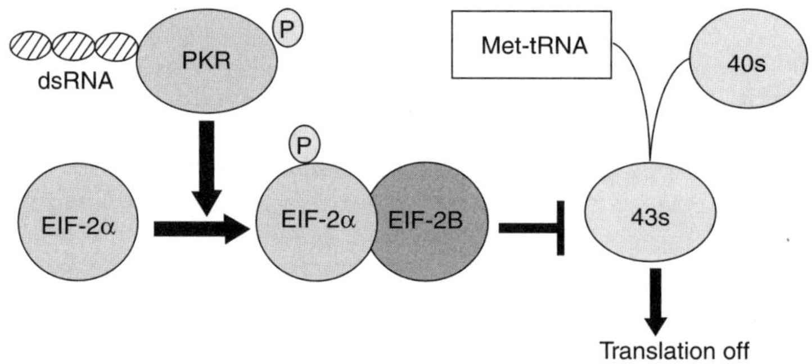

Figure 8.4 PKR activation by reovirus infection leads to viral translation inhibition

not participate in the formation of the 43S pre-initiation complex, translation initiation is halted (Figure 8.4).

Although many data support the hypothesis that PKR is responsible for the inhibition of cellular and viral protein synthesis during reovirus infection, other investigations have recently demonstrated that some strains are also able to inhibit translation in a PKR-independent manner. Whereas infection with some strains (such as Jones) leads to a dramatic or moderate inhibition of translation when PKR is expressed, other strains such as type 3 Dearing are minimally inhibited when PKR is expressed suggesting the involvement of alternative pathways. Although PKR activation is most commonly associated with its ability to phosphorylate eIF-2α, there are data indicating that PKR can also impact translation through its effects on other signalling pathways including NF-κB, signal transducer and activator of transcription (STAT)-1 and -3, mitogen-activated protein kinase kinases (MAPKK) 4/7 and 3/6 and protein phosphatase 2A (Figure 8.5) (Williams, 2001). This may explain why some

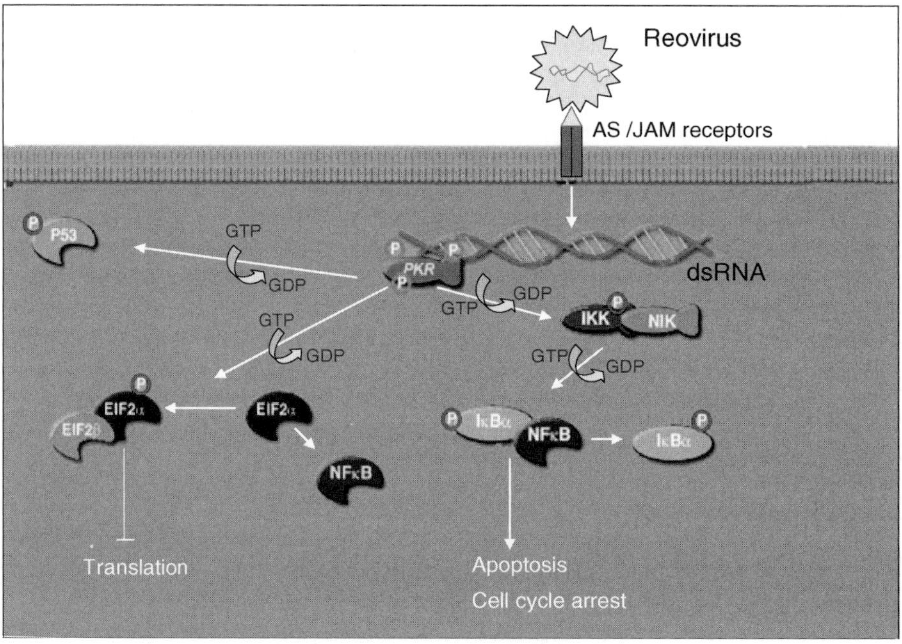

Figure 8.5 PKR signalling pathway

strains can inhibit cellular translation when eIF-2α is not phosphorylated. Additionally, other IFN-regulated gene products such as RNase L can down-regulate cellular protein synthesis in reovirus-infected cells in the absence of PKR. RNase L is a component of the interferon antiviral system that, when activated, cleaves single strand RNA and interferes with translation by degrading mRNA and/or RNA (Smith *et al.*, 2005). Preclinical data have, however, also shown that some reovirus strains can replicate despite the presence of PKR and RNAase L (Smith *et al.*, 2005).

The mechanisms developed by reovirus to overcome the IFN-stimulated antiviral system have yet to be elucidated. It is unclear if by decreasing cellular protein synthesis, the activity of these IFN-regulated gene products creates an environment that favours the translation of reovirus transcripts, thus increasing progeny virion production.

8.6 Mechanisms of reovirus-induced cytotoxicity

8.6.1 Apoptosis

To induce apoptosis, reovirus may require the activation of cell surface death receptors (DR), specifically DR4 or 5, by their ligand, tumour necrosis factor (TNF)-related apoptosis-inducing ligand (TRAIL). Preclinical data have demonstrated that reovirus can induce apoptosis in some cancer cell lines and that this apoptosis can be inhibited by TRAIL receptor activation blockade (Clarke *et al.*, 2001). Reovirus can also sensitize cancer cells to TRAIL-induced apoptosis (Clarke *et al.*, 2002). These findings may, however, be reovirus strain and cell line specific. Reovirus-infected cells can also release TRAIL, which can bind DR in an autocrine fashion to induce apoptosis through the activation of the initiator caspase and its downstream effectors (Kominsky *et al.*, 2002). Reovirus-induced apoptosis may require the cooperation of the NF-κB complex, which upregulates the expression of pro-apoptotic genes (Clark *et al.*, 2003). NF-κB regulates both TRAIL and its receptors.

Early in the reovirus replication cycle, the initial binding of reovirus to both cell-surface sialic acid (SA) and the JAM receptors has also been reported to be essential for apoptosis induction. In these studies, reovirus infection induced by SA-binding alone is not, however, sufficient to induce apoptosis in the absence of JAM-induced signalling (Barton *et al.*, 2001) because it is the direct interaction between reovirus $\sigma 1$ and JAM that initiates the signalling cascade that leads to NF-κB activation. Overall, reovirus strains (particularly serotype T3) which bind to both JAM-1 and SA on the host cell surface induce higher levels of tumour cell apoptosis than other strains which only bind to JAM-1 (Conolly *et al.*, 2001, Weiner *et al.*, 1988)

8.6.2 Cell cycle disruption

The earliest effect of infection by all reovirus serotypes is the inhibition of DNA synthesis. Reoviruses also inhibit proliferation by inducing G2/M cell cycle arrest. Some reovirus strains (including serotype 3 Dearing and Abney) result in greater inhibition of proliferation than other serotypes. It has been reported that this ability to inhibit DNA synthesis and induce G2/M phase cell cycle arrest is related to the s1 gene (Pogglioli *et al.*, 2000); $\sigma 1$ also inhibits cdc2 kinase activity which is responsible for the transition from G2 to M, although the mechanism for this effect is still unclear. Although the $\sigma 1$ protein is required for reovirus-induced G2/M arrest, the latter occurs in the absence of apoptosis, suggesting that reovirus-induced inhibition of proliferation involves a different pathway (Poggioli *et al.*, 2001). Serotype 3 reovirus infection also alters microtubule function *in vitro* (Babiss *et al.*, 1979).

8.7 Preclinical experience

Reovirus has been shown to induce oncolysis in a wide range of human cancer cell lines including colorectal, pancreatic, malignant glioma, ovarian and breast carcinomas known to express activated Ras either by mutation or by activated upstream signaling. *In vivo* studies have also shown that reovirus induces tumour regression in animal xenograft models (Hirasawa *et al.*, 2003). In initial xenograft studies with v-erb transformed NIH-3T3 cells, single intratumoral administration of reovirus resulted in more than 80 per cent tumour regression in approximately 75 per cent of the xenografts (Coffey *et al.*, 1998). The U87 human malignant glioma cell line, which possesses high levels of activated RAS due to PDGFR overexpression, has been used to study reovirus antitumour activity. Severe combined immunodeficient (SCID) mice were implanted with U87 cells subcutaneously in the hind limb and received a single intratumoral injection of reovirus. Tumour regression was observed in 80 per cent of mice in this study, with no remaining viable tumour at the injection site or in other tissues (Coffey *et al.*, 1998). Further preclinical *in vivo* studies have also demonstrated the inhibition of tumour invasion and metastases following intracerebral injection with no evidence of detectable neurotoxicity (Yang *et al.*, 2003). Morphological correlation between areas of viral infection and the presence of apoptosis have been reported after intracerebral reovirus administration (Oberhaus *et al.*, 1997). Reovirus also induces an oncolytic effect in immunocompetent and syngeneic models of liver metastases from pancreatic cancer xenografts (Himeno *et al.*, 2005).

Reovirus also has antitumour activity in hematopoietic tumour models *in vivo*. Single doses of reovirus injected intravenously and intratumorally in SCID mice induced tumour regression in different primary lymphoid tumour xenografts (Alain *et al.*, 2002). Additionally, *in vitro* and *ex vivo* data have demonstrated that reovirus can purge haematopoietic stem cell harvests of contaminating tumour cells with no detectable toxicity to normal human stem cells (Thirukkumaran *et al.*, 2003).

8.8 Immunogenicity

Systemic administration of reovirus has been shown to induce a brisk host immune response both *in vitro* and *in vivo* in xenograft models. The development of neutralizing antireovirus antibodies following reovirus administration increases viral clearance and negatively impacts viral delivery to tumour.

To evaluate the effect that the immune response has on reovirus oncolytic activity, immunocompetent mice were treated with an intramuscular injection of reovirus. This resulted in detectable antibody after two weeks, after which the mice were implanted with C3 allografts. Two weeks later, immunized and non-immunized mice were treated with repeated intratumoral administration of reovirus. Complete regressions were reported at the site of injection, with no sign of toxicity. There was no difference in antitumour activity between immunized and non-immunized mice (Coffey *et al.*, 1998). However, the efficacy of reovirus decreased when administered intravenously to immunocompetent mice. When these immunocompetent mice were treated with reovirus and concomitant immunosuppressant therapy, a decrease in tumour burden and improved survival was observed (Hirasawa *et al.*, 2002). These data are relevant to ongoing clinical trials since data from the first systemic phase I trial have demonstrated that all patients developed high levels of neutralizing anti-reovirus antibody titres following reovirus intravenous infusion (Vidal *et al.*, 2005). The abrogation of this reovirus immune response by immunosuppression may expose tumour to a higher effective viral dose, increasing reovirus replication and enhancing oncolytic activity.

8.9 Clinical experience

To date, three clinical trials with this oncolytic virus have been completed. The first was a phase I clinical trial evaluating the intralesional administration of reovirus serotype 3 Dearing strain (Reolysin®) to patients with histologically confirmed cutaneous metastases from advanced cancer. The dose ranges tested were single injections of 1×10^7 to 1×10^{10} tissue culture infectious dose ($TCID_{50}$) with no dose-limiting toxicity being observed in 19 patients. The toxicities observed were grade 2 or less and comprised occasional transient flu like symptoms and headache. Evidence of antitumour activity was seen in several patients, including one complete regression and one partial regression of injected lesions (Morris *et al.*, 2002). The second study involved direct intratumoral injection of 5×10^9 $TCID_{50}$ into previously untreated prostatic cancers (T2) in patients scheduled to have total prostatectomies three weeks later. All six treated patients tolerated therapy well with no delay in time to prostatectomy for any patient. Toxicity was minimal and involved flu-like symptoms such as fever, rhinorrhoea and myalgias.

There are four further phase I trials currently ongoing: two administering reovirus intratumorally and two trials with reovirus delivered intravenously to patients with metastatic solid tumours. The first involves intratumoral injection of Reolysin directly into recurrent intracerebral malignant gliomas. Twelve patients have completed treatment to date with no dose-limiting toxicity described.

A trial of intratumoral reovirus injections in combination with local palliative radiotherapy is also ongoing. To date, seven patients with a variety of tumour types have been treated with increasing doses of reovirus from 1×10^8 to 1×10^{10} $TCID_{50}$ on days 2 and 4, with concomitant radiotherapy at a dose of 20 Gy/fraction. Minimal toxicity has been observed during reovirus treatment. There has been no induction of neutralizing anti-reovirus antibodies thus so far. Local responses and a systemic response outside the radiation field have been observed (Vidal *et al.*, 2006).

Two separate phase I trials are exploring the intravenous administration of reovirus to determine the feasibility, safety, interaction with the immune system and oncolytic activity of this agent. In the first trial, 29 patients have been treated to date (Vidal *et al.*, 2005). Doses from 1×10^8 to 3×10^{10} $TCID_{50}$ have been administered for 1, 3 and 5 consecutive days every 28 days (defined as a cycle). No dose-limiting toxicity (DLT) has been observed and the maximum tolerated dose (MTD) has not been defined. Toxicity has been mild with fever, chills, flu-like symptoms, fatigue, rash and asymptomatic lymphopenia. No evidence of reovirus excretion has been observed in blood, urine, saliva and stool by reverse transcriptase polymerase chain reaction. Early and robust induction of neutralizing antireovirus antibodies has been observed on day 7 in all patients with these titres remaining high during subsequent cycles of treatment. A patient with hormone refractory prostate cancer had a 50 per cent fall in prostate specific antigen associated with radiological evidence of tumour necrosis. Tumour necrosis and a large number of reovirus particles were observed in biopsy material from a metastatic node. Further decrements in carcinoembryonic antigen in two patients with colorectal cancer have been observed with disease stabilization for 3 and 6 months.

8.10 Conclusions

Reovirus is a selectively cytotoxic wild type virus that seems to fulfil many of the criteria for an oncolytic agent. Preclinical data have clearly shown that reovirus selectively replicates and kills tumour cells with an activated Ras signalling pathway. Clinical trials have demonstrated that reovirus is well tolerated when administered to cancer patients, either intratumorally or intravenously. Evidence of reovirus replication in tumour cells has been acquired from tumour biopsies from patients treated with systemic reovirus. Finally, preliminary evidence of clinical antitumour activity has been observed.

However, the utility of this potential anticancer agent needs to be further optimized. The robust antiviral tumoral immune response observed after systemic reovirus administration may compromise reovirus delivery to tumour and limit its oncolytic activity. Concomitant treatment with immunosuppressants may suppress these innate antiviral responses (Ikeda *et al.*, 1999) and decrease plasma levels of circulating neutralising antireovirus antibodies. Recent data have shown, however, that the cellular immune response mediated by regulatory T cells following the death of normal cells could also restrain the host autoimmune response against tumour antigens. Strategies to modulate the immune response should be taken into consideration when using viral combinations with immunosuppressive agents to maximize antitumour activity.

Finally, tumour types with activated Ras such as pancreas, colorectal cancer or glioblastoma multiforme represent attractive targets for reovirus therapy. It is likely, however, that reovirus will be most effective when administered in combination with standard cytotoxic agents to circumvent cross-resistance and increase therapeutic effect. Such combinations now need to be explored in the clinic

References

Adams DJ, Spendelove JC, Spendelove RS, *et al.* (1982). Aerosol stability of infectious and potentially infectious reovirus particles. *Appl Environ Microbiol* **44**, 903–908.

Alain T, Hirasawa K, Pon KJ, *et al.* (2002). Reovirus therapy of lymphoid malignancies. *Blood* **100**, 4146–4153.

Babiss LE, Luftig RB, Weatherbee JA, *et al.* (1979). Reovirus serotypes 1 and 3 differ in their in vitro association with microtubules. *J Virol* **30**, 863–874.

Barton ES, Forrest JC, Connolly JL, *et al.* (2001). Junction adhesion molecule is a receptor for reovirus. *Cell* **104**(3), 441–451.

Ben-Levy R, Paterson HF, Marshall CJ, *et al.* (1994). A single autophosphorylation site confers oncolgenicity to the neu/ErbB-2 receptor and enables coupling to the MAP kinase pathway. *EMBO J* **13**(14), 3302–3311.

Borsa J, Sargent MD, Lievaart PA, *et al.* (1981). Reovirus: evidence for a second step in the intracellular uncoating and transcriptase activation process. *Virology* **111**(1), 191–200.

Bos JL (1989). Ras oncogenes in human cancer: a review. *Cancer Res* **49**, 4682–4689.

Clark P, Maintzer S, Moffitt L, *et al.* (2003). Two distinct phases of virus-induced nuclear factor kB regulation enhance tumor necrosis factor-related apoptosis inducing ligand-mediated apoptosis in virus-infected cell. *J Bio Chem* **278**, 18092–18100.

Clarke P, Meintzer S, Gibson S (2001). Reovirus-induced apoptosis is mediated by TRIAL. *J Virol* **74**, 8135–8139.

Clarke P, Meintzer S, Spalding A, *et al.* (2002). Caspase-8 dependent sensitization of cancer cells to TRIAL-induced apoptosis following reovirus-infection. *Oncogene* **20**, 6910–6919

Coffey MC, Strong JE, Forsyth P, *et al.* (1998). Reovirus as an agent against tumours with activated Ras Pathway. *Science* **282**, 1332–1334.

Conolly JL, Barton ES, Dermody TS (2001). Reovirus binding to cell surface sialic acid potentiates virus-induced apoptosis. *J Virol* **75**, 4029–4039.

Coombs KM, Mak SC and LD Petrycky-Cox (1994). Studies of the major reovirus core protein sigma 2: reversion of the assembly-defective mutant tsC447 is an intragenic process and involves back mutation of Asp-383 to Asn. *J Virol* **68**(1), 177–186

Dryden KA, Wang G, Yeager M, *et al.* (1993). Early steps in reovirus infection are associated with dramatic changes in supramolecular structure and protein conformation: analysis of virions and subviral particles by cryoelectron microscopy and image reconstruction. *J Cell Biol* **122**(5), 1023–1041.

Fraser RD, Furlong DB, Trus BL, *et al.* (1990). Molecular structure of the cell-attachment protein of reovirus: correlation of computer-processed electron micrographs with sequence-based predictions. *J Virol* **64**(6), 2990–3000.

Furlong DB, Nibert ML, Fields BN (1988). Sigma 1 protein of mammalian reoviruses extends from the surfaces of viral particles. *J Virol* **62**(1), 246–56.

Glaser JH, Balistrei WF, Morecki R (1984). Role of reovirus type 3 in persistent infantile cholestasis. *J Pediatr* **105**, 912–915

Hashiro G, Loh PC, Yau CT (1977). The preferential cytotoxicity of reovirus for certain transformed cell lines. *Arch Virol* **54**(4), 307–315.

Hermanson M, Funa K, Koopmann J, *et al.* (1996). Association of loss of heterozygosity on chromosome 17p with high platelet-derived growth factor alpha receptor expression in human malignant gliomas. *Cancer Res* **56**(1), 164–171.

Himeno Y, Etoh T, Matsuto T, *et al.* (2005). Efficacy of oncolytic reovirus againts liver metastases from pancreatic cancer in immunocompetent models. *Int J Oncol* **27**, 901–906

Hirasawa K, Hirasawa K, Nishikawa SG, *et al.* (2002). Oncolytic reovirus against ovarian and colon cancer. *Cancer Res* **62**, 1696–1671.

Hirasawa K, Nishikawa SG, Norman KL, *et al.* (2003). Systemic reovirus therapy of metastatic cancer in immune-competent mice. *Cancer Res* **63**, 348–353.

Jackson GG, Muldoon RL (1973). Virus causing common respiratory infection in man IV reoviruses and adenoviruses. *J Infect Dis* **128**(6), 811–823.

Ikeda K, Ichikawa T, Wakimoto H, *et al.* (1999). Oncolytic virus therapy of multiple tumors in the brain requires suppression of innate and elicited antiviral responses. *Nat Med* **8**, 881–887.

Kominsky DJ, Bickel RJ, Tayler KL (2002). Reovirus induced apoptosis requires both death receptor- and mitochondrial-mediated caspase-dependent pathways of cell death. *Cell Death Differ* **9**, 926–933.

Liu Y, Nusrat A, Schnell FJ, *et al.* (2000). Human junction adhesion molecule regulates tight junction resealing in epithelia. *J Cell Sci* **113**, 1–11.

Meurs E, Chong K, Galabro J *et al.* (1990). Molecular cloning and characterization of the human double-stranded RNA-activated protein kinase induced by interferon. *Cell* **2**, 379–390.

Morecki R, Glaser JH, Cho S, *et al.* (1982). Biliary atresia and reovirus type 3 infection. *N Engl J Med* **19**, **307**(8), 481–484.

Morris DG, Forsyth PA, Paterson AH, *et al.* (2002). A phase I clinical trial evaluating intralesional Reolysin (reovirus) in histologically confirmed malignancies. *Proc Am Soc Clin Oncol* 24a:92.

Nirbert ML, Chappell JD and Dermody TS (1995). Infectious subviron particles of reovirus type 3 Dearing exhibit a loss in infectivity and contain a cleaved σ1 protein. *J Virol* **69**, 5057–5059.

Nirbert ML, Schiff LA, Fields BN (1996). *Reovirus and their replication*. In: Fields BN, Knipe DM, Howley PM (eds), *Fundamental Virology*. Lippincott-Raven, Philadelphia, PA, pp. 691–730.

Norman KL, Hirasawa K, Yang AD, *et al.* (2004). Reovirus oncolysis: the Ras/RalGEF/p38 pathway dictates host cell permissiveness to reovirus infection. *Proc Natl Acad Sci U S A* **101**, 11099–11104.

Oberhaus SM, Smith RL, Clayton GH, Dermody TS, Tyler KL (1997). Reovirus infection and tissue injury in the mouse central nervous system are associated with apoptosis. *J Virol* **17**, 2100–2106.

Paul RW, Choi AH, Lee PW (1989). The alpha-anomeric form of sialic acid is the minimal receptor determinant recognized by reovirus. *Virology* **172**(1), 382–385.

Pogglioli G, Keefer C, Connolly J, *et al.* (2000). Reovirus-induced G2/M cell cycle arrest requires δ1s and occurs in the absence of the apoptosis. *J Virol* **74**, 9562–9570.

Pogglioli GJ, Dermody TS, Tyler KL (2001). Reovirus-induced δ1-dependent G2/M phase cell cycle arrest is associated with inhibition of p34(cdc2). *J Virol* **75** (16), 7429–7434.

Reinisch KM, Nirbert ML, Harrison SC (2000). Structure of the reovirus core at 3.6 Å resolution. *Nature* **404**, 960–967.

Ridinger DN, Spendelove RS, Barnett BB (1982). Evaluation of cell lines and immunofluorescence and plaque assay procedures for qualifying reovirus in sewage. *Appl Environ Microbiol* **43**, 740–746.

Rodriguez WJ, Kim HW, Arrobio JO, *et al.* (1977). Clinical features of acute gastroenteritis associated with human reovirus-like agent in infants and young children. *J Pediatr* **91**(2), 188–193.

Rogers SJ, Harrington KJ, Rhys-Evans P, *et al.* (2005). Biological significance of c-erbB family oncogenes in head and neck cancer. *Cancer Met Rev* **24**(1), 47–69.

Rosen L (1963). Reovirus infections in human volunteers. *Am J Hyg* **77**, 29–37.

Smith JA, Schmechel SC, Williams BRG, *et al.* (2005). Involvement of the interferon-regulated antiviral proteins PKR and RNase L in reovirus-induced shutoff of cellular translation. *J Virol* **79**(4), 2240–2250.

Stanley NF (1974). In: Reovirus for the prevention of neoplasia. In: Kurstak E, Kurstak K (eds), *Comparative Diagnosis of Viral Diseases*. Academic Press, New York, pp. 385–421.

Stanley NF, (1974). *Comparative Diagnosis of Viral Diseases*. Academic Press, New York.

Strong JE, Tang D, Lee PWK (1993). Evidence that the epidermal growth factor receptor on host cells confers reovirus infection efficiency. *Virology* **197**(1), 405–411.

Strong JE, Coffey MC, Tang D, *et al.* (1998). The molecular basis of viral oncolysis; usurpation of the Ras signalling pathway by reovirus. *EMBO J* **17**, 3351–3362.

Tao Y, Farsetta DL, Nibert ML, *et al.* (2002). RNA synthesis in a cage-structural studies of reovirus polymerase lambda3. *Cell* **111**(5), 733–745.

Thirukkumaran CM, Luider JM, Stewart DA, *et al.* (2003). reovirus oncolysis as a novel purging strategy for autologous stem cell transplantation. *Blood* **102**, 377–387.

Tosteson MT, Nibert ML, Fields BN (1993). Ion channels induced in lipid bilayers by subvirion particles of the nonenveloped mammalian reoviruses. *Proc Natl Acad Sci U S A* **90**(22), 10549–10552.

Tyler KL, Fields BN (1996). Reoviruses. In: Fields BN, Knipe DM, Howley PM (eds), *Fields Virology*, 3rd edition. Lippincott-Raven, Philadelphia, PA, pp. 1597–1623

Tyler KL, Squier MK, Rodgers SE, *et al.* (1995). Differences in the capacity of reovirus strains to induce apoptosis are determined by the viral attachment protein sigma 1. *J Virol* **69**(11), 6972–6979.

Tyler KL, Barton ES, Ibach ML, *et al.* (2004). Isolation and molecular characterization of a novel type 3 reovirus from a child with meningitis. *J Infect Dis* **189**(9), 1664–1675.

Vidal L, Pandha HS, Harrington KJ, *et al.* (2005). A phase I study of wild-type reovirus (Reolysin) given intravenously to patients with advanced malignancies. *Proc Am Assoc Cancer Res* **204**, C31.

Vidal L, Twigger K, White CL, *et al.* (2006). Phase I trial of intratumoral administration of reovirus type 3

(Reolysin) in combination with radiation in patients with advanced malignancies *Pro Am Assoc Cancer Res* **171**, 3998.

Vorbuger SA, Pataer A, Swiher SG, *et al.* (2004). Genetically targeted cancer: therapy tumor destruction by PKR activation. *Am J Pharmacogenom* **4**(3), 180–198.

Weiner DB, Girard K, Williams WV, *et al.* (1988). Reovirus type 1 and type 3 differ in their binding to isolated intestinal epithelial cells. *Micro Pathol* **5** (1), 29–40.

Williams BR (2001). Signal integration via PKR. *Sci STKE* **89**, RE2.

Wilson GA, Morrison LA, Fields BN (1994). Association of the reovirus S1 gene with serotype 3-induced biliary atresia in mice. *J Virol* **68**(10), 6458–6465.

Yang WQ, Senger D, Muzik H, *et al.* (2003). Reovirus prolongs survival and reduces the frequency of spinal and leptomeningeal metastases from medulloblastoma. *Cancer Res* **63**, 3162–3172.

9

Oncolytic vaccinia

M. Firdos Ziauddin and **David L. Bartlett**

9.1 Introduction

Vaccinia virus has a long history of use in humans. In 1776, Edward Jenner inoculated a young boy with cowpox isolated from the hand of a milkmaid. He then demonstrated that this boy was protected from smallpox (Moss, 1996). Vaccination against smallpox was eventually widely adopted. In the 1930s, it was determined that the strain of virus being used for vaccination was distinct from cowpox. This strain was later identified as vaccinia. As a result of the smallpox vaccination programme, smallpox was eradicated worldwide by 1977.

In the decades following the eradication of smallpox, vaccinia has had continued utility in research and vaccine development. With the development of recombinant DNA technology, it became possible to genetically engineer the virus for use as an expression vector in these applications.

Vaccinia possesses several characteristics that make it an effective expression vector. It has a large genome, into which 25 kb of foreign DNA can be inserted without deletion of any viral sequence. It has strong promoters to drive high levels of transgene expression. It has a wide host range, and can infect almost all cell types. Vaccinia has a quick, efficient life cycle, and spreads efficiently from cell to cell. As an expression vector, vaccinia has found clinical applicability in development of vaccines against unrelated infectious diseases and cancer.

It has also been recognized that vaccinia has several characteristics that make it a good candidate for use as an oncolytic virus. Cells infected by vaccinia are rapidly destroyed. Vaccinia has a natural tumour tropism and can be made very tumour-selective by deletion of certain genes. It has a well-known safety profile. In addition, its power as an expression vector can be harnessed to make it a more effective anticancer therapeutic. The focus of this chapter is the development of vaccinia as a tumour-selective oncolytic virus.

9.2 Biology of vaccinia virus

Vaccinia virus is a member of the Poxviridae family. Based on host range, sequence homology, and antigenicity, poxviruses are further classified into two major subfamilies, Chordopox (vertebrate pox) and Entomopox (insect pox). Many strains of vaccinia developed during the era of worldwide smallpox vaccination. Regional differences in production and maintenance of the virus resulted in various strains with different characteristics. The New York City Board of Health (NYCBH) strain was obtained from England in 1856 and originally used for smallpox vaccination in the US (Fenner *et al.*, 1989). The Western Reserve (WR) strain is a

Viral Therapy of Cancer Edited by Kevin J. Harrington, Richard G. Vile and Hardev S. Pandha

laboratory derivative of this strain and is one of the more virulent strains in laboratory animals and non-human primates. It has not yet been utilized in humans. Another derivative of the NYCBH strain, the Wyeth strain, produced by Wyeth Laboratories as a smallpox vaccine, has been the backbone commonly used for experimental vaccines in clinical trials. The modified vaccinia Ankara (MVA) strain was developed through multiple rounds of infection in avian cells. This strain is highly attenuated and does not replicate in human and other mammalian cells (Sutter and Moss, 1995). The attenuated strains usually result from genetic mutations, such as deletion of viral genes, and so have genomes of smaller sizes in comparison with the WR strain.

The genome of vaccinia virus has been fully sequenced and found to contain double-stranded DNA with inverted terminal repeats and a terminal hairpin loop that mimics a large circular single-stranded DNA. The genome consists of 191 636 bp and encodes for 2063 proteins of 65 or more amino acids (Goebel *et al.*, 1990; Antoine *et al.*, 1998).

Like other poxviruses, vaccinia is a large, brick-shaped particle (350 nm in diameter), consisting of outer lipoprotein membranes surrounding a complex core structure (Moss, 1996). The core structure contains the viral DNA and several virally encoded enzymes, including a multisubunit DNA-dependent RNA polymerase, transcription factor, capping and methylating enzymes, and poly(A) polymerase. These enzymes enable synthesis of viral mRNAs.

Vaccinia, like other poxviruses, spends its entire life cycle in the cytoplasm, and does not integrate into the host genome (Figure 9.1) The virus has very few interactions with host cellular proteins, allowing for rapid, efficient replication without negative effects from host cell defences (Moss and Earl, 1998; Guerra *et al.*, 2003)

There are two infectious forms of vaccinia, the intracellular mature virion (IMV) and the extracellular enveloped virion (EEV). The IMV is released when infected cells are lysed and is the form harvested for use in the laboratory. The EEV is the form released from the cell by membrane

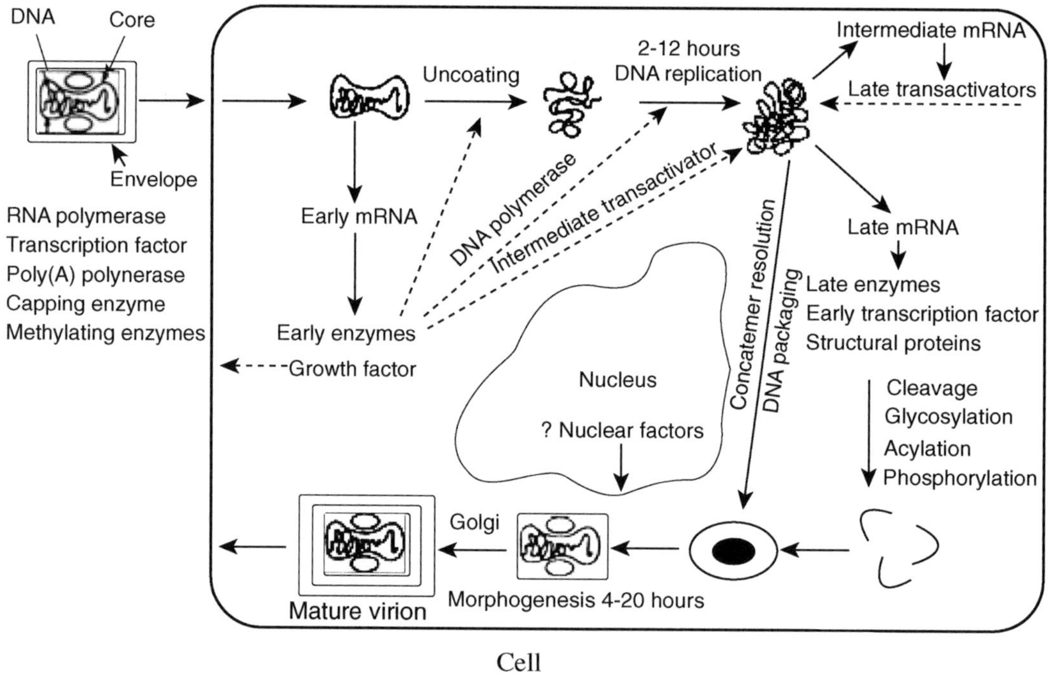

Figure 9.1 Vaccinia life cycle. Reproduced from *Science* 1991;252:1662–1667 courtesy of: American Association for the Advancement of Science

fusion and is responsible for cell-to-cell spread. The EEV is too fragile to withstand the purification process in the laboratory (Smith and Vanderplasschen, 1998).

Vaccinia attachment and uptake likely involve the A27L and D8L proteins found in the IMV membrane. These proteins seem to bind heparin sulfate and chondroitin sulfate on the cell surface (Chung *et al.*, 1998; Hsiao *et al.*, 1999). Viral fusion with the cell is quickly followed by release of the viral transcriptional enzymes that transcribe early viral mRNA. Early mRNA typically encodes for proteins involved with uncoating of the viral DNA and transcriptional factors for intermediate mRNA production. Intermediate mRNA encodes for late transactivators that lead to late mRNA synthesis. Late proteins include viral structural proteins and early transcriptional factors to be incorporated into the mature virion. Viral DNA replication occurs, forming concatemers, which are then resolved into individual genomes and assembled into mature virions. The mature virion contains three membranes after assembly. The outer membrane fuses with the cell membrane, leading to release of the double-membrane viral particle (the EEV). The EEV remains attached to the cell surface through the A34R gene product, facilitating cell-to-cell spread of the virus without release into the bloodstream (McIntosh and Smith, 1996).

Vaccinia exhibits efficient replication and rapid cell-to-cell spread. It induces a profound cytopathic effect soon after viral entry, as early viral enzymes inhibit host cell function. Four to 6 h after infection, there is almost complete inhibition of host protein synthesis, allowing very efficient expression of viral genes and viral replication. $\sim$10 000 copies of the viral genome are produced within 12 h of infection. Half of these copies are incorporated into mature virions and released.

9.3 Tumour selectivity and antitumour effect

Vaccinia has a known ability to induce tissue destruction by replicative necrosis. It produces vaccinia necrosum, a spreading necrotic ulcer, in immunosuppressed hosts. The WR strain is more efficient in this process than the vaccine used clinically. After intradermal injection of 10^6 plaque-forming units (p.f.u.) in rhesus macaques, a spreading necrotic ulcer of 108 cm^2 develops around the injection site in only 8 days, without systemic spread of the virus. This ability to rapidly spread and cause tissue destruction makes vaccinia unique among current oncolytic viral vectors in clinical and preclinical development. The goal in developing vaccinia as an anticancer therapeutic is to direct its tissue-destructive ability selectively to tumour cells.

Vaccinia has a natural tumour tropism. After intravenous injection of the WR strain, the highest amount of virus is found in the tumour, followed by the ovary, with minimal to no virus in other tissues (Whitman *et al.*, 1994). The mechanism of this natural tropism has not been established.

Vaccinia is a large virus and may require a leaky vasculature for access to tissue. Historically, smallpox virus has been known to have tropism for injured and irritated skin (Ricketts, 1996). The virus may have relatively easy access to this inflamed tissue because of leaky vasculature due to histamine release. The tropism of the virus to tumour and ovarian follicles may be related to leaky vasculature in these tissues, which are known to be sites of vascular endothelial growth factor (VEGF) production (Neeman *et al.*, 1997).

Hyperthermia has been studied as a condition that can increase the permeability of the vasculature to vaccinia and augment vaccinia delivery to tumours after systemic injection (Chang *et al.*, 2005). Hyperthermia increases the permeability of the vascular endothelium to nanoparticles. It has been demonstrated that hyperthermia does not alter tumour cells' susceptibility to the intrinsic cytopathogenicity of the virus compared to normothermia, nor does it change viral infectivity or levels of viral marker gene expression. In an *in vitro* model of endothelial cell monolayer permeability, hyperthermia was shown to reversibly increase the permeability of the monolayer to vaccinia.

The effects of hyperthermia on viral delivery, marker gene activity, and antitumour response were tested in an *in vivo* model of subcutaneous MC38 flank tumours in immunocompetent mice. Vaccinia was administered systemically under

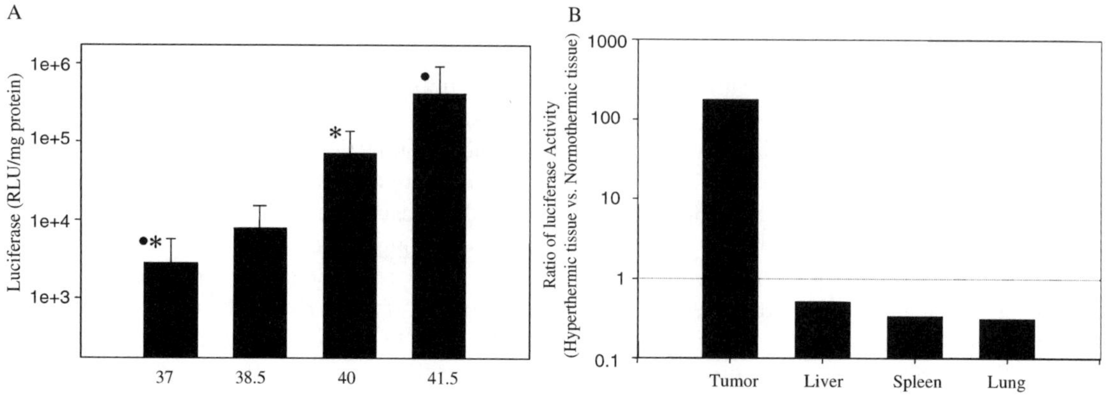

Figure 9.2 Enhancement by hyperthermia of transgene expression and vaccinia extravasation into tumour tissue. Reproduced from *Hum Gene Ther* 2005;16(4):435–444, with permission from Mary Ann Liebert, Inc

conditions of local tumour hyperthermia (41.5°C for 30 min). Following this treatment, there were significantly higher levels of vaccinia marker gene activity (>100-fold) in tumours treated under hyperthermic conditions than those treated under normothermic conditions. This effect was tumour-specific (Figure 9.2).

A greater antitumour effect was also demonstrated. Fifty per cent of mice with 1-cm subcutaneous tumours treated with systemic vaccinia and hyperthermia had complete tumour regression. Thirteen days after treatment, the mean tumour volume in mice treated under hyperthermic conditions was 110 mm^3, compared to 3169 mm^3 in mice treated under normothermic conditions. Mice treated with hyperthermia alone were included as controls and did not have the tumour regression demonstrated in the group treated with both vaccinia and hyperthermia. These results point to the importance of viral extravasation for successful treatment (Chang *et al.*, 2005).

Another characteristic of tumours which may make them more suitable for viral replication compared to normal tissue is the relatively greater fraction of dividing cells. Dividing cells have an accessible pool of nucleotides from which viruses may draw for construction of new virions. In fact, poxviruses encode for a protein, vaccinia growth factor (VGF), which is secreted by infected cells and binds growth factor receptors on surrounding resting cells, stimulating them to proliferate in order to prime them for vaccinia infection (Buller *et al.*, 1988). Given this favorable milieu in dividing cells, one might then expect viral infection and toxicity in other dividing cells *in vivo*, e.g. bone marrow-derived cells and gastrointestinal mucosa. There is toxicity to these normal tissues by chemotherapy which targets dividing cells. Vaccinia, however, when given systemically in murine, rat, rabbit, or primate studies, is not toxic to gastrointestinal mucosa nor to bone marrow-derived cells. Viral gene expression is not evident and there is no viral recovery from gastrointestinal mucosa. Vaccinia is recovered from bone marrow, but the significance of this finding is unclear. Animals dying of vaccinia infection do not have neutropenia or thrombocytopenia (McCart *et al.*, 2000). The presence of dividing cells in a tumour seems to support viral replication, but it does not completely explain the viral tropism to the tumour.

Deleting genes which are required for vaccinia virus replication in non-dividing cells may make the virus more specific for tumour cells. One such gene is the thymidine kinase (TK) gene which is essential for the synthesis of deoxythymidine monophosphate (dTMP) and deoxyuridine monophosphate (dUMP) for DNA and RNA synthesis. In normal cells, the expression of this gene has been found to be dependent on the phase of the cell cycle, peaking during S phase. In transformed cells, however, TK activity is greater than that in normal cells during all phases of the cell cycle (Hengtschlager

Table 9.1 Tissue luciferase activity (RLU/mg protein) after intravenous delivery of 10^6 vaccinia luciferase Reproduced from *Monographs in Virology*. 2001. Volume 22, pp. 130–159, with permission from S. Karger AG (Basel).

Tumour model	Tumour	Ovary	Liver	Lung
Adenocarcinoma liver metastases in immunocompetent mice	46 000 000	1450	—	600
Subcutaneous sarcoma in rat	4337	0.74	0.023	0.056
VX-2 liver metastases in rabbit	2103	132	9	7
Human melanoma in athymic mice	558 000	78 000	215	963

et al., 1994). Deletion of the TK gene results in viral dependence on host cell nucleotides, which are more readily available in dividing cells than in resting cells. A TK-deleted vaccinia virus has been shown to be tumour selective in several tumour models, including murine colon cancer and melanoma, rat sarcoma, human colon cancer in nude mice, and rabbit kidney cancer (Gnant *et al.*, 1999a; Puhlmann *et al.*, 2000; Bartlett *et al.*, 2001) (Table 9.1) In subcutaneous tumour models, it has been shown to replicate in the tumour and lead to an antitumor response (McCart *et al.*, 2000). The antitumour response is enhanced in athymic, nude mice, indicating that the response is directly related to intratumoral vaccinia replication and not to a bystander inflammatory response (Gnant *et al.*, 1999b).

In an effort to further increase tumour specific viral replication, a vaccinia virus was constructed with the vaccinia growth factor (VGF) gene deleted in parallel with the TK gene. As discussed above, VGF is encoded by vaccinia and stimulates proliferation of surrounding cells. During infection with a TK-deleted virus, this cell proliferation with its attendant increase in available nucleotides may compensate for the loss of viral TK. Furthermore, cell signalling initiated by VGF has been shown to be important for viral replication and pathogenicity. VGF binds to growth factor receptors, such as EGFR, and initiates signaling through the Ras/MAPK/ERK pathway. This signaling is necessary for viral replication (de Magalhaes *et al.*, 2001). Inhibition of this pathway blocks vaccinia pathogenicity (Yang *et al.*, 2005). In tumour cells, however, a dysregulated EGFR pathway may allow viral replication.

It has been found that this double-deleted virus (vvDD) does indeed have markedly enhanced tumour specificity (McCart *et al.*, 2001); 10^9 p.f.u. given systemically to a nude mouse caused no toxicity, while causing significant regression of established subcutaneous tumours (Figure 9.3) Mice given the same dose of wild type virus had a median survival of 5 days. vvDD was not pathogenic when 10^9 p.f.u. were administered IV to rhesus macaques (Zeh *et al.*, 2002). Unlike the WR strain, 10^6 p.f.u. of vvDD inoculated intradermally to rhesus macaques did not cause replicative necrosis. While the virulence of the vvDD was diminished relative to the WR strain, replication in tumours was not reduced, and tumour-specificity was increased. Eight days after intravenous delivery, similar titres of the WR strain and vvDD could be recovered from subcutaneous tumours in mice, but there were markedly reduced titres of vvDD compared to the wild type in normal tissues (Table 9.2) The tumour selectivity, antitumor activity, and low virulence of this double-deleted vector suggest that it is a promising candidate for an anticancer therapeutic.

Other strategies to achieve tumour-selective replication of vaccinia have been investigated.

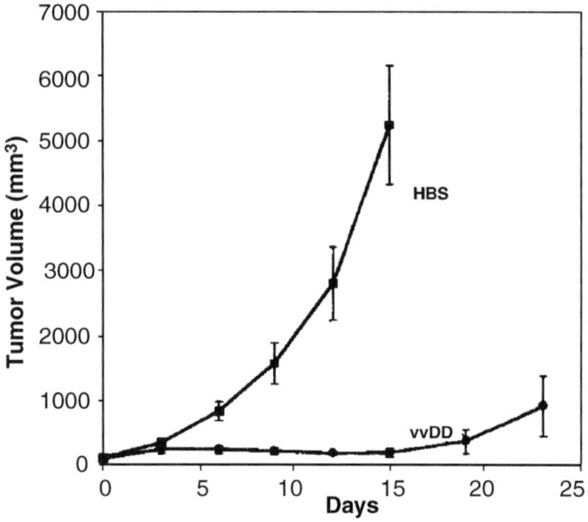

Figure 9.3 Mean tumor volume after intraperitoneal treatment of subcutaneous MC38 in nude mice. On day 0, 8 days after inoculation with tumour, mice were injected with 10^9 p.f.u. of replicating vvDD-GFP ($n = 10$) or Hanks' balanced salt solution (HBSS) control ($n = 10$). Control HBSS had no effect on tumours. Reproduced from *Cancer Res* 2001; 61(24):8751–8757, with permission from the American Association for Cancer Research

One of these strategies involves deletion of host range genes so that the virus would replicate in tumour cells but not in normal cells. In a sense, it is surprising that host range genes have been identified in vaccinia, since this virus seems to have minimal dependence and interaction with host cellular factors and has a wide host range. At least five host range genes have been described in poxviruses, including Chinese hamster ovary (CHO)hr, K1L, E3L, and SPI-1 (Perkus *et al.*, 1990; Wyatt *et al.*, 1998).

The products of these host range genes interact with the cell in some selective way to allow for cell-specific replication. Some host range genes are thought to function as inhibitors of apoptosis in some cells but not in others. A general host cellular response to viral infection is apoptosis, defensively stopping host cell processes that may be essential for viral replication. Many viruses have evolved genes which encode for proteins that inhibit cellular apoptosis. Deletion of these genes may result in viral replication selectively in tumour cells, which may have intrinsic defects in apoptotic pathways which compensate for loss of the anti-apoptotic genes. An example is the adenovirus E1B protein, which binds to p53 and inhibits its function. E1B-deleted adenovirus has diminished ability to replicate in normal host cells, but preserved replication in tumour cells (Bischoff *et al.*, 1996). Some poxvirus host range genes are known to inhibit apoptosis. The cowpox CrmA gene inhibits interleukin-1B-converting enzyme, a downstream mediator of apoptosis (Ray *et al.*, 1992). The vaccinia homologue of this gene encodes for a serine protease inhibitor (serpin),

Table 9.2 Median viral recovery from nude mouse tissues. Median (range) viral p.f.u./mg total protein of tissues 8 days after infection with WT (F13L+), TK- (VJS6), VGF- (VSC20), or vvDD-GFP

	WT	TK-	VGF-	vvDD-GFP
Brain	2.8 (2.4–4.9) $\times 10^4$*	1.3 (0.21–20) $\times 10^2$	1.5 (0.76–4.3) $\times 10^2$	0 (0–8)*
Liver	3 (0.8–11)	7 (0.6–13)	1 (0.24–1.1)	0.1 (0–0.2)
Spleen	5.1 (0.59–21) $\times 10^2$	12 (6–16)	23 (16–308)	8(0–16)
Testes[a]	54 (0.4–2800)	12 (0.13–24) $\times 10^2$	0.6 (0.4–0.8)	6.8 (0.7–28)
Bone marrow	1.0 (0.08–10) $\times 10^4$	3.0 (0.75–7.6) $\times 10^3$	1.1 (0.41–2100) $\times 10^3$	5.0 (0–12) $\times 10^2$
Ovary	7.1 (2.6–9.7) $\times 10^6$	9.3 (2.3–15) $\times 10^6$	2.1 (0.41–3.9) $\times 10^7$	8.6 (0.6–172) $\times 10^6$
Tumour	17.0 (1.2–14) $\times 10^6$	4.6 (0.3–6.6) $\times 10^6$	2.3 (0.5–2.6) $\times 10^7$	6.5 (0.4–6.5) $\times 10^6$

Nude mice ($n = 3$–5) were infected with 10^7 p.f.u. of virus. On day 8 after infection, tissues were harvested and homogenized, and a standard plaque assay was performed. Reprinted from McCart JA, *et al.* Systemic cancer therapy with a tumor-selective vaccinia virus mutant lacking thymidine kinase and vaccinia growth factor genes. *Cancer Res* 2001;61(24):8751–8757, with permission from the American Association for Cancer Research.

*$P = 0.011$

[a]Testes samples obtained in a separate experiment.

SPI-2. Vaccinia encodes three serpins – SPI-1, SPI-2, and SPI-3 (Kotwal and Moss, 1989; Law and Smith, 1992; Zhou *et al.*, 1992). SPI-3 inhibits virus-induced cell-cell fusion but does not have a known pro- or anti-apoptotic function. SPI-1 and SPI-2 inhibit apoptosis and have been deemed host range genes. SPI-1 inhibits apoptosis by binding cathepsin G and exerting effects on the mitochondrial apoptotic pathway (Moon *et al.*, 1999). As noted above, SPI-2 inhibits ICE. SPI-2 also inhibits granzyme B. It also inhibits apoptosis initiated by some death ligands through death receptors such as the Fas receptor or the type I tumour necrosis factor (TNF) receptor (Dobbelstein and Shenk, 1996; Macen *et al.*, 1996; Kettle *et al.*, 1997).

The role of apoptosis in vaccinia replication has been the subject of some controversy. Intentional induction of apoptosis in transformed cell lines was shown to not alter vaccinia replication (Kettle *et al.*, 1997). Vaccinia has a rapid life cycle and does not rely on host cell proteins. Furthermore, vaccinia inhibits host cellular function very early during infection, similar to the situation with apoptosis. Much of what has been reported in this regard has been from *in vitro* studies using transformed cell lines, which likely have inherent resistance to apoptosis. There are few *in vivo* studies. One *in vivo* study called into question the significance of the inhibition of apoptosis in terms of viral replication, demonstrating that an SPI-2-deleted mutant was not attenuated *in vivo* (Kettle *et al.*, 1995). It may be that a difference in efficiency was not seen because of the mode of delivery of the virus, which was administered intranasally in this study.

A recent study has examined a WR vaccinia virus with deletions of SPI-1 and SPI-2 (vSP) (Guo *et al.*, 2005). *In vitro*, this virus replicated preferentially in transformed or p53-null cells compared with normal cells. The mechanism and kinetics of cell death after viral infection was also examined. The hypothesis was that deletion of the antiapoptotic genes would allow infected normal cells to die by apoptosis. This hypothesis was not upheld by the data; however some interesting observations were made. When assessed at 18 h after viral infection, normal cells (normal human fibroblasts, NHF) were seen to die with faster kinetics than cancer cells (H460, human non-small cell lung cancer). Furthermore, at this 18 h timepoint, the vSP seemed to induce more normal cells to die via necrosis compared with the wild type virus (Figure 9.4) The vSP had reduced pathogenicity compared with the wild type virus. After intraperitoneal (i.p.) injection of 10^7 p.f.u. of vSP into nude mice, the median survival of these animals was 32 days, compared with 13 days for nude mice who received the same dose of the wild type virus. No toxicity was seen in immunocompetent C57BL/6 mice given 10^8 p.f.u. of vSP i.p., whereas 8/10 of these mice given the same dose of wild type virus died within 7 days. Enhanced tumour selectivity of the vSP compared to the wild type was also seen *in vivo*. Biodistribution of the vSP and wild type virus in nude mice with established subcutaneous MC38 tumours was evaluated by marker gene expression (lacZ) and viral titres. The mice were treated with 10^7 p.f.u. of either vSP or wild type virus intraperitoneally. There was similarly high titre and gene expression of both viruses in the tumour. However, there were significantly reduced levels of vSP in normal tissues compared with the wild type virus. Both viruses had similarly low titres in the brain. vSP was also compared to the double-deleted virus in terms of biodistribution and was found to have even greater tumour selectivity than this virus (Figure 9.5).

The vSP virus was also found to be superior to the wild type in terms of an antitumour effect. In a subcutaneous MC38 tumour model in nude mice, the vSP demonstrated significantly greater antitumour effect and significantly less pathogenicity. Furthermore, in a subcutaneous MC38 tumour model in an immunocompetent host (C57BL/6 mice), an antitumour effect of the vSP virus was demonstrated (Figure 9.6). In summary, deletion of host range genes resulted in a vaccinia virus with improved tumour selectivity, less pathogenicity, and enhanced antitumour effect. The antitumor effect apparently resulted from viral replication and not an immunological response against the tumour, as the antitumour effect was more pronounced in immunodeficient nude mice.

Another strategy for targeting vaccinia selectively to tumours is to alter viral surface proteins

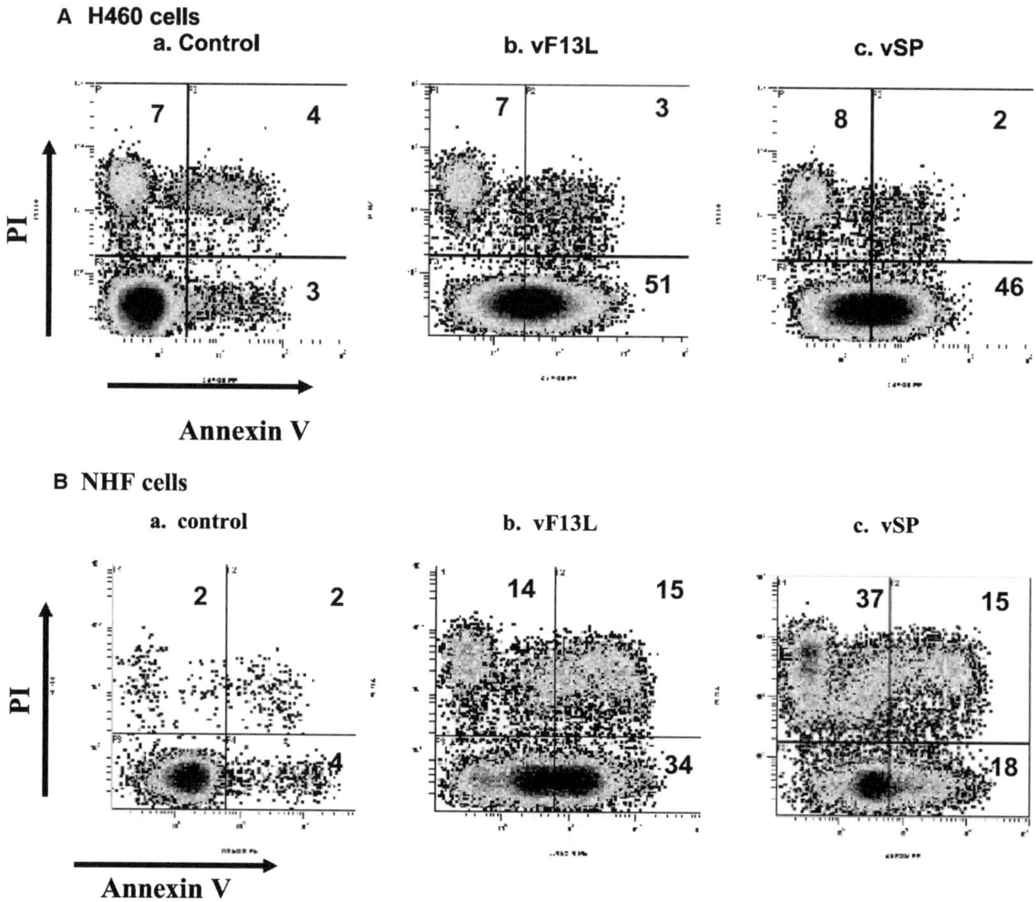

Figure 9.4 Comparable apoptosis of various constructs tested in vaccinia-infected cells. [Human cancer H460 cells (A) and normal human primary fibroblasts (NHF; B)] were infected with vaccinia viruses at a MOI of 5 for over 1 h in 1 ml medium with 2% fetal bovine serum. Virus was aspirated and the cells were washed with phosphate-buffered saline once before complete growth medium was added. The cells were harvested at 18 h after infection. Cells were stained with Annexin V–phycoerythrin and –propidium iodide (PI) by using apoptosis kits under the conditions provided by the manufacturers. The stained cells were analysed by flow cytometry. Reproduced from *Cancer Res* 2005;65(21):9991–9998, with permission from the American Association for Cancer Research

that mediate viral binding to cells so that the virus binds selectively to tumour cells. While this principle has been shown to be effective with other viruses such as adenovirus, there may be difficulty in applying the same method to vaccinia. The mechanism of vaccinia uptake into cells is not clearly understood. There is no defined cell-surface receptor for vaccinia. Furthermore, the IMV and EEV forms of the virus are antigenically distinct (Boulter, 1969). Altering the surface proteins of the IMV form which is available in the lab may have no significance for targeting, as it is the EEV form which spreads from cell to cell (see above).

Nevertheless, this strategy has been under study. Numerous proteins have been identified on the surface of the EEV. Investigators have been able to produce stable expression of proteins on the EEV by fusion with particular EEV proteins. For example, an HIV glycoprotein has been expressed on the surface of the EEV by fusion with the EEV

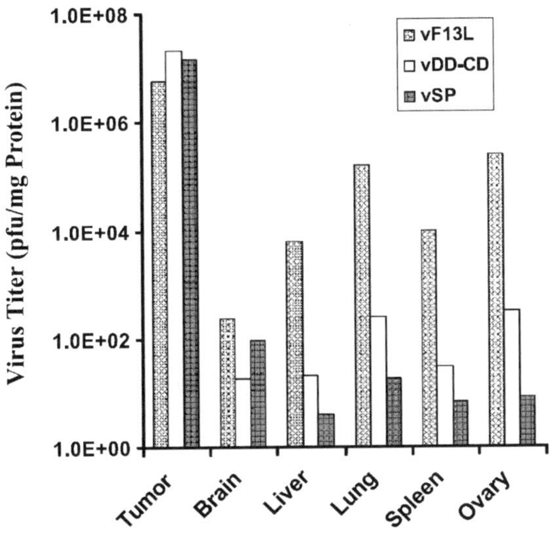

Figure 9.5 Enhanced recovery of vSP vaccinia from tumour when compared with normal tissues. MC38 cells (2.5×10^5) were injected s.c. into athymic nude mice. When the tumours reached ~75–125 mm^3 (in ~9 days), 1×10^7 p.f.u. of specified vaccinia virus were injected i. p. into each mouse. Eight days after virus administration, tumour and normal tissues were collected. The median values of virus titres (p.f.u./mg protein) were determined by titration of the virus in tissue lysates on CV-1 cells. $P < 0.05$ for normal tissues except brain between the groups treated with vF13L+ and vSP. Reproduced from *Cancer Res* 2005;65(21):9991–9998, with permission from the American Association for Cancer Research

B5R protein (Katz *et al.*, 1997). This virus did not have altered tropism, however. In another study, another EEV membrane protein, the viral haemagglutinin protein (A56R), was fused to a scFv against tumour-associated antigen ErbB-2 (Galmiche *et al.*, 1997). This virus was able to bind the antigen ErbB-2 in solution but would not bind to ErbB-2 expressing cells. Again, there was no altered tropism.

Another EEV protein, the A34R protein, appears to play a significant role in viral release from cells as well viral infection of neighbouring cells (McIntosh and Smith, 1996). Deletion of this protein results in failure of the virus to form normal-sized plaques *in vitro*. *In vivo* activity is attenuated, also. In an effort to use this surface

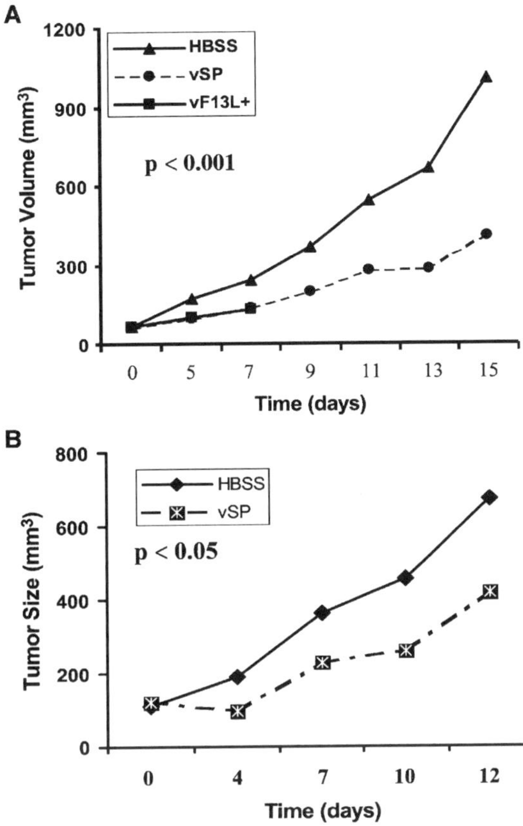

Figure 9.6 Diminished MC38 tumour growth in mice treated with vSP virus. Mice (nude and C57BL/6) were injected s.c. with 2.5×10^5 MC38 tumour cells. At the median tumour volume of 75–125 mm^3, mice were injected i.p. with 1.0×0^8 p.f.u. of vaccinia virus vSP, vF13L+, or saline HBSS. The tumour sizes and health of mice were monitored. A, MC38 tumour in nude mice. In the group of mice treated with vF13L+ (wt), all mice died between days 8 to 16 and, therefore, no further tumour measurement was considered after day 8. Data are representative of two independent experiments ($P < 0.001$). B, MC38 tumour in immunocompetent C57BL/6 mice. Tumour-bearing mice were injected with 2.0×10^8 p.f.u. of viruses i.p. Data are representative of five independent experiments ($P < 0.05$). Reprinted from Guo ZS, *et al.* The enhanced tumor selectivity of an oncolytic vaccinia lacking the host range and antiapoptosis genes SPI-1 and SPI-2. *Cancer Res* 2005;65(21): 9991–9998, with permission from the American Association for Cancer Research

protein to target vaccinia selectively in an ovarian tumour model, vaccinia shuttle plasmids were generated creating fusion genes with a full-length A34R gene or a truncated (48 nucleotides) form of the gene fused to a scFv specific for ovarian cells (MOvλ specific for folate-binding protein, FBP). The truncated form lacked the lectin-binding domain, which is an essential component of the A34R protein in terms of facilitating cell-to-cell spread of the virus. These fusion genes were recombined into vaccinia such that the native A34R gene was replaced with the fusion genes. The wild type, A34R deletion mutant, A34R-MOv full-length (M168), and A34r-MOv truncated (M148) were then compared in terms of infectivity, plaque morphology, growth curves, and immunofluorescence in different cell lines, some of which expressed the FBP and some of which did not. M148 was attenuated in cells lacking the FBP and exhibited a similar phenotype to the A34R-deleted virus. M168 had a similar phenotype to the wild type virus, with no apparent advantage for replication in FBP-positive cells. While the MOv antibody constructs were stable, there was no specific binding to FBP-expressing cells, FBP bound to plates, or a MOv anti-idiotype antibody. The fusion was, therefore, nonfunctional (Zeh and Bartlett, 2002). Clearly, better understanding of the vaccinia–host cell interaction and the viral cell entry process is needed before this aspect of the viral life cycle can be manipulated to improve viral tropism for tumour cells.

9.4 Improving antitumour effects through bystander effects

It is unlikely that 100 per cent of tumour cells would be infected with vaccinia *in vivo* after local or systemic administration. Perhaps the antitumour effect of the virus would be improved by incorporating into the virus a mechanism to kill tumour cells which have not been infected – a bystander effect.

One strategy to induce bystander cytotoxicity is to take advantage of the strong viral promoters to drive expression of a transgene, the product of which is toxic to cells surrounding the infected cells. This strategy has been studied using both the purine nucleoside phosphorylase gene and cytosine deaminase (CD) gene (Gnant *et al.*, 1999; Puhlmann *et al.*, 1999). These genes encode for enzymes which convert prodrugs to their cytotoxic form. The prodrugs are administered following the virus. Since the genes are expressed in infected tumour cells, the prodrug is converted to its toxic form in the tumour and can exert its cytotoxic effect on the tumour cells. Significant antitumour effects have been demonstrated with each of these systems in a murine model of hepatic metastases.

There exists a complex interaction between the oncolytic effect of the replicating virus and enzyme–prodrug effect. While the virus can directly infect and kill tumour cells, the cytotoxic effect of the enzyme–prodrug system can directly decrease viral replication (McCart *et al.*, 2000). This relationship has been explored in the vaccinia–CD system to better define the advantageous and detrimental interactions between the replicating oncolytic effect and the enzyme–prodrug therapeutic effect. *In vitro* infection of cancer cells with vaccinia–CD at a high multiplicity of infection (MOI > 0.1) led to cell death secondary to a viral cytopathic effect. The addition of the prodrug had no added effect. At a low MOI, there was no viral cytopathic effect alone, but the addition of the prodrug (5-FC) caused significant cell death. Cell lysates demonstrated 300-fold reduced viral recovery from cells treated with both vaccinia–CD and 5-FC, compared to vaccinia treatment alone. It appears that the converted prodrug (5-FU) has inhibitory effects on viral replication, either by direct interference with viral DNA synthesis or by killing surrounding cells prior to infection by vaccinia virus.

This interaction was further explored *in vivo* in a subcutaneous MC38 tumour model in nude mice. The mice were administered 10^8 p.f.u. of vaccinia–CD. One group of mice received the prodrug, 5-FC, while another group did not receive the prodrug. Both of these groups had smaller tumours than the untreated controls, suggesting that replicating vaccinia alone is cytotoxic to tumours *in vivo* after systemic injection. Addition of 5-FC improved the antitumour response, with 50 per cent complete regressions, when a low dose of virus (10^7 p.f.u.) was injected intratumorally. Interestingly, there was prolonged survival from viral-mediated death in the group that was treated with 5-FC. Together, these results suggest that addition of an enzyme–prodrug system to a replicating vaccinia virus can improve the antitumour response and decrease viral pathogenicity.

9.5 Immune response to vaccinia and vaccinia immune evasion strategies

Immune clearance of vaccinia is an impediment to its efficiency as an oncolytic virus. As a result of the smallpox vaccination programme most older cancer patients have been immunized to vaccinia. Since smallpox immunizations were terminated in the 1970s younger patients do not have preformed immunity to this virus. Vaccination of this younger population may become necessary, however, due to the potential use of smallpox as a biological weapon.

Cellular immunity to vaccinia is quite potent and seems to be more important than humoral immunity in viral clearance. Progressive vaccinia infection correlates with a defect in cell-mediated immunity (Fulginiti *et al.*, 2003). In athymic nude mice with established tumours, vaccinia is able to replicate in tumour cells for $>$30 days (McCart *et al.*, 2000). In immunocompetent hosts, however, gene expression lasts for only $\sim$8 days (Puhlmann *et al.*, 2000).

Vaccinia has evolved immune evasion strategies. The EEV form of the virus has host-derived complement control proteins in the outer envelope – CD46, CD55, CD59 (Vanderplasschen *et al.*, 1998). The EEV is resistant to antibody neutralization (Smith *et al.*, 2002). Vaccinia encodes for proteins that suppress both innate immunity and the T helper (Th)-1 immune response (Smith *et al.*, 1997).

Vaccinia virus has at least three genes which encode for proteins that can block the function of interferon (IFN) family members IFN-α and IFN-β (Alcami *et al.*, 1998; Mahalingan and Karupiah, 2000; Seet and McFadden, 2002). These IFNs are secreted by a variety of cells in response to innate danger signals. They can induce an antiviral state and upregulate adaptive immune functions. Vaccinia also carries genes for multiple inhibitors of chemokines, some of the earliest substances produced during the initiation of an immune response (Alcami *et al.*, 1998; Mahalingan and Karupiah, 2000; Seet and McFadden, 2002). In addition to actively suppressing early innate responses to viral infections, vaccinia also produces several factors which inhibit the development of a Th1 immune response. Vaccinia encodes for at least three factors that can directly block the function of IFN-λ, one of the most potent Th1 cytokines (Symons *et al.*, 1995; Najarro *et al.*, 2001). Vaccinia also encodes for interleukin (IL)-18-binding protein (IL-18BP), a naturally produced soluble factor that blocks the binding of IL-18 to its cognate receptor. IL-18BP has been shown to be one of the most potent inhibitors of the development of a Th1 immune response (Novick *et al.*, 1999). Vaccinia encodes for several other immunosuppressive factors including factors to block complement activation, IL-1β soluble factor, and soluble TNF receptor antagonist (Howard *et al.*, 1991; Engelstad *et al.*, 1992; Bowie *et al.*, 2000). These observations suggest that subverting the early innate immune response and slowing the development of Th1 responses are important for the efficacy of oncolytic viral therapy.

These inferences have been confirmed in several different models of vaccinia infection. For example, a particular strain of vaccinia, the modified vaccinia Ankara (MVA) strain has markedly low virulence associated with deletion of immune-evasion genes. This strain was created by serial passage through chick embryo fibroblasts. It replicates poorly in mammalian cells (Sutter and Moss, 1995). The genome of this strain has been sequenced and found to be missing genes that encode for proteins that inhibit IFN-α, IFN-β, IFN-λ, TNF, and chemokines (Blanchard *et al.*, 1998). The importance of innate immunity in viral clearance is further evidenced by the observation that IFN-α and IFN-β knockout mice demonstrate markedly enhanced susceptibility to vaccinia viral infection (Deonarain *et al.*, 2000).

Other studies have confirmed the critical role of Th1 response to clearance of vaccinia viral infection. The Th1 (IFN-λ, IL-12) and Th2 (IL-4, IL-10) cytokine balance in clearance of vaccinia has been examined using cytokine knockouts (van den Broek *et al.*, 2000). Viral replication was enhanced in IL-12 and IFN-λ knockout mice, with IL-$12^{-/-}$ demonstrating greater susceptibility to infection than IFN-λ-deficient mice. Consistent with this finding was that development of anti-vaccinia CTL was completely abrogated in IL-12 knockout mice, but remained normal in IFN-λ knockout mice. In contrast, IL-4- and IL-10 knockout mice showed marked enhancement of vaccinia viral clearance, suggesting that these cytokines naturally suppress the host response to vaccinia. IL-$10^{-/-}$mice exhibited greater inhibition of viral replication than IL-$4^{-/-}$ mice. When the effects of each of these cytokines on vaccinia infection was examined in recombinant viral constructs, local expression of IL-4 showed a much greater inhibition of host responses. In fact, whereas the absence of IL-10 resulted in improved clearance of vaccinia virus that was mediated by increased levels of IL-6 and IL-1, the local expression of IL-10 had little to no effect on viral clearance. In another study, IL-12 and IL-18 were seen to act in synergy to clear vaccinia infection. Natural killer (NK) and T cells were involved in virus clearance (Gherardi *et al.*, 2003).

While vaccinia has some mechanisms in place for immune evasion, it is clear that early clearance of the virus by the immune response in immunocompetent hosts limits the antitumor efficacy. Strategies must be employed to regulate the immune response to vaccinia in order to develop it as a viable option as a cancer therapeutic. Investigators have designed at least four strategies to circumvent the premature immune clearance of the virus *in vivo*. One strategy is to mutate the viral coat proteins so that the virus evades recognition by the immune system. However, vaccinia, like other poxviruses, is antigenically very complex, and it is unlikely that one or two mutations in the viral membrane proteins could significantly alter antibody recognition. Furthermore, any mutations in the viral envelope may decrease the infectivity of the virus.

Another strategy would be to develop other poxviruses that do not cross-react with vaccinia and are able to selectively infect and lyse human tumour cells. Viruses from the Yatapox genus infect monkeys and have secondarily infected monkey caretakers (Grace and Mirand, 1965). The yaba-like disease (YLD) virus has been isolated from skin lesions in monkeys. It replicates in human cells and causes limited skin lesions similar to vaccinia inoculations, but this virus does not cross-react serologically with vaccinia. The YLD virus was studied as another replicating poxvirus for antitumour therapy (Hu *et al.*, 2001). The virus was characterized in terms of plaque morphology, host range, and replication efficiency. The YLD virus could be grown in high titres, like vaccinia, under normal conditions using CV-1 cells. However, the host range of the YLD virus was more restrictive than that of vaccinia. Whereas the YLD virus replicated well in monkey and human cell lines, it did not replicate in rodent cell lines. Infectivity and plaque formation were slower and less efficient than seen with vaccinia, even in permissive cells. Enzyme-linked immunosorbent assay on YLD-coated plates using vaccinia antiserum showed no reactivity, verifying the supposition that no antibody cross-reactivity exists between the two viruses. Experiments were also done using a plasmid with the vaccinia synthetic promoter expressing β-galactosidase. YLD-infected cells were transfected with this plasmid, demonstrating that the YLD polymerase recognized the vaccinia promoters.

A YLD shuttle plasmid was created and used to recombine GFP as a marker gene into the TK locus of the YLD virus. (Lack of sequence homology prohibited use of a vaccinia shuttle plasmid.) Using this viral construct, it was demonstrated that the YLD-GFP virus could mediate highly efficient gene transfer and expression in monkey and human cells. The efficiency was not as great as with the vaccinia virus constructs, but much more efficient that other vectors investigated, including adenovirus and liposome-mediated gene delivery. The YLD virus was tested *in vivo*. A2780 human ovarian peritoneal carcinomatosis was established in nude mice, which were then administered YLD intraperitoneally at 10^8 p.f.u.. The tumour was sampled every 4 days to obtain cells for fluorescence-activated cell sorting analysis (FACS) for GFP expression. By 12 days, 20 per cent of tumour cells expressed GFP. So, YLD is potentially useful for vaccination or local injection of tumours. While it is more efficient than some other vectors being studied, it remains much less efficient than vaccinia.

A third strategy is to create a viral recombinant that actively suppresses host cellular immune responses. Several groups have reported that insertion of Th2-like cytokines, such as IL-4 or IL-10, into vaccinia virus increases *in vivo* viral replication and slows host clearance of infection (Sharma *et al.*, 1996; Alvarez-Vallina *et al.*, 1997; van den Broek *et al.*, 2000). There are serious safety concerns, though, regarding creation of a potentially pathogenic virus that is not immunologically cleared. The threat of bioterrorism increases the concern.

A fourth strategy is to transiently suppress the host immune system. Multiple immunosuppressive agents are available due to their development for use in solid organ transplantation. These agents can very specifically target specific pathways of the host immune response. Knowledge of the mechanisms of action of these agents combined with our growing understanding of the immune response to vaccinia virus should allow us to reversibly slow the immune response to this vector. The slowed immune response will theoretically allow for more efficient *in vivo* viral replication in the tumour, higher transgene expression and greater oncolysis. The authors hypothesize that this will be a feasible approach with the double-deleted vaccinia mutant because of the remarkable tumour selectivity that it exhibits. As noted above, this vector was not pathogenic in nude mice when injected systemically at doses up to 10^8 p.f.u.

This approach of transient immunosuppression is under investigation (McCart *et al.*, 2001). Early studies focused on depletion of immune cells in order to better understand which effector cells are critical for allowing efficient *in vivo* infection of tumour cells. Antibodies to CD4 and CD8 murine T cells were administered to tumour-bearing mice (MC38 subcutaneous tumours), individually or combined, to deplete both populations of T cells. T-cell depletion was verified by FACS. TK-deleted vaccinia expressing β-galactosidase was then delivered intraperitoneally and β-galactosidase expression in the subcutaneous tumour was analysed. T-cell depleted animals demonstrated a higher peak gene expression level compared to non-depleted animals. Intact animals depleted the virus and β-galactosidase expression within 10 days, whereas depleted mice demonstrated prolonged expression throughout the duration of the study. CD4- and CD8-expressing T cells contributed equally to immune clearance of vaccinia. The most prolonged expression was achieved with combined depletion. Immunoglobulin G (IgG) and IgM formation against vaccinia was measured, and as expected, there was no isotype switch to IgG in the T-cell depleted host. The absence of the isotype switch may allow for repeat administration of the virus. Finally, tumour responses were examined after delivery of 10^9 p.f.u. of vaccinia intraperitoneally. There was significantly improved inhibition of tumour growth in T-cell depleted animals, compared to intact animals (Zeh and Bartlett, 2002).

Successful reinfection of a preimmunized host appears to be possible through transient immunosuppression. Rhesus macaques were preimmunized with wild type WR strain vaccinia. When they were reinfected with the same virus, no live virus could be recovered from the animal. The same animals (with high IgG against vaccinia) were then pretreated with FK-506, prednisone, cyclophosphamide, and complement inhibitor, then reinfected with wild type virus. The virus spread quickly and was recoverable from blood and other organs (Zeh and Bartlett, 2002).

9.6 Virus-driven antitumour immune response

A potential mechanism for inducing bystander cytotoxicity is by expression of factors that drive an immunologic response to the uninfected tumour cells.

The author's (D.B.) experience has suggested that viral-induced oncolysis alone does not drive an efficient antitumour immune response. This observation is not surprising, given that vaccinia possesses immune evasion mechanisms. A hypothesis which has been tested in several studies is that to create an effective *in vivo* tumour oncolysis vaccine, the vector must induce expression of a cytokine capable of driving a Th1/Tc1 immune response (Karupiah *et al.*, 1991; Sambhi *et al.*, 1991; Ruby *et al.*, 1995; Gherardi *et al.*, 1999 Ramirez *et al.*, 2000; Perera *et al.*, 2001). However, one of the chief limitations to efficient *in vivo* viral replication and high tumour infection rates is premature immune clearance of the virus. Insertion of cytokines would be expected to exacerbate this problem. Preliminary observations have indicated that insertion of FAS-L into vaccinia paradoxically decreases its efficacy in immunocompetent animals, presumably by promoting clearance of the virus before a significant percentage of tumour infection has occurred (McCart *et al.*, 2001). This decrease in viral replication efficiency has been observed in other recombinant vaccinia systems using cytokine transgenes. A recombinant vaccinia containing IL-2 or IL-15 demonstrates markedly reduced replication efficacy *in vivo*. (Perera *et al.*, 2001). Similarly, others have shown that recombinant vaccinia expressing various cytokines results in marked impairment of *in vivo* replication efficiency (Karupiah *et al.*, 1991; Sambhi *et al.*, 1991; Ruby *et al.*, 1995).

It is clear that premature clearance of the virus is harmful to *in vivo* viral replication and consequently to the direct oncolytic properties of the virus, but it is not as clear that it will diminish the immunostimulatory/vaccination capabilities of the virus (Ramirez *et al.*, 2000). A recombinant VV expressing both the IL-12 gene and the human immunodeficiency virus (HIV) env gene, showed that expression of IL-12 in the replication-competent vector directly decreased the *in vivo* replication efficiency; however, when lower titres of virus were used to inoculate the animal, IL-12 transgene expression resulted in augmentation of the anti-HIV-env response (Gherardi *et al.*, 1999).

Perhaps this problem of premature clearance of virus due to cytokine expression can be circumvented by creation of an inducible expression system. Such a system would make expression of the cytokine dependent on a controlled exogenous signal, allowing efficient viral replication prior to induction of cytokine expression. This type of system may allow efficient use of both the oncolytic and immunostimulatory aspects of the virus.

9.7 Imaging

If recombinant vaccinia vectors are studied in clinical trials, it will be important to noninvasively track the biodistribution and persistence of these vectors. As with the delivery of therapeutic genes, it is possible to deliver transgenes that allow imaging of vaccinia-infected cells. One such system which has been reported incorporates the gene for the human somatostatin receptor type 2 (SSTR2) (McCart *et al.*, 2004). Somatostatin and its synthetic analogues bind this receptor. A recombinant double-deleted vaccinia vector into which the gene for this receptor was inserted was shown to express the SSTR2 on infected cells. This vector was systemically delivered to nude mice bearing subcutaneous MC38 tumours. Six days later the mice were given an intravenous injection of a radiolabeled long-acting somatostatin analogue, ^{111}In-pentetreotide. Imaging with a gamma camera demonstrated tumour-specific labelling by ^{111}In-pentetreotide (Figure 9.7). Furthermore, this localization on imaging correlated with the biodistribution measured by tissue viral titres (Table 9.3). This system allowed repeated imaging over 3 weeks.

Other investigators have used luciferase expression to image vaccinia-infected cells (Yu *et al.*, 2004). Vaccinia virus of the Lister Institute for Viral Preparations (LIVP) strain was used to construct a recombinant vaccinia virus expressing *Renilla reniformis* luciferase–*Aequorea victoria* green fluorescent protein (GFP) fusion protein (RUC-GFP) – rVV-*RUC-GFP*. This virus did not have deletions of the TK or VGF genes. Imaging demonstrated localization of this vector to tumours and metastases in immunocompromised and immunocompetent mice.

These reports demonstrate that delivery of transgenes by vaccinia allow for non-invasive tracking of the virus *in vivo*. Furthermore, localizing the virus may aid in identifying tumours and metastases. The size of the vaccinia genome and the ability to insert large sequences of DNA would allow for the construction of recombinant vaccinia vectors for the expression of therapeutic as well as imaging transgenes.

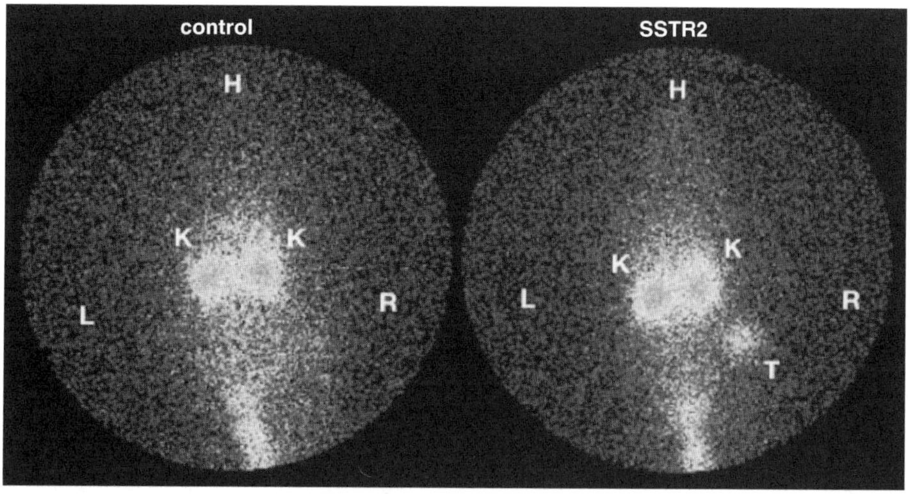

Figure 9.7 Posterior whole-body images of tumour-bearing athymic mice 1 week after i.p. injection with vaccinia virus and 24 h after i.v. (tail vein) injection with ^{111}In-pentetreotide. Tumour (T) is visible on the right flank of vvDD-SSTR2-injected mouse (right) but not in the control vvDD-GFP-injected mouse (left). Prominent visualization of both kidneys (K) is noted in all animals, as well as the tail (site of injection). The right (R) and left (L) sides of the mouse as well as the head (H) are indicated. Reproduced from *Mol Ther* 2004;10(3):553–561, with permission from Elsevier

9.8 Current and potential clinical applications

To date, no clinical trials have examined the use of vaccinia as an oncolytic virus. There have been several trials, however, utilizing vaccinia as a vector to induce an immune response against cancer or infectious diseases such as HIV. Clinical trials conducted in the 1960s and 1990s, which involved intratumoral injection of vaccinia in patients with melanoma or intravesical instillation of the virus in patients with bladder cancer, showed a correlation between an immune response at the site of vaccinia inoculation and an antitumor response (Thorne and Kirn, 2004). Based on this finding, clinical trials of vaccinia have largely focused on using vaccinia to drive an immunological antitumour response. Vaccinia oncolysates and recombinant vaccinia virus expressing tumour-associated antigens have been used for vaccination. In addition to encoding for

Table 9.3 Correlation of vaccinia/SSTR2 biodistribution with ^{111}In-pentetreotide biodistribution

Tissue	Tumour		Liver		Kidney	
Virus	hSSTR2	EGFP	hSSTR2	EGFP	hSSTR2	EGFP
RT-PCR (SSTR2)	+	–	–	–	–	–
Viral titres (mean p.f.u./mg)	5.1×10^8	ND	1.9×10^2	ND	1.2×10^2	ND
ROI analysis (mean counts/pixel)	0.46*	0.26*	0.91	1.02	5.54	5.51
Biodistribution (%ID/g)	0.94**	0.18**	0.37	0.37	6.91	7.53

ND, not done; ROI, region of interest; EGFP, enhanced green fluorescent protein; RT-PCR, reverse transcriptase–polymerase chain reaction.
*$P = 0.04$;
**$P < 0.0001$.

the tumour antigens, vaccinia has been constructed to encode for costimulatory molecules, such as MUC-1 and IL-2, to enhance the immune response (Scholl *et al.*, 2000). It has been hypothesized that these vaccinia constructs may be more effective in inducing an anti-tumour immune response if they are injected intralesionally, thus acting as an immune adjuvant in the tumour microenvironment (Thorne and Kirn, 2004; Thorne *et al.*, 2005). This approach has shown some promise in a trial in which a vaccinia–GM-CSF recombinant virus was administered intralesionally in patients with recurrent and/or refractory melanoma (Mastrangelo *et al.*, 1999).

While these trials have begun to show the utility of vaccinia in inducing an immune response to cancer, the preclinical studies described in this chapter show the potential of vaccinia as an oncolytic virus. It is tumour-selective and can be made more tumour-selective by gene deletions. Based on the tumour selectivity and safety demonstrated by the double-deleted virus, there is a proposed clinical trial in our institution using this virus to treat patients with cutaneous malignancies.

In summary, vaccinia appears to have great potential as a targeted therapy for cancer. It has been shown to be safe through an extensive clinical experience in smallpox vaccination and tumour vaccine trials. It is tumour-selective. It has strong promoters and can induce expression of genes to produce a bystander effect. It can induce an immune response against tumours. Perhaps an inducible expression system would allow application of all these antitumour effects at once by allowing efficient replication before induction of toxic transgenes or cytokine expression. It appears that regulation of the immune response to vaccinia will be very important to the success of its application as an oncolytic virus.

References

Alcami A, Symons JA, Collins PD, Williams TJ, Smith GL (1998). Blockade of chemokine activity by a soluble chemokine binding protein from vaccinia virus. *J Immunol* **160**(2), 624–633.

Alvarez-Vallina L, Agha-Mohammadi S, Hawkins RE, Russell SJ (1997). Pharmacological control of antigen responsiveness in genetically modified T lymphocytes. *J Immunol* **159**(12), 5889–95.

Antoine G, Scheiflinger F Dorner F, Falkner FG (1998). The complete genomic sequence of the modified vaccinia Ankara strain: comparison with other orthopoxviruses. *Virology* **242**, 365–396.

Bartlett DL (2001). Vaccinia virus. In Hernaiz Driever P, Rabkin SD (eds), *Replication-Competent Viruses for Cancer Therapy.* Monographs in Virology, Volume 22, pp. 130–159. S. Karger AG (Basel).

Bischoff JR, Kirn DH, Williams A, *et al.* (1996). An adenovirus mutant that replicates selectively in p53-deficient human tumor cells. *Science* **274**(5286): 373–376.

Blanchard TJ, Alcami A, Andrea P, Smith GL (1998). Modified vaccinia virus Ankara undergoes limited replication in human cells and lacks several immunomodulatory proteins: implications for use as a human vaccine. *J Gen Virol* **79**(Pt 5): 1159–67.

Boulter EA (1969). Protection against poxviruses. *Proc R Soc Med* **62**(3), 295–297.

Bowie A, Kiss-Toth E, Symons JA, Smith GL, Dower SK, O'Neill LA (2000). A46R and A52R from vaccinia virus are antagonists of host IL-1 and toll-like receptor signaling. *Proc Natl Acad Sci U S A* **97** (18), 10162–7.

Buller RML, Chakrabarti S, Cooper JA, Twardzik DR, Moss B (1988). Deletion of the vaccinia virus growth factor gene reduces virus virulence. *J Virol* **62**(3), 866–874.

Chang E, Chalikonda S, Friedl J, *et al.* (2005). Targeting vaccinia to solid tumors with local hyperthermia. *Hum Gene Ther* **16**(4), 435–444.

Chung CS, Hsiao JC, Chang YS, Chang W (1998). A27L protein mediates vaccinia virus interaction with cell surface heparan sulfate. *J Virol* **72**, 1577–1585.

de Magalhaes JC, Andrade AA, Silva PN, *et al.* (2001). A mitogenic signal triggered at an early stage of vaccinia virus infection: implication of MEK/ERK and protein kinase A in virus multiplication. *J Biol Chem* **276**(42), 38353–38360.

Deonarain R, Alcami A, Alexiou M, Dallman MJ, Gewert DR, Porter AC (2000). Impaired antiviral response and alpha/beta interferon induction in mice lacking beta interferon. *J Virol* **74**(7), 3404–9.

Dobbelstein M, Shenk T (1996). Protection against apoptosis by the vacinnia virus SPI-2 (B13R) gene product. *J Virol* **70**(9), 6479–6485.

Engelstad M, Howard ST, Smith GL (1992). A constitutively expressed vaccinia gene encodes a 42-kDa glycoprotein related to complement control factors that forms part of the extracellular virus envelope. *Virology* **188**(2), 801–810.

Fenner F, Wittek R, Dumbell KR (1989). Vaccinia virus: the tool for smallpox eradication. In: Fenner F, Wittek R, Dumbell KR (eds), *The Orthopoxviruses*. Academic Press, Inc. New York, pp. 143–170.

Fulginiti VA, Papier A, Lane JM, Neff JM, Henderson DA (2003). Smallpox vaccination: a review, part II. Adverse events. *Clin Infect Dis* **37**(2), 251–271.

Galmiche MC, Rindisbacher L, Wels W, Wittek R, Buchegger F (1997). Expression of a functional single chain antibody on the surface of extracellular enveloped vaccinia virus as a step towards selective tumour cell targeting. *J Gen Virol* **78**(11), 3019–3027.

Gherardi MM, Ramirez JC, Esteban M (2003). IL-12 and IL-18 act in synergy to clear vaccinia virus infection: involvement of innate and adaptive components of the immune system. *J Gen Virol* **84**(Pt 8): 1961–1972.

Gherardi MM, Ramirez JC, Rodriguez D, *et al.* (1999). IL-12 delivery from recombinant vaccinia virus attenuates the vector and enhances the cellular immune response against HIV-1 Env in a dose-dependent manner. *J Immunol* **162**(11), 6724–33.

Gnant MF, Noll LA, Irvine KR, *et al.* (1999a). Tumor-specific gene delivery using recombinant vaccinia virus in a rabbit model of liver metastases. *J Natl Cancer Inst* **91**(20), 1744–1750.

Gnant MF, Puhlmann M, Alexander HR, Bartlett DL (1999b). Systemic administration of a recombinant vaccinia virus expressing the cytosine deaminase gene and subsequent treatment with 5-fluorocytosine leads to tumor specific gene expression and prolongation of survival in mice. *Cancer Res* **59**, 3396–3404.

Goebel SJ, Johnson GP, Perkus ME, Davis SW, Winslow JP, Paoletti E (1990). The complete DNA sequence of vaccinia virus. *Virology* **179**, 247–266.

Grace JT Jr, Mirand EA (1965). Yaba virus infection in humans. *Exp Med Surg* **23**(2), 213–216.

Guerra S, Lopez-Fernandez LA, Pascual-Montano A, Munoz M, Harshman K, Esteban M (2003). Cellular gene expression survey of vaccinia virus infection of human HeLa cells. *J Virol* **77**(11), 6493–6506.

Guo ZS, Naik A, O'Malley ME, *et al.* (2005). The enhanced tumor selectivity of an oncolytic vaccinia lacking the host range and antiapoptosis genes SPI-1 and SPI-2. *Cancer Res* **65**(21), 9991–9998.

Hengstschlager M, Knofler M, Mullner EW, Ogris E, Wintersberger E, Wawra E (1994). Different regulation of thymidine kinase during the cell cycle of normal versus DNA tumor virus-transformed cells. *J Biol Chem* **269**(19), 13836–13842.

Howard ST, Chan YS, Smith GL (1991). Vaccinia virus homologues of the Shope fibroma virus inverted terminal repeat proteins and a discontinuous ORF related to the tumor necrosis factor receptor family. *Virology* **180**(2), 633–647.

Hsiao JC, Chung CS, Chang W (1999). Vaccinia virus envelope D8L protein binds to cell surface chondroitin sulfate and mediates the adsorption of intracellular mature virions to cells. *J Virol* **73**, 8750–8761.

Hu Y, Lee J, McCart JA, *et al.* (2001). Yaba like disease virus: and alternating replicating pox vector for cancer gene therapy. *J Virol* **75**(21), 10300–10308.

Karupiah G, Woodhams CE, Blanden RV, Ramshaw IA (1991). Immunobiology of infection with recombinant vaccinia virus encoding murine IL-2. Mechanisms of rapid viral clearance in immunocompetent mice. *J Immunol* **147**(12), 4327–4332.

Katz E, Wolffe EJ, Moss B (1997). The cytoplasmic and transmembrane domains of the vaccinia virus B5R protein target a chimeric human immunodeficiency virus type 1 glycoprotein to the outer envelope of nascent vaccinia virions. *J Virol* **71**(4), 3178–87.

Kettle S, Blake NW, Law KM, Smith GL (1995). Vaccinia virus serpins B13R(SPI-2) and B22R (SPI-1) encode M_r 38.5 and 40K, intracellular polypeptides that do not affect virus virulence in a murine intranasal model. *Virology* **206**(1), 136–147.

Kettle S, Alcami A, Khanna A, Ehret R, Jassoy C, Smith GL (1997). Vaccinia virus serpin B13R (SPI-2) inhibits interleukin-1 beta-converting enzyme and protects virus-infected cells from TNF- and Fas-mediated apoptosis, but does not prevent Il-1 beta-induced fever. *J Gen Virol* **78**(Pt 3):677–685.

Kotwal GJ, Moss B (1989). Vaccinia virus encodes two proteins that are structurally related to members of the plasma serine protease inhibitor superfamily. *J Virol* **63**(2), 600–606.

Law KM, Smith GL (1992). A vaccinia serine protease inhibitor which prevents virus-induced cell-fusion. *J Gen Virol* **73** (Pt 3):549–557.

Macen JL, Garner RS, Musy PY, *et al.* (1996). Differential inhibition of the Fas- and granule-mediated cytolysis pathways by the orthopoxvirus cytokine response modifier A/SPI-2 and SPI-1 protein. *Proc Natl Acad Sci U S A* **93**(17), 9108–9113.

Mahalingam S, Karupiah G (2000). Modulation of chemokines by poxvirus infections. *Curr Opin Immunol* **12**(4), 409–412.

Mastrangelo MJ, Maguire HC Jr, Eisenlohr LC, *et al.* (1999). Intratumoral recombinant GM-CSF-encoding virus as gene therapy in patients with cutaneous melanoma. *Cancer Gene Ther* **6**(5), 409–22.

McCart JA, Mehta N, Scollard D, *et al.* (2004). Oncolytic vaccinia virus expressing the human somatostatin receptor SSTR2: molecular imaging after systemic delivery using ^{111}In-Pentetreotide. *Mol Ther* **10**(3), 553–561.

McCart JA, Puhlmann M, Lee J, *et al.* (2000). Complex interactions between the replicating oncolytic effect and the enzyme/prodrug effect of vaccinia-mediated tumor regression. *Gene Ther* **7**, 1217–1223.

McCart JA, Ward JM, Lee J, *et al.* (2001). Systemic cancer therapy with a tumor-selective vaccinia virus mutant lacking thymidine kinase and vaccinia growth factor genes. *Cancer Res* **61**(24), 8751–8757.

McIntosh AA, Smith GL (1996). Vaccinia virus glycoprotein A34R is required for infectivity of extracellular enveloped virus. *J Virol* **70**(1), 272–281.

Moon KB, Turner PC, Moyer RW (1999). SPI-1-dependent host range of rabbitpox virus and complex formation with cathepsin G is associated with serpin motifs. *J Virol* **73**(11), 8999–9010.

Moss B (1996). Genetically engineered poxviruses for recombinant gene expression, vaccination, and safety. *Proc Natl Acad Sci U S A* **93**, 11341–11348.

Moss B, Earl PL (1998). Expression of proteins in mammalian cells using vaccinia viral vectors. *Curr Prot Mol Biol* **43**, 16.

Najarro P, Traktman P, Lewis JA (2001). Vaccinia virus blocks gamma interferon signal transduction: viral VH1 phosphatase reverses Stat1 activation. *J Virol* **75**(7), 3185–3196.

Neeman M, Abramovitch R, Schiffenbauer YS, Tempel C (1997). Regulation of angiogenesis by hypoxic stress: from solid tumors to the ovarian follicle. *Int J Exp Pathol* **78**(2), 57–70.

Novick D, Kim SH, Fantuzzi G, Reznikov LL, Dinarello CA, Rubinstein M (1999). Interleukin-18 binding protein: a novel modulator of the Th1 cytokine response. *Immunity* **10**(1), 127–136.

Perera LP, Goldman CK, Waldmann TA (2001). Comparative assessment of virulence of recombinant vaccinia viruses expressing IL-2 and IL-15 in immunodeficient mice. *Proc Natl Acad Sci U S A* **98**(9), 5146–5151.

Perkus ME, Goebel SJ, Davis SW, *et al.* (1990). Vaccinia virus host range genes. *Virology* **179**(1), 276–286.

Puhlman M, Gnant M, Brown CK, Alexander HR, Bartlett DL (1999). Thymidine kinase-deleted vaccinia virus expressing purine nucleoside phosphorylase as a vector for tumor-directed gene therapy. *Hum Gene Ther* **10**(4), 649–657.

Puhlmann M, Brown CK, Gnant M, *et al.* (2000). Vaccinia as a vector for tumor-directed gene therapy: biodistribution of a thymidine kinase-deleted mutant. *Cancer Gene Ther* **7**(1), 66–73.

Ramirez JC, Gherardi MM, Rodriguez D, Esteban M (2000). Attenuated modified vaccinia virus Ankara can be used as an immunizing agent under conditions of preexisting immunity to the vector. *J Virol* **74**(16), 7651–7655.

Ray C, Black RA, Kronheim SR, *et al.* (1992). Viral inhibition of inflammation: cowpox virus encodes an inhibitor of the interleukin-1 beta converting enzyme. *Cell* **69**(4), 597–604.

Ricketts TF (1966). *Diagnosis of Smallpox*. Cassell and Company, London, pp. 6–13.

Ruby J, Bluethmann H, Aguet M, Ramshaw IA (1995). CD40 ligand has potent antiviral activity. *Nat Med* **1**(5), 437–441.

Sambhi SK, Kohonen-Corish MR, Ramshaw IA (1991). Local production of tumor necrosis factor encoded by recombinant vaccinia virus is effective in controlling viral replication *in vivo*. *Proc Natl Acad Sci USA* **88**(9), 4025–4029.

Scholl SM, Balloul JM, Le Goc G, *et al.* (2000). Recombinant vaccinia virus encoding human MUC1 and IL2 as immunotherapy in patients with breast cancer. *J Immunother* **23**(5), 570–580.

Seet BT, McFadden G (2002). Viral chemokine-binding proteins. *J Leukoc Biol* **72**(1), 24–34.

Sharma DP, Ramsay AJ, Maguire DJ, Rolph MS, Ramshaw IA (1996). Interleukin-4 mediates down regulation of antiviral cytokine expression and cytotoxic T-lymphocyte responses and exacerbates vaccinia virus infection *in vivo*. *J Virol* **70**(10), 7103–7107.

Smith GL, Vanderplasschen A (1998). Extracellular enveloped virus: entry, egress, and evasion. In: Enjuanes A (ed.), *Coronaviruses and Arteriviruses*. Plenum Press, New York, pp. 395–414.

Smith GL, Symons JA, Khanna A, Vanderplasschen A, Alcami A (1997). Vaccinia virus immune evasion. *Immunol Rev* **159**, 137–154.

Smith GL, Vanderplasschen A, Law M (2002). The formation and function of extracellular enveloped vaccinia virus. *J Gen Virol* **83** (Pt 12): 2915–2931.

Sutter G, Moss B (1995). Novel vaccinia vector derived from the host range restricted and highly attenuated MVA strain of vaccinia virus. *Dev Biol Stand* **84**, 195–200.

Symons JA, Alcami A, Smith GL (1995). Vaccinia virus encodes a soluble type I interferon receptor of novel structure and broad species specificity. *Cell* **81**(4), 551–560.

Thorne SH, Kirn DH (2004). Future directions for the field of oncolytic virotherapy: a perspective on the use of vaccinia virus. *Exp Opin Biol Ther* **4**(8), 1307–1321.

Thorne SH, Hwang TH, Kirn DH (2005). Vaccinia virus and oncolytic virotherapy of cancer. *Curr Opin Mol Ther* **7**(4), 359–365.

van Den Broek M, Bachmann MF, Kohler G, *et al.* (2000). IL-4 and IL-10 antagonize IL-12-mediated protection against acute vaccinia virus infection with a limited role of IFN-gamma and nitric oxide synthetase 2. *J Immunol* **164**(1), 371–378.

Vanderplasschen A, Mathew E, Hollionshead M, Sim RB, Smith GL (1998). Extracellular enveloped vaccinia virus is resistant to complement because of incorporation of host complement control proteins into its envelope. *Proc Natl Acad Sci U S A* **95**(13), 7544–7549.

Whitman ED, Tsung K, Paxson J, Norton JA (1994). *In vitro* and *in vivo* kinetics of recombinant vaccinia virus cancer-gene therapy. *Surgery* **116**, 183–188.

Wyatt LS, Carroll MW, Czerny C-P, Merchlinsky M, Sisler JR, Moss B (1998). Marker rescue of the host range restriction defects of modified vaccinia virus Ankara. *Virology* **251**(2), 334–342.

Yang H, Kim SK, Kim M, *et al.* (2005). Antiviral chemotherapy facilitates control of poxvirus infections through inhibition of cellular signal transduction. *J Clin Invest* **115**(2), 379–387.

Yu YA, Shabahang S, Timiryasova TM, *et al.* (2004). Visualization of tumors and metastases in live animals with bacteria and vaccinia virus encoding light-emitting proteins. *Nat Biotechnol* **22**(3), 313–320.

Zeh HJ, Bartlett DL (2002). Development of a replication-selective, oncolytic poxvirus for the treatment of human cancers. *Cancer Gene Ther* **9**(12), 1001–1012.

Zhou J, Sun XY, Fernando GJ, Frazer IH (1992). The vaccinia virus K2L gene encodes a serine protease inhibitor which inhibits cell-cell fusion. *Virology* **189**(2), 678–686.

10

Newcastle Disease Virus: a promising vector for viral therapy of cancer

Volker Schirrmacher and **Philippe Fournier**

10.1 Introduction

In recent years a large variety of viruses have been tested as vectors for oncolytic cancer therapy (Parato *et al.*, 2005), for gene therapy (Chlichlia *et al.*, 2005) or for immune stimulation (Schirrmacher, 2005a). RNA viruses are rapidly emerging as particularly promising agents for virotherapy of cancer. Integral to the lifecycle of all RNA viruses is the formation of double-stranded RNA (dsRNA), which activates a spectrum of cellular defence mechanisms involving interferons (IFN)-α and β. Mutations in tumour cells often cripple the IFN system to allow uninhibited proliferation and to provide resistance to apoptosis. Therefore tumours provide a relatively permissive substrate for the propagation of RNA viruses (Russell, 2002). The most promising ones are attenuated strains of mumps virus, Newcastle disease virus (NDV), measles virus (MV), vesicular stomatitis virus (VSV), human reovirus, poliovirus and influenza virus.

The avian paramyxovirus NDV (Alexander, 1997) is one of five species of viruses that are under clinical evaluation (Aghi and Martzua, 2005). It is a negative strand RNA virus with interesting antineoplastic (Cassel and Garret, 1965; Reichard *et al.*, 1991, Lorence *et al.* 1994) and immune-stimulating properties (Lorence *et al.*, 1988; Zeng *et al.*, 2002). Most remarkable is its capacity to induce strong type I interferon responses by viral protein (Zeng *et al.*, 2002) and RNA (Fournier *et al.*, 2003). Detection of foreign RNA induces an innate antiviral programme that initiates the transcription of RNA-responsive genes. The responses involve a multimodal machinery of gene regulation by the Interferon Regulatory Factor (IRF) family of transcription factors (Taniguchi and Takaoka; 2002) and link innate and adaptive immunity (LeBon and Tough, 2002).

William A. Cassel and Joseph G. Sinkovics were the two main pioneers of translational research in the 1960s and 1970s on oncolytic NDV strains (Sinkovics and Horvath, 2000) and on oncolysate mediated immunotherapy (Cassel *et al.*, 2005). Early success with the viral vaccine in humans was first reported in the USA in the early 1980s, but it is only recently that interest in NDV has resurfaced both in academia and private industry (Nelson, 1999).

10.2 Structure, taxonomy, pathogenicity and oncolytic properties of NDV

NDV is an enveloped virus with a negative sense, single-stranded RNA genome of 15 186 nucleotides. These code for a large (L) protein, haemagglutinin-neuraminidase (HN) protein, fusion (F)

Viral Therapy of Cancer Edited by Kevin J. Harrington, Richard G. Vile and Hardev S. Pandha

protein, matrix (M) protein, phosphoprotein (P) and nucleoprotein (NP) in the 5′–3′ direction (Alexander, 1997). The RNA-dependent RNA polymerase involves the proteins L, P and NP which are translated in infected cells at free ribosomes in the cytoplasm (Alexander, 1997; Wise *et al.*, 2004a).

During 1993 the International Committee on the Taxonomy of Viruses rearranged the paramyxoviruses and placed NDV within the Rubulavirus genus among the Paramyxovirinae subfamily. More recently, a new genus, Avulavirus, has been created for the avian-specific Paramyxovirinae (De Leeuw and Peeters, 1999). This distinction is supported by comparative nucleotide sequence analysis of the M, N and L genes (De Leeuw and Peeters, 1999; Seal *et al.*, 2002; Wise *et al.*, 2004a).

Isolates of NDV are categorized into three pathotypes depending on the severity of disease (Alexander, 1997). Lentogenic NDV isolates do not usually cause overt clinical signs in adult birds and are considered of low virulence. Viruses of intermediate virulence that cause respiratory disease, but are not usually fatal, are termed mesogenic. Among the highly virulent velogenic NDV isolates, there are viscerotropic forms marked by lesions of the digestive tract, while neurotropic forms are characterized clinically by respiratory and neurological signs. NDV has a wide host range with at least 27 of 50 orders of birds susceptible to infection, although there is wide variation in clinical response, even among species of the same genus (Kaleta and Baldauf, 1988).

The more virulent NDV strains have a furin cleavage site in their F protein (Peeters *et al.*, 1999; Römer-Oberdörfer *et al.*, 2003), which allows its activation in a proteolytic environment such as the tumour microenvironment. This allows multicyclic viral replication and cross-infection from one tumour cell to another. Cytopathic effects can be seen be a plaque assay in tumour monolayers. The possession of hydrophobic fusion peptides within the viral envelope promotes syncytium formation between infected tumour cells whereby the virus spreads without an extracellular phase leaving an oncolytic plaque. Lytic NDV strains have a high capacity for killing tumour cells. One infectious particle leads to death of at least 10 000 cancer cells in 2–3 days.

10.3 Human application and safety

NDV vaccines are used to treat human cancer and contain either the virus alone or virus-infected tumour cells from the patient's own tumour or tumour cell lines grown in the laboratory. Three cellular mechanisms have been proposed. First, oncolytic strains may simply kill the tumour cells directly. Second, for non-lytic strains, the viral proteins inserted into the tumour cell membrane after infection may enhance an immune response. Finally, the virus itself may stimulate the host to produce cytokines such as interferons or tumour necrosis factor (TNF) which in turn lead to the activation of natural killer (NK) cells, macrophages and sensitized T cells. When applied to humans, this avian virus usually induces only mild fever for a day. Severe adverse effects have not been reported in spite of applications in several thousands of people over two decades in Europe and the US (Nelson, 1999).

Certain NDV strains have been applied in the past to prepare oncolysates or were given directly systemically (Table 10.1). Different studies were reported in the nineties on the treatment of patients with oncolysates obtained with the NDV strains 73-T and Italian. These non-randomized and not prospective studies suggested benefits to the treated patients. Some case reports have been published on the systemic use of NDV strains in patients as direct anticancer reagent: NDV strain Hickman in myelogenous leukaemia (Wheelock and Dingle, 1964) and NDV strain MTH-68/H in glioblastoma (Csatary and Bakacs, 1999; Csatary *et al.* 2004). Concomitant treatments made the interpretation of tumour responses difficult. A phase II study performed in Hungary based on inhalation of MTH-68/H by 33 advanced chemo-refractory patients suggested a better survival among these patients (Csatary *et al.*, 1993). In recent years, the systemic application of NDV has been studied more systematically. The NDV strain HUJ (replication-defective, lentogenic; Theravir,

Table 10.1 Clinical studies on cancer using different NDV strains

NDV Strain	Treatment	Disease (route)	Clinical phase	Clinical observations (n)	References
Immunotherapy with NDV oncolysates					
73-T	Oncolysate (i.d.)	Malignant melanoma (stage II) ($n = 83$)	Phase II ($n = 110$)	Over 60% OS and DFS(10 years)	Cassel and Murray, 1992
		Head and neck cancer ($n = 23$)		Unusual DFS (10 years)	
		Cerebral metastases ($n = 6$)		Unusual DFS (10 years)	
	Oncolysate + IL-2 + IFN2a (i.d.)	Advanced renal cancer	Phase II ($n = 208$)	Improved DFS	Anton *et al.*, 1996 Kirchner *et al.*, 1995
Italien	Oncolysate (i.d.)	Breast or ovarian metastatic	Phase II (n > 100)	Increased OS	Mallmann *et al.*, 1991 Mallmann *et al.*, 1992
			Phase II ($n = 22$)	9 complete/partial responses	Mallmann, 1993
Ulster	Oncolysate[a] (i.d.)	Glioblastoma multiforme	Phase II (n = 11)	Only a peripheral immune response	Schneider *et al.*, 2001
Systemic application of NDV					
Hickman	Virus (i.v.)	Acute myeloid myeloma	Case report (n = 1)	Partial response	Wheelock and Dingle, 1964
MTH-68	Virus (i.v.)	Glioblastoma multiforme	Case series (n = 4)	Survival rates of at least 5–9 years	Csatary *et al.*, 2004 Csatary and Bakacs, 1999
	Virus (inhal.)	Advanced chemorefractory cancer	Phase II (n = 33)	18 responses (2 complete) OS (2 years) 7/33 (versus 0/26 in the control group) Improved quality of life	Csatary *et al.*, 1993
HUJ	Virus (i.v.)	Glioblastoma	Phase I/II (n = 14)	1 complete (but not durable) remission, 4 responses	Freeman *et al*, 2006
PV701 (MK107)	Virus (i.v. bolus)	Advanced chemorefractory cancer	Phase I (n = 79)	2 major responses 4 year survival in a patient with mesothelioma	Pecora *et al.*, 2002
	Virus (i.v. with desensitisation)	Advanced chemorefractory cancer	Phase I (n = 16)	5 responses	Laurie *et al.*, 2006
	Virus (i.v. with low infusion rate)	Advanced chemorefractory cancer	Phase I (n = 18)	6 responses (4 major; 2 minor) 6 survivals for at least 2 years	Hotte *et al.*, 2007

OS, overall survival; DFS, disease free survival; i.v., intravenous; i.d., intradermal; inhal., inhalation; n, number of treated patients.
[a]Obtained after addition 3 h after NDV infection of the cytotoxic agent cisplatinum at a concentration of 500 μg/ml during 16 h.

Jerusalem, Israel) was administered intravenously (i.v.) to 14 glioblastoma patients using intrapatient dose escalation (Freeman *et al.*, 2006). A complete response lasting 3 months was reported. Replication-independent mechanisms may cause efficacy with this attenuated strain.

Even replication competent, oncolytic NDV *(PV 701)* is well tolerated in patients with advanced solid cancer in doses of at least 3×10^9 infectious particles by the i.v. route and of at least 4×10^{12} infectious particles by the intratumoral route (Pecora *et al.*, 2002; Lorence *et al.*, 2003). Dose-limiting toxicities included dyspnoea, diarrhoea, and dehydration (Pecora *et al.*, 2002). When patients were desensitized with a lower initial dose, the maximum tolerated dose (MTD) was increased 10-fold. Two objective responses were reported, and six patients had stable disease for 6 months. Virus was shown in only one tumour biopsy, however. A subsequent two-step i.v. desensitization trial reported improved patient tolerance and a single lymph node metastasis response (Laurie *et al,*. 2006). Another phase I study (Hotte *et al.*, 2007) suggests that toxicity can be significantly modulated through altering the infusion rate. Slowing the rate of i.v. infusion and a desensitization protocol also reduced toxicity and increased the MTD of PV701.

What is remarkable is that systemic applications of very high doses of NDV have been extremely well tolerated. NDV was associated with transient thrombocytopenia and diffuse vascular leak. Out of all the patients of these clinical studies of i.v. treated patients, only one possibly treatment-related death (in a terminal patient) was reported. This death was associated with rapid tumour lysis in the lungs by PV701. This compares favourably with safety problems of other phase I oncology studies.

10.4 Tumour-selective replication of NDV

We recently compared several tumorigenic human cell lines to non-tumorigenic human cells from the blood for sensitivity to become infected by a recombinant lentogenic strain of NDV with incorporated transgene EGFP (Fiola *et al.*, 2006). While fluorescence signals in non-tumorigenic cells were only weak or missing completely, a massive and long-lasting transgene expression was observed in all tumour cell lines. The majority of tumour cells could be infected and viral replication was associated with an increase in the cell surface density of viral antigens. A further analysis revealed several defects of tumour cells in their antiviral defence responses: they showed no response to UV inactivated NDV while non-tumorigenic cells reacted with induction of high levels of the antiviral enzymes PKR (Clemens and Elia, 1997), a dsRNA-responsive protein kinase, and MxA (Haller and Kochs, 2002), a dynamin-like GTPase with antiviral activity. Upon co-incubation with live NDV, tumour cells showed a delayed response in increased expression of the antiviral enzymes in comparison to peripheral blood mononuclear cells (PBMC). While in non-tumorigenic cells the replication cycle of NDV stopped after production of positive strand RNA (Fiola *et al.*, 2006), tumour cells continued in the replication cycle and copied viral genomes 10–50 h after infection. Thus, NDV can replicate up to 10 000 times better in human cancer cells than in most normal human cells.

10.5 Virally based cancer immunotherapy and danger signals

The first published 'danger model' of immunity (Matzinger, 1994) proposed only one mechanism for immune recognition of danger: that perceived by dendritic cells (DC) upon release of cellular contents following necrosis of a diseased cell in its neighbourhood. This model predicts a superior effect of a lytic as opposed to a non-lytic virus in the treatment of tumours, because tumour cells necrotically destroyed by the virus would be phagocytosed and perceived as dangerous by DCs. In such a process these professional antigen presenting cells (APC) would (i) process tumour-associated antigens (TAA), (ii) become activated and (iii) present processed TAA peptides to T cells for cognate interaction and immune response induction (Matzinger, 2002).

Recently, a modified medical hypothesis has been formulated which suggests that T lymphocytes themselves correlate danger signals to antigen (Forden, 2004). This model associates danger also with non-lytic viruses if these are upregulating danger signals in their host cells. Such an event will quickly cause its host cells to be killed by the immune system, probably in a variety of ways. Killed infected tumour cells are likely to result in TAA being presented by DC along with potent costimulation. Recently, it was shown that dsRNA in the apoptotic bodies of virus-infected dead cells is recognized by CD8 α^+ DCs that have high expression of Toll-like receptor 3 (TLR-3) (Kawai and Akira, 2005; Schulz *et al.*, 2005). This promotes cross-priming of T cells to virus-infected cells (Schulz *et al.*, 2005). Since the immune system as a whole must correlate antigens to disease, it makes great sense that T cells should gather danger information, both because of their antigen specificity and because they continually contact host cells throughout most tissues in the body as they scan them for the presence of their cognate antigens. In response to danger signals, both myeloid and plasmacytoid DC precursors (Liu, 2005) are rapidly mobilized into the circulation. After accumulating in the lymph nodes through distinct trafficking pathways (Yoneyama *et al.*, 2005), DC interact with lymphocytes temporally and spatially to establish effective immune responses. This involves cognate APC–T-cell interactions and interferons.

We have recently proposed that one might be able to 'condition' the immune system to associate recognition of TAA with danger (Schirrmacher, 2005b). This hypothesis is based on conditioning experiments involving immune responses in animal models. Training could be performed by repeated vaccination with vaccine expressing TAA and danger signals. This may 'educate' DC (Colonna *et al.*, 2005) and memory T cells (MTC) (Kaech *et al.*, 2002) to integrate the multiple signals derived from the virus and the tumour cells to direct and 'fine-tune' the immune response accordingly. Once the immune system's MTC are trained, possibly also via involving neuronal networks to associate TAA recognition with danger (Remondini *et al.*, 2003), they may not require the danger signal any longer and can react to TAA without being tolerized (Matzinger, 1994).

10.6 NDV: a danger signal inducing vector

Features of viruses that can enhance their potency as vaccine vectors include their ability to induce immunological danger signals at the site of infection. Features of viruses that can diminish their potency as vaccine vectors include the presence of virally encoded inhibitors of immunity such as TAP inhibitors (Gewurz *et al.*, 2001), cytokine decoys (Hengel *et al.*, 2005) and viral proteins that block TLR- and RIG-1-Mda5-dependent signalling pathways to antagonize type I IFN induction (Horvath *et al.*, 2004; Hengel *et al.*, 2005). The retinoic acid inducible gene RIG-1 (Yoneyama *et al.*, 2004) is a cytoplasmic dsRNA receptor capable of stimulating activation of IRF-3 thereby mediating its nuclear translocation and induction of IFN-α. The evolutionary distance of NDV as an avian virus from mammalian cells has prevented its adaptation to the latter. It thus did not develop any known mechanism to evade immune responses in mammals including man. The release of high amounts of IFN-α by plasmacytoid DC from rodents or humans upon contact with NDV indicates that these cells of the innate immune system have sensed danger. It has only recently become clear that IFN-α has an important adjuvant function in the immune response. It activates DCs (LeBon and Tough, 2002), induces TRAIL in NK cells (Sato *et al.*, 2001) and monocytes (Washburn *et al.*, 2003) and the interleukin (IL)-12 receptor β chain in T cells (Rogge *et al.*, 1997). Together with IL-12, IFN-α polarizes the T cell towards a cell-mediated T helper 1 (Th1) response characterized by delayed-type hypersensitivity (DTH) and cytotoxic T lymphocyte (CTL) activity. In addition, IFN-α induces the upregulation of molecules which are important for antigen recognition (e.g. HLA (Washburn and Schirrmacher, 2002), cell–cell interaction (e.g. cell adhesion molecules, CAM (Washburn and Schirrmacher, 2002) and cytotoxicity (e.g. TRAIL (Sato *et al.*, 2001, Washburn *et al.*, 2003).

Interferons triggered by microbial components, for instance via Toll-like receptors (Kawai and Akira, 2005), initiate a widespread cellular resistance response via induction of antiviral proteins (Taniguchi and Takaoka, 2002). In its natural host, the bird, NDV has developed, like other viruses (Hengel *et al.*, 2005, Horvath, 2004), immune escape mechanisms. NDV can evade IFN-mediated responses via the V-protein which interferes with STAT mediated interferon signals (Horvath, 2004). This immune evasion mechanism of NDV, however, is species restricted (Park *et al.*, 2003). Therefore, in rodent and mammalian non-malignant cells, the NDV induced interferon response is capable of preventing viral replication (Fiola *et al.*, 2006).

10.7 The human cancer vaccine ATV-NDV

The virus-modified tumour vaccine for human application developed by us consists of virus infected intact viable and irradiated autologous tumour cells. This strategy is based on preclinical studies in metastatic animal tumours. Antimetastatic effects were observed after local postoperative vaccination with NDV-infected autologous tumour cells (Heicappell *et al.*, 1986). The vaccination activated a tumour-line specific T-cell mediated immune response, which also protected against a second challenge with the same tumour line (Schirrmacher and Heicappell, 1987).

10.7.1 Reasons for the use of autologous tumour cells

A tumour vaccine normally consists of a specific component containing one or more TAA and a non-specific component with an adjuvant function to augment the immune response against TAA. The question of how to define and select TAA that are suitable as targets for immunotherapy of metastases and for inducing long-term protective immunity is important but still unresolved. A major aspect relates to true tumour rejection antigens (TRA) (Coggin *et al.*, 2005) and to the choice between common (shared) or unique (individual) TAA (Lewis, 2004). The latter may be more important for tumour rejection responses because corresponding T-cell receptors are expected to have higher affinities and because T cells with specificity for a unique mutant peptide were found to dominate the immune response in comparison to T cells recognizing shared antigens. The logical extrapolation from this is to use an individualized approach for tumour vaccine generation. The promise of cancer vaccines has been discussed (Gilboa, 1997). The specific components of the vaccine that we developed are patient-derived (autologous) live tumour cells (ATV).

10.7.2 Introduction of foreign viral genes and of 'danger signals'

Hiroshi Kobayashi was first to introduce the concept of viral xenogenization of intact tumour cells (Kobayashi, 1997) as a means to increase the tumour cells' immunogenicity. Virus infection of tumour cells is meanwhile a frequently used procedure in human gene therapy for transferring virus incorporated cloned therapeutic genes. Such infection, however, is often limited by the number of tumour cells that can reliably be infected as well as by issues of selectivity and safety. In the case of infection with NDV, a large variety of human tumour cells, either freshly isolated from operation specimens or derived from cell culture, were shown to be efficiently infectable (Schirrmacher *et al.*, 1999). Viral replication was found to be independent of tumour cell proliferation which can be explained by the cytoplasmic replication cycle of this RNA virus. 4–48 h after infection with the lentogenic strain NDV *Ulster*, two of the six viral genes, e.g. HN and F, modify the tumour cell surface where they strongly increase in cell surface density (Schirrmacher *et al.*, 1997) and introduce new cell adhesion molecules (e.g. HN) for lymphocyte interactions (Schirrmacher *et al.*, 1997) and T cell co-stimulation (Ertel *et al.*, 1993). In addition, human tumour cell infection by NDV leads to upregulation of human leukocyte antigen and intracellular adhesion molecule-1 and to induction of interferons, chemokines and finally apoptosis (Washburn and Schirrmacher, 2002). dsRNA, a by-product of viral replication, can activate cytoplasmic PKR (Clemens and Elia, 1997) as well as the cytoplasmic RNA-dependent helicase RIG-1 (Kato *et al.*, 2005) and also TLR-3 (Alexopoulou *et al.*, 2001).

Thus, the rationale of this ATV-NDV vaccine is to link multiple TAA from individual patient derived tumour cells with multiple danger signals derived from the infection by NDV (dsRNA, IFN-α, HN cell surface protein (Zeng *et al.*, 2002). This allows activation of multiple innate immune responses (monocytes (Washburn *et al.*, 2003), dendritic cells (Bai *et al.*, 2002) and NK cells (Sato *et al.*, 2001)) as well as adaptive immune responses (CD4 and CD8 T cells (Ertel *et al.*, 1993, Von Hoegen *et al.*, 1990) (summarized in Schirrmacher *et al.*, 1998, Schirrmacher, 2005b).

10.8 Pre-existing antitumour memory T cells from cancer patients and their activation by antitumour vaccination with ATV-NDV

Antigen-specific MTC could be an ideal source for effective immunotherapy since MTC show higher frequencies and exert stronger immune responses than naïve T cells (Kaech *et al.*, 2002). Previous studies revealed that tumour-or pathogen-specific MTCs are enriched in the bone marrow (BM) of mice and humans (Feuerer *et al.*, 2001a, b). Such cells have 'central' or 'effector' memory phenotype and can be re-stimulated *in vitro* by autologous DC loaded with TAA to produce IFN-γ and to become cytotoxic (Beckhove *et al.*, 2004). Re-stimulated human MTCs but not naïve T cells infiltrated autologous tumour but not normal skin transplants and caused tumour regression after transfer into tumour-xenotransplanted non-obese diabetic–severe combined immunodeficient mice (Feuerer *et al.*, 2001b, Beckhove *et al.*, 2004). The therapeutic efficiency of MTC was augmented upon co-transfer of TAA loaded DC (Bai *et al.*, 2003).

Bone Marrow (BM) is known as a primary lymphoid organ involved in haemato- and lymphopoiesis. Surprisingly, we were able recently to demonstrate that naïve antigen-specific T cells migrated to the BM, where they were primed by resident antigen-presenting DC suggesting that BM can exert also secondary immune functions (Feuerer *et al.*, 2003). Activation of naïve CD8 T cells requires at least three signals which are derived from (i) antigen, (ii) costimulation and (iii) Th1 type cytokines (IL-12, IFN-α) (Curtsinger *et al.*, 2003).

Pre-existing antitumour memory T cells from cancer patients could apparently be activated *in situ* by the tumour vaccine ATV-NDV as seen by augmentation of antitumour DTH memory responses (Schirrmacher, 2005b). The conclusion that ATV-NDV vaccine can present TAA directly to MTC and stimulate them is supported by the following results: (i) in a coculture with a TAA-specific memory T-cell clone, ATV-NDV stimulated T-cell proliferation and IL-2 production, while ATV without NDV infection induced tolerance (Termeer *et al.*, 2000); (ii) viability of the irradiated vaccine was important for CTL activation (Schirrmacher and von Hoegen, 1993) and for clinical efficacy (Ahlert *et al.*, 1997); and (iii) APC transfected with the viral HN cDNA showed increased CTL stimulatory capacity (Ertel *et al.*, 1993). The same was true for NDV infection of tumour-stimulatory cells (Von Hoegen *et al.*, 1990). A strong increase of antitumour DTH reactivity (>5 mm) to tumour challenge after vaccination was observed and correlated with a survival advantage (Pomer *et al.*, 1995).

10.9 Clinical trials of antitumour vaccination with ATV-NDV

10.9.1 Phase I studies

After having optimized a technical procedure for isolating live tumour cells from freshly operated tumour specimens and having calculated average yields and stability parameters we started to perform phase I clinical studies. In the ESb animal tumour model we had described that an optimal vaccine composition which yielded 50 per cent survival benefit after a single inoculation (Heicappell *et al.*, 1986) was composed of 10^7 irradiated tumour cells infected by 32 haemagglutinating units (HU) of NDV *Ulster*. NDV

Ulster is first adsorbed to the isolated human tumour cells *in vitro* (1-h binding). Then the virus-modified tumour vaccine is injected intradermally thus allowing for virus replication *in vivo* at the site of vaccine application. NDV *Ulster* has a monocyclic abortive replication cycle in tumour cells (Washburn *et al.*, 2002) which takes about 6–40 h. This time is sufficient to generate anti-tumour DTH skin responses, which are dependent on pre-existing TAA-specific memory T cells. We selected the non-lytic strain *Ulster* for reasons of safety during application in cancer patients and also because we intended to develop a whole cell cancer vaccine consisting of virus-infected intact viable irradiated cancer cells.

The first systematic optimization studies were performed in breast carcinoma (Ahlert *et al.*, 1997), colorectal carcinoma (Lehner *et al.*, 1990) and in renal carcinoma patients (Pomer *et al.*, 1995). Two to 3 weeks after primary tumour operation, the vaccine, inactivated by 200 Gy γ-irradiation, was applied intradermally at the upper thigh. Optimal skin reactions were observed with 1×10^7 tumour cells infected with 32 HU NDV *Ulster*. With this vaccine formulation, 85 per cent of colorectal carcinoma patients and 90 % of renal carcinoma patients showed about 7–11 mm skin indurations at the vaccination site. We then evaluated skin responsiveness in patients not only to the vaccine ATV-NDV but also to autologous tumour cells without virus (ATV) and to various controls. The DTH responsiveness to ATV could be distinguished from recall antigen responses of the MÕrieux test and correlated with *in vitro* enzyme-linked immunospot tests (ELISPOT) thus demonstrating distinct respective memory responses to TAA in individual patients (Bai *et al.*, 2003). With the ELISPOT assay, we also found out that DC, when pulsed with viral oncolysates from the ATV-NDV vaccine, stimulate antitumour MTC responses from cancer patients more strongly than when pulsed with ATV-derived tumour lysate (Bai *et al.*, 2002). Supernatants from co-cultures of MTC and viral oncolysate pulsed DC contained increased titres of IFN-α and IL-15 (Bai *et al.*, 2002).

We were able to potentiate antitumour DTH reactivity by repeated vaccinations with ATV-NDV. Among 264 patients tested, 44 per cent showed significantly increased DTH reactivity to ATV after a course of three vaccinations. In additional 23 per cent of patients we induced *de novo* DTH immune reactivity to ATV (Schirrmacher, 2005b, Schirrmacher, 2005c). The intradermal vaccinations were well tolerated and could be repeated many times without causing serious problems. A few patients developed mild fever and/or mild headache for 1–2 days. There was no evidence of autoimmune phenomena, such as vasculitis, rheumatoid arthritis or lymphatic disorders (Schirrmacher *et al.*, 1998, Schirrmacher, 2005c)

10.9.2 Phase II studies

A variety of phase II studies were performed over the last 10 years to evaluate the efficacy of post-operative antitumour vaccination with ATV-NDV (Table 10.2). The results have recently been summarized (Schirrmacher *et al.*, 1998, Schirrmacher, 2005c). In a phase II study of primary operated breast cancer patients, the 5-year survival rate in vaccinated patients was more than 30 per cent higher than in a comparable control group (Ahlert *et al.*, 1997, Schirrmacher, 2005c). In two recent studies, patients received a well defined ATV-NDV vaccine derived from 10^7 cell culture adapted autologous tumour cells. In the head and neck squamous cell carcinoma study (Karcher *et al.*, 2004) 61 per cent of the 20 vaccinated patients were still alive after 5 years, which compares favourably with the expected 38 per cent under standard therapy. Our last phase II clinical study was performed with patients suffering from Glioblastoma multiforme (Steiner *et al.*, 2004), a most devastating brain tumour disease. The median overall survival of 23 vaccinated patients was twice as long as that of a non-vaccinated control group involving 87 patients from the same hospital. There was one complete remission of remaining brain tumour after operation and several long-term surviving patients who had developed a specific antitumoral long-term memory (Steiner *et al.*, 2004).

Table 10.2 Clinical studies done with ATV-NDV

Disease	Clinical phase (n)	Clinical observations	References
Colorectal carcinoma (locally advanced)	Phase II (n = 57)	Improved OS and DFS (2 and 5 years)	Ockert *et al.*, 1996 Schirrmacher *et al.*, 1998
Colorectal carcinoma (RO resected liver metastasis)	Phase II (n = 23)	Improved OS and DFS (2 years)	Schlag *et al.*, 1992
Malignant melanoma	Phase II (n = 21)	Improved OS and DFS (2 years)	Schirrmacher *et al.*, 1998
Glioblastoma multiforme	Phase II (*n* = 25)	Improved OS and DFS (2 years)	Steiner *et al.*, 2004
Breast carcinoma (locally advanced)	Phase II (*n* = 32)	Improved OS and DFS (5 years)	Ahlert *et al.*, 1997
Head and neck squamous cell carcinoma (stage III + IV)	Phase II (n = 18)	Improved OS and DFS (5 years)[a]	Karcher *et al.*, 2004
Ovarian carcinoma	Phase II (*n* = 82)	Improved OS and DFS	Möbus *et al.*, 1993
Renal cell carcinoma (advanced)	Phase II (n = 40)	Improved OS and DFS (4 years)°	Pomer *et al.*, 1995
Pancreatic carcinoma (stage G3)	Phase II (n = 53)	Improved OS and DFS	Schirrmacher *et al.*, 1998

OS, overall survival; DFS, disease free survival; *n*, number of treated patients.
[a]IL-2 and IFN-alpha 2b were added to the ATV-NDV vaccine just before application.

10.9.3 A unifying hypothesis about the underlying mechanism

We propose in a unifying hypothesis that cancer patients contain pre-existing tumour-reactive memory T cells which rest in particular reservoirs such as the bone marrow. These can be re-activated by antitumour vaccination with ATV-NDV and mobilized to circulate in the blood and to infiltrate tumour tissues thereby causing antitumour effects.

By studying a 'tumour dormancy' situation in detail in an animal model, we found that MTC can control residual tumour cells (Schirrmacher *et al.*, 2001). Conversely, TAA produced by such dormant tumour cells has an influence on the memory cells. This was found to be important for the maintenance of long-term tumour-reactive memory (Mahnke *et al.*, 2005). Active specific immunotherapy (ASI) by ATV-NDV appears to re-activate tumour-reactive memory cells and to elevate the level of circulating TAA specific MTC (Karcher *et al.*, 2004, Steiner *et al.*, 2004). We propose that a certain threshold of tumour-reactive immune memory is necessary for the control of residual tumour cells that remain after most therapies and also for long-term survival of treated cancer patients (Schirrmacher, 2005b). Provided there is sufficient memory, a situation of a balance between immune control and residual tumour may persist in patients for varying time periods. A recent study on circulating tumour cells in patients with breast cancer dormancy suggested a balance between tumour replication and cell death for as long as 22 years (Meng *et al.*, 2004). If this hypothesis about the importance of a threshold of antitumour memory for the control of residual cancer is correct, a change of paradigm has to occur in clinical oncology as described in more detail elsewhere (Schirrmacher, 2005b).

10.10 NDV-specific recombinant bispecific antibodies to augment antitumour immune responses

To further increase the immunogenicity of the human tumour cell vaccine ATV-NDV, we have added a second step after the virus infection: attachment of one or two immunostimulatory molecules to the vaccine. While virus infection introduces danger signals to activate innate immune responses (see 10.6, above), the new bispecific T-cell stimulatory molecules are designed to activate adaptive immune responses (Haas *et al.*, 2005a). When bound via one arm to HN molecules of the ATV-NDV vaccine, the other arm (either anti CD3 or anti CD28) bound to T cells and caused a polyclonal T-cell response. Such response could be useful to overcome the potential various T-cell suppressive mechanisms of tumour cells. We demonstrated the induction of strong antitumour activity in human lymphocytes upon coincubation with a virus modified tumour vaccine containing anti-CD3 and anti-CD28 bispecific antibodies. Peripheral blood mononuclear cells or purified T cells that were coincubated with such a tumour vaccine for 3 days were able to destroy monolayers of human breast carcinoma or other carcinoma cells (Haas *et al.*, 2005b). Serial transfer to new tumour cell monolayers revealed antitumour cytotoxic activity in such effector cells that lasted as long as about 10 days. Non-tumour target cells appeared to be much less sensitive to the activated effector cells. Antitumour activity of the activated effector cells was mediated through soluble factors as well as through direct cell contact of effector cells with the non-targeted bystander tumour cells (Haas *et al.*, 2005b). The new reagents can not only be combined with the ATV-NDV vaccine but also with NDV based systemic oncolytic virus therapy. They could potentially boost post-oncolytic anti-tumour immunity which appears to be very important for cancer patients' survival.

10.11 NDV-binding bispecific fusion proteins to improve cancer specific virus targeting

NDV binds to ubiquitously expressed sialic acid containing cellular receptors (Suzuki *et al.*, 1985, Ferreira *et al.*, 2004). In the case of systemic administration of NDV, its binding to normal cells, including large amounts of erythrocytes, could prevent it from reaching the tumour tissues and could cause undesired side effects. Since efficient distribution at the tumour site may be one of the most critical parameters for tumour selective gene delivery and for antitumour efficacy of oncolytic virotherapy (Demers *et al.*, 2003), we have developed adaptor molecules that redirect the virus to tumour tissue (Bian *et al.*, 2005a). The targeting molecule used, αHN-IL-2, contains an *scFv* antibody cloned from a neutralizing HN specific hybridoma linked to the human cytokine IL-2. The targeted gene delivery to IL-2 receptor positive tumour cells was blocked by the target ligand human IL-2 (Bian *et al.*, 2005a). Selective virus entry was observed *in vitro* in a mixture of target positive and target negative human tumour cell lines (Bian *et al.*, 2005a and b). Retargeted virus infection of tumour cells required specific binding via the bispecific fusion protein and membrane fusion via the viral F-protein (Bian *et al.*, 2005b). When this approach was tested *in vivo* after systemic virus inoculation into tumour-bearing mice, the modification of NDV by the bispecific protein did not compromise the efficiency of gene delivery into target positive tumours (Bian *et al.*, 2005c) but greatly reduced viral gene expression in target negative tumours and in normal tissues (liver, spleen, kidney and lung) and thus reduced side effects (Bian *et al.*, 2006).

10.12 Recombinant NDV as a new vector for vaccination and gene therapy

Attenuated live NDV vaccines have been widely used to control Newcastle Disease for more than 50 years and their use has tremendously contributed to the world wide development of today's highly efficient poultry industry (Senne *et al.*, 2004). A new rapid diagnostic RT-PCR test was recently developed (Wise *et al.*, 2004b), which also differentiates low virulent from virulent NDV, thus minimizing the disadvantage of live virus vaccines in the face of an outbreak. Despite the availability of live virus vaccines of good potency, the intrinsic ability of attenuated strains to revert in virulence makes control of this disease by vaccination difficult. Through the use of recombinant DNA technology, it is now possible to generate recombinant strains from non-segmented negative-sense RNA viruses (Huang *et al.*, 2003). By introducing multiple gene deletion mutations in the genome, it is possible to generate a new class of attenuated vaccines that are safe and may not result in reversion to virulence. Reverse genetic systems were established to recover recombinant NDV (rNDV) entirely from cloned cDNA (Peeters *et al.*, 1999). This unique molecular genetic methodology provides the means not only to investigate the functions of various virus encoded genes but also to allow the use of these viruses to express heterologous foreign genes. In this way, NDV could be more specifically designed and its efficacy enhanced.

Certain characteristics of NDV suggest that rNDV expressing a foreign protein would be very good vaccine candidates. NDV grows to very high titres in many cell lines and eggs and it elicits strong humoral and cellular immune responses *in vivo*. NDV naturally infects via respiratory and alimentary tract mucosal surfaces, so it is especially useful to deliver protective antigens of respiratory disease pathogens. The foreign protein is expressed together with only a few NDV proteins. In contrast, pox and herpes virus vectors express a large number of additional proteins from their large-size genomes. NDV replicates in the cytoplasm of the infected cells without a DNA phase which eliminates the problem of integration of the viral genome into the host cell DNA. The virus does not undergo detectable genetic recombination (Toyoda *et al.*, 1989), which make this expression vector stable and safe.

Recent studies have demonstrated the potential of NDV as a vaccine vector. The results showed that the expression levels of foreign proteins [e.g. CAT (Huang *et al.*, 2001) or influenza virus haemagglutinin (Nakaya *et al.*, 2001)] are quite high and the foreign genes are very stable after many passages *in vitro* and *in vivo*. rNDV with reduced V protein expression was found to lack pathogenicity for chicken embryos (Mebatsion *et al.*, 2001) and rNDV expressing foreign HN from human parainfluenza virus type 3 (Bureyev *et al.*, 2005) were found to be highly immunogenic in primates.

10.13 Conclusion

The modular nature of transcription, undetectable rate of recombination and the lack of a DNA phase in the replication cycle make NDV a suitable candidate for the rational design of a safe live attenuated vaccine and gene therapy vector. A deeper understanding of NDV molecular biology may be subsequently applied to the control of Newcastle Disease and other avian diseases including *IBDV* (Huang, *et al.*, 2004) and avian influenza H5N1. There has been a recent claim from the Chinese newspaper *China Daily* of the successful construction of rNDV expressing avian influenza H5.

For gene therapy of cancer, one can envisage rNDV with heterologous genes coding for cytokines, chemokines or co-stimulatory ligands in order to increase antitumour immune responses. Alternatively, the new gene products could be targeted against oncogenes, growth factors or against angiogenesis. Besides being a vector for polyvalent vaccines, NDV can be engineered as a surrogate virus in which the viral envelope can be completely replaced with other viral envelope proteins or by chimeric envelope proteins. This will enable manipulation of host range and cell tropism. In this way, NDV can be engineered to target specific tumour tissues.

In conclusion, the advantages of using NDV as vector are its safety and tolerability in cancer patients, and its efficient and selective replication in cancer tissue. Oncolytic strains of NDV kill cancer cells efficiently. The viral vector is able to augment autologous antitumour immune responses. There may be additional mechanisms *in vivo* such as cell fusion and syncytium formation which allow viral escape from neutralizing antibodies. The general human population is seronegative when tested against NDV antigens. The viral vector is not able to lead to cellular transformation and a robust virus production and manufacturing system is available. NDV is thus a promising vector for viral therapy of cancer. Its inherent anti-neoplastic properties can now be explained by direct oncolytic effects as well as by indirect host immune system mediated effects. The effectiveness of a combined oncolytic/

immunostimulatory tumour therapy approach was recently demonstrated (Bernt *et al.*, 2005).

Note added in proof

Recent results of a prospective randomized trial of adjuvant, active, specific immunization after resection of liver metastases, revealed that the ATV-NDV vaccine was effective in colon carcinoma, but not in rectum carcinoma patients (Schulze *et al.*, *Cancer Immunology Immunotherapy*, in press, 2008).

References

Aghi M, Martzua RL (2005). Oncolytic viral therapies – the clinical experience. *Oncogene* **24**, 7802–7815.

Ahlert T, Sauerbrei W, Bastert G, *et al.* (1997). Tumor cell number and viability as quality and efficacy parameters of autologous virus modified cancer vaccines. *J Clin Oncol* **15**, 1354.

Alexander DJ (1997). Newcastle disease and other Paramyxoviridae infections. In: Calnek, BW, Barnes HJ, Beard CW, McDougald, L, Saif JYM (eds), *Diseases of Poultry*, 10th edition. Iowa State University, Ames IA, pp. 541–569.

Alexopoulou L, Holt AC, Medzhitov R, Flavell RA (2001). Recognition of double-stranded RNA and activation of NF-kappa B by toll-like receptor. *Nature* **413**, 732.

Anton P, Kirchner H, Jonas U *et al.* (1996). Cytokines and tumor vaccination. *Cancer Biother Radiopharm* **11**(5), 315–318.

Bai L, Koopmann J, Fiola C, Fournier P, Schirrmacher V (2002). Dendritic cells pulsed with viral oncolysates potently stimulate autologous T cells from cancer patients. *Int J Oncol* **21**, 685–694.

Bai L, Beckhove P, Feuerer M, *et al.* (2003). Cognate interactions between memory T cells and tumor antigen presenting dendritic cells from bone marrow of breast cancer patients: bi-directional cell stimulation, survival and anti-tumor activity *in vivo. Int J Cancer* **103**, 73–83.

Beckhove P, Feuerer M, Dolenc M, *et al.* (2004). Specifically activated memory T cell subsets from cancer patients recognize and reject xenotransplanted autologous tumors. *J Clin Invest* **114**, 67–76.

Bernt KM, Ni S, Tieu A-T, Lieber A (2005). Assessment of a combined, adenovirus-mediated oncolytic and immunostimulatory tumor therapy. *Cancer Res* **65** (10), 4343–4352.

Bian H, Fournier P, Moormann R, Peeters B, Schirrmacher V (2005a). Selective gene transfer *in vitro* to tumor cells via recombinant Newcastle Disease Virus. *Cancer Gene Ther* **12**, 295–303.

Bian H, Fournier P, Moormann R, Peeters B, Schirrmacher V (2005b). Selective gene transfer to tumor cells by recombinant Newcastle Disease Virus via a bispecific fusion protein. *Int J Oncol* **26**, 431–439.

Bian H, Fournier P, Peeters B, Schirrmacher V (2005c). Tumor-targeted gene transfer *in vivo* via recombinant Newcastle Disease Virus modified by a bispecific fusion protein. *Int J Oncol* **27**, 377–384.

Bian H, Wilden H, Fournier P, Peeters B, Schirrmacher V (2006). In vivo efficacy of systemic tumor targeting of a viral RNA vector with oncolytic properties using a bispecific adapter protein. *Int J Oncol* **29**(6), 1359-69.

Bureyev A, Huang Z, Yang L, *et al.* (2005). Recombinant Newcastle Disease Virus expressing a foreign viral antigen is attenuated and highly immunogenic in primates. *J Virol* **79**, 13275–13284.

Cassel WA, Garret RE (1965). Newcastle disease virus as an antineoplastic agent. *Cancer* **7**, 863–868.

Cassel WA, Murray DR (1992). A ten-year follow-up on stage II malignant melanoma patients treated post-surgically with Newcastle disease virus oncolysate. *Med Oncol Tumor Pharmacother* **9**(4), 169–171.

Cassel WA, Murray DR, Olkowski ZL. Newcastle Disease Virus oncolysate in the management of stage III malignant melanoma. In: Sinkovics JG, Horvatz JC (eds), *Viral Therapy of Human Cancer.* Marcel Dekker, New York, 2005. pp. 577–689.

Clemens MJ, Elia A (1997). The double-stranded RNA-dependent protein kinase PKR: structure and function. *J. Interferon Cytokine Res* **17**(9), 503–524.

Chlichlia K, Schirrmacher V, Sandaltzopoulos R (2005). Cancer immunotherapy: battling tumors with gene vaccines. *Current Medical Chemistry – Anti-inflammatory and anti-allergy agents* **4**, 353–365.

Coggin IH, Barsoum AL, Rohrer IW, Thurnher M, Zeiss M (2005). Contemporary definitions of tumor specific antigens, immunogens and markers as related to the adaptive response of the cancer-bearing host. *Anticancer Res* **25**, 2345–2355.

Colonna M, Pulendran B, Iwasaki A (2005). Dendritic cells at the host-pathogen interface. *Nat Immunol* **7**(2), 117–120.

Csatary LK, Eckhardt S, Bukosza I *et al.* (1993). Attenuated veterinary virus vaccine for the treatment of cancer. *Cancer Detect Prev* **17**(6), 619–627.

Csatary LK, Bakacs T (1999). Use of Newcastle disease virus vaccine (MTH-68/H) in a patient with high-grade glioblastoma. *JAMA* **281**(17), 1588–1589.

Csatary LK, Gosztonyi G, Szeberenyi J, *et al.* (2004). MTH-68/H oncolytic viral treatment in human high-grade gliomas. *J Neurooncol* **67**, 83–93.

Curtsinger JM, Lins DC, Mescher MF (2003). Signal 3 determines tolerance versus full activation of naive CD8 T cells: dissociating proliferation and development of effector function. *J Exp Med* **197**, 1141–1151.

De Leeuw O, Peeters B (1999). Complete nucleotide sequence of Newcastle disease virus: evidence for the existence of a new genus within the subfamily paramyxovirinae. *J Gen Virol* **80**, 131–136.

Demers GW, Johnson DE, Tsai V, *et al.* (2003). Pharmacologic indicators of antitumor efficacy for oncolytic virotherapy. *Cancer Res* **63**, 4003–4008.

Ertel C, Millar NS, Emmerson PT, Schirrmacher V, von Hoegen P (1993). Viral hemagglutinin augments peptide specific cytotoxic T-cell responses. *Eur J Immunol* **23**, 2592.

Ferreira L, Villar E, Munoz-Barroso I (2004). Ganliosides and N-glycoproteins function as Newcastle disease virus receptors. Int J Biochem Cell Biol **36**, 2344–2356.

Feuerer M, Beckhove P, Bai L, *et al.* (2001a). Therapy of human tumors in NOD/SCID mice with patient derived re-activated memory T cells from bone marrow. *Nat Med* **7**(4), 452–458.

Feuerer M, Rocha M, Bai L, *et al.* (2001b). Enrichment of memory T cells and other profound immunological changes in the bone marrow from untreated breast cancer patients. *Int J Cancer* **92**(1), 96–105.

Feuerer M, Beckhove P, Garbi N, *et al.* (2003). Bone marrow as a priming site for T cell responses to blood-borne antigen. *Nat Med* **9**, 1151–1157.

Fiola C, Peeters B, Fournier P, Arnold A, Bucur M, Schirrmacher V (2006). Tumor selective replication of Newcastle Disease Virus: association with defects of tumor cells in antiviral defence. *Int J Cancer* **119**(2), 328–338.

Forden C (2004). Do T lymphocytes correlate danger signals to antigen? *Medical Hypothesis* **62**, 989–906.

Fournier P, Zeng J, Schirrmacher V (2003). Two ways to induce innate immune responses in human PBMCs: paracrine stimulation of IFN-α responses by viral protein or dsRNA. *Int J Oncol* **23**, 673–680.

Freeman AI, Zakay-Rones Z, Gomori JM *et al.* (2006). Phase I/II trial of intravenous NDV-HUJ oncolytic virus in recurrent glioblastoma multiforme. *Mol Ther* **13**(1), 221–228.

Gewurz BE, Gaudet R, Tortorella D, Wang EW, Ploegh HL (2001). Virus subversion of immunity: A structural perspective. *Curr Opin Immunol* **3**(4), 442–450.

Gilboa E (1997). The promise of cancer vaccines. *Nat Rev Cancer* **4**, 401–411.

Haas C, Lulei M, Fournier P, Arnold A, Schirrmacher V (2005a). T-cell triggering by CD3- and CD28-binding molecules linked to a human virus-modified tumor cell vaccine. *Vaccine* **23**, 2439–2453.

Haas C, Lulei M, Fournier P, Arnold A, Schirrmacher V (2005b). A tumor vaccine containing anti-CD3 and anti-CD28 bispecific antibodies triggers strong and durable anti-tumor activity in human lymphocytes. *Int J Cancer* **118**, 658–667.

Haller O, Kochs G (2002). Interferon-induced mx proteins: dynamin-like GTPases with antiviral activity. *Traffic* **3**(10), 710–717.

Heicappell R, Schirrmacher V, von Hoegen P, *et al.* (1986). Prevention of metastatic spread by postoperative immunotherapy with virally modified autologous tumor cells. I: parameters for optimal therapeutic effects. *Int J Cancer* **37**, 569.

Hengel H, Koszinowski UH, Conzelmann KK (2005). Viruses know it all: new insights into IFN networks. *Trends Immunol* **26**, 396–401.

Horvath CM (2004). Weapons of STAT destruction. Interferon evasion by paramyxovirus V protein. *Eur J Biochem* **271**, 4621–4628.

Hotte SJ, Lorence RM, Hirte HW *et al.* (2007). An optimized clinical regimen for the oncolytic virus PV701. *Clin Cancer Res* **13**(3), 977–985.

Huang Z, Krishnamurthy S, Panda A, Samal SK (2001). High-level expression of a foreign gene from the most 3′-proximal locus of a recombinant Newcastle Disease Virus. *J Gen Virol* **82**, 1729–1736.

Huang Z, Elankumaran S, Panda A, Samal S (2003). Recombinant Newcastle disease virus as a vaccine vector. *Poultry Sci* **82**, 899–906.

Huang Z, Elankumaran S, Yunus AS, Samal SK (2004). A recombinant Newcastle Disease Virus (NDV) expressing VP2 protein of infectious bursal disease virus (IBDV) protects against NDV and IBDV. *J Virol* **78**, 10054–10063.

Kaech SM, Wherry EJ, Ahmed R (2002). Effector and memory T cell differentiation: implications for vaccine development. *Nat Rev Immunol* **2**, 251–262.

Kaleta EF, Baldauf C (1988). Newcastle disease in free-living and pet birds. In: Alexander DJ (ed.), *Newcastle Disease*. Kluwer Academic Publishers, Boston, MA, pp 197–256.

Karcher J, Dyckhoff G, Beckhove P, *et al.* (2004). Antitumor vaccination with HNSCC with autologous virus-modified tumor cells. *Cancer Res* **64**(21), 8057–8061.

Kato H, Sato S, Yoneyama M, *et al.* (2005). Cell type-specific involvement of RIG-I in antiviral response. *Immunity* **23**, 19–28.

Kawai T, Akira S (2005). Pathogen recognition with Toll-like receptors. *Curr Opin Immunol* **17**, 338–344.

Kirchner HH, Anton P, Atzpodien J (1995). Adjuvant treatment of locally advanced renal cancer with autologous virus-modified tumor vaccines. *World J Urol* **13**(3), 171–173.

Kobayashi H (1997). Viral xenogenization of intact tumor cells. *Adv Cancer Res* **30**, 279–299.

Laurie SA, Bell JC, Atkins HL, *et al.* (2006). A phase 1 clinical study of intravenous administration of PV701, an oncolytic virus, using two-step desensitization. *Clin Cancer Res* **12**(8), 2555–2562.

LeBon A, Tough DF. (2002) Links between innate and adaptive immunity via type I interferon. *Curr Opin Immunol* **14**, 432–436.

Lehner B, Schlag P, Liebrich W, Schirrmacher V (1990). Post-operative active specific immunization in curatively resected colorectal cancer patients with virus-modified autologous tumor cell vaccine. *Cancer Immunol Immunother* **32**, 173.

Lewis JJ (2004). Therapeutic cancer vaccines: using unique antigens. *Proc Natl Acad Sci U S A* **101**(Suppl. 2): 14653–14656.

Liu Y-J (2005). IPC: Professional type 1 interferon-producing cells and plasmacytoid dendritic cell precursors. *Annu Rev Immunol* **23**, 275–306.

Lorence RM, Rood PA, Kelly KW (1988). Newcastle disease virus as an antineoplastic agent: induction of tumor necrosis factor-α and augmentation of its cytotoxicity. *J Natl Cancer Inst* **80**, 1305–1312.

Lorence RM, Reichard KW, Katubig BB *et al.* (1994). Complete regression of human neuroblastoma xenografts in athymic mice after local Newcastle disease virus therapy. *J Natl Cancer Inst* **86**, 1228–1233.

Lorence RM, Pecora AL, Major PP *et al.* (2003). Overview of phase I studies of intravenous administration of PV701, an oncolytic virus. *Curr Opin. Mol Ther* **5** (6), 618–624.

Mahnke YD, Schwendemann J, Beckhove P, Schirrmacher V (2005). Maintenance of long-term tumour-specific T-cell memory by residual dormant tumor cells. *Immunology* **115**, 325–336.

Mallmann P (1993). Autologous tumor-cell vaccination and lymphokine-activated tumor-infiltrating lymphocytes (LAK-TIL). *Hybridoma* **12**(5), 559–566.

Mallmann P, Eis-Hübinger AM, Spiegel G, Krebs D. Erste Erfahrungen einer Immuntherapie mit autologen virusmodifizierten Tumorzellen bei Patientinnen mit fortgeschrittenen gynäkologischen Tumoren. In: Melchert F, Neises M, Wischnik A (eds), *Klinische Tumorimmunologie in der Gynäkologie,*. Serie *Aktuelle Onkologie, W. Zuckschwerdt Verlag München, Bern, Wien, San Francisco, 1991*, pp 233–241.

Mallmann P, Eis-Hübinger AM, Krebs D (1992). Lymphokine-activated tumor-infiltrating lymphocytes and autologous tumor vaccine in breast ovarian cancer. *Onkologie* **15**, 490–496.

Matzinger P (1994). Tolerance, danger, and the extended family. *Annu Rev Immunol* **12**, 991–1045.

Matzinger P (2002). The danger model: a renewed sense of self. *Science* **296**, 301–305.

Mebatsion T, Verstegen S, De Vaan LT, *et al.* (2001). A recombinant Newcastle Disease Virus with low-level V protein expression is immunogenic and lacks pathogenicity for chicken embryos. *J Virol* **75**, 420–428.

Meng S, Tripathy D, Frenkel EP, *et al.* (2004). Circulating tumor cells in patients with breast cancer dormancy. *Clin Cancer Res* **10**, 8152–8162.

Möbus V, Horn S, Stöck M, Schirrmacher V (1993). Tumor cell vaccination for gynecological tumors. *Hybridoma* **12**(5), 543–696.

Nakaya T, Cros J, Park MS *et al.* (2001). Recombinant Newcastle Disease Virus as a vaccine vector. *J Virol* **75**, 11868–11873.

Nelson NJ (1999). Scientific interest in Newcastle Disease Virus is reviving. *J Natl Cancer Inst* **91**, 1708–1710.

Ockert D, Schirrmacher V, Beck N, *et al.* (1996). Newcastle disease virus-infected intact autologous tumor cell vaccine for adjuvant active specific immunotherapy of resected colorectal carcinoma. *Clin Cancer Res* **2**(1), 21–28.

Parato KA Senger D, Forsyth PAJ, Bell JC (2005). Recent progress in the battle between oncolytic viruses and tumours. *Nat Rev Cancer* AOP published online Nov. 2005; doi:10.1038/nrc1750.

Park MS, Garcia-Sastre A, Cros JF, Basler CF, Palese P (2003). Newcastle disease virus V protein is a determinant of host range restriction. *J Virol* **77**(17), 9522–9532.

Pecora AL, Rizvi N, Cohen GI, *et al.* (2002). Phase I trial of intravenous administration of PV701, an oncolytic virus, in patients with advanced solid cancers. *J Clin Oncol* **20**(9), 2251–2266.

Peeters BP, de Leeuw OS, Koch G, Gielkens AL (1999). Rescue of Newcastle disease virus from cloned cDNA: evidence that cleavability of the fusion protein is a major determinant for virulence. *J Virol* **73**, 5001–9.

Pomer S, Schirrmacher V, Thiele R, Löhrke H, Staehler G (1995). Tumor response and 4 year survival data of patients with advanced renal cell carcinoma treated with autologous tumor vaccine and subcutaneous r-IL-2 and IFN-Alpha 2b. *Int J Oncol* **6**, 947.

Reichard KW, Lorence RM, Cascino CJ *et al.* (1991). Newcastle disease virus selectively kills human cancer cells. *J Surg Res* **52**, 448–453.

Remondini D, Bazzani A, Franceschi C, Bersani F, Verondini E, Castellani G (2003). Role of connectivity in immune and neural network models: memory development and aging. *Riv Biol.* **96**(2), 225–39.

Römer-Oberdörfer A, Werner O, Veits J, Mebatsion T, Mettenleiter TC (2003). Contribution of the length of the HN protein and the sequence of the F protein cleavage site to Newcastle disease virus pathogenicity. *J Gen Virol* **84**, 3121–3129.

Rogge L, Barberis-Maino L, Biffi M, *et al.* (1997). Selective expression of an interleukin-12 receptor component by human T helper 1 cells. *J Exp Med* **185**, 825–831.

Russell SJ (2002). RNA viruses as virotherapy agents. Review. *Cancer Gene Ther* **9**, 961–966.

Sato K, Hida S, Takayanagi H, *et al.* (2001). Antiviral response by natural killer cells through TRAIL gene induction by IFN-alpha/beta. *Eur J Immunol* **31**, 3138–3146.

Schirrmacher V (2001). T cell immunity in the induction and maintenance of a tumor dormant state. *Semin Cancer Biol* **11**, 285.

Schirrmacher V (2005a). T cell mediated immunotherapy of Metastases: State of the art in (2005). *Exp Opin Biol Ther* **5**(8), 1051–1068.

Schirrmacher V (2005b). Anti-tumor immune memory and its activation for control of residual tumor cells and improvement of patient survival. In: Sinkovics J, Horvath J (eds), *Virus Therapy of Human Cancers.* Marcel Decker Monograph, Marcel Dekker, New York, pp. 481–531.

Schirrmacher V (2005c). Clinical trials of antitumor vaccination with an autologous tumor cell vaccine modified by virus infection: improvement of patient survival based on improved anti-tumor immune memory. *Cancer Immunol Immunother* **54/6**, 587–598.

Schirrmacher V, Heicappell R (1987). Prevention of metastatic spread by postperative immunotherapy with virally modified autologous tumor cells. II. Establishment of specific systemic anti tumor immunity. *Clin Exp Met* **5**(2), 147–156.

Schirrmacher V, Von Hoegen P (1993). Importance of tumor cell membrane integrity and viability for CTL activation by cancer vaccines. *Vaccine Res* **2**, 183.

Schirrmacher V, Haas C, Bonifer R, Ertel C (1997). Virus potentiation of tumor vaccine T cell stimulatory capacity requires cell surface binding but not infection. *Clin Cancer Res* **3**, 1135–1148.

Schirrmacher V, Ahlert T, Pröbstle T, *et al.* (1998). Immunization with virus modified tumor cells. *Semin Oncol* **25**(6), 677–696.

Schirrmacher V, Haas C, Bonife, R, Ahlert T, Gerhards R, Ertel C (1999). Human tumor cell modification by virus infection: an efficient and safe way to produce cancer vaccine with pleiotropic immune stimulatory properties when using Newcastle Disease Virus. *Gene Ther* **6**, 63–73.

Schirrmacher V, Griesbach A, Ahlert Th (2001). Antitumor effects of Newcastle Disease Virus *in vivo*: local versus systemic effects. *Int J Oncol* **18**, 945–952.

Schlag P, Manasterski M, Gerneth T, *et al.* (1992). Active specific immunotherapy with Newcastle-disease-virus-modified autologous tumor cells following resection of liver metastases in colorectal cancer. First evaluation of clinical response of a phase II-trial. *Cancer Immunol Immunother* **35**(5), 325–330.

Schneider T, Gerhards R, Kircher E, Firsching R (2001). Preliminary results of active specific immunization with modified tumor cell vaccine in glioblastoma multiforme. *J Neurooncol* **53**(1), 39–46.

Schulz O, Diebold SS, Chen M, *et al.* (2005). Toll-like receptor 3 promotes cross-priming to virus-infected cells. *Nature* **453**, 887–892.

Seal BS, Crawford JM, Sellers HS, Locek DP, King DJ (2002). Nucleotide sequence analysis of the Newcastle disease virus nucleocapsid protein gene and phylogenetic relationship among the paramyxoviridae. *Virus Res* **83**, 119–129.

Senne AD, King DJ, Kapcynski DR (2004). Control of Newcastle Disease by vaccination. In: Schudel A. Lombard M (eds), *Control of Infectious Animal Diseases by Vaccination.* Dev. Biol. 119. Karger, Basel, pp. 165–170.

Sinkovics JG, Horvath JC (2000). Newcastle disease Virus (NDV): brief history of its oncolytic strains. *J Clin Virol* **16**(1), 1–15.

Steiner HH, Bonsanto MM, Beckhove P, *et al.* (2004). Anti-tumor vaccination of patients with glioblastoma multiforme in a case-control study: feasibility, safety and clinical benefit. *J Clin Oncol* **22**(21), 4272–4281.

Suzuki Y, Suzuki T, Matsunaga M, Matsumoto M (1985). Gangliosides as paramyxovirus receptor. Structural requirement of sialo-oligosaccharides in receptors for hemagglutinating virus of Japan (Sendai Virus) and Newcastle Disease Virus. *J Biol Chem* **97**, 1189–1199.

Taniguchi T, Takaoka A (2002). The interferon-alpha/beta system in antiviral responses: a multimodal machinery of gene regulation by the IRF family of transcription factors. *Curr Opin Immunol* **14**, 111–116.

Termeer CC, Schirrmacher V, Bröcker EB, Becker JC (2000). Newcastle Disease Virus infection induces B7–1/B7–2 independent T-cell costimulatory activity in human melanoma cells. *Cancer Gene Ther* **7**, 316.

Toyoda T, Sakaguchi T, Hirota H, *et al.* (1989). Newcastle disease virus evolution. II Lack of gene recombination in generating virulent and avirulent strains. *Virology* **169**(2), 273–282.

Von Hoegen P, Zawatzky R, Schirrmacher V (1990). Modification of tumor cells by a low dose of Newcastle Disease Virus, III: potentiation of tumor specific cytolytic T cell activity via induction of interferon α,β. *Cell Immunol* **126**, 80.

Washburn B, Schirrmacher V (2002). Human tumor cell infection by Newcastle Disease Virus leads to upregulation of HLA and cell adhesion molecules and to induction of interferons, chemokines and finally apoptosis. *Int J Oncol* **21**, 85–93.

Washburn B, Weigand MA, Grosse-Wilde A, *et al.* (2003). TNF-related apoptosis-inducing ligand mediates tumoricidal activity of human monocytes stimulated by Newcastle Disease Virus. *J Immunol* **170**(4), 1814–1821.

Wise MG, Sellers HS, Alvarez R, Seal BS (2004a). RNA-dependent RNA polymerase gene analysis of worldwide Newcastle disease virus isolates representing different virulence types and their phylogenetic relationship with other members of the paramyxoviridae. *Virus Res* **104**, 71–80.

Wise MG, Suarez DL, Seal BS, *et al.* (2004b). Development of a real-time reverse-transcription PCR for detection of Newcastle Disease Virus RNA in clinical samples. *J Clin Microbiol* **42**, 329–338.

Wheelock EF, Dingle JH (1964). Observations on the repeated administration of viruses to a patient with acute leukemia. A preliminary report. *N Engl J Med* **271**(13), 645–651.

Yoneyama M, Kikuchi M, Natsukawa T, Shinobu N, Imaizumi T, Miyagishi M, Taira K, Akira S, Fujita T (2004). The RNA helicase RIG-I has an essential function in double-stranded RNA-induced innate antiviral responses. *Nat Immunol.* **5**(7), 730-7.

Yoneyama H, Matsuno K, Matsushima K (2005). Migration of dendritic cells. *Int J Hematol* **81**, 204–207.

Zeng J, Fournier P, Schirrmacher V (2002). Induction of interferon α and tumor necrosis factor-related apoptosis-inducing ligand in blood mononuclear cells by hemagglutinin-neuraminidase but not F protein of Newcastle Disease Virus. *Virology* **297**, 19–30.

11

Vesicular Stomatitis Virus

John Bell, Kelly Parato and **Harold Atkins**

11.1 Introduction

Over the last several years of advancement in the field of oncolytic virotherapy of cancer, many viruses have been put forth as potential therapeutic candidates, and are at various stages of clinical development (Parato *et al.*, 2005). This chapter will discuss vesicular stomatitis virus (VSV), a small bullet-shaped RNA virus, which has shown significant anti-tumour activity *in vitro* and in numerous preclinical models. The mode of tumour selectivity, preclinical experience with attenuated variants, and the interaction of VSV with the host immune system will be the focus of this chapter, highlighting some of the challenges facing this and other oncolytic viruses, and will uncover new opportunities for therapeutic targeting of VSV to tumours *in vivo*.

11.2 VSV: genomic organization and life cycle

VSV is a member of the Rhabdoviridae family, and a relative of rabies virus. It is a monomorphic, enveloped, bullet-shaped virus that ranges from 100 to 430 nm in length and 80 nm diameter. Its nucleocapsid core contains a single copy of the ~11 200 nucleotide negative-sense RNA genome. The core is surrounded by matrix and enveloped in a lipid bilayer studded with viral glycoprotein surface spikes (Management, 2006). The five viral genes are linearly arranged along the genome. The genomic organization and VSV gene function are detailed in Figure 11.1. Each gene has a polyadenylation signal and is separated from its neighbours by small intragenic regions. The coding sequences are sandwiched between a 51 bp 3′ leader sequence and a 36 bp 5′ untranslated region that are important *cis*-acting elements for encapsidation and replication. These are the minimum elements required for the formation of transmissible defective-interfering particles (Letchworth *et al.*, 1999).

Infection follows VSV attachment to the cell membrane which initiates viral endocytosis through clathrin-coated pits (Sun *et al.*, 2005). Specific cellular receptors for VSV have not been identified and the exact cellular components responsible for attachment remain uncertain. Within the lysosomes, pH-dependent changes trigger G protein-mediated fusion of the virus to the cellular membrane (Fredericksen and Whitt, 1995), releasing the nucleocapsid into the cytoplasm. Transcription and replication of the viral genome by the viral RNA-dependent RNA polymerase complex occurs in the cytoplasm. It is unclear how the viral polymerase balances transcription and replication but site-dependent phosphorylation of VSV-P (Das and Pattnaik, 2004), subunit composition of the polymerase (Gupta *et al.*, 2003) or its interaction with cellular proteins influences this life-cycle decision. Genomes are

Viral Therapy of Cancer Edited by Kevin J. Harrington, Richard G. Vile and Hardev S. Pandha

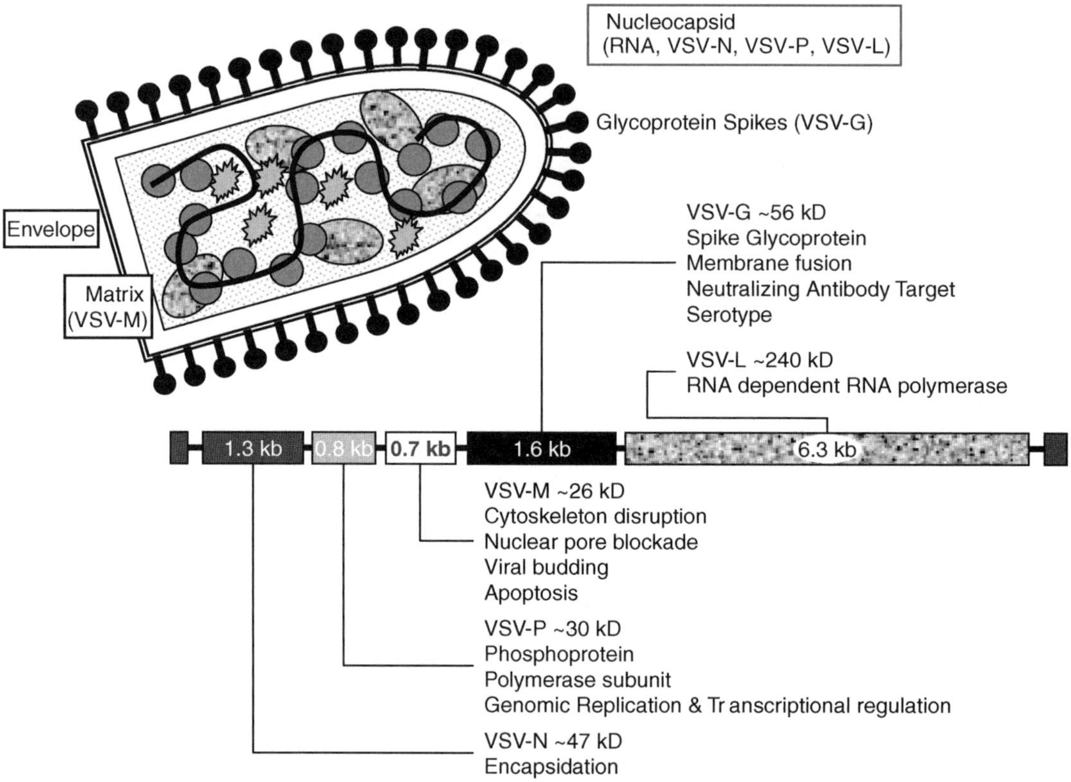

Figure 11.1 Schematic representation of vesicular stomatitis virus. VSV is a single stranded negative sense RNA virus, with a characteristic bullet-like morphology. The nucleocapsid consists of the RNA genome surrounded by M (matrix), and viral nucleocapsid (N), phosphoprotein (P), and large (L) proteins which comprise the RNA-dependent RNA-polymerase necessary for viral replication and gene expression. The virion is enveloped, and covered with glycoprotein (G) spikes for viral binding and entry into target cells. Depicted are the complete assembled virion, and the genomic organization of the VSV genome

encapsidated by newly synthesized N protein. Matrix protein (VSV-M) alters many cellular processes. VSV-M plays a role in the inhibition of cellular protein synthesis by blocking transport of cellular mRNA through nuclear pores (von Kobbe *et al.*, 2000), initiation of apoptosis through interactions with the mitochondria (Gadaleta *et al.*, 2005), cell rounding and detachment by inducing cytoskeleton depolymerization (Lyles and McKenzie, 1997) and directing viral assembly and budding (Irie *et al.*, 2004) at the inner leaflet of the plasma membrane. The viral glycoprotein (VSV-G) is synthesized in the endoplasmic reticulum, glycosylated in the Golgi and transported to the cell's plasma membrane where it associates into microdomains (Brown and Lyles, 2003) at the site of viral budding.

11.3 Host range and pathogenesis of VSV infection

VSV grows in a variety of arthropods, including sand flies (Comer *et al.*, 1990), black flies (Howerth *et al.*, 2002), mosquitos (Liu and Zee, 1976) and others (Nunamaker *et al.*, 2003). The virus also infects mammals and evidence of infection has been found in most types of wild mammals in the Americas (Letchworth *et al.*, 1999). VSV causes an illness indistinguishable from foot and mouth disease in domestic farm animals.

Cows, horses and pigs develop painful blisters and ulcers on their tongues, oral mucous membranes, feet and teats. The mouth sores last 2 or 3 weeks and may reduce the animal's feed intake leading to weight loss. Subclinical infections can occur. Infection from the arthropod vector likely results in transmission of the virus in the wild, but VSV is highly contagious and aerosol transmission from the saliva of infected animals results in the rapid spread through animal herds. VSV occurs mainly in the Western hemisphere in a range between the mid USA to the temperate regions of South America. The virus is endemic in Mexico and Central America with evidence of seropositivity in more than 50 per cent of animals tested (Rodriguez *et al.*, 1990). At the edges of its range, VSV infections occur in an epidemic fashion (Rodriguez, 2002).

Many strains of the two predominant serotypes, VSV-Indiana and VSV-New Jersey, have been identified. About 80 per cent of infections are caused by VSV-New Jersey. Three subtypes of VSV-Indiana have been identified and are evolutionarily related. The structure of the genome and the sequence of regulatory regions are conserved across VSV strains but inter-strain sequence variations exist in the coding and intragenic regions. The strains tend to be geographically localized and phylogeny tends to follow geographical proximity (Rodriguez, 2002).

Evidence of widespread VSV infection in humans has been inferred from high prevalence of seropositivity in endemic regions (Tesh *et al.*, 1969) and in workers with occupational exposure (Reif *et al.*, 1987). VSV can cause symptomatic infections in humans. Infections are related to contact with infected animals or occupational exposures. It is thought that aerosols from the sneeze of infected animals lead to VSV inoculation in the eye (Reif *et al.*, 1987). Conjunctivitis develops 1 to 3 days after infection. This is followed by an acute flu-like illness lasting 3 to 6 days and characterized by fever, chills, pharyngitis, nausea, vomiting, myalgias and headaches (Letchworth *et al.*, 1999). Oral vesicles may occur. Specific treatment for VSV infection does not exist. While there have been two case reports of encephalitis associated with VSV infection (Quiroz *et al.*, 1988), infection is self-limiting and there are generally no lasting or serious complications.

11.4 Control of VSV infection by the innate type I interferon response

Viral infection of an immune competent host activates both the innate and adaptive arms of the host immune response. The innate immune component of antiviral immunity is primarily governed by activation of type I interferon (IFN-α/β), while the adaptive component comprises production of virus-specific antibodies and cytotoxic T-cell activation (see below). The type I IFN response is an antigen non-specific defence mechanism that is triggered immediately following viral detection and does not develop immunological memory for antigens. Cells have evolved sophisticated networks to sense invading viruses early in the infection process, through intracellular sensors of viral nucleic acid, including protein kinase receptor (PKR), Toll-like receptor-3 (TLR3), double-stranded RNA (dsRNA), Toll-like receptor 7/8 (TLR7/8), single-stranded RNA (ssRNA), and retinoic acid inducible gene I (RIG-I), leading to the rapid induction of type I interferons (reviewed in (Gale and Katze, 1998; Samuel, 2001; Hertzog *et al.*, 2003; Kaempfer, 2003; Yoneyama *et al.*, 2004; Bowie and Haga, 2005) and associated antiviral responses. Specifically, viral dsRNA activates: (1) 2–5 oligoadenylate synthetase and RNase L expression, leading to viral mRNA degradation; (2) PKR activation and concomitant eIF2α phosphorylation and translational arrest; and (3) activation of nuclear factor (NF)-κB and downstream production of IFN-β. Independent of PKR or 2–5 OAS, ssRNA or dsRNA can be detected by TLRs or RIG-I, leading to induction of IFN-β transcription. Once IFN-β binds to the type I IFN receptor on the infected or adjacent cells, transcription of many gene products (de Veer *et al.*, 2001; Chawla-Sarkar *et al.*, 2003; Khabar *et al.*, 2004; Sana *et al.*, 2005) ensues, which are collectively called interferon stimulated genes (ISG), and which include multiple isoforms of IFN-α, and numerous genes involved in downstream antiviral activities including cell cycle arrest, apoptosis, and immune stimulation. Most viruses elicit type I IFN induction during the viral life cycle, however VSV above most other viruses, is exquisitely sensitive to the collective antiviral actions of type I IFNs.

11.5 Cancer cells are insensitive to type I interferon

The biological properties of the interferons and ISG are not only antiviral; they also have physiological consequences that are incompatible with efficient tumour growth (de Veer *et al.*, 2001; Chawla-Sarkar *et al.*, 2003, Ikeda *et al.*, 2002; Khabar *et al.*, 2004; Dunn *et al.*, 2005). As tumours progress, they become non-responsive to interferons and lose expression of key ISG such as the major histocompatibility complex (MHC) genes, the protein products of which are required for antigen presentation to the adaptive immune system. As a result, the tumour becomes invisible to the host immune system (Ikeda *et al.*, 2002; Dunn *et al.*, 2005). Other ISG promote apoptosis, halt cell growth or are anti-angiogenic, so it is not surprising to find that many different kinds of tumour cells have acquired defects in the ability to respond to interferons (Stojdl *et al.*, 2000b, 2003). Therefore, it seems likely that an effective strategy to tailor a viral therapeutic for replication in tumour cells is to select or design a virus that is especially sensitive to the antiviral properties of interferons and, in this regard, VSV is an ideal candidate. VSV replication is strongly suppressed in interferon-responsive normal tissues but is still able to flourish in interferon-nonresponsive tumour cells (Stojdl *et al.*, 2003) (Figure 11.2). Most viruses carry genes whose products are dedicated to evading the antiviral activity of interferons, and VSV is no exception. As a result of this sensitivity to IFN, the matrix (M) protein of wild-type VSV has evolved to inhibit IFN induction through inhibition of nucleocytoplasmic export of host mRNAs, including IFN mRNAs, thereby allowing viral replication to proceed uninterrupted. Manipulation of this function of matrix (discussed below) is a key factor in the therapeutic benefit offered by VSV during oncolytic virotherapy.

To this end, several investigators have demonstrated the concept that: (1) tumour cells are infected and killed by VSV; (2) killing by virus can be restricted to tumour cells, while sparing normal cells; and (3) this selectivity is likely attributable to IFN sensitivity of normal versus tumour cells, albeit not to the exclusion of other cancer cell defects rendering tumour cells VSV-sensitive, including ras or myc activation, or p53 deficiency.

11.6 VSV preferentially replicates in and lyses tumour cells *in vitro*

Based on defects in the IFN pathway in tumour cells, which confers a growth advantage by rendering tumour cells refractory to the anti-proliferative and pro-apoptotic effects of type I IFN, VSV preferentially infects, replicates in, and lyses a variety of tumour cell types *in vitro* (Balachandran and Barber, 2000; Stojdl *et al.*, 2000b). Lysis of human tumour cells occurs even in the presence of IFN that protects normal primary cell cultures (Stojdl *et al.*, 2000b). Given the complexity of the IFN response pathway, any number of defects could be at play in tumour cells. While it is unclear which points of IFN regulation are implicated in various tumour types assessed to date, it is well established that PKR deficiency imparts extreme sensitivity to VSV infection (Stojdl *et al.*, 2000a). In addition to IFN defects, VSV oncolysis is effective against tumours with aberrant p53, ras, or myc status, and induces multiple caspase-dependent apoptotic pathways, even in cells with normal PKR activity (Balachandran *et al.*, 2001).

As VSV appears to efficiently replicate in tumour cells with a variety of genetic defects, a wide range of tumour cell lines and *in vivo* models have demonstrated efficient oncolysis and tumour control by this therapeutic. VSV exhibits oncolytic activity against primary human human T lymphotropic virus-1 (HTLV-1) infected T lymphocytes from adult T-cell leukaemia patients, but not against non-leukaemic cells from HTLV-1 associated myelopathy/tropical spastic paraparesis, nor naïve $CD4^+$ T cells or chronic lymphocytic leukamia patients (Cesaire *et al.*, 2006). VSV efficiently kills most leukaemia cell lines and leukaemic cells in primary multiple myeloma patients, and can purge bone marrow of leukaemic cells, but had no effect against normal clonogenic bone marrow progenitor cells or peripheral blood leukocytes (Lichty *et al.*, 2004).

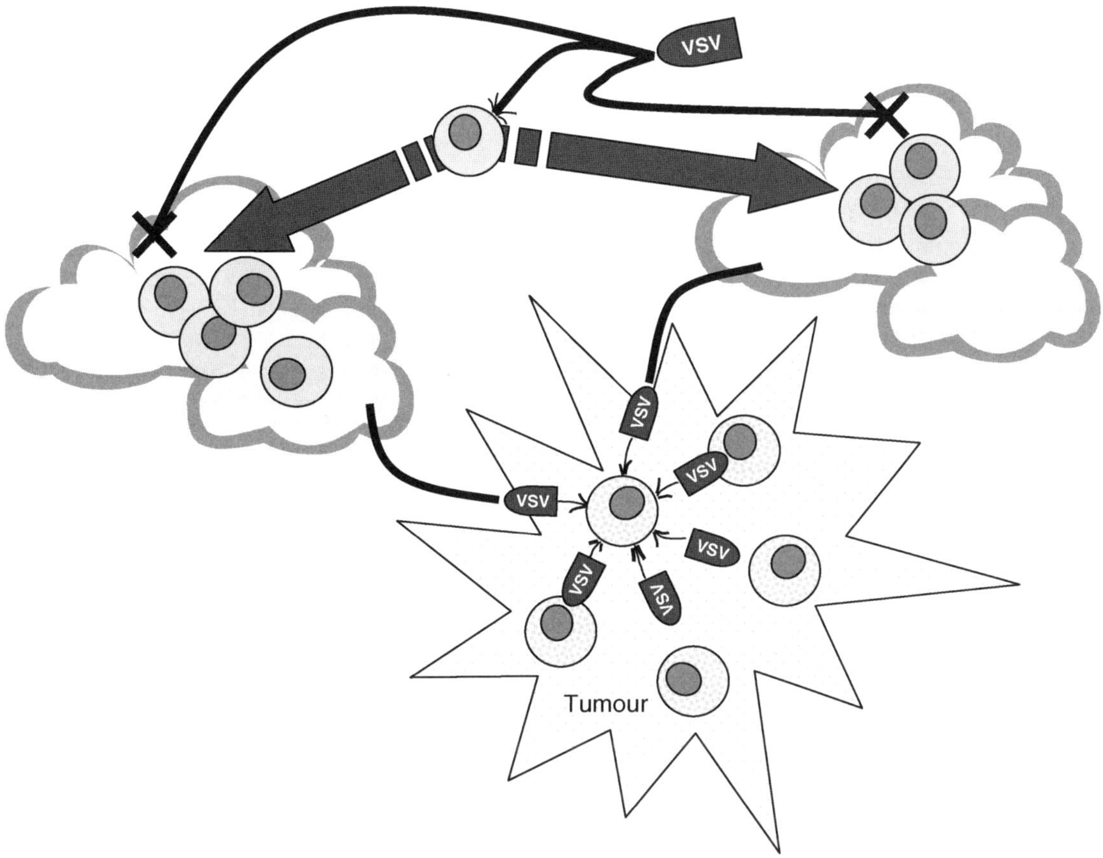

Figure 11.2 Interferon protects normal but not tumour cells from VSV infection. Upon infection of a normal cell with VSV, the type I IFN response is initiated, leading to secretion of IFN which acts in an autocrine and paracrine manner, to prevent further replication and infection of normal cells (block arrows). However, tumour cells with defects in either induction of the IFN response, or in responsiveness to the antiviral effects of IFN, are not protected, and VSV infection spreads, leading to cytolysis of infected tumour cells

A549 and Lewis lung carcinomas (LLC), and LNCaP human prostate carcinomas are effectively killed by VSV *in vitro* (Ahmed *et al.*, 2004; Li *et al.*, 2004). U87MG and M0595 glioblastoma cells are susceptible to VSV oncolysis, and this sensitivity can be enhanced by serial passage on glioblastoma cells *in vitro* (Wollmann *et al.*, 2005). In an organotypic brain tissue slice glioma co-culture model, rat C6 or human U87 glioma cells were eliminated by VSV, albeit with significant collateral damage to neurons in co-culture, even in the presence of IFN-β which restricted infection of normal cells. Upon removal of the glycoprotein (G) from the VSV genome, the virus was restricted to single cycle infection and was unable to spread, and this modification allowed VSV to eliminate C6 and U87 glioma cells with no collateral neuronal damage (Duntsch *et al.*, 2004). This points to at least one opportunity to attenuate VSV to restrict cytotoxicity which may occur independent of direct viral infection. Several publications have demonstrated VSV's lytic capacity on numerous human cancer cell lines derived from lung, ovary, prostate, colon, breast, brain, and melanoma (Balachandran *et al.*, 2001; Stojdl *et al.*, 2000b, 2003) and, as such, VSV represents a widely applicable cancer therapeutic.

11.7 VSV attenuation: enhanced tumour selectivity and therapeutic index

The matrix (M) protein of VSV plays a critical role in shutoff of host gene expression during viral replication. Specifically, the M protein disrupts the Rae1/mrnp41 mRNA nuclear export pathway, thereby restricting nuclear export of host cell mRNAs, including IFN mRNAs, while viral mRNAs are efficiently transcribed and translated in the cytoplasm (Faria *et al.*, 2005). Therefore, while VSV can be sensed by PKR or other host cell RNA sensors leading to induction of the IFN response, wild type VSV prevents the manifestation of this response by sequestering host cell transcripts from translation machinery, even in normal cells with an otherwise intact IFN pathway. To this end, several attempts have been made to attenuate wild-type VSV. Small plaque mutants of VSV (AV1, AV2) derived by serial passage on IFN-responsive cells were confirmed to induce 20 to 50 times more IFN-α than the parental wild-type VSV in PC3 prostate carcinoma and CAKI-1 renal carcinoma cells. As a consequence of this attenuation, these mutants were safe to administer to BALB/c mice at much higher doses than wild type VSV. Sequencing of AV1 and AV2 revealed minor amino acid substitutions in the matrix protein, specifically M51R in AV1, and V221F and S226R in AV2. These mutations resulted in the failure of matrix to block nuclear export of IFN-β transcripts, thereby allowing the IFN response to progress in cells with an intact IFN pathway (Stojdl *et al.*, 2003). The new inherent IFN-inducing capability of AV1 and AV2 provided new utility to VSV as a prospective therapeutic vector, eliminating the requirement for exogenous IFN addition in order to confer cancer-killing and replication specificity. Not only does the IFN-inducing capability of AV1/AV2 heighten the therapeutic index of the virus by enhancing tumour-directed specificity and increasing the maximum virus dose tolerated in mice, but IFN induced by AV1/AV2 protects PKR$^{-/-}$ mice from lethal challenge with wild type VSV (Stojdl *et al.*, 2003). This observation suggests that in the unlikely event of a reversion of AV1/AV2 to the wild type phenotype during replication the safety of the therapeutic vector will not be compromised, in the face of some retention of the AV1/2 IFN-inducing phenotype.

11.8 Engineered/recombinant VSV

With the advent of a VSV cloning and rescue system by Rose's group (Lawson *et al.*, 1995), investigators have been able to modify the VSV genome in attempts to improve safety and to enhance the therapeutic index of VSV against tumours *in vivo*.

The *in vitro* selected small plaque mutant of VSV, AV1, with the M51R mutation in the matrix protein, rendered this virus unable to block host cell gene expression including IFN, thereby enhancing selectivity of VSV replication for tumor cells without the need for exogenous IFN addition (Stojdl *et al.*, 2003). This mutation has been re-engineered as a deletion of methionine at amino acid position 51 in the M protein (ΔM51) and rescued into VSV, which retains its interferon inducing status and inherent tumour cell selectivity based on IFN response defects in tumour cells, and is therefore highly attenuated. This variant has proven highly selective in killing numerous glioma cell lines and primary human glioma cultures from surgical specimens, while normal cells (HS68, NIH3T3) remained resistant to VSV$^{\Delta M51}$. Intratumoral administration of VSV$^{\Delta M51}$ to U87 and U118 glioblastoma cell lines in nude mice resulted in marked tumour regression, while systemic administration to U87 orthotopic tumours showed significantly enhanced survival over dead-virus treated controls, and exhibited localization to multifocal gliomas and invasive glioma cells (Lun *et al.*, 2006). This same effect could be recapitulated by inclusion of murine IFN-β in the VSV genome, allowing enhanced replication in IFN-unresponsive tumour cells over normal cells, and significantly attenuated the virus *in vivo* (Obuchi *et al.*, 2003).

Inclusion of immunotherapeutic or suicide genes into VSV is in its infancy experimentally, but holds great promise for enhancing the therapeutic benefit of VSV during oncolytic virotherapy.

VSV encoding cytosine deaminase/uracil phosphoribosyl transferase (CD/UPRT) which converts 5-fluorocytosine to 5-fluorouracil caused decreased A20 or TS/A tumour burden during intratumoral administration, and facilitated the emergence of IFN-γ expressing $CD8^+$T cells in spleens of treated mice (Porosnicu *et al.*, 2003). VSV encorporating thymidine kinase or interleukin-4 selectively kills cancer cells *in vitro* in the presence of exogenous IFN, and during intratumoral administration exhibits enhanced oncolytic activity against breast carcinoma or melanoma, relative to a GFP-encoding VSV control. This activity was also associated with antitumour cytotoxic T lymphocyte (CTL) responses, and showed efficacy during systemic administration in the metastatic TS/A model of breast cancer, and prolonged survival (Fernandez *et al.*, 2002).

In lieu of deriving or engineering attenuated variants of VSV, an alternative strategy to ensure cancer selectivity is to retarget or restrict infection of VSV to a limited set of cells, and this has been elegantly demonstrated by Ira Bergman. Two strategies have been developed to retarget VSV to breast cancer cells. The first is to replace the VSV glycoprotein (G) with the Sindbis virus glycoprotein, modified to reduce its native binding function, and which also contains the immunoglobulin Fc binding domain of protein A. VSV virus particles coated with this chimeric Sindbis glycoprotein are then conjugated to a monoclonal antibody, 4D5, directed against the Her2/neu receptor to retarget VSV binding and infection of Her2 neu positive breast cancer cells (Bergman *et al.*, 2003). Alternatively the Sindbis glycoprotein is modified to contain a single chain antibody against Her2 neu, and VSV expressing this modification demonstrated preferential replication of erbB2-expressing cells compared to erbB2-null cells (Bergman *et al.*, 2004). Thus, retargeting VSV from its broad host cell tropism is an alternate attenuation strategy conferring tumour-directed oncolysis and should be further explored.

11.9 VSV effectively eradicates tumours *in vivo*

In vivo efficacy of VSV has aptly been demonstrated in numerous animal models, both human xenografts and immune competent syngeneic systems. Wild type VSV in a subcutaneous xenograft melanoma model administered intratumorally demonstrated decreased tumour burden relative to untreated control mice (Stojdl *et al.*, 2000b). The IFN-inducing attenuated mutant AV1 was also effective in OVCAR2 xenografts and CT26 colon adenocarcinoma lung metastasis model in immune competent Balb/c mice during systemic viral administration (Stojdl *et al.*, 2003). Although IFN-inducing mutants of VSV are extremely effective at infecting and killing CT26 cells *in vitro*, one intravenous dose of virus was insufficient to eradicate CT26 lung metastases *in vivo* (unpublished observations), indicating that perhaps previously unappreciated physiological barriers obstruct the *in vivo* application of VSV and oncolytic viruses as a whole. However, 6 doses of AV1 or AV2 variants of VSV given intravenously every 48 h at 5×10^8 plaque-forming units (p.f.u.) per dose very effectively eradicated CT26 tumours in immune competent mice, and established durable cures beyond 100 days (Stojdl *et al.*, 2003). Wild type VSV also inhibits growth of C6 glioblastomas by intravenous administration and is also effective in controlling BALB/3T3Myc, BALB/3T3 Ras, and CH3Ag104 sarcoma *in vivo* during intratumoral treatment (Balachandran *et al.*, 2001). Inhibition of p53 null C6 glioblastoma tumours *in vivo* occurred without infecting and replicating in normal tissues (Balachandran and Barber, 2000). In a head-to-head comparison of wild type and M51R IFN-inducing VSV, both viruses were equally effective in controlling tumour growth of LNCaP prostate tumours; however wild type virus killed 50–71 per cent of mice (Ahmed *et al.*, 2004). A variant of VSV, VSV-rp30 selected for enhanced lytic activity on U87MG *in vitro* also could spread and kill subcutaneous U87MG tumours in CB.17 severe combined immunodeficient mice *in vivo* (Wollmann *et al.*, 2005).

The laboratory of Savio Woo has performed extensive analyses of the *in vivo* application of VSV to treat animal models of multi-focal colorectal cancer metastases or hepatocellular carcinoma (HCC). In MCA26 colorectal carcinoma in the livers of BALB/c mice, a single intratumoral injection of VSV prolonged survival (Huang *et al.*,

2003), while human and rat HCC cells *in vitro* showed preferential replication of VSV over normal liver cells, and one intratumoral dose in rats showed selective replication in tumours within the liver, and prolonged survival (Ebert *et al.*, 2003). A modification of VSV to create a fusogenic variant expressing a fusion protein from Newcastle disease virus (NDV) (L289A F protein) showed syncytia formation between tumour cells via membrane fusion, and more prolonged survival in a rat model of HCC treated with fusogenic VSV by hepatic arterial infusion (Ebert *et al.*, 2004). In multifocal colorectal carcinoma in the livers of rats (chemically-induced rat colorectal carcinoma (LMCR) cells introduced by the ileocaecal vein), VSV gained access to multiple colorectal carcinoma lesions by hepatic arterial infusion and demonstrated enhanced survival (Shinozaki *et al.*, 2004, 2005c), and the antitumour response was further enhanced by repeated dosing with hepatic arterial infusion (Shinozaki *et al.*, 2005b). In line with VSV selectivity for IFN defective tumour cells, treatment of a rat model of HCC with VSV and prophylactic IFN-α enhanced the therapeutic index, increased the maximum tolerated dose by ½ log, and prolonged survival of tumour-bearing rats (Shinozaki *et al.*, 2005a).

While VSV shows efficacy against numerous cancer models *in vivo* on its own, combination therapy warrants consideration, and has begun to be explored. Combination of VSV (5×10^4–10^8 p.f.u.) daily for five days with 5–125 mg/kg/day every third day for a total of four doses of gemcitabine, showed enhanced lung carcinoma regression over either therapy alone (Li *et al.*, 2004). Therefore, while VSV holds great promise as a single therapeutic agent, its potential may be greatly enhanced in combination with traditional chemotherapy or radiation therapy regimens, and should be explored further.

VSV shows significant infectivity and cytolysis of tumour cells *in vitro*, and has potent antitumour activity *in vivo* in numerous cancer models. While VSV infection and spread *in vitro* is rapid, clearly the need for multiple therapeutic doses of virus to elicit cures *in vivo* (Stojdl *et al.*, 2003) suggests that significant barriers exist to delivery and/or spread of virus within tumours *in vivo*. These barriers warrant future investigation in order to enhance the therapeutic index of wild type or recombinant VSVs, and may represent novel targets for cancer treatment by VSV or other oncolytic viruses.

11.10 VSV and the host immune response

VSV has been extensively studied for decades as a prototypical IFN-restricted, acute viral infection. In addition to the first-line innate control of VSV infection by the IFN response, it has also become clear that the adaptive immune response represents a significant potential barrier to the application of VSV as a therapeutic vector, in both aspects of delivery, and spread or persistence within a tumour, once initial infection has been achieved. This is a particular threat since multiple doses of VSV seem to be required to elicit the most potent anti-tumour responses or cures in immune competent hosts.

While neutralizing antibodies, cytotoxic T lymphocytes, and T helper cell proliferation and function are activated by VSV infection (Seiler *et al.*, 1998), there is a very clear role for the antibody response in protection against VSV infection. This is particularly evident in IFN-α/β receptor knockout mice, which are highly susceptible to VSV infection due to unchecked virus replication. Passive transfer of neutralizing antibodies, but not T cells, are protective against VSV pathology in type I IFN receptor$^{-/-}$ mice (Steinhoff *et al.*, 1995). Protection of mice against VSV infection by antibody is independent of the immunoglobulin subclass/isotype, avidity, rate of neutralization, and *in vitro* neutralization activity, as long as a minimum threshold serum concentration of antibody is reached (Bachmann *et al.*, 1997). The lack of antibody class specificity in protection of host versus VSV implies that the antibody protective activity comprises a mix of neutralization and complement fixation/opsonization of virus particles. The viral glycoprotein G is the major antigenic target of neutralizing and non-neutralizing antibodies generated in response to VSV infection, and neutralizing epitopes against G define serotype specificity, while non-neutralizing epitopes are cross-reactive among serotypes (Lefrancois and Lyles, 1983). Generation of neutralizing IgM antibodies against G is largely T cell independent, and is largely dependent on antigen trapping by complement receptors on follicular dendritic cells (Freer *et al.*, 1994; Ochsenbein *et al.*, 1999).

While VSV G has been identified as a target of $CD4^+$ lytic cells (Browning *et al.*, 1990), N (nucleoprotein) is the predominant antigen recognized by VSV specific $CD8^+$ cytotoxic T lymphocytes (CTL) (Puddington *et al.*, 1986), and recognition of N-derived or G-derived CTL epitopes by CTL is cross-reactive among serotypes. In addition, while $CD4^+$ T-cell activity is detected in response to VSV infection, CTL priming can be independent of CD4 cells but is dependent on CD40 ligand and CD28/B7 interactions, as is the case for antibody class switching (Andreasen *et al.*, 2000; McAdam *et al.*, 2000). In addition, peripheral blood lymphocytes (PBL) from VSV infected mice show evidence for enhanced natural killer (NK) lytic activity *ex vivo* on Yac-1 target cells (Stitz *et al.*, 1985), indicating that NK cells are also activated during VSV infection of mice. The complement cascade is also induced or activated during VSV infection, leading to the production of complement component C5a, a potent anaphylotoxin and perhaps the most potent inflammatory mediator known (Chen and Reiss, 2002). Taken together, various aspects of innate and adaptive immunity are induced following VSV infection *in vivo*, yet it appears that type I IFN and to a lesser extent, antibody against VSV are of paramount importance in restricting VSV replication and pathogenesis. Although T lymphocytes are not apparently vital for control of VSV infection *in vivo*, it remains to be determined whether the T-cell compartment, in addition to antibodies, restricts the utility of VSV as a cancer therapeutic in immune competent hosts.

11.11 Host immunity vs. therapeutic efficacy

During VSV treatment of advanced multifocal HCC in rats, administering a low virus dose via hepatic arterial infusion allowed sustained tumour selective viral replication, which was restricted once antibodies were generated (Shinozaki *et al.*, 2005b). Yet despite neutralizing antibody responses, oncolytic viruses such as VSV, adenovirus and NDV have shown efficacy in immune competent animals or in human patients (Lorence *et al.*, 2003; Stojdl *et al.*, 2003). However, multiple doses of VSV were needed to cure mice with CT26 lung tumours, and these doses were given over a period of time coincident with antibody generation against VSV. Given that multiple doses of VSV can be efficacious, yet there is little data illustrating that all doses result in augmented delivery/replication in tumour, this opens the door to the possibility that oncolytic virotherapy leads to some amount of anti-tumour immunity, and if so, this can be explored and exploited. The potential emergence of antitumour adaptive immunity during oncolytic virotherapy therapy may contribute significantly not only to rejection of a primary tumour mass, but may also aid in the establishment of prolonged tumour cures by inducing immunological memory to relevant tumour antigens to prevent recurrence, and may also aid in the control and elimination of secondary metastases in the absence of direct infection with oncolytic viruses. Uncovering whether VSV possesses adjuvant activity, or can be armed with therapeutic transgenes to facilitate antitumour immunity may open new therapeutic opportunities for VSV, and in addition to facilitating tumour clearance may also establish anti-tumour immunological memory and concomitant durable cures.

11.12 VSV is a potent vaccine

Viral vectors have previously been exploited for use as gene therapy and vaccine vectors in the cancer setting and in treatment or manipulation of numerous other disease conditions. Viruses have proven particularly useful in vaccine applications due to the capacity to introduce antigens and immunomodulatory genes of choice into vector concurrent with the adjuvant/inflammatory properties inherent to viruses. While there is abundant literature about the prior use of some oncolytic viruses as tumour vaccines independent of direct in situ tumour oncolysis, notably NDV and adenovirus, little has been explored regarding in situ tumour vaccination by VSV.

The notion that VSV may evoke antitumour immunity as an in situ vaccine during oncolytic virotherapy has particular merit considering the recent history of VSV application as a vaccine vector for a number of infectious diseases. Initially, incorporation of influenza HA (haemagglutinin) into VSV was shown to be efficacious against lethal influenza challenge and bronchial pneumonia (Roberts *et al.*, 1998, 1999). Since then, VSV has also been proven effective as a vaccine against cottontail rabbit papillomavirus (L1 major capsid protein) (Reuter *et al.*, 2002, Roberts *et al.*, 2004) and severe acute respiratory syndrome (SARS-coronavirus spike protein) (Kapadia *et al.*, 2005), and is receiving significant attention as an human immunodeficiency virus (HIV) vaccine vector. Inclusion of HIV env or gag into VSV elicits long-term specific memory in the T lymphocyte compartment and excellent recall responses in mice (Haglund *et al.*, 2002b, Haglund *et al.*, 2002a). Additionally, primary responses generated in rhesus monkeys can be boosted in heterologous vaccine vectors such as vaccinia, and somewhat protects against lethal SHIV89.6P challenge in delaying the onset of acquired immune deficiency syndrome and maintaining significantly longer periods of viral load-free status (Rose *et al.*, 2001; Ramsburg *et al.*, 2004). Efforts to improve the utility of VSV as a vaccine vector have revealed that including foreign antigens further upstream in the VSV genome confers more potent immunization (Roberts *et al.*, 2004), and that VSV pathogenesis is reduced by truncating the cytoplasmic domain of G from 29 to 9 amino acids, with comparable immunization against an included foreign transgene (Publicover *et al.*, 2004). As well, single cycle infectious VSV is as effective in generating an anti-env $CD8^+$ T cell and antibody response in mice compared to wild-type virus (Publicover *et al.*, 2005), and inclusion of granulocyte–macrophage colony-stimulating factor (GM-CSF) in VSV enhances memory $CD8^+$ T-cell activity and is highly attenuated with respect to the wild type vector (Ramsburg *et al.*, 2005).

11.13 Innate sensing of VSV and the antitumour response

T lymphocyte responses against VSV are largely primed by dendritic cells (Ciavarra *et al.*, 2000), and in infected cells, VSV is detected or sensed by several dsRNA sensing molecules that additionally interplay with immune cell activation. While the prototypic dsRNA sensor is considered to be the dsRNA-dependent protein kinase PKR, which has been shown to be involved in detection of VSV (Stojdl *et al.*, 2000a), recent work has uncovered additional dsRNA sensory pathways. FADD (Fas-associated protein with death domain) is involved in a PKR- and TLR3-independent dsRNA sensing pathway that involves TBK-1, IRF-1 activation, and receptor interacting protein-1 (Balachandran *et al.*, 2004). While TLR3 has been identified as a receptor for dsRNA *in vitro*, TLR3 is not universally required for antiviral responses, as $TLR3^{-/-}$ mice show no impairment in their ability to generate an adaptive immune response against VSV, and show no alteration in viral pathogenesis (Edelmann *et al.*, 2004). Alternatively, TLR7 which recognizes ssRNA viruses including VSV and influenza is vital for host defense against VSV. $TLR7^{-/-}$ mice or $MyD88^{-/-}$ mice show a reduced response to infection against VSV in terms of reduced IFN-α production (Lund *et al.*, 2004). Additional support for a role of TLR sensing of VSV comes from evidence that fibroblasts deficient in TRAF3, a critical link between TLR adapters and IFR-3/7 kinases that are important for IRF activation and the IFN response, are defective in their type I IFN responses to VSV infection (Oganesyan *et al.*, 2006). Viral replication is alternatively sensed by the recently identified cytosolic DExD/H box RNA helicase RIG-I (Yoneyama *et al.*, 2004). While RIG-I-mediated dsRNA-dependent signaling leads to NF-κB, IRF-3 activation and type I IFN induction, and has been shown to sense viral infection by hepatitis C (Foy *et al.*, 2005), the flaviviruses Japanese encephalitis virus (JEV) and dengue virus serotype 2 (DEN-2) (Chang *et al.*, 2006), and the paramyxovirus Sendai virus (Melchjorsen *et al.*, 2005), no association has yet been made between RIG-I and induction of type I IFN by VSV. Regardless of the intracellular dsRNA sensor, pathogen-associated danger signals sensed by antigen presenting cells trigger adaptive immune responses against the pathogen. Recent work has illustrated that the combination of dsRNA recognition by TLR-3 in dendritic cells, along with viral antigen phagocytosed from virus infected cells undergoing apoptosis was sufficient to prime cytotoxic T cells against viral antigen, even though the antigen-presenting cell itself was not infected (Schulz *et al.*, 2005). It is therefore conceivable that dsRNA present in virally-infected tumour cells during oncolysis may stimulate antigen presentation and T-cell priming against viral and tumour antigen in a similar manner, following phagocytosis by antigen presenting cells.

11.14 So what is a good oncolytic virus?

VSV is not a human pathogen, but will readily infect human cells. As such, there is limited chance of pre-existing anti-VSV immunity in the prospective human cancer patient population. Considering that VSV elicits a rapid and robust neutralizing antibody and cytotoxic T-cell response in immune competent mice, the existence of immunity to VSV would be problematic for the therapeutic utility of VSV against human cancer, as it has been shown to be in mouse models of numerous oncolytic viruses [reviewed in (Parato *et al.*, 2005)].

In addition, an oncolytic virus suitable for use as a cancer therapeutic should be selective for replication in tumour cells, with limited or no toxicity in normal tissues. Clearly wild type VSV in the presence of type I IFN, or IFN-inducing variants of VSV show this level of selectivity *in vitro* and *in vivo* in a variety of human cancers and preclinical animal models. *In vivo*, the selectivity of viral infection of lung tumour metastases is as precise as excluding adjacent normal tissue within the tumour-affected organ (Figure 11.3).

Ideally an oncolytic virus should be amenable to genetic manipulation either to facilitate tumour specificity or attenuation, or to arm the vector with therapeutic or suicide genes, to enhance the therapeutic index. Alternatively, employing imaging/reporter transgenes into the oncolytic vector can facilitate monitoring of delivery, spread and persistence or elimination of the vector *in vivo*, which in preclinical or early clinical trial settings may be particularly instructive. VSV readily accepts transgenes into its genome, which can be rescued into infectious virus as described above.

The viral life cycle of an oncolytic virus should ideally include rapid replication, cytolysis and spread, in order to amplify the therapeutic inoculum *in situ*, allowing the virus to spread rapidly particularly in an immune competent host. A virus that spreads particularly well cell to cell or within the tumour microenvironment would theoretically minimize exposure of the virus to the host immune system systemically once it accesses the tumour itself. Of course, a virus with a rapid replicative cycle culminating in lysis rather than establishing a chronic infection would maximize target cell destruction.

A virus that can be given to patients systemically would be beneficial in the setting of treating metastatic or inaccessible tumours. While the majority of early clinical trials of oncolytic viruses are principally intratumoral and show safety and limited immune involvement, when delivered locally, a virus that is efficacious and safe systemically is ideal, and warrants significant exploration. Clearly preclinical data relating to systemic VSV application to solid or metastatic tumours show significant efficacy in mice; whether this holds true in humans is another concern. Recent studies have demonstrated that lentiviral vectors pseudotyped with VSV-G packaged in murine cells are inactivated by human serum, and packaging in a human cell line may help render virus resistant to inactivation (Ory *et al.*, 1996). Yet other groups maintain that virus packaged in human cell lines still is sensitive to inactivation by human serum (DePolo *et al.*, 2000). Vector modifications such as polyethylene glycol (PEG) conjugation increases plasma vector

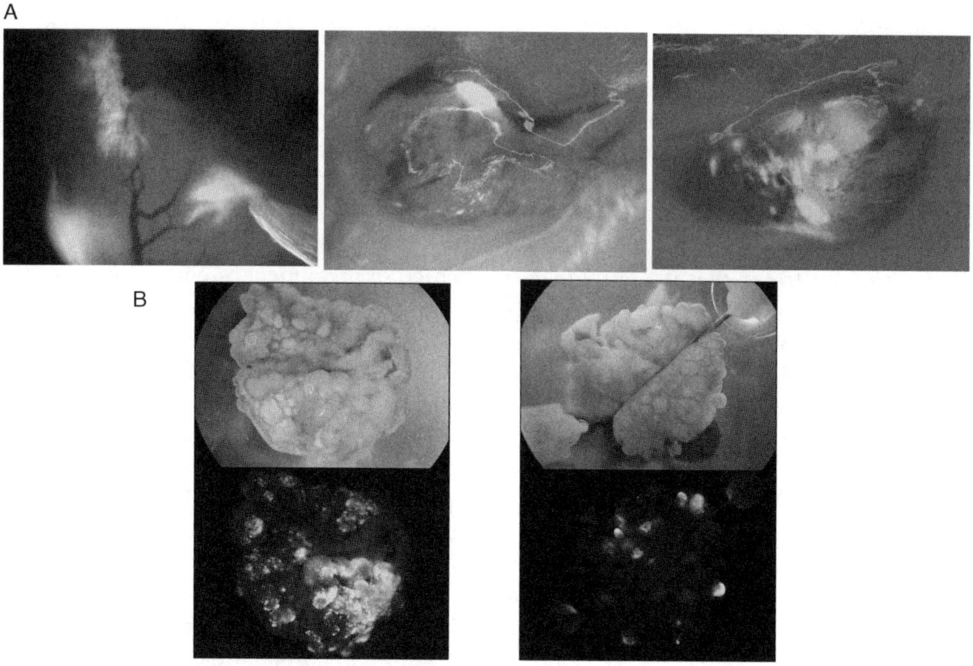

Figure 11.3 IFN-inducing VSVΔM51-GFP replicates specifically in tumour tissue, but not surrounding normal tissue. An IFN-inducing mutant of VSV expressing GFP was administered systemically to BALB/c mice bearing (A) subcutaneous, or (B) lung CT26 tumours. In (A), after 24 h post-treatment, VSV infection appears to be most concentrated around areas of significant vasculature. (B) In lung tumours, VSV infection appears rampant at 11 h post-infection, but tapers off by 72 h. In both the lung and subcutaneous tumours, viral infection and replication are only observed in tumour tissue, but not in adjacent normal tissue

half life (Croyle *et al.*, 2004), and additionally shields the virus from antibody neutralization.

For safety purposes, an oncolytic virus that does not pose a risk for integration into the host genome, and possible downstream deleterious genetic events is most desirable. Replication of VSV occurs strictly in the cytoplasm and does not entail a DNA intermediate, and therefore no significant risk of genomic integration exists. In addition, while not necessarily vital for an oncolytic virus candidate, viruses that elicit antitumour immunity during the process of tumour treatment would have the added benefit of helping tumour regression via a secondary mechanism, harnessing the power and adaptability of the immune system to eliminate tumour cells, and keep metastatic spread and new tumour growth at bay. Among other viruses, VSV induces antigen presenting cell activation via TLR3 or -7 and, therefore, has the potential to act as an adjuvant in vivo, yet this point remains to be proven. VSV replicates well in hypoxic tumours, which gives it an advantage over traditional cancer therapeutics which are often limited by hypoxic regions (Connor *et al.*, 2004). Last but not least, VSV has a broad host cell range and can therefore, bind and infect a diverse array of cancers from multiple species.

11.15 Future challenges for VSV

While VSV satisfies numerous criteria putting it forth as an excellent oncolytic virus candidate, several challenges remain in adapting this agent for maximal therapeutic utility. The robust neutralizing antibody response that occurs shortly after infection is a significant barrier to subsequent VSV treatments. Several strategies have been employed in other oncolytic virus settings including epitope/serotype modifications, polymer coating viruses to shield antigens, liposomal encapsulation, and immune suppression [reviewed in (Parato *et al.*, 2005)]. Whether any of these methodologies will help VSV evade the antibody response has yet to be determined. However caution needs to be considered: in the absence of IFN restriction of VSV, the lack of antibody/or evading antibodies could pose a safety risk. Like other viruses, VSV is subject to uptake/absorption by the liver, and is poorly delivered to the tumour environment. Factors determining access of VSV to tumours such as leaky vasculature or infection of vascular endothelium have not been fully elucidated, but should be considered in efforts to improve delivery and efficacy of VSV early in treatment. Finally, wild type VSV is well documented to cause encephalopathy in mice. Therefore, to ensure safety of a VSV-based therapeutic, the virus should be sufficiently attenuated or delivered in a route/dosing regimen to minimize this risk of central nervous system infection.

Despite these obstacles, in numerous preclinical models, VSV remains an excellent candidate oncolytic virus with substantial efficacy during systemic administration. Future efforts to arm VSV with new immunoregulatory genes may facilitate the evolution of antitumour immunity during oncolytic virotherapy, especially considering VSV's considerable utility as a vaccine vector in the infectious disease arena. Inclusion of imaging genes that allow non-invasive monitoring of virus replication in vivo in the next generation of VSV vectors should facilitate further understanding of the biology of the virus, delivery and spread throughout tumours. The scientific and clinical community awaits phase I trials of this agent with interest.

References

Ahmed M, Cramer SD, Lyles DS (2004). Sensitivity of prostate tumors to wild type and M protein mutant vesicular stomatitis viruses. *Virology* **330**, 34–49.

Andreasen SO, Christensen JE, Marker O, Thomsen AR (2000). Role of CD40 ligand and CD28 in induction and maintenance of antiviral CD8+ effector T cell responses. *J Immunol* **164**, 3689–3697.

Bachmann MF, Kalinke U, Althage A, *et al.* (1997). The role of antibody concentration and avidity in antiviral protection. *Science* **276**, 2024–2027.

Balachandran S, Barber GN (2000). Vesicular stomatitis virus (VSV;therapy of tumors. *IUBMB Life* **50**, 135–138.

Balachandran S., Porosnicu M, Barber GN (2001). Oncolytic activity of vesicular stomatitis virus is effective against tumors exhibiting aberrant p53, Ras, or myc function and involves the induction of apoptosis. *J Virol* **75**, 3474–9.

Balachandran S, Thomas E, Barber GN (2004). A FADD-dependent innate immune mechanism in mammalian cells. *Nature* **432**, 401–405.

Bergman I, Whitaker-Dowling P, Gao Y, Griffin JA (2004). Preferential targeting of vesicular stomatitis virus to breast cancer cells. *Virology* **330**, 24–33.

Bergman I, Whitaker-Dowling P, Gao Y, Griffin JA, Watkins SC (2003). Vesicular stomatitis virus expressing a chimeric Sindbis glycoprotein containing an Fc antibody binding domain targets to Her2/neu over-expressing breast cancer cells. *Virology* **316**, 337–47.

Bowie AG, Haga IR (2005). The role of Toll-like receptors in the host response to viruses. *Mol Immunol* **42**, 859–67.

Brown EL, Lyles DS (2003). A novel method for analysis of membrane microdomains: vesicular stomatitis virus glycoprotein microdomains change in size during infection, and those outside of budding sites resemble sites of virus budding. *Virology* **310**, 343–358.

Browning M, Reiss CS, Huang AS (1990). The soluble viral glycoprotein of vesicular stomatitis virus efficiently sensitizes target cells for lysis by CD4+ T lymphocytes. *J Virol,* **64**, 3810–3816.

Cesaire R, Oliere S, Sharif-Askari E, *et al.* (2006). Oncolytic activity of vesicular stomatitis virus in primary adult T-cell leukemia. *Oncogene* **25**, 349–358.

Chang TH, Liao CL, Lin YL (2006). Flavivirus induces interferon-beta gene expression through a pathway involving RIG-I-dependent IRF-3 and PI3K-dependent NF-kappaB activation. *Microbes Infect* **8**, 157–171.

Chawla-Sarkar M, Lindner DJ, Liu, YF, *et al.* (2003). Apoptosis and interferons: role of interferon-stimulated genes as mediators of apoptosis. *Apoptosis* **8**, 237–249.

Chen N, Reiss CS (2002). Innate immunity in viral encephalitis: role of C5. *Viral Immunol* **15**, 365–372.

Ciavarra RP, Greene AR, Horeth DR, Buhrer K, Van Rooijen N, Tedeschi B (2000). Antigen processing of vesicular stomatitis virus in situ. Interdigitating dendritic cells present viral antigens independent of marginal dendritic cells but fail to prime CD4 +;and CD8 +;T cells. *Immunology* **101**, 512–520.

Comer JA, Tesh RB, Modi GB, Corn JL, Nettles VF (1990). Vesicular stomatitis virus, New Jersey serotype: replication in and transmission by *Lutzomyia shannoni* (Diptera: Psychodidae). *Am J Trop Med Hyg* **42**, 483–490.

Connor JH, Naczki C, Koumenis C, Lyles DS (2004). Replication and cytopathic effect of oncolytic vesicular stomatitis virus in hypoxic tumor cells *in vitro* and *in vivo*. *J Virol* **78**, 8960–8970.

Croyle MA, Callahan SM, Auricchio A, *et al.* (2004). PEGylation of a vesicular stomatitis virus G pseudotyped lentivirus vector prevents inactivation in serum. *J Virol* **78**, 912–921.

Das SC, Pattnaik AK (2004). Phosphorylation of vesicular stomatitis virus phosphoprotein P is indispensable for virus growth. *J Virol* **78**, 6420–30.

De Veer MJ, Holko M, Frevel M, *et al.* (2001). Functional classification of interferon-stimulated genes identified using microarrays. *J Leukoc Biol* **69**, 912–20.

DePolo NJ, Reed JD, Sheridan PL, *et al.* (2000). VSV-G pseudotyped lentiviral vector particles produced in human cells are inactivated by human serum. *Mol Ther* **2**, 218–22.

Dunn GP, Bruce AT, Sheehan KC, *et al.* (2005). A critical function for type I interferons in cancer immunoediting. *Nat Immunol* **6**, 722–729.

Duntsch CD, Zhou Q, Jayakar HR, *et al.* (2004). Recombinant vesicular stomatitis virus vectors as oncolytic agents in the treatment of high-grade gliomas in an organotypic brain tissue slice-glioma coculture model. *J Neurosurg* **100**, 1049–1059.

Ebert O, Shinozaki K, Huang TG, Savontaus MJ, Garcia-Sastre A, Woo SL (2003). Oncolytic vesicular stomatitis virus for treatment of orthotopic hepatocellular carcinoma in immune-competent rats. *Cancer Res* **63**, 3605–3611.

Ebert O, Shinozaki K, Kournioti C, Park MS, Garcia-Sastre A, Woo SL (2004). Syncytia induction enhances the oncolytic potential of vesicular stomatitis virus in virotherapy for cancer. *Cancer Res* **64**, 3265–3270.

Edelmann KH, Richardson-Burns S, Alexopoulou L, Tyler KL, Flavell RA, Oldstone MB (2004). Does Toll-like receptor 3 play a biological role in virus infections? *Virology* **322**, 231–238.

Faria PA, Chakraborty P, Levay A, *et al.* (2005). VSV disrupts the Rae1/mrnp41 mRNA nuclear export pathway. *Mol Cell* **17**, 93–102.

Fernandez M, Porosnicu M, Markovic D, Barber GN (2002). Genetically engineered vesicular stomatitis virus in gene therapy: application for treatment of malignant disease. *J Virol* **76**, 895–904.

Foy E, Li K, Sumpter R, Jr, *et al.* (2005). Control of antiviral defenses through hepatitis C virus disruption of retinoic acid-inducible gene-I signaling. *Proc Natl Acad Sci U S A* **102**, 2986–2991.

Fredericksen BL, Whitt MA (1995). Vesicular stomatitis virus glycoprotein mutations that affect membrane fusion activity and abolish virus infectivity. *J Virol* **69**, 1435–1443.

Freer G, Burkhart C, Ciernik I, Bachmann, MF, Hengartner H, Zinkernagel RM (1994). Vesicular stomatitis virus Indiana glycoprotein as a T-cell-dependent and -independent antigen. *J Virol* **68**, 3650–3655.

Gadaleta P, Perfetti X, Mersich S, Coulombie F (2005). Early activation of the mitochondrial apoptotic pathway in vesicular stomatitis virus-infected cells. *Virus Res* **109**, 65–69.

Gale M Jr, Katze MG (1998). Molecular mechanisms of interferon resistance mediated by viral-directed inhibition of PKR, the interferon-induced protein kinase. *Pharmacol Ther* **78**, 29–46.

Gupta AK, Shaji D, Banerjee AK (2003). Identification of a novel tripartite complex involved in replication of vesicular stomatitis virus genome RNA. *J Virol* **77**, 732–738.

Haglund K, Leiner I, Kerksiek K, Buonocore L, Pamer E, Rose JK (2002a). High-level primary CD8+ T-cell response to human immunodeficiency virus type 1 gag and env generated by vaccination with recombinant vesicular stomatitis viruses. *J Virol* **76**, 2730–2738.

Haglund K, Leiner I, Kerksiek K, Buonocore L, Pamer E, Rose JK (2002b). Robust recall and long-term memory T-cell responses induced by prime-boost regimens with heterologous live viral vectors expressing human immunodeficiency virus type 1 Gag and Env proteins. *J Virol* **76**, 7506–7517.

Hertzog PJ, O'Neill LA, Hamilton JA (2003). The interferon in TLR signaling: more than just antiviral. *Trends Immunol* **24**, 534–539.

Howerth EW, Mead DG, Stallknecht DE (2002). Immunolocalization of vesicular stomatitis virus in black flies (Simulium vittatum). *Ann N Y Acad Sci* **969**, 340–345.

Huang TG, Ebert O, Shinozaki K, Garcia-Sastre A, Woo SL (2003). Oncolysis of hepatic metastasis of colorectal cancer by recombinant vesicular stomatitis virus in immune-competent mice. *Mol Ther* **8**, 434–440.

Ikeda H, Old LJ, Schreiber RD (2002). The roles of IFN gamma in protection against tumor development and cancer immunoediting. *Cytokine Growth Factor Rev* **13**, 95–109.

Irie T, Licata JM, Jayakar HR, Whitt MA, Bell P, Harty RN (2004). Functional analysis of late-budding domain activity associated with the PSAP motif within the vesicular stomatitis virus M protein. *J Virol* **78**, 7823–7827.

Kaempfer R (2003). RNA sensors: novel regulators of gene expression. *EMBO Rep* **4**, 1043–1047.

Kapadia SU, Rose JK, Lamirande E, Vogel L, Subbarao K, Roberts A (2005). Long-term protection from SARS coronavirus infection conferred by a single immunization with an attenuated VSV-based vaccine. *Virology* **340**, 174–182.

Khabar KS, Al-Haj L, Al-Zoghaibi F, *et al.* (2004). Expressed gene clusters associated with cellular sensitivity and resistance towards anti-viral and anti-proliferative actions of interferon. *J Mol Biol* **342**, 833–846.

Lawson ND, Stillman EA, Whitt MA, Rose JK (1995). Recombinant vesicular stomatitis viruses from DNA. *Proc Natl Acad Sci U S A* **92**, 4477–4781.

Lefrancois L, Lyles DS (1983). Antigenic determinants of vesicular stomatitis virus: analysis with antigenic variants. *J Immunol* **130**, 394–398.

Letchworth GJ, Rodriguez LL, Del Cbarrera J (1999). Vesicular stomatitis. *Vet J* **157**, 239–260.

Li Q, Wei YQ, Wen YJ, *et al.* (2004). Induction of apoptosis and tumor regression by vesicular stomatitis virus in the presence of gemcitabine in lung cancer. *Int J Cancer* **112**, 143–149.

Lichty BD, Stojdl DF, Taylor RA, *et al.* (2004). Vesicular stomatitis virus: a potential therapeutic virus for the treatment of hematologic malignancy. *Hum Gene Ther* **15**, 821–831.

Liu IK, Zee YC (1976). The pathogenesis of vesicular stomatitis virus, serotype Indiana, in Aedes aegypti mosquitoes. I. Intrathoracic injection. *Am J Trop Med Hyg* **25**, 177–185.

Lorence RM, Pecora AL, Major PP, *et al.* (2003). Overview of phase I studies of intravenous administration of PV701, an oncolytic virus. *Curr Opin Mol Ther* **5**, 618–624.

Lun X, Senger D, Alain T, *et al.* (2006). Effects of intravenously administered recombinant vesicular stomatitis virus (VSVDeltaM51) on multi-focal and invasive gliomas. *J Natl Cancer Inst* **98**(21): 1546–1557.

Lund JM, Alexopoulou L, Sato A, *et al.* (2004). Recognition of single-stranded RNA viruses by Toll-like receptor 7. *Proc Natl Acad Sci USA* **101**, 5598–5603.

Lyles DS, McKenzie MO (1997). Activity of vesicular stomatitis virus M protein mutants in cell rounding is correlated with the ability to inhibit host gene expression and is not correlated with virus assembly function. *Virology* **229**, 77–89.

ICTVdB Management (2006) 01.062.0.01.001. Vesicular stomatitis Indiana virus. In: ICTVdB – The Universal Virus Database, version 4. Bchen-Osmond, C. (Ed), Columbia University, New York, USA ICTVdB – The Universal Virus Database, ver 4; http://www.ncbi.nlm.nih.gov/ICTVdb/ICTVdB/

McAdam AJ, Farkash EA, Gewurz BE, Sharpe AH (2000). B7 costimulation is critical for antibody class switching and CD8(+;cytotoxic T-lymphocyte generation in the host response to vesicular stomatitis virus. *J Virol* **74**, 203–208.

Melchjorsen J, Jensen SB, Malmgaard L, *et al.* (2005). Activation of innate defense against a paramyxovirus is mediated by RIG-I and TLR7 and TLR8 in a cell-type-specific manner. *J Virol* **79**, 12944–12951.

Nunamaker RA, Lockwood JA, Stith CE, *et al.* (2003). Grasshoppers (Orthoptera: Acrididae;could serve as reservoirs and vectors of vesicular stomatitis virus. *J Med Entomol* **40**, 957–563.

Obuchi M, Fernandez M, Barber GN (2003). Development of recombinant vesicular stomatitis viruses that exploit defects in host defense to augment specific oncolytic activity. *J Virol* **77**, 8843–8856.

Ochsenbein AF, Pinschewer DD, Odermatt B, Carroll MC, Hengartner H, Zinkernagel RM (1999). Protective T cell-independent antiviral antibody responses are dependent on complement. *J Exp Med* **190**, 1165–1174.

Oganesyan G, Saha SK, Guo B, *et al.* (2006). Critical role of TRAF3 in the Toll-like receptor-dependent and independent antiviral response. *Nature* **439**, 208–211.

Ory DS, Neugeboren BA, Mulligan RC (1996). A stable human-derived packaging cell line for production of high titer retrovirus/vesicular stomatitis virus G pseudotypes. *Proc Natl Acad Sci U S A* **93**, 11400–11406.

Parato KA, Senger D, Forsyth PA, Bell JC (2005). Recent progress in the battle between oncolytic viruses and tumours. *Nat Rev Cancer* **5**, 965–76.

Porosnicu M, Mian A, Barber GN (2003). The oncolytic effect of recombinant vesicular stomatitis virus is enhanced by expression of the fusion cytosine deaminase/uracil phosphoribosyltransferase suicide gene. *Cancer Res* **63**, 8366–8376.

Publicover J, Ramsburg E, Rose JK (2004). Characterization of nonpathogenic, live, viral vaccine vectors inducing potent cellular immune responses. *J Virol* **78**, 9317–9324.

Publicover J, Ramsburg E, Rose JK (2005). A single-cycle vaccine vector based on vesicular stomatitis virus can induce immune responses comparable to those generated by a replication-competent vector. *J Virol* **79**, 13231–8.

Puddington L, Bevan MJ, Rose JK, Lefrancois L (1986). N protein is the predominant antigen recognized by vesicular stomatitis virus-specific cytotoxic T cells. *J Virol* **60**, 708–717.

Quiroz E, Moreno N, Peralta PH, Tesh RB (1988). A human case of encephalitis associated with vesicular stomatitis virus (Indiana serotype;infection. *Am J Trop Med Hyg* **39**, 312–314.

Ramsburg E, Publicover J, Buonocore L, *et al.* (2005). A vesicular stomatitis virus recombinant expressing granulocyte-macrophage colony-stimulating factor induces enhanced T-cell responses and is highly attenuated for replication in animals. *J Virol* **79**, 15043–15053.

Ramsburg E, Rose NF, Marx PA, *et al.* (2004). Highly effective control of an AIDS virus challenge in macaques by using vesicular stomatitis virus and modified vaccinia virus Ankara vaccine vectors in a single-boost protocol. *J Virol* **78**, 3930–3940.

Reif JS, Webb PA., Monath TP, *et al.* (1987). Epizootic vesicular stomatitis in Colorado, 1982: infection in occupational risk groups. *Am J Trop Med Hyg* **36**, 177–182.

Reuter JD, Vivas-Gonzalez BE, Gomez D, *et al.* (2002). Intranasal vaccination with a recombinant vesicular stomatitis virus expressing cottontail rabbit papillomavirus L1 protein provides complete protection against papillomavirus-induced disease. *J Virol* **76**, 8900–8909.

Roberts A, Buonocore L, Price R, Forman J, Rose JK (1999). Attenuated vesicular stomatitis viruses as vaccine vectors. *J Virol* **73**, 3723–3732.

Roberts A, Kretzschmar E, Perkins AS, *et al.* (1998). Vaccination with a recombinant vesicular stomatitis virus expressing an influenza virus hemagglutinin provides complete protection from influenza virus challenge. *J Virol* **72**, 4704–4711.

Roberts A, Reuter JD, Wilson JH, Baldwin S, Rose JK (2004). Complete protection from papillomavirus challenge after a single vaccination with a vesicular stomatitis virus vector expressing high levels of L1 protein. *J Virol* **78**, 3196–3199.

Rodriguez LL (2002). Emergence and re-emergence of vesicular stomatitis in the United States. *Virus Res* **85**, 211–219.

Rodriguez LL, Vernon S, Morales AI, Letchworth GJ (1990). Serological monitoring of vesicular stomatitis New Jersey virus in enzootic regions of Costa Rica. *Am J Trop Med Hyg* **42**, 272–281.

Rose NF, Marx PA, Luckay A, *et al.* (2001). An effective AIDS vaccine based on live attenuated vesicular stomatitis virus recombinants. *Cell* **106**, 539–549.

Samuel CE (2001). Antiviral actions of interferons. *Clin Microbiol Rev* **14**, 778–809, table of contents.

Sana TR, Janatpour MJ, Sathe M, McEvoy LM, McClanahan TK (2005). Microarray analysis of primary endothelial cells challenged with different inflammatory and immune cytokines. *Cytokine* **29**, 256–269.

Schulz O, Diebold SS, Chen M, *et al.* (2005). Toll-like receptor 3 promotes cross-priming to virus-infected cells. *Nature* **433**, 887–892.

Seiler P, Brundler MA, Zimmermann C, *et al.* (1998). Induction of protective cytotoxic T cell responses in the presence of high titers of virus-neutralizing antibodies: implications for passive and active immunization. *J Exp Med* **187**, 649–654.

Shinozaki K, Ebert O, Kournioti C, Tai YS, Woo SL (2004). Oncolysis of multifocal hepatocellular carcinoma in the rat liver by hepatic artery infusion of vesicular stomatitis virus. *Mol Ther* **9**, 368–376.

Shinozaki K, Ebert O, Suriawinata A, Thung SN, Woo SL (2005a). Prophylactic alpha interferon treatment increases the therapeutic index of oncolytic vesicular stomatitis virus virotherapy for advanced hepatocellular carcinoma in immune-competent rats. *J Virol* **79**, 13705–13713.

Shinozaki K, Ebert O, Woo SL (2005b). Eradication of advanced hepatocellular carcinoma in rats via repeated hepatic arterial infusions of recombinant VSV. *Hepatology* **41**, 196–203.

Shinozaki K, Ebert O, Woo SL (2005c). Treatment of multi-focal colorectal carcinoma metastatic to the liver of immune-competent and syngeneic rats by hepatic artery infusion of oncolytic vesicular stomatitis virus. *Int J Cancer* **114**, 659–664.

Steinhoff U, Muller U, Schertler A, Hengartner H, Aguet M, Zinkernagel RM (1995). Antiviral protection by vesicular stomatitis virus-specific antibodies in alpha/beta interferon receptor-deficient mice. *J Virol* **69**, 2153–2158.

Stitz L, Althage A, Hengartner H, Zinkernagel R (1985). Natural killer cells vs cytotoxic T cells in the peripheral blood of virus-infected mice. *J Immunol* **134**, 598–602.

Stojdl DF, Abraham N, Knowles S, *et al.* (2000a). The murine double-stranded RNA-dependent protein kinase PKR is required for resistance to vesicular stomatitis virus. *J Virol* **74**, 9580–958

Stojdl DF, Lichty B, Knowles S, *et al.* (2000b). Exploiting tumor-specific defects in the interferon pathway with a previously unknown oncolytic virus. *Nat Med* **6**, 821–825.

Stojdl DF, Lichty BD, Tenoever BR, *et al.* (2003). VSV strains with defects in their ability to shutdown innate immunity are potent systemic anti-cancer agents. *Cancer Cell* **4**, 263–75.

Sun X, Yau VK, Briggs BJ, Whittaker GR (2005). Role of clathrin-mediated endocytosis during vesicular stomatitis virus entry into host cells. *Virology* **338**, 53–60.

Tesh RB, Peralta PH, Johnson KM (1969). Ecologic studies of vesicular stomatitis virus. I. Prevalence of infection among animals and humans living in an area of endemic VSV activity. *Am J Epidemiol* **90**, 255–261.

Von Kobbe C, Van Deursen JM, Rodrigues JP, *et al.* (2000). Vesicular stomatitis virus matrix protein inhibits host cell gene expression by targeting the nucleoporin Nup98. *Mol Cell* **6**, 1243–1252.

Wollmann G, Tattersall P, Van Den Pol AN (2005). Targeting human glioblastoma cells: comparison of nine viruses with oncolytic potential. *J Virol* **79**, 6005–6022.

Yoneyama M, Kikuchi M, Natsukawa T, *et al.* (2004). The RNA helicase RIG-I has an essential function in double-stranded RNA-induced innate antiviral responses. *Nat Immunol* **5**, 730–737.

12

Measles as an Oncolytic Virus

Adele Fielding

12.1 Introduction

The vaccine strain of measles virus (MV) is now being intensively investigated as a viral agent for the therapy of cancer. Initial support for the concept was provided by several case reports of spontaneous regression of haematological malignancies after wild-type measles infection (Bluming and Ziegler, 1971; Pasquinucci, 1971; Zygiert, 1971; Mota, 1973). Within the past decade there has been rapid development and translation into early phase clinical studies. The use of replicating attenuated MV for cancer therapy is facilitated by the ready ability to genetically engineer MV and rescue the virus from cloned DNA (Radecke *et al.*, 1995; Schneider *et al.*, 1997). Administration of large doses of genetically modified MV to patients with cancer would be very unlikely to have significant untoward effects at a population level since there is no human or animal reservoir of MV in the environment and most individual are immune to MV. Wild type MV is known to be extremely genetically stable (Rima *et al.*, 1997) and this appears to be the case with attenuated genetically engineered MV. A considerable understanding of the biology of MV has informed and guided the goal-orientated research and development described in this chapter. An outline of some of the salient features of measles biology is provided here as a prerequisite to understanding its potential uses and limitations as a cancer therapeutic.

12.2 Measles virus and the consequences of natural infection

MV is a negative strand RNA virus of the family Paramyxoviridae. In susceptible populations, infection with wild type MV causes the childhood illness, measles (Katz, 1995; Griffin and Bellini, 1996). MV normally enters the human host via the respiratory route, assumed to be via epithelial cells. It is generally accepted that there is extensive replication in the draining lymph nodes, followed by a primary cell-associated viraemia. Subsequent replication occurs within multiple reticuloendothelial sites (Katz, 1995). MV replication in lymphoid tissue was recognized in early studies by its characteristic cytopathic effect where gross cell–cell fusion leads to the presence of multinucleated giant cells within infected lymph nodes (Warthin, 1931). A secondary viraemia then occurs, leading to infection of endothelial and epithelial cells, at which point the immune response leads to the classic signs of the exanthemous illness 'measles'. The infection is usually self-limited and spontaneous recovery is the rule. However, there is a considerable mortality after MV infection, particularly in the developing world. This largely relates to the immune suppression following natural MV infection, which generates an increased risk of development of secondary infections.

In the immune response against wild type MV-infection, anti-MV antibodies are detected

Viral Therapy of Cancer Edited by Kevin J. Harrington, Richard G. Vile and Hardev S. Pandha

coincidentally with the onset of the rash. There is an initial immunoglobulin M (IgM) response, followed by IgG1 and IgG4, with antibodies against the N, F and H proteins being detected in the largest quantity (Griffin, 1995). Anti-MV antibody titres have a close correlation with protective immunity. Indeed, neutralizing antibody solely against either the H or F proteins generated by DNA immunization can protect rhesus macaques against measles challenge (Polack *et al.*, 2000). However, cellular immunity is also important in and is responsible for eventual elimination of MV following initial control of viraemia (Permar *et al.*, 2004; Pan *et al.*, 2005).

MV is also able to subvert an appropriate immune response in a variety of ways. A range of immune defects have been demonstrated after infection with wild type MV and include decreased delayed type hypersensitivity response, decreased T-cell proliferation in response to mitogens, depletion of circulating activated T cells, altered of interleukin-12 (IL-12) and interferon responses and impaired natural killer cell function (Griffin, 1995; Gans *et al.*, 1999; Nanan *et al.*, 2000). Detailed mechanistic aspects of MV-related immunosuppression are beyond the scope of this chapter but the interactions between various MV proteins and the cellular receptors CD46 and SLAM are particularly important. The subject is well reviewed by Kerdiles *et al.* (2006).

Canine distemper virus (CDV) is closely related to MV. A recently developed ferret model of CDV infection may add to our understanding of the spread of wild type MV in humans; in this model, the spread of a recombinant CDV expressing GFP was carefully tracked following intranasal infection. A dramatic initial infection of circulating lymphocytes was demonstrated, followed by secondary lymphoid organs. Epithelial cells did not become infected until late in the illness (von Messling *et al.*, 2004). The widespread infection of lymphocytes was contributory to the immune suppression in this model. In humans natural MV infection results in suppression of T-cell function and profound lymphopenia; however, this not thought to result from direct massive infection of T cells but is attributed to contact inhibition (Schneider-Schaulies and Dittmer, 2006).

12.3 MV vaccine

A live attenuated strain of the virus has been used as a vaccine for over 40 years. All current MV vaccines are based on the Enders attenuated strain which was isolated from a patient named David Edmonston (Enders and Peebles, 1954). The vaccine strains were originally developed by blind-passage on human and non-human cell lines. There are a now a number of different MV vaccine strains available and careful sequence analysis has revealed the similarities and differences between them in both coding and non-coding regions (Parks *et al.*, 2001a, b). Nonetheless, the molecular basis for attenuation has not been fully elucidated. MV vaccine has been administered to millions of individuals worldwide with an excellent safety record. A small number of reports in the literature document atypical measles infections in the months following vaccination in very severely immunocompromised individuals (Mitus *et al.*, 1962, Mawhinney *et al.*, 1971, Monafo *et al.*, 1994). The vaccine strain of MV was strongly suspected but not confirmed at the causative agent of what became a fatal illness in these patients. However, the most recent report confirmed that the measles-like illness was due to uncontrolled proliferation of the vaccine strain virus, which had been administered to a young man with human immunodeficiency virus (HIV) infection and undetectable level T cells (Angel *et al.*, 1998). At the current time, the vaccine is not recommended for patients who are significantly immune suppressed. For patients with HIV infection, a CD4 T-cell count of >200 cells/μl is recommended for safe vaccination (Watson *et al.*, 1998). The vaccine strain of MV at typical vaccine doses is not a pathogen in normal healthy individuals.

12.4 MV genetics and engineering

The MV genome, approximately 16 kb is encapsidated by a nucleocapsid protein (N), encodes six structural protein products. A polymerase (L) and its co-factor, phosphoprotein, (P) associate with the RNA and N protein to form a ribonucleoprotein complex. This complex is surrounded by matrix (M) protein. Two oligomeric glycoproteins, haemagglutinin (H) and fusion (F) form the viral envelope. The two MV envelope glycoproteins H and F work in concert to elicit virus–cell membrane fusion. Attachment, mediated via H is followed by membrane fusion mediated via F. Cells which express MV F and H proteins at their surface become highly fusogenic (Norrby and Oxman, 1990). The P gene also encodes the non-structural proteins V and C, which are dispensable for viral growth in cultivated cells but are necessary for host invasion *in-vivo* (Nagai and Kato, 2004).

The first reverse genetic system for the rescue of an attenuated Edmonston-B strain of measles virus from cloned DNA was described (Radecke *et al.*, 1995) in 1995. Subsequently, a number of foreign proteins have been efficiently expressed from additional transcription units of MV including marker genes such as β-galactosidase (Cornu, 1997), green fluorescent protein (GFP) (Duprex *et al.*, 1999). A number of potentially therapeutic proteins have also been expressed, including human IL-12 (Singh *et al.*, 1999), hepatitis-B surface antibody (Singh and Billeter, 1999) and granulocyte–macrophage colony-stimulating factor (Grote *et al.*, 2003). MV particles are pleomorphic and there is no known size constraint on the length of genome which can be encapsidated.

12.5 MV receptors

Two receptors for MV have so far been identified to date (Dhiman *et al.*, 2004). Signalling lymphocyte activation molecule (SLAM, CD150) (Tatsuo *et al.*, 2000) belongs to a family of receptors with a crucial role in normal immune reactions in both innate and adaptive immunity (Veillette, 2006) and its expression is confined to immune cells. CD46 (Dorig *et al.*, 1993; Naniche *et al.*, 1993), (membrane co-factor protein) is a regulator of complement-mediated cell death whose expression is ubiquitous. For the tissue culture adapted vaccines strain of MV, both receptors are used, whilst for wild type, the relevance of CD46 use is not clear, since most wild-type strains cannot enter cells via CD46. CD46 is also a receptor for several other pathogens, both viruses and bacteria (Cattaneo, 2004).

12.6 Animal models for the study of MV pathogenesis and oncolysis

MV only replicates within primate cells and its only natural host is the human. Measles can be transmitted experimentally to primates; hence primates have been used in studies of MV vaccination and immunity. Considerable effort has been invested to create relevant small animal models for *in vivo* study. It is worth reviewing these briefly, since they have relevance to interpreting preclinical *in vivo* results of MV-oncolysis. There are now a number of transgenic rodent models in which some or all of the murine tissues express human CD46 or SLAM and recently, both (Shingai *et al.*, 2005; Welstead *et al.*, 2005). The vaccine strains of MV only replicate well within CD46 transgenic models if another immune defect, such as a defective interferon receptor (Mrkic *et al.*, 1998, 2000; Roscic-Mrkic *et al.*, 2001),or Rag-1 knockout (Oldstone *et al.*, 2005) is also present. After intranasal inoculation, replication is terminated within a few days by an immune response and the virus is eliminated. Wild type MV strains typically do not replicate within CD46 transgenic models. None of the rodent models recapitulate all of the hallmarks of human measles and may not be entirely relevant to study aspects of the human illness 'measles'. However, in context they have added insight into the biology of MV. Models in which some aspects of the pathogenesis of MV can be inferred from the behaviour of a closely related virus, such as CDV, also have some relevance.

MV oncolysis has predominantly been studied in immune-deficient mice in which local or disseminated tumours derived from human cell lines have been established. MV has been administered both locally and systemically. In these models, the virus cannot replicate in any tissue other than the tumour tissue and the mouse is unable to mount an immune response. Thus, no implications about pharmacokinetic, biodistribution or potential toxicity can be made. Nor can any positive or negative contribution from the immune system be evaluated. Thus, these studies can only confirm proof of principle of oncolytic activity of the virus within the tumour cell line under investigation.

Biodistribution and pharmacokinetic studies have typically been carried out in CD46 transgenic mice (Peng *et al.*, 2003b, 2006). The vaccine strain of MV is not pathogenic in these mice – as indeed it is not pathogenic in humans – no significant toxicities from in excess of 10^8 plaque-forming units have been seen in these murine studies. Worst-case scenario toxicity studies are typically carried out in non-human primates and these may be required prior to clinical studies. There is currently no single model allowing the study of MV oncolysis within an immune-competent, tumour-bearing animal all of whose cells are potentially permissive for MV infection.

12.7 Oncolytic activity of MV

MV has been shown to have anti-tumour activity in a number of models of malignancy in immune deficient mice. All of the models employ xenografted cell lines in established subcutaneous, orthotopic or disseminated tumours. The replicating attenuated MV has typically been compared to a control, UV-irradiated preparation. Due to the natural tropism of MV, B-cell malignancies were among the first investigated, In models of B-cell non-Hodgkin lymphoma (Grote *et al.*, 2001) and myeloma (Peng *et al.*, 2001) large established tumour xenografts regressed substantially or completely after administration of MV both intratumourally and intravenously. Other routes of administration relevant to anatomical sites of naturally occurring tumour have also been used. Intraperitoneal MV was administered to mice with advanced intraperitoneal tumours derived from a transformed ovarian epithelial cell line (SKOV-3.p.1) and enhanced the median survival of these mice. Intratumoural injection of MV into orthotopic human glioma xenografts likewise enhanced survival compared to mice injected with control (Phuong *et al.*, 2003). A recent publication details similar activity of MV against xenografted breast cancer cell lines (McDonald *et al.*, 2006). Various techniques have been used to confirm that virus replicates within the tumour, for example *in-situ* hybridization of MV specific RNA (Grote *et al.*, 2001; Peng *et al.*, 2001) or expression of a soluble marker peptide (Peng *et al.*, 2002a; Phuong *et al.*, 2002). In general, a total threshold dose between 10^6 and 10^8 plaque forming units is needed. If local injection is used, divided dosing is typically needed whereas, at least in the most sensitive tumour cells, a single intravenous dose can be sufficient for tumour regression (Peng *et al.*, 2001). Virus replicates within tumour cells and causes direct cell death, possibly via apoptosis (Phuong *et al.*, 2003; McDonald *et al.*, 2006). Although the virus does spread within the tumours and multinucleated syncytia can be found, clear evidence of virally-generated pathology can sometimes be elusive. It is likely that, even in SCID and nude mice, a host-response contributes to tumour regression (Grote *et al.*, 2003).

12.8 Mechanism of specificity

Whilst for some oncolytic RNA viruses, the mechanism of specificity is relatively clear (Stojdl *et al.*, 2000; Marcato *et al.*, 2005), this is not yet so for MV. There are some in-vitro data to suggest that MV-related lysis has greater specificity for tumour cells than their normal counterparts. Several studies have compared replication of attenuated MV within tumour cell lines or primary cells with that in non-transformed cells from the same tissues and concluded that replication and cell death is limited to the transformed cells (Peng *et al.*, 2001, 2002a; McDonald *et al.*, 2006). To date, the only published mechanistic explanation concerns the density of the MV receptor, CD46. It was determined that the absence of MV-induced cytopathic effect (CPE) in infected non-transformed cells was not due to lack of production of viral proteins. However, when non-permissive Chinese hamster ovary cells were engineered to express human CD46 at range of densities, cell–cell fusion, responsible for the MV CPE was minimal at low receptor densities but increased considerably above a threshold density (Anderson *et al.*, 2004). There are some indications that CD46 is expressed at higher levels on tumour cells compared to non-transformed counterparts; the expression of CD46 was much greater on multiple myeloma cells than on normal haematopoietic cells of various lineages (Ong *et al.*, 2006).

12.9 Targeting MV entry

Despite the suggestions that MV CPE is relatively specific for transformed cells, CD46 is ubiquitously distributed and all CD46-expressing cells are permissive for infection. Although non-transformed cells may not be directly subjected to a MV CPE, infection and signalling through CD46 is unlikely to be completely without consequences, in particular immune consequences. Therefore, targeting MV entry to specific cell types may, on a practical level, help avoid some of the potential safety concerns stemming from administering a replicating virus systemically. To what extent targeted entry, *per se*, will prove to be of real value in the clinical exploitation of measles as an oncolytic virus is not yet clear However, targeting virus entry has long been an intellectually engaging challenge for groups working on a variety of viral vectors and interestingly, measles has proved to be one of the few viruses/viral vectors where true re-targeted entry to specific cell types has actually been achieved. Furthermore, this achievement has now been robustly replicated through a number of different entry targets, and merits a detailed consideration of the concepts.

The first proof of principle demonstrations that measles virus was able to enter cells through non-native receptors was published in 2002 (Schneider *et al.*, 2000). The H protein, a type II transmembrane glycoprotein was modified by the addition of epidermal growth factor (EGF) or insulin-like growth factor (IGF-1) at the carboxy (extracellular) terminus. Recombinant virus expressing the modified H was rescued readily and replicated to titres approaching that of the parental strain. The modified virus was able to enter non-permissive rodent cells which had been engineered to express human EGF-R and IGF-R. Subsequently, using the same approach was used to demonstrate that MV entry could be targeted through a number of different receptors by addition of single chain antibodies against carcinoembryonic antigen (CEA) (Hammond *et al.*, 2001), CD20 (Bucheit *et al.*, 2003) and CD38 (Peng *et al.*, 2003a). Extending the concept a little further, tumour vasculature was targeted by the expression of a disintegrin M28L echistatin, which binds with high affinity to alpha(v)beta(3) integrin. In this study, xenografts from myeloma cell lines, which were resistant to MV oncolysis with unmodified MV, regressed after administration of the virus targeted to the vasculature (Hallak *et al.*, 2005).

Another interesting concept in targeting extended the repertoire of targets to specific peptide–major histocompatibility complex (MHC) ligands by display of a high-affinity single-chain T-cell receptor (scTCR) which can recognize and bind to specific peptide–MHC complex (Peng *et al.*, 2004). Taken together, these studies demonstrated that MV could enter non-permissive cells specifically via an interaction between a ligand displayed on the H glycoprotein of MV, a single chain antibody, a growth factor or even a scTCR. Viruses modified in this way appeared to be stable, as evidenced by no loss of the displayed domain during in-vitro passage. Although natural MV receptors are single transmembrane domain proteins, targeting could occur via multiple transmembrane domain receptors, nor is there a need for antigen internalization upon ligand binding for re-targeted entry to proceed. Data presented suggest that the interaction between the displayed domain and the targeted receptor are likely to be sufficient to activate necessary conformational change to the MV F protein thus activating the fusion capability of the MV envelope.

Whilst these studies confirmed the proof of possibility of re-targeting MV entry, all of the viruses generated were still capable also of binding to and entering permissive cells via the native receptors. Therefore, the next step was to find a strategy to ablate binding to the native receptors. A careful mutagenesis study by Vongpungsawad et al. (2004) allowed identification of residues within MV H which were important for both CD46 and SLAM-induced fusion. When these mutations were transferred into genomic MV cDNAs, viruses were rescued which could replicate selectively in either

CD46 or SLAM-expressing cells. Further work, using different combinations of point mutations to ablate binding to both SLAM and CD46 added EGF and scFVantiCD38 domains to these effectively receptor-blind H and demonstrated targeted cell-cell fusion of human cells by adenoviral vectors expressing the targeted H proteins (Nakamura *et al.*, 2004).

The final demonstration of truly re-targeted MVs resulted from finding a way in which retargeted replicating viruses could be rescued and propagated. A 'pseudoreceptor' system, in which MVs with ablated binding to their native receptors were targeted to CD38 or EGF-R expressing cells was established, to overcome the fact that ablation of the native measles virus receptor interactions was incompatible with virus rescue and growth. Vero cells, in which MV is usually propagated, were engineered to express a single-chain antibody that recognizes a six-histidine peptide. Viruses incorporating an H6 peptide at the C terminus of their ablated H proteins could be rescued and propagated on the Vero-alpha His cells expressing this pseudoreceptor. Fully re-targeted infection was demonstrated in vitro and in vivo after systemic administration.

12.10 Enhancing the oncolytic activity of MV

We have already seen than MV readily tolerates the insertion of foreign genes with little or no loss of replicative ability. Unsurprisingly, genes to enhance activity or to allow enhanced detection of activity are being actively investigated. The soluble peptides CEA and human chorionic gonadotropin (B-chain) were cloned into the MV genome upstream of N. Infected cells secreted the peptides allowing measurement of the concentrations of virally encoded marker peptides in culture supernatants or in serum. Kinetics of marker peptide expression correlated with therapeutic outcome in murine xenograft models (Peng *et al.*, 2002b).

A replication competent attenuated MV engineered to express the thyroidal sodium iodide symporter NIS (MV NIS) provided a theoretically elegant and practically attractive method of both tracking MV replication and increasing its therapeutic efficacy. NIS is a membrane ion channel with a normal physiological role is of concentrating iodine within the thyroid gland. This property has previously been ingeniously exploited for cancer therapeutic purposes. Experimental tumours transduced with a gene encoding NIS will concentrate systemically administered radioiodine, which is toxic and results in regression (Spitzweg and Morris, 2002). When MV is engineered to express NIS as an additional transcription unit, it retains the oncolytic potential of the parent virus as well as inducing NIS expression in infected cells. In *in vivo* tumour models, the distribution and expression of MV can be monitored non-invasively over time by gamma-camera imaging of ^{123}I uptake (Dingli *et al.*, 2004). Furthermore, the expression of NIS enhances the therapeutic effect of MV. A relatively small dose of virus followed by ^{131}I can lead to the eradication of tumours which are normally resistant to MV-mediated lysis. Further contributions to tracking MV activity using this system were made by examining ^{124}I positron emission tomography (PET)/computerized tomography (CT) imaging and comparing the findings to ^{123}I gamma camera imaging. Combined modality imaging using PET/CT allowed accurate and non-invasive imaging of the gene expression, suggesting that this could be a useful clinical tool (Dingli *et al.*, 2006b). In determining how best to plan the use of this complex system clinically, a mathematical model of 'radiovirotherapy' was developed, to add to what has been determined experimentally and suggest what additional data might be needed to determine the virus dose, and dosing and timing of radio-iodine (Dingli *et al.*, 2006a). This 'radiovirotherapy' system holds particular promise for radiosensitive tumours such as myeloma and is being developed for clinical trials in this disease.

12.11 Interactions with the immune system

Cancer is a systemic disease in most cases and requires systemic therapy. Most adults are immune to measles. Hence, systemic administration of MV would be expected to generate a robust antibody response. Although there is evidence that MV replication can occur in previously immune individuals without causing disease, it is highly likely that immune-mediated diminution or obliteration of systemically administered therapeutic virus will occur. A considerable focus of effort is needed to address this issue. The administration of virotherapy will ultimately be in conjunction with chemotherapy, radiotherapy or antibody therapy, all of which are able to temporarily suppress the immune response and this may be sufficient in some cases. Co-administration of MV with immune-suppressive agents is being explored (Myers *et al.*, 2004).

Administration of systemic viruses within cell-carriers is a more promising direction to follow, since deliberate immune suppression in accompaniment to replicating virus therapy would, raise several safety concerns and need to proceed very cautiously. In additional, certain cell types may localize to tumours and this could be used advantageously to assist in directing delivery of virus to appropriate sites. A recent publication details very promising relevant results with an oncolytic vaccinia virus in an immunocompetent mouse tumour model: administration of pre-infected cytokine-induced killer cells resulted in a prolonged eclipse phase in which the vaccinia was sheltered from the immune system until trafficking and to infiltration of the tumour resulted in tumour regression (Thorne *et al.*, 2006). On-going studies, published only in abstract form, are now exploring the use of carrier cells specifically to deliver systemic MV.

12.12 Potential specific toxicities of clinical use of replicating attenuated MV

Attenuated MV is not a human pathogen and has an excellent safety record when used as a vaccine. In severely immune-suppressed patients, the possibility of uncontrolled replication resulting in fatal atypical measles infection exists, as discussed earlier in the chapter. In patients with normal immune systems, administration of large dose of attenuated MV could theoretically result in immune suppression, although the clinico-pathological implications of the known immune suppressive activity of wild type MV for the use of vaccine strains of MV for oncolysis are unknown. A number of candidate mechanisms for MV-mediated immune suppression have been elucidated, many of them are mediated by specific ligation of the known MV receptors, CD46 and SLAM (Kerdiles *et al.*, 2006). A recent study employing engineered CDV in a ferret model suggested that SLAM recognition is necessary for morbillivirus virulence (von Messling *et al.*, 2006). Hence, ablated binding to SLAM, even outside the context of MV targeting could have a positive impact in this context. Furthermore, the immune suppressive profile of MV may still compare favourably with that of current cancer therapies. However, it is important that clinical studies of MV oncolysis should all include a thorough and careful scientific evaluation of the immune status of study subjects.

12.13 Clinical trials

An early phase I clinical trial of the intralesional injection of commercially available Edmonston-Zagreb (E-Z) vaccine strain in patients with cutaneous T-cell lymphoma has been carried out in Europe (Heinzerling *et al.*, 2005). Five patients were treated with escalating doses of E-Z MV. In each case, this was preceded by systemic alpha interferon administration. The injections, a maximum dose of 1000 $TCID_{50}$, were administered over eight cycles of two injections each and were well tolerated. There were some apparent reductions in size and thickness of the injected lesions in four of five patients but no evidence of response in distant lesions. Antibody staining of excised lesions showed reactivity for MV N protein and some change in the inflammatory cell infiltrate. All patients had an increase in anti-MV antibody titre. This very preliminary study does not allow any firm conclusions about the oncolytic activity of MV in this context. The very small dose was given for both pragmatic (it is the commercially available vaccine dose) and safety reasons.

A second phase I clinical trial in patients with locally advanced ovarian cancer is now on-going in the USA. The study employs the ‘trackable’ MV-CEA, and, extrapolating likely effective dosage from murine data, has a potentially more clinically relevant dose escalation design with a target dose of 10^9 $TCID_{50}$. The higher titres of virus are permitted by specific good manufacturing practice (GMP) manufacture of the study virus and delivery via the intraperitoneal route, which permits relatively large volumes of administration. The study reached the 10^6 $TCID_{50}$ dose level with only minor toxicity to date. No CEA in serum has yet been detected.

12.14 Conclusions

Attenuated MV is already a well-studied virus and has a long history of safe human use. A number of recent studies have demonstrated that it has properties of highly significant interest to the field of oncolytic virus therapy. Technological progress in preclinical evaluation, targeting virus entry, virus modification and GMP-manufacture has been rapid. Clinical studies with good correlative science are underway and are the only way to answer with certainty the many remaining questions.

References

Anderson BD, Nakamura T, Russell SJ, Peng KW (2004). High CD46 receptor density determines preferential killing of tumor cells by oncolytic measles virus. *Cancer Res* **64**, 4919–4926.

Angel JB, Walpita P, Lerch RA, *et al.* (1998). Vaccine-associated measles pneumonitis in an adult with AIDS. *Ann Intern Med* **129**, 104–106.

Bluming AZ, Ziegler JL (1971). Regression of Burkitt's lymphoma in association with measles infection. *Lancet* **2**, 105–106.

Bucheit AD, Kumar S, Grote DM, *et al.* (2003). An oncolytic measles virus engineered to enter cells through the CD20 antigen. *Mol Ther* **7**, 62–72.

Cattaneo R (2004). Four viruses, two bacteria, and one receptor: membrane cofactor protein (CD46) as pathogens' magnet. *J Virol* **78**, 4385–4388.

Cornu T. Use of measles virus as a vector: stable expression of a supplementary large transcription unit or additional foreign viral glycoproteins [MSc thesis].University of Zurich, Zurich, 1997.

Dhiman N, Jacobson RM, Poland GA (2004). Measles virus receptors: SLAM and CD46. *Rev Med Virol* **14**, 217–229.

Dingli D, Peng KW, Harvey ME, *et al.* (2004). Image-guided radiovirotherapy for multiple myeloma using a recombinant measles virus expressing the thyroidal sodium iodide symporter. *Blood* **103**, 1641–1646.

Dingli D, Cascino MD, Josic K, Russell SJ, Bajzer Z (2006a). Mathematical modeling of cancer radiovirotherapy. *Math Biosci* **199**, 55–78.

Dingli D, Kemp BJ, O'Connor MK, Morris JC, Russell SJ, Lowe VJ (2006b). Combined I-124 positron emission tomography/computed tomography imaging of NIS gene expression in animal models of stably transfected and intravenously transfected tumor. *Mol Imaging Biol* **8**, 16–23.

Dorig RE, Marcil A, Chopra A, Richardson CD (1993). The human CD46 molecule is a receptor for measles virus (Edmonston strain). *Cell* **75**, 295–305.

Duprex WP, McQuaid S, Hangartner L, Billeter MA, Rima BK (1999). Observation of measles virus cell-to-cell spread in astrocytoma cells by using a green fluorescent protein-expressing recombinant virus. *J Virol* **73**, 9568–9575.

Enders J, Peebles T (1954). Propagation in tissue culture of cytopathic agents from patients with measles. *Proc Soc Exp Biol Med* **86**, 277–286.

Gans HA, Maldonado Y, Yasukawa LL, *et al.* (1999). IL-12, IFN-gamma, and T cell proliferation to measles in immunized infants. *J Immunol* **162**, 5569–5575.

Griffin D, Bellini W (1996). Measles virus. In: PM, H. (ed.), *Fields Virology.* Lippincott-Raven, Philadelphia, pp. 1267–1313.

Griffin DE (1995). Immune responses during measles virus infection. In: Billeter MA (ed.), *Measles Virus.* Springer-Verlag, Berlin, pp. 117–134.

Grote D, Russell SJ, Cornu T, *et al.* (2001). Live attenuated measles virus induces regression of human lymphoma xenografts in immunodeficient mice. *Blood* **97**, 3746–3754.

Grote D, Cattaneo R, Fielding AK (2003). Neutrophils contribute to the measles virus-induced antitumor effect: enhancement by granulocyte macrophage colony-stimulating factor expression. *Cancer Res* **63**, 6463–6468.

Hallak LK, Merchan JR, Storgard CM, Loftus JC, Russell SJ (2005). Targeted measles virus vector displaying echistatin infects endothelial cells via alpha(v)beta3 and leads to tumor regression. *Cancer Res* **65**, 5292–5300.

Hammond AL, Plemper RK, Zhang J, Schneider U, Russell SJ, Cattaneo R (2001). Single-chain antibody displayed on a recombinant measles virus confers entry through the tumor-associated carcinoembryonic antigen. *J Virol* **75**, 2087–96.

Heinzerling L, Kunzi V, Oberholzer PA, Kundig T, Naim H, Dummer R (2005). Oncolytic measles virus in cutaneous T-cell lymphomas mounts antitumor immune responses in vivo and targets interferon-resistant tumor cells. *Blood* **106**, 2287–2294.

Katz M (1995). Clinical spectrum of measles. *Curr Top Microbiol Immunol* **191**, 1–12.

Kerdiles YM, Sellin CI, Druelle J, Horvat B (2006). Immunosuppression caused by measles virus: role of viral proteins. *Rev Med Virol* **16**, 49–63.

Marcato P, Shmulevitz M, Lee PW (2005). Connecting reovirus oncolysis and Ras signaling. *Cell Cycle* **4**, 556–559.

Mawhinney H, Allen IV, Beare JM, *et al.* (1971). Dysgammaglobulinaemia complicated by disseminated measles. *Br Med J* **2**, 380–381.

McDonald CJ, Erlichman C, Ingle JN, *et al.* (2006). A measles virus vaccine strain derivative as a novel oncolytic agent against breast cancer. *Breast Cancer Res Treat* **99**(2): 177–184.

Mitus A, Holloway A, Evans AE, Enders JF (1962). Attenuated measles vaccine in children with acute leukemia. *Am J Dis Child* **103**, 413–418.

Monafo WJ, Haslam DB, Roberts RL, Zaki SR, Bellini WJ, Coffin CM (1994). Disseminated measles infection after vaccination in a child with a congenital immunodeficiency. *J Pediatr* **124**, 273–276.

Mota HC (1973). Infantile Hodgkin's disease: remission after measles. *Br Med J* **2**, 421.

Mrkic B, Pavlovic J, Rulicke T, *et al.* (1998). Measles virus spread and pathogenesis in genetically modified mice. *J Virol* **72**, 7420–7427.

Mrkic B, Odermatt B, Klein MA, Billeter MA, Pavlovic J, Cattaneo R (2000). Lymphatic dissemination and comparative pathology of recombinant measles viruses in genetically modified mice. *J Virol* **74**, 1364–1372.

Myers R, Harvey M, Soeffker D, Greiner S, Peng KW, Russell SJ (2004). Modulating the immune response to oncolytic measles virus. *Mol Ther* **9**, S108.

Nagai Y, Kato A (2004). Accessory genes of the paramyxoviridae, a large family of nonsegmented negative-strand RNA viruses, as a focus of active investigation by reverse genetics. *Curr Top Microbiol Immunol* **283**, 197–248.

Nakamura T, Peng KW, Vongpunsawad S, *et al.* (2004). Antibody-targeted cell fusion. *Nat Biotechnol* **22**, 331–336.

Nanan R, Rauch A, Kampgen E, Niewiesk S, Kreth HW (2000). A novel sensitive approach for frequency analysis of measles virus-specific memory T-lymphocytes in healthy adults with a childhood history of natural measles. *J Gen Virol* **81**(Pt 5): 1313–1319.

Naniche D, Varior-Krishnan G, Cervoni F, *et al.* (1993). Human membrane cofactor protein (CD46) acts as a cellular receptor for measles virus. *J Virol* **67**, 6025–6032.

Norrby E, Oxman MN (1990). Measles virus. In: Fields BN (ed.), *Virology*. Raven Press, Ltd, New York, NY, pp. 1013–1044.

Oldstone MB, Dales S, Tishon A, Lewicki H, Martin L (2005). A role for dual viral hits in causation of subacute sclerosing panencephalitis. *J Exp Med* **202**, 1185–1190.

Ong HT, Timm MM, Greipp PR, *et al.* (2006). Oncolytic measles virus targets high CD46 expression on multiple myeloma cells. *Exp Hematol* **34**, 713–720.

Pan CH, Valsamakis A, Colella T, *et al.* (2005). Inaugural Article: Modulation of disease, T cell responses, and measles virus clearance in monkeys vaccinated with H-encoding alphavirus replicon particles. *Proc Natl Acad Sci U S A* **102**, 11581–11588.

Parks CL, Lerch RA, Walpita P, Wang HP, Sidhu MS, Udem SA (2001a). Analysis of the noncoding regions of measles virus strains in the Edmonston vaccine lineage. *J Virol* **75**, 921–933.

Parks CL, Lerch RA, Walpita P, Wang HP, Sidhu MS, Udem SA (2001b). Comparison of predicted amino acid sequences of measles virus strains in the Edmonston vaccine lineage. *J Virol* **75**, 910–920.

Pasquinucci G (1971). Possible effect of measles on leukamia. *Lancet* **1**, 136.

Peng KW, Ahmann GJ, Pham L, Greipp PR, Cattaneo R, Russell SJ (2001). Systemic therapy of myeloma xenografts by an attenuated measles virus. *Blood* **98**, 2002–2007.

Peng K-W, TenEyck C, Galanis E, Kalli K, Hartmann L, Russell S (2002a). Intraperitoneal therapy of ovarian cancer using an engineered measles virus. *Mol Ther* **5**, S444.

Peng KW, Facteau S, Wegman T, O'Kane D, Russell SJ (2002b). Non-invasive *in vivo* monitoring of trackable viruses expressing soluble marker peptides. *Nat Med* **8**, 527–531.

Peng KW, Donovan KA, Schneider U, Cattaneo R, Lust JA, Russell SJ (2003a). Oncolytic measles viruses displaying a single-chain antibody against CD38, a myeloma cell marker.*Blood* **101**, 2557–2562.

Peng KW, Frenzke M, Myers R *et al.* (2003b). Biodistribution of oncolytic measles virus after intraperitoneal administration into Ifnar-CD46Ge transgenic mice. *Hum Gene Ther* **14**, 1565–1577.

Peng KW, Holler PD, Orr BA, Kranz DM, Russell SJ (2004). Targeting virus entry and membrane fusion through specific peptide/MHC complexes using a high-affinity T-cell receptor. *Gene Ther* **11**, 1234–1239.

Peng KW, Hadac EM, Anderson BD, *et al.* (2006). Pharmacokinetics of oncolytic measles virotherapy: eventual equilibrium between virus and tumor in an ovarian cancer xenograft model. *Cancer Gene Ther* **13**, 732–738.

Permar SR, Klumpp SA, Mansfield KG, *et al.* (2004). Limited contribution of humoral immunity to the clearance of measles viremia in rhesus monkeys. *J Infect Dis* **190**, 998–1005.

Phuong L, Peng K-W, Allen C, *et al.* (2002). Use of a vaccine strain of measles virus genetically engineered to produce CEA as a novel therapeutic agent against glioblastoma multiforme. *Mol Ther* **5**, S302.

Phuong LK, Allen C, Peng KW, *et al.* (2003). Use of a vaccine strain of measles virus genetically engineered to produce carcinoembryonic antigen as a novel therapeutic agent against glioblastoma multiforme. *Cancer Res* **63**(10): 2462–2469.

Polack FP, Lee SH, Permar S, *et al.* (2000). Successful DNA immunization against measles: Neutralizing antibody against either the hemagglutinin or fusion glycoprotein protects rhesus macaques without evidence of atypical measles. *Nat Med* **6**, 776–781.

Radecke F, Spielhofer P, Schneider H, *et al.* (1995). Rescue of measles viruses from cloned DNA. *EMBO J* **14**, 5773–5784.

Rima BK, Earle JA, Baczko K, *et al.* (1997). Sequence divergence of measles virus haemagglutinin during natural evolution and adaptation to cell culture *J Gen Virol* **78**, 97–106.

Roscic-Mrkic B, Schwendener R, Odermatt B, *et al.* (2001). The roles of macrophages in measles virus infection of genetically modified mice. *J Virol* **75**, 3343–3351.

Schneider H, Spielhofer P, Kaelin K, *et al.* (1997). Rescue of measles virus using a replication-deficient vaccinia-T7 vector. *J Virol Methods* **64**, 57–64.

Schneider U, Bullough F, Vongpunsawad S, Russell SJ, Cattaneo R (2000). Recombinant measles viruses efficiently entering cells through targeted receptors. *J Virol* **74**, 9928–9936.

Schneider-Schaulies S, Dittmer U (2006). Silencing T cells or T-cell silencing: concepts in virus-induced immunosuppression. *J Gen Virol* **87**, 1423–1438.

Shingai M, Inoue N, Okuno T, *et al.* (2005). Wild-type measles virus infection in human CD46/CD150-transgenic mice: CD11c-positive dendritic cells establish systemic viral infection. *J Immunol* **175**, 3252–3261.

Singh M, Billeter M (1999). A recombinant measles virus expressing biologically active human interleukin-12. *J Gen Virol* **80**, 101–106.

Singh M, Cattaneo R, Billeter MA (1999). A recombinant measles virus expressing hepatitis B virus surface antigen induces humoral immune responses in genetically modified mice. *J Virol* **73**, 4823–4828.

Spitzweg C, Morris JC (2002). The sodium iodide symporter: its pathophysiological and therapeutic implications. *Clin Endocrinol (Oxf)* **57**, 559–574.

Stojdl D, Lichty B, Knowles S, *et al.* (2000). Exploiting tumor-specific defect in the interferon pathway with a previously unknown oncolytic virus. *Nat Med* **6**, 821–825.

Tatsuo H, Ono N, Tanaka K, Yanagi Y (2000). SLAM (CDw150) is a cellular receptor for measles virus. *Nature* **406**, 893–897.

Thorne SH, Negrin RS, Contag CH (2006). Synergistic antitumor effects of immune cell-viral biotherapy. *Science* **311**, 1780–1784.

Veillette A (2006). Immune regulation by SLAM family receptors and SAP-related adaptors. *Nat Rev Immunol* **6**, 56–66.

von Messling V, Milosevic D, Cattaneo R (2004). Tropism illuminated: lymphocyte-based pathways blazed by lethal morbillivirus through the host immune system. *Proc Natl Acad Sci U S A* **101**, 14216–14221.

von Messling V, Svitek N, Cattaneo R (2006). Receptor (SLAM [CD150]) recognition and the V protein sustain swift lymphocyte-based invasion of mucosal tissue and lymphatic organs by a morbillivirus. *J Virol* **80**, 6084–6092.

Vongpunsawad S, Oezgun N, Braun W, Cattaneo R (2004). Selectively receptor-blind measles viruses: Identification of residues necessary for SLAM- or CD46-induced fusion and their localization on a new hemagglutinin structural model. *J Virol* **78**, 302–313.

Warthin A (1931). Occurrence of numerous large giant cells in tonsils and pharyngeal mucosa in the prodromal stage of measles *Arch Pathol* **11**, 864–874.

Watson JC, Hadler SC, Dykewicz CA, Reef S, Phillips L (1998). Measles, mumps, and rubella–vaccine use and strategies for elimination of measles, rubella, and congenital rubella syndrome and control of mumps: recommendations of the Advisory Committee on Immunization Practices (ACIP). *MMWR Recomm Rep* **47**, 1–57.

Welstead GG, Iorio C, Draker R, *et al.* (2005). Measles virus replication in lymphatic cells and organs of CD150 (SLAM) transgenic mice. *Proc Natl Acad Sci U S A* **102**, 16415–16420.

Zygiert Z (1971). Hodgkin's disease: remission after measles. *Lancet* **1**, 593.

13 Alphaviruses

Ryuya Yamanaka

13.1 Introduction

Several viral and non-viral vector systems have been developed for efficient gene delivery and high heterologous gene expression. Viral vectors commonly used for therapeutic DNA delivery include replication-deficient forms of retroviruses, adenoviruses and adeno-associated viruses. Although retroviruses have the advantage of mediating stable gene transfer by integrating into the host genome and have a low potential for immunogenicity, this vector delivery system has some problems with respect to therapeutic use including difficulties in producing high titres of retrovirus, the fact that only actively dividing cells are capable of being infected and the possibility of insertional mutagenesis (Mulligan, 1993). One SCID-X1-treated patients developed a leukaemia-like condition, possibly as a result of random integration into the host genome (Check, 2002). The adenovirus vector system, while capable of delivering genes with high efficiency to a wide spectrum of non-dividing cells *in vivo* (Englehardt *et al.*, 1993a), unfortunately, has produced only transient expression with different gene products. Transient expression may result from a strong immune response from host cells against the adenovirus (Engelhardt *et al.*, 1993b). Adeno-associated virus (AAV) can produce long-term and efficient transgene expression in various cell types (Rabinowitz and Samulski, 1998). However, there are several disadvantages of AAV, such as restricted packaging capacity, difficulty in scaling up production, pre-existing immunity to human AAV vectors and integration into the host genome is random (Rabinowitz and Samulski, 1998). Alphaviruses have received considerable attention for use as virus-based expression vectors. Alphavirus expression systems differ from currently available viral delivery systems in that they are RNA viruses, known to generate high levels of protein expression *in vitro*. Semliki Forest virus (SFV), a member of the alphaviruses is being developed as a vector for expression of heterologous genes and has many advantages as an expression vector system. In SFV systems, because helper RNA does not contain a packaging signal, it will not form a defective interfering particle or be packaged with recombinant RNA. Furthermore, replication occurs entirely in the cytoplasm of the infected cells as an RNA molecule, without a DNA intermediate (Strauss and Strauss, 1994). This is in contrast to retroviruses, which must enter the nucleus and integrate into the host genome for initiation of vector activity. Thus, retrovirus vectors have applications in long-term expression of foreign genes, while SFV is useful primarily for transient high level expression. Furthermore, although adenovirus vectors can express high levels of foreign genes, these systems are more complex than SFV and express many highly antigenic virus-specific gene products

Viral Therapy of Cancer Edited by Kevin J. Harrington, Richard G. Vile and Hardev S. Pandha

including structural proteins (Rosenfeld *et al.*, 1991). In contrast, current SFV systems only express the four viral replicase proteins (nsP1–4) required for RNA amplification in the transduced cells. This system has been found to express significant quantities of heterologous proteins *in vitro* (Levis *et al.*, 1990) and *in vivo* (Piper *et al.*, 1994). It is possible to introduce at least 7 kb, which means that several genes, either under separate subgenomic promoters (Zhang *et al.*, 1997) or Internal Ribosomal Entry Site (IRES) sequences, can be inserted. The generation of recombinant SFV particles is extremely rapid. High-titre virus stocks (10^9–10^{10} infectious particles/ml) can be produced within a few days and no further purification or concentration is required. This is an advantage compared to the more time consuming and labour-intensive methods and the relatively low yields obtained for retrovirus and AAV vectors.

The rapid high-titre production and high level gene expression capacity have made alphavirus vectors attractive for cancer gene therapy strategies. Replication-deficient forms of alphavirus vectors such as SFV (Liljestrom and Garoff, 1991), Sindbis virus (SIN) (Xiong *et al.*, 1989) and Venezuelan equine encephalitis virus (VEE) (Davis *et al.*, 1989) have been engineered. This review discusses the background and application of these alphavirus vector systems for cancer gene therapy.

13.2 RNA viruses as gene expression vectors

RNA viruses infect a wide range of organisms, such as prokaryotics and eukaryotics. Many of them produce high levels of viral proteins, which makes them good candidates for virus vectors as heterologous gene expression systems (Huang *et al.*, 1989). The RNA genome is first converted into genomic complementary DNA (cDNA) which is placed downstream of a promoter for a DNA dependent RNA polymerase. Heterologous genes that encode the proteins of interest should be introduced into the cDNA downstream from a promoter. The whole cDNA is transcribed *in vitro* into the genomic viral RNA that can be transfected into cultured cells or animals. Many of the RNA viruses that infect eukaryotic organisms replicate exclusively in the cytoplasm and expression of their genes would be independent of host nuclear programmes.

RNA viruses have a different mode of replication. The main feature dividing RNA viruses into two categories is the polarity of the RNA genome. RNA genomes with positive polarity are those in which the genomic RNA functions as mRNA. The genome of viruses such as togaviruses, picornaviruses and flaviviruses is a single strand of RNA of positive polarity. RNA genomes in which the sequence is complementary to the mRNA must be transcribed into mRNA before translation will yield protein production. Other characteristics which vary among the different RNA-virus families are whether the genome exists as a single molecule of RNA or as several molecules of different structure.

13.3 The biology of alphaviruses

Alphaviruses are the major genus of the Togavirus family (Schlesinger and Schlesinger, 1996) that infect many types of hosts, ranging from mosquitos to avian and mammalian species (Strauss and Strauss, 1994). They include SFV and SIN, being developed as vectors for the expression of heterologous genes. The genome is a single-stranded RNA of positive polarity, that is capped at the 5′ terminus and polyadenylated at the 3′ terminus (Strauss and Strauss, 1986; Schlesinger and Schlesinger, 1996). Alphavirus virions are small (60 nm in diameter), spherical and possess a lipid envelope through which an arrangement of 80 glycoprotein spikes project. The alphavirus particle contains a single genomic RNA complex with 240 molecules of a basic capsid protein surrounded by a lipid bilayer containing 240 E1E2 envelope glycoprotein heterodimers.

13.3.1 Genomic organization

The single-stranded, around 12 kb long genomes of SFV and SIN are divided into two open reading frames (ORFs) (Strauss and Strauss, 1994). The first ORF encodes four non-structural proteins, designated nsP1 to nsP4, responsible for transcription and replication of viral RNA. The nonstructural proteins are translated from the genomic RNA and function to transcribe negative sense as well as positive sense viral RNA. The second ORF located in the 3′ one-third of the genome, under the control of a 26S subgenomic promoter, codes for the structural proteins required for the encapsidation of the viral genome and their proper assembly into enveloped particles. They include the capsid protein, the glycoproteins E1, E2 and E3 and the 6K protein. These glycoproteins form stable, non-covalently bound heterodimers that trimerize to form a functional subunit and spike on the virus surface. The E1 glycoprotein is highly conserved among alphaviruses and is involved in cell attachment, membrane fusion and entry. The E2 glycoprotein contains the most potent epitopes, eliciting neutralizing antibodies. The structural proteins are not necessary for viral replication, but are required for virus propagation, together with the packaging signal located in the coding region of nsP2 in SFV and of nsP1 in SIN. The genomic and subgenomic RNAs are both capped and polyadenylated and there are UTRs located at the 5′ and 3′ termini.

The non-structural proteins are required for replication of the genomic RNA and transcription of subgenomic RNA from the full-length negative sense RNA. nsP1 is the viral capping enzyme and is believed to play a major role in binding and assembly of the alphavirus replication complex at intracellular membranes (Ahola *et al.*, 1999, 2000). nsP2 protein is an RNA binding protein that has NTPase activity and likely functions as an RNA helicase to unwind duplex RNA (Rikkonen *et al.*, 1994). nsP2 also functions as a protease that is required for post-translational processing of the nonstructural polyproteins (Ding and Schlesinger, 1989). nsP3 is a phosphoprotein with two major regions; an amino terminal region that is highly conserved among the alphaviruses and a carboxy-terminal region that is neither conserved nor required for viral replication (Li *et al.*, 1990; Lastarza *et al.*, 1994). nsP4 has been identified as the viral polymerase on the basis of sequence homology with other RNA-dependent RNA polymerases (Hahn *et al.*, 1989; Sawicki *et al.*, 1990).

Certain features of the genome are essential for replication of alphavirus RNA. The 5′ UTR contains a stem-loop structure that is conserved among alphaviruses (Strauss and Strauss, 1994). A conserved sequence element (CSE) capable of forming a stem-loop structure has been identified in the 5′ coding region of the nsP1 protein (Niesters and Strauss, 1990). The 3′ UTR of alphaviruses, contains a 19 nt CSE immediately upstream of the poly A tail. This sequence is believed to be the promoter for minus strand genomic RNA synthesis (Strauss and Strauss, 1994).

Another sequence element conserved among alphaviruses is located within the junction region between the nonstructural and structural protein ORF (Strauss and Strauss, 1994). This 21 nt sequence is included in the minimal promoter required for synthesis of subgenomic RNA. The structural proteins, along with a packaging signal are required for the encapsidation of viral genomic RNA into nucleocapsids, are located in the coding region of either nsP1 or nsP2 proteins (Frovola *et al.*, 1997).

13.3.2 Alphavirus life cycle

The replication cycle of the alphavirus is shown in Figure 13.1. The virus enters the cell by receptor-mediated endocytosis, such as major histocompatibility complex I (MHC I) molecules and high-affinity laminin receptors mediated by the E2 portion of the glycoprotein spike (Helenius *et al.*, 1978; Wang *et al.*, 1992). Fusion of the virus membrane with the endosomal membrane releases the viral nucleocapsid into the cytoplasm, where translation of the genomic RNA occurs. The genomic RNA initially serves as an mRNA for translation of the viral non-structural proteins required for initiation of viral RNA amplification. Only the non-structural viral proteins are translated from the genomic RNA. Four polypeptides are generated after post-translational cleavage by the nsP2 protease (Hardy and Strauss, 1989). These function together in a replication complex which is required for the synthesis of the negative strand RNA. RNA replication occurs via synthesis of a full-length minus-strand intermediate that is

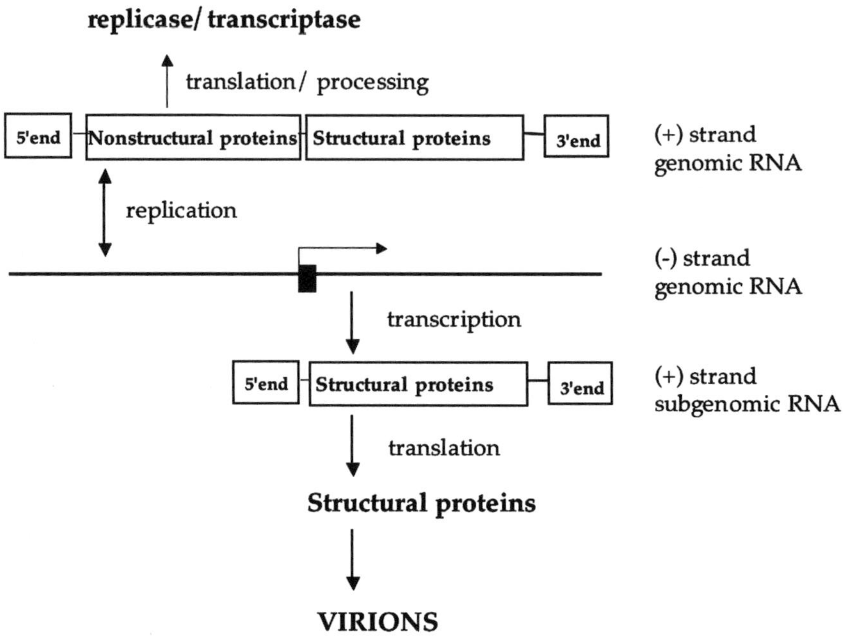

Figure 13.1 Alphavirus replication cycle

used as a template for synthesis of additional genome-length RNAs and for transcription of a plus-strand subgenomic RNA from an internal promoter. The synthesis of minus, plus and subgenomic RNAs is regulated via proteolytic processing of non-structural polyprotein replicase components. The viral structural proteins translated from 26S RNA are synthesized as a polyprotein with the N-terminal capsid protein functioning as an auto-protease. The capsid proteins subunits interact with the genomic RNA to form a nucleocapsid that matures by budding through the plasma membrane, acquiring a lipid bilayer envelope with embedded viral glycoproteins. Virus budding through the plasma membrane occurs and the infectious virus is released from the host-cell membrane.

13.4 Heterologous gene expression using alphavirus vectors

Alphaviruses, which have a single stranded RNA genome of positive polarity, have some additional features that make them attractive tools for gene expression and gene therapy applications. Levis *et al.* developed a defective interfering RNA capable of expressing a reporter gene in cells infected with helper SIN virus (Levis *et al.*, 1987). Xiong *et al.* (1989) developed a self-replicating RNA expression vector by modifying a SIN virus infectious clone by replacing the structural protein gene region with the reporter gene. The replicon vector lacked the structural protein-coding region, so a helper SIN virus was used to supply the structural proteins to package the replicon RNA into particles.

The basic strategy for alphavirus expression of heterologous genes has been to construct cDNAs of the alphavirus genome in which the heterologous gene is placed downstream from a promoter used to transcribe a subgenomic RNA. The alphavirus genome contains four nonstructural genes, nsP1–4, which are responsible for highly efficient RNA replication in infected cells (Lundstrom, 2003). The nsP1–4 generates the replicase complex, which construct a minus strand template from which an estimated 200 000 copies of RNA are made (Strauss and Strauss, 1994). Capped RNA transcripts, produced by *in vitro* transcription

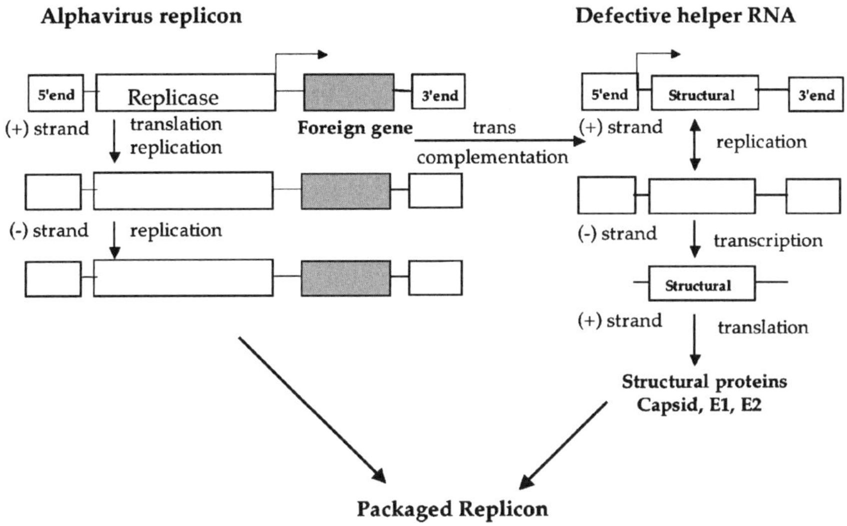

Figure 13.2 Packaging of replicons by co-transfection of helper RNA expressing the alphavirus structural proteins

with SP6 or T7 polymerase, are typically used to transfect tissue culture cells, usually baby hamster kidney cell line (BHK) or chicken embryo fibroblasts. The RNA transcript is transfected into cells either by lipofection (Felgner and Ringold, 1989) or electroporation (Liljestrom and Garoff, 1991). These RNA molecules function as mRNA. The subgenomic RNA synthesized in the transfected cells is translated into the heterologous protein. Alphavirus infection causes apoptosis, which results in a rapid onset of host cell death, leading to take-over of host cell protein synthesis. Several alphaviruses, such as SFV, SIN and VEE have been engineered as foreign gene expression vectors. Replication-deficient vector systems contain the replicon and the foreign gene in an expression vector and the structural genes are separately provided in a helper vector (Figure 13.2).

13.5 Cancer gene therapy strategies using alphavirus vectors

13.5.1 Direct tumour killing or antiangiogenic therapy

SFVs used as transient RNA expression vectors, which have been demonstrated to induce apoptosis in transfected cells by inhibition of host macromolecular synthesis (Lundstrom *et al.*, 1997), have also an inherent p53-independent apoptosis-inducing property. This suggests the attractive possibility that they can be applied to cancer gene therapy. Several therapeutic genes were introduced into tumour cells using SFV vectors with efficient antitumor effect such as the herpes simplex virus thymidine kinase (HSV-TK) (Loimas *et al.*, 2001), interleukin-12 (IL-12) (Asselin-Paturel, *et al.*, 1999; Yamanaka *et al.*, 2000), granulocyte–macrophage colony-stimulating factor (GM-CSF) (Klimp *et al.*, 2001), and Bax (Murphy *et al.*, 2001). Tseng *et al.* (2004) showed that the systemic delivery of SIN-luciferase vectors specifically targets primary and metastatic tumour cells, inducing tumour suppression and eradication. They proposed that specific targeting was achieved due to the inherent property of tumour cells to express excess, unoccupied high-affinity laminin receptors on their surface.

Antiangiogenic therapy using SFV carrying the endostatin gene (Yamanaka *et al.*, 2001a), IL-12 (Asselin-Paturel, 1999) was investigated to improve therapeutic efficacy. A marked reduction of intratumoral vascularization was seen in the tumour sections from the SFV treated animals. Gene therapy with antiangiogenic genes delivered via SFV may be a candidate for the development of new cancer therapy. Targeting of SFV vectors should be further investigated because of the broad host range.

13.5.2 Cancer vaccine

Several features of alphaviruses make them useful for vaccine development: (1) alphaviruses infect a broad range of animals, including humans, often with no symptoms; (2) the seroprevalence of alphaviruses is low, thus reducing the probability of interference with immune responses to the vectors; (3) alphaviruses have lymph node tropism that results in effective antigen presentation and induction of a strong immune response.

Alphavirus vectors have been utilized to generate tumour-specific cytotoxic T cells by immunization of the host. The therapy resulted in long-lasting immunity. Immune responses and protection against tumour challenges have been achieved. rSFV treatment resulted in long-term immunity as observed by the lack of tumour recurrence in the majority of tumour-regressing mice after rechallenge with the tumour (Colmenero *et al.*, 2002). Several therapeutic genes were introduced to immunize animals such as P1A (Colmenero *et al.*, 1999, 2002, human papillomavirus HPV oncoproteins E6, E7 (Daemen *et al.*, 2002, 2003, 2004), IL-12 (Colmenero *et al.*, 2002, Yamanaka *et al.*, 2002a; Rodriguez-Madoz *et al.*, 2005; Chikkanna-Gowda *et al.*, 2005) and IL-18 (Yamanaka *et al.*, 2003). Vaccination with total tumour-derived materials has also been reported using dendritic cells pulsed with SFV mediated tumour cell cDNA for experimental brain tumours and showed prolonged survival of tumour-bearing animals, and induction of a CTL response (Yamanaka *et al.*, 2001b, 2002b). SIN expressing the human papillomavirus type 16 (HPV-16) E7 antigen linked to the transmembrane and cytoplasmic regions of the lysosome-associated membrane protein 1 (LAMP-1) had a significant increase of E7 specific $CD4^+$, $CD8^+$ T-cell and antismog responses (Cheng *et al.*, 2002). E7/LAMP-1 RNA replicon-transfected apoptotic cells can be taken up by dendritic cells and presented efficiently through the MHC class I pathway (Cheng *et al.*, 2002). Double-stranded RNA (dsRNA) are produced during replicon amplification and may further enhance the immune response through mechanisms that involve cellular responses to dsRNA such as increased expression of class I self antigen presentation and activation of dendritic cells (Cella *et al.*, 1999). It was demonstrated that a single intramuscular injection of SFV-LacZ RNA prolonged the survival time of mice with established tumours and even protected mice from tumour challenge (Ying *et al.*, 1999). An advantage to using alphavirus-based RNA vaccines, as compared to inactivated virus vaccines, is that cytoplasmic replication and antigen expression allows for efficient processing and presentation by MHC class I molecules to induce cellular immune responses. Another mechanism that may enhance the potency of alphavirus-mediated immunization is the induction of apoptosis following replicon expression, and the resultant cross-priming of antigen presenting cells (Lundstrom, 2003). A phase I clinical trial on advanced melanoma and renal cell carcinoma using encapsulated SFV particles expressing IL-12 showed no liposome or SFV-related toxicity (Strauss and Strauss, 1994). Also, liposome-encapsulated SFV-IL-12 is being tested in a phase I/II clinical trial on glioma (Ren *et al.*, 2003).

VEE virus appears to be unique in its natural cell tropism compared with either SIN or SFV. VEE virus is naturally lymphotropic (Walker *et al.*, 1976; McDonald and Johnston, 2000), while SIN and SFV are not. The lymphotropic nature of VEE has benefited VEE-based vaccine vectors as they have been shown to elicit strong cellular and humoral immune responses in vaccinated animals at very low immunization doses (Hevey *et al.*, 1998); Pushko *et al.*, 2000). Immunization of a murine E7 tumour model with VEE replicons expressing the HPV E7 protein resulted in a strong MHC class I-restricted CD8 T-cell response and prevention of tumour development following challenge of mice with tumour cells (Velders *et al.*, 2001).

13.5.3 Replication-competent alphaviruses

Recently, replication-competent viruses have been studied as anticancer agents. The replication-competent SIN AR339 strain was evaluated for cervical and ovarian cancer therapy (Unno *et al.*, 2005). SIN AR339 infection was able to induce cytopathic effects and apoptosis in several cervical and ovarian cancer cells but not in normal keratinocytes. In vivo studies revealed that systemic treatment with the single injection of SIN induces necrosis within tumours at a remote site. The SIN AR339 strain demonstrated the possibility of cancer-specific cytotoxicity and targeting (Unno *et al.*, 2005).

13.6 Alphavirus vector development for gene therapy application (Table 13.1)

Alphavirus vectors have turned out to be efficient for many gene therapy applications. However, there have been some limitations, particularly related to broad host range, host cell toxicity and the transient nature of expression. To address these questions, novel vectors have been constructed for both SIN and SFV. Several efforts have been conducted to develop vectors with cell specific targeting. Incorporation of heterologous ligands or receptors into the virion envelope may allow targeting of engineered alphavirus RNAs to specific cell types. London *et al.* (1992) first introduced the possibility that alphaviruses could be modified to alter cell tropism. Modification of a full-length clone of SIN to include a neutralization epitope from RVFV in the SIN E2 gene resulted in SIN particles with the RVFV epitope on the surface of recombinant viruses (London *et al.*, 1992). The SIN envelope has been engineered to reduce the infection of normal cells by introduction of IgG binding domains from protein A into the E2 region (Ohno *et al.*, 1997). The presence of the protein A domains in the SIN envelope allowed efficient infection of host cells treated with a monoclonal antibody against a surface marker protein. Targeted infection of host cells could also be achieved by the introduction of α-and β-human chorionic gondotropin gene sequences into the SIN envelope, where no infection of BHK cells or human cancer cells lacking LH/CG receptors occurred, while choriocarcinoma cells showed high infection rates (Sawai and Meruelo, 1998). SIN vectors have also been used to *in vitro* transcribe biotinylated and self-replicating SIN genomic RNA with streptavidin-protein A fusion protein and mAbs and could in the presence of cationic liposomes result in specific transfection of cancer cells in a monoclonal antibody dose-dependent manner (Sawai *et al.*, 1998). SIN vectors with a single point mutation at residue 160 in the envelope glycoprotein E2 resulted in increased infection of human dendritic cells (Gardner *et al.*, 2000).

Table 1. Alphavirus vector development

Alphavirus vectors	References
Replication-deficient vectors	
SFV	11
SIN	12
VEE	13
Targeted vectors	
SIN	63,64,65,66,67
Non-cytopathogenic vectors	
SFV	72
SIN	70
Temperature-sensitive vectors	
SFV	73
Hybrid vectors	
SFV	75
Helper vectors	
SFV	80
DNA vectors	
SFV	83,84,85,88
SIN	81,82,87,88
Replication-competent vectors	
SIN	62

Using the SFV1 vector system, continual reporter gene expression was evident in neuronal cell bodies for up to 3 weeks post-inoculation, but with time cell degeneration and axonopathy, neuronal loss was increasingly apparent (Graham *et al.*, 2005). In an adult rat model of intranasal infection with SFV4 strains, apoptosis was documented in the rostral migratory stream, where neuronal precursor cells migrate from the proliferating subependymal layer into the olfactory bulb. In addition, the SFV4 strain provokes extensive areas of necrosis in the superficial layers of the olfactory bulb and in cortical, thalamic and hippocampal neurons (Sammin *et al.*, 1999). The SFV1 and SFV4 vector systems are limited in their potential for CNS gene therapy by neurotoxicity. Several mutant alphavirus vectors have reduced vector cytotoxicity. SIN and SFV vectors with point mutations in the non-structural genes, nsP2 and nsP4, generated novel vectors with reduced cytotoxicity and temperature-sensitive expression (Agapov *et al.*, 1998; Boorsma *et al.*, 2000; Lundstrom *et al.*, 2003). The introduction of the nsP2-L713P point mutation into the SFV-PD vector resulted in a novel vector with low cytotoxicity and long-term gene expression (Lundstrom *et al.*, 2003). Engineering expression vectors based on the avirulent SFV strain A7(74) resulted in novel vectors, SFVnsPA7 vector showed astrocyte spe- cific gene expression at 37°C and neuron specific expression at 31°C in hippocampal slice cultures (Ehrengruber *et al.*, 2003). The non-cytopathogenic phenotype was not restricted to a limited number of cell lines, but present in all cell lines tested, so far as well as in primary hippocampal neurons in culture. A replication-competent SFV-VA7 vector was generated by introduction of a second subgenomic promoter in the viral 3′ non-translated region (Vaha-Koskela *et al.*, 2003). The SFV-VA7 vector efficiently transduced pyramidal cells, neurons and glia cells in hippocampal slice cultures.

Using a SFV-retrovirus hybrid vector system, functional retrovirus particles were generated, in vitro transcribing retrovirus vector RNA from the SFV 26S subgenomic promoter (Wahlfors *et al.*, 1997). Furthermore, minigene-containing retroviral vectors were produced using an alphavirus/retrovirus hybrid vector system (Wahlfors and Morgan, 1999). When the Phoenix retroviral packaging cell lines were infected with alphavirus/retrovirus particles, cytoplasmically producing factor η minigene-containing retroviral vectors were generated.

The second generation helper vector pSFV-Helper2 increased the safety, since three point mutations in the SFV structural genes prevent any amplification of replication-proficient SFV particles generation through expression and helper vector recombination (Berglund, 1993). The generation of a package cell line for alphavirus production added further improvements (Polo *et al.*, 1999). In another approach, the structural genes are split onto two separate helper vectors (Frolov *et al.*, 1997; Smerdou and Liljestrom, 1999).

Although less efficient than transfection of full-length RNAs, alphavirus replication can also be initiated by transfection of plasmid DNA (Dubensky *et al.*, 1996; Herweijer *et al.*, 1995; Kohno *et al.*, 1998; Leitner *et al.*, 2000; Ni *et al.*, 2004; Yamanaka and Xanthopolous, 2004, 2005; Leitner *et al.*, 2000). In this case, full-length 5′-capped RNAs are transcribed in the nucleus using a polymerase II promoter and transported to the cytoplasm, the site of primary translation and RNA amplification. They have been reported to be more efficient than other conventional DNA vectors, requiring 100- to 1000-fold less DNA per immunization (Leitner *et al.*, 2000; Berglund *et al.*, 1998). The induction of apoptosis by replicase-based nucleic acid vaccines not only represents a safety feature, but also seems to be critical for the activation of antigen presenting cells (Leitner *et al.*, 2003). There is no risk of stable integration into the chromosome, which ensures that there is no long-term expression of antigen that might induce tolerance. This strategy offers the highest safety, owing to the absence of the genes coding for the structural proteins, which eliminates the risk of producing replication-proficient viruses. The induction of apoptotic mechanisms is an important safety feature associated with DNA based alphavirus vaccines, since it would eliminate the risk associated with integration of DNA into the host cell genome. These plasmid vectors are most adequate for *in vivo* application, where transient, high-level protein expression is necessary, such as for vaccines.

13.7 Conclusions

The studies with SIN and SFV suggest that both viruses are promising as vectors for heterologous gene expression. The attractive characteristics of alphaviruses are the rapid high-titre virus production and high gene expression. Infection of cultured cells with the SFV vectors shuts off the synthesis of host-cell proteins. The inhibition of host-cell protein synthesis may be due to the expression of a particular viral gene product or it could be a consequence of the very high level of expression of the heterologous gene. SFV vectors do not integrate into the host genome, limiting their use to short-term expression strategies. The transient nature of alphavirus-mediated gene expression has often been considered as a disadvantage. Indeed, this is a problem if long-term expression is desired. It needs to be analyzed, whether the novel replication-competent SIN and SFV vectors can substantially extend the duration of expression and allow the use of alphavirus vectors also for long-term expression. On the other hand, it is under certain circumstances advantageous to obtain transient expression. Using vectors with a short half-life and no capacity for chromosomal integration increases the safety but, there is one report of a fatal infection by SFV (Willems *et al.*, 1979). However, the vectors derived from SFV that are unable to provide any viral structural proteins should be safe, although the possibility of recombination and the production of segmented genome particles has not been fully resolved. Most strains of SIN are not pathogenic and so they may be preferred for many purposes. The safety issues of alphavirus vectors should be further investigated for gene therapy applications. In the future, continued work on these vectors should provide us with improved expression systems. To use alphavirus vectors in clinical trials, development of packaging cell lines and large-scale GMP grade material production should be further investigated.

References

Agapov EV, Frolov I, Lindenbach BD, Pragai BM, Schlesinger S, Rice CM (1998). 'Non-cytopathogenic Sindbis RNA vectors for heterologous gene expression', *Proc Natl Acad Sci U S A* **95**(22), 12989–12994.

Ahola T, Lampio A, Auvinen P, Kaariainen L (1999). Semliki Forest virus mRNA capping enzyme requires association with anionic membrane phospholipids for activity. *EMBO J* **18**(11), 3164–3172.

Ahola T, Kujala P, Tuittila M, *et al.* (2000). Effects of palmitoylation of replicase protein nsP1 on alphavirus infection. *J Virol* **74**(15), 6725–6733.

Asselin-Paturel C, Lassau U, Guinebretiere J-M, *et al.* (1999). Transfer of the murine interleukine-12 gene *in vivo* by a Semliki Forest virus vector induces B16 tumor regression through inhibition of tumor blood vessel formation monitored by Doppler ultrasonography. *Gene Ther* **6**(4), 606–615.

Berglund P, Smerdou C, Fleeton MN, Tubulekas I, Liljestrom P (1998). Enhancing immune responses using suicidal DNA vaccines. *Nat Biotechnol* **16**(6), 562–565.

Berglund P, Sjoberg M, Garoff H, Atkins GJ, Sheahan BJ, Liljestrom P (1993). Semliki Forest virus expression system: production of conditionally infectious recombinant particles.*Bio/Technology* **11**(8), 916–920.

Boorsma M, Nieba L, Koller D, Bachmann MF, Bailey JE, Renner WA (2000). A temperature-regulated replicon-based DNA expression system. *Nat Biotechnol* **18** 429–432.

Cella M, Salio M, Sakakibara Y, Langen H, Julkunen I, Lanzavecchia A (1999). Maturation, activation, and protection of dendritic cells induced by double-stranded RNA. *J Exp Med* **189**(5), 821–829.

Check E (2002). Gene therapy: a tragic setback. *Nature* **420**(6912): 116–118.

Cheng WF, Hung CF, Hsu KF, *et al.* (2002). Cancer immunotherapy using Sindbis virus replicon particles encoding a VP22-antigen fusion. *Hum Gene Ther* **13** (4), 553–568.

Chikkanna-Gowda CP, Sheahan BJ, Fleeton MN, Atkins GJ (2005). Regression of mouse tumours and inhibition of metastases following administration of a Semliki Forest virus vector with enhanced expression of IL-12. *Gene Ther* **12**(16), 1253–1263.

Colmenero P, Lilijestrom P, Jondal M (1999). Induction of P185 tumor immunity by recombinant Semliki Forest virus expressing the P1A gene. *Gene Ther* **6** (10), 1728–1733.

Colmenero P, Chen M, Castanos-Velez E, Liljestrom P, Jondal M (2002). Immunotherapy with recombinant SFV-replicons expressing the P815A tumor antigen or IL-12 induces tumor regression. *Int J Cancer* **98**(4), 554–560.

Daemen T, Regts J, Holtrop M, Wilschut J (2002). Immunization strategy against cervical cancer involving an alphavirus vectors expressing high levels of a stable fusion protein of human papillomavirus 16 E6 and E7. *Gene Ther* **9**(2), 85–94.

Daemen T, Riezebos-Brilman A, Bungener L, Regts J, Dontje B, Wilschut J (2003). Eradication of established HPV16-transformed tumours after immunisation with recombinant Semliki Forest virus expressing a fusion protein of E6 and E7. *Vaccine* **21**(11–12): 1082–1088.

Daemen T, Riezebos-Brilman A, Regts J, Dontje B, van der Zee A, Wilschut J (2004). Superior therapeutic efficacy of alphavirus-mediated immunization against human papilloma virus type 16 antigens in a murine tumour model: effects of the route of immunization. *Antivir Ther* **9**(5), 733–742.

Davis NL, Willis LV, Smith JF, Johnston RE (1989). *In vitro* synthesis of infectious venezuelan equine encephalitis virus RNA from a cDNA clone: analysis of a viable deletion mutant. *Virology* **171**(1), 189–204.

Ding MX, Schlesinger MJ (1989). Evidence that Sindbis virus NSP2 is an autoprotease which processes the virus nonstructural polyprotein. *Virology* **171**(1), 280–284.

Dubensky TW Jr, Driver DA, Polo JM, *et al.* (1996). Sindbis virus DNA-based expression vectors: utility for *in vitro* and *in vivo* gene transfer. *J Virol* **70**(1), 508–519.

Ehrengruber MU, Renggli M, Raineteau O, *et al.* (2003). Semliki Forest virus A7 (74) transduces hippocampal neurons and glial cells in a temperature-dependent dual manner. *J Neurovirol* **9**(1), 16–28.

Engelhardt JF, Yang Y, Stratford-Perricaudet LD, *et al.* (1993a). Direct gene transfer of human CFTR into human bronchial epithelia of xenografts with E1-deleted adenoviruses. *Nat Genet* **4**(1), 27–34.

Engelhardt JF, Simon RH, Yang Y, *et al.* (1993b). Adenovirus-mediated transfer of the CFTR gene to lung of nonhuman primates: biological efficacy study. *Hum Gene Ther* **4**(6), 759–769.

Felgner PL, Ringold GM (1989). Cationic liposome-mediated transfection. *Nature* **337**(6205): 387–388.

Frolov I, Frolova E, Schlesinger S (1997). Sindbis virus replicons and Sindbis virus: assembly of chimeras and of particles deficient in virus RNA. *J Virol* **71**(4), 2819–2829.

Frolova E, Frolov I, Schlesinger S (1997). Packaging signals in alphaviruses. *J Virol* **7**(1), 248–258.

Gardner JP, Frolov I, Perri S, *et al.* (2000). Infection of human dendritic cells by a sindbis virus replicon vector is determined by a single amino acid substitution in the E2 glycoprotein. *J Virol* **74**(24), 11849–11857.

Graham A, Walker R, Baird P, Hahn CN, Fazakerley JK (2005). CNS gene therapy applications of the Semliki Forest Virus 1 vector are limited by neurotoxicity. *Mol Ther* **13**(3), 631–635.

Hahn YS, Grakoui A, Rice CM, Strauss EG, Strauss JH (1989). Mapping of RNA-temperature-sensitive mutants of Sindbis virus: complementation group F mutants have lesions in nsP4. *J Virol* **63**(3), 1194–1202.

Hardy WR, Strauss JH (1989). Processing the nonstructural polyproteins of Sindbis virus: nonstructural proteinase is in the C-terminal half of nsP2 and functions both *in cis* and *in trans*. *J Virol* **63**(11), 4653–4664.

Helenius A, Morein B, Fries E, *et al.* (1978). Human (HLA-A and HLA-B) and murine (H-2K and H-2D) histocompatibility antigens are cell surface receptors for Semliki Forest virus. *Proc Natl Acad Sci USA* **75** (8), 3846–3850.

Herweijer H, Latendresse JS, Williams P, *et al.* (1995). A plasmid-based self-amplifying Sindbis virus vector. *Hum Gene Ther* **6**(9), 1161–1167.

Hevey M, Negley D, Pushko P, Smith J, Schmaljohn A (1998). Marburg virus vaccines based upon alphavirus replicons protect guinea pigs and nonhuman primates. *Virology* **251**(1), 28–37.

Huang HV, Rice CM, Xiong C, Schlesinger S (1989). RNA viruses as gene expression vectors. *Virus Genes* **3** (1), 85–91.

Klimp AH, van der Vaart E, Lansink PO, *et al.* (2001). Activation of peritoneal cells upon *in vivo* transfection with a recombinant alphavirus expressing GM-CSF. *Gene Ther* **8**(4), 300–307.

Kohno A, Emi N, Kasai M, Tanimoto M, Saito H (1998). Semliki Forest virus-based DNA expression vector: transient protein production followed by cell death. *Gene Ther* **5**(3), 415–418.

Lastarza MW, Grakoui A, Rice CM (1994). Deletion and duplication mutations in the C-terminal nonconserved region of Sindbis virus nsP3: effects on phosphorylation and on virus replication in vertebrate and invertebrate cells. *Virology* **202**(1), 224–232.

Leitner WW, Hwang LN, deVeer MJ, *et al.* (2003). Alphavirus-based DNA vaccine breaks immunological tolerance by activating innate antiviral pathways. *Nat Med* **9**(1):33–39.

Leitner WW, Ying H, Driver DA, Dubensky TW, Restifo NP (2000). Enhancement of tumor-specific immune response with plasmid DNA replicon vectors. *Cancer Res* **60**(1), 51–55.

Levis R, Huang H, Schlesinger S (1987). Engineered defective interfering RNAs of Sindbis virus express bacterial chloramphenicol acetyltransferase in avian cells. *Proc Natl Acad Sci U S A* **84**(14), 4811–4815.

Levis R, Schlesinger S, Huang HV (1990). Promoter for Sindbis virus RNA-dependent subgenomic RNA transcription. *J Virol* **64**(4), 1726–1733.

Li GP, La Starza MW, Hardy WR, Strauss JH, Rice CM (1990). Phosphorylation of Sindbis virus nsP3 *in vivo* and *in vitro*. *Virology* **179**, 416–427.

Liljestrom P, Garoff H (1991). A new generation of animal cell expression vectors based on the Semliki Forest virus replicon. *Biotechnology (NY)* **9**(12), 1356–1361.

Loimas S, Toppinen MR, Visakorpi T, Janne J, Wahlfors J (2001). Human prostate carcinoma cells as targets for herpes simplex virus thymidine kinase-mediated suicide gene therapy. *Cancer Gene Ther.* **8**(2), 137–144.

London SD, Schmaljohn AL, Dalrymple JM, Rice CM (1992). Infectious enveloped RNA virus antigenic chimeras. *Proc Natl Acad Sci U S A* **89**(1), 207–211.

Lundstrom K (2003). Latest development of viral vectors for gene therapy. *Trends Biotechnol* **21**(3), 117–122.

Lundstrom K, Pralong W, Martinou J-C (1997). Anti-apoptotic effect of Bcl-2 overexpression in RIN cells infected with Semliki Forest virus. *Apoptosis* **2**(2), 189–191.

Lundstrom K, Abenavoli A, Malgaroli A, Ehrengruber MU (2003). Novel Semliki Forest virus vectors with reduced cytotoxicity and temperature sensitivity for long-term enhancement of transgene expression. *Mol Ther* **7**(2), 202–209.

MacDonald GH, Johnston RE (2000). Role of dendritic cell targeting in Venezuelan equine encephalitis virus pathogenesis. *J Virol* **74**(2), 914–922.

Mulligan RC (1993). The basic science of gene therapy. *Science* **260**(5110): 926–932.

Murphy AM, Sheahan BJ, Atkins G (2001). Induction of apoptosis in Bcl-2 expressing rat prostate cancer cells using the Semliki Forest virus vector. *Int J Cancer* **94** (4), 572–578.

Ni B, Lin Z, Zhou L, *et al.* (2004). Induction of P815 tumor immunity by DNA-based recombinant Semliki Forest virus or replicon DNA expressing the P1A gene. *Cancer Detect Prev* **28**(6), 418–425.

Niesters HG, Strauss JH (1990). Mutagenesis of the conserved 51-nucleotide region of Sindbis virus. *J Virol* **64**(4), 1639–1647.

Ohno K, Sawai K, Iijima Y, Levin B, Meruelo D (1997). Cell specific targeting of Sindbis virus vectors displaying IgG binding domains of protein A. *Nat Biotechnol* **15**(8), 763–767.

Piper RC, Slot JW, Li G, Stahl PD, James DE (1994). Recombinant Sindbis virus as an expression system for cell biology. *Methods Cell Biol* **43**, 55–78.

Polo JM, Belli BA, Driver DA, *et al.* (1999). Stable alphavirus packaging cell lines for Sindbis virus- and Semliki Forest virus-derived vectors. *Proc Natl Acad Sci U S A* **96**(8), 4598–4603.

Pushko P, Bray M, Ludwig G.V, *et al.* (2000). Recombinant RNA replicons derived from attenuated Venezuelan equine encephalitis virus protect guinea pigs and mice from Ebola hemorrhagic fever virus. *Vaccine* **19**(1), 142–153.

Rabinowitz JE, Samulski J (1998). Adeno-associated virus expression systems for gene transfer (Review). *Curr Opin Biotechnol* **9**(5), 470–475.

Ren H, Boulikas T, Lundstrom K, Soling A, Warnke PC, Rainov NG (2003). Immunogene therapy of recurrent glioblastoma multiforme with a liposomally encapsulated replication-incompetent Semliki forest virus vector carrying the human interleukin-12 gene – a phase I/II clinical protocol. *J Neurooncol* **64**(1–2): 147–154.

Rikkonen M, Peranen J, Kaariainen L (1994). ATPase and GTPase activities associated with Semliki Forest virus nonstructural protein nsP2. *J Virol* **68**(9), 5804–5810.

Rodriguez-Madoz JR, Prieto J, Smerdou C (2005). Semliki forest virus vectors engineered to express higher IL-12 levels induce efficient elimination of murine colon adenocarcinomas. *Mol Ther* **12**(1), 153–163.

Rosenfeld MA, Siegfried W, Yoshimura K, *et al.* (1991). Adenovirus-mediated transfer of a recombinant alpha 1-antitrypsin gene to the lung epithelium *in vivo*. *Science* **252**(5004): 431–434.

Sammin DJ, Butler D, Atkins GJ, Sheahan BJ (1999). Cell death mechanisms in the olfactory bulb of rats infected intranasally with Semliki forest virus. *Neuropathol Appl Neurobiol* **25**(3), 236–243.

Sawai K, Meruelo D (1998). Cell-specific transfection of choriocarcinoma cells by using Sindbis virus hCG expressing chimeric vector. *Biochem Biophys Res Commun* **248**(2), 315–323.

Sawai K, Ohno K, Iijima Y, Levin B, Meruelo D (1998). A novel method of cell-specific mRNA transfection. *Mol Genet Metab* **64**(1), 44–51.

Sawicki D, Barkhimer DB, Sawicki SG, Rice CM, Schlesinger S (1990). Temperature sensitive shut-off of alphavirus minus strand RNA synthesis maps to a nonstructural protein, nsP4. *Virology* **174**(1), 43–52.

Schlesinger S, Schlesinger MJ. Togaviridae:the viruses and their replication. In Fields BN, Knipe PM (eds), *Fields Virology*, 3rd edition. Lippincott-Raven, Philadelphia, 1996.

Smerdou C, Liljestrom P (1999). Two-helper RNA system for production of recombinant Semliki forest virus particles. *J Virol* **73**, 1092–1098.

Strauss EG., Strauss JH. Structure and replication of the alphavirus genome. In: Schlesinger S, Schlesinger MJ (eds), *The Togaviridae and Flaviviridae*. Plenum Press, New York, 1986.

Strauss JH, Strauss EG (1994). The alphaviruses: gene expression, replication, and evolution. *Microbiol Rev* **58**(4), 491–562.

Tseng JC, Levin B, Hurtado A, *et al.* (2004). Systemic tumor targeting and killing by Sindbis viral vectors. *Nat Biotechnol* **22**(1), 70–77.

Unno Y, Shino Y, Kondo F, *et al.* (2005). Oncolytic viral therapy for cervical and ovarian cancer cells by Sindbis virus AR339 strain. *Clin Cancer Res* (11) **12**, 4553–4560.

Vaha-Koskela MJ, Tuittila MT, Nygardas PT, *et al.* (2003). A novel neurotropic expression vector based on the avirulent A7 (74) strain of Semliki Forest virus. *J Neurovirol* **9**(1), 1–15.

Velders MP, McElhiney S, Cassetti MC, *et al.* (2001). Eradication of established tumors by vaccination with Venezuelan equine encephalitis virus replicon particles delivering human papillomavirus 16 E7 RNA. *Cancer Res* **61**(21), 7861–7867.

Wahlfors JJ, Morgan RA (1999). Production of minigene-containing retroviral vectors using an alphavirus/retrovirus hybrid vector system. *Hum Gene Ther* **10** (7), 1197–1206.

Wahlfors JJ, Xanthopoulos KG (1997). Morgan RA. Semliki Forest virus-mediated production of retroviral vector RNA in retroviral packaging cells. *Hum Gene Ther* **8**(17), 2031–2041.

Walker DH, Harrison A, Murphy K, Flemister M, Murphy FA (1976). Lymphoreticular and myeloid pathogenesis of Venezuelan equine encephalitis in hamsters. *Am J Pathol* **84**(2), 351–370.

Wang KS, Kuhn RJ, Strauss EG, Ou S, Strauss JH (1992). High-affinity laminin receptor is a receptor for Sindbis virus in mammalian cells. *J Virol* **66**(8), 4992–5001.

Willems WR, Kaluza G, Boschek CB, *et al.* (1979). Semliki forest virus: cause of a fatal case of human encephalitis. *Science* **203**(4385): 1127–1129.

Xiong C, Levis R, Shen P, Schlesinger S, Rice CM, Huang HV (1989). Sindbis virus: an efficient, broad host range vector for gene expression in animal cells. *Science* **243**(4895): 1188–1191.

Yamanaka R, Xanthopoulos KG (2005). Induction of antigen-specific immune responses against malignant brain tumors by intramuscular injection of sindbis DNA encoding gp100 and IL-18. *DNA Cell Biol* **24**(5), 317–324.

Yamanaka R, Xanthopoulos KG (2004). Development of improved Sindbis virus-based DNA expression vector. *DNA Cell Biol* **23**(2), 75–80.

Yamanaka R, Zullo SA, Tanaka R, Ramsey J, Blaese M, Xanthopoulos KG (2000). Induction of a therapeutic antitumor immunological response by intratumoral injection of genetically engineered Semliki Forest virus to produce interleukin 12. *Neurosurg Focus* **9**, Article 7.

Yamanaka R, Zullo SA, Ramsey J, *et al.* (2001a). Induction of therapeutic antitumor antiangiogenesis by intratumoral injection of genetically engineered endostatin-producing Semliki Forest virus. *Cancer Gene Ther* **8**(10), 796–802.

Yamanaka R, Zullo SA, Tanaka R, Blaese M, Xanthopoulos KG. (2001b). Enhancement of antitumor immune response in glioma models in mice by genetically modified dendritic cells pulsed with Semliki Forest virus – mediated by complementary DNA. *J Neurosurg* **94**(3), 474–481.

Yamanaka R., Zullo SA, Ramsey J, *et al.* (2002a). Marked enhancement of antitumor immune responses in mouse brain tumor models by genetically modified dendritic cells producing Semliki Forest virus-mediated interleukin-12. *J Neurosurg* **97**(3), 611–618.

Yamanaka R, Yajima N, Tsuchiya N, *et al.* (2002b). Administration of interleukin-12 and -18 enhancing the antitumor immunity of genetically modified dendritic cells that had been pulsed with Semliki forest virus-mediated tumor complementary DNA. *J Neurosurg* **97**(5), 1184–1190.

Yamanaka R, Tsuchiya N, Yajima N, *et al.* (2003). Induction of antitumor immunologic response by intratumoral injection of genetically engineered Semliki Forest Virus to produce IL-18 combined with systemic administration of IL-12. *J Neurosurg* **99**(4), 746–753.

Ying H, Zaks TZ, Wang RF, *et al.* (1999). Cancer therapy using a self-replicating RNA vaccine. *Nat Med* **5**(7), 823–827.

Zhang J, Asselin-Paturel C, Bex F, *et al.* (1997). Cloning of human IL-12 p40 and p35 DNA into the Semliki Forest virus vector: expression of IL-12 in human tumor cells. *Gene Ther* **4**(4), 367–374.

14

Tumour-suppressor Gene Therapy

Bingliang Fang and **Jack A Roth**

14.1 Tumour-suppressor genes

14.1.1 Discovery of tumour-suppressor genes

A gene whose loss-of-function mutations facilitate development of tumours is called a tumour-suppressor gene. The first evidence of the presence of tumour-suppressor genes came from cell hybrid experiments. Hybrid cells derived by fusion of normal cells with tumour cells lost their tumorigenicity or immortality (Stanbridge, 1976; Pereira-Smith and Smith, 1983). This finding suggested that normal tumour-suppressor genes in normal cells can compensate for the inactivation of tumour-suppressor genes in tumour cells and thus reverse the tumorigenic phenotype. Suppression of tumorigenicity also was observed in hybrids between different tumour cell lines (Weissman and Stanbridge, 1983). Yet the tumorigenicity of those fusion cells could be restored when some chromosomes were lost (Harris, 1988; Marshall, 1991), indicating that tumorigenicity was suppressed in the hybrids by a genetic mechanism and that different tumours might have different tumour-suppressor genes. Subsequent findings showed that introduction of a single chromosome by microcells is sufficient to suppress the tumorigenic phenotypes. For example, introduction of a single copy of human chromosome 11 is sufficient to completely suppress the tumorigenic phenotype of HeLa cells, suggesting the presence of tumour-suppressor gene(s) on this chromosome (Saxon *et al.*, 1986).

Cloning of the retinoblastoma susceptibility gene (*RB*) provided direct evidence that recessive mutations on two alleles of the same tumour-suppressor gene are involved in development of retinoblastoma (Friend *et al.*, 1986; Lee *et al.*, 1987), a malignant tumour arising in the retina of the eye. Inherited retinoblastoma showed an autosomal dominant inheritance of early onset of multiple tumors in both retinas. Although every retinal cell inherits the *RB* gene, only a very few develop retinoblastoma, suggesting that another molecular event besides an *RB* mutation contributes to the disease. This phenomenon led to development of the 'two-hit' hypothesis to explain that an autosomal dominant form of inheritance of familial cancers could be accounted for by recessive loss-of function mutations (Knudson, 1971). An affected individual can get a mutant inactive allele by inheritance and another by somatic mutation. Indeed, the first cloning of the *RB* gene showed that, in retinoblastoma samples, this gene was either homozygously deleted or its mRNA shortened (Lee *et al.*, 1987).

The somatic inactivation of an allele often results from loss of some genetic material inherited from one of the parents and can be detected by loss of heterozygosity analysis (Ponder, 1988). This method has been used extensively in searching

Viral Therapy of Cancer Edited by Kevin J. Harrington, Richard G. Vile and Hardev S. Pandha

for candidate tumour-suppressor genes in both familial and sporadic forms of cancers. For example, loss of heterozygosity of markers on chromosome 17p was detected at high frequency in various cancers, including breast, lung, and colon cancers. The refining of common deletions in 17p in colon carcinomas is confined to a site that harbours the *p53* gene (Baker *et al.*, 1989). Introducing a wild type *p53* (wt-*p53*) gene to cancer cells suppresses their growth *in vitro*. However, a mutation resulting in a single conservative amino acid substitution in the *p53* gene product abrogates this suppressive ability (Baker *et al.*, 1990), demonstrating that *p53* is a tumour-suppressor gene. It is now known that alterations in the *p53* gene occur frequently in human cancers.

14.1.2 Biological function of tumour-suppressor genes

In addition to the *RB* and *p53* genes, numerous other tumour-suppressor genes have been identified (Vogelstein and Kinzler, 2004). Those gene are classified on the basis of their biological functions as gatekeepers or caretakers (Kinzler and Vogelstein, 1997).

Gatekeepers are genes that directly regulate cellular functions involved in cell growth, differentiation, and cell death. *PTEN*, *p53*, *p73*, *Fhit*, *RB*, von Hippel–Lindau, neurofibromatosis type 1, and adenomatous polyposis coli (*APC*) genes are in this category. Mutations in gatekeepers directly predispose to cancers by allowing uncontrolled cell proliferation and disrupting apoptosis. In contrast, caretakers control cellular processes that repair genetic alterations and maintain genomic integrity. These genes include *ATM* (ataxia telangiectasia mutated), *ATR* (ATM and Rad3-related), *BRCA1*, *BRCA2*, and mismatch-repair genes. Their inactivation leads to genetic instabilities that result in increased mutation of all genes, including gatekeepers. Genes such as *p53* can function both as gatekeepers and caretakers. As a gatekeeper, p53 protein downregulates expression of *Bcl2* and directly activates expression of *Bax* (Miyashita *et al.*,1994), Noxa (Oda *et al.*, 2000a), *p53*-regulated apoptosis-inducing protein 1 (*p53AIP1*) (Oda *et al.*, 2000b), *p53*-upregulated modulator of apoptosis (*PUMA*) (Nakano and Vousden, 2001; Yu *et al.*, 2001), *Fas* (Owen-Schaub *et al.*,1995), and death receptor killer/DR5 (Wu *et al.*, 1997), igniting both mitochondrion- and membrane-mediated apoptotic pathways. *p53* also transactivates the *PIG3* gene, which encodes a homologue of reduced nicotinamide adenine dinucleotide phosphate-quinone oxidoreductase, a potent generator of reactive oxygen species, which are powerful inducers of apoptosis (Polyak *et al.*,1997). As a caretaker, *p53* transactivates p21 (el Deiry *et al.*, 1993) and 14-3-3σ (Hermeking *et al.*, 1997; Chan *et al.*, 1999), which mediate G_1 and G_2 arrest, preventing replication of damaged DNA or transmission of damaged DNA to the next generation. Moreover, *p53* activates transcription of genes directly involved in DNA repair, such as *p53R2* (Tanaka *et al.*, 2000) and proliferating cell nuclear antigen (*PCNA*) (Morris *et al.*, 1996).

Nevertheless, the functions of most tumour-suppressor genes are not yet completely understood. For example, a number of transcriptional targets that can be regulated by p53 protein have been identified (for review, see Xu and Raafat el-Gewely, 2001). Yet, the list of genes regulated by p53 is still expanding. In fact, by using oligonucleotide arrays containing 6000 human genes, Zhao *et al.* (2000) identified 107 p53-induced and 54 p53-repressed genes that fall into functional categories of apoptosis, growth arrest, cytoskeletal functions, growth factors and their inhibitors, extracellular matrix, and adhesion. Moreover, tumour-suppressor genes often interweave with other tumour-suppressor genes or with oncogenes, forming networks that govern cell growth, apoptosis, differentiation, and genome integrity. One such example is p53-mediated upregulation of p21, which binds to a number of cyclins and cyclin-dependent kinases (cdk), including cyclin D1-cdk4, leading to inhibition of cyclin D1-cdk-4, whose major target is retinoblastoma protein (Rb). The Rb protein binds to a number of the E2F family proteins, forming complexes that inhibit transcription of the genes regulated by E2F transcription factors, including dihydrofolate reductase and *PCNA*. Phosphorylation of RB by cyclin D1-cdk4 releases Rb protein from the E2F complex, leading to E2F-mediated transcription and cell-cycle progression.

14.2 Use of tumour-suppressing genes for cancer therapy

14.2.1 Conceptual development

A final proof for a candidate gene to be considered a tumour-suppressor gene is that restoring its function by introducing the normal gene into cancer cells can suppress some malignant phenotype. Indeed, retrovirus-mediated gene transfer of the *RB* gene into retinoblastoma or osteosarcoma cells significantly suppressed cell growth and tumorigenicity in nude mice (Huang *et al.*, 1988). Similarly, reintroducing the wt-*p53* gene to colorectal cancer cells or glioblastoma cells resulted in cell-cycle arrest and growth suppression of those cells (Baker *et al.*, 1990; Mercer *et al.*, 1990).

It is now known, however, that no single gene defect can cause cancer (Vogelstein and Kinzler, 2004). Development of malignant phenotypes is caused by malfunction of multiple genes. The facts that cancer cells in a tumour mass are often heterogeneous and that genetic defects in a tumour mass are never completely characterized make repair of genetic defects in cancer cells for cancer therapy a daunting task. Nevertheless, the goal of cancer gene therapy is to eliminate, not to repair, malignant cells. Thus, the observation that expression of the wt-*p53* gene triggers apoptosis in some cancer cells stimulated interest in using such tumour-suppressor genes for cancer therapy (Shaw *et al.*, 1992; Fujiwara *et al.*, 1993).

Evidence has shown that activation or overexpression of a single oncogene such as *myc* or *ras* in normal human cells induces only apoptosis or senescence (Evan *et al.*, 1992; Serrano *et al.*, 1997), suggesting that mammalian cells have safeguards against carcinogenesis. Evidence also has shown that apoptosis induction by *myc* or *ras* oncogenes requires the presence of wt-*p53* (Hermeking and Eick, 1994; Tanaka *et al.*, 1994; Wagner *et al.*, 1994; Serrano *et al.*, 1997). Moreover, p53 promotes apoptosis when *E2F1* expression is dysregulated or when the *RB* gene is deficient (Morgenbesser *et al.*, 1994; Qin *et al.*, 1994; Symonds *et al.*, 1994; Wu and Levine, 1994). Thus, p53-mediated apoptosis may serve as an important safeguard mechanism to prevent carcinogenesis by oncogene activation or mutations of other tumour-suppressor genes, and restoration of p53 functions could be an attractive approach for cancer therapy. In addition to *p53*, many other tumour-suppressor genes have been investigated in preclinical and clinical trials of cancer therapy, including *BRCA1* (Shao *et al.*, 1996; Tait *et al.*, 1999), *p16* (Schrump *et al.*,1996; Sandig *et al.*,1997), *FHit* (Dumon *et al.*, 2001; Ramesh *et al.*, 2001; Nishizaki *et al.*, 2004) and *RB* (Demers *et al.*, 1998; Claudio *et al.*, 1999; Zhang *et al.*, 2003).

14.2.2 Technology development

Initial experiments with retrovirus-mediated *p53* gene transfer demonstrated that *p53*-expressing retroviral vector can penetrate human lung cancer spheroids in culture and inhibit their growth by inducing apoptosis (Fujiwara *et al.*, 1993). Direct administration of *p53*-expressing retroviral vector to orthotopic lung cancers with abnormal *p53* expression inhibited local tumour growth *in vivo* (Fujiwara *et al.*, 1994a). Those studies led to clinical trials of retrovirus-mediated wt-*p53* gene therapy in patients with lung cancer (Roth *et al.*, 1996). Meanwhile, it was found that adenovectors can be easily manipulated and produced in high titers, and adenovector-mediated *p53* gene transfer led to high *in vivo* transduction efficiency and high levels of *p53* expression. *In vivo* administration of an adenovector expressing *p53* led to induction of apoptosis in tumour cells and suppression of tumour growth (Liu *et al.*, 1994; Zhang *et al.*, 1994). In most subsequent preclinical and clinical studies, therefore, adenovector has been used for *p53* gene therapy.

Because tumour-suppressor genes are expressed in normal cells, they are expected be nontoxic to normal cells. Indeed, overexpression of *p53* and *FHIT* at pharmaceutical levels resulted in no toxic effects in normal bronchial epithelial cells (Zhang *et al.*, 1995; Ji *et al.*, 1999). Furthermore, targeting the pharmaceutical effects of a therapeutic gene to cancer cells is feasible by using tumour-selective expression. For example, the toxicity of a therapeutic gene can be prevented by using tumour-specific promoters such as human telomerase reverse transcriptase (hTERT) promoter (Gu *et al.*, 2000, 2002) or other tissue-specific promoters.

A major issue in cancer gene therapy is efficiency of gene delivery. In contrast to chemotherapeutic agents, gene-based medicines are usually macromolecules, and their distribution *in vivo* may be limited by their size. Thus far, no technology is available that can deliver a therapeutic gene to 100 per cent of cancer cells. Interestingly, evidence has shown that 100 per cent transduction of tumour cells *in vivo* may not be necessary to elicit complete tumour regression. Non-transduced tumour cells can be killed by 'bystander effects' that may be associated with various mechanisms. For example, treatment with the *p53* gene may elicit bystander effects by inhibiting the angiogenic response. Transduction of cancer cells with adenoviral *p53* (Ad-*p53)* resulted in significant reduction of vascular endothelial growth factor production by cancer cells. Moreover, intratumoral injection of Ad-*p53* inhibited tumour cell-induced angiogenesis *in vivo* (Bouvet *et al.*, 1998). Suppression of angiogenesis also may result from *p53*-mediated upregulation of thrombospondin, a potent angiogenesis inhibitor (Dameron *et al.*, 1994). Alternatively, Ad-*p53* gene transfer may induce bystander effects by upregulating expression of Fas, DR5, and insulin-like growth factor (IGF) 1 binding protein 3 (IGF1BP3) (Backbinder *et al.*, 1995). IGF1BP3 is known to block IGF action and inhibit cell growth by sequestering free IGF. IGF1BP3 also can induce apoptosis in an IGF-independent manner (Raiah *et al.*, 1997; Grimberg, 2000). Finally, a bystander effect may be elicited by local inflammatory or immune responses triggered by transgene products (Abina *et al.*, 1996) or by viral proteins in viral vectors (Zoltick *et al.*, 2001; Bowen *et al.*, 2002).

14.3 Clinical trials of *p53* gene replacement

The first clinical trial of *p53* gene therapy was carried out with a retroviral vector expressing wt-*p53* under control of the beta-actin promoter (Roth *et al.*, 1996). The vector was injected directly into the tumours of nine patients with unresectable non-small cell lung cancer (NSCLC) already proven resistant to other interventions. Laboratory analysis revealed that apoptosis was more frequent in tissue biopsies taken after treatment than in those taken before treatment. Tumour regression was noted in three patients, and tumour growth stabilized in three other patients. No clinically significant vector-related toxic effects were noted up to 5 months after treatment, demonstrating the feasibility and safety of gene therapy (Roth *et al.*, 1996).

Subsequent clinical trials utilized a replication-defective adenovector expressing the wt-*p53* gene under control of a cytomegalovirus promoter. A phase I trial enrolled 28 NSCLC patients whose cancers had not responded to conventional treatments, and successful vector delivery into tumour was demonstrated in 80 per cent of evaluable patients (Swisher *et al.*, 1999). Gene expression was detected in 46 per cent, and apoptosis was demonstrated in all but one of the patients expressing the gene. Despite repeated injections (up to six) vector-related toxicity was minimal. Of 25 evaluable patients, two (8 per cent) had reduction of tumour size by more than a 50 per cent and 16 (64 per cent) had disease stabilization for 2–14 months. These results demonstrated that repeated intratumoral injections of Ad-*p53* are well tolerated and can mediate antitumour activity in a subset of patients with advanced NSCLC.

Authors of a phase I study of 33 patients with head and neck squamous cell carcinoma (HNSCC) also reached the conclusion that transfer of the Ad-*p53* construct caused little toxicity and, once again, significant clinical responses were observed – this time in 9 of 18 clinically evaluable patients Clayman *et al.*, 1998) A subsequent phase II clinical trial of Ad-*p53* in over 200 patients with recurrent or refractory HNSCC resulted in demonstration of complete or partial responses in approximately 10 per cent of patients, and some evidence of antitumor activity was observed in 60 per cent of patients (Goodwin *et al.*, 1999).

Phase I clinical studies with Ad-*p53* also have been performed in patients with ovarian cancer (intraperitoneal delivery) or glioma (intracerebral delivery) (Lang et al., 2003; Wolf *et al.*, 2004). In the study of intraperitoneal delivery to patients with ovarian cancer, Ad-*p53* was given daily for 5 days every 3 weeks at doses ranging from 3×10^{10} to 3×10^{12} viral particles. Fifteen (88 per cent) patients were evaluable for toxicity, and no dose-limiting toxic effects were observed (Wolf *et al.*, 2004). Therapeutic response or disease stabilization was observed in 36 per cent of evaluable patients. Intracerebral delivery of Ad-*p53* to tumour of patients with recurrent glioma at doses ranging from 3×10^{10} to 3×10^{12} viral particles also was found to be safe. A maximum tolerated dose was not reached. However, effective transduction occurred only within a short distance (5 mm) of the injection site (Lang *et al.*, 2003).

An adenovector expressing wt-*p53* gene driven by a Rous sarcoma virus promoter has been evaluated for treatment of patients with laryngeal cancer or HNSCC (Peng, 2005). It was approved for the treatment of HNSCC by the State Food and Drug Administration of China (SFDA) in 2003 and was trademarked as Gendicine.

14.4 Tumour-suppressor gene therapy in multimodality anticancer treatment

14.4.1 Preclinical studies

Using *p53*-deficient mouse embryonic fibroblasts, Lowe *et al.* (1993) demonstrated that *p53* function is required for efficient induction of apoptosis by ionizing radiation and various anticancer chemotherapeutics. This observation suggested that loss of *p53* function due to gene mutation or deletion, as seen in more than 60 percent of human cancers, may account for cross-resistance of cancer cells to anticancer agents. Subsequently, combination of *p53* gene therapy with chemotherapy or radiotherapy has been tested zealously by several groups. Introducing the wt-*p53* gene into a *p53*-deleted human NSCLC cell line by adenovector-mediated gene transfer markedly increased cellular sensitivity to the chemotherapeutic drug cisplatin (Fujiwara *et al.*, 1994b). Moreover, direct injection of Ad-*p53* into *p53*-null tumors, followed by administration of cisplatin, induced massive apoptotic destruction of the tumors. Furthermore, Nguyen *et al.* (1996) reported that exposing cancer cells to cisplatin enhanced transgene expression from adenovirus-mediated gene transfer. Sequential administration of cisplatin and Ad-*p53* yielded significantly greater apoptosis and tumour growth suppression in an animal model of lung cancer than did administration of either agent alone. This observation led to a phase I clinical trial of Ad-*p53* gene therapy in sequence with cisplatin for treatment of lung cancer (Nemunaitis *et al.*, 2000). Similarly, Li *et al.* (2002) reported that docetaxel and paclitaxel enhanced adenovirus-mediated transgene expression. Thus far, enhancements of the antitumor activity of the *p53* gene by combinations with cisplatin, doxorubicin, 5-fluorouracil, methotrexate, or etoposide have been reported (Nguyen *et al.*, 1996; Dorigo *et al.*, 1998; Jones *et al.*, 1998; Nielsen et al., 1998; Gurnani *et al.*, 1999; Nishizaki *et al.*, 2001; Lebedeva *et al.*, 2001). For example, Lebedeva *et al.* (2001) reported that systemic administration of Ad-*p53* plus doxorubicin led to a significant reduction in the incidence of experimental lung metastasis from breast cancers when compared with Ad-*p53* or doxorubicin alone.

Combination of *p53* gene therapy with radiotherapy has been tested by several groups. Spitz *et al.* (1996) demonstrated that introducing wt-*p53* into cancer cells by adenovector-mediated gene transfer sensitized cancer cells to chemotherapy and radiotherapy. Similarly, Kawabe *et al.* (2001) showed that adenovector-mediated *p53* gene expression sensitized both *p53*-mutant and wt-*p53* NSCLC cells to radiotherapy, but not normal human lung fibroblasts. Immunoblot analysis showed that Bax expression was increased in NSCLC cells treated with the combination therapy but remained unchanged in normal cells, suggesting that treatment with Ad-*p53* radiosensitizes cancer cells but not normal cells. Using two-dimensional and three-dimensional isobologram modelling and rigorous statistical analysis, Nishizaki *et al.* (2001) demonstrated synergistic antitumour activity of Ad-*p53* gene transfer combined with docetaxel and radiation therapy in various human lung cancer cell lines. *In vitro* and *in vivo* studies showed that combination of these three therapeutic agents (*p53* gene, radiotherapy, and docetaxel) exhibited synergistic inhibitory effects on tumour cell growth in all four cell lines tested. Together, the results of these studies suggest that combined *p53* gene therapy, chemotherapy, and radiotherapy may have advantages over single-modality therapy. Some of those observations have led to clinical trials of combinations of *p53* gene plus chemotherapy or *p53* plus ionizing radiation.

p53 gene therapy plus antiangiogenesis therapy is expected to be more effective than either of these therapies alone, because tumour cells deficient in *p53* have been reported to display a diminished rate of apoptosis under hypoxic conditions (Yu *et al.*, 2002). In mice, tumors derived from *p53*(−/−) human colorectal cancer cells were less responsive to antiangiogenic combination therapy than isogenic *p53*(+/+) tumors. These results suggest that even though antiangiogenic therapy targets genetically stable endothelial cells in the tumour vasculature, genetic alterations that decrease the vascular dependence of tumour cells can influence the therapeutic response of tumours to this therapy (Yu *et al.*, 2002).

14.4.2 Clinical studies

A phase I trial of *p53* gene transfer in combination with cisplatin therapy was performed in 24 patients with NSCLC previously unresponsive to conventional treatments (Nemunaitis *et al.*, 2000). Seventy-five percent of the patients had previously experienced tumour progression on cisplatin- or carboplatin-containing regimens. Intravenous cisplatin was administered, and 3 days later *p53* was delivered by intratumoral injection; patients received as many as six monthly courses. The result showed that the combination therapy was well tolerated and yielded evidence of clinical activity. Of the 23 evaluable patients, 17 had stable disease for at least 2 months, two achieved partial responses, and four had progressive disease. Intratumor transgene mRNA was identified in 43 per cent of assessable patients.

Schuler and coworkers (2001) reported on 25 patients with nonresectable NSCLC who were enrolled in an open-label, multicentre phase II study of three cycles of either carboplatin plus paclitaxel or cisplatin plus vinorelbine in combination with intratumoral injection of Ad-*p53* on day 1. Since all patients had disseminated cancer and only a single lesion was injected in each patient, the gene therapy was not expected to affect survival. The size of the Ad-*p53*-injected lesion was compared to that of a control lesion. Calculation of the areas of the Ad-*p53*-treated lesions and the comparator lesions at the end of each treatment cycle revealed significant differences in tumour regression between the Ad-*p53*-treated lesions and the comparator lesions of all study patients after the second cycle.

Buller and coworkers treated 36 patients with recurrent ovarian carcinoma with either a single or multiple intraperitoneal doses of Ad-*p53* over multiple cycles in combination with platinum-based agents (Buller *et al.*, 2002). The results demonstrated safety and clinical benefit for both regimens. The treatment was well tolerated. Median survival of individuals who received multiple-dose Ad-*p53* with chemotherapy was 12–13.0 months, while that of those treated with single-dose Ad-*p53* was only 5 months. Ten patients who received multiple-dose treatment were long-term survivors (more than 20 months), whereas only two who received single-dose treatment were long-term survivors.

On the basis of preclinical observations that the *p53* gene can increase the sensitivity of tumours to ionizing radiation, a phase II trial of Ad-*p53* plus radiotherapy was performed in patients with non-metastatic, locoregionally advanced NSCLC who could not tolerate chemoradiation and resection. The patients received three intratumoral injections of Ad-*p53* in doses as high as 3×10^{12} viral particles/injection on days 1, 18, and 32, in conjunction with radiation therapy (60 Gy). Again, the treatment was performed on an outpatient basis and was well tolerated. Eleven (62 per cent) patients underwent biopsies, and specimens from eight were negative on pathological examination. The 1-year progression-free survival rate was 45.5 per cent. Among 13 evaluable patients, five (39 per cent) had a complete response and three (23 per cent) had a partial response or disease stabilization (Roth *et al.*, 2001; Swisher *et al.*, 2002). Most of the treatment failures were due to metastatic disease, not local progression.

A multicentre, randomized clinical trial with Gendicine was conducted in 135 patients with HNSCC. The patients received radiation therapy (70 Gy over 8 weeks) in combination with intratumoral injection of Ad-*p53* (10^{12} viral particles/dose/week for a total of 8 weeks) or radiation therapy alone. Complete remission was seen in 64 per cent of patients who received Ad-*p53* combined with radiation therapy but in only 19 per cent of patients who received radiation therapy alone, a difference that was highly significant statistically (Peng, 2005). This clinical trial formed the basis for approval of Ad-*p53* for head and neck cancer by the China State Food and Drug Administration, thus making Ad-*p53* the first gene therapy approved for human use.

14.5 Future prospects

Results of clinical trials with the *p53* gene have demonstrated that direct intratumour injection of a tumour-suppressor gene is well tolerated. Favourable clinical responses are observed even when therapy comprises replacement with only a single tumour-suppressor gene, despite the fact that a wide diversity of genetic lesions is present in cancer cells. Expression of the therapeutic gene has been documented even in the presence of a humoral immune response against gene delivery vehicles. Combination of tumour-suppressor gene therapy with convention chemotherapy or radiotherapy led to synergistic therapeutic effects or overcoming of resistance.

It is clear, however, that the current approach to tumour-suppressive cancer gene therapy can be improved further. A major issue in cancer gene therapy is vector technology. Because of their high *in vivo* transduction efficiencies and easy production, adenovectors are the most widely used vehicles in cancer gene therapy. Intratumoral injection of adenovectors often leads to only limited transduction, however, typically within a short distance of the injection site. Use of conditionally replicative adenoviruses might promote vector spread in tumour sites because replication and release of the vector will lead to another round of transduction of neighbouring cells. Moreover, vector replication inside tumour cells will dramatically increase the copy number of vector genome and expression of the therapeutic protein. It was reported that incorporating the p53-expressing cassette into an oncolytic viral vector, AdDelta24, increased the antitumor activity of the vector by more than 100-fold (Van Beusechem *et al.*, 2002). Thus, delivery of tumour-suppressor genes by oncolytic vectors could be a promising approach for cancer therapy. Technologies for efficient systemic gene delivery are needed, especially for treatment of metastatic diseases. Recently, nanoscale synthetic particles with encapsulated plasmid DNA have been development for systemic gene delivery of p53 and other tumour suppressor genes. Preclinical study showed that systemic administration of the FUS1-expressing plasmid vector complexed with DOTAP:cholesterol (DOTAP:Chol) liposome, termed FUS1 nanoparticle, elicited a decrease in the number of metastatic tumour nodules and prolonged survival of animals with lung metastasis tumors (Ito *et al.*, 2004; Uno *et al.*,. 2004). A phase I clinical trial with systemic administration of FUS1-nanoparticles is now underway in stage IV lung cancer patients at the University of Texas M.D. Anderson Cancer Center in Houston, Texas. Technologies in vector targeting or tumour-selective expression will be helpful in increasing efficacy and reducing possible systemic toxicity. Further explorations of bystander effects, antiangiogenesis, immune responses, and inflammatory responses at tumour sites will improve delivery of the therapeutic effects to untransduced cancer cells. Thus, improvement in gene delivery systems, enhanced induction of bystander effects, and further adjuvant use of gene therapy with conventional chemotherapy, radiation therapy, and surgery will all lead to better clinical responses and enhance the antitumour activities of tumour-suppressor genes.

Acknowledgements

We thank Kathryn Hale for editorial review and Debbie Smith for assistance in preparation of this manuscript. This work was supported in part by grants from the National Cancer Institute (RO1

CA98582-01A1 to B.F. and SPORE in Lung Cancer, P50CA70907to J.A.R.).

References

Abina MA, Lee MG, Descamps V, *et al.* (1996). *LacZ* gene transfer into tumor cells abrogates tumorigenicity and protects mice against the development of further tumors. *Gene Ther* **3**, 212–216.

Baker SJ, Fearon ER, Nigro JM, *et al.* (1989). Chromosome 17 deletions and *p53* gene mutations in colorectal carcinomas. *Science* **244**, 217–221.

Baker SJ, Markowitz S, Fearon ER, Willson JK, Vogelstein B (1990). Suppression of human colorectal carcinoma cell growth by wild-type p53. *Science* **249**, 912–915.

Bouvet M, Ellis LM, Nishizaki M, *et al.* (1998). Adenovirus-mediated wild-type *p53* gene transfer down-regulates vascular endothelial growth factor expression and inhibits angiogenesis in human colon cancer. *Cancer Res* **58**, 2288–2292.

Bowen GP, Borgland SL, Lam M, Libermann TA, Wong NC, Muruve DA (2002). Adenovirus vector-induced inflammation: capsid-dependent induction of the C-C chemokine RANTES requires NF-kappa B. *Human Gene Ther* **13**, 367–379.

Buckbinder L, Talbott R, Velasco-Miguel S, *et al.* (1995). Induction of the growth inhibitor IGF-binding protein 3 by p53. *Nature* **377**, 646–649.

Buller RE, Shahin MS, Horowitz JA, *et al.* (2002). Long-term follow-up of patients with recurrent ovarian cancer after Ad*p53* gene replacement with SCH58500. *Cancer Gene Ther* **9**, 567–572.

Chan TA, Hermeking H, Lengauer C, Kinzler KW, Vogelstein B (1999). 14-3-3Sigma is required to prevent mitotic catastrophe after DNA damage. *Nature* **401**, 616–620.

Claudio PP, Fratta L, Farina F, *et al.* (1999). Adenoviral *RB2/p130* gene transfer inhibits smooth muscle cell proliferation and prevents restenosis after angioplasty. *Circ Res* **85**, 1032–1039.

Clayman GL, el Naggar AK, Lippman SM, *et al.* (1998). Adenovirus-mediated *p53* gene transfer in patients with advanced recurrent head and neck squamous cell carcinoma. *J Clin Oncol* **16**, 2221–2232.

Dameron KM, Volpert OV, Tainsky MA, Bouck N (1994). The p53 tumor suppressor gene inhibits angiogenesis by stimulating the production of thrombospondin. *Cold Spring Harbor Symp Quant Biol* **59**, 483–489.

Demers GW, Harris MP, Wen SF, Engler H, Nielsen LL, Maneval DC (1998). A recombinant adenoviral vector expressing full-length human retinoblastoma susceptibility gene inhibits human tumor cell growth. *Cancer Gene Ther* **5**, 207–214.

Dorigo O, Turla ST, Lebedeva S, Gjerset RA (1998). Sensitization of rat glioblastoma multiforme to cisplatin *in vivo* following restoration of wild-type p53 function. *J Neurosurg* **88**, 535–540.

Dumon KR, Ishii H, Fong LY, *et al.* (2001). *FHIT* gene therapy prevents tumor development in Fhit-deficient mice. *Proc Natl Acad Sci U S A* **98**, 3346–3351.

el Deiry WS, Tokino T, Velculescu VE, *et al.* (1993). WAF1, a potential mediator of p53 tumor suppression. *Cell* **75**, 817–825.

Evan GI, Wyllie AH, Gilbert CS, *et al.* (1992). Induction of apoptosis in fibroblasts by c-myc protein. *Cell* **69**, 119–128.

Friend SH, Bernards R, Rogelj S, *et al.* (1986). A human DNA segment with properties of the gene that predisposes to retinoblastoma and osteosarcoma. *Nature* **323**, 643–646.

Fujiwara T, Grimm EA, Mukhopadhyay T, Cai DW, Owen-Schaub LB, Roth JA (1993). A retroviral wild-type *p53* expression vector penetrates human lung cancer spheroids and inhibits growth by inducing apoptosis. *Cancer Res* **53**, 4129–4133.

Fujiwara T, Cai DW, Georges RN, Mukhopadhyay T, Grimm EA, Roth JA (1994a). Therapeutic effect of a retroviral wild-type p53 expression vector in an orthotopic lung cancer model. *J Natl Cancer Inst* **86**, 1458–1462.

Fujiwara T, Grimm EA, Mukhopadhyay T, Zhang WW, Owen-Schaub LB, Roth JA (1994b). Induction of chemosensitivity in human lung cancer cells *in vivo* by adenovirus-mediated transfer of the wild-type *p53* gene. *Cancer Res* **54**, 2287–2291.

Goodwin WJ, Esser D, Clayman GL, Nemunaitis J, Yver A, Dreiling LK (1999). Randomized phase II study of intratumoral injection of two dosing schedules using a replication-deficient adenovirus carrying the *p53* gene (AD5CMV-P53) in patients with recurrent/refractory head and neck cancer. *Proc Am Soc Clin Oncol* **19**, 445a.

Grimberg A (2000). P53 and IGFBP-3: apoptosis and cancer protection. *Mol Genet Metab* **70**, 85–98.

Gu J, Kagawa S, Takakura M, *et al.* (2000). Tumor-specific transgene expression from the human telomerase reverse transcriptase promoter enables targeting of the therapeutic effects of the *Bax* gene to cancers. *Cancer Res* **60**, 5359–5364.

Gu J, Andreeff M, Roth JA, Fang B (2002). hTERT promoter induces tumor-specific *Bax* gene expression and cell killing in syngenic mouse tumor model and prevents systemic toxicity. *Gene Ther* **9**, 30–37.

Gurnani M, Lipari P, Dell J, Shi B, Nielsen LL (1999). Adenovirus-mediated *p53* gene therapy has greater efficacy when combined with chemotherapy against human head and neck, ovarian, prostate, and breast cancer. *Cancer Chemother Pharmacol* **44**, 143–151.

Harris H (1988). The analysis of malignancy by cell fusion: the position in 1988. *Cancer Res* **48**, 3302–3306.

Hermeking H, Eick D (1994). Mediation of c-Myc-induced apoptosis by p53. *Science* **265**, 2091–2093.

Hermeking H, Lengauer C, Polyak K, *et al.* (1997). 14-3-3 sigma is a p53-regulated inhibitor of G_2/M progression. *Mol Cell* **1**, 3–11.

Huang HJ, Yee JK, Shew JY, *et al.* (1988). Suppression of the neoplastic phenotype by replacement of the *RB* gene in human cancer cells. *Science* **242**, 1563–1566.

Ito I, Ji L, Tanaka F, *et al.* (2004). Liposomal vector mediated delivery of the 3p FUS1 gene demonstrates potent antitumor activity against human lung cancer *in vivo*. *Cancer Gene Ther* **11**, 733–739.

Ji L, Fang B, Yen N, Fong K, Minna JD, Roth JA (1999). Induction of apoptosis and inhibition of tumorigenicity and tumor growth by adenovirus vector-mediated fragile histidine triad (FHIT) gene overexpression. *Cancer Res* **59**, 3333–3339.

Jones NA, Turner J, McIlwrath AJ, Brown R, Dive C (1998). Cisplatin- and paclitaxel-induced apoptosis of ovarian carcinoma cells and the relationship between bax and bak up-regulation and the functional status of p53. *Mol Pharmacol* **53**, 819–826.

Kawabe S, Munshi A, Zumstein LA, Wilson DR, Roth JA, Meyn RE (2001). Adenovirus-mediated wild-type *p53* gene expression radiosensitizes non-small cell lung cancer cells but not normal lung fibroblasts. *Int J Radiat Biol* **77**, 185–194.

Kinzler KW, Vogelstein B (1997). Cancer-susceptibility genes. Gatekeepers and caretakers. *Nature* **386**, 761–763.

Knudson AG, Jr (1971). Mutation and cancer: statistical study of retinoblastoma. *Proc Natl Acad Sci U S A* **68**, 820–823.

Lang FF, Bruner JM, Fuller GN, *et al.* (2003). Phase I trial of adenovirus-mediated *p53* gene therapy for recurrent glioma: biological and clinical results. *J Clin Oncol* **21**, 2508–2518.

Lebedeva S, Bagdasarova S, Tyler T, Mu X, Wilson DR, Gjerset RA (2001). Tumor suppression and therapy sensitization of localized and metastatic breast cancer by adenovirus *p53*. *Hum Gene Ther* **12**, 763–772.

Lee WH, Bookstein R, Hong F, Young LJ, Shew JY, Lee EY (1987). Human retinoblastoma susceptibility gene: cloning, identification, and sequence. *Science* **235**, 1394–1399,.

Li Y, Okegawa T, Lombardi DP, Frenkel EP, Hsieh JT (2002). Enhanced transgene expression in androgen independent prostate cancer gene therapy by taxane chemotherapeutic agents. *J Urol* **167**, 339–346.

Liu TJ, Zhang WW, Taylor DL, Roth JA, Goepfert H, Clayman GL (1994). Growth suppression of human head and neck cancer cells by the introduction of a wild-type *p53* gene via a recombinant adenovirus. *Cancer Res* **54**, 3662–3667.

Lowe SW, Ruley HE, Jacks T, Housman DE (1993). p53-dependent apoptosis modulates the cytotoxicity of anticancer agents. *Cell* **74**, 957–967.

Marshall CJ (1991). Tumor suppressor genes. *Cell* **64**, 313–326.

Mercer WE, Shields MT, Amin M, *et al.* (1990). Negative growth regulation in a glioblastoma tumor cell line that conditionally expresses human wild-type *p53*. *Proc Natl Acad SciU S A* **87**, 6166–6170.

Miyashita T, Krajewski S, Krajewska M, *et al.* (1994). Tumor suppressor p53 is a regulator of bcl-2 and bax gene expression *in vitro* and *in vivo*. *Oncogene* **9**, 1799–1805.

Morgenbesser SD, Williams BO, Jacks T, DePinho RA (1994). p53-dependent apoptosis produced by Rb deficiency in the developing mouse lens. *Nature* **371**, 72–74.

Morris GF, Bischoff JR, Mathews MB (1996). Transcriptional activation of the human proliferating-cell nuclear antigen promoter by p53. *Proc Natl Acad Sci U S A* **93**, 895–899.

Nakano K, Vousden KH (2001). *PUMA*, a novel proapoptotic gene, is induced by p53. *Mol Cell* **7**, 683–694.

Nemunaitis J, Swisher S G, Timmons T, *et al.* (2000). Adenovirus-mediated *p53* gene transfer in sequence with cisplatin to tumors of patients with non-small-cell lung cancer. *J Clin Oncol* **18**, 609–622.

Nguyen DM, Spitz FR, Yen N, Cristiano RJ, Roth JA (1996). Gene therapy for lung cancer: enhancement of tumor suppression by a combination of sequential systemic cisplatin and adenovirus-mediated *p53* gene transfer. *J Thorac Cardiovasc Surg* **112**, 1372–1376.

Nielsen LL, Lipari P, Dell J, Gurnani M, Hajian G (1998). Adenovirus-mediated *p53* gene therapy and paclitaxel have synergistic efficacy in models of human head and neck, ovarian, prostate, and breast cancer. *Clin Cancer Res* **4**, 835–846.

Nishizaki M, Meyn RE, Levy LB, *et al.* (2001). Synergistic inhibition of human lung cancer cell growth by adenovirus-mediated wild-type *p53* gene transfer in combination with docetaxel and radiation therapeutics *in vitro* and *in vivo*. *Clin Cancer Res* **7**, 2887–2897.

Nishizaki M, Sasaki J, Fang B, *et al.* (2004). Synergistic tumor suppression by coexpression of FHIT and p53 coincides with FHIT-mediated MDM2 inactivation and p53 stabilization in human non-small cell lung cancer cells. *Cancer Res* **64**, 5745–5752.

Oda E, Ohki R, Murasawa H, *et al.* (2000a). Noxa, a BH3-only member of the Bcl-2 family and candidate mediator of p53-induced apoptosis. *Science* **288**, 1053–1058.

Oda K, Arakawa H, Tanaka T, *et al.* (2000b). p53AIP1, a potential mediator of p53-dependent apoptosis, and its regulation by Ser-46-phosphorylated p53. *Cell* **102**, 849–862.

Owen-Schaub LB, Zhang W, Cusack JC, *et al.* (1995). Wild-type human p53 and a temperature-sensitive mutant induce Fas/APO-1 expression. *Mol Cell Biol* **15**, 3032–3040.

Peng Z (2005). Current status of gendicine in China: recombinant human Ad-p53 agent for treatment of cancers. *Hum Gene Ther* **16**, 1016–1027.

Pereira-Smith OM, Smith JR (1983). Evidence for the recessive nature of cellular immortality. *Science*, **221**, 964–966.

Polyak K, Xia Y, Zweier JL, Kinzler KW, Vogelstein B (1997). A model for p53-induced apoptosis. *Nature* **389**, 300–305.

Ponder B (1988). Cancer. Gene losses in human tumours. *Nature* **335**, 400–402.

Qin XQ, Livingston DM, Kaelin WG, Jr, Adams PD (1994). Deregulated transcription factor E2F-1 expression leads to S-phase entry and p53-mediated apoptosis. *Proc Natl Acad Sci U S A* **91**, 10918–10922.

Rajah R, Valentinis B, Cohen P (1997). Insulin-like growth factor (IGF)-binding protein-3 induces apoptosis and mediates the effects of transforming growth factor-beta1 on programmed cell death through a p53- and IGF-independent mechanism. *J Biol Chem* **272**, 12181–12188.

Ramesh R, Saeki T, Templeton NS, *et al.* (2001). Successful treatment of primary and disseminated human lung cancers by systemic delivery of tumor suppressor genes using an improved liposome vector. *Mol Ther* **3**, 337–350.

Roth JA, Nguyen D, Lawrence DD, *et al.* (1996). Retrovirus-mediated wild-type *p53* gene transfer to tumors of patients with lung cancer. *Nat Med* **2**, 985–991.

Roth JA, Grammer SF, Swisher SG, *et al.* (2001). Gene therapy approaches for the management of non-small cell lung cancer. *Semin Oncol* **28**, 50–56.

Sandig V, Brand K, Herwig S, Lukas J, Bartek J, Strauss M (1997). Adenovirally transferred *p16INK4/CDKN2* and *p53* genes cooperate to induce apoptotic tumor cell death. *Nat Med* **3**, 313–319.

Saxon PJ, Srivatsan ES, Stanbridge EJ (1986). Introduction of human chromosome 11 via microcell transfer controls tumorigenic expression of HeLa cells. *EMBO J* **5**, 3461–3466.

Schrump DS, Chen GA, Consuli U, Jin X, Roth JA (1996). Inhibition of esophageal cancer proliferation by adenovirally mediated delivery of p16INK4. *Cancer Gene Ther* **3**, 357–364.

Schuler M, Herrmann R, De Greve JL, *et al.* (2001). Adenovirus-mediated wild-type *p53* gene transfer in patients receiving chemotherapy for advanced non-small-cell lung cancer: results of a multicenter phase II study. *J Clin Oncol* **19**, 1750–1758.

Serrano M, Lin AW, McCurrach ME, Beach D, Lowe SW (1997). Oncogenic *ras* provokes premature cell senescence associated with accumulation of p53 and p16INK4a. *Cell* **88**, 593–602.

Shao N, Chai YL, Shyam E, Reddy P, Rao VN (1996). Induction of apoptosis by the tumor suppressor protein BRCA1. *Oncogene* **13**, 1–7.

Shaw P, Bovey R, Tardy S, Sahli R, Sordat B, Costa J (1992). Induction of apoptosis by wild-type p53 in a human colon tumor-derived cell line. *Proc Natl Acad Sci U S A* **89**, 4495–4499.

Spitz FR, Nguyen D, Skibber JM, Meyn RE, Cristiano RJ, Roth JA (1996). Adenoviral-mediated wild-type *p53* gene expression sensitizes colorectal cancer cells to ionizing radiation. *Clin Cancer Res* **2**, 1665–1671.

Stanbridge EJ (1976). Suppression of malignancy in human cells. *Nature* **260**, 17–20.

Swisher SG, Roth JA, Carbone DP (2002). Genetic and immunologic therapies for lung cancer. *Semin Oncol* **29**, 95–101.

Swisher SG, Roth JA, Nemunaitis J, *et al.* (1999). K. Adenovirus-mediated *p53* gene transfer in advanced non-small-cell lung cancer. *J Natl Cancer Inst* **91**, 763–771.

Symonds H, Krall L, Remington L, *et al.* (1994). p53-dependent apoptosis suppresses tumor growth and progression *in vivo*. *Cell* **78**, 703–711.

Tait DL, Obermiller PS, Hatmaker AR, Redlin-Frazier S, Holt JT (1999). Ovarian cancer *BRCA1* gene therapy: phase I and II trial differences in immune response and vector stability. *Clin Cancer Res* **5**, 1708–1714.

Tanaka H, Arakawa H, Yamaguchi T, *et al.* (2000). A ribonucleotide reductase gene involved in a p53-dependent cell-cycle checkpoint for DNA damage. *Nature* **404**, 42–49.

Tanaka N, Ishihara M, Kitagawa M, *et al.* (1994). Cellular commitment to oncogene-induced transformation or apoptosis is dependent on the transcription factor IRF-1. *Cell* **77**, 829–839.

Uno F, Sasaki J, Nishizaki M, *et al.* (2004). Myristoylation of the fus1 protein is required for tumor suppression in human lung cancer cells. *Cancer Res* **64**, 2969–2976.

Van Beusechem VW, van den Doel PB, Grill J Pinedo HM, Gerritsen WR (2002). Conditionally replicative adenovirus expressing p53 exhibits enhanced oncolytic potency. *Cancer Res* **62**, 6165–6171.

Vogelstein B, Kinzler KW (2004). Cancer genes and the pathways they control. *Nat Med* **10**, 789–799.

Wagner AJ, Kokontis JM, Hay N (1994). Myc-mediated apoptosis requires wild-type p53 in a manner independent of cell cycle arrest and the ability of p53 to induce p21waf1/cip1. *Genes Dev* **8**, 2817–2830.

Weissman BE, Stanbridge EJ (1983). Complementation of the tumorigenic phenotype in human cell hybrids. *J Natl Cancer Inst* **70**, 667–672.

Wolf JK, Bodurka DC, Gano JB, *et al.* (2004). A phase I study of Ad*p53* (INGN 201; ADVEXIN) for patients with platinum- and paclitaxel-resistant epithelial ovarian cancer. *Gynecol Oncol* **94**, 442–448.

Wu GS, Burns TF, McDonald ER III, *et al.* (1997). *KILLER/DR5* is a DNA damage-inducible p53-regulated death receptor gene. *Nat Genet* **17**, 141–143.

Wu X, Levine AJ (1994). p53 and E2F-1 cooperate to mediate apoptosis. *Proc Natl Acad Sci U S A* **91**, 3602–3606.

Xu H, Raafat el-Gewely M (2001). P53-responsive genes and the potential for cancer diagnostics and therapeutics development. *Biotechnol Annu Rev* **7**, 131–164.

Yu J, Zhang L, Hwang PM, Kinzler KW, Vogelstein B (2001). *PUMA* induces the rapid apoptosis of colorectal cancer cells. *Mol Cell* **7**, 673–682.

Yu JL, Rak JW, Coomber BL, Hicklin DJ, Kerbel RS (2002). Effect of p53 status on tumor response to antiangiogenic therapy. *Science* **295**, 1526–1528.

Zhang WW, Fang X, Mazur W, French BA, Georges RN, Roth JA (1994). High-efficiency gene transfer and high-level expression of wild-type p53 in human lung cancer cells mediated by recombinant adenovirus. *Cancer Gene Ther* **1**, 5–13.

Zhang WW, Alemany R, Wang J, Koch PE, Ordonez NG, Roth JA (1995). Safety evaluation of Ad5CMV-*p53 in vitro* and *in vivo*. *Hum Gene Ther* **6**, 155–164.

Zhang X, Multani AS, Zhou JH, *et al.* (2003). Adenoviral-mediated retinoblastoma 94 produces rapid telomere erosion, chromosomal crisis, and caspase-dependent apoptosis in bladder cancer and immortalized human urothelial cells but not in normal urothelial cells. *Cancer Res* **63**, 760–765.

Zhao R, Gish K, Murphy M, *et al.* (2000). Analysis of p53-regulated gene expression patterns using oligonucleotide arrays. *Genes Dev* **14**, 981–993.

Zoltick PW, Chirmule N, Schnell MA, Gao GP, Hughes JV, Wilson JM (2001). Biology of E1-deleted adenovirus vectors in nonhuman primate muscle. *J Virol* **75**, 5222–5229.

15

RNA Interference and Dominant Negative Approaches

Charlotte Moss and **Nick Lemoine**

15.1 Introduction

This chapter describes the mechanisms underlying dominant negative, antisense and RNA interference for anticancer therapy. Concentrating on RNAi, we explore some of the challenges faced in translating this technology into a clinical setting and using recent examples, examine viral vectors used as their delivery vehicles.

Dominant negative approaches inhibit the function of a cancer-associated protein, by the expression of a structurally similar protein that has lost its function and thus acts as a competitive inhibitor of the normal protein. Antisense techniques prevent the transcription and translation of cancer-associated genes using oligonucleotides. The ability of antisense oligonucleotides to inhibit gene expression by sequence-specific binding of complementary target mRNA was discovered in the 1970s. Despite continued efforts, these agents are not in widespread clinical use. The more recently discovered RNA interference pathway has rekindled optimism about the value of these approaches for therapeutic use. Experience gained from the development of antisense technology has informed the rapid progress seen in RNAi.

15.2 Oligonucleotide agents

There are three categories of oligonucleotide agents (illustrated in Figure 15.1):

1. Antisense oligonucleotides that recruit RNAse H to cleave target mRNA are termed enzyme-dependent antisense agents. This includes single-stranded DNA, RNA and phosphorothioate antisense oligos. Enzyme-independent antisense sequences cause steric hindrance of translation or splicing upon target binding. Steric blocking antisense includes 2′-*O*-alkyl (usually in chimeras with RNase-H dependent antisense), peptide nucleic acid (PNA), locked nucleic acid (LNA) and Morpholino antisense oligos. By definition, antisense molecules work in a 1:1 stoichiometric relationship to their target.

2. Ribozymes or deoxyribozymes are catalytically active nucleic acids which induce RNA cleavage of the phosphodiester backbone of the RNA target by transesterification. The hammerhead ribozyme is the most widely used in molecular biology. These molecules have the advantage that they can process and destroy a higher number of target molecules

Viral Therapy of Cancer Edited by Kevin J. Harrington, Richard G. Vile and Hardev S. Pandha

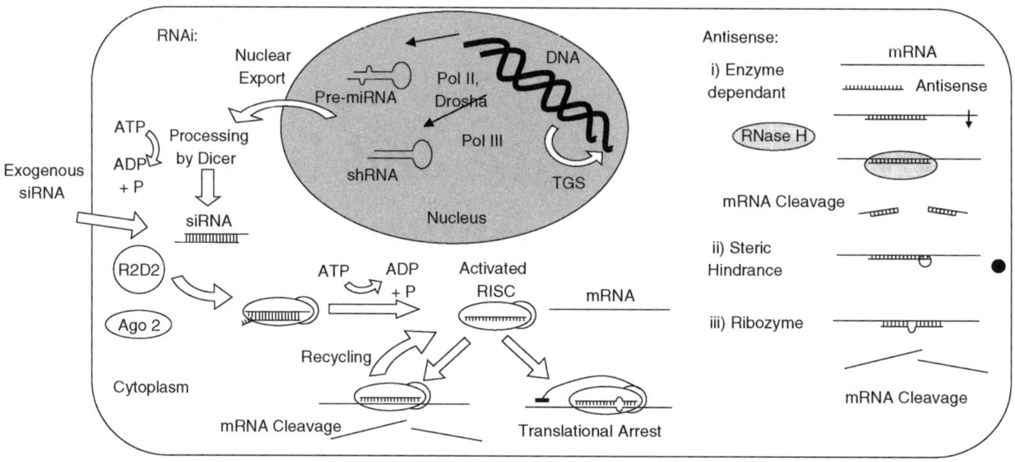

Figure 15.1 Mechanisms of NA mediated gene silencing: RNAi: microRNAs are transcribed from host DNA by Pol II as Pri-miRNA and processed by the nuclear RNase III enzyme Drosha to 60nt Pre-miRNA. Short hairpin RNA, (shRNA) is transcribed from vector DNA by pol III. The RNAi intermediates are exported from the nucleus by Exportin 5 and processed by the cytoplasmic Rnase III Dicer into 21nt duplexes. siRNAs from exogenous or endogenous sources are then bound by the dsRNA binding protein R2D2 forming RISC (RNA induced silencing complex). RISC activation involves unwinding the siRNA and requires ATP. The "guide" strand of the siRNA remains incorporated in the activated RISC, which probes mRNA for complementary sequences. Upon identification of a target, the degree of complementarity of the guide strand to the target determines the fate of the target strand. In the nucleus, transcriptional gene silencing (TGS) may also occur via an RNAi related mechanism, leading to RNA directed DNA methylation and histone modification. Antisense: i) Antisense oligonucleotide binds to the target mRNA which is then degraded by RNAse H. ii) Chemically modified antisense binds to target mRNA and blocks translation by steric hindrance iii) Ribozymes have intrinsic catalytic activity and upon binding cleave the target mRNA

per molecule of ribozyme. However, they require specific sequences for RNA cleavage, which limits their potential targets.

3. Short interfering RNA, (siRNA) are double-stranded RNAs that utilize the RNA interference pathway to induce gene silencing. RNAi is an endogenous pathway, evolutionarily highly conserved in eukaryotes. It appears to have developed as a defence against viral infection, and is now recognized as a key player in physiological post-transcriptional gene regulation. This natural pathway has been exploited by molecular biologists using either synthetic RNA duplexes, termed short interfering RNAs (siRNA) or vectors expressing short hairpin RNAs (shRNA) that mimic intermediates in the RNAi pathway resulting in efficient, specific gene silencing. Viral vectors present a naturally evolved, efficient and flexible mechanism of delivery for shRNA to mammalian cells. They also have the potential to provide the means for long term gene silencing through RNAi.

15.3 Mechanism of RNAi

RNAi is initiated when a cell encounters double-stranded RNA (dsRNA). This may originate from exogenous sources: either transcribed from an invading virus, or experimentally introduced synthetic RNA. Endogenous sources of dsRNA include non-coding microRNA (miRNA) genes or transposons. Primary miRNA (pri-miRNA) transcripts, >100 nucleotides (nt) in length, are processed in the nucleus into smaller (~60nt), stem loop structures called precursor miRNA, (pre-miRNA) by Drosha before being exported to the cytoplasm by Exportin 5. Short hairpin RNA, (shRNA) are siRNA-precursors that mimic ~50-mer pre-miRNA and are processed by DICER before entering RISC complex. shRNAs are encoded by exogenous vectors that may be integrated into the host cell genome.

In the cytoplasm, DICER (an RNAse III enzyme) cleaves dsRNA >30 basepairs into 21–23nt duplexes with a 2nt overhang at the 3′ end. Such duplexes are termed short interfering RNA (siRNA), or mature miRNA when derived from pre-miRNA. siRNAs, regardless of their origin, subsequently bind to a multiprotein nuclease complex to form the RNA-induced silencing complex or RISC. ATP-dependent unwinding of the siRNA duplex is required for activation of RISC, allowing either strand of the siRNA to act as a guide for recognition of its complementary target mRNA.

The fate of the target mRNA depends on the degree of complementarity of the guide strand to the target. siRNAs that are exactly complementary to their target cause site-specific cleavage of the mRNA target by the RNA endonuclease Ago2 (also known as 'slicer'). The cleaved mRNA is then released and degraded, leaving the activated RISC available to locate and cleave further target mRNA (Elbashir *et al.*, 2001).

Exogenous sequences are designed to target a unique sequence in the coding region of mRNA and are perfectly complementary to their target. Endogenous miRNAs tend to bind to the 3′ untranslated region (UTR) of their target mRNA. They bind with imperfect complementarity and repress gene expression by blocking translation of the target mRNA. Thus, both siRNA and miRNA assemble with and function via RISC; their mechanism of gene silencing is interchangeable, depending on the degree of base pairing to their target (Table 15.1).

Certain siRNA sequences induce transcriptional gene silencing (TGS) in yeast and plants, via the formation and maintenance of heterochromatin. Recently it has been shown that TGS can be induced in mammalian cells by dsRNA in a similar way (Morris, 2006). TGS occurs in the nucleus and is mediated by dsRNA sequences directed against gene promoter regions. This cause RNA- directed DNA methylation of the homologous promoter and/or covalent histone modification of the chromatin, resulting in long term gene suppression (Matzke and Birchler, 2005). The role of transcriptional gene silencing in mammalian gene regulation is not fully understood at present. However, the potential for utilizing this mechanism for long-term or even heritable gene suppression through siRNA-mediated epigenetic changes could be another exciting potential application of this technology (Lemoine, 2005).

RNAi was successfully used to induce targeted gene silencing in *Caenorhabditis elegans* and *Drosophila*, using long dsRNA (Fire *et al.*, 1998). However, early studies in mammalian cells were hampered by the interferon response, which is triggered by dsRNA duplexes >30 bp. The interferon response activates a cascade of interferon-stimulated genes (ISGs), which results in inhibition of protein synthesis and cell death. It can be evaded *in vitro* by reducing the dsRNA duplexes to 19–25 bp, first demonstrated in 2001 (Elbashir *et al.*, 2001).

15.4 RNAi and antisense compared

RNAi has been widely adopted as a standard technique for gene silencing; this has led to excitement regarding its potential for therapeutic use. It is favoured because of its superior potency, specificity and efficiency in comparison to previous antisense technologies. Lower concentrations of nucleic acid are required for effective gene silencing by RNAi. It has been estimated that the half maximal inhibition levels (IC_{50}) of siRNAs are 100–1000-fold lower than an optimised antisense oligonucleotide directed against the same target. More durable gene silencing (up to 10 mammalian cell divisions) can be achieved following transient transfection with siRNA (Dorsett and Tuschl, 2004).

Table 15.1 Contrasting miRNA and siRNA mediated gene silencing

	miRNA	siRNA
Source	Endogenous (genomically encoded)	Exogenous (viral/experimentally introduced)
Precursor	Pre-miRNAs (~70mers)	None or shRNA (short hairpin RNA ~ 50-mers)
Complementarity to target	Incomplete	Perfect/near perfect
Binding to target	Usually at 3′ UTR	Anywhere on target – usually in coding region
Effect	Block translation	Site specific cleavage of target mRNA by Slicer/Ago 2

Antisense strategies are also limited by difficulty in finding effective sequences. This largely depends on the secondary structure of the target mRNA; efficient binding of antisense RNA can also be blocked by binding of proteins to the mRNA, which can be difficult to predict. Thus, identification of an active antisense is a somewhat trial and error process, often requiring screening of 20–40 antisense oligonucleotides to find an active sequence (Morgan *et al.*, 1993; Lebedeva and Stein, 2001). Hopefully, computational design tools can improve this and a recently founded AS database and design tool is freely available via http://www.bioit.org.cn/ao/aobase (Bo *et al.*, 2006).

RNAi has a higher success rate in finding a potent siRNA sequence, (about one in four) and allows silencing of a wider range of genes than antisense methods. The knockdown of multiple genes is also readily achieved with RNAi, facilitating the dissection of networks of gene interactions.

Compared to antisense methods, RNAi also has advantages in specificity. The efficiency and robustness of RNAi technology has enabled its development for functional genomic screening, where libraries of RNAi molecules directed against thousands of genes are used to identify novel components in a given pathway (Berns *et al.*, 2004; Ngo *et al.*, 2006). Such libraries are also used to screen for 'synthetic lethal' gene–gene interactions, to identify genes that become essential for the survival of cancer cells, as new drug targets (Ngo *et al.*, 2006).

15.5 siRNA design

siRNAs consist of 21bp dsRNA with symmetrical 3′ hydroxyl groups and 5′ phosphates at each end. The thermodynamic stability of the first few base pairs of either end determines which end unwinds more easily and which strand incorporates into the RISC complex. For efficient, specific silencing, it is critical that the siRNA antisense strand is incorporated into RISC and acts as the guide to target cleavage of sense mRNA. Therefore low thermodynamic stability at the 5′ end of the antisense strand is preferred (Figure 15.2).

The 5′ end of the antisense siRNA strand is known as the 'seed region', this area directs the specificity of the siRNA; in some circumstances only seven contiguous nucleotides are sufficient for recognition of a homologous target for gene silencing (Jackson *et al.*, 2006). Cleavage of the target mRNA is carried out by Ago2/Slicer, and occurs between nt positions 10 and 11 on the guide strand, (counting from the 5′end of the guide). A single base pair change here is not tolerated and results in loss of silencing effect (Brummelkamp *et al.*, 2002b).

The following web-based tool allows free siRNA design following registration: http://jura.wi.mit.edu/bioc/siRNAext/ (Yuan *et al.*, 2004). Some companies (e.g. http://www.dharmacon.com) have designed siRNA sequences for the entire genome and guarantee gene silencing by their reagents. However, validation of the efficacy of the sequences in model systems remains an essential step.

The folding of the target mRNA is crucial in antisense design and the secondary structure of the RNA target can also influence the efficacy of an siRNA. Westerhout *et al.* demonstrated that a variant of the human immunodeficiency virus (HIV)-1 virus was able to escape the effects of a previously potent siRNA by evolving a mutation upstream of the target sequence. The mutation stabilized an alternative secondary structure in the target region, rendering it inaccessible to the siRNA (Westerhout *et al.*, 2005). Predictions of RNA secondary structure can be obtained from the mfold website, http://bioweb.pasteur.fr/seqanal/interfaces/mfold-simple.html (Zuker, 2003). Such algorithms may in the future be incorporated into the rational design of siRNAs and antisense oligonucleotides.

15.6 Off-target effects

Investigation of antisense techniques have shown non-specific silencing effects are encountered following antisense mediated gene silencing. This is largely due to 'irrelevant cleavage' by RNAse H, a relatively low stringency enzyme that is able to cleave mRNA with as few as 4/5 contiguous base pairs of homology to the intended target (Monia, 1997). Furthermore, application of microarray technology to examine the specificity of antisense

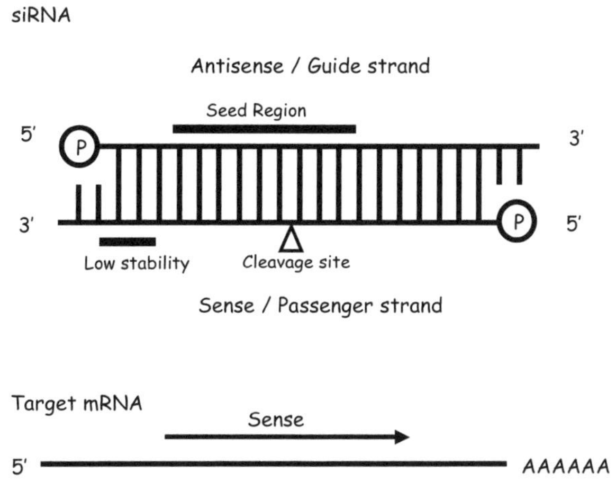

Figure 15.2 Anatomy of an siRNA: siRNAs consist of 21bp dsRNA with symmetrical 3′ hydroxyl groups and 5′ phosphates at each end. The thermodynamic stability of the first few base pairs of either end determines which end unwinds more easily and which strand incorporates into the RISC complex. For efficient, specific silencing, it is critical that the siRNA antisense strand is incorporated into RISC and acts as the guide to target cleavage of sense mRNA. Therefore low thermodynamic stability at the 5′ end of the antisense strand is preferred. The 5′ end of the antisense strand is known as the "seed region", this area directs the specificity of the siRNA; in some circumstances only 7 contiguous nucleotides are sufficient for recognition of a homologous target for gene silencing, (Jackson et al.,2006). Cleavage of the target mRNA is carried out by Ago2 (Argonaut 2 / Slicer), and occurs between nt positions 10 and 11 on the guide strand, (counting from the 5′end of the guide). A single base pair change here is not tolerated and results in loss of silencing effect, (Brummelkamp *et al.*, 2002). However, minor alterations to the siRNA such as chemical modification of some nucleotides is tolerated elsewhere, this can be exploited for example to alter the siRNA stability; prolonging its effect, without altering the specificity of gene silencing

oligonucleotide treatments revealed significant changes in expression in ~2 per cent of genes tested of comparable magnitude to the downregulation of the intended target mRNA. (Fisher *et al.*, 2002)

Similarly, closer examination of siRNA effects has revealed unintentional silencing of genes with partial homology to the intended target. Given the interchangeable nature of siRNA and miRNA, it is not surprising that genome-wide microarray analysis of gene expression following siRNA showed off-target effects. Jackson et al found that off-target gene silencing can occur in mRNAs with as few as seven contiguous matches to the siRNA seed region at the 5′ end of the antisense strand (Jackson *et al.*, 2006). It appears that the off-target effects on gene expression observed are less pronounced than the silencing of the intended target and can be partially abrogated by minimizing the dose of siRNA. These findings have led to further refinements in siRNA design criteria with more sophisticated sequence comparison tools to filter and exclude candidate siRNAs containing sequences homologous to non-target mRNA. These are based on the NCBI BLAST website (http://ncbi.nih.gov/BLAST/) (Jackson *et al.*, 2003, Reynolds *et al.*, 2004).

For functional genetic studies, two independent siRNA sequences targeting the same gene are often used to control for unforeseen off-target effects. It is also essential that low doses of potent siRNA sequences be used (e.g. ≤20 nM; Semizarov *et al.*, 2003). A number of studies have elegantly controlled for off target effects by validating the specificity of their knockdown phenotype by re-introducing the target gene in an siRNA-resistant form and demonstrating a recovery of the wild type phenotype despite ongoing siRNA expression. Such 'functional controls' are deemed the gold standard for RNAi experiments. At present, few *in vivo* studies have employed more than one siRNA against the same target as a control for off target effects.

Although the phenomenon of off target effects due to unintended activity of siRNAs was demonstrated by expression microarray, if siRNAs can induce miRNA-like translational suppression, the full ramifications of this would only be detectable by a proteomics approach. One study has reported unexpected off target effects seen more dramatically at the protein level than at mRNA level. These appeared to be unrelated to the degree of target gene silencing (Scacheri *et al.*, 2004).

The recent findings that siRNAs directed against gene promoter sequences can induce transcriptional gene silencing (TGS), raises the possibility that it might also be prudent to screen and exclude further siRNA designs bearing homology to genomic DNA sequences rather than simply exclude homologous cDNA sequences through a BLASTn search when designing siRNAs, although this is not done routinely at present.

In addition to predictable off target effects due to mRNA sequences bearing partial homology to the intended target, it has been proposed that introduction of siRNAs could saturate the RNAi machinery and antagonise the function of endogenous miRNA resulting in up-regulation of genes usually controlled by miRNA. This would be expected to produce dose-dependent changes in gene expression rather than sequence-dependent effects. Although this is not thought to be a prominent feature in cell culture systems where extremely low doses of siRNA are used, such effects could potentially be more relevant in whole animal studies and clinical applications because of the wider variation in the siRNA dose delivered to different tissues and variation in the extent to which different cell types rely on miRNA-mediated gene regulation. The effects of saturating doses of siRNA may also be more relevant in clinical studies if a cocktail of siRNAs against multiple targets is used in a bid to avoid the emergence of resistance, increase the efficacy of suppression or target multiple pathways simultaneously.

A recent report found significant morbidity and mortality in mice due to saturation of the shRNA/miRNA pathway following intravenous injection of an optimized shRNA delivery vector based on dsDNA containing adeno-associated virus serotype 8 (AAV8). The optimized vector was engineered to consist of a dsAAV2 genome pseudotyped with AAV8 capsids. One AAV2 DNA packaging signal was replaced with one from AAV4, yielding a 'stabilized double-stranded' (sds) vector. The shRNA sequences were expressed from a human U6 promoter (Grimm *et al.*, 2006).

Sustained high-level expression of shRNA resulted in dose-dependent hepatotoxicity in 36/49 (73 per cent) of mice tested; this led to death within 2 months in 23 mice (47 per cent) due to liver failure. These effects were not dependent on the shRNA sequence used since 49 different vectors against 6 targets were used. Greater toxicity was noted from expression of longer stem sequences in the shRNA as follows; 25-mer > 23-mer > 21-mer > 19-mer. No evidence of activation of the interferon pathway was found following shRNA expression in the affected mice. A direct correlation was found between high levels of shRNA and toxicity. Morbidity was associated with a specific reduction in levels of liver microRNAs indicating possible competition for limiting, shared cellular co-factors. Nuclear Exportin 5 was identified as a limiting, shared component of the shRNA/miRNA pathway and overexpression of Exportin 5 was shown in this and other studies to enhance shRNA silencing (Grimm *et al.*, 2006, Yi *et al.*, 2005).

More encouragingly, when the less toxic, shorter shRNA sequences were delivered at a lower dose, below the threshold for saturation of the shRNA/miRNA saturation, long-term gene silencing was achieved without toxic effects. Notably, one 19-mer sequence directed against human α-1 anti-trypsin (hAAT), administered at a dose of 10^{11} particles gave efficient and persistent hAAT knockdown in mice for over a year (Grimm *et al.*, 2006).

15.7 Induction of innate immunity

As previously stated, long dsRNA sequences trigger an interferon 1 response in mammalian cells. Delivery of siRNAs under 30 bp avoids interferon secretion via OAS (2′5′oligoadenylate synthetase) and prevents this. However subsequent studies using microarray expression analysis revealed that under certain circumstances siRNAs of 21 bp can induce the expression of a subset of ISG genes. This partial interferon response seems to be dose dependent and sequence dependent and does not cause cell death (Bridge *et al.*, 2003; Sledz *et al.*, 2003). It appears that further receptors of the innate immune system are capable of recognizing siRNAs, these include: Toll-like receptors, (TLR3, TLR7, TLR8 interact with RNA) dsRNA protein kinase (PKR) and retinoic acid-inducible gene 1 (RIG1).

The Toll-like receptors exist in the endosomal/lysosomal compartments of dendritic cells, where they respond to sequence-specific triggers, recognising short nucleotide motifs that resemble viral sequences. Transfection of siRNA with lipid complexes delivers them to the endosomal compartment. It has been shown that liposomal transfection of siRNA is more immunostimulatory in mouse models than injection of naked siRNA and this may be mediated by TLR signalling (Judge *et al.*, 2005).

Sledz *et al.* found a dose-dependent activation of the Jak-Stat pathway mediated by the dsRNA dependent protein kinase (PKR) on examining the effects of two independent siRNA sequences (Sledz *et al.*, 2003).

More recently, blunt ended 21–27 nt dsRNA oligonucleotides were shown to activate components of the interferon system via RIG1, leading to cell toxicity. This response was avoided by transfection of dsRNA sequences containing 2nt 3′ overhangs an effect that depends on the terminal structure of the siRNA rather than the sequence. Interestingly, it was also noted that certain cancer cell lines (293T, HT1080) appear to be deficient in parts of the RIG1 pathway; however the response was restored in these cells by over-expression of RIG1 (Marques *et al.*, 2006).

Since certain sequence motifs are known to enhance triggering of the interferon cascade via TLRs, these sequences can be avoided at the siRNA design stage to evade the host response. It has also been reported that the occurrence of non-specific effects can be reduced by the introduction of a central 'bulge' in the siRNA duplex by introducing a single nucleotide mutation into the passenger/sense strand. This strategy is thought to avoid activation of dsRNA-responsive cell signalling by making the duplex more closely resemble native miRNA (Cullen, 2006).

An alternative approach to evade the host cell interferon response has been developed, aimed at blocking the nuclear export of vector-mediated expression of long dsDNA. A novel RNA polymerase II based vector was made, (pDECAP) which produces a several hundred base pair dsRNA transcript lacking a 5′ cap and poly A tail, which prevents export of the dsRNA to the cytoplasm. Injection of embryonic stem cells constructs targeting the transcriptional co-repressor Ski exhibited a similar phenotype to Ski deficient embryos (Shinagawa and Ishii, 2003).

It remains important to control carefully for siRNA effects before drawing conclusions regarding a gene knockdown phenotype. Therefore it is preferable to compare the effects of more than one independent siRNA/antisense sequence directed against the same target. Given that sequence-independent triggering of interferon may also occur, it may be difficult to define a truly negative control siRNA sequence, so serum levels of cytokines and interferon should be checked in whole animal experiments. A number of *in vivo* studies have included measurements of serum cytokines in their analyses with encouraging results.

15.8 Methods of delivery

15.8.1 Synthetic siRNA duplexes

The biggest challenge facing RNAi technology as a therapeutic agent is *in vivo* delivery of siRNA. RNAs do not readily cross the cell membrane because of their high molecular mass and high negative charge. The main delivery strategies employed use either chemically synthesized siRNA or DNA vectors expressing shRNA. A variety of methods have been used to introduce 21bp synthetic siRNAs into cells. The standard method for cells in culture is via transfection of RNA mixed with lipoprotein complexes, e.g. Oligofectamine. The efficiency of transfection is cell line-dependent. Other methods such as electroporation are also employed in difficult to transfect cell lines.

In whole animal studies, local administration of naked siRNA duplexes has shown some encouraging results. For systemic use, further strategies are in development to increase the circulating half-life of siRNA and target them appropriately. The obvious drawback to the use of transient transfection of siRNA as an anticancer strategy is that the effects are short lived. The duration of silencing following transfection depends partly on degradation of siRNA by endonucleases present in serum but also on the rate of cell division. It is thought that once in the cell, binding to the RISC complex protects siRNA from nuclease attack, allowing greater persistence of gene silencing than antisense sequences. Typically, siRNA gene silencing will last 5–7 days in cultured cells. Rapidly dividing cells dilute the pool of activated RISC until it is no longer at an effective dose, which implies that cancer cells may naturally clear siRNAs faster than normal cells.

15.8.2 Short hairpin RNA/vector-based methods

The duration of RNAi effect may be prolonged through the stable transfection of DNA vectors, which direct the expression of siRNA or shRNA sequences. Short hairpin RNA, (shRNA) are siRNA-precursors that are processed by DICER before entering the RISC complex (see Figure 15.1). Commonly used promoters are RNA Polymerase II or III (Pol II or Pol III). Relatively little cloning capacity is required since both the promoter and transcripts are short, meaning that a wide range of viral vectors are suitable for such purposes and perhaps paving the way for expression of multiple shRNA sequences from the same vector.

The prototype short hairpin vector, pSUPER was developed in 2002; it directs the expression of RNA sequences, which spontaneously form stem loop structures, or shRNA. Using an RNA polymerase III promoter, a 49-nt precursor transcript is synthesized consisting of two 19-nt 'target' sequences separated by a short spacer and followed by five thymidines (T5) as a termination signal. The stem-loop precursor RNA mimics endogenous Pre-miRNA and is processed in the cell by Dicer to siRNA into an effective 21nt siRNA containing the sense and antisense target sequences. An antibiotic resistance marker in the vector allows the selection of successfully transfected cells and stable expression of the shRNA can be maintained (Brummelkamp *et al.*, 2002b).

DNA vectors are delivered to cells by transfection or by electroporation, although this is relatively toxic to cells. Here we will describe some of the viral approaches of shRNA vector delivery.

15.8.3 Viral delivery of vector-based shRNA

Given that viruses have evolved as machines for delivery of nucleic acids into cells it seems natural that attempts should be made to exploit their potential as a platform for effective therapeutic RNAi (Dallas and Vlassov, 2006). Viral systems may offer solutions to many of the problems surrounding the therapeutic use of AS/RNAi, such as efficient transduction in difficult to transfect cell types (e.g. primary cell lines and non-dividing cells) and delivery of RNA to the cytoplasm rather than the endosome. The use of conditional or cell selective promoters, conditionally replicating viruses and viral tropism offers flexibility and the prospect of targeting RNA to selected cells. Integrating viruses offer the potential for stable gene silencing.

RNAi is a good candidate for a new form of gene therapy as the transgenes are short and potent. However, the toxicity and long term effects of RNAi mediated gene silencing are unknown and the possibility of the emergence of resistance has also been raised.

The viral delivery systems that have been designed for RNAi gene silencing include: adenoviral (AdV), AAV and lentiviral vectors (LV), commonly using the Pol III promoters U6 and HI. These lie upstream of the transcript and are ubiquitously expressed and constitutively active (Raoul *et al.*, 2006).

Viral vectors usually drive the expression a shRNA transcript which self anneals into a hairpin configuration (Figure 15.3i), alternatively two promoters are used in tandem to drive expression of the sense and antisense siRNA strands separately (Figure 3ii). The tandem expression of independent siRNA strands results in lower efficiency of gene silencing due to less efficient annealing of the siRNA and export to the cytoplasm (Miyagishi *et al.*, 2004).

A reporter gene expression cassette can be added to allow detection of transduced cells in which silencing occurs, enabling the comparison of transduction rates between cells infected with control or active constructs (Figure 15.3iii). HIV-lentiviral and -adenoviral vectors were used to target Skp-2 in human small cell lung carcinoma cell lines that overexpress Skp-2, a

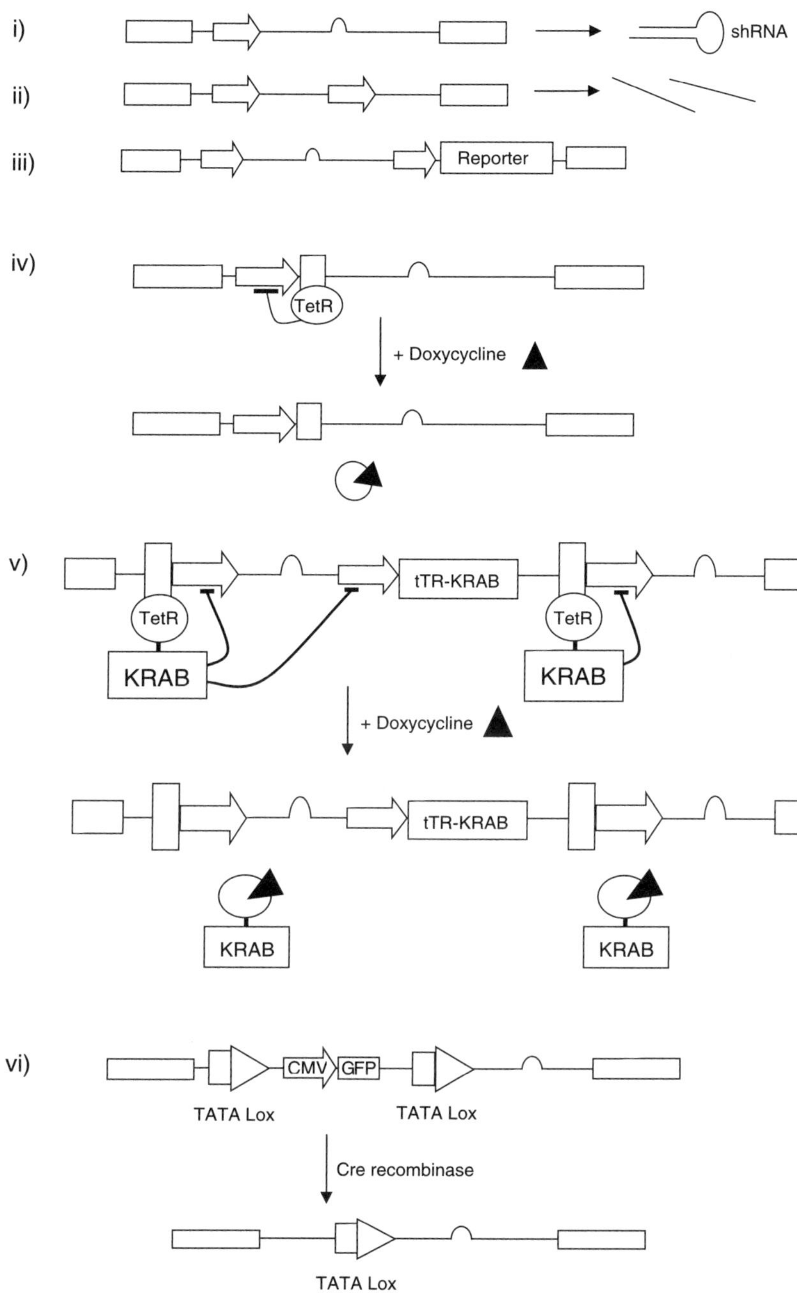

Figure 15.3 Viral vectors for RNAi. This scheme illustrates the various forms of viral RNAi vectors described in the text. (i) shRNA expressed in its simplest form under the control of a Pol III or Pol II promoter. (ii) siRNA strands arranged in tandem for separate expression. (iii) A reporter gene is added. (iv) Conditional expression via the Tet inducible system. (v) Conditional single LV vector platform using the KRAB-TetR fusion protein. (vi) Cre-Lox dependant integration and activation of shRNA expression (LV)

substrate-recognition subunit of the stem cell factor (SCF) ubiquitin–protein ligase complex involved in $p27^{kip1}$ degradation, which is frequently over-expressed in small cell lung cancer. HIV vectors were constructed from an HIV-U6i–green fluorescent protein (GFP) plasmid containing two expression units: in the first an shRNA expression cassette under the human U6 promoter, in the second GFP was transcribed from the cytomegalovirus (CMV) promoter. For the adenoviral vectors in the study the vector plasmid pAdF35 and the shuttle vector plasmid pHMCMV-GFP1 were used to create a construct consisting of a CMV promoter driving GFP with the bovine growth hormone (BGH) poly (A) signal. Downstream from this, the shRNA expression unit was located. At an multiplicity of infection (MOI) of 100, HIV-lentivirus mediated-RNAi for Skp-2 resulted in efficient inhibition of the *in vitro* cell growth of cancer cells with increased Skp-2 (ACC-LC-172) but no significant effect on the growth of cells without high Skp-2 expression (SBC-1). Intratumoral injection (1×10^8 IFU administered every 2 days, a total of three times) of adenovirus siRNA vector for Skp-2 efficiently inhibited growth of established sub-cutaneous tumour on non-obese diabetic/severe combined immunodeficient mice (Sumimoto *et al.*, 2005).

Shuttling the shRNA insert between different vectors allows variation of the promoter used to drive expression. Inducible systems of shRNA expression in cells have been developed to allow conditional knockdown of target genes. The first report of this approach used a doxycycline-regulated form of the Pol III H1 promoter to drive expression of shRNA targeting β-Catenin in colorectal cancer cells (Figure 15.3iv). The tetracycline operator (TetO) sequence was situated between the H1 promoter and shRNA sequence. Cells stably transfected with plasmids expressing the Tet repressor were subsequently transfected with pTER-B-catenin constructs. Specific inhibition of β-catenin ativity was rapidly induced by addition of doxycycline in these cells (van de Wetering *et al.*, 2003). An almost identical design was subsequently used with an adenoviral vector (Ad-H1tetO plus Ad-TR) to produce regulated silencing of p53 and c-Myc in A549 cells (Hosono *et al.*, 2004).

A novel lentiviral vector has also been used to produce a Dox-dependent system of shRNA expression. The tetracycline repressor (tetR) was fused to the vertebrate repressor Kruppel associated box domain, (KRAB) to produce tTRKRAB fusion protein, (Figure 15.3v). KRAB is a zinc finger DNA binding protein that induces epigenetic repression over an adjacent 2–3Kb region. An shRNA expression cassette (H1-shRNA), inserted downstream of a TetO sequence was placed adjacent to the KRAB based repressor, and cloned into the 3′LTR of a self-inactivated lentivirus, to create a Dox- inducible single lentiviral vector platform. Following genomic integration, tightly controlled, reversible regulation of endogenous TP53 in MCF7 cells and in a mouse xenograft model was demonstrated on addition of doxycycline (Szulc *et al.*, 2006).

A cell-specific promoter-based RNAi system has been constructed using an alveolar epithelial type II cell-specific promoter for surfactant protein-C (SP-C). Adenoviral vectors were constructed based on pA/PL-DEST vector. shRNA sequences were cloned into a pENTR/D-Topo vector construct in between the SP-C promoter and poly A tail to produce three adenoviral vectors containing shRNAs targeting lamin A/C, Annexin 2, or GFP. These were shown to specifically down-regulate expression of targeted genes in alveolar type II cells using cell and organ culture *in vitro* and *in vivo* in rats, without silencing genes in other pneumocytes. It was suggested that similarly, a cancer-associated promoter (for example human telomerase RT, hTERT) could be used for cancer cell-specific shRNA-mediated oncogene silencing (Gou *et al.*, 2004).

Conditionally replicating adenoviruses, (CRAds) which selectively replicate in neoplastic cells have been developed as delivery vehicles to target the expression of shRNA to cancer cells. A U6-Ffl1 shRNA expression cassette targeting exogenous Firefly luciferase 1 was inserted into the genome of Ad5-24E3, a CRAd whose replication is restricted to cancer cells with a disrupted Rb pathway. This was tested in four different cell lines (A549, MCF7s, HeLa and SaOs-2 cells) using an MOI of 500, 40–70% silencing was achieved, demonstrating that shRNAs expressed from

CRAds are able to suppress the expression of the target gene *in vitro* (Carette *et al.*, 2004).

Certain viral-based DNA vectors allow permanent expression of shRNA in mammalian cells since LV DNA expression cassettes integrate into the host cell genome and their expression will then be maintained indefinitely. The first demonstration of this was using pRETRO-SUPER, based on a retroviral vector (Brummelkamp *et al.*, 2002a). The Cre-LoxP system has been used to allow irreversible conditional activation of shRNA synthesis (Figure 15.3vi). In an LV vector, (pSico) the U6 promoter TATA box was replaced by a LoxP site, a second LoxP site was placed upstream of the shRNA. Following Cre recombinase action, GFP reporter activity is lost, one TATA Lox site remains, enabling U6 to drive expression of the shRNA. Cre-mediated recombination was demonstrated in MEFs by conditional knockdown of p53, nucleophosmin and DNA methyltransferase 1. Conditional, tissue-specific RNA interference in Cre-expressing transgenic mice was also shown (Ventura *et al.*, 2004).

However, translation of these techniques from *in vitro* to *in vivo* experiments raises all the risks and anxieties regarding the oncogenic potential of integration of DNA into the host genome. A possible solution to this may emerge through the development of AAV vector genomes engineered to increase persistence as episomal forms. However, since the effects of long-term expression of shRNA are unknown, anxieties will doubtless remain, making extensive testing of these therapies mandatory.

In addition to the anxieties surrounding gene therapy and genomic integration, viral based vectors have some other drawbacks. For example, it is more difficult to control the dosage of siRNA at a cellular level(s) this may result in greater or more variable toxicity, and with increased dosage there may be increased potential for non-specific effects and the triggering of innate immune responses.

15.8.4 Virosomes as siRNA delivery vehicles

Modified virus envelopes or virosomes have been employed as delivery vehicles for siRNA. Influenza virus envelopes were used to deliver fluorescent dye-labelled siRNAs by intraperitoneal injection into mice. Virosomes are vesicles that bear the influenza virus spike protein haemagglutinin (HA) in their membrane, which mediates the binding of the particles to cells. The virosomes are taken up by receptor-mediated endocytosis, and fuse with the endosomal membrane to release their contents into the cytoplasm (de Jonge *et al.*, 2006).

SV40 pseudovirions have also been shown to be capable of delivering siRNA to 45 human lymphoblastoid cells with greater efficiency than lipid transfection. VP1, a capsid protein from the simian virus 40 was used for *in vitro* packaging of pseudovirions containing siRNA (Kimchi-Sarfaty *et al.*, 2005).

15.9 Antisense

A number of studies have been carried out using viral vectors for delivery of antisense sequences. A recombinant adeno-associated virus (rAAV) vector expressing antisense HPV16E7 (AAV-HPV16E7AS), using a CMV promoter, was shown to inhibit cell proliferation *in vitro* and *in vivo* in HPV16/18 positive cervical cancer CaSki cells (Wu *et al.*, 2006).

An oncolytic adenovirus armed with an antisense sequence directed against chk2 it has shown promising results. M3, a novel *E1A CR-2* deleted Ad5 was engineered with a chk2 antisense sequence inserted into the deleted 6.7K/gp19K region. The virus preferentially replicates in a variety of cancer cell lines and silenced cancer-associated chk2. Systemic administration of M3 (2×10^5 plaque-forming units (p.f.u. for 5 days i.v.) followed by cisplatin (3 mg/kg/day i.p. for 4 days) proved superior to its parent virus in cisplatin-refractory hepatic xenograft mouse models. Complete response was seen in 9/12 (75 percent) mice treated with M3 + cisplatin, compared to 3/12 (25 per cent) CR in mice given the parent virus (Ad5/dE1A + cisplatin) (Chen *et al.*, 2006).

A number of AS ODNs are currently being tested in clinical trials, including AS ODN targeting *Bcl-2*, *XIAP* (X-linked inhibitor of apoptosis protein) and *TGF-beta-2* (transforming growth factor beta-2). None of these studies are currently using viral delivery systems. However, they have served to demonstrate that antisense strategies appear to be well tolerated in human subjects and can be safely combined with conventional chemotherapy, radiotherapy and newer targeted therapies (Lacasse *et al.*, 2005; Moore *et al.*, 2006).

The results of the phase III trial of oblimersen (Genasense), an antisense to *Bcl-2* were disappointing; 241 patients who had failed standard treatment for chronic lymphocytic leukaemia (CLL) were randomized to receive chemotherapy with fludarabine and cyclophosphamide with or without Genasense. The addition of Genasense significantly increased the proportion of patients with a major response (19 patients (16 per cent) who were treated with Genasense plus Flu/Cy achieved a complete remission CR or nodular partial remission (nPR), (11 and 8 patients, respectively), compared with 8 patients (7 per cent) who were treated with chemotherapy alone (3 CRs, 5 nPRs) ($P = 0.039$). However when partial responders (PRs) were included, the overall response rate was similar for both treatment arms (41 per cent for Genasense/Flu/Cy and 45 per cent for Flu/Cy; $P = \mathrm{NS}$) http://www.genta.com.

The only current FDA-approved antisense drug for human therapeutic use is a phosphorothioate antisense oligo, fomivirsen (Vitravene) targetting CMV IE2 for intravitreal injection to combat CMV retinitis in patients with HIV (2002).

15.10 Dominant negative approaches

Examples of virus-mediated dominant negative anti-cancer approaches include two studies using truncated insulin-like growth factor-I (IGF-I) receptors (IGF-Ir/dn) cloned into recombinant adenoviruses in murine models of human pancreatic cancer and gastric cancer. The viruses were constructed with the truncated IGF-Ir sequence under control of a CMV early enhancer and promoter and followed by the SV40 poly(A) site using the pAC shuttle plasmid and the vector plasmid pJM17 to create Ad-IGF-Ir/482st.

IGF-Ir/dn expression suppressed tumorigenicity *in vivo* and increased chemotherapy/radiation-induced apoptosis. Infection with Ad-IGF-Ir/482st blocked both IGF-I and IGF-II induced activation of *Akt-1*. Intratumoral injections of the adenovirus (1×10^8 p.f.u. on 5 successive days) given in combination with i.p. 5-FU (50 mg/kg) completely suppressed growth of established tumours. When given alone intraperitoneally to mice bearing intraperitoneal nodules of AsPC-1 or MKN45 cells, the number and size of tumour nodules was reduced by the adenovirus and survival was significantly increased in the treated mice (Min *et al.*, 2003, 2005).

15.11 Research applications of siRNA

RNAi has already demonstrated its value as a revolutionary tool in research for defining gene function and dissecting complex networks of regulation and gene function. High-throughput RNAi screens are now emerging in mammalian systems which will continue to accelerate progress in increasing our basic biological knowledge.

siRNA design has evolved so that sequences increasingly resemble native miRNA sequences, (for example by the introduction of mismatches in the passenger strand sequence and targeting the 3′ UTR). These subtle changes appear to increase efficiency and reduce non-specific effects of the siRNA, by evading host cell defences.

15.12 Therapeutic applications of siRNA

RNAi shows great potential as a novel therapeutic agent in a variety of diseases. Because of its huge potential and wide applicability, this new technology has generated lots of enthusiasm. Many pharmaceutical companies are investing in research aimed at finding techniques to improve delivery, targeting and durability of RNAi response.

Cancer cells present a fertile ground for siRNA therapy, since they contain many well-characterized gene targets, many of which would be applicable in multiple tumour types. RNAi has been used in preclinical studies targeting numerous cancer-related genes involved in oncogenesis, in tumour–host interactions and genes important for treatment resistance, (reviewed in Pai *et al.*, 2006).

The diversity of genetic lesions in cancer cells may present problems; making it difficult to find

RNAi targets with broad effectiveness in heterogeneous cells populations. The exquisite specificity of RNAi and the genetic instability characteristic of cancer cells leads to the prospect that *in vivo* RNAi will be a strategy prone to the development of resistance, since small changes in the target sequence may render a previously potent siRNA ineffective (Westerhout *et al.*, 2005). To combat the emergence of resistance, it has been proposed that RNAi therapy should aim to target multiple sequences in multiple genes. However, this strategy would not avoid the possibility that genes of the RNAi machinery could also become mutated and limit the effectiveness of RNA interference. There is evidence that a number of viruses have evolved anti-RNAi mechanisms for example; wild type adenoviruses block RNAi by expressing RNA at such high levels that it saturates the RNAi processing machinery, HIV tat encodes a suppressor of DICER (Dallas and Vlassov, 2006). It is conceivable that cancer cells could develop similar RNAi resistance mechanisms.

Although RNAi has been shown to be capable of not only aiding in the identification of new targets; and silencing these genes has produced encouraging results both *in vitro* and in preclinical animal models, the translation of these findings into the clinic raises multiple new challenges and it remains to be seen whether these can be overcome. Certainly, no time is being wasted in the drive to realize the therapeutic potential of RNAi. At present, the most promising results *in vivo* have been seen following localized administration of siRNA. The first phase I clinical trial of intravitreal siRNA injection targeting vascular endothelial growth factor receptor 1 (VEGFR1), to treat age-related macular degeneration is currently underway (Campochiaro, 2006).

As with any new therapy, siRNA will need rigorous testing to examine potential side effects, with particular vigilance for off-target effects and triggering of host immune responses, which are intrinsic properties of the RNAi pathway. Development will no doubt take time but the good news is that prior research in antisense technology (AS and ribozymes) may be of value in solving some of the problems common to oligonucleotide strategies (Lemoine, 2005).

References

(2002) *Am J Ophthalmol*, **133**, 467–74.

(2003) *Nat Cell Biol*, **5**, 489–90.

Berns K, Hijmans EM, Mullenders J, *et al.* (2004) *Nature* **428**, 431–7.

Bo X, Lou S, Sun D, Yang J, Wang S (2006). *Nucl. Acids Res* **34**, D664–667.

Bridge AJ, Pebernard S, Ducraux A, Nicoulaz AL, Iggo R (2003). *Nat Genet* **34**, 263–264.

Brummelkamp TR, Bernards R, Agami R (2002a). *Cancer Cell* **2**, 243–247.

Brummelkamp TR, Bernards R, Agami R (2002b). *Science* **296**, 550–553.

Campochiaro PA (2006). *Gene Ther* **13**, 559–562.

Carette JE, Overmeer RM, Schagen FHE, *et al.* (2004) *Cancer Res* **64**, 2663–2667.

Chen G, Zhou J, Gao Q, *et al.* (2006). Oncolytic adenovirus-mediated transfer of the antisense chk2 selectively inhibits tumor growth *in vitro* and *in vivo*. *Cancer Gene Ther* **13**10: 930–939.

Cullen BR (2006). *Gene Ther* **13**, 503–508.

Dallas A, Vlassov AV (2006). *Med Sci Monit* **12**, RA 67–74.

de Jonge J, Holtrop M, Wilschut J, Huckriede A (2006). *Gene Ther* **13**, 400–411.

Dorsett Y, Tuschl T (2004). *Nat Rev Drug Discov* **3**, 318–329.

Elbashir SM, Harborth J, Lendeckel W, Yalcin A, Weber K, Tuschl T (2001). *Nature* **411**, 494–498.

Fire A, Xu S, Montgomery MK, Kostas SA, Driver SE, Mello CC (1998). *Nature* **391**, 806–811.

Fisher AA, Ye D, Sergueev DS, Fisher MH, Shaw BR, Juliano RL (2002). *J Biol Chem* **277**, 22980–22984.

Gou D, Narasaraju T, Chintagari NR, Jin N, Wang P, Liu L (2004). *Nucleic Acids Res* **32**, e134.

Grimm D, Streetz KL, Jopling CL, *et al.* (2006). *Nature* **441**, 537–541.

Hosono T, Mizuguchi H, Katayama K, *et al.* (2004). *Hum Gene Ther* **15**, 813–819.

Jackson AL, Bartz SR, Schelter J, *et al.* (2003). *Nat Biotechnol* **21**, 635–637.

Jackson AL, Burchard J, Schelter J, *et al.* (2006). Widespread siRNA 'off-target' transcript silencing mediated by seed region sequence complementarity. *RNA* **12**7: 1179–1187.

Judge AD, Sood V, Shaw JR, Fang D, McClintock K, MacLachlan I (2005). *Nat Biotechnol* **23**, 457–462.

Kimchi-Sarfaty C, Brittain S, Garfield S, Caplen NJ, Tang Q, Gottesman MM (2005). *Hum Gene Ther* **16**, 1110–1115.

Lacasse EC, Kandimalla ER, Winocour P, *et al.* (2005) *Ann NY Acad Sci* **1058**, 215–234.

Lebedeva I, Stein CA (2001). *Annu Rev Pharmacol Toxicol* **41**, 403–419.

Lemoine NR (2005). *Gut* **54**, 1215–1216.

Marques JT, Devosse T, Wang D, *et al.* (2006). *Nat Biotechnol* **24**, 559–565.

Matzke MA, Birchler JA (2005). *Nat Rev Genet* **6**, 24–35.

Min Y, Adachi Y, Yamamoto H, *et al.* (2003). *Cancer Res* **63**, 6432–6441.

Min Y, Adachi Y, Yamamoto H, *et al.* (2005). *Gut* **54**, 591–600.

Miyagishi M, Sumimoto H, Miyoshi H, Kawakami Y, Taira K (2004). *J Gene Med* **6**, 715–723.

Monia BP (1997). *Anticancer Drug Des* **12**, 327–339.

Moore J, Seiter K, Kolitz J, *et al.* (2006). *Leuk Res* **30**, 777–783.

Morgan R, Edgem, Colman A (1993). *Nucleic Acids Res.* **21**, 4615–4620.

Morris KV (2006). *Biotechniques* Suppl, 7–13.

Ngo VN, Davis RE, Lamy L, *et al.* (2006). *Nature* **441**, 106–110.

Pai SI, Lin YY, Macaes B, Meneshian A, Hung CF, Wu TC (2006). *Gene Ther* **13**, 464–477.

Raoul C, Barker SD, Aebischer P (2006). *Gene Ther* **13**, 487–495.

Reynolds A, Leake D, Boese Q, Scaringe S, Marshall WS, Khvorova A (2004). *Nat Biotechnol* **22**, 326–330.

Scacheri PC, Rozenblatt-Rosen O, Caplen NJ, *et al.* (2004). *Proc Natl Acad Sci U S A* **101**, 1892–1897.

Semizarov D, Frost L, Sarthy A, Kroeger P, Halbert DN, Fesik SW (2003). *Proc Natl Acad Sci U S A* **100**, 6347–6352.

Shinagawa T, Ishii S (2003). *Genes Dev* **17**, 1340–1345.

Sledz CA, Holko M, de Veer MJ, Silverman RH, Williams BR (2003). *Nat Cell Biol* **5**, 834–839.

Sumimoto H, Yamagata S, Shimizu A, *et al.* (2005). *Gene Ther* **12**, 95–100.

Szulc J, Wiznerowicz M, Sauvain MO, Trono D, Aebischer P (2006). *Nat Methods* **3**, 109–116.

van de Wetering M, Oving I, Muncan V, *et al.* (2003). *EMBO Rep* **4**, 609–615.

Ventura A, Meissner A, Dillon CP, *et al.* (2004). *Proc Natl Acad Sci U S A* **101**, 10380–10385.

Westerhout EM, Ooms M, Vink M, Das AT, Berkhout B (2005). *Nucleic Acids Res* **33**, 796–804.

Wu S, Meng L, Wang S, *et al.* (2006). *Clin Cancer Res* **12**, 2032–2037.

Yi R, Doehle BP, Qin Y, *et al.* (2004). *Nucleic Acids Res* **32**, W130–134.

Zuker M (2003). *Nucleic Acids Res* **31**, 3406–3415.

16

Gene-directed Enzyme Prodrug Therapy

Silke Schepelmann, Douglas Hedley, Lesley M. Ogilvie and **Caroline J. Springer**

16.1 Introduction

Conventional chemotherapy treatments for cancer usually lack selectivity for tumour cells and it is often difficult to deliver sufficient levels of cytotoxic drug to malignant cells without adverse effects to normal tissue. Gene-directed enzyme prodrug therapy (GDEPT) is a suicide gene therapy approach that aims to improve the selectivity of chemotherapy by targeting cytotoxic drugs to tumour cells selectively, thus protecting normal cells from damage (Bridgewater *et al.*, 1995). GDEPT is a two-stage therapy (Figure 16.1). In the first step, a gene delivery vector is used to target an exogenous prodrug-activating enzyme to tumour cells. The aim is to express the enzyme efficiently and selectively in the tumour cells before administration of a prodrug, which is relatively non-toxic until converted into a cytotoxic drug by the enzyme. If the gene for the enzyme has been targeted successfully, the activated prodrug will be produced in the tumour cells only, killing them and sparing normal tissue from damage (Figure 16.1). Thus, higher concentrations of cytotoxic drug can be delivered to the tumour cells compared to conventional chemotherapy. GDEPT systems that use viral vectors are sometimes referred to as VDEPT (virus-directed enzyme prodrug therapy) (Huber *et al.*, 1991).

Current gene therapy vectors cannot target all cells within a tumour. Even a few surviving cancer cells are sufficient for re-growth of the tumour and therefore, a 'bystander effect' is required, in which the cells that express the prodrug-activating enzyme are able to lead to killing of the non-expressing neighbouring cells (Figure 16.1). The bystander effect results from the intercellular transfer of the activated prodrug, either by passive or active mechanisms (Greco and Dachs, 2001; Springer, 2004). The bystander effect is responsible for killing the uninfected cells within a tumour and the tumour-supporting cells, such as the endothelium of tumour blood vessels. Thus, 100 per cent gene transfer efficiencies are not required for success in GDEPT therapies.

16.2 Enzyme-prodrug systems for GDEPT

A large number of enzyme-prodrug combinations have been described (Greco and Dachs, 2001; Niculescu-Duvaz and Springer, 2004), many of which have been assessed in preclinical GDEPT studies (Table 16.1) or in clinical trials (Table 16.2). Examples of some of these combinations are described below.

16.2.1 Ganciclovir activated by thymidine kinase

Ganciclovir (2-amino-9-(1,3-dihydroxypropan-2-yloxymethyl)-3H-purin-6-one; GCV) is a guanosine analogue that is phosphorylated by the Herpes

Viral Therapy of Cancer Edited by Kevin J. Harrington, Richard G. Vile and Hardev S. Pandha

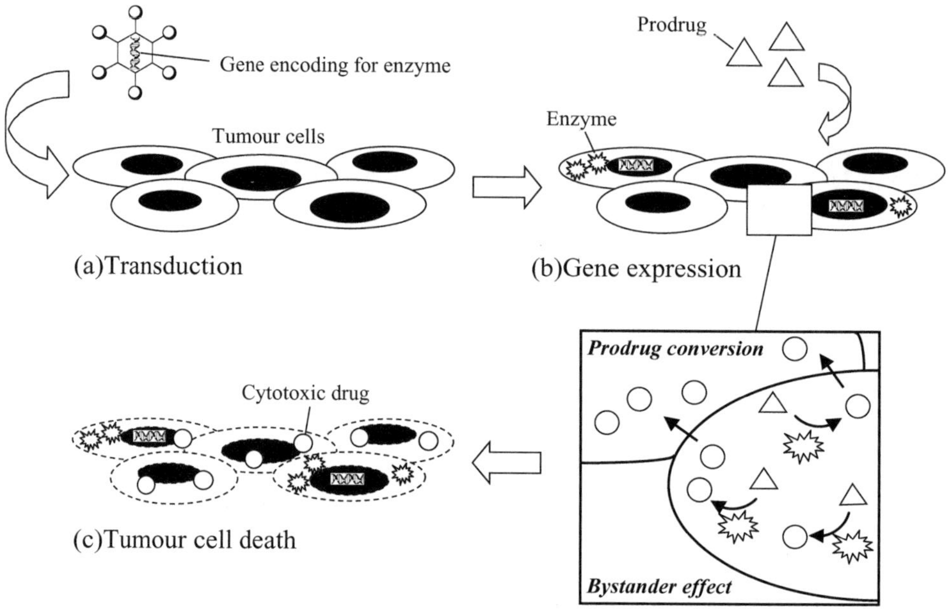

Figure 16.1 Schematic diagram of gene-directed enzyme prodrug therapy (GDEPT). (a) In the first step, the gene encoding for a prodrug-converting enzyme is delivered to the tumour cells by a viral vector. (b) Once sufficient levels of enzyme are present in the transduced cells, a non-toxic prodrug is administered. (c) The enzyme catalyses the conversion of the prodrug into an active, cytotoxic drug, which spreads to untransduced cells that do not express the enzyme ('bystander effect'), killing both expressing and non-expressing tumour cells

simplex type 1 (HSV)-thymidine kinase (TK). The resulting ganciclovir monophosphate is further phosphorylated by intracellular kinases to the triphosphate (GCV-TP), which is a potent cytotoxic agent (Rooseboom *et al.*, 2004). During the S-phase, GCV-TP is incorporated into the newly synthesized DNA, resulting in inhibition of DNA polymerase (Mar *et al.*, 1985) and DNA fragmentation (Hamzeh and Lietman, 1991). Cells treated with GCV exhibit irreversible G_2-M checkpoint arrest, possibly due to inactivation of the cdc2/cyclin B complex (Halloran and Fenton, 1998). There is evidence that the HSV-TK/GCV combination requires the presence of p53 to induce apoptosis (Krohne *et al.*, 2001) and gap-junctional intercellular communication is needed to produce a bystander effect in neighbouring uninfected cells (Fick *et al.*, 1995). Endogenous thymidine kinases exist in mammalian cells, however, the *Herpes simplex* enzyme catalyses the phosphorylation of GCV more efficiently (Moolten, 1986).

16.2.2 5-Fluorocytosine activated by cytosine deaminase

The prodrug 5-fluorocytosine (5-FC) is modified by cytosine deaminase (CD) to 5-fluorouracil (5-FU), which is a commonly used chemotherapy drug for a range of cancers. 5-FU is converted by nucleotide salvage pathways in the host cell to 5-fluorodeoxyuridylate (5-FdUMP), an irreversible inhibitor of the cellular thymidylate synthetase that is involved in the synthesis of DNA. Other cytotoxic metabolites of 5-FU are 5-fluorouridine 5′-triphosphate (5-FUTP) and 5-fluoro-2-deoxyuridine 5′-triphosphate (5-FdUTP), which inhibit RNA and DNA synthesis, respectively (Knox, 1997). The 5-FC/CD combination arrests dividing cells in the G_1 and early S phase (Nishiyama *et al.*, 1982). Moreover, Huber *et al.* (1994) observed a potent bystander effect of the treatment *in vitro*. CD is not expressed by mammalian cells and bacterial CD (*Escherichia coli* CD) or CD from

Table 16.1 Preclinical *in vivo* VDEPT studies

Enzyme	Vector	Prodrug	Tumour	Tumour type	Reference
TK	Replication-defective retrovirus	GCV	C6, 9L, BT4C	Rat glioma	(Ezzeddine *et al.*, 1991, Barba *et al.*, 1993, Sandmair *et al.*, 2000)
			MDAMB361, SW620 CaCo2	Human breast or colon carcinoma	(Mavria and Porter, 2001)
			ARO	Human thyroid carcinoma	(Barzon *et al.*, 2002)
			MCF-7	Human breast carcinoma	(Zeng *et al.*, 2006)
	Lentivirus	GCV	Diethylnitrosamine-induced rat liver carcinoma		(Gerolami *et al.*, 2004)
	Replication-competent retrovirus	GCV	CNS-1	Human glioma	(Solly *et al.*, 2003)
	Replication-defective adenovirus	GCV	SK-Hep-1	Human liver carcinoma	(Kaneko *et al.*, 1995)
			Diethylnitrosamine-induced rat liver carcinoma		(Gerolami *et al.*, 2000)
			RCM-1	Human colorectal carcinoma	(Okabe *et al.*, 2003)
	Oncolytic adenovirus	GCV	LS180, HT29	Human colon carcinoma	(Wildner *et al.*, 1999)
			HEp-2	Human head and neck carcinoma	(Morris and Wildner, 2000)
			A549, MDAH2774	Human lung or ovarian carcinoma	(Wildner and Morris, 2000)
			U-87	Human glioma	(Nanda *et al.*, 2001)
			A549	Human lung carcinoma	(Lambright *et al.*, 2001)
	Replication-defective HSV	GCV	U-87	Human glioma	(Miyatake *et al.*, 1997, Moriuchi *et al.*, 2005)
			CT26	Murine colon carcinoma	(Toda *et al.*, 2001)
	Oncolytic HSV	GCV	9L	Rat glioma	(Boviatsis *et al.*, 1994)
	Vesicular stomatitis virus	GCV	B16, D1-DMBA3	Murine melanoma or breast carcinoma	(Fernandez *et al.*, 2002)
	Sindbis virus	GCV	BHK	Transformed hamster kidney cells	(Tseng *et al.*, 2006)
	Adeno-associated virus	GCV	U-251SP	Human glioma	(Okada *et al.*, 1996)

(Continued)

Table 16.1 (*Continued*)

Enzyme	Vector	Prodrug	Tumour	Tumour type	Reference
			MeWo	Human melanoma	(Schoensiegel *et al.*, 2004)
CD	Replication-defective retrovirus	5-FC	K12	Rat colon carcinoma	(Humphreys *et al.*, 2001)
	Repication-competent retrovirus	5-FC	U87	Human glioma	(Wang *et al.*, 2003, Tai *et al.*, 2005)
	Replication-defective adenovirus	5-FC	MKN-45	Human gastric carcinoma	(Lan *et al.*, 1997)
			LoVo	Human colon carcinoma	(Zhang *et al.*, 2003)
	Oncolytic adenovirus	5-FC	LNCaP	Human prostate carcinoma	(Zhan *et al.*, 2005)
			SK-Mel-28	Human melanoma	(Liu *et al.*, 2006a)
			B16, LLC	Murine melanoma or lung carcinoma	(Liu *et al.*, 2006b)
	Vaccinia	5-FC	MC38	Murine colon carcinoma	(Gnant *et al.*, 1999)
	Oncolytic Herpes simplex virus	5-FC	MC26	Murine colon carcinoma	(Nakamura *et al.*, 2001)
CD-UPRT	Replication-defective adenovirus	5-FC	SW480	Human colon carcinoma	(Chung-Faye *et al.*, 2001)
	Vesicular stomatitis virus	5-FC	A20, TSA	Murine lymphoma or breast carcinoma	(Porosnicu *et al.*, 2003)
CD-TK	Oncolytic adenovirus	5-FC/GCV	C33-A	Human cervical carcinoma	(Rogulski *et al.*, 2000)
CD and TK	Replication-defective Herpes simplex virus	5-FC/GCV	9L	Rat glioma	(Moriuchi *et al.*, 2002)
NR	Replication-defective adenovirus	CB1954	SW480	Human colorectal carcinoma	(Lipinski *et al.*, 2001)
			C33-A, A2780	Human cervical or ovarian carcinoma	(Bilsland *et al.*, 2003)
	Oncolytic adenovirus	CB1954	SW480	Human colorectal carcinoma	(Chen *et al.*, 2004)
			SW620	Human colorectal carcinoma	(Lukashev *et al.*, 2005)
CPG2	Oncolytic adenovirus	ZD2767P	HepG2, Hep3B	Human hepatocarcinoma	(Schepelmann *et al.*, 2005)
CYP2B1	Oncolytic Herpes simplex virus	CPA	MC26	Murine colon carcinoma	(Pawlik *et al.*, 2002)

Table 16.1 (*Continued*)

Enzyme	Vector	Prodrug	Tumour	Tumour type	Reference
CYP2B6	Replication-defective retrovirus	CPA	HT29, MDAMB231, MDAMB468	Human colon or breast carcinoma	(Kan *et al.*, 2001)
CYP2B6 and P450R	Replication-defective adenovirus	CPA	A549, PC-3	Human lung or prostate carcinoma	(Jounaidi and Waxman, 2004)
P450R	Replication-defective adenovirus	Tirapazamine	HT1080	Human fibrosarcoma	(Cowen *et al.*, 2004)
PNP	Replication-defective adenovirus	Fludarabine	HuH-7	Human hepatocarcinoma	(Mohr *et al.*, 2000)
PNP	Ovine atadenovirus	Fludarabine	RM-1	Murine prostate carcinoma	(Martiniello-Wilks *et al.*, 2004)
PNP	Vaccinia	6-MPDR	MC-38	Murine colon carcinoma	(Puhlmann *et al.*, 1999)
Methioninase	Replication-defective adenovirus	SeMet	N1S1	Rat hepatocarcinoma	(Miki *et al.*, 2001)
CE	Replication-defective adenovirus	CPT-11	MG-63	Human osteosarcoma	(Oosterhoff *et al.*, 2003)
CE	Oncolytic adenovirus	CPT-11	C33-A	Human cervical carcinoma	(Stubdal *et al.*, 2003)
CE and CYP2B1	Oncolytic Herpes simplex virus	CPA, CPT-11	Gli36ΔEGFR	Human glioma	(Tyminski *et al.*, 2005)

Table 16.2 Clinical VDEPT trials

Enzyme	Vector	Prodrug	Tumour	Phase	Reference
TK	Replication-defective retrovirus	GCV	Glioblastoma multiforme	I/II	(Klatzmann *et al.*, 1998, Shand *et al.*, 1999, Prados *et al.*, 2003)
				III	(Rainov, 2000)
			Melanoma, breast, lung or sarcoma	I	(Singh *et al.*, 2001)
	Replication-defective adenovirus	GCV	Melanoma	I	(Morris *et al.*, 2000)
			Prostate	I	(Shalev *et al.*, 2000)
		ACV	Ovarian	I	(Hasenburg *et al.*, 2001)
		GCV	Colon	I	(Sung *et al.*, 2001)
		Valacyclovir	Prostate	I	(Kubo *et al.*, 2003)
		GCV	Malignant glioma	I	(Immonen *et al.*, 2004)
		GCV	Retinoblastoma	I	(Chevez-Barrios *et al.*, 2005)
		GCV	Mesothelioma	I	(Sterman *et al.*, 2005)
CD	Replication-defective adenovirus	5-FC	Colon	I	(Crystal *et al.*, 1997)
CD-TK	Oncolytic adenovirus	5-FC/GCV	Prostate	I	(Freytag *et al.*, 2002, 2003)
CYP2B6	Replication-defective retrovirus	CPA	Breast, melanoma	I	(Braybrooke *et al.*, 2005)

yeast (yCD) is used in GDEPT. However, yCD converts 5-FC more efficiently into 5-FU than *E. coli* CD (Hamstra *et al.*, 1999). Co-expression of CD and uracil phosphoribosyltransferase (UPRT) as a fusion protein (CD-UPRT) has been reported to increase the sensitivity of tumour xenografts to 5-FC (Chung-Faye *et al.*, 2001). UPRT is an enzyme that catalyses the conversion of 5-FU directly into 5-dUMP, bypassing the rate-limiting reactions controlled by the cellular enzymes.

16.2.3 CB1954 activated by nitroreductase

In the presence of the cofactor reduced nicotinamide adenine dinucleotide phosphate (NADPH), the minor FMN-dependent nitroreductase from *E. coli* (NR) reduces the monofunctional alkylating agent CB1954 (5-(arizidin-1-yl)-2,4-dinitrobenzamide) to the cognate 2- and 4-hydroxylamino compounds. The latter reacts with intracellular alkylthioesters to produce a bifunctional alkylating agent that cross-links DNA strands (Knox *et al.*, 1988). The only mammalian enzyme known to activate CB1954 is human DT-diaphorase but this enzyme has a much lower k_{cat} for the prodrug than NR from *E. coli* (4 min^{-1} as opposed to 360 min^{-1}) (Anlezark *et al.*, 1992), providing a rationale for the use of NR with CB1954 in VDEPT. NR/CB1954 mounts a marked bystander effect both *in vitro* (Anlezark *et al.*, 1992) and *in vivo* (Djeha *et al.*, 2000).

16.2.4 Nitrogen mustard prodrugs activated by carboxypeptidase G2

The *Pseudomonas* RS16 enzyme carboxypeptidase G2 (CPG2) hydrolyses the nitrogen mustard prodrugs CMDA (4-[(2-chloroethyl)(2-mesyloxyethyl) amino]benzoyl-L-glutamic acid) (Springer, 1993) and ZD2767P (4-[*N,N*-bis(2-iodoethyl)amino] phenoxycarbonyl L-glutamic acid) (Springer *et al.*, 1995), releasing glutamic acid and the cognate cytotoxic mustard drugs. The mustard drugs are bifunctional alkylating agents. There is no human equivalent of CPG2; thus, endogenous prodrug activation does not occur. Another advantage of the CPG2 enzyme is that it does not depend on cofactors and that it converts prodrugs directly to cytotoxic drugs without requiring host enzymes that could be absent or limited in tumour cells. Furthermore, the cytotoxic drugs kill both cycling and non-dividing cells; thus, quiescent tumour cells are unlikely to survive and re-grow. CPG2 mounts a potent bystander effect, as exemplified with CMDA *in vitro* (Stribbling *et al.*, 2000) and ZD2767P *in vivo* (Friedlos *et al.*, 2002; Schepelmann *et al.*, 2007). CPG2 has been expressed on the surface of tumour cells (Marais *et al.*, 1997; Spooner *et al.*, 2000; Cowen *et al.*, 2002). Extracellular prodrug conversion is thought to enhance the bystander effect since the activated drug will be formed in the tumour interstitial spaces rather than intracellularly (Oosterhoff *et al.*, 2003).

16.2.5 Oxazaphosphorines activated by cytochrome P450

Cyclophosphamide (2-[bis(2-chloroethyl)amino] tetrahydro-2*H*-1,3,2-oxazaphosphorine-2-oxide; CPA) and ifosfamide (3-(2-chloroethyl)-2-[(2-chloroethyl) amino]tetrahydro-2*H*-1,3,2-oxazaphosphorine 2-oxide; IFA) are oxazaphosphorine prodrugs that are metabolized by two alternative cytochrome P450 pathways; drug activation by 4-hydroxylation and drug inactivation by *N*-dechloroethylation, which generates the toxic compound chloroacetaldehyde. The cytochrome P450 enzymes CYP2B6, CYP2C and CYP3A4 catalyse the oxidation of CPA to its 4-hydroxy product, which then spontaneously rearranges with the release of acrolein to produce an oxazaphosphorine mustard drug. IFA is activated by CYP3A4, 2A6, 2B6, 3A5 and 2C9/18/19. The activated forms of the prodrugs are bifunctional alkylating agents (Rooseboom *et al.*, 2004). In addition, the release of the active metabolite acrolein may contribute to the cytotoxicity of CFA as well as the strong bystander effect generated by the mustard (Knox, 1997). The flavoenzyme NADPH-P450 reductase (P450R) increases the activity of P450 (Jounaidi and Waxman, 2004). For VDEPT, CYP2B6 has been expressed as a CYP2B6/NADPH-cytochrome P450 reductase fusion protein (CYP2B6/RED). The RED donates electrons to CYP and enhances the efficacy of the system (Tychopoulos *et al.*, 2005). Endogenous CYP enzymes are expressed in the liver, not in other tissues or in tumour cells. Thus, viral targeting of the enzymes to tumour cells followed by intratumoral injection of prodrug is thought to spare non-tumour tissues, although there will be non-GDEPT activation of prodrug leaking from the tumour by endogenous CYPs in the liver.

16.2.6 Fludarabine activated by purine nucleoside phosphorylase

The *E. coli* enzyme purine nucleoside phosphorylase (PNP) is used in combination with prodrug nucleosides, such as fludarabine (2-fluoroadenine-9-β-D-arabinofuranoside). The exact mechanism by which the PNP/fludarabine combination exerts its cytotoxic effect is not fully understood, however, fludarabine is converted to the free nucleoside 9-β-D-arabinosyl-2-fluoroadenine (F-ara-A), which is subsequently phosphorylated to the toxic 5′-triphosphate (Rooseboom *et al.*, 2004). The mechanism also appears to be p53-independent and exerts a bystander effect (Hong *et al.*, 2004). Mammalian analogues of *E. coli* PNP exist; however, fludarabine is metabolized to 2-fluoroadenine more efficiently by the bacterial PNP than by the mammalian enzymes. Another prodrug that is activated by *E. coli* PNP is 9-(β-2-deoxy-erythropentofuranosyl)-6-methylpurine (MeP-dR), which is activated to the active metabolite 6-methylpurine. Mammalian PNPs cannot catalyse this reaction (Sorscher *et al.*, 1994).

16.2.7 Selenomethionine activated by methioninase

Selenomethionine (SeMET) is metabolized by the *Pseudomonas putida* enzyme methionine α, γ-lyase (methioninase). The resulting products are α-ketobutyrate, ammonia and the toxic compound methylselenol, which generates superoxide anion radicals. The superoxide causes mitochondrial damage, cytochrome C release and induces apoptosis (Rooseboom *et al.*, 2004). SeMET cannot be converted to methylselenol by mammalian cells (Rooseboom *et al.*, 2004). The antitumour efficacy of the methioninase/SeMET combination has been demonstrated (Miki *et al.*, 2001). Furthermore, this enzyme prodrug system exerts a potent bystander effect (Miki *et al.*, 2001), which was also observed in cells overexpressing the mitochondrial antiapoptosis protein bcl-2, a condition known to inhibit other enzyme prodrug therapies (Yamamoto *et al.*, 2003).

16.2.8 CPT-11 activated by carboxylesterase

The camptothecin-derivative CPT-11 (irinotecan, 7-ethyl-10-[4-(1-piperidino)-1-piperidino]carboxyloxy -comptothecin) is hydrolysed by carboxylesterases (CE) to produce SN-38, a potent topoisomerase-1 inhibitor (Tanizawa *et al.*, 1994). Camptothecins prevent supercoiled DNA cleavage and relaxation, resulting in double-strand breaks in DNA during replication and disabling cell division. CEs are predominantly found in liver microsomes in humans; however, CPT-11 is a relatively poor substrate for human CE. In contrast, rabbit CE is 100 to 1000 times more efficient at converting CPT-11 to SN-28 and has been used in GDEPT protocols (Stubdal *et al.*, 2003). Furthermore, a secreted form of CE has been developed for VDEPT (Oosterhoff *et al.*, 2003).

16.3 Gene delivery vectors for GDEPT

A number of viral and non-viral gene delivery systems have been developed for enzyme prodrug therapies. Non-viral vectors include bacterial vectors (BDEPT) (Pawelek *et al.*, 1997), genetically modified cells (Salmons *et al.*, 2003), liposomes (Schatzlein, 2001) or naked DNA (Herweijer and Wolff, 2003). Non-viral vectors present a number of advantages, especially in terms of safety following their administration to humans. However, viruses are more efficient at delivering genes into cells. Consequently, the majority of current gene therapy clinical trials use viral vectors and non-viral gene transfer has not been studied to the same extent in humans as viral gene therapy (http://www.wiley.co.uk/genmed/clinical/).

Viral vectors for GDEPT can be administered locally at the tumour site or systemically in an approach that targets metastasized cancer cells as well as solid tumours. However, whilst systemic virus administration is likely to be more effective, there is a risk that extratumoral expression of the prodrug-converting enzyme may occur, leading to systemic toxicity upon prodrug administration. Consequently, in all clinical GDEPT studies so far, the viral vectors have been administered locally or near the tumour site. In any gene therapy approach, it is important that the vectors are safe to administer to humans and that the dose required to achieve a therapeutic effect is well below the dose that produces adverse side effects. Initial viral vectors for cancer gene therapy were engineered to be replication-defective (Kirn *et al.*, 2001), but replication-selective oncolytic viruses (virotherapy) have been developed that replicate in and destroy cancer cells (Kirn and McCormick, 1996). These vectors have the advantage that each virus particle can infect a tumour cell that generates progeny capable of spreading to other cells. They achieve higher efficiencies of gene delivery compared with replication-defective viruses (Ichikawa and Chiocca, 2001). They are also oncolytic and have intrinsic antitumour activity. In addition to oncolytic viruses, replicating retroviral vectors have been developed that are not intrinsically cytolytic (Solly *et al.*, 2003). Many GDEPT systems are currently being developed, and here we discuss those that have already been tested in preclinical *in vivo* models or clinical trials.

16.3.1 Retroviruses

Retroviruses can be divided into those derived from murine leukaemia virus (MLV) or lentiviral vectors, such as the human immunodeficiency virus (HIV). Both vector types integrate into the host genome following their transcription into DNA. However, MLV viruses can only replicate in dividing cells, whilst cell proliferation is not essential for HIV-based vectors (Naldini *et al.*, 1996). An advantage of retroviral vectors is that infected cells pass on the stably integrated transgene to their progeny, thus achieving long-term expression. However, in GDEPT, transient gene transfer is usually sufficient as the aim is to destroy the target cells shortly after infection. Furthermore, retroviral vector integration has been associated with insertional oncogenesis (Hacein-Bey-Abina *et al.*, 2003). Lentiviral GDEPT vectors are replication-defective, but MLV-based vectors can be either non-replicating or replicating (replication-competent retroviruses, onco-retroviruses) (VandenDriessche *et al.*, 2003).

Replication-defective retroviruses

Lentiviruses have been developed for GDEPT (De Palma *et al.*, 2003; Gerolami *et al.*, 2004); however, they are in their relative infancy and have not been assessed yet in clinical GDEPT applications. By contrast, there is a substantial amount of preclinical and clinical data for MLV vectors. Because of their intrinsic selectivity for dividing cells, MLV-based viruses are suitable candidates for brain cancer GDEPT, where the only dividing cells are those within a tumour (Rainov and Ren, 2003).

A number of preclinical studies have been performed using glioma xenograft models *in vivo* (Ezzeddine *et al.*, 1991; Barba *et al.*, 1993; Sandmair *et al.*, 2000). Replication-defective MLV viruses have also been used to deliver prodrug-converting enzymes to tumour endothelial cells (Mavria *et al.*, 2005) or to thyroid (Barzon *et al.*, 2002), colon (Kan *et al.*, 2001), breast (Kan *et al.*, 2001) or liver cancer (Humphreys *et al.*, 2001) xenografts. Clinical GDEPT studies have been conducted in patients with breast cancer or melanoma (Braybrooke *et al.*, 2005) or glioblastoma multiforme (Klatzmann *et al.*, 1998; Shand *et al.*, 1999; Prados *et al.*, 2003). There has been one phase III, controlled trial of retroviral GDEPT therapy for glioblastoma multiforme (Rainov, 2000). However, 4 years of follow-up has demonstrated no benefit of the GDEPT treatment, which was probably due to poor transduction efficiencies (Rainov, 2000).

Replication-competent retroviruses

Replication-competent retroviruses have been developed as a means to improve retroviral gene delivery. In contrast to other replicating viruses such as adenoviruses or vaccinia, replication-competent retroviruses do not lyse infected cells. Although replication-competent retroviruses were shown to be efficient in GDEPT therapies for experimental gliomas (Solly *et al.*, 2003; Wang *et al.*, 2003; Tai *et al.*, 2005), thus far, they have not been tested in humans. Retroviral vectors have been associated with insertional mutagenesis and oncogenesis (Hacein-Bey-Abina *et al.*, 2003). Therefore, tumour-selective targeting of replication-competent retroviruses is crucial and tissue-selective promoters have been used to confine viral replication to tumour cells (Logg *et al.*, 2002).

16.3.2 Adenoviruses

Retroviruses and adenoviruses are amongst the most common gene therapy vectors in clinical trials (Edelstein *et al.*, 2004). There is a vast clinical experience for adenoviruses. Adenovirus vaccines have been used safely and efficiently (Lichtenstein and Wold, 2004) and the most common serotypes, Ad2 and Ad5, have been developed as vectors for gene therapy. These vectors achieve high transduction and transgene expression efficiencies in a broad range of dividing and non-dividing cells. Another advantage of adenoviruses is that the adenoviral genome remains episomal in infected cells and does not usually integrate into the host DNA, minimizing the risk of insertional mutagenesis. Adenovirus-mediated gene expression is relatively short-term, but can last for up to several weeks. For GDEPT approaches, this is usually sufficient as the aim is to destroy the cancer cells shortly after infection, which is in contrast to gene replacement strategies where long-term expression is required.

Replication-defective adenoviruses

Both replication-defective and replicating adenoviruses have been used for GDEPT. Replication-defective adenoviruses have been assessed in numerous phase I clinical GDEPT trials, most of which used the TK/GCV or the CD/5-FC enzyme prodrug system. In these studies, various types of cancer have been treated, including malignant mesothelioma (Sterman *et al.*, 2005), malignant glioma (Immonen *et al.*, 2004), retinoblastoma (Chevez-Barrios *et al.*, 2005), melanoma (Morris *et al.*, 2000), metastatic colorectal liver carcinoma (Crystal *et al.*, 1997; Sung *et al.*, 2001), ovarian (Hasenburg *et al.*, 2001) and prostate cancer (Kubo *et al.*, 2003). In another trial, an NR-expressing adenovirus was tested as a single agent without prodrug in patients with primary and secondary liver cancer (Palmer *et al.*, 2004). In all of these studies, the vectors were delivered by direct injection into the tumours or near the tumour site. Adenoviruses are not intrinsically tumour-selective and for systemic GDEPT treatments, strategies are required that target the expression of the prodrug-converting enzyme to the tumour cells. Even after intratumoral vector administration, leakage of a non-targeted virus from the tumour site may cause adverse effects when the prodrug is administered.

Consequently, in a range of preclinical studies with replication-defective adenoviruses, tumour-selective promoters have been used for transcriptional targeting to liver cancer (TK, α-fetoprotein promoter) (Kaneko *et al.*, 1995; Gerolami *et al.*, 2000), melanoma (TK, tyrosinase promoter) (Siders *et al.*, 1998), ovarian or cervical (NR, human telomerase promoter) (Bilsland *et al.*, 2003), colon [TK, carcinoembryonic antigen (CEA) promoter (Okabe *et al.*, 2003); CD, CEA promoter (Zhang *et al.*, 2003); NR, CTP-1 (synthetic, beta-catenin-dependent) promoter (Lipinski *et al.*, 2001)] and gastric carcinoma (CD, CEA promoter) (Lan *et al.*, 1997). Similarly, P450R has been expressed under the control of a hypoxia-responsive promoter with the aim to sensitize hypoxic tumour cells to the bioreductive drug tirapazamine (Cowen *et al.*, 2004). However, other studies have used adenoviral GDEPT vectors that expressed prodrug-converting enzymes under the control of the ubiquitous cytomegalovirus (CMV) promoter (Miki *et al.*, 2001; Oosterhoff *et al.*, 2003; Jounaidi and Waxman, 2004; Palmer *et al.*, 2004). In a direct *in vivo* comparison of a CMV-driven and a transcriptionally targeted NR-adenovirus, systemic exposure to the non-targeted vector resulted in toxicity when the prodrug was given, which was due to liver damage that was associated with high levels of hepatic NR expression (Lipinski *et al.*, 2001). These findings underline the importance of tumour targeting and suggest that non-targeted viral vectors may not be suitable for systemic GDEPT applications.

Oncolytic adenoviruses

Replication-selective adenoviruses are the most widely studied oncolytic viruses in the clinic. The first adenovirus mutant that selectively targeted viral replication to tumour cells (*dl*1520, Onyx-015) was described a decade ago (Bischoff *et al.*, 1996). There are three strategies to restrict the replication of adenoviruses to cancer cells. One involves the use of tumour-selective promoters to drive expression of one or more viral genes that regulate viral replication (Kurihara *et al.*, 2000; Brunori *et al.*, 2001; Jakubczak *et al.*, 2003; Ryan *et al.*, 2004; Schepelmann *et al.*, 2005). In an alternative approach, selective transcription factor-binding sites have been inserted into early adenoviral promoters (Lukashev *et al.*, 2005). Finally, viral genes can be deleted that are essential for replication in normal but dispensable in tumour cells (Bischoff *et al.*, 1996; Fueyo *et al.*, 2000; Heise *et al.*, 2000; Chen *et al.*, 2004; Wang *et al.*, 2005). The tumour-selectivity and safety of oncolytic viruses as single agents has been demonstrated in a number of clinical studies (Ganly *et al.*, 2000; Nemunaitis *et al.*, 2000, 2003). However, antitumour efficacy could only be achieved in trials that combined the viral treatment with conventional chemotherapy (Khuri *et al.*, 2000; Reid *et al.*, 2002; Hecht *et al.*, 2003). Consequently, oncolytic adenoviruses have been 'armed' with prodrug-converting GDEPT enzymes to increase their efficacy. This approach has a multifaceted way of attacking tumours. First, there are successive rounds of virus-mediated cancer cell killing. Second, replicating vectors should deliver relatively long-lived expression of the GDEPT enzymes, which are then able to convert a larger number of prodrug molecules, thus enhancing the amplification effect. By spreading throughout the tumours, the viruses should also improve enzyme delivery and expression levels compared with replication-defective vectors. Third, the GDEPT-mediated bystander effect will kill uninfected cells within the tumours. Furthermore, virus/prodrug combinations tailored to act synergistically may enhance the efficacy of either monotherapy (Bernt *et al.*, 2002). Finally, oncolytic viruses and GDEPT kill tumour cells by different mechanisms, making it less likely for resistance to the treatment to develop.

The incorporation of the *tk* gene into the genomes of oncolytic adenoviruses enhanced the antitumour effect in models of malignant glioma, colon, lung and ovarian cancer when the virus was given in combination with GCV compared with virus alone (Wildner *et al.*, 1999; Wildner and Morris, 2000; Nanda *et al.*, 2001). However, in other studies, GCV did not improve the efficacy of TK-expressing oncolytic adenoviruses (Morris and Wildner, 2000; Wildner and Morris, 2000; Lambright *et al.*, 2001), probably due to inhibition of adenoviral replication by the prodrug metabolites (Post *et al.*, 2003). Oncolytic adenoviruses have also been engineered to express CD (Zhan *et al.*, 2005; Liu *et al.*, 2006a; Liu and Deisseroth, 2006), NR (Chen *et al.*, 2004; Lukashev *et al.*, 2005), CPG2 (Schepelmann *et al.*, 2005) or CE (Stubdal *et al.*, 2003). In two of these approaches, the fibre proteins of the viruses were modified in order to increase the transduction efficiency for tumour cells that express low levels of the coxsackievirus and adenovirus receptor (CAR) by enabling the vectors to use the integrins $\alpha_V\beta_3$ and $\alpha_V\beta_5$, which are abundant on tumour cells, as alternative receptors (Liu *et al.*, 2006a; Liu and Deisseroth, 2006).

Onyx-015 has been armed with a TK/CD fusion protein for double suicide gene therapy (Freytag *et al.*, 1998). The resulting virus, Ad5-FGNR or Ad5-CD/TKrep, is the best-characterized oncolytic GDEPT vector so far (Paielli *et al.*, 2000; Rogulski *et al.*, 2000) and a clinical trial was conducted for the treatment of prostate cancer (Freytag *et al.*, 2002). This was the first reported trial that used an oncolytic adenovirus to deliver therapeutic genes to humans. The treatment was found to be safe with no dose-limiting toxicity and evidence of some tumour responses (Freytag *et al.*, 2002). In preclinical studies, the TK/CD GDEPT system was shown to enhance the efficacy of radiotherapy (Rogulski *et al.*, 2000). Therefore, a second clinical trial was conducted, combining Ad5-CD/TKrep and the two prodrugs with radiotherapy (Freytag *et al.*, 2003). Results from this trial showed that the combination therapy was safe. There was also evidence of a possible interaction between the GDEPT and the radiation therapy (Freytag *et al.*, 2003). Subsequently, Ad5-CD/TKrep was further modified to express an improved yCD/mutTK$_{SR39}$ fusion enzyme and the adenovirus death protein (ADP), which has been shown to enhance the cytolytic activity of oncolytic adenoviruses *in vitro*. Relative to the parental Ad5-CD/TKrep adenovirus, Ad5-yCD/mutTK(SR39)rep-ADP did not cause increased *in vivo* toxicity (Barton *et al.*, 2006). Furthermore, Ad5-yCD/mutTK(SR39)rep-ADP demonstrated greater tumor cell kill *in vitro* and enhanced anti-tumour efficacy in preclinical models of human cancer (Barton *et al.*, 2006).

16.3.3 Vaccinia

Vaccinia virus is a cytolytic virus. Its safety has been assessed in humans for longer than any other virus due to its worldwide use as the vaccine against smallpox (Fenner, 1996) and more recently, as a vaccine in clinical trials for cancer immunotherapy (Kwak *et al.*, 2003). Vaccinia has a very good safety record and causes no known human disease, although it can be associated with eczema vaccination in immunocompromised patients. The virus spends its entire life cycle in the cytoplasm of infected cells and has never been shown to integrate (Zeh and Bartlett, 2002). Many laboratory animals and cell types can be infected with vaccinia, allowing the use of cell and animal models that are relevant to human disease (Shen and Nemunaitis, 2005). Despite this broad host range, vaccinia inherently targets tumour after systemic administration. It is thought that this natural tropism is due to the size of the virus (>200 nm) and that only leaky vasculature (such as in tumours) allows extravasation of the virus from the blood vessels (Peplinski *et al.*, 1996; Chang *et al.*, 2005).

Approaches have been described to enhance further the tumour-selectivity of vaccinia. Transcriptional targeting strategies cannot be employed due to the fact that viral replication occurs in the cytoplasm and independently of the host DNA synthesis machinery. However, vaccinia can be targeted to tumour cells by deletion or modification of genes that are necessary for replication in normal, non-dividing cells (McCart *et al.*, 2001). Vaccinia has been armed with the enzymes CD (Gnant *et al.*, 1999) or PNP (Puhlmann *et al.*, 1999). In both cases, GDEPT had antitumour activity, however, there was also evidence that the activated forms of the prodrugs interfered with viral replication (Gnant *et al.*, 1999; Puhlmann *et al.*, 1999), which may diminish the therapeutic potential of replicating viruses (Post *et al.*, 2003). These findings highlight that oncolytic GDEPT vectors require enzyme prodrug systems that do not inhibit their replication (Bernt *et al.*, 2002; Porosnicu *et al.*, 2003; Schepelmann *et al.*, 2005).

16.3.4 Herpes simplex virus

Herpes simplex virus type 1 (HSV-1) virus has a natural tropism for neuronal cells. However, the virus can also infect a broad range of other cell types, both dividing and non-dividing (Latchman, 2002; Shen and Nemunaitis, 2006). HSV vectors are efficient gene delivery vehicles. In addition, they do not integrate into the cellular genome, precluding insertional mutagenesis (Varghese and Rabkin, 2002).

Replication-defective HSV

Replication-defective HSV vectors can be engineered by the deletion or disruption of essential viral genes (Shen and Nemunaitis, 2006). The neurotropism of HSV renders the virus suitable for the treatment of cancers of the central nervous system. Furthermore, HSV naturally expresses TK, which makes it an attractive candidate for TK-mediated GDEPT. In combination with prodrug, TK generates cytotoxic metabolites that are incorporated into replicating DNA. Thus, HSV-mediated TK-GDEPT therapies for the brain target only the replicating cells, but not the neurons or quiescent glia (Shen and Nemunaitis, 2006). Consequently, HSV vectors have been developed for the delivery of TK to glioma models (Miyatake *et al.*, 1997; Moriuchi *et al.*, 2002, 2005). Replication-defective HSV viruses have also been used to deliver TK in combination with other therapeutic proteins (Toda *et al.*, 2001; Moriuchi *et al.*, 2005) and in addition, a replication-defective HSV virus has been used to co-express TK and CD in models of gliosarcoma for double suicide gene therapy (Moriuchi *et al.*, 2002).

Oncolytic HSV

A direct comparison between a replication-defective and an oncolytic HSV vector showed that transgene delivery and expression lasted longer when the oncolytic virus was used and that transgene distribution was improved in human tumour xenografts (Ichikawa and Chiocca, 2001). A variety of oncolytic HSV viruses have been developed for non-GDEPT cancer gene therapy applications and some have safely completed phase I clinical trials (Markert *et al.*, 2000; Rampling *et al.*, 2000; Fong Y, 2002). HSV does not naturally target cancer cells and in order to achieve tumour-selectivity, oncolytic HSV vectors have to be genetically engineered using gene deletion/modification or transcriptional targeting. Notably, oncolytic HSV was the first virus designed to replicate selectively in dividing cells due to a deletion in the *tk* gene (Martuza *et al.*, 1991). However, for clinical use, it is desirable to retain the viral TK-activity as it provides an inherent safety mechanism against uncontrolled viral replication (Shen and Nemunaitis, 2006).

Endogenous viral TK expression has been used for GDEPT treatment of glioma models (Boviatsis *et al.*, 1994), however, it has been shown that the TK/GCV system inhibits not only tumour cell proliferation but also HSV replication (Nakamura *et al.*, 2001; Pawlik *et al.*, 2002). Therefore, oncolytic HSV vectors have been developed for GDEPT combinations, in which the cytotoxic compounds affect HSV replication to a lesser extent than TK/GCV (yCD/5-FC (Nakamura *et al.*, 2001), CD-UPRT/5-FC (Simpson *et al.*, 2006), CYP2B1/CPA (Pawlik *et al.*, 2002) or CYP2B1 and secreted CE for double suicide gene therapy in combination with CPA and CPT-11 (Tyminski *et al.*, 2005)). It has been suggested that the spread of oncolytic HSV vectors within tumours can be enhanced by co-injection of collagenase (McKee *et al.*, 2006). Thus far, replicating HSV viruses have not been tested in clinical GDEPT trials.

16.3.5 Oncolytic RNA viruses

Many RNA viruses are naturally oncolytic (Russell, 2002) and some have safely completed phase I clinical studies as single agents (Pecora *et al.*, 2002; Csatary *et al.*, 2004). RNA virus engineering for gene delivery is complicated by the fact that the genomes of RNA viruses can only be manipulated in their DNA forms. However, 'reverse genetics' systems have recently been developed for the rescue of several positive- and negative-strand RNA viruses (Russell, 2002). Vesicular stomatitis virus (VSV) is the only oncolytic RNA virus so far that has been used as a GDEPT delivery vector. VSV replicates efficiently in human tumour cells, whilst propagation is inhibited in non-transformed human cells. It is a non-integrating virus that replicates in the cytoplasm and has no known transforming properties (Porosnicu *et al.*, 2003). VSV has been engineered to express TK (Fernandez *et al.*, 2002) or CD-UPRT (Porosnicu *et al.*, 2003) for the conversion of GCV or 5-FC, respectively. Interestingly, the prodrug metabolite 5-FU did not significantly interfere with the viral replication (Porosnicu *et al.*, 2003). This may be due to the fact that VSV is an RNA virus whose cytoplasmic replication may be less affected by the inhibitory actions of 5-FU than that of DNA viruses, making VSV particularly suitable for CD-mediated GDEPT (Porosnicu *et al.*, 2003).

16.3.6 Sindbis virus

Sindbis virus is a blood-borne, replication-defective RNA virus that infects tumour cells selectively and induces apoptosis (Zrachia *et al.*, 2002; Tseng *et al.*, 2004). Recombinant Sindbis vectors retain the blood-borne attribute and can be delivered systemically (Tseng *et al.*, 2004). They are also efficient gene delivery vectors and suitable vectors for GDEPT, as exemplified by a study that used a TK-expressing Sindbis vector in combination with GCV in an ovarian cancer model (Tseng *et al.*, 2006).

16.3.7 Adeno-associated virus

Adeno-associated virus (AAV) is naturally replication defective. AAV vectors can infect both dividing and quiescent populations of different cell types. They transduce their genomes into stable episomal isoforms and/or integrate it into the host genome, mediating long-term expression. However, AAV-mediated vector integration is relatively uncommon (compared with retroviruses) and so far, there is no evidence that AAV insertional mutagenesis causes cancer (Kay and Nakai,

2003). AAV vectors have been used less extensively for GDEPT than, for example, adenoviruses. They do not have a natural tropism for tumours and targeting strategies are required to limit the expression of the prodrug-activating enzymes to cancer cells.

Following transcriptional targeting approaches, TK-expressing AAV vectors has been developed for the treatment of hepatoma or melanoma models (Su *et al.*, 2000, Schoensiegel *et al.*, 2004). However, AAV transduction and transgene expression levels are often low. This is due to the fact that at the beginning of the AAV life cycle, the single-stranded virus genome has to be converted into its double-stranded form. This viral second strand synthesis is the rate-limiting step and the underlying reason for inefficient transduction and gene expression (Ferrari *et al.*, 1996). In different attempts to improve the potency of AAV-mediated GDEPT, the AAV-TK system has been combined with immune stimulators (Okada *et al.*, 1996; Su *et al.*, 2000; Janouskova *et al.*, 2003). However, high transduction and enzyme expression levels are crucial for cancer gene therapy and further improvements are required to enhance the efficacy of AAV vectors for GDEPT.

16.4 Conclusions

Viral vectors are rapidly emerging as delivery vectors for GDEPT. Different viruses vary in the size of the transgenes they can accommodate; however, in GDEPT, this is not usually an issue as the genes encoding for the most commonly used enzymes are relatively short. The current data suggest that replicating, oncolytic viruses are superior compared to replication-defective vectors.

GDEPT is more complex than conventional chemotherapy as it utilizes two components, the virus and the prodrug. The dosing and the timing of vector and prodrug administration are crucial for a successful therapy. Preclinical dose–response and time course experiments in animal models are useful tools to mimic the clinical situation, however, the resulting data cannot always be extrapolated to humans because of species differences (Russell, 2002; Raper *et al.*, 2003). For clinical GDEPT study designs, it is difficult to predict how long it will take to achieve sufficient levels of transgene expression in the tumours after virus administration. Targeted, replicating viruses are designed to spread throughout the tumours, whilst they are cleared from healthy tissues, creating a differential of enzyme expression between tumours and normal tissues. Adverse events arising from systemic GDEPT can be greatly reduced by waiting for this differential to develop before the prodrug is given. In the ideal clinical setting, the timing of virus and prodrug should be individually tailored for each patient. Novel techniques are required that monitor tumour transduction, viral gene expression and virus elimination (Peng *et al.*, 2002). In animal models, optical imaging technologies allow real-time, quantitative and three-dimensional monitoring (Yu *et al.*, 2004; Tseng *et al.*, 2004; Lee *et al.*, 2006; Hurtado *et al.*, 2006). However, these techniques have the drawback of attenuation of the signal by overlaying tissues, which becomes even more significant in patients. Cost-effective imaging methods are required for routine clinical use.

Another issue for clinical GDEPT applications is the role of the immune system. Pre-existing immunity against viral vectors can be a beneficial safeguard against uncontrolled virus spread. However, it can also be a risk factor (Raper *et al.*, 2003) or compromise the efficacy of the treatment, especially when the virus is delivered systemically. Immunosuppressive therapy could potentially be deployed to counteract antiviral immunity. Notably, many cancer patients are immunocompromised due to their disease or previous treatments. On the other hand, the generation of antibodies against the virus and/or the prodrug-converting enzyme could be beneficial as they should direct the immune response of the patient against the transduced tumour cells, enhancing the antitumour effect of the therapy. GDEPT therapies are usually tested in immunodeficient mice bearing human xenograft tumours. These models are of limited significance for the clinic and novel, immunocompetent animal models may more accurately reflect the situation in patients (Hallden *et al.*, 2003; Thomas *et al.*, 2006). Nevertheless, preclinical and clinical GDEPT data so far have been encouraging and more GDEPT trials can be expected in the near future.

References

Anlezark GM, Melton RG, Sherwood RF, Coles B, Friedlos F, Knox RJ (1992). The bioactivation of 5-(aziridin-1-yl)-2,4-dinitrobenzamide (CB1954)-I. Puri fication and properties of a nitroreductase enzyme from *Escherichia coli* – A potential enzyme for antibody-directed enzyme prodrug therapy (ADEPT) *Biochem Pharmacol* **44**, 2289–2295.

Barba D, Hardin J, Ray J, Gage FH (1993). Thymidine kinase-mediated killing of rat brain tumors. *J Neurosurg* **79**, 729–735.

Barton KN, Paielli D, Zhang Y, *et al.* (2006). Second-generation replication-competent oncolytic adenovirus armed with improved suicide genes and ADP gene demonstrates greater efficacy without increased toxicity. *Mol Ther* **13**, 347–356.

Barzon L, Bonaguro R, Castagliuolo I, *et al.* (2002). Transcriptionally targeted retroviral vector for combined suicide and immunomodulating gene therapy of thyroid cancer. *J Clin Endocrinol Metab* **87**, 5304–5311.

Bernt KM, Steinwaerder DS, Ni S, Li ZY, Roffler SR, Lieber A (2002). Enzyme-activated prodrug therapy enhances tumor-specific replication of adenovirus vectors. *Cancer Res* **62**, 6089–6098.

Bilsland AE, Anderson CJ, Fletcher-Monaghan AJ, *et al.* (2003). Selective ablation of human cancer cells by telomerase-specific adenoviral suicide gene therapy vectors expressing bacterial nitroreductase. *Oncogene* **22**, 370–380.

Bischoff JR, Kirn DH, Williams A, *et al.* (1996). An adenovirus mutant that replicates selectively in p53-deficient human tumor cells. *Science* **274**, 373–376.

Boviatsis EJ, Park JS, Sena-Esteves M, *et al.* (1994). Long-term survival of rats harboring brain neoplasms treated with ganciclovir and a herpes simplex virus vector that retains an intact thymidine kinase gene. *Cancer Res* **54**, 5745–5751.

Braybrooke JP, Slade A, Deplanque G, *et al.* (2005). Phase I study of MetXia-P450 gene therapy and oral cyclophosphamide for patients with advanced breast cancer or melanoma. *Clin Cancer Res* **11**, 1512–1520.

Bridgewater JA, Springer CJ, Knox RJ, Minton NP, Michael NP, Collins MK (1995). Expression of the bacterial nitroreductase enzyme in mammalian cells renders them selectively sensitive to killing by the prodrug CB1954. *Eur J Cancer* **31A**, 2362–70.

Brunori M, Malerba M, Kashiwazaki H, Iggo R (2001). Replicating adenoviruses that target tumors with constitutive activation of the wnt signaling pathway. *J Virol* **75**, 2857–2865.

Chang E, Chalikonda S, Friedl J, *et al.* (2005). Targeting vaccinia to solid tumors with local hyperthermia. *Hum Gene Ther* **16**, 435–444.

Chen MJ, Green NK, Reynolds GM, *et al.* (2004). Enhanced efficacy of *Escherichia coli* nitroreductase/CB1954 prodrug activation gene therapy using an E1B-55K-deleted oncolytic adenovirus vector. *Gene Ther* **11**, 1126–1136.

Chevez-Barrios P, Chintagumpala M, Mieler W, *et al.* (2005). Response of retinoblastoma with vitreous tumor seeding to adenovirus-mediated delivery of thymidine kinase followed by ganciclovir. *J Clin Oncol* **23**, 7927–7935.

Chung-Faye GA, Chen MJ, Green NK, *et al.* (2001). *In vivo* gene therapy for colon cancer using adenovirus-mediated, transfer of the fusion gene cytosine deaminase and uracil phosphoribosyltransferase. *Gene Ther* **8**, 1547–1554.

Cowen RL, Williams JC, Emery S, *et al.* (2002). Adenovirus vector-mediated delivery of the prodrug-converting enzyme carboxypeptidase G2 in a secreted or GPI-anchored form: High-level expression of this active conditional cytotoxic enzyme at the plasma membrane. *Cancer Gene Ther* **9**, 897–907.

Cowen RL, Williams KJ, Chinje EC, *et al.* (2004). Hypoxia targeted gene therapy to increase the efficacy of tirapazamine as an adjuvant to radiotherapy: reversing tumor radioresistance and effecting cure. *Cancer Res* **64**, 1396–1402.

Crystal RG, Hirschowitz E, Lieberman M, *et al.* (1997). Phase I study of direct administration of a replication deficient adenovirus vector containing the *E. coli* cytosine deaminase gene to metastatic colon carcinoma of the liver in association with the oral administration of the pro-drug 5-fluorocytosine. *Hum Gene Ther* **8**, 985–1001.

Csatary LK, Gosztonyi G, Szeberenyi J, *et al.* (2004). MTH-68/H oncolytic viral treatment in human high-grade gliomas. *J Neurooncol* **67**, 83–93.

De Palma M, Venneri MA, Naldini L (2003). *In vivo* targeting of tumor endothelial cells by systemic delivery of lentiviral vectors. *Hum Gene Ther* **14**, 1193–1206.

Djeha AH, Hulme A, Dexter MT, *et al.* (2000). Expression of *Escherichia coli* B nitroreductase in established human tumor xenografts in mice results in potent antitumoral and bystander effects upon systemic administration of the prodrug CB1954. *Cancer Gene Ther* **7**, 721–731.

Edelstein ML, Abedi MR, Wixon J, Edelstein RM (2004). Gene therapy clinical trials worldwide 1989–2004-an overview. *J Gene Med* **6**, 597–602.

Ezzeddine ZD, Martuza RL, Platika D, *et al.* (1991). Selective killing of glioma cells in culture and *in vivo* by retrovirus transfer of the herpes simplex virus thymidine kinase gene. *New Biol* **3**, 608–614.

Fenner F (1996). Poxviruses. In: Fields BN, Knipe DM, Howley PM (eds), *Fields Virology*, third edition. Lippincott-Raven Publishers, Philadelphia.

Fernandez M, Porosnicu M, Markovic D, Barber GN (2002). Genetically engineered vesicular stomatitis virus in gene therapy: application for treatment of malignant disease. *J Virol* **76**, 895–904.

Ferrari FK, Samulski T, Shenk T, Samulski RJ (1996). Second-strand synthesis is a rate-limiting step for efficient transduction by recombinant adeno-associated virus vectors. *J Virol* **70**, 3227–3234.

Fick J, Barker II, FG, Dazin P, Westphale EM, Beyer EC, Israel MA (1995). The extent of heterocellular communication mediated by gap junctions is predictive of bystander tumor cytotoxicity *in vitro*. *Proc Natl Acad Sci U S A* **92**, 11071–11075

Fong YKN, Jarnagin W, Stanziale S, *et al.* (2002). Phase 1 study of a replication-competent herpes simplex oncolytic virus for treatment of hepatic colorectal metastases. American Society of Clinical Oncologists Annual Meeting, Orlando, FL, USA.

Freytag SO, Rogulski KR, Paielli DL, Gilbert JD, Kim JH (1998). A novel three-pronged approach to kill cancer cells selectively: concomitant viral, double suicide gene, and radiotherapy. *Hum Gene Ther* **9**, 1323–1333.

Freytag SO, Khil M, Stricker H, *et al.* (2002). Phase I study of replication-competent adenovirus-mediated double suicide gene therapy for the treatment of locally recurrent prostate cancer. *Cancer Res* **62**, 4968–4976.

Freytag SO, Stricker H, Pegg J, *et al.* (2003). Phase I study of replication-competent adenovirus-mediated double-suicide gene therapy in combination with conventional-dose three-dimensional conformal radiation therapy for the treatment of newly diagnosed, intermediate- to high-risk prostate cancer. *Cancer Res* **63**, 7497–7506.

Friedlos F, Davies L, Scanlon I, *et al.* (2002). Three new prodrugs for suicide gene therapy using carboxypeptidase G2 elicit bystander efficacy in two xenograft models. *Cancer Res* **62**, 1724–1729.

Fueyo J, Gomez-Manzano C, Alemany R, *et al.* (2000). A mutant oncolytic adenovirus targeting the Rb pathway produces anti-glioma effect *in vivo*. *Oncogene* **19**, 2–12.

Ganly I, Kirn D, Eckhardt G, *et al.* (2000). A phase I study of Onyx-015, an E1B attenuated adenovirus, administered intratumorally to patients with recurrent head and neck cancer. *Clin Cancer Res* **6**, 798–806.

Gerolami R, Cardoso J, Lewin M, *et al.* (2000). Evaluation of HSV-tk gene therapy in a rat model of chemically induced hepatocellular carcinoma by intratumoral and intrahepatic artery routes. *Cancer Res* **60**, 993–1001.

Gerolami R, Uch R, Faivre J, *et al.* (2004). Herpes simplex virus thymidine kinase-mediated suicide gene therapy for hepatocellular carcinoma using HIV-1-derived lentiviral vectors. *J Hepatol* **40**, 291–297.

Gnant MF, Puhlmann M, Alexander HR Jr, Bartlett DL (1999). Systemic administration of a recombinant vaccinia virus expressing the cytosine deaminase gene and subsequent treatment with 5-fluorocytosine leads to tumor-specific gene expression and prolongation of survival in mice. *Cancer Res* **59**, 3396–3403.

Greco O, Dachs GU (2001). Gene directed enzyme/prodrug therapy of cancer: historical appraisal and future prospectives. *J Cell Physiol* **187**, 22–36.

Hacein-Bey-Abina S, Von Kalle C, Schmidt M, *et al.* (2003). LMO2-associated clonal T cell proliferation in two patients after gene therapy for SCID-X1. *Science* **302**, 415–419.

Hallden G, Hill R, Wang, Y, *et al.* (2003). Novel immunocompetent murine tumor models for the assessment of replication-competent oncolytic adenovirus efficacy. *Mol Ther* **8**, 412–424.

Halloran PJ, Fenton RG (1998). Irreversible G2-M arrest and cytoskeletal reorganization induced by cytotoxic nucleoside analogues. *Cancer Res* **58**, 3855–3865.

Hamstra DA, Rice DJ, Fahmy S, Ross BD, Rehemtulla A (1999). Enzyme/prodrug therapy for head and neck cancer using a catalytically superior cytosine deaminase. *Hum Gene Ther* **10**, 1993–2003.

Hamzeh FM, Lietman PS (1991). Intranuclear accumulation of subgenomic noninfectious human cytomegalovirus DNA in infected cells in the presence of ganciclovir. *Antimicrob Agents Chemother* **35**, 1818–1823.

Hasenburg A, Tong XW, Fischer DC, *et al.* (2001). Adenovirus-mediated thymidine kinase gene therapy in combination with topotecan for patients with recurrent ovarian cancer: 2.5-year follow-up. *Gynecol Oncol* **83**, 549–554.

Hecht JR, Bedford R, Abbruzzese JL, *et al.* (2003). A phase I/II trial of intratumoral endoscopic ultrasound injection of ONYX-015 with intravenous gemcitabine in unresectable pancreatic carcinoma. *Clin Cancer Res* **9**, 555–561.

Heise C, Hermiston T, Johnson L, *et al.* (2000). An adenovirus E1A mutant that demonstrates potent and selective systemic anti-tumoral efficacy. *Nat Med* **6**, 1134–1139.

Herweijer H, Wolff JA (2003). Progress and prospects: naked DNA gene transfer and therapy. *Gene Ther* **10**, 453–458.

Hong JS, Waud WR, Levasseur DN, *et al.* (2004). Excellent *in vivo* bystander activity of fludarabine phosphate against human glioma xenografts that express the *Escherichia coli* purine nucleoside phosphorylase gene. *Cancer Res* **64**, 6610–6615.

Huber BE, Richards CA, Krenitsky TA (1991). Retroviral-mediated gene therapy for the treatment of hepatocellular carcinoma: an innovative approach for cancer therapy. *Proc Natl Acad Sci U S A* **88**, 8039–8043.

Huber BE, Austin EA, Richards CA, Davis ST, Good SS (1994). Metabolism of 5-fluorocytosine to 5-fluorouracil in human colorectal tumor cells transduced with the cytosine deaminase gene: Significant antitumor effects when only a small percentage of tumor cells express cytosine deaminase. *Proc Natl Acad Sci U S A* **91**, 8302–8306

Humphreys MJ, Ghaneh P, Greenhalf W, *et al.* (2001). Hepatic intra-arterial delivery of a retroviral vector expressing the cytosine deaminase gene, controlled by the CEA promoter and intraperitoneal treatment with 5-fluorocytosine suppresses growth of colorectal liver metastases. *Gene Ther* **8**, 1241–1247.

Hurtado A, Tseng JC, Meruelo D (2006). Gene therapy that safely targets and kills tumor cells throughout the body. *Rejuvenation Res* **9**, 36–44.

Ichikawa T, Chiocca EA (2001). Comparative analyses of transgene delivery and expression in tumors inoculated with a replication-conditional or -defective viral vector. *Cancer Res* **61**, 5336–5339.

Immonen A, Vapalahti M, Tyynela K, *et al.* (2004). AdvHSV-tk gene therapy with intravenous ganciclovir improves survival in human malignant glioma: a randomised, controlled study. *Mol Ther* **10**, 967–972.

Jakubczak JL, Ryan P, Gorziglia M, *et al.* (2003). An oncolytic adenovirus selective for retinoblastoma tumor suppressor protein pathway-defective tumors: dependence on E1A, the E2F-1 promoter, and viral replication for selectivity and efficacy. *Cancer Res* **63**, 1490–1499.

Janouskova O, Sima P, Kunke D (2003). Combined suicide gene and immunostimulatory gene therapy using AAV-mediated gene transfer to HPV-16 transformed mouse cell: decrease of oncogenicity and induction of protection. *Int J Oncol* **22**, 569–577.

Jounaidi Y, Waxman DJ (2004). Use of replication-conditional adenovirus as a helper system to enhance delivery of P450 prodrug-activation genes for cancer therapy. *Cancer Res* **64**, 292–303.

Kan O, Griffiths L, Baban D, *et al.* (2001). Direct retroviral delivery of human cytochrome P450 2B6 for gene-directed enzyme prodrug therapy of cancer. *Cancer Gene Ther* **8**, 473–482.

Kaneko S, Hallenbeck P, Kotani T, *et al.* (1995). Adenovirus-mediated gene therapy of hepatocellular carcinoma using cancer-specific gene expression. *Cancer Res* **55**, 5283–5287.

Kay MA, Nakai H (2003). Looking into the safety of AAV vectors. *Nature* **424**, 251.

Khuri FR, Nemunaitis J, Ganly I, *et al.* (2000). A controlled trial of intratumoral ONYX-015, a selectively-replicating adenovirus, in combination with cisplatin and 5-fluorouracil in patients with recurrent head and neck cancer. *Nat Med* **6**, 879–885.

Kirn DH, McCormick F (1996). Replicating viruses as selective cancer therapeutics. *Mol Med Today* **2**, 519–527.

Kirn D, Martuza RL, Zwiebel J (2001). Replication-selective virotherapy for cancer: Biological principles, risk management and future directions. *Nat Med* **7**, 781–787.

Klatzmann D, Valery CA, Bensimon G, *et al.* (1998). A phase I/II study of herpes simplex virus type 1 thymidine kinase 'suicide' gene therapy for recurrent glioblastoma. Study Group on Gene Therapy for Glioblastoma. *Hum Gene Ther* **9**, 2595–2604.

Knox RJ, Friedlos F, Jarman M, Roberts JJ (1988). A new cytotoxic, DNA interstrand crosslinking agent, 5-(aziridin-1-yl)-4-hydroxylamino-2-nitrobenzamide, is formed from 5-(aziridin-1-yl)-2,4-dinitrobenzamide (CB 1954) by a nitroreductase enzyme in Walker carcinoma cells. *Biochem Pharmacol* **37**, 4661–4669.

Knox RJ, Connors TA (1997). Prodrugs in cancer chemotherapy. *Pathol Oncol Res* **3**, 309–324.

Krohne TU, Shankara S, Geissler M, *et al.* (2001). Mechanisms of cell death induced by suicide genes encoding purine nucleoside phosphorylase and thymidine kinase in human hepatocellular carcinoma cells in vitro. *Hepatology* **34**, 511–518

Kubo H, Gardner TA, Wada Y, *et al.* (2003). Phase I dose escalation clinical trial of adenovirus vector carrying osteocalcin promoter-driven herpes simplex virus thymidine kinase in localized and metastatic hormone-refractory prostate cancer. *Hum Gene Ther* **14**, 227–241.

Kurihara T, Brough DE, Kovesdi I, Kufe DW (2000). Selectivity of a replication-competent adenovirus for human breast carcinoma cells expressing the MUC1 antigen. *J Clin Invest* **106**, 763–771.

Kwak H, Horig H, Kaufman HL (2003). Poxviruses as vectors for cancer immunotherapy. *Curr Opin Drug Discov Devel* **6**, 161–168.

Lambright ES, Amin K, Wiewrodt R, *et al.* (2001). Inclusion of the herpes simplex thymidine kinase gene in a replicating adenovirus does not augment antitumor efficacy. *Gene Ther* **8**, 946–953.

Lan KH, Kanai F, Shiratori Y, *et al.* (1997). *In vivo* selective gene expression and therapy mediated by adenoviral vectors for human carcinoembryonic antigen-producing gastric carcinoma. *Cancer Res* **57**, 4279–4284.

Latchman DS (2002). Herpes simplex virus vectors for gene delivery to a variety of different cell types. *Curr Gene Ther* **2**, 415–426.

Lee CT, Lee YJ, Kwon SY, *et al.* (2006). *In vivo* imaging of adenovirus transduction and enhanced therapeutic efficacy of combination therapy with conditionally replicating adenovirus and adenovirus-p27. *Cancer Res* **66**, 372–377.

Lichtenstein DL, Wold WS (2004). Experimental infections of humans with wild-type adenoviruses and with replication-competent adenovirus vectors: replication, safety, and transmission. *Cancer Gene Ther* **11**, 819–829.

Lipinski KS, Djeha AH, Ismail T, Mountain A, Young LS, Wrighton CJ (2001). High-level, beta-catenin/TCF-dependent transgene expression in secondary colorectal cancer tissue. *Mol Ther* **4**, 365–371.

Liu Y, Deisseroth A (2006). Tumor vascular targeting therapy with viral vectors. *Blood,* **107**, 3027–3033.

Liu Y, Ye T, Maynard J, Akbulut H, Deisseroth A (2006a). Engineering conditionally replication-competent adenoviral vectors carrying the cytosine deaminase gene increases the infectivity and therapeutic effect for breast cancer gene therapy. *Cancer Gene Ther* **13**, 346–356.

Liu Y, Ye T, Sun D, Maynard J, Deisseroth A (2006b). Tumor-specific therapeutic effect induced by an oncolytic adenoviral vector containing heat shock protein 70 and prodrug activation genes. *Gene Ther* **13**, 1235–1243.

Logg CR, Logg A, Matusik RJ, Bochner BH, Kasahara N (2002). Tissue-specific transcriptional targeting of a replication-competent retroviral vector. *J Virol* **76**, 12783–12791.

Lukashev AN, Fuerer C, Chen MJ, Searle P, Iggo R (2005). Late expression of nitroreductase in an oncolytic adenovirus sensitizes colon cancer cells to the prodrug CB1954. *Hum Gene Ther* **16**, 1473–1483.

Mar EC, Chiou JF, Cheng YC, Huang ES (1985). Inhibition of cellular DNA polymerase α and human cytomegalovirus-induced DNA polymerase by the triphosphates of 9-(2-hydroxyethoxymethyl) guanine and 9-(1,3-dihydroxy-2-propoxymethyl)guanine. *J Virol* **53**, 776–780.

Marais R, Spooner RA, Stribbling SM, Light Y, Martin J, Springer CJ (1997). A cell surface tethered enzyme improves efficiency in gene-directed enzyme prodrug therapy. *Nat Biotechnol* **15**, 1373–1377.

Markert JM, Medlock MD, Rabkin SD, *et al.* (2000). Conditionally replicating herpes simplex virus mutant, G207 for the treatment of malignant glioma: results of a phase I trial. *Gene Ther* **7**, 867–874.

Martiniello-Wilks R, Wang XY, Voeks DJ, *et al.* (2004). Purine nucleoside phosphorylase and fludarabine phosphate gene-directed enzyme prodrug therapy suppresses primary tumour growth and pseudo-metastases in a mouse model of prostate cancer. *J Gene Med* **6**, 1343–1357.

Martuza RL, Malick A, Markert JM, Ruffner KL, Coen DM (1991). Experimental therapy of human glioma by means of a genetically engineered virus mutant. *Science* **252**, 854–856.

Mavria G, Porter CD (2001). Reduced growth in response to ganciclovir treatment of subcutaneous xenografts expressing HSV-tk in the vascular compartment. *Gene Ther* **8**, 913–920.

Mavria G, Harrington KJ, Marshall CJ, Porter CD (2005). *In vivo* efficacy of HSV-TK transcriptionally targeted to the tumour vasculature is augmented by combination with cytotoxic chemotherapy. *J Gene Med* **7**, 263–275.

McCart JA, Ward JM, Lee J, *et al.* (2001). Systemic cancer therapy with a tumor-selective vaccinia virus mutant lacking thymidine kinase and vaccinia growth factor genes. *Cancer Res* **61**, 8751–8757.

McKee TD, Grandi P, Mok W, *et al.* (2006). Degradation of fibrillar collagen in a human melanoma xenograft improves the efficacy of an oncolytic herpes simplex virus vector. *Cancer Res* **66**, 2509–2513.

Miki K, Xu M, Gupta A, *et al.* (2001). Methioninase cancer gene therapy with selenomethionine as suicide prodrug substrate. *Cancer Res* **61**, 6805–6810.

Miyatake S, Martuza RL, Rabkin SD (1997). Defective herpes simplex virus vectors expressing thymidine kinase for the treatment of malignant glioma. *Cancer Gene Ther* **4**, 222–228.

Mohr L, Shankara S, Yoon SK, *et al.* (2000). Gene therapy of hepatocellular carcinoma *in vitro* and *in vivo* in nude mice by adenoviral transfer of the *Escherichia coli* purine nucleoside phosphorylase gene. *Hepatology* **31**, 606–614.

Moolten FL (1986). Tumor chemosensitivity conferred by inserted herpes thymidine kinase genes: paradigm for a prospective cancer control strategy. *Cancer Res* **46**, 5276–5281.

Moriuchi S, Wolfe D, Tamura M, *et al.* (2002). Double suicide gene therapy using a replication defective herpes simplex virus vector reveals reciprocal interference in a malignant glioma model. *Gene Ther* **9**, 584–591.

Moriuchi S, Glorioso JC, Maruno M, *et al.* (2005). Combination gene therapy for glioblastoma involving herpes simplex virus vector-mediated codelivery of mutant IkappaBalpha and HSV thymidine kinase. *Cancer Gene Ther* **12**, 487–496.

Morris JC, Ramsey WJ, Wildner O, Muslow HA, Aguilar-Cordova E, Blaese RM (2000). A phase I study of intralesional administration of an adenovirus vector expressing the HSV-1 thymidine kinase gene (AdV.RSV-TK) in combination with escalating doses of ganciclovir in patients with cutaneous metastatic malignant melanoma. *Hum Gene Ther* **11**, 487–503.

Morris JC, Wildner O (2000). Therapy of head and neck squamous cell carcinoma with an oncolytic adenovirus expressing HSV-tk. *Mol Ther* **1**, 56–62.

Nakamura H, Mullen JT, Chandrasekhar S, Pawlik TM, Yoon SS, Tanabe KK (2001). Multimodality therapy with a replication-conditional herpes simplex virus 1 mutant that expresses yeast cytosine deaminase for intratumoral conversion of 5-fluorocytosine to 5-fluorouracil. *Cancer Res* **61**, 5447–5452.

Naldini L, Blomer U, Gallay P, *et al.* (1996). *In vivo* gene delivery and stable transduction of nondividing cells by a lentiviral vector. *Science* **272**, 263–267.

Nanda D, Vogels R, Havenga M, Avezaat CJ, Bout A, Smitt PS (2001). Treatment of malignant gliomas with a replicating adenoviral vector expressing herpes simplex virus-thymidine kinase. *Cancer Res* **61**, 8743–8750.

Nemunaitis J, Ganly I, Khuri F, *et al.* (2000). Selective replication and oncolysis in p53 mutant tumors with ONYX-015, an E1B-55kD gene-deleted adenovirus, in patients with advanced head and neck cancer: a phase II trial. *Cancer Res* **60**, 6359–6366.

Nemunaitis J, Cunningham C, Tong AW, *et al.* (2003). Pilot trial of intravenous infusion of a replication-selective adenovirus (ONYX-015) in combination with chemotherapy or IL-2 treatment in refractory cancer patients. *Cancer Gene Ther* **10**, 341–352.

Niculescu-Duvaz I, Springer CJ (2004). Introduction to the background, principles, and state of the art in suicide gene therapy. *Methods Mol Med* **90**, 1–28.

Nishiyama T, Kawamura Y, Kawamoto K (1982). Antineoplastic effect of 5-fluorocytosine and cytosine deaminase on brain tumor. *Neurol Med Chir (Tokyo)* **22**, 344–352.

Okabe S, Arai T, Yamashita H, Sugihara K (2003). Adenovirus-mediated prodrug-enzyme therapy for CEA-producing colorectal cancer cells. *J Cancer Res Clin Oncol* **129**, 367–373.

Okada H, Miyamura K, Itoh T, *et al.* (1996). Gene therapy against an experimental glioma using adeno-associated virus vectors. *Gene Ther* **3**, 957–964.

Oosterhoff D, Witlox MA, Van Beusechem VW, *et al.* (2003). Gene-directed enzyme prodrug therapy for osteosarcoma: sensitization to CPT-11 *in vitro* and *in vivo* by adenoviral delivery of a gene encoding secreted carboxylesterase-2. *Mol Cancer Ther* **2**, 765–771.

Paielli DL, Wing MS, Rogulski KR, *et al.* (2000). Evaluation of the biodistribution, persistence, toxicity, and potential of germ-line transmission of a replication-competent human adenovirus following intraprostatic administration in the mouse. *Mol Ther* **1**, 263–274.

Palmer DH, Mautner V, Mirza D, *et al.* (2004). Virus-directed enzyme prodrug therapy: intratumoral administration of a replication-deficient adenovirus encoding nitroreductase to patients with resectable liver cancer. *J Clin Oncol* **22**, 1546–1552.

Pawelek JM, Low KB, Bermudes D (1997). Tumor-targeted Salmonella as a novel anticancer vector. *Cancer Res* **57**, 4537–4544.

Pawlik TM, Nakamura H, Mullen JT, *et al.* (2002). Prodrug bioactivation and oncolysis of diffuse liver metastases by a herpes simplex virus 1 mutant that expresses the CYP2B1 transgene. *Cancer* **95**, 1171–1181.

Pecora AL, Rizvi N, Cohen GI, *et al.* (2002). Phase I trial of intravenous administration of PV701, an oncolytic virus, in patients with advanced solid cancers. *J Clin Oncol* **20**, 2251–2266.

Peng KW, Facteau S, Wegman T, O'Kane D, Russell SJ (2002). Non-invasive *in vivo* monitoring of trackable viruses expressing soluble marker peptides. *Nat Med* **8**, 527–531.

Peplinski GR, Tsung K, Casey MJ, *et al.* (1996). *In vivo* murine tumor gene delivery and expression by systemic recombinant vaccinia virus encoding interleukin-1beta. *Cancer J Sci Am* **2**, 21.

Porosnicu M, Mian A, Barber GN (2003). The oncolytic effect of recombinant vesicular stomatitis virus is enhanced by expression of the fusion cytosine deaminase/uracil phosphoribosyltransferase suicide gene. *Cancer Res* **63**, 8366–8376.

Post DE, Khuri FR, Simons JW, Van Meir EG (2003). Replicative oncolytic adenoviruses in multimodal cancer regimens. *Hum Gene Ther* **14**, 933–946.

Prados MD, McDermott M, Chang SM, *et al.* (2003). Treatment of progressive or recurrent glioblastoma multiforme in adults with herpes simplex virus thymidine kinase gene vector-producer cells followed by intravenous ganciclovir administration: a phase I/II multi-institutional trial. *J Neurooncol* **65**, 269–278.

Puhlmann M, Gnant M, Brown CK, Alexander HR, Bartlett DL (1999). Thymidine kinase-deleted vaccinia virus expressing purine nucleoside phosphorylase as a vector for tumor-directed gene therapy. *Hum Gene Ther* **10**, 649–657.

Rainov NG (2000). A phase III clinical evaluation of herpes simplex virus type 1 thymidine kinase and ganciclovir gene therapy as an adjuvant to surgical resection and radiation in adults with previously untreated glioblastoma multiforme. *Hum Gene Ther* **11**, 2389–2401.

Rainov NG, Ren H (2003). Oncolytic viruses for treatment of malignant brain tumours. *Acta Neurochir Suppl* **88**, 113–123.

Rampling R, Cruickshank G, Papanastassiou V, *et al.* (2000). Toxicity evaluation of replication-competent herpes simplex virus (ICP 34.5 null mutant 1716) in patients with recurrent malignant glioma. *Gene Ther* **7**, 859–866.

Raper SE, Chirmule N, Lee FS, *et al.* (2003). Fatal systemic inflammatory response syndrome in a ornithine transcarbamylase deficient patient following adenoviral gene transfer. *Mol Genet Metab* **80**, 148–158.

Reid T, Galanis E, Abbruzzese J, *et al.* (2002). Hepatic arterial infusion of a replication-selective oncolytic adenovirus (dl1520): phase II viral, immunologic, and clinical endpoints. *Cancer Res* **62**, 6070–6079.

Rogulski KR, Wing MS, Paielli DL, Gilbert JD, Kim JH, Freytag SO (2000). Double suicide gene therapy augments the antitumor activity of a replication-competent lytic adenovirus through enhanced cytotoxicity and radiosensitization. *Hum Gene Ther* **11**, 67–76.

Rooseboom M, Commandeur JNM, Vermeulen NPE (2004). Enzyme-catalyzed activation of anticancer prodrugs. *Pharmacol Rev* **56**, 53–102

Russell SJ (2002). RNA viruses as virotherapy agents. *Cancer Gene Ther* **9**, 961–966.

Ryan PC, Jakubczak JL, Stewart DA, *et al.* (2004). Antitumor efficacy and tumor-selective replication with a single intravenous injection of OAS403, an oncolytic adenovirus dependent on two prevalent alterations in human cancer. *Cancer Gene Ther* **11**, 555–569.

Salmons B, Lohr M, Gunzburg WH (2003). Treatment of inoperable pancreatic carcinoma using a cell-based local chemotherapy: results of a phase I/II clinical trial. *J Gastroenterol* **38** (Suppl 15), 78–84.

Sandmair AM, Turunen M, Tyynela K, *et al.* (2000). Herpes simplex virus thymidine kinase gene therapy in experimental rat BT4C glioma model: effect of the percentage of thymidine kinase-positive glioma cells on treatment effect, survival time, and tissue reactions. *Cancer Gene Ther* **7**, 413–421.

Schatzlein AG (2001). Non-viral vectors in cancer gene therapy: principles and progress. *Anticancer Drugs* **12**, 275–304.

Schepelmann S, Hallenbeck P, Ogilvie LM, *et al.* (2005). Systemic gene-directed enzyme prodrug therapy of hepatocellular carcinoma using a targeted adenovirus armed with carboxypeptidase G2. *Cancer Res* **65**, 5003–5008.

Schepelmann S, Ogilvie LM, Hedley D *et al.* (2007). Suicide gene therapy of human colon carcinoma xenografts using an armed oncolytic adenovirus expressing carboxypeptidase G2. *Cancer Res* **67**, 4949–4955.

Schoensiegel F, Paschen A, Sieger S, *et al.* (2004). MIA (melanoma inhibitory activity) promoter mediated tissue-specific suicide gene therapy of malignant melanoma. *Cancer Gene Ther* **11**, 408–418.

Shalev M, Miles BJ, Thompson TC, *et al.* (2000). Suicide gene therapy for prostate cancer using a replication-deficient adenovirus containing the herpesvirus thymidine kinase gene. *World J Urol* **18**, 125–129.

Shand N, Weber F, Mariani L, *et al.* (1999). A phase 1–2 clinical trial of gene therapy for recurrent glioblastoma multiforme by tumor transduction with the herpes simplex thymidine kinase gene followed by ganciclovir. GLI328 European-Canadian Study Group. *Hum Gene Ther* **10**, 2325–2335.

Shen Y, Nemunaitis J (2005). Fighting cancer with vaccinia virus: teaching new tricks to an old dog. *Mol Ther* **11**, 180–195.

Shen Y, Nemunaitis J (2006). Herpes simplex virus 1 (HSV-1) for cancer treatment. *Cancer Gene Ther* **13**, 975–992.

Siders WM, Halloran PJ, Fenton RG (1998). Melanoma-specific cytotoxicity induced by a tyrosinase promoter-enhancer/herpes simplex virus thymidine kinase adenovirus. *Cancer Gene Ther* **5**, 281–291.

Simpson GR, Han Z, Liu B, Wang Y, Campbell G, Coffin RS (2006). Combination of a fusogenic glycoprotein, prodrug activation, and oncolytic herpes simplex virus for enhanced local tumor control. *Cancer Res* **66**, 4835–4842.

Singh S, Cunningham C, Buchanan A, Jolly DJ, Nemunaitis J (2001). Toxicity assessment of intratumoral injection of the herpes simplex type I thymidine kinase gene delivered by retrovirus in patients with refractory cancer. *Mol Ther* **4**, 157–160.

Solly SK, Trajcevski S, Frisen C, *et al.* (2003). Replicative retroviral vectors for cancer gene therapy. *Cancer Gene Ther* **10**, 30–39.

Sorscher EJ, Peng S, Bebok Z, Allan PW, Bennett LL Jr, Parker WB (1994). Tumor cell bystander killing in colonic carcinoma utilizing the *Escherichia coli* DeoD gene to generate toxic purines. *Gene Ther* **1**, 233–238.

Spooner RA, Martin J, Friedlos F, Marais R, Springer CJ (2000). In suicide gene therapy, the site of subcellular localization of the activating enzyme is more important than the rate at which it activates prodrug. *Cancer Gene Ther* **7**, 1348–1356.

Springer CJ (1993). CMDA, an antineoplastic prodrug. *Drugs Future* **18**, 212–215.

Springer CJ (2004). Introduction to vectors for suicide gene therapy. *Methods Mol Med* **90**, 29–46.

Springer CJ, Dowell R, Burke PJ, *et al.* (1995). Optimization of alkylating agent prodrugs derived from phenol and aniline mustards: a new clinical candidate prodrug (ZD2767) for antibody-directed enzyme prodrug therapy (ADEPT). *J Med Chem* **38**, 5051–5065.

Sterman DH, Recio A, Vachani A, *et al.* (2005). Long-term follow-up of patients with malignant pleural mesothelioma receiving high-dose adenovirus herpes simplex thymidine kinase/ganciclovir suicide gene therapy. *Clin Cancer Res* **11**, 7444–7453.

Stribbling SM, Friedlos F, Martin J, *et al.* (2000). Regressions of established breast carcinoma xenografts by carboxypeptidase G2 suicide gene therapy and the prodrug CMDA are due to a bystander effect. *Hum Gene Ther* **11**, 285–292.

Stubdal H, Perin N, Lemmon M, *et al.* (2003). A prodrug strategy using ONYX-015-based replicating adenoviruses to deliver rabbit carboxylesterase to tumor cells for conversion of CPT-11 to SN-38. *Cancer Res* **63**, 6900–6908.

Su H, Lu R, Ding R, Kan YW (2000). Adeno-associated viral-mediated gene transfer to hepatoma: thymidine kinase/interleukin 2 is more effective in tumor killing in non-ganciclovir (GCV)-treated than in GCV-treated animals. *Mol Ther* **1**, 509–515.

Sung MW, Yeh HC, Thung SN, *et al.* (2001). Intratumoral adenovirus-mediated suicide gene transfer for hepatic metastases from colorectal adenocarcinoma: results of a phase I clinical trial. *Mol Ther* **4**, 182–191.

Tai CK, Wang WJ, Chen TC, Kasahara N (2005). Single-shot, multicycle suicide gene therapy by replication-competent retrovirus vectors achieves long-term survival benefit in experimental glioma. *Mol Ther* **12**, 842–851.

Tanizawa A, Fujimori A, Fujimori Y, Pommier Y (1994). Comparison of topoisomerase I inhibition, DNA damage, and cytotoxicity of camptothecin derivatives presently in clinical trials. *J Natl Cancer Inst* **86**, 836–842

Thomas MA, Spencer JF, La Regina MC, *et al.* (2006). Syrian hamster as a permissive immunocompetent animal model for the study of oncolytic adenovirus vectors. *Cancer Res* **66**, 1270–1276.

Toda M, Martuza RL, Rabkin SD (2001). Combination suicide/cytokine gene therapy as adjuvants to a defective herpes simplex virus-based cancer vaccine. *Gene Ther* **8**, 332–339.

Tseng JC, Levin B, Hurtado A, *et al.* (2004). Systemic tumor targeting and killing by Sindbis viral vectors. *Nat Biotechnol* **22**, 70–77.

Tseng JC, Zanzonico PB, Levin B, Finn R, Larson SM, Meruelo D (2006). Tumor-specific *in vivo* transfection with HSV-1 thymidine kinase gene using a Sindbis viral vector as a basis for prodrug ganciclovir activation and PET. *J Nucl Med* **47**, 1136–1143.

Tychopoulos M, Corcos L, Genne P, Beaune P, De Waziers I (2005). A virus-directed enzyme prodrug therapy (VDEPT) strategy for lung cancer using a CYP2B6/NADPH-cytochrome P450 reductase fusion protein. *Cancer Gene Ther* **12**, 497–508

Tyminski E, Leroy S, Terada K, *et al.* (2005). Brain tumor oncolysis with replication-conditional herpes simplex virus type 1 expressing the prodrug-activating genes, CYP2B1 and secreted human intestinal carboxylesterase, in combination with cyclophosphamide and irinotecan. *Cancer Res* **65**, 6850–6857.

VandenDriessche T, Collen D, Chuah MK (2003). Biosafety of onco-retroviral vectors. *Curr Gene Ther* **3**, 501–515.

Varghese S, Rabkin SD (2002). Oncolytic herpes simplex virus vectors for cancer virotherapy. *Cancer Gene Ther* **9**, 967–978.

Wang WJ, Tai CK, Kasahara N, Chen TC (2003). Highly efficient and tumor-restricted gene transfer to malignant gliomas by replication-competent retroviral vectors. *Hum Gene Ther* **14**, 117–127.

Wang Y, Xue SA, Hallden G, *et al.* (2005). Virus-associated RNA I-deleted adenovirus, a potential oncolytic agent targeting EBV-associated tumors. *Cancer Res* **65**, 1523–1531.

Wildner O, Morris JC (2000). The role of the E1B 55 kDa gene product in oncolytic adenoviral vectors expressing herpes simplex virus-tk: assessment of antitumor efficacy and toxicity. *Cancer Res* **60**, 4167–4174.

Wildner O, Blaese RM, Morris JC (1999). Therapy of colon cancer with oncolytic adenovirus is enhanced by the addition of herpes simplex virus-thymidine kinase. *Cancer Res* **59**, 410–413.

Yamamoto N, Gupta A, Xu M, *et al.* (2004). Methioninase gene therapy with selenomethionine induces apoptosis in bcl-2-overproducing lung cancer cells. *Cancer Gene Ther* **10**, 445–450.

Yu YA, Shabahang S, Timiryasova TM, *et al.* (2004). Visualization of tumors and metastases in live animals with bacteria and vaccinia virus encoding light-emitting proteins. *Nat Biotechnol* **22**, 313–320.

Zeh HJ, Bartlett DL (2002). Development of a replication-selective, oncolytic poxvirus for the treatment of human cancers. *Cancer Gene Ther* **9**, 1001–1012.

Zeng ZJ, Li ZB, Luo SQ, Hu WX (2006). Retrovirus-mediated tk gene therapy of implanted human breast cancer in nude mice under the regulation of Tet-On. *Cancer Gene Ther* **13**, 290–297.

Zhan J, Gao Y, Wang W, *et al.* (2005). Tumor-specific intravenous gene delivery using oncolytic adenoviruses. *Cancer Gene Ther* **12**, 19–25.

Zhang M, Li S, Nyati MK, *et al.* (2003). Regional delivery and selective expression of a high-activity yeast cytosine deaminase in an intrahepatic colon cancer model. *Cancer Res* **63**, 658–663.

Zrachia A, Dobroslav M, Blass M, *et al.* (2002). Infection of glioma cells with Sindbis virus induces selective activation and tyrosine phosphorylation of protein kinase C delta. Implications for Sindbis virus-induced apoptosis. *J Biol Chem* **277**, 23693–23701.

17

Immunomodulatory Gene Therapy

Denise Boulanger and **Andrew Bateman**

17.1 Introduction

The fact that the immune system can to some extent fight against tumours has been suggested since the early 1900s (Coley, 1896). In the intervening years huge progress has been made in our understanding of fundamental immunology and tumour immunology. We now know that virtually all cancers will express tumour-specific antigens, and we know the molecular identity of many (van der Bruggen, 2007). We have a greater appreciation of the mechanism required to generate a de novo immune response and the crucial role of antigen-presenting cells (APC); and dendritic cells (DC) in particular (Banchereau and Steinman, 1998). More recently we have obtained greater awareness about the negative regulators of immune activation and subversion by tumours (Yamaguchi and Sakaguchi, 2006).

The presence of tumour infiltrating T lymphocytes (TILs) has also been recognized as an important prognostic marker, particularly in melanoma, ovarian and colon cancer (Galon *et al.*, 2006). This suggests that the host immune system can have a definite impact on cancer outcome. In addition we have seen the regular use of immunotherapies in the clinic: bacille Calmette–Guérin (BCG) in superficial bladder cancer, cytokines in renal cancer and melanoma, and an increasing range of targeted monoclonal antibodies. Direct confirmation of the potential of highly specific tumour – reactive T cells has been seen in adoptive transfer protocols in patients with metastatic melanoma (Dudley *et al.*, 2002).

Over the same period researchers have attempted to use viruses in cancer therapy; initially live virus inoculation, after reports of tumour regression coinciding with natural viral infections (Nelson, 1999). Subsequent molecular biology advances have allowed the manipulation and refinement of recombinant viral vectors. So that today, with only a few exceptions e.g. Reovirus, recombinant viral vectors are being developed and taken forward into the clinic. Strategies incorporating recombinant viral vectors specifically to promote immune responses will be the focus of this chapter.

17.2 Immunotherapy strategies using viral vectors

17.2.1 Tumour antigen-specific immunotherapy

The existence of tumour antigens recognized by T lymphocytes and their importance in the immune response was first described in mouse models. These are reviewed by Boon *et al.* (1994) who describe the identification of mouse and human tumour associated antigens and the different approaches that lead to their identification.

Viral Therapy of Cancer Edited by Kevin J. Harrington, Richard G. Vile and Hardev S. Pandha

Table 17.1 Classification of tumour-associated antigens

Category	Description	Examples
Unique antigens	Viral antigens	E6, E7 (HPV)
	Immunoglobulin/T-cell receptor idiotype	Lymphoma, leukaemia
	Mutated antigens	BCR-ABL, K-ras, casp-5
Cancer-germline (cancer-testis) antigens	Proteins only normally expressed in placental trophoblast and testicular germ cells – both MHC class 1 negative	MAGE, BAGE, GAGENY-ESO-1
Differentiation antigens	Normal proteins expressed in the tissue of origin of the tumour	Tyrosinase, gp100, TRP-2PSA, CEA
Overexpressed antigens	Ubiquitous proteins overexpressed in tumour cells	HER-2/neu, MUC-1, p53 Telomerase, survivin

Induction of auto-immunity should not occur if targeting unique antigens as these are specific for the tumour. However, apart from viral antigens, the remaining antigens will most likely be unique to the individual patient; requiring development of a personalized vaccine.

The remaining antigens are shared to some extent between normal tissues and tumour cells. The risk of inducing auto-immunity following successful vaccination will be greatest for differentiation antigens and overexpressed antigens. This may be acceptable in vitiligo, but potentially more significant with other antigens e.g. CEA.

Unless derived from oncogenic viruses (human papillomaviruses, Epstein–Barr virus), tumour associated antigens (TAA) are self-antigens, either mutated, incorrectly glycosylated (MUC-1), expressed at a higher level than usual (CEA, PSA, MUC-1) or expressed at a different developmental stage than in normal cells (5T4) (Table 17.1). Immunizing against a specific TAA can induce production of TAA-specific CD4 or CD8 T cells and/or TAA-specific antibodies. The tumour cells can be killed either directly by cytotoxic T cells, by antibody- or complement-dependent cellular cytotoxicity. As a result of tumour cell death new antigens can be presented to the immune system (epitope spreading or antigen cascade resulting in a bystander effect) and also a local inflammation can be generated, creating a more immunogenic environment.

Vaccination against TAAs has been the focus of the vast majority of clinical trials of recombinant viral vector immunotherapy (see below).

17.2.2 Adoptive transfer of virally transduced APC

Active immunization relies on efficient antigen delivery to APC. This can require the administration of a large dose of virus; possibly resulting in toxicity due to the infection of non-target cells. In order to increase efficiency of antigen delivery to professional antigen presenting cells and/or reduce toxicity, DC can be harvested from the patients, infected or transduced *in vitro* and then readministered to the patients. Viral gene delivery into DC has been achieved using poxviruses ($CD34^+$ derived DC transduced with MVA-hTyr (Di Nicola *et al.*, 2004), herpesviruses, adenoviruses, retroviruses, lentiviruses and adeno-associated viruses, reviewed in (Wysocki *et al.*, 2002).

17.2.3 Delivery of immunomodulators

Naive T cells require more than one signal for activation and subsequent proliferation into effector cells. Signal 1 is antigen specific, generated by the interaction between the major histocompatibility complex (MHC) class I or II molecule bearing a specific peptide on the antigen presenting cell and the T-cell receptor (TCR)–CD3 complex on the CD8, or CD4, T cell respectively. Signal 2 is non-antigen specific and can be provided by a series of costimulatory molecules on the APC, such as the B7 (B7.1/CD80 and B7.2/CD86) family of costimulatory molecules interacting with the CD28 ligand on the T cell, CD40 interacting with CD154 or CD40 ligand, CD54

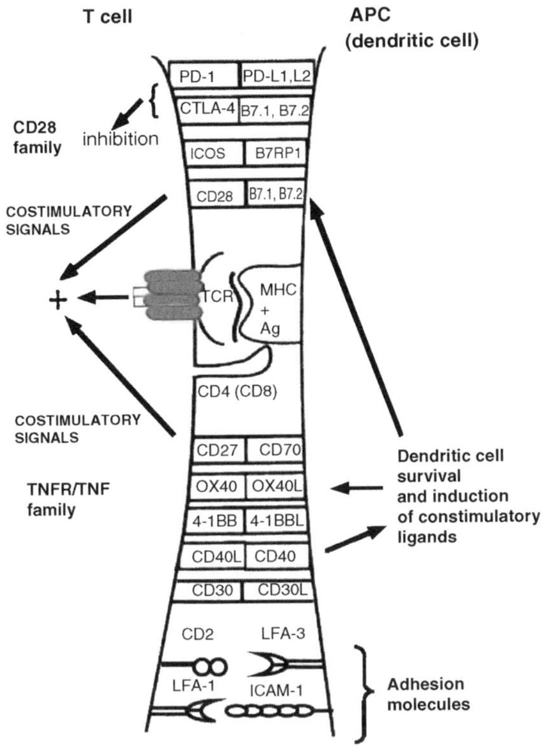

Figure 17.1 Schematic diagram of co-stimulatory receptors and adhesion molecules involved in T cell–dendritic cell interaction. Adapted from Bertram *et al.*, (2004)

(ICAM-1, intercellular adhesion molecule-1) interacting with the CD11a/CD18 complex (leukocyte function-associated antigen-1 (LFA-1)/β2-integrin) and CD58 (LFA-3) interacting with CD2 (Figure 17.1).

Another ligand for the B7 molecules, also present on T cells, is cytotoxic T lymphocyte antigen-4 (CTLA-4), which is structurally similar to CD28 but has a higher avidity. CD28 is expressed on both resting and activated T cells and provides a positive costimulatory signal. Whereas CTLA-4 is expressed only on activated T cells, appears within 24 h of stimulation, and creates a negative feedback mechanism to moderate T-cell proliferation, ensuring T-cell homeostasis. In the absence of signal 2 a naive T cell will be tolerized. Likewise, a resting T cell receiving signal 1, in the absence of signal 2, will become anergic. As part of a range of immune escape mechanisms tumour cells can loose signal 1 by downregulating expression of the MHC I molecules, or of different molecules involved in antigen processing and presentation (such as tapasin, TAP, β2- microglobulin). Additionally, tumour cells usually lack intrinsic expression of signal 2 costimulatory molecules, which are only expressed by APC. A strategy to induce an immune response against tumour cells is to deliver tumour antigens in combination with costimulatory molecules (see below), or to administer anti-CTLA-4 blocking antibodies. Addition of anti-CTLA-4 monoclonal antibody enhanced the immune response generated after vaccination with a recombinant vaccinia MVA recombinant vector expressing p53, resulting in rejection of palpable p53-overexpressing tumour cells and in lasting tumour immunity in mice (Espenschied *et al.*, 2003). Those therapies have been recently tested in patients: treatment with anti-CTLA-4 antibody and interleukin-2 (IL-2) induced objective tumour response but also autoimmune toxicity in patients with metastatic melanoma (Maker *et al.*, 2005). Likewise, melanoma peptide vaccination combined with anti-CTLA-4 antibody treatment induces durable objective response in some patients but also induced autoimmunity (Attia *et al.*, 2005).

The results of several clinical trials showed that systemic administration of cytokines or blocking antibodies often results in high toxicity. The recent clinical trial, testing a superagonist monoclonal antibody directed against CD28 (TGN1412), on six healthy individuals who developed rapid grade-IV toxicity with multi-organ failure illustrates the risks associated with those approaches (Farzaneh *et al.*, 2007). Safer options include local administration by intratumoral injection of cytokines or costimulatory molecules expressing viral vectors e.g. adenovirus-interferon-γ (IFN-γ) (Dummer *et al.*, 2004), recombinant vaccinia-TRICOM (<u>tri</u>ad of <u>co</u>stimulatory <u>m</u>olecules (B7.1, ICAM-1 and LFA-3 = TRICOM; Kaufman *et al.*, 2004) or vaccination using viral vectors coexpressing a tumour antigen and one or more cytokines or costimulatory molecules. Depending on the viral vector used a sustained delivery over several days or even weeks can be achieved.

Recombinant poxvirus vectors expressing costimulatory molecules have been progressively improved. (Hodge *et al.*, 1999) showed that the degree of T-cell activation using a vector containing TRICOM was far greater than the sum of the effect of each vector containing only one costimulatory molecule. Tumour cells infected with TRICOM expressing recombinants as stimulator cells induced *in vitro* a higher proliferation of naive CD4 and CD8 cells, a higher production of IL-2 by CD4 and of IFN-γ by CD8 T cells. Also, vaccination of mice with rV-CEA-TRICOM induced a higher T-cell response associated with a higher survival rate after tumour challenge.

Murine DC and splenocytes infected with vaccinia or fowlpox recombinants expressing TRICOM increased the level of expression of costimulatory molecules which enhanced their efficacy to activate T cells, requiring less signal 1 and less APC (Hodge *et al.*, 2000).

As indicated above, as well as costimulatory molecules, cytokines have been incorporated into viral vectors. The choice of cytokine has generally been one to aid DC maturation and presentation (granulocyte–macrophage colony-stimulating factor (GM-CSF), IL-12) or promote T-cell proliferation (IL-2, IFN-γ).

However with any strategy the clinical benefit for patients needs to be evaluated carefully: intratumoral injection of melanoma patients with canarypox vectors expressing B7.1 and IL-12 (ALVAC B7.1 + ALVAC-IL-12) induced the production of factors (VEGF, IL-10) that may even suppress the antitumour immune response (Triozzi *et al.*, 2005).

17.2.4 Suicide gene therapy/*viro*therapy

As opposed to the previous strategies, viral vectors used for suicide gene therapy or virotherapy target and kill the tumour cells. As a result of tumour cell death, and presence of foreign (viral) antigens, local inflammation can ensue; promoting a suitable environment for uptake of tumour antigens by DC and the subsequent development of a tumour specific immune response. These therapies are considered elsewhere in this book.

17.3 Viruses used as viral vectors in cancer immunotherapy

The aim of cancer immunotherapy is to generate a specific antitumour immune response of sufficient size, quality and duration to produce a clinically meaningful effect. Viral vectors are an attractive choice of antigen delivery system for this to occur. These vectors are capable of sustained TAA +/− immunomodulator expression in vivo. In addition they mimic a natural infection and provide potent danger signals; conditions that favour APC activation and generation of an adaptive immune response.

Numerous viral vector systems have been developed; we will focus on those which have been tested in clinical trials or hold the greatest promise for future development in this setting.

17.3.1 Poxviruses

The Poxviridae family includes a number of genera classified according to genome, antigenicity, morphology and host range. The orthopoxvirus genus infects mammals and includes the best known poxviruses: variola virus responsible for smallpox, cowpox virus and vaccinia virus, close relatives that have been successfully used to vaccinate humans against smallpox. Another genus with interest in the context of viral vectors is the avipoxvirus genus, infecting mainly birds. Two members of this genus, canarypox virus and fowlpox virus have been developed as viral vectors.

Recombinant poxviruses, especially vaccinia virus recombinants, were first developed as research tools, reviewed in (Mackett *et al.*, 1984). They were later on developed as successful and safe live vaccines against a number of pathogens in different hosts, reviewed by Paoletti (1996) and by Moss (1991, 1996), and as therapeutic vaccines against cancer in experimental mouse models (Lathe *et al.*, 1987). A number of clinical trials have now been carried out, some of which we will describe below.

Poxviruses have the ability to express large inserts spread in different insertion sites (at least 25 000 base pairs of foreign DNA can be inserted in the TK insertion site of the vaccinia virus genome, lack nuclear integration, are easy to construct without the need for complex packaging cell lines and are very stable. They also have strong adjuvant properties resulting from their intrinsic expression of MHC II restricted viral epitopes and to their ability to activate innate immunity through the activation of NK cells (Brooks *et al.*, 2006; Chisholm and Reyburn, 2006) and subsequent maturation of DC (Ryan *et al.*, 2007).

Vaccinia virus

Vaccinia vaccines have an extensive safety record due to the fact that vaccinia virus has been administered to millions of individuals during the smallpox eradication campaign. The first vaccine strains used for this purpose were replication competent and caused some adverse effects in a minority of people, the most serious effect being encephalitis. Attenuated strains, such as NYVAC, were therefore generated. The highly attenuated but still immunogenic NYVAC strain was generated from the replication competent Copenhagen strain by precise deletion of 18 open reading frames, including host range genes (Tartaglia *et al.*, 1992). Modified Vaccinia virus Ankara (MVA) was generated by extensive serial passages on chicken embryo fibroblasts of the parental pathogenic Ankara strain (Mayr, 1999). This treatment resulted in 6 large deletions (total of 31 kbp) including at least two host-range genes (Meyer *et al.*, 1991). Due to those deletions MVA is severely host restricted in most mammalian cells but is still able to induce a strong immune response. Early protein expression, DNA replication and late protein expression occur but viral morphogenesis is blocked (Sutter and Moss, 1992). MVA was specially developed for immunizing high-risk patients (nervous system, skin disorders or allergies) during the smallpox vaccination campaign. It has been administered safely to over 120 000 people at the end of the vaccination campaign (early 1970s); including people at risk for receiving the earlier strains of vaccinia virus such as immunocompromised people. It has therefore an excellent safety record. In addition, high levels of gene expression can be obtained and high titer stocks can be produced.

This chapter will focus on recombinants that have been tested in humans.

Vaccinia replication competent strains (Copenhagen and Wyeth, the licensed smallpox vaccine, derived from the original New York City Board of Health NYCBH strain, which is associated with the lowest incidence of clinical complications) have a short life cycle, rapid cell to cell spread, wide host cell range and strong lytic ability with inherent systemic tumour targeting (Whitman *et al.*, 1994). This makes them ideal for virotherapy targeted to tumour cells. Different strategies have been developed to increase further their specificity toward cancer cells (Zeh and Bartlett, 2002). Thymidine kinase (TK) knock-out mutants grow preferentially in dividing cells, especially in cancer cells where TK activity is constitutively high (Puhlmann *et al.*, 2000). A high tumour selectivity has been achieved by double TK/viral growth factor (VGF) deletion mutants (McCart *et al.*, 2001).

In addition to their use in virotherapy the large cloning capacity of vaccinia vectors make them ideal for vaccination purposes allowing insertion of several transgenes.

Pre-existing immunity against vaccinia virus can, however, reduce the efficiency of the vaccines. Although one vaccination with such vectors seemed to prime a response against the transgene, even in patients with prior smallpox vaccination at childhood, further vaccinations with the same vector have limited efficiency due to a robust neutralizing response against the vaccinia vector and lower response against the transgene. The increased immune response mounted against the virus is shown by the decrease in inoculation site reactions over successive inoculations (Gulley *et al.*, 2002) and by analysing the anti-vaccinia virus antibody and CTL response (Smith *et al.*, 2005).

Avipoxviruses

Avipoxvirus vectors do not replicate in mammalian cells. Virus replication is blocked at different steps depending on the cell type but early gene expression occurs (Taylor *et al.*, 1988; Somogyi *et al.*, 1993). Avipoxviruses, first developed and commercialized as vaccine vectors against poultry diseases (Boyle and Coupar, 1988), have therefore been considered as a safe alternative, or complementing vectors for vaccinating mammals, reviewed in (Limbach and Paoletti, 1996). The utility of fowlpox virus vectors in immunizing non-avian species was demonstrated in 1988 by the inoculation of different mammal species with a fowlpox recombinant expressing the rabies virus glycoprotein. The vaccination protected them against a live rabies virus challenge (Taylor *et al.*, 1988). The efficiency of canarypox virus as a rabies recombinant was tested shortly after (Taylor *et al.*, 1991). The first reported use of a fowlpox recombinant in the prevention and treatment of a murine experimental cancer was published in 1995 (Wang *et al.*, 1995). Late proteins, against which neutralizing antibodies are usually produced, are usually not expressed or at a reduced level depending on the cell type. This could explain why avipoxvirus vectors, as opposed to vaccinia vectors, can be administered several times inducing increasing immune response against the transgene over successive injections: canarypox expressing CEA could be given up to eight times with continued increases in CEA T-cell precursors (Marshall *et al.*, 2000). In addition, as avipoxviruses do not usually infect mammals, antibodies are rarely seen in humans: (Rosenberg *et al.*, 2003) detected anti-fowlpox antibodies (1: 1600 titre) only in 1 out of 38 patients.

Another advantage of fowlpox virus vectors is their sustained transgene expression in infected cells for up to 3 weeks in contrast to replication competent vaccinia expressing antigens from 1 to 2 days before cell death.

17.3.2 Adenoviruses

There has been considerable interest in the use of adenoviruses for cancer therapy. This has been predominantly focused on conditionally replicating adenoviruses and is dealt with in detail in Chapter 1.

The ability to generate recombinant replication incompetent vectors of choice relatively easily has meant adenoviruses have been used in a number of immunomodulatory strategies: vaccination against TAA (Rosenberg *et al.*, 1998b), delivery of anti-angiogenesis agents [Adenovirus-sFLT-1, (Mahasreshti *et al.*, 2003)] and delivery of cytokines have all been examined. In these studies it was confirmed that adenoviral vectors are highly immunogenic and not suitable for strategies where multiple boosts are required. However *ex vivo* they have been used to transduce human DC with TAA prior to re-infusion into patients. This is due to the efficient transduction of DC and ability not to affect DC maturation and function (Rea *et al.*, 1999).

Promising results have also been seen following local administration of adenovirus vectors in a number of different clinical trials: An adenovirus recombinant expressing IL-2 (adeno-IL2, TG1024) has been tested in 20 patients with metastatic melanoma or other solid tumours. The vector was injected intratumorally. The treatment was well tolerated and several disease stabilizations were observed (Liu *et al.*, 2004). The same team also tested an intratumoral injection of an adenovirus–IFN-γ recombinant (TG1042) in phase I and I/II clinical trials on 13 patients with advanced cutaneous lymphoma. Some adverse effects were noted but compensated by the benefit of the treatment; with an overall response rate of 60 per cent (4 complete and 2 partial responses out of 10 evaluable patients). Some of the patients showed a systemic complete response with clearance of other non-injected skin lesions (Dummer *et al.*, 2004, Liu *et al.*, 2004).

17.3.3 Alphaviruses

Alphaviruses (genus of the Togaviridae family) are small (50–70 nm diameter) enveloped viruses with a positive single-stranded RNA genome. The viral replication occurs in the cytoplasm so there is no risk of integration of foreign sequences within the host genome. The use of alphaviruses in cancer therapy, reviewed in (Atkins *et al.*, 2004) is still in its infancy and very few clinical trials have been performed. A strength appears to be that following infection with alphaviral vectors cells undergo apoptosis. Apoptotic bodies can be taken up by APCs, allowing antigenic epitopes to be cross-presented to T cells.

Semliki Forest virus (SFV)

Semliki Forest virus was first isolated from mosquitoes in Uganda. Its natural hosts are unknown. Although serological studies indicate that human infection is relatively common, SFV has been linked to human disease on only two occasions (Mathiot *et al.*, 1990). However acute lethal encephalitis can be induced in experimental rodent models. There are ongoing clinical trials in patients with glioblastoma multiforme; receiving a genetically modified replication disabled SFV vector expressing IL-12 and encapsulated in cationic liposomes (Ren *et al.*, 2003). This is also being explored in patients with melanoma and renal cell carcinoma.

Sindbis virus (SIN)

Sindbis virus is widespread, maintained primarily in birds and transmitted by mosquitoes, and can infect humans (Sindbis fever; associated with a rash and arthralgia). Sindbis virus has high gene transfer efficiency in human cells and replicates very efficiently. Entry into the cell is mediated by interaction with the laminin receptor upregulated in many human cancers, including breast cancer and colorectal cancer, conferring preferential targeting to tumour cells (Tseng *et al.*, 2002). Impressive results have been obtained in a mouse model after systemic delivery of a replication defective SIN vector. The vector specifically targeted tumours growing subcutaneously, intrapancreatically, intraperitoneally and in the lungs, and could induce complete tumour regression (Tseng *et al.*, 2004). The efficacy of the vector has been enhanced further by the incorporation of cytokine genes such as IL-12 and IL-15, which may enhance the induction of an immune response against antigens expressed by the tumour. No clinical trial has been performed so far to our knowledge.

17.3.4 Herpesviruses

Herpesviruses are large (120–200 nm diameter) enveloped viruses containing a linear dsDNA genome. Replication and transcription occur in the nucleus. Herpesviruses infect a very wide range of animals, including mammals, birds, reptiles, amphibians and fish. Many herpesviruses remain latent in the infected host, either in the central nervous system or in cells from the immune system and can be reactivated during immunodepression episodes.

Herpes simplex virus (HSV)

HSV has so far been mainly developed as an oncolytic agent, as detailed in Chapter 7. However HSV vectors have been used in the delivery of immunomodulatory molecules such as IL-12, GM-CSF and vaccination against TAA such as gp100, MART-1 and Tyrosinase. HSV encoding GM-CSF (OncoVEX$^{\text{GM-CSF}}$) has been tested in a phase I clinical trial in a range of solid tumours with encouraging results (Hu *et al.*, 2006).

17.4 Clinical trials against specific TAA

A single injection with a viral vector as part of an immunomodulatory vaccine strategy is highly unlikely to be sufficient to control or eliminate a tumour. Repetitive injections will more likely be required. Pre-existing immunity against the vector has already been mentioned as a main limitation for the use of viral vectors. Likewise, repetitive injections of a vector lead to the development of an immune response directed against the viral vector, and this has been shown to limit the level of immune response against the therapeutic insert. Heterologous prime-boost strategies using different poxviruses (vaccinia, fowlpox or canarypox), or different families of viruses (adenovirus and poxviruses) have proven to be more efficient.

Most clinical trials using viral vectors have been phase I trials evaluating safety of the vaccines. Most of them have been well tolerated with only mild or moderate reactions such as local transient inflammatory reactions at the site of injection.

This chapter can not review all clinical trials that have been performed. It aims at giving examples of strategies, focusing on some strategies that show promising results, and showing how the vectors and/or strategies have been improved over time. This process comes not only as a result of accumulating observations from clinical trials but also of the ever increasing understanding of the different arms of the immune response, how to improve the efficiency of antigen presentation and how to counteract immune evasion mechanisms.

17.4.1 Carcinoembryonic antigen (CEA)

The first clinical trial using a vaccinia (Wyeth) recombinant expressing CEA demonstrated that patients with metastatic carcinoma could elicit CEA specific T-cell response after vaccination. Although the patients received three injections it was clear from the inoculation lesion size that an anti-vector response after the first injection limited the boosting efficiency against the tumour antigen (Tsang *et al.*, 1995).

In a later phase I clinical trial patients with advanced carcinomas were vaccinated i.m. three times at monthly intervals with a canarypox recombinant virus expressing CEA (ALVAC-CEA). No objective antitumour effect was observed but seven of nine of the HLA-A2 patients had a significant increase in CEA specific precursors (Marshall *et al.*, 1999).

The results of the first phase I clinical trial using heterologous poxvirus prime/boost regimens was published in 2000. In this trial, 18 patients with advanced tumours expressing CEA received either recombinant vaccinia (rV-CEA) for priming, followed by three vaccinations with canarypox ALVAC-CEA, or a priming injection with the ALVAC-CEA (three times) followed by one vaccination with rV-CEA, and then subsequent injections with the ALVAC-CEA recombinant. The first strategy (VAAA) was more efficient to generate CEA-specific T-cell responses and resulted in longer survival than treatment with AAAV. ALVAC-CEA could be given up to eight times with continued increases in CEA T-cell precursors. Further increase was seen when GM-CSF and low doses of IL-2 were given with the last injections (Marshall *et al.*, 2000).

In a pilot study, patients with CEA expressing tumours received a canarypox recombinant expressing CEA and B7.1 (ALVAC-CEA B7.1), intradermally (i.d.) every other week four times and then boosted monthly. Leukocytic infiltration and CEA expression were observed in biopsies of vaccine sites. Six out of 39 (15 per cent) patients had a reduction in their serum CEA and had stable disease after four vaccinations. HLA-A2 positive patients demonstrated an increase in CEA-specific CTLs. The number of prior chemotherapy regimens was inversely correlated with the ability to generate a T-cell response (von Mehren *et al.*, 2000).

In another trial patients with advanced metastatic CEA expressing carcinoma were vaccinated with ALVAC-CEA B7.1 with or without GM-CSF. GM-CSF was given s.c. into the region of vaccination on 5 consecutive days, with vaccination i.d. on day 3. In preclinical models this strategy had brought significant therapeutic benefit. However the vaccination was associated with the induction of a CEA specific T-cell response in patients treated with vaccine alone, but not with the vaccine and GM-CSF. In responders disease stabilization was noted for up to 13 months (von Mehren *et al.*, 2001). Once more the number of prior chemotherapy regimens was negatively correlated with the generation of a T-cell response, whereas there was a positive correlation between the number of months from the last chemotherapy regimen and the T-cell response.

In an attempt to improve responses, CEA vectors were generated encoding a modified CEA molecule, CEA(6D), which contains an agonist epitope of CEA605–613 (Asn610 replaced by Asp) to enhance recognition by T cells. The vectors also expressed TRICOM and the strategy involved local GM-CSF administration. Fifty-eight patients with advanced CEA-expressing cancers were vaccinated with vaccinia (Wyeth strain) and/or fowlpox virus CEA(6D)-TRICOM recombinants in different regimens. The strategy generated a significant immune response and had clinical benefit in some patients with advanced cancer: 40 per cent of the patients had a stable disease after four monthly vaccinations, including 24 per cent stable for more than 6 months. Nineteen per cent of the patients had decreasing or stable CEA and one patient had a pathologically complete response. The trend showed that patients who received only the fowlpox recombinant did not do as well as patients who received a primary vaccination with the vaccinia recombinant and then booster vaccinations with the fowlpox vector. Also, patients who received GM-CSF with the vaccines seemed to do better than those who did not receive it. Twenty-four per cent of the patients who had stable disease after six vaccinations then received vaccinations every 3 months. It is interesting to note that the majority of them on progression reverted back to the monthly vaccination schedule and half of them restabilized their disease. (Marshall *et al.*, 2005).

In another phase I trial 14 patients (11 with colorectal cancer and 3 with non-small cell lung cancer) received one or two cycles of injections with *ex vivo* generated DC infected with a recombinant fowlpox vector encoding the modified CEA (CEA(6D) and TRICOM (rF-CEA(6D)-TRICOM). One patient had a decrease in the CEA level and minor regression in adenopathy; five patients were stable for at least 3 months. Ten patients showed an increase in CEA-specific T cells (Morse *et al.*, 2005).

The same team extended the previous study and analysed the NK cell responses in nine of the previous patients (five colorectal cancer, three lung cancer, one urachal adenocarcinoma). All patients with increased NK activity after vaccination had a better prognosis. The authors suggest therefore that NK response may correlate more closely to the clinical outcome than T-cell response and would be worth monitoring routinely in clinical trials (Osada *et al.*, 2006).

17.4.2 Melanoma

The majority of recognized melanoma tumour antigens (gp100, MART-1, TRP-1, TRP-2, tyrosinase) are differentiation antigens and are therefore expressed on normal melanocytes. Most self-antigens are poorly immunogenic and peptides derived from these proteins have intermediate to low binding affinity to MHC molecules; consistent with the hypothesis that T cells recognizing high-affinity self-peptides would be eliminated by negative selection in the thymus. Immunization against these antigens therefore requires the breaking of immune tolerance or activation of anergized T cells. Additional melanoma antigens are cancer-germline antigens (MAGE, BAGE, GAGE, NY-ESO-1 gene families) and are attractive candidates for T-cell mediated immunotherapy.

MAGE antigen

Forty patients with advanced cancer (37 melanoma patients) were sequentially injected four times with a recombinant canarypox virus containing a minigene encoding antigenic peptides MAGE-3 (168–176) and MAGE-1 (161–169) (ALVAC mini-MAGE-1/3). After viral injections patients were boosted 3 times with the MAGE-3 and MAGE-1 peptides. Tumour regression with a detectable CTL response was detected in a minority of patients (van Baren *et al.*, 2005).

Tyrosinase (hTyr)

In a phase I clinical trial 20 patients with stage II melanoma received three injections of 5×10^8 IU of MVA-hTyr at 4-week intervals. A strong response to the viral vector (CD4, CD8 and antibody responses) was observed but no T-cell or antibody response was observed against hTyr. This included no increase in one patient who had a detectable pre-existing T-cell response against tyrosinase peptides. The anti-vector response was thought to have played a role in the lack of hTyr responses, but additionally this strategy was thought insufficient to break tolerance (Meyer *et al.*, 2005).

In a phase II clinical trial 64 patients with advanced metastatic melanoma received a heterologous prime/boost vaccination with recombinant vaccinia and fowlpox vectors encoding hTyr alone, or combined with IL-2. Whereas the prime/boost immunization enhanced anti-tyrosinase immunity in a minority of patients with metastatic melanoma, it was ineffective alone in mediating clinical benefit. A substantial number of mixed responses and minor responses were seen but overall the combination with IL-2 did not result in significant clinical benefit over IL-2 alone. Again failure to break tolerance and the development of some antigen loss variants were considered to be the main impediment to successful therapy (Lindsey *et al.*, 2006).

In a phase I trial six patients with stage IV melanoma were vaccinated with autologous $CD34^+$-derived DC transduced with MVA recombinant expressing human tyrosinase (MVA-hTyr). MVA-hTyr expressing DC (1×10^8) were given at each vaccination: the first intravenously (i.v.) then three subsequent injections subcutaneously (s.c.), with 14 days between each. One patient had a partial response with shrinkage of a s.c. nodule, later surgically removed, and remained disease free for more than 2 years. Two patients developed vitiligo. Significant long-lasting increase in the frequency of hTyr-specific CTLs, with an effector memory/T terminally differentiate stages CCR7-CD45RA$^{-/+}$, was observed in most patients. (Di Nicola *et al.*, 2004).

gp100

The gp100 melanoma antigen is a 661 amino-acid melanosomal matrix protein involved in melanin synthesis.

In a phase I trial 54 patients with metastatic melanoma were vaccinated with a recombinant adenovirus expressing either gp100 or MART-1 alone, or followed by the administration of IL-2. One of 16 patients who received adenovirus MART-1 alone had a complete response after four injections. Of the 20 patients who received adenovirus MART-1 + IL-2, two had a complete response and two a partial one. No response was observed in 6 patients who received the adenovirus gp100 alone. One complete response was observed in 12 patients who received adenovirus gp100 + IL-2. Thus, except for the single patient who responded after vaccination with the adenovirus MART-1 vaccine alone, there was no evidence that the adenoviral immunization enhanced the anti-tumour effects seen with IL-2 administration alone. High levels of neutralizing antibodies might have impaired antigen delivery: 52 of 53 patients had substantial preimmunization titre of anti-adenovirus neutralizing antibodies and more than 75 per cent of them showed a significant titer increase after a single immunization (Rosenberg *et al.*, 1998b).

In a pilot study patients with metastatic melanoma were immunized three times with a recombinant fowlpox vaccine (POXVAC-TC vaccine strain) expressing either the native gp100, a modified gp100 in which 2 aa have been substituted to increase binding avidity to the MHC molecule (Thr210 - Met, Ala288 - Val), or a gp100 minigene including an endoplasmic reticulum (ER) signal sequence upstream of the modified gp100:209–217(210M) peptide; with the aim of facilitating transport of the gp100 peptide into the ER, where MHC I loading occurs. One out of seven patients developed specific CTLs after vaccination with native gp100, 18 out of 28 with the modified gp100 and 12 out of 16 with the minigene. Modifying individual anchor amino-acids to create epitopes with higher binding affinity for MHC I molecules can therefore substantially increase the immunogenicity of gp100. Only one patient who had received the modified gp100 vaccine showed a partial clinical response after vaccine alone treatment. Once patients showed evidence of progressive disease they received IL-2. None of the 13 patients who received native gp100 or modified gp100 showed any clinical benefit. Six out of 12 patients who had received the minigene showed objective cancer regression after subsequent IL-2 treatment; including three complete responders, showing no evidence of disease at 12–24 months (Rosenberg *et al.*, 2003). These results are quite encouraging when compared to treatment with IL-2 only, which can induce clinical response in 15 per cent of treated patients (Rosenberg *et al.*, 1998a).

The route of administration was studied in a report of 42 patients with high risk Melanoma treated using six different protocols. The differing strategies included gp100 peptide vaccination only, priming with ALVAC-modified gp100 (gp100m) and boosting with peptide, and the co-administration of tetanus toxoid to provide CD4 help. Peptide or ALVAC were given s.c. or by intranodal injection. Intranodal injection of ALVAC-gp100 resulted in earlier CTL response and higher anti-ALVAC antibody production (Spaner *et al.*, 2006). Peptide only gave no gp100-reactive CTL responses versus 8/18 for the ALVAC-gp100. However, even in CTL responders levels were not maintained and became unrecordable 3 months post vaccination.

17.4.3 Solid tumours expressing MUC1

MUC1 is a highly glycosylated mucin protein normally found on the surface of epithelial cells in many types of tissue including breast, prostate, lung, pancreas, stomach, ovaries, fallopian tubes, intestine. MUC1 is a transmembrane protein containing a small intracellular tail. The extracellular domain consists of a large number (30–100) of 20 amino acid tandem repeats. The peptide core is highly glycosylated, conferring a rigid rod-like structure that can extend several hundred nanometres from the apical cell surface into the lumen of ducts and glands. MUC1 is often overexpressed in epithelial tumour cells, overexpression being a bad prognosis. MUC1 expressed in tumour cells is heavily underglycosylated, revealing new peptide and carbohydrate epitopes, otherwise cryptic, recognized either by antibodies or CTL. Non-MHC restricted cytotoxic T-cell activity has also been described (Barnd *et al.*, 1989, Wajchman *et al.*, 2004). Circulating MUC1 glycoprotein (tumour marker CA15-3) is found in advanced stage cancer patients.

A recombinant vaccinia (Copenhagen) vector expressing MUC1 and IL-2 (VV-MUC1-IL-2, TG1031) was tested in a phase I/II clinical trial on nine patients with advanced inoperable breast cancer recurrences to the chest wall. Patients were treated with one or two injections of the recombinant vaccine intramuscularly. Specific T cells and increased NK activity were reported but limited clinical benefit was observed (Scholl *et al.*, 2000). A subsequent phase II trial involved 31 patients who received repeated injections with the same TG1031 vector. Two patients achieved partial responses with clinical regression of axillary lymph nodes, seen after the first injection, or of liver metastases after three injections, and 15 patients had some disease stabilization. A rise in circulating CD4 T lymphocytes was seen in some patients and in particular in the two responding patients. Mild autoimmunity effects (anti-DNA or anti-TPO antibodies) were seen in three patients including one of the partial responders (Scholl *et al.*, 2003, Liu *et al.*, 2004).

The same vector was also tested in a phase I study on patients with prostate cancer. Some of them showed stable disease (stable or decreased level of PSA) for 1–2 years (Pantuck *et al.*, 2004).

TG1031, based on a replication competent vaccinia, suffered regulatory issues and was subsequently replaced by TG4010 based on vaccinia MVA (MVA-MUC1-IL-2). At the same time as increasing safety, expression of the MUC1 sequence was improved by stabilization of the tandem-repeat portion and by using a stronger promoter. This new vaccine has been tested in different phase II clinical trials in patients affected with different types of cancer. No objective clinical responses were observed in patients with advanced metastatic breast cancer, but more promising results were observed in patients with advanced lung cancer; with 25 per cent or more of the patients showing disease stabilization (Liu *et al.*, 2004).

17.4.4 Colorectal cancer – 5T4

An MVA recombinant expressing the tumour antigen 5T4 (TroVax) has been evaluated in a phase I/II trial in 22 metastatic colorectal cancer patients. Of 17 evaluable patients, 16 showed 5T4 specific cellular responses, whereas 14 developed a 5T4 specific antibody response. TroVax was able to boost the immune response in the presence of MVA neutralizing antibodies. Periods of disease stabilization ranging from 3 to 18 months were observed in five patients, all of whom mounted a 5T4 specific immune response (Harrop *et al.*, 2006). Trovax is also currently in trial in renal cell carcinoma, prostate cancer and breast cancer (http://www.oxfordbiomedica.co.uk/trovax.htm).

17.4.5 Prostate cancer

Forty per cent of patients diagnosed with prostate cancer will die of this disease, accounting for 10 000 deaths in the UK each year. Novel therapies are required and of these there is considerable enthusiasm for immunotherapy approaches. This is due to a number of factors: firstly the prostate is a non-essential organ and therefore inducing an immune reaction within the gland will not cause major problems; and secondly, a number of specific prostate antigens have been identified. Prostate-specific antigen (PSA) is the most targeted to date.

PSA

In a phase I clinical trial, 33 patients with advanced prostate cancer with rising PSA levels after prostatectomy, radiation therapy or with metastatic disease, were vaccinated 3 times at 4-week intervals with a vaccinia (Wyeth strain) recombinant vector expressing PSA (rV-PSA) with or without GM-CSF. 14/33 patients had stable disease for at least 6 months and 9 of them remained stable for 11–25 months. The greatest increase in PSA-specific precursor frequency was observed after the first vaccination. Subsequent vaccinations did not produce substantial additional increases (Eder *et al.*, 2000).

A phase I clinical trial was then conducted using the same vaccine (rV-PSA or PROSTVAC: Vaccinia Wyeth (NYCBH)) in patients with metastatic androgen-independent prostate cancer. Forty-two patients were given a total of three vaccinations at 4-week intervals. There was no objective tumour response even in selected patients who showed an immunologic response (Gulley *et al.*, 2002).

In a phase II trial a heterologous prime/boost strategy was tested using vaccinia (Wyeth) and fowlpox recombinant viruses expressing PSA (rV and rF-PSA respectively). 64 men with biochemical progression after local therapy for prostate cancer were studied. The trial was conducted in three arms: (a) rF-PSA ×4, (b) rF-PSA ×3 then rV-PSA ×1, (c) rV-PSA ×1 then rF-PSA ×3. Vaccination was well tolerated and there was no evidence of autoimmunity, nor the development of anti-PSA antibodies in any of the patients. Forty-five per cent of patients remained free of PSA progression at 19 months of follow up, with 46 per cent of HLA-A2 patients indicating an increase in PSA-specific T-cell precursors. The study was not powered to detect differences between arms but there was a trend to improved outcome following priming with rV-PSA (Kaufman *et al.*, 2004).

As indicated previously, many patients with clinically localized prostate cancer subsequently progress systemically. Therefore a strategy was developed to include patients at an earlier stage of the disease. A randomized phase II study looked to examine the combination of local therapy: external beam radiation therapy, with heterologous prime/boost vaccination against PSA. Thirteen patients received standard radiation therapy alone. Seventeen patients received a prime vaccination using two recombinant vaccinia (Wyeth) vaccines expressing PSA and the B7.1 costimulatory molecule. They then received seven injections monthly using a recombinant fowlpox vector expressing PSA. The patients received a s.c. injection of GM-CSF as well at the site of vaccination and a s.c. injection of IL-2 in the abdomen. Standard external beam radiation therapy was given between the fourth and the sixth vaccinations.

Although this study did not follow the patients long enough to determine if the strategy had any clinical benefit, it showed that some patients developed CTLs specific not only for PSA but also to other prostate cancer antigens, such as PSMA, PAP, PSCA and MUC1. This phenomenon known as epitope spreading or antigen cascade suggests that PSA-specific CTL generated by the vaccine have been responsible for tumour cell death and presentation of new TAA to the immune system (this phenomenon can also be observed after radiotherapy). The study showed as well that although the level of circulating PSA-specific CTL decreased in some patients after local radiation (but increased in others), it increased again later. This suggests that radiation and vaccination therapies can be combined and will hopefully result in positive clinical benefit (Gulley *et al.*, 2005).

Another combined therapy strategy conducted by the same group used the same vectors combined with antiandrogen therapy (nilutamide) (Arlen *et al.*, 2005).

In a phase II trial, 28 patients with metastatic androgen independent prostate cancer received a vaccine regime composed of rV-PSA admixed with rV-B7.1 as a prime, and sequential boosts with rF-PSA alone or with weekly administration of the chemotherapeutic agent docetaxel. The results of the trial showed that docetaxel (and ‘short bursts’ of steroids) can be administered safely with immunotherapy without inhibiting T cell responses. Furthermore, vaccinated patients transferred to docetaxel treatment may respond longer to the drug compared to a historical control of patients who received docetaxel alone (Arlen *et al.*, 2006).

To improve efficacy of the vaccine strategy TRICOM corresponding vectors have been developed. Safety of rV-PSA-TRICOM (PROSTVAC-V) and rF-PSA-TRICOM (PROSTVAC-F) has been recently tested in an heterologous prime-boost therapeutic vaccine (PROSTVAC-VF/TRICOM) (Dipaola *et al.*, 2006). Phase I/II clinical studies with these vectors, including a study incorporating an anti-CTLA-4 antibody, are on-going.

17.4.6 NY-ESO-1

Following encouraging results after vaccination with a recombinant NY-ESO-1 protein in a saponin-based adjuvant, recombinant vaccinia and fowlpox vectors were developed. Vaccination outcome in a very diverse patient cohort seemed to show some benefit that might appear clearer in future clinical trials (Jager *et al.*, 2006).

17.4.7 HPV-associated cancers.

Human papillomaviruses (HPV) have been associated with a variety of epithelial proliferative diseases. The best known consequence of HPV infection is cervical cancer. HPV infection (mostly HPV16) is followed by a long latency before the appearance of precancerous epithelial changes that can develop within 10 to 15 years in invasive carcinomas. Although there is now a commercially available papilloma prophylactic vaccine, women already infected must rely on early detection followed by treatment by surgery and/or radiotherapy. Therapeutic vaccination could be an attractive alternative.

HPV E6 and E7 proteins are involved in the induction of neoplasia resulting from their interaction with the tumour repressor genes *p53* and *Rb* respectively. HPV E2 downregulates the expression of E6 and E7, and can also promote cell arrest and apoptosis.

Rosales *et al.* (2000) showed in a rabbit experimental model that intratumoral injection of an MVA recombinant expressing the bovine papillomavirus E2 protein could result in impressive regressions. No T cell cytotoxicity was observed but E2 antibodies were produced and activated macrophage antibody-dependent cytotoxicity. Based on those very promising results, a phase I/II clinical trial was conducted on 36 women with cervical intraepithelial neoplasia (CIN) 1 to 3 grade lesions. They received an MVA-human HPV E2 vaccine injected directly into the cervix. All patients developed antibodies and specific cytotoxicity; with 34 of the patients (94 per cent) achieving complete elimination of precancerous lesions. In the remaining two patients precancerous lesions were reduced from grade CIN 3 to CIN 1 (Corona Gutierrez *et al.*, 2004). The same vaccine strategy was then applied in a phase II trial to 34 women with high grade lesions (CIN 2 and 3). 20 showed total elimination, 11 had a 50 per cent reduction in lesion size, 2 had a reduction to grade CIN 2 and 1 a reduction to CIN 1 (Garcia-Hernandez *et al.*, 2006).

An MVA recombinant expressing modified HPV16 E6 and E7 proteins and IL-2 has also been developed (MVA-HPV-IL-2, TG8042). The E6 and E7 sequences inserted in the MVA vector were mutated to inhibit interaction with p53 and pRb, and fused with an ER signal sequence and a transmembrane domain to improve antigen presentation. Some encouraging results have been observed in phase II clinical trials depending on the HPV-related disease and its stage (Liu *et al.*, 2004).

17.5 Conclusions and future prospects

To date viral vector delivery of immunologically relevant molecules has resulted in very little benefit to patients (Rosenberg *et al.*, 2004). This is despite compelling results from preclinical tumour models. A major issue is that the vast majority of patients entered in to the studies described above were late stage and had received numerous rounds of prior therapies. These are unlikely to be the optimum target population for immunotherapy due to (i) tumour effects of immunoselection and immunosubversion (Zitvogel *et al.*, 2006) which are only likely to have increased during tumour progression, and (ii) the general physiological and immunological condition of these patients. Another major issue is the limitation of the efficiency of viral vectors due to the development of both humoral and cellular immunity against viral epitopes rather than against the transgene. The efficiency of a prime injection, using for instance, a common adenovirus or herpesvirus vector, against which most individuals have antibodies, or a vaccinia vector in patients who have been vaccinated against smallpox, can already be limited. Repeated injections will become increasingly inefficient. Several strategies have been developed to circumvent this problem: local delivery of the viral vector (intratumoral, intraocular, intracerebral) avoiding exposure to neutralizing antibodies, use of adenovirus serotypes rarely infecting humans (Barouch *et al.*, 2004) or RGD retargeted adenovirus vectors less sensitive to neutralization or the use of replication deficient poxvirus vectors or viral particles coated with cationic liposomes (Naito *et al.*, 2007). Heterologous prime-boost vaccinations will not only limit the antibody response against the viral vector but also reduce T-cell immunity against immunodominant vector epitopes, and have been shown to enhance both the antibody response and T-cell immunity against the transgene.

What we do know from these studies is that demonstrable immune responses to virally delivered target tumour antigens can be demonstrated in a number of patients. The belief is that if these T-cell responses can reach as yet undefined quantitative and qualitative levels, then clinical responses will be witnessed (Boon and Van den Eynde, 2003). How to generate these improvements in T-cell responses is the focus of the on-going research effort. This includes: (a) defining the optimal antigen/molecule to deliver; (b) defining the optimal prime-boost combinations of recombinant viral vectors and the number of boost injections required; (c) defining the optimum route; whether directly into the patient or by *ex vivo* DC transduction and then delivery; (d) timing, particularly with reference to other anticancer treatment, i.e radiation, chemotherapy, etc. It is likely that immunomodulatory gene therapy will be incorporated into treatment strategies earlier in the clinical course and be combined with treatments aimed at diminishing the immunosuppressive environment in cancer patients. In this way the expectation remains that recombinant viral vectors delivering tumour antigens will find a definite role in cancer treatment.

References

Arlen PM, Gulley JL, Todd N, *et al.* (2005). Antiandrogen, vaccine and combination therapy in patients with nonmetastatic hormone refractory prostate cancer. *J Urol* **174**, 539–546.

Arlen PM, Gulley JL, Parker C, *et al.* (2006). A randomized phase II study of concurrent docetaxel plus vaccine versus vaccine alone in metastatic androgen-independent prostate cancer. *Clin Cancer Res* **12**, 1260–1269.

Atkins GJ, Smyth JW, Fleeton MN, Galbraith SE, Sheahan BJ (2004). Alphaviruses and their derived vectors as anti-tumor agents. *Curr Cancer Drug Targets* **4**, 597–607.

Attia P, Phan GQ, Maker AV, *et al.* (2005). Autoimmunity correlates with tumor regression in patients with metastatic melanoma treated with anti-cytotoxic T-lymphocyte antigen-4. *J Clin Oncol* **23**, 6043–6053.

Banchereau J, Steinman RM (1998). Dendritic cells and the control of immunity. *Nature* **392**, 245–252.

Barnd DL, Lan MS, Metzgar RS, Finn OJ (1989). Specific, major histocompatibility complex-unrestricted recognition of tumor-associated mucins by human cytotoxic T cells. *Proc Natl Acad Sci USA* **86**, 7159–7163.

Barouch DH, Pau MG, Custers JH, *et al.* (2004). Immunogenicity of recombinant adenovirus serotype 35 vaccine in the presence of pre-existing anti-Ad5 immunity. *J Immunol* **172**, 6290–6297.

Bertram EM, Dawicki W, Watts TH (2004). Role of T cell costimulation in anti-viral immunity. *Semin Immunol* **16**, 185–196.

Boon T, Cerottini JC, Van Den Eynde B, Van Der Bruggen P & Van Pel A (1994). Tumor antigens recognized by T lymphocytes. *Annu Rev Immunol* **12**, 337–365.

Boon T, Van Den Eynde B (2003). Tumour immunology. *Curr Opin Immunol* **15**, 129–130.

Boyle DB, Coupar BE (1988). Construction of recombinant fowlpox viruses as vectors for poultry vaccines. *Virus Res* **10**, 343–356.

Brooks CR, Elliott T, Parham P, Khakoo SI (2006). The inhibitory receptor NKG2A determines lysis of vaccinia virus-infected autologous targets by NK cells. *J Immunol* **176**, 1141–1147.

Chisholm SE, Reyburn HT (2006). Recognition of vaccinia virus-infected cells by human natural killer cells depends on natural cytotoxicity receptors. *J Virol* **80**, 2225–2233.

Coley W (1896). Further observations upon the treatment of malignant tumors with the toxins of erysipelas and *Bacillus prodigiosus* with a report of 160 cases. *Bull Johns Hopkins Hosp* **7**, 157.

Corona Gutierrez CM, Tinoco A, Navarro T, *et al.* (2004). Therapeutic vaccination with MVA E2 can eliminate precancerous lesions (CIN 1, CIN 2, and CIN 3) associated with infection by oncogenic human papillomavirus. *Hum Gene Ther* **15**, 421–431.

Di Nicola M, Carlo-Stella C, Mortarini R, *et al.* (2004). Boosting T cell-mediated immunity to tyrosinase by vaccinia virus-transduced, CD34(+)-derived dendritic cell vaccination: a phase I trial in metastatic melanoma. *Clin Cancer Res* **10**, 5381–5390.

Dipaola R, Plante M, Kaufman H, *et al.* (2006). A phase I trial of Pox PSA vaccines (PROSTVAC(R)-VF) with B7-1, ICAM-1, and LFA-3 co-stimulatory molecules (TRICOMtrade mark) in patients with prostate cancer. *J Transl Med* **4**, 1.

Dudley ME, Wunderlich JR, Robbins PF, *et al.* (2002). Cancer regression and autoimmunity in patients after clonal repopulation with antitumor lymphocytes. *Science* **298**, 850–854.

Dummer R, Hassel JC, Fellenberg F, *et al.* (2004). Adenovirus-mediated intralesional interferon-gamma gene transfer induces tumor regressions in cutaneous lymphomas. *Blood* **104**, 1631–1638.

Eder JP, Kantoff PW, Roper K, *et al.* (2000). A phase I trial of a recombinant vaccinia virus expressing prostate-specific antigen in advanced prostate cancer. *Clin Cancer Res* **6**, 1632–1638.

Espenschied J, Lamont J, Longmate J, *et al.* (2003). CTLA-4 blockade enhances the therapeutic effect of an attenuated poxvirus vaccine targeting p53 in an established murine tumor model. *J Immunol* **170**, 3401–3407.

Farzaneh L, Kasahara N, Farzaneh F (2007). The strange case of TGN1412. *Cancer Immunol Immunother* **56**, 129–134.

Galon J, Costes A, Sanchez-Cabo F, *et al.* (2006). Type, density, and location of immune cells within human colorectal tumors predict clinical outcome. *Science* **313**, 1960–1964.

Garcia-Hernandez E, Gonzalez-Sanchez JL, Andrade-Manzano A, *et al.* (2006). Regression of papilloma high-grade lesions (CIN 2 and CIN 3) is stimulated by therapeutic vaccination with MVA E2 recombinant vaccine. *Cancer Gene Ther* **13**, 592–597.

Gulley J, Chen AP, Dahut W, *et al.* (2002). Phase I study of a vaccine using recombinant vaccinia virus expressing PSA (rV-PSA) in patients with metastatic androgen-independent prostate cancer. *Prostate* **53**, 109–117.

Gulley JL, Arlen PM, Bastian A, *et al.* (2005). Combining a recombinant cancer vaccine with standard definitive radiotherapy in patients with localized prostate cancer. *Clin Cancer Res* **11**, 3353–3362.

Harrop R, Connolly N, Redchenko I, *et al.* (2006). Vaccination of colorectal cancer patients with modified vaccinia Ankara delivering the tumor antigen 5T4 (TroVax) induces immune responses which correlate with disease control: a phase I/II trial. *Clin Cancer Res* **12**, 3416–3424.

Hodge JW, Rad AN, Grosenbach DW, *et al.* (2000). Enhanced activation of T cells by dendritic cells engineered to hyperexpress a triad of costimulatory molecules. *J Natl Cancer Inst* **92**, 1228–1239.

Hodge JW, Sabzevari H, Yafal AG, Gritz L, Lorenz MG, Schlom J (1999). A triad of costimulatory molecules synergize to amplify T-cell activation. *Cancer Res* **59**, 5800–5807.

Hu JC, Coffin RS, Davis CJ, *et al.* (2006). A phase I study of OncoVEXGM-CSF, a second-generation oncolytic herpes simplex virus expressing granulocyte macrophage colony-stimulating factor. *Clin Cancer Res* **12**, 6737–6747.

Jager E, Karbach J, Gnjatic S, *et al.* (2006). Recombinant vaccinia/fowlpox NY-ESO-1 vaccines induce both humoral and cellular NY-ESO-1-specific immune responses in cancer patients. *Proc Natl Acad Sci U S A* **103**, 14453–14458.

Kaufman HL, Wang W, Manola J, *et al.* (2004). Phase II randomized study of vaccine treatment of advanced prostate cancer (E7897): a trial of the Eastern Cooperative Oncology Group. *J Clin Oncol* **22**, 122–132.

Lathe R, Kieny MP, Gerlinger P, *et al.* (1987). Tumour prevention and rejection with recombinant vaccinia. *Nature* **326**, 878–880.

Limbach KJ, Paoletti E (1996). Non-replicating expression vectors: applications in vaccine development and gene therapy. *Epidemiol Infect* **116**, 241–256.

Lindsey KR, Gritz L, Sherry R, *et al.* (2006). Evaluation of prime/boost regimens using recombinant poxvirus/tyrosinase vaccines for the treatment of patients with metastatic melanoma. *Clin Cancer Res* **12**, 2526–2537.

Liu M, Acres B, Balloul JM, *et al.* (2004). Gene-based vaccines and immunotherapeutics. *Proc Natl Acad Sci U S A* **101**(Suppl 2): 14567–14571.

Mackett M, Smith GL, Moss B (1984). General method for production and selection of infectious vaccinia virus recombinants expressing foreign genes. *J Virol* **49**, 857–864.

Mahasreshti PJ, Kataram M, Wang MH, *et al.* (2003). Intravenous delivery of adenovirus-mediated soluble

FLT-1 results in liver toxicity. *Clin Cancer Res* **9**, 2701–2710.

Maker AV, Phan GQ, Attia P, *et al.* (2005). Tumor regression and autoimmunity in patients treated with cytotoxic T lymphocyte-associated antigen 4 blockade and interleukin 2: a phase I/II study. *Ann Surg Oncol* **12**, 1005–1016.

Marshall JL, Hawkins MJ, Tsang KY, *et al.* (1999). Phase I study in cancer patients of a replication-defective avipox recombinant vaccine that expresses human carcinoembryonic antigen. *J Clin Oncol* **17**, 332–337.

Marshall JL, Hoyer RJ, Toomey MA, *et al.* (2000). Phase I study in advanced cancer patients of a diversified prime-and-boost vaccination protocol using recombinant vaccinia virus and recombinant nonreplicating avipox virus to elicit anti-carcinoembryonic antigen immune responses. *J Clin Oncol* **18**, 3964–3973.

Marshall JL, Gulley JL, Arlen PM, *et al.* (2005). Phase I study of sequential vaccinations with fowlpox-CEA (6D)-TRICOM alone and sequentially with vaccinia-CEA(6D)-TRICOM, with and without granulocyte-macrophage colony-stimulating factor, in patients with carcinoembryonic antigen-expressing carcinomas. *J Clin Oncol* **23**, 720–731.

Mathiot CC, Grimaud G, Garry P, *et al.* (1990). An outbreak of human Semliki Forest virus infections in Central African Republic. *Am J Trop Med Hyg* **42**, 386–393.

Mayr A (1999). [Historical review of smallpox, the eradication of smallpox and the attenuated smallpox MVA vaccine]. *Berl Munch Tierarztl Wochenschr* **112**, 322–328.

McCart JA, Ward JM, Lee J, *et al.* (2001). Systemic cancer therapy with a tumor-selective vaccinia virus mutant lacking thymidine kinase and vaccinia growth factor genes. *Cancer Res* **61**, 8751–8757.

Meyer H, Sutter G, Mayr A (1991). Mapping of deletions in the genome of the highly attenuated vaccinia virus MVA and their influence on virulence. *J Gen Virol* **72** (Pt 5): 1031–1038.

Meyer RG, Britten CM, Siepmann U, *et al.* (2005). A phase I vaccination study with tyrosinase in patients with stage II melanoma using recombinant modified vaccinia virus Ankara (MVA-hTyr). *Cancer Immunol Immunother* **54**, 453–467.

Morse MA, Clay TM, Hobeika AC, *et al.* (2005). Phase I study of immunization with dendritic cells modified with fowlpox encoding carcinoembryonic antigen and costimulatory molecules. *Clin Cancer Res* **11**, 3017–3024.

Moss B (1991). Vaccinia virus: a tool for research and vaccine development. *Science* **252**, 1662–1667.

Moss B (1996). Genetically engineered poxviruses for recombinant gene expression, vaccination, and safety. *Proc Natl Acad Sci U S A* **93**, 11341–11348.

Naito T, Kaneko Y, Kozbor D (2007). Oral vaccination with modified vaccinia virus Ankara attached covalently to TMPEG-modified cationic liposomes overcomes pre-existing poxvirus immunity from recombinant vaccinia immunization. *J Gen Virol* **88**, 61–70.

Nelson NJ (1999). Viruses and cancer. *J Natl Cancer Inst* **91**, 1709.

Osada T, Clay T, Hobeika A, Lyerly HK, Morse MA (2006). NK cell activation by dendritic cell vaccine: a mechanism of action for clinical activity. *Cancer Immunol Immunother* **55**, 1122–1131.

Pantuck AJ, Van Ophoven A, Gitlitz BJ, *et al.* (2004). Phase I trial of antigen-specific gene therapy using a recombinant vaccinia virus encoding MUC-1 and IL-2 in MUC-1-positive patients with advanced prostate cancer. *J Immunother* **27**, 240–253.

Paoletti E (1996). Applications of pox virus vectors to vaccination: an update. *Proc Natl Acad Sci U S A* **93**, 11349–11353.

Puhlmann M, Brown CK, Gnant M, *et al.* (2000). Vaccinia as a vector for tumor-directed gene therapy: biodistribution of a thymidine kinase-deleted mutant. *Cancer Gene Ther* **7**, 66–73.

Rea D, Schagen FH, Hoeben RC, *et al.* (1999). Adenoviruses activate human dendritic cells without polarization toward a T-helper type 1-inducing subset. *J Virol* **73**, 10245–10253.

Ren H, Boulikas T, Lundstrom K, Soling A, Warnke PC, Rainov NG (2003). Immunogene therapy of recurrent glioblastoma multiforme with a liposomally encapsulated replication-incompetent Semliki forest virus vector carrying the human interleukin-12 gene–a phase I/II clinical protocol. *J Neurooncol* **64**, 147–154.

Rosales C, Graham VV, Rosas GA, Merchant H, Rosales R (2000). A recombinant vaccinia virus containing the papilloma E2 protein promotes tumor regression by stimulating macrophage antibody-dependent cytotoxicity. *Cancer Immunol Immunother* **49**, 347–360.

Rosenberg SA, Yang JC, White DE, Steinberg SM (1998a). Durability of complete responses in patients with metastatic cancer treated with high-dose interleukin-2: identification of the antigens mediating response. *Ann Surg* **228**, 307–319.

Rosenberg SA, Zhai Y, Yang JC, *et al.* (1998b). Immunizing patients with metastatic melanoma using recombinant adenoviruses encoding MART-1 or gp100 melanoma antigens. *J Natl Cancer Inst* **90**, 1894–1900.

Rosenberg SA, Yang JC, Schwartzentruber DJ, *et al.* (2003). Recombinant fowlpox viruses encoding the anchor-modified gp100 melanoma antigen can generate antitumor immune responses in patients with metastatic melanoma. *Clin Cancer Res* **9**, 2973–2980.

Rosenberg SA, Yang JC, Restifo NP (2004). Cancer immunotherapy: moving beyond current vaccines. *Nat Med* **10**, 909–915.

Ryan EJ, Harenberg A, Burdin N (2007). The canarypox-virus vaccine vector ALVAC triggers the release of IFN-gamma by natural killer (NK) cells enhancing Th1 polarization. *Vaccine* **25**, 3380–3390.

Scholl SM, Balloul JM, Le Goc G, *et al.* (2000). Recombinant vaccinia virus encoding human MUC1 and IL2 as immunotherapy in patients with breast cancer. *J Immunother* **23**, 570–580.

Scholl S, Squiban P, Bizouarne N, *et al.* (2003). Metastatic breast tumour regression following treatment by a gene-modified vaccinia virus expressing MUC1 and IL-2. *J Biomed Biotechnol* **2003**, 194–201.

Smith CL, Mirza F, Pasquetto V, *et al.* (2005). Immunodominance of poxviral-specific CTL in a human trial of recombinant-modified vaccinia Ankara. *J Immunol* **175**, 8431–8437.

Somogyi P, Frazier J, Skinner MA (1993). Fowlpox virus host range restriction: gene expression, DNA replication, and morphogenesis in nonpermissive mammalian cells. *Virology* **197**, 439–444.

Spaner DE, Astsaturov I, Vogel T, *et al.* (2006). Enhanced viral and tumor immunity with intranodal injection of canary pox viruses expressing the melanoma antigen, gp100. *Cancer* **106**, 890–899.

Sutter G, Moss B (1992). Nonreplicating vaccinia vector efficiently expresses recombinant genes. *Proc Natl Acad Sci U S A* **89**, 10847–108451.

Tartaglia J, Perkus ME, Taylor J, *et al.* (1992). NYVAC: a highly attenuated strain of vaccinia virus. *Virology* **188**, 217–232.

Taylor J, Trimarchi C, Weinberg R, *et al.* (1991). Efficacy studies on a canarypox-rabies recombinant virus. *Vaccine* **9**, 190–193.

Taylor J, Weinberg R, Languet B, Desmettre P, Paoletti E (1988). Recombinant fowlpox virus inducing protective immunity in non-avian species. *Vaccine* **6**, 497–503.

Triozzi PL, Allen KO, Carlisle RR, Craig M, Lobuglio AF, Conry RM (2005). Phase I study of the intratumoral administration of recombinant canarypox viruses expressing B7.1 and interleukin 12 in patients with metastatic melanoma. *Clin Cancer Res* **11**, 4168–4175.

Tsang KY, Zaremba S, Nieroda CA, Zhu MZ, Hamilton JM, Schlom J (1995). Generation of human cytotoxic T cells specific for human carcinoembryonic antigen epitopes from patients immunized with recombinant vaccinia-CEA vaccine. *J Natl Cancer Inst* **87**, 982–990.

Tseng JC, Levin B, Hirano T, Yee H, Pampeno C, Meruelo D (2002). *In vivo* antitumor activity of Sindbis viral vectors. *J Natl Cancer Inst* **94**, 1790–1802.

Tseng JC, Levin B, Hurtado A, *et al.* (2004). Systemic tumor targeting and killing by Sindbis viral vectors. *Nat Biotechnol* **22**, 70–77.

van Baren N, Bonnet MC, Dreno B, *et al.* (2005). Tumoral and immunologic response after vaccination of melanoma patients with an ALVAC virus encoding MAGE antigens recognized by T cells. *J Clin Oncol* **23**, 9008–9021.

van der Bruggen P, Van Pel VS, Van Den Eynde AB (2007). www.cancerimmunity.org/peptidedatabase/Tcellepitopes.htm.

von Mehren M, Arlen P, Tsang KY, *et al.* (2000). Pilot study of a dual gene recombinant avipox vaccine containing both carcinoembryonic antigen (CEA) and B7.1 transgenes in patients with recurrent CEA-expressing adenocarcinomas. *Clin Cancer Res,* **6**, 2219–2228.

von Mehren M, Arlen P, Gulley J, *et al.* (2001). The influence of granulocyte macrophage colony-stimulating factor and prior chemotherapy on the immunological response to a vaccine (ALVAC-CEA B7.1) in patients with metastatic carcinoma. *Clin Cancer Res* **7**, 1181–1191.

Wajchman HJ, Pierce CW, Varma VA, Issa MM, Petros J, Dombrowski KE (2004). Ex vivo expansion of CD8+CD56+ and CD8+CD56- natural killer T cells specific for MUC1 mucin. *Cancer Res* **64**, 1171–1180.

Wang M, Bronte V, Chen PW, *et al.* (1995). Active immunotherapy of cancer with a nonreplicating recombinant fowlpox virus encoding a model tumor-associated antigen. *J Immunol* **154**, 4685–4692.

Whitman ED, Tsung K, Paxson J, Norton JA (1994). *In vitro* and *in vivo* kinetics of recombinant vaccinia virus cancer-gene therapy. *Surgery* **116**, 183–188.

Wysocki PJ, Grabarczyk P, Mackiewicz-Wysocka M, Kowalczyk DW, Mackiewicz A (2002). Genetically modified dendritic cells–a new, promising cancer treatment strategy? *Exp Opin Biol Ther* **2**, 835–845.

Yamaguchi T, Sakaguchi S (2006). Regulatory T cells in immune surveillance and treatment of cancer. *Semin Cancer Biol* **16**, 115–123.

Zeh HJ, Bartlett DL (2002). Development of a replication-selective, oncolytic poxvirus for the treatment of human cancers. *Cancer Gene Ther* **9**, 1001–1012.

Zitvogel L, Tesniere A, Kroemer G (2006). Cancer despite immunosurveillance: immunoselection and immunosubversion. *Nat Rev Immunol* **6**, 715–727.

18

Antiangiogenic Gene Delivery

Anita T. Tandle and **Steven K. Libutti**

18.1 Angiogenesis: role in tumour development and metastasis

Angiogenesis is a multistep process of formation of new blood vessels from pre-existing vasculature. It can be divided into two phases: a phase of activation (sprouting) and a phase of resolution (Risau, 1997). The phase of activation encompasses (i) increased vascular permeability and extravascular fibrin deposition, (ii) vessel wall disassembly, (iii) basement membrane degradation, (iv) cell migration and extracellular matrix invasion and (v) endothelial cell (EC) proliferation and capillary lumen formation. The phase of resolution includes inhibition of EC proliferation, cessation of cell migration, basement membrane reconstitution, junctional complex maturation, and vessel wall assembly, including recruitment and differentiation of smooth muscle cells and pericytes (Risau, 1997). The phase of resolution results in the establishment of blood flow in the newly formed vessel.

Physiological angiogenesis occurs during wound healing, the female endometrial cycle and during pregnancy. Under these conditions, it is a very tightly regulated process, controlled by both positive and negative regulators of angiogenesis. Positive regulators of angiogenesis include, vascular endothelial growth factor (VEGF), fibroblast growth factor (FGF), platelet derived growth factor (PDGF), angiopoietins, families of cytokines and their tyrosine kinase receptors (Felmeden *et al.*, 2003). The important angiogenic growth factors and their mechanism of action are listed in Table 18.1. Interaction of growth factors with their receptors initiates signalling cascades that regulate expression of additional growth factors, cell cycle progression, cell–cell and cell–matrix interactions, synthesis of matrix-remodelling proteases and extracellular matrix components (Gagne *et al.*, 2004). A number of endogenous negative regulators of angiogenesis have been described, including endostatin, angiostatin, thrombospondin (TSP)-1 and -2, and the fragments of collagen IV, canstatin and tumstatin. The negative regulators of angiogenesis have been listed in Table 18.2. Any imbalance between these regulators can give rise to abnormal angiogenesis, which is important in many diseases including cancer.

The malignant transformation of normal cells is a complex and multistep process, involving several genetic changes in the key regulatory elements of the cell cycle. At this stage, the primary tumour is still very small and supported by pre-existing blood vessels for nutritional and oxygen requirements. However, once the tumour grows beyond 1–3 mm, it is starved of nutrients and oxygen, unless new vasculature is established (Folkman, 1990). In order to obtain sufficient oxygen and nutrients, growing tumours need to establish an additional blood supply by stimulating angiogenesis. At this point, the angiogenic switch occurs, the net balance of endogenous inhibitors and

Viral Therapy of Cancer Edited by Kevin J. Harrington, Richard G. Vile and Hardev S. Pandha

Table 18.1 Angiogenic growth factors

Growth factor	Mechanism of action
VEGF family	Increases EC proliferation, survival, and vascular permeability, lymphangiogenesis, upregulates proteases for matrix organization (Jain, 2005)
FGF	Angiogenesis (Mignatti and Rifkin, 1996)
Angl	EC sprouting and vessel maturation (Sato *et al.*, 1995)
Matrix metalloproteinases	Control degradation of basement membrane and ECM and cause release of other angiogenic factors (Jain, 2003).
PDGF	EC proliferation and tube formation (Battegay *et al.*, 1994)
Integrins	Neovascularization (Lode *et al.*, 2002)
Cox2	Increase expression of VEGF, inhibition of EC apoptosis (Gallo *et al.*, 2001, Liu *et al.*, 2000)
TGF-β	Capillary formation (Roberts *et al.*, 1986)
Angiotropin	EC migration (Hockel *et al.*, 1987)
IL-1	Modulation of VEGF expression (Salven *et al.*, 2002)
IL-6	VEGF expression (Huang *et al.*, 2004)
IL-8	EC survival, proliferation and MMP synthesis (Li *et al.*, 2003)
TNF-α	At low concentration, it releases angiogenic growth factors (Yoshida *et al.*, 1997)
PlGF	Vessel growth, synergistic to VEGF (Carmeliet *et al.*, 2001)

VEGF, vascular endothelial growth factor; FGF, fibroblast growth factor; Angl, angiopoietin 1; PDGF, platelet derived growth factor; Cox2, cyclooxygenase 2; TGF-β, transforming growth factor-beta; IL, interleukin; TNF-α, tumour necrosis factor-alpha; PIGF, placental growth factor; EC, endothelial cells; ECM, extracellular matrix.

Table 18.2 Negative regulators of angiogenesis

Regulator	Mechanism of action	Reference
Endostatin	Cleavage product of ECM, inhibits EC migration and promotes apoptosis	(Shichiri and Hirata, 2001; Abdollahi *et al.*, 2004)
Angiostatin	Internal fragment of plasminogen, inhibits EC migration, proliferationand induces apoptosis	(Holmgren *et al.*, 1995; (Claesson-Welsh *et al.*, 1998))
Tumstatin/canstatin	Cleavage product of ECM, inhibits EC proliferation, migration and induces apoptosis	(Mundel and Kalluri, 2007)
Thrombospondin-1	Platelet derived protein, inhibits angiogenesis	(de Fraipont *et al.*, 2001)
TIMPs	Inhibits activity of metalloproteinases	(Baker *et al.*, 2002)
EMAP-II	Cytokine, inhibits EC proliferation, induces apoptosis, sensitizes vasculature to TNF-α	(Kayton and Libutti, 2001)
TNF-α	Cytokine, induces EC apoptosis and vascular damage	(Lin *et al.*, 2007)
Interleukins 12 and 18	Cytokines, antiangiogenic eggect via induction of interferon gamma	(Trinchieri, 2003; Cao *et al.*, 1999)
Platelet factor-4	Chemokine, inhibits EC proliferation and migration	(Bikfalvi and Gimenez-Gallego, 2004)
Interferon-inducible protein-10	Chemokine, potent immunomodulatory and antiangiogenic activity	(Neville *et al.*, 1997)

EC, endothelial cell; ECM, extracellular matrix; TIMPs, tissue inhibitors of metalloproteinases; EMAP-II, endothelial monocyte activating polypeptide-II; TNF-α, tumour necrosis factor alpha.

proangiogenic growth factors shifts in favour of the angiogenic growth factors. In a transgenic mouse model of pancreatic islet carcinogenesis, it has been shown that a certain proportion of hyperplastic islets become angiogenic by switching on angiogenesis in the normal quiescent islet capillaries (Casanovas *et al.*, 2005). In tumours, the angiogenic switch is associated with a marked decrease (three- to four-fold) in overall tumour cell apoptosis (Holmgren *et al.*, 1995). It is partly due to the release of anti-apoptotic factors from the ECs, and mutations in oncogenes and loss and or mutations in tumour suppressor genes (Folkman, 2003). The production of angiogenic growth factors by tumour cells may be continuous and is increased by tumour hypoxia. Every increase in the tumour cell population must be preceded by an increase in new capillaries that converge upon the tumour (Folkman and Klagsbrun, 1987). The first compelling evidence of this was described using the non-vascularized cornea of rabbits to show that tumours are angiogenesis dependent (Gimbrone *et al.*, 1972). On the other hand, experimental models have shown that, under the influence of endogenous angiogenesis inhibitors, metastases remain dormant and tumour cell proliferation is balanced by an equivalent rate of cell death (Holmgren *et al.*, 1995). More recently, accumulating evidence has linked angiogenesis to the pathophysiology of various haematopoietic malignancies (Bertolini *et al.*, 2000; Ribatti, 2004). Increased levels of the angiogenic growth factors VEGF and FGF have been show in acute and chronic lymphocytic leukaemias to be correlated with a poor clinical outcome (Perez-Atayde *et al.*, 1997; Chen *et al.*, 2000b; Duensing and Atzpodien, 1995; Molica *et al.*, 1999).

These early discoveries, followed by subsequent studies, led to the identification and isolation of a number of angiogenic growth factors. In addition to tumour derived factors, the extracellular matrix and different cell types are also involved in tumour-induced angiogenesis (Verheul *et al.*, 2004). Tumour-associated macrophages, pericytes, and fibroblasts are all involved in tumour-induced angiogenesis. A subpopulation of fibroblasts, so called cancer-associated fibroblasts (CAFs) are important promoters of tumour growth and progression (Kalluri and Zeisberg, 2006). In breast carcinomas 80 per cent of stromal fibroblasts seem to acquire this phenotype (Sappino *et al.*, 1988). Also fibroblasts and inflammatory cells are the principal source of host-derived VEGF (Fukumura *et al.*, 1998).

Another recently studied cell type involved in angiogenesis is the circulating endothelial cells (CEC), which shed from the tumour itself or are derived from bone marrow. Previously, it was thought that all new ECs in a growing blood vessel were derived from the division of pre-existing differentiated ECs. However, studies have shown the existence of circulating endothelial progenitor cells (CEP), which could be mobilized out of the marrow compartment, enter the peripheral circulation and be incorporated into newly forming vessels (Asahara *et al.*, 1997). These peripheral blood cells may not only contribute to angiogenesis, but are a likely target of many antiangiogenic drugs (Bertolini *et al.*, 2003; Schuch *et al.*, 2003; Shaked *et al.*, 2005).

Most of the human solid tumours and haematopoietic malignancies are angiogenesis dependent (Bertolini *et al.*, 2000). The degree of vascularization of the primary tumour correlates directly with the presence of bone marrow micrometastases at diagnosis in breast cancer patients (Fox *et al.*, 1997). Intratumoral vascularization has prognostic value for cancers of the breast, colon, cervix, lung, melanoma and others (Kitadai *et al.*, 2004). Thus, inhibition of angiogenesis could prove beneficial across a wide variety of human cancers. However, development of effective cancer therapy will require simultaneous targeting of multiple steps in tumour development.

18.2 Targeting tumour vasculature as an approach for cancer treatment

The promise of antiangiogenic therapies stems from low resistance, low toxicity, potential for long term maintenance therapy, and possible combination with conventional chemotherapy (Sharma and Odunsi, 2005). Antiangiogenic treatment looks very promising for various reasons: (a) a single vessel provides nutrition for and facilitates removal of waste products of metabolism from hundreds of thousands of tumour cells, and has to be damaged at only one point to block blood flow, (b) EC killing is not required, a change of shape or local initiation of blood coagulation may be sufficient, (c) the EC is adjacent to the blood supply ensuring adequate drug delivery (Thorpe *et al.*, 2003). As tumour EC come into constant contact with circulating blood, it makes a unique target for antiangiogenic therapies. It is generally accepted that ECs in tumour vessels have a higher rate of proliferation and are mostly activated (Sedlacek, 2001). ECs in the angiogenic vessels within solid tumours express several proteins that are absent in established vessels (Folkman, 1995, Arap *et al.*, 1998). Given such inherent molecular diversity of the normal and tumour microvasculature, resident receptor proteins for selective targeting of diagnostic and therapeutic agents to specific vascular beds can be identified and exploited (Yao *et al.*, 2006).

The identification of effective targets on EC is important. With the advances in techniques that allow full genome analysis, in particular, microarray analysis combined with bioinformatics and mRNA-based serial analysis of gene expression (SAGE) a wealth of expression data is now available in the public domain (Neri and Bicknell, 2005). Using *in-vivo* phage peptide libraries, specific peptides that home selectively to the vasculature of specific organs have been identified (Pasqualini and Ruoslahti, 1996, Rajotte *et al.*, 1998). Injection of phage peptide libraries into the circulation of nude mice bearing human breast carcinomas, identified three main peptide motifs that targeted phage to the tumour vasculature (Arap *et al.*, 1998). One motif contained the sequence Arg-Gly-Asp (RGD) that binds to cell surface integrin receptors (Koivunen *et al.*, 1995). A second motif contained the sequence Asn-Gly-Arg (NGR), which has been identified as a cell adhesion motif (Koivunen *et al.*, 1994). The third motif was Gly-Ser-Leu (GSL), frequently present in breast carcinoma, Kaposi's sarcoma and malignant melanoma (Arap *et al.*, 1998). An RGD/NGR coupled doxorubicin was effective in breast carcinoma xenografts at much lower concentrations and inhibited both primary tumours and metastases (Arap *et al.*, 1998). Thus, ECs in tumour vasculature present a different signature and hence the possibility of vascular targeting. There are also many phenotypic differences between tumour and normal vessels including levels of the VEGF, Eph and Tie2 receptors (Ruoslahti, 2002), adhesion molecules and integrin receptors (Brooks *et al.*, 1994). Angiogenic vessels express elevated levels of the tyrosine kinase receptors for VEGF and Eph receptors (Ruoslahti, 2002). The $\alpha v\beta 3$ and $\alpha v\beta 5$ integrins are upregulated in endothelial cells undergoing angiogenesis, and their level of expression in tumour vasculature correlates with the grade of malignancy in certain cancers (Ruoslahti, 2002). Integrin $\alpha v\beta 3$ is required for the survival and maturation of newly forming blood vessels, an event essential for the proliferation of tumours (Brooks *et al.*, 1994). Oh and colleagues identified several endothelial surface markers induced by tumour microenvironment, which are not present in the normal tissue microenvironment (Oh, 2004). Also, normal vessels maintain their vascular integrity by surrounding pericytes, which are absent in newly formed immature tumour vessels. This makes them dependent of VEGF for survival (Benjamin, 1999; Kerbel, 2000).

Several lessons learned from early clinical trials in antiangiogenic therapy would seem to support a role for antiangiogenic gene transfer strategies over other treatment modalities (Tandle *et al.*, 2004). Agents that have so far been tested clinically include endogenous inhibitors such as angiostatin, endostatin and PF-4 (Sim *et al.*, 2000) and agents that can neutralize angiogenic peptides or their receptors such as antibodies to FGF, VEGF and matrix metalloproteinase (MMP) inhibitors (Ziche *et al.*, 2004).

Angiostatin, endostatin and TNP-470 which inhibit EC proliferation are in phase I/II clinical trials for advanced solid tumours as well as some haematological malignancies (Ziche *et al.*, 2004). Another way to inhibit angiogenesis is to inhibit angiogenic growth factors and growth factor receptors. A monoclonal antibody that inactivates VEGF, Avastin, has been recently tried in phase III trial for breast, prostate, colorectal and renal cancers (Shih, 2006) Tyrosine kinase inhibitors, such as SU-5416, SU-6668 and PTK787 which inhibit growth factor receptor signalling are also at different stages of clinical trials in various malignancies showing promising results (Ziche *et al.*, 2004). The other category of inhibitors include agents that inhibit activity of matrix metalloproteases, such as Marimastat, Primostat and Neovastat which are mostly in phase III clinical trials for breast, lung, pancreas, gliomas and brain cancers (Ziche *et al.*, 2004)

However, gene therapy targeting ECs has several advantages compared to systemic treatments; lower risk of inhibiting normal physiological angiogenesis and accumulation and persistence of antiangiogenic agents in tumour tissues (Sedlacek, 2001).

18.3 Viral vectors to deliver antiangiogenic gene products

Viral vectors have so far proved most popular for gene therapy applications. For further information on specific properties of viral vectors please refer to specific chapters. Preclinical studies using viral vectors to deliver antiangiogenic molecules or target angiogenic growth factors are listed in Table 18.3.

18.3.1 Adenovirus (Ad)

Adenoviruses have been developed as gene delivery vectors since the early 1980s (Berkner, 1988). They are non-enveloped, linear, double stranded DNA virus. The entry of adenovirus into susceptible cell requires Ad binding to the primary cellular receptor called coxsackie receptor (CAR) through the globular knob domain of the fibre capsid protein (Louis, 1994). Subsequent internalization of the virus by receptor-mediated endocytosis requires the interaction of Arg-Gly-Asp (RGD) peptide sequences in the penton base protein with integrin receptors on the host cell (Wickham, 1993). Ad have been widely used as a gene therapy vectors for the following properties; although short term, they mediate very high levels of gene expression of the transgene, the expressed proteins undergo a complex post-translational modifications required for the correct function, they can infect a wide range of dividing and non-dividing mammalian cells, and the production and purification of virus is relatively easy (Douglas, 2007). However, Ad has a number of limitations as a gene therapy vector. The inability of the vector to infect tumour cells expressing low levels of CAR receptor (Kim, 2002). The use of first generation Ad vectors induces both an innate and an acquired immune response in the host (Young, 2001), and the later eliminates the transducer cell. In addition to cellular immunity, a humoral antibody response is generated to the adenoviral vector. This leads to a reduction in adenoviral-mediated gene delivery upon repeat vector administration. Moreover, sometimes the first dose may also be inefficient in patients who possess neutralizing antibodies to the commonly used Ad vectors, as a result of prior exposure to the parental vector (Douglas, 2007). Intravenous administration of adenoviral vectors leads to preferential transduction of the liver in rodents (Ghosh *et al.*, 2000). Adenovirus rapidly binds to circulating platelets, which causes their activation/aggregation and subsequent entrapment in liver sinusoids. Further, virus-platelet aggregates are taken up by Kupffer cells and degraded (Stone *et al.*, 2007). However, in spite of some disadvantages with Ads, the improved second and third generation vectors are more efficient as a delivery tool.

TSP-1 is a multifunctional extracellular glycoprotein regulating cell proliferation, migration and apoptosis in a variety of physiological and pathological conditions (Chen *et al.*, 2000a). As the first protein to be recognized as a potential endogenous suppressor of angiogenesis, TSP-1 has been shown to inhibit angiogenesis in multiple *in vitro* and *in-vivo* assays (Tolsma *et al.*, 1993; Dawson *et al.*, 1999). Adenovirus expressing TSP-1 has been used to treat K562 tumours in a mouse xenograft model (Liu *et al.*, 2003). The Ad-TSP-1 treated tumours showed a statistically significant reduction in microvessel density and tumour volume compared to controls.

Tie2 is an endothelium-specific receptor tyrosine kinase known to play a role in tumour angiogenesis (Lin, 1997[A1]). A replication-deficient adenoviral vector was used for systemic delivery of soluble Tie2 (AdExTek) capable of blocking Tie2 activation in several models of primary and metastatic cancer (Lin, 1998). Administration of AdExTek to mice

Table 18.3 Antiangiogenic gene delivery

Vector	Target/antiangiogenic inhibitor	Findings (reference)
Adenovirus	sTie2 receptor	Inhibition of vascularization, primary tumour growth and lung metastasis (Lin *et al.*, 1998)
Adenovirus	Angiostatin	Inhibition of angiogenesis, increase in apoptosis and inhibition of primary tumour growth (Griscelli *et al.*, 1998)
Adenovirus	Antisense VEGF	Decrease in MVD, inhibition of breast cancer xenograft (Im *et al.*, 2001)
Adenovirus	Endostatin	Inhibition of adenocarcinoma xenograft (Feldman *et al.*, 2000)
Adenovirus	Platelet derived factor 4	Inhibition of angiogenesis and head and neck tumour xenograft (Wu *et al.*, 2003)
Adenovirus	Thrombospondin-1	Inhibition of MVD and leukaemia xenograft (Liu *et al.*, 2003)
Adenovirus/oncolytic virus	Soluble VEGF receptor	Enhanced anti-tumour and anti-angiogenic effects in colon and prostate xenografts (Thorne *et al.*, 2006)
Adenovirus	TNF-α+radiation	Tumour vessel thrombosis and complete tumour regression in 71% malignant glioma xenografts (Staba *et al.*, 1998)
Retrovirus	Endostatin	Sustained and significant inhibition of tumour growth (Feldman *et al.*, 2001)
Retrovirus	TIMP-2	Inhibition of local invasion and tumour growth (Imren *et al.*, 1996)
AAV	Angiostatin+endostatin	Complete protection from tumour development and survival in combination treatment (Ponnazhagan *et al.*, 2004)
AAV	Angiostatin+endostatin	Increased tumour cell apoptosis, decreased blood vasculature and increased tumour free survival (Isayeva *et al.*, 2005)
AAV	sVEGFR1/R2	reduces overall tumour volume and increases median survival time in glioblastoma multiforme xenograft (Harding *et al.*, 2006)
HSV	IL-12	Decrease level of angiogenesis, inhibition of progression of tumour growth and increase in median survival in prostate cancer (Varghese *et al.*, 2006)
Measles virus	$\alpha v\beta 3$	Significant regression of resistant multiple myeloma xenografts (Hallak, 2005)

MVD, microvessel density; TIMP-2, Tissue inhibitor of metalloproteinases-2; AAV, Adeno-associated virus; HSV, herpes simplex virus; IL-12, interleukin-12.

with two different well established primary tumours, a murine mammary carcinoma or a murine melanoma, significantly inhibited the growth rate of both tumours. Administration of AdExTek also inhibited tumour metastasis when delivered at the time of surgical excision of primary tumours. More recently, it has been shown that an intratumoral, combination therapy employing Ad expressing a fusion protein between endostatin–angiostatin and Ad expressing soluble Tie2, could achieve a complete regression of the injected, as well as the contralateral uninjected tumour and prolonged the tumour-free survival in 80 per cent of the animals with prostate xenografts (Raikwar, 2005).

An antiangiogenic adenovirus vector, Ad-Flk1-Fc, which expresses a soluble VEGF receptor capable of inhibiting tumour angiogenesis and growth has recently been described (Thorne *et al.*, 2006).

Previously an oncolytic adenovirus, dl922/947, whose replication and subsequent cytotoxicity is restricted to cancer cells with a loss of the G_1–S cell cycle checkpoint was also developed. Which methods for combining these therapies are also being developed, yield significantly greater anti-tumour effects than the respective monotherapies (Thorne *et al.*, 2006).

18.3.2 Adeno-associated virus (AAV)

AAV is a single-strand, non-enveloped DNA virus with a 4.7-kb genome (Srivastava *et al.*, 1983). The AAV type 2 receptor is a membrane-associated heparin sulfate proteoglycan, which is present on many cell surfaces, thus explaining the broad host infectivity of this virus (Summerford and Samulski, 1998). It also uses $\alpha_v\beta_5$ integrin, fibroblast growth factor receptor type 1, the hepatocyte growth factor receptor c-Met, or laminin receptor as co-receptors to enable internalization (White *et al.*, 2007). AAV integrates in the host genome giving long-term gene expression. All viral-encoded genes, approximately 96% of the viral genome, can be replaced with foreign DNA of choice and packaged into an AAV virion. The current design of AAV vectors provides an efficient DNA carrier system, free from the possibility of recombination with wild type virus (Guha *et al.*, 2000[A2]). Unlike retroviruses, AAV vectors can infect non-dividing cells, which extends their usefulness to the treatment of slow-growing tumours, although the rate of transduction of non-cycling cells is much lower than that for dividing cells in culture (Russell *et al.*, 1994). However, the single-strand viral genome necessitates second-strand synthesis to generate a transcriptionally active double-strand intermediate, which may be rate-limiting in transduction of cells by AAV (Ferrari et al, 1996). Also, vascular ECs are poorly transduced by AAV (Nicklin *et al.*, 2001). Hence, AAV vectors with enhanced tropism for EC will be useful for diverse gene therapeutics targeted at the vasculature. AAV has been used to deliver antiangiogenic targets in preclinical animal models.

Long-term inhibition of VEGF signalling using AAV expressing the soluble truncated form of VEGFR-2 has been achieved (Davidoff *et al.*, 2002). The intraportal injection of this vector generated a systemic state of angiogenesis inhibition. Tumour development was prevented in 67 per cent of mice, with significant growth reduction of tumours in the remaining mice, indicating feasibility of using AAV for antiangiogenic gene delivery. This approach might prove beneficial in treating distant metastasis.

18.3.3 Retrovirus

Retroviruses are a class of enveloped viruses containing a single stranded RNA molecule as the genome. They are able to efficiently integrate permanently into the human genome, where they provide the basis for permanent expression of foreign DNA. Retroviral vectors efficiently infect proliferating cells and provide long-term transgene expression as their genome integrates into the host DNA (Daly and Chernajovsky, 2000). However, they cannot infect non-dividing cells as they cannot cross the nuclear membrane (Miller *et al.*, 1990).

Retrovirus vectors have been used to deliver antiangiogenic genes in many preclinical animal models. Wang and co-workers used retrovirus to transfer endostatin to human liver carcinoma cells. Only in three out of five mice were tumours formed and the mean size of flank tumours was 94.5 per cent smaller than that from the control cell inoculation (Wang *et al.*, 2002). Similarly, Chen *et al.* showed that in a colon carcinoma xenograft model, there was an 86 per cent reduction in tumour size in the endostatin-transduced group, accompanied by a reduction in vessels, compared to the control group (Chen *et al.*, 2003). Thus, retroviruses can allow functional expression of the endostatin gene in human tumours, showing promise for an antitumour strategy using antiangiogenesis.

18.3.4 Lentivirus

Lentiviruses are members of the retrovirus family. Lentiviruses have the unique ability amongst retroviruses of being able to infect non-cycling cells. Vectors derived from lentiviruses have provided a huge advancement in technology and seemingly offer the means to achieve significant levels of gene transfer *in vivo*. Gene transfer can be achieved in quiescent cells, non-dividing or terminally differentiated cells such as neurons by their ability to translocate into the nucleus. Lentiviral vectors are especially useful in transducing cells which lack receptors for adenoviruses. However, lentivirus has low transduction efficiency for ECs and may result in significant vector-associated cytotoxicity (Shichinohe *et al.*, 2001). The generation of replication-competent lentivirus can be minimized by deleting a major portion of the parental genome into the insertion cassette. Vectors that are based on HIV retain <5 per cent of the parental genome (Thomas *et al.*, 2003). Lentiviral vectors have been engineered to include genes such as MMP-2, angiostatin and endostatin (Pfeifer *et al.*, 2000; Shichinohe *et al.*, 2001). One of the first reports using lentivirus to express endostatin and angiostatin showed significant inhibition of cellular proliferation on coculture of transduced tumour cells with endothelial cells (Shichinohe *et al.*, 2001). Pfeifer and co-workers used lentivirus to deliver a non-catalytic fragment of MMP2 in different angiogenesis models (Pfeifer *et al.*, 2000). Lentiviral delivery of MMP2 blocked basic fibroblast growth factor-induced matrix metalloproteinase 2 activation and angiogenesis on chicken chorioallantoic membranes. It also inhibited tumour-induced angiogenesis and tumour growth in a nude mouse model. Thus, lentiviral vectors can deliver sufficient quantities of antiangiogenic substances to achieve therapeutic effects *in vivo*.

18.3.5 Herpes simplex virus-1 (HSV-1)

Herpes simplex virus-1 is an enveloped double-stranded DNA virus. To initiate infection, the virus first attaches to cell surface receptors. The initial attachment involves the interaction of viral envelope glycoproteins with the glycosaminoglycan moieties of cell surface heparin sulfate (Shen and Nemunaitis, 2006). Thus, it can infect a wide array of cell types in various species, is able to transduce non-dividing as well as dividing cells, can accommodate multiple transgenes, and lacks insertional mutagenesis (Shen and Nemunaitis, 2006).

HSV has a natural tropism for neuronal tissue and has shown promise in the treatment of brain tumours. However, a mutant HSV has been shown efficacious and safe for use in localized human malignancies of non-neuronal origin such as malignant mesothelioma (Kucharczuk *et al.*, 1997).

An oncolytic strain of HSVtk, HSVtk-1716, has been shown to infect tumour endothelium, thus exerting direct antiangiogenic activity in tumours (Benencia *et al.*, 2005).

Recently, a hybrid vector between AAV and bacteriophage (AAVP) carrying the HSV-tk gene cassette, was shown to be very effective in targeting tumour vasculature *in vivo* (Hajitou *et al.*, 2006). Nude mice bearing human prostate cancer tumours showed significant reduction in tumour viability and tumour volumes after systemic injection of AAVP-HSKtk virus.

18.3.6 Bacteriophage vectors

Bacteriophage vectors infect and multiply inside bacteria by making use of some or all of the host biosynthetic machinery. Bacteriophages can be used to display up to tens of billions of peptides and proteins, including human antibodies and enzymes, on their surface. This technique allows screening of a large number of peptide and antibody libraries. The phage display peptide libraries have been used to identify peptides that home to tumours through the circulation and that specifically bind to the tumour ECs or lymphatic cells (Arap *et al.*, 1998, Laakkonen *et al.*, 2002). The bacteriophage has been used previously to target mammalian cell surface receptors for gene delivery either using a specific ligand or an antibody. FGF2 was attached to the bacteriophage coat protein and demonstrated long term transgene expression in FGF receptor expressing cells (Larocca et al., 1998, 1999). A bacteriophage displaying the adenoviral penton base and carrying a transgene has also been reported (Di Giovine *et al.*, 2001). However, the reported transduction efficiency for these vectors is very low ($\sim$1–9 per cent) in spite of using multivalent phagemid particles (Larocca et al., 1998, 1999, 2001).

18.3.7 Adeno-associated viral phage vectors (AAVP)

Because any one virus lack all the suitable properties to be the ideal gene delivery vector, many researchers are combining the best suited properties of two or more vectors in order to obtain better vector systems. One of the most interesting hybrid couples is the site-specific integration machinery of wild type AAV with the efficient internalization and nuclear targeting properties of adenovirus (Thomas, 2003).

Recently, a hybrid vector was introduced containing *cis*-elements from AAV and single-stranded bacteriophage called AAVP. This vector was engineered to target integrin receptors (Hajitou *et al.*, 2006). AAVP provides superior tumour transduction over phage alone and the incorporation of inverted terminal repeats from AAV is associated with an improved fate of the delivered transgene. Bacteriophage containing cyclic RGD as the targeting motif has a 40- to 80-fold greater selectivity for tumour blood vessels *in vivo* (Pasqualini *et al.*, 1997).

Using AAVP, the physiologic distribution of tumour-targeting has been evaluated using immunofluorescence microscopy in blood vessels of the RIP-Tag2 pancreatic islet tumour mouse model (Hanahan, 1985; Yao *et al.*, 2006). RIPTag2 mice have been particularly instructive in the study of the parameters of angiogenesis. By virtue of expressing the SV40 virus oncoproteins in the pancreatic islet β cells, RIPTag2 mice develop islet carcinomas in multiple stages of hyperplastic/dysplastic islet, angiogenic dysplastic islets, solid tumours and invasive carcinomas (Hanahan, 1985).

Unlike the non-targeted AAVP, targeted virus homed to vascular ECs in a dose dependent fashion. The distribution of phage was similar to $\alpha v\beta 3$ and $\alpha 5\beta 1$ integrin expression. Blood vessels that survived treatment with AG-013736, a small molecule inhibitor of VEGF and PDGF receptors, had only 4 per cent as much binding of AAVP compared with vessels in untreated tumours. Cellular distribution of viral particles in surviving tumour vessels matched the $\alpha 5\beta 1$ integrin expression. The authors speculated that the reduction in integrin expression on tumour vessels after antiangiogenic therapy raises the possibility that integrin-targeted delivery of diagnostics or therapeutics may be compromised (Yao *et al.*, 2006).

18.4 Viral targeting

In the case of most DNA viruses and retroviruses, the viral genome is ultimately translocated into the nucleus of the host cell in order to be expressed and replicated, or integrated into the host genome. Thus, efficient cell targeting, endosomal escape from degradation, and nuclear targeting are desirable elements of an efficient cancer-specific targeting viral vector (Glover *et al.*, 2005). A major drawback of current approaches to antiangiogenic gene therapy is the lack of tissue/tumour-specific targeting. In theory, tumour targeting can be achieved either by transcriptional targeting or transductional targeting.

18.4.1 Transcriptional targeting

Transcriptional targeting involves genetically limiting the expression of the introduced gene to specific tissues through the use of a tissue specific promoter (TSP). One of the earliest TSPs explored for cancer was the carcinoembryonic antigen (CEA) promoter, expressed in most gastric, pancreatic, colon and lung cancers (Bauerschmitz *et al.*, 2002). *In vitro* and *in vivo* models have shown higher endothelial gene expression by using the endothelial cell specific promoters, such as murine preproendothelin-1 (PPE-1) promoter, DF-3/MUC-1 promoter, von-Willebrand factor, intracellular adhesion molecule-2 (ICAM-2), tie 2, VEGFR-2 and E-selectin (Varda-Bloom *et al.*, 2001).

The activation of proapoptotic caspases mediates apoptosis of neovascular endothelial cells, and overcomes the prosurvival effect of vascular endothelial growth factor or basic fibroblast growth factor (Nor *et al.*, 2002). A transcriptionally targeted adenoviral vector that mediates expression of inducible caspase-9 under human VEGFR2 promoter (Ad-hVEGFR2-icaspase-9) specifically in endothelial cells has been described (Song *et al.*, 2005). In the severe combined immunodeficient mouse model of human angiogenesis, the local delivery of Ad-hVEGFR2-icaspase-9 followed by intraperitoneal injection of a cell-permeable dimerizer drug, AP20187, resulted in endothelial cell apoptosis and local ablation of microvessels. Thus, a transcriptionally targeted antiangiogenic adenoviral vector can mediate neovascular disruption upon activation of a caspase-based artificial death switch.

Using the concept of transcriptional targeting, Richardson *et al.* replaced viral transcriptional control elements within the long terminal repeat of retrovirus to impart the desired specificity (Richardson *et al.*, 2004). The authors have developed such hybrid LTR retroviruses, incorporating sequences from human promoters for VEGF receptors and ICAM-2 as a cancer gene therapy strategy targeting tumour endothelial cells. This may provide transcriptional specificity as well as vector safety with respect to prevention of gene activation at sites of proviral integration (Richardson *et al.*, 2004).

The murine preproendothelin-1 (PPE-1) promoter

Endothelins are a family of peptides produced by vascular endothelium in response to injury, hypoxia, ischaemia and shear stress (Varda-Bloom *et al.*, 2001). Since endothelin is produced and activates mitogenic signalling in several human cancer cells, and interacts with VEGF, its promoter can be used to direct gene expression specifically in the vasculature (Matsuura *et al.*, 1998, Bagnato *et al.*, 1997). The murine PPE-1 promoter is highly specific for vascular ECs due to the presence of the *cis* element ETC/D/E in the promoter, which is absent in the human ET-1 promoter. An adenovirus expressing luciferase under the control of the PPE-1 promoter has been described (Varda-Bloom *et al.*, 2001). Systemic administration of the Ad vector, in mice bearing Lewis lung carcinoma, resulted in high and specific activity of PPE-1 in the neovasculature of primary tumours and lung metastasis.

In subsequent studies, the authors designed adenovirus expressing a chimeric death receptor derived from the modified PPE-1 promoter to trigger EC specific apoptosis (Greenberger *et al.*, 2004). Expression of a chimeric death receptor, composed of Fas and tumour necrosis factor (TNF) receptor 1, resulted in specific apoptosis of ECs *in vitro* and sensitization of cells to the proapoptotic effects of TNF-α. The antitumoral activity of the vectors was assayed in two mouse models. In the B16 melanoma model, a single systemic injection of virus to the tail vein caused growth retardation of tumour and reduction of tumour mass with central tumour necrosis. When the Lewis lung carcinoma lung-metastasis model was utilized, systemic injection of the vector resulted in reduction of lung-metastases, via an antiangiogenic mechanism. Moreover, by application of the PPE-1-based transcriptional control, a humoral immune response against the transgene was avoided (Greenberger *et al.*, 2004). Thus, the data provide evidence that transcriptionally controlled angiogenesis-targeted gene therapy is feasible.

The DF3 promoter

DF3/MUC1 is a tumour-associated antigen that is over expressed on many human carcinomas, including breast, pancreatic, and colon cancer (Kufe *et al.*, 1984; Friedman *et al.*, 1986; Abe and Kufe, 1987). More than 75 per cent of human epithelial ovarian carcinomas express DF3 protein (Friedman *et al.*, 1986). DF3/MUC1 gene expression is regulated by sequences between positions −598 and −485 bp upstream from the transcription start site. Studies have demonstrated that this DF3/MUC1 promoter/enhancer region confers selective expression of diverse transgenes in MUC1-positive breast cancer cells. Also, this promoter/enhancer has been used to regulate expression of E1A in an adenoviral mutant, Ad-DF3-E1, which replicates preferentially in DF3/MUC1-positive cancer cells (Kurihara *et al.*, 2000). It has been shown that infection with adenovirus expressing TNF-α under the control of a DF3/CMV promoter was associated with selective replication and production of TNF-α in cells that express MUC1. Moreover, treatment of MUC1-positive, but not MUC1-negative, xenografts with a single injection of Ad.DF3-E1-TNF-α was effective in inducing stable tumour regression (Kurihara *et al.*, 2000).

Treatment responsive promoters

Another strategy for cancer gene therapy involves restricting gene expression with a conventional form of treatment, such as chemotherapy or radiation (Bauerschmitz *et al.*, 2002). Therapeutic ionizing radiation can cause DNA strand break and DNA nucleoprotein conformation change, which may alter transduction signals resulting in activation of early growth response genes including

Egr-1 whose gene products then activate other genes, which are important in the cellular response to radiation injury. EGR-1 regulates transcription of genes involved in growth inhibition and apoptosis. Radiation-induced apoptotic function of Egr-1 is directly mediated by its target genes p53, Rb and TNF-α (Das *et al.*, 2001; Wu *et al.*, 2006). Recently, Egr-1 has been used in clinical trials to deliver adenovirus expressing TNF-α after ionizing radiation to allow temporal and spatial control of TNF-α release (Senzer *et al.*, 2004, McLoughlin *et al.*, 2005).

18.4.2 Transductional targeting

Transductional targeting involves the chemical or genetic modification of vectors, redirecting its tropism from the native receptor, to a new one preferentially expressed on target cells (Bauerschmitz *et al.*, 2002). It can be done by incorporating either a ligand or bispecific antibody that block virus receptor and direct to EC receptor. Singly ablated (CAR ablated) or doubly ablated (both CAR and integrin binding ablated) Ad has been studied for its liver transduction, and bloodstream persistence after intraperitoneal administration (Akiyama *et al.*, 2004). The doubly ablated vector showed lower liver transduction with significantly improved delivery to cancers located in the peritoneal cavity, as well as to metastatic tumours located throughout the body by virtue of its enhanced bloodstream persistence. It has been shown that combining both the transductional and transcriptional targeting improves the specificity of transgene expression *in vivo* (Reynolds *et al.*, 2001). The combination of transductional targeting to a pulmonary endothelial marker, angiotensin-converting enzyme and an endothelial-specific promoter VEGFR-1 resulted in a synergistic, 300 000-fold improvement in the selectivity of transgene expression for lung versus the usual site of vector sequestration, the liver (Reynolds *et al.*, 2001).

TNFerade™

TNF-α is a multifunctional cytokine with potent antitumour and antivascular properties. However severe systemic toxicity has limited its use to isolated limb perfusions for melanoma and sarcoma patients (Alexander *et al.*, 1998). A gene therapy approach, using intratumoral delivery is one way to reduce the systemic toxicity. TNFerade is a second generation (E1-, partial E3-, and E-4 deleted) adenoviral vector, expressing the TNF-α gene driven by a radiation-inducible promoter, Egr-1 (Senzer *et al.*, 2004; McLoughlin *et al.*, 2005). The approach capitalizes on the known therapeutic synergy between radiation and TNF-α (Hallahan *et al.*, 1995). The activity of TNFerade in combination with radiation has been evaluated in a number of different human xenografts. The combined effect of TNFerade and radiation was found to significantly enhance the effect of either modality alone in these models (Senzer *et al.*, 2004). The first phase I clinical trial was conducted recently in order to evaluate the safety, tolerance and feasibility of intratumoral administration of TNFerade in conjunction with radiation in patients with various solid tumours. The clinical trial showed that repeated intratumoral injection of TNFerade in combination with external beam radiation was safe (Senzer *et al.*, 2004). Overall, 70 per cent patients demonstrated objective tumour responses.

RGD targeting

Integrins are heterodimeric glycoproteins consisting of an α-subunit and a β-subunit. They are an important class of transmembrane molecules involved in cell-cell and cell-matrix interaction (Ruoslahti and Pierschbacher, 1987). Compared with quiescent established blood vessels, ECs in angiogenic blood vessels express additional proteins, such as the $\alpha v\beta 3$, $\alpha v\beta 5$ (Ruoslahti, 2002). Integrin $\alpha v\beta 3$ is expressed on proliferating but not on quiescent ECs and expressed by various tumours. It mediates cellular adhesion to vitronection, fibrinogen, laminin, collagen and von Willebrand factor through their exposed RGD amino acid sequence (Plow *et al.*, 2000). Many researchers have exploited this ligand-receptor binding to target viral vectors to tumour endothelium. The double cyclic RGD peptide binds with a 200-fold greater *in vitro* affinity to $\alpha v\beta 3$ and $\alpha v\beta 5$ integrins and a 50-fold greater affinity to the $\alpha 5\beta 1$ integrin than the linear peptide (Koivunen *et al.*, 1995). RGD-targeted, integrin expressing adenoviral vectors have been used to effectively transfer genes to human melanoma (Nakamura *et al.*, 2002). The RGD-targeted adenovirus infected cells independent of CAR receptor expression.

18.5 Concluding remarks

There is growing enthusiasm for antiangiogenic approaches to treat human cancers (Yang *et al.*, 2003). A number of human malignancies both solid as well as haematopoietic are considered candidates for antiangiogenic therapeutic intervention and this has considerably broadened the scope of potential antiangiogenic gene therapy. The inherent difficulties with production and the cost of producing purified recombinant antiangiogenic inhibitors, coupled with the need for the long term administration, makes gene therapy an exciting alternative strategy to circumvent these difficulties.

However, development of second and third generation viral vectors, advances in high titre purification methods and selective cell/tissue targeting can circumvent some of the problems associated with viral vectors. The development of more selective anticancer targets, which can discriminate between tumour cells and normal cells, should be the goal of current anticancer research. Also, the optimal antiangiogenic treatment strategy including optimal dosing, timing of treatment and proper evaluation of tumour response has not been finalized. Recently, the potential and limitations of a wide range of techniques used for imaging of tumour vasculature including positron emission tomography, X-ray computed tomography, magnetic resonance imaging, ultrasound and optical imaging has been reviewed. The authors discussed which methods are likely to have the sensitivity and robustness for monitoring responses to cancer therapy and described ways in which imaging has been used in clinical trials to date. Such techniques will be important additional tools for assessing the clinical efficacy of the new drugs being developed including antiangiogenic gene therapy (Miller *et al.*, 2005). The clinical studies so far indicate that angiogenesis is not solely regulated by any one factor but it is regulated on many levels. Hence, inhibition of any single factor is not going to be very beneficial. As better vectors are developed, combination strategies continue to evolve, and increased understanding of the complex role that endogenous angiogenesis inhibitors play in tumour growth and progression takes place; antiangiogenic gene therapy will certainly be evaluated in future clinical trials.

References

Abdollahi A, Hahnfeldt P, Maercker C, *et al.* (2004). Endostatin's antiangiogenic signaling network. *Mol Cell* **13**(5), 649–663.

Abe M, Kufe DW (1987). Identification of a family of high molecular weight tumour-associated glycoproteins. *J Immunol* **139**, 257–261.

Akiyama M, Thorne S, Kirn D, *et al.* (2004). Ablating CAR and integrin binding in adenovirus vectors reduces nontarget organ transduction and permits sustained bloodstream persistence following intraperitoneal administration. *Mol Ther* **9**, 218–230.

Alexander HR Jr, Bartlett DL, Libutti SK, Fraker DL, Moser T, Rosenberg SA (1998). Isolated hepatic perfusion with tumor necrosis factor and melphalan for unresectable cancers confined to the liver. *J Clin Oncol* **16**, 1479–1489.

Arap W, Pasqualini R, Ruoslahti E (1998). Cancer treatment by targeted drug delivery to tumor vasculature in a mouse model. *Science* **279**, 377–380.

Asahara T, Murohara T, Sullivan A, *et al.* (1997). Isolation of putative progenitor endothelial cells for angiogenesis. *Science* **275**, 964–967.

Bagnato A, Tecce R, Di Castro V, Catt KJ (1997). Activation of mitogenic signaling by endothelin 1 in ovarian carcinoma cells. *Cancer Res* **57**, 1306–1311.

Baker AH, Edwards DR, Murphy G (2002). Metalloproteinase inhibitors: biological actions and therapeutic opportunities. *J Cell Sci* **115**(Pt 19), 3719–3727.

Battegay EJ, Rupp J, Iruela-Arispe L, Sage EH, Pech M (1994). PDGF-BB modulates endothelial proliferation and angiogenesis *in vitro* via PDGF beta-receptors. *J Cell Biol* **125**, 917–928.

Bauerschmitz GJ, Barker SD, Hemminki A (2002). Adenoviral gene therapy for cancer: from vectors to targeted and replication competent agents (review). *Int J Oncol* **21**, 1161–1174.

Belotti D, Vergani V, Drudis T, *et al.* (1996). The microtubule-affecting drug paclitaxel has antiangiogenic activity. *Clin Cancer Res* **2**, 1843–1849.

Benencia F, Courreges MC, Conejo-Garcia JR, *et al.* (2005). Oncolytic HSV exerts direct antiangiogenic activity in ovarian carcinoma. *Hum Gene Ther* **16**, 765–778.

Benjamin L, Golijanin E, Itin D, Pode A, Keshet DE (1999). Selective ablation of immature blood vessels in established human tumors follows vascular endothelial growth factor withdrawal. *J Clin Invest* **103**(2), 159–165.

Berkner KL (1988). Development of adenovirus vectors for the expression of heterologous genes. *Biotechniques* **67**: 616–629.

Bertolini F, Mancuso P, Gobbi A, Pruneri G (2000). The thin red line: angiogenesis in normal and malignant hematopoiesis. *Exp Hematol* **28**, 993–1000.

Bertolini F, Paul S, Mancuso P, *et al.* (2003). Maximum tolerable dose and low-dose metronomic chemotherapy have opposite effects on the mobilization and viability of circulating endothelial progenitor cells. *Cancer Res* **63**, 4342–4346.

Bevilacqua MP, Stengelin S, Gimbrone MA Jr, Seed B (1989). Endothelial leukocyte adhesion molecule 1: an inducible receptor for neutrophils related to complement regulatory proteins and lectins. *Science* **243**, 1160–1165.

Bikfalvi A, Gimenez-Gallego G (2004). The control of angiogenesis and tumor invasion by platelet factor-4 and platelet factor-4-derived molecules. *Semin Thromb Hemost* **30**, 1137–1144.

Brooks PC, Montgomery AM, Rosenfeld M, *et al.* (1994). Integrin alpha v betz 3 antagonists promote tumour regression by inducing apoptosis of angiogenic blood vessels. *Cell* **79**, 1157–1164.

Browder T, Butterfield CE, Kraling BM, *et al.* (2000). Antiangiogenic scheduling of chemotherapy improves efficacy against experimental drug-resistant cancer. *Cancer Res* **60**, 1878–1886.

Cao R, Farnebo J, Kurimoto M, Cao Y (1999). Interleukin-18 acts as an angiogenesis and tumor suppressor. *FASEB J* **13**, 152195–152202.

Carmeliet P, Moons L, Luttun A, *et al.* (2001). Synergism between vascular endothelial growth factor and placental growth factor contributes to angiogenesis and plasma extravasation in pathological conditions. *Nat Med* **7**, 575–583.

Casanovas O, Hicklin DJ, Bergers G, Hanahan D (2005). Drug resistance by evasion of antiangiogenic targeting of VEGF signaling in late-stage pancreatic islet tumors. *Cancer Cell* **8**, 299–309.

Chen H, Herndon ME, Lawler J (2000a). The cell biology of thrombospondin-1. *Matrix Biol* **19**, 597–614.

Chen H, Treweeke AT, West DC, *et al.* (2000b). *In vitro* and *in vivo* production of vascular endothelial growth factor by chronic lymphocytic leukemia cells. *Blood* **96**, 3181–3187.

Chen W, Fu J, Liu Q, Ruan C, Xiao S (2003). Retroviral endostatin gene transfer inhibits human colon cancer cell growth *in vivo*. *Chin Med J (Engl)* **116**(10), 1582–1584.

Claesson-Welsh L, Welsh M, Ito N, *et al.* (1998). Angiostatin induces endothelial cell apoptosis and activation of focal adhesion kinase independently of the integrin-binding motif RGD. *Proc Natl Acad Sci U S A* **95**, 105579–105583.

Daly G, Chernajovsky Y (2000). Recent developments in retroviral-mediated gene transduction. *Mol Ther* **2**, 5423–5434.

Das A, Chendil D, Dey S, *et al.* (2001). Ionizing radiation down-regulates p53 protein in primary Egr-1−/− mouse embryonic fibroblast cells causing enhanced resistance to apoptosis. *J Biol Chem* **276**, 3279–3286.

Davidoff AM, Nathwani AC, Spurbeck WW, Ng CY, Zhou J, Vanin EF (2002). rAAV-mediated long-term liver-generated expression of an angiogenesis inhibitor can restrict renal tumor growth in mice. *Cancer Res* **62**(11), 3077–3083.

Dawson DW, Volpert OV, Pearce SF, *et al.* (1999). Three distinct D-amino acid substitutions confer potent antiangiogenic activity on an inactive peptide derived from a thrombospondin-1 type 1 repeat, *Mol Pharmacol* **55**, 332–338.

de Fraipont F, Nicholson AC, Feige JJ, Van Meir EG (2001). Thrombospondins and tumor angiogenesis. *Trends Mol Med* **79**: 401–407.

Di Giovine M, Salone B, Martina Y, *et al.* (2001). Binding properties, cell delivery, and gene transfer of adenoviral penton base displaying bacteriophage. *Virology* **282**, 102–112.

Dias S, Hattori K, Zhu Z, *et al.* (2000). Autocrine stimulation of VEGFR-2 activates human leukemic cell growth and migration, *J Clin Invest* **106**, 511–521.

Douglas JT (2007). Adenoviral vectors for gene therapy. *Mol Biotechnol* **36**(1), 71–80.

Duensing S, Atzpodien J (1995). Increased intracellular and plasma levels of basic fibroblast growth factor in B-cell chronic lymphocytic leukemia. *Blood* **85**, 1978–1980.

During MJ, Naegele JR, O'Malley KL, Geller AI (1994). Long-term behavioral recovery in parkinsonian rats by an HSV vector expressing tyrosine hydroxylase. *Science* **266**, 1399–1403.

Feldman AL, Alexander HR, Hewitt SM, *et al.* (2001). Effect of retroviral endostatin gene transfer on subcutaneous and intraperitoneal growth of murine tumors. *J Natl Cancer Inst* **93**, 1014–10-20.

Feldman AL, Friedl J, Lans TE, *et al.* (2002). Retroviral gene transfer of interferon-inducible protein 10 inhibits growth of human melanoma xenografts. *Int J Cancer* **99**, 149–153.

Feldman AL, Restifo NP, Alexander HR, *et al.* (2000). Antiangiogenic gene therapy of cancer utilizing a recombinant adenovirus to elevate systemic endostatin levels in mice. *Cancer Res* **60**, 1503–1506.

Felmeden DC, Blann AD, Lip GY (2003). Angiogenesis: basic pathophysiology and implications for disease. *Eur Heart J* **24**, 586–603.

Ferrari FK, Samulski T, Shenk T, Samulski RJ (1996). Second-strand synthesis is a rate-limiting step for efficient transduction by recombinant adeno-associated virus vectors. *J Virol* **70**(5), 3227–3234.

Folkman J, Klagsbrun M (1987). Angiogenic factors, *Science* **235**, 442–447.

Folkman J (2003). Angiogenesis and apoptosis. *Semin Cancer Biol* **13**, 159–167.

Folkman J (1995). Angiogenesis in cancer, vascular, rheumatoid and other disease. *Nat Med* **1**, 27–31.

Folkman J (1990). What is the evidence that tumors are angiogenesis dependent? *J Natl Cancer Inst* **82**, 4–6.

Fox SB, Leek RD, Bliss J, *et al.* (1997). Association of tumor angiogenesis with bone marrow micrometastases in breast cancer patients. *J Natl Cancer Inst* **89**, 1044–1049.

Fox SB, Turner GD, Gatter KC, Harris AL (1995). The increased expression of adhesion molecules ICAM-3, E- and P-selectins on breast cancer endothelium, *J Pathol* **177**, 369–376.

Friedman EL, Hayes DF, Kufe DW (1986). Reactivity of monoclonal antibody DF3 with a high molecular weight antigen expressed in human ovarian carcinomas, *Cancer Res* **46**, 5189–5194.

Fukumura D, Xavier R, Sugiura T, *et al.* (1998). Tumor induction of VEGF promoter activity in stromal cells. *Cell* **94**, 715–725.

Gagne P, Akalu A, Brooks PC (2004). Challenges facing antiangiogenic therapy for cancer: impact of the tumor extracellular environment. *Expert Rev Anticancer Ther* **4**, 129–140.

Gallo O, Franchi A, Magnelli L, *et al.* (2001). Cyclooxygenase-2 pathway correlates with VEGF expression in head and neck cancer. Implications for tumor angiogenesis and metastasis. *Neoplasia* **3**, 53–61.

Ghosh SS, Takahashi M, Thummala NR, Parashar B, Chowdhury NR, Chowdhury JR (2000). Liver-directed gene therapy: promises, problems and prospects at the turn of the century. *J Hepatol* **32** (1 Suppl2), 38–52.

Gimbrone MA Jr, Leapman SB, Cotran RS, Folkman J (1972). Tumor dormancy *in vivo* by prevention of neovascularization. *J Exp Med* **136**, 261–276.

Glover DJ, Lipps HJ, Jans DA (2005). Towards safe, non-viral therapeutic gene expression in humans, *Nat Rev Genet* **6**, 299–310.

Greenberger S, Shaish A, Varda-Bloom N, *et al.* (2004). Transcription-controlled gene therapy against tumor angiogenesis. *J Clin Invest* **113**, 1017–1024.

Griscelli F, Li H, Bennaceur-Griscelli A, *et al.* (1998). Angiostatin gene transfer: inhibition of tumor growth *in vivo* by blockage of endothelial cell proliferation associated with a mitosis arrest. *Proc Natl Acad Sci U S A* **95**, 6367–6372.

Hajitou A, Trepel M, Lilley CE, *et al.* (2006). A hybrid vector for ligand-directed tumor targeting and molecular imaging. *Cell* **125**(2), 385–398.

Hallahan DE, Vokes EE, Rubin SJ, *et al.* (1995). Phase I dose-escalation study of tumor necrosis factor-alpha and concomitant radiation therapy. *Cancer J Sci Am* **1**, 204.

Hanahan D (1985). Heritable formation of pancreatic beta-cell tumours in transgenic mice expressing recombinant insulin/simian virus 40 oncogenes. *Nature* **315**, 115–122.

Harding TC, Lalani AS, Roberts BN, *et al.* (2006). AAV serotype 8-mediated gene delivery of a soluble VEGF receptor to the CNS for the treatment of glioblastoma. *Mol Ther* **13**, 956–966.

Hockel M, Sasse J, Wissler JH (1987). Purified monocyte-derived angiogenic substance (angiotropin) stimulates migration, phenotypic changes, and 'tube formation' but not proliferation of capillary endothelial cells *in vitro*. *J Cell Physiol* **133**, 1–13.

Holmgren L, O'Reilly MS, Folkman J (1995). Dormancy of micrometastases: balanced proliferation and apoptosis in the presence of angiogenesis suppression. *Nat Med* **1**, 149–153.

Holmgren L, O'Reilly MS, Folkman J (1995). Dormancy of micromctastases: balanced proliferation and apoptosis in the presence of angiogenesis suppression. *Nat Med* **1**, 27149–27153.

Huang SP, Wu MS, Shun CT, *et al.* (2004). Interleukin-6 increases vascular endothelial growth factor and angiogenesis in gastric carcinoma. *J Biomed Sci* **11**, 517–527.

Im SA, Kim JS, Gomez-Manzano C, *et al.* (2001). Inhibition of breast cancer growth *in vivo* by anti-angiogenesis gene therapy with adenovirus-mediated antisense-VEGF. *Br J Cancer* **84**, 1252–1257.

Imren S, Kohn DB, Shimada H, Blavier L, DeClerck YA (1996). Overexpression of tissue inhibitor of metalloproteinases-2 retroviral-mediated gene transfer *in vivo* inhibits tumor growth and invasion. *Cancer Res* **56**, 2891–2895.

Isayeva T, Ren C, Ponnazhagan S (2005). Recombinant adeno-associated virus 2-mediated antiangiogenic prevention in a mouse model of intraperitoneal ovarian cancer. *Clin Cancer Res* **11**, 1342–1347.

Jain RK (2005). Antiangiogenic therapy for cancer: current and emerging concepts. *Oncology (Williston Park)* **19**, 7–16.

Jain RK (2003). Molecular regulation of vessel maturation. *Nat Med* **9**, 685–693.

Kalluri R, Zeisberg M (2006). Fibroblasts in cancer, *Nat Rev Cancer* **6**, 392–401.

Kansas GS (1996). Selectins and their ligands: current concepts and controversies, *Blood* **88**, 3259–3287.

Kayton ML, Libutti SK (2001). Endothelial monocyte activating polypeptide II (EMAP II) enhances the effect of TNF on tumor-associated vasculature. *Curr Opin Investig Drugs* **2**(1), 136–138.

Kerbel RS (2000). Tumor angiogenesis: past, present and the near future. *Carcinogenesis* **21**3: 505–515.

Kim M, Zinn K, Barnett R, *et al.* (2002). The therapeutic efficacy of adenoviral vectors for cancer gene therapy is limited by a low level of primary adenovirus receptors on tumour cells. *Eur J Cancer* 1917–26.

Kitadai Y, Onogawa S, Kuwai T, *et al.* (2004). Angiogenic switch occurs during the precancerous stage of human esophageal squamous cell carcinoma. *Oncol Rep* **11**, 315–319.

Koivunen E, Wang B, Ruoslahti E (1994). Isolation of a highly specific ligand for the alpha 5 beta 1 integrin from a phage display library. *J Cell Biol* **124**, 373–380.

Koivunen E, Wang B, Ruoslahti E (1995). Phage libraries displaying cyclic peptides with different ring sizes: ligand specificities of the RGD-directed integrins. *Biotechnology (N Y)* **13**, 265–270.

Kucharczuk JC, Randazzo B, Chang MY, *et al.* (1997). Use of a 'replication-restricted' herpes virus to treat experimental human malignant mesothelioma. *Cancer Res* **57**, 466–471.

Kufe D, Inghirami G, Abe M, Hayes D, Justi-Wheeler H, Schlom J (1984). Differential reactivity of a novel monoclonal antibody (DF3) with human malignant versus benign breast tumors. *Hybridoma* **3**, 223–232.

Kurihara T, Brough DE, Kovesdi I, Kufe DW (2000). Selectivity of a replication-competent adenovirus for human breast carcinoma cells expressing the MUC1 antigen. *J Clin Invest* **106**, 763–771.

Kuzu I, Beksac M, Arat M, Celebi H, Elhan AH, Erekul S (2004). Bone marrow microvessel density (MVD) in adult acute myeloid leukemia (AML): therapy induced changes and effects on survival. *Leuk Lymphoma* **45**, 1185–1190.

Laakkonen P, Porkka K, Hoffman JA, Ruoslahti E (2002). A tumor-homing peptide with a targeting specificity related to lymphatic vessels. *Nat Med* **8**, 751–755.

Larocca D, Burg MA, Jensen-Pergakes K, Ravey EP, Gonzalez AM, Baird A (2002). Evolving phage vectors for cell targeted gene delivery. *Curr Pharm Biotechnol* **6**(3), 145–157.

Larocca D, Jensen-Pergakes K, Burg MA, Baird A (2001). Receptor-targeted gene delivery using multivalent phagemid particles. *Mol Ther* **3**, 476–484.

Larocca D, Kassner PD, Witte A, Ladner RC, Pierce GF, Baird A (1999). Gene transfer to mammalian cells using genetically targeted filamentous bacteriophage. *FASEB J* **13**, 727–734.

Larocca D, Witte A, Johnson W, Pierce GF, Baird A (1998). Targeting bacteriophage to mammalian cell surface receptors for gene delivery. *Hum Gene Ther* **9**, 162393–162399.

Li A, Dubey S, Varney ML, Dave BJ, Singh RK (2003). IL-8 directly enhanced endothelial cell survival, proliferation, and matrix metalloproteinases production and regulated angiogenesis. *J Immunol* **170**, 3369–3376.

Lin J, Ziring D, Desai S, *et al.* (2007). TNFalpha blockade in human diseases: An overview of efficacy and safety. *Clin Immunol* **1521**, 6616.

Lin P, Buxton JA, Acheson A, *et al.* (1998). Antiangiogenic gene therapy targeting the endothelium-specific receptor tyrosine kinase Tie2. *Proc Natl Acad Sci U S A* **95**, 8829–8834.

Liu P, Wang Y, Li YH, *et al.* (2003). Adenovirus-mediated gene therapy with an antiangiogenic fragment of thrombospondin-1 inhibits human leukemia xenograft growth in nude mice. *Leuk Res* **27**, 701–708.

Liu XH, Kirschenbaum A, Yao S, Lee R, Holland JF, Levine AC (2000). Inhibition of cyclooxygenase-2 suppresses angiogenesis and the growth of prostate cancer *in vivo*. *J Urol* **164**, 820–825.

Lode HN, Wenkel J, Gaedicke G (2002). Angiogenesis, views and news. The 93rd annual meeting of the American Association for Cancer Research. April 6–10 2002, Moscone Convention Center, San Francisco, USA. *Exp Opin Biol Ther* **2**, 671–676.

Louis N, Fender P, Barge A, Kitts P, Chroboczek J (1994). Cell-binding domain of adenovirus serotype 2 fiber. *J Virol* **68**6: 4104–4106.

Matsuura A, Yamochi W, Hirata K, Kawashima S, Yokoyama M (1998). Stimulatory interaction between vascular endothelial growth factor and endothelin-1 on each gene expression. *Hypertension* **32**, 89–95.

McLoughlin JM, McCarty TM, Cunningham C, *et al.* (2005). TNFerade, an adenovector carrying the transgene for human tumor necrosis factor alpha, for patients with advanced solid tumors: surgical experience and long-term follow-up. *Ann Surg Oncol* **12**, 825–830.

Mignatti P, Rifkin DB (1996). Plasminogen activators and matrix metalloproteinases in angiogenesis. *Enzyme Protein* **49**, 117–137.

Miller DG, Adam MA, Miller AD (1990). Gene transfer by retrovirus vectors occurs only in cells that are actively replicating at the time of infection. *Mol Cell Biol* **10**, 84239–84242.

Miller JC, Pien HH, Sahani D, Sorensen AG, Thrall JH (2005). Imaging angiogenesis: applications and potential for drug development. *J Natl Cancer Inst* **97**, 172–187.

Modlich U, Pugh CW, Bicknell R (2000). Increasing endothelial cell specific expression by the use of heterologous hypoxic and cytokine-inducible enhancers. *Gene Ther* **7**, 896–902.

Molica S, Vitelli G, Levato D, Gandolfo GM, Liso V (1999). Increased serum levels of vascular endothelial growth factor predict risk of progression in early B-cell chronic lymphocytic leukaemia. *Br J Haematol* **107**, 605–610.

Mundel TM, Kalluri, R (2007). Type IV collagen-derived angiogenesis inhibitors. Microvasc Res **74**, 85–89.

Nakamura T, Sato K, Hamada H (2002). Effective gene transfer to human melanomas via integrin-targeted adenoviral vectors. *Hum Gene Ther* **13**, 613–626.

Neri D, Bicknell R (2005). Tumour vascular targeting. *Nat Rev Cancer* **5**, 436–446.

Neville LF, Mathiak G, Bagasra O (1997). The immunobiology of interferon-gamma inducible protein 10 kD (IP-10): a novel, pleiotropic member of the C-X-C chemokine superfamily. *Cytokine Growth Factor Rev* **8**, 207–219.

Nicklin SA, Buening H, Dishart KL, *et al.* (2001). Efficient and selective AAV2-mediated gene transfer directed to human vascular endothelial cells. *Mol Ther* **4**(3), 174–181.

Nor JE, Hu Y, Song W, Spencer DM, Nunez G (2002). Ablation of microvessels *in vivo* upon dimerization of iCaspase-9. *Gene Ther* **9**(7), 444–451.

Noro T, Miyake K, Suzuki-Miyake N, *et al.* (2004). Adeno-associated viral vector-mediated expression of endostatin inhibits tumor growth and metastasis in an orthotropic pancreatic cancer model in hamsters, *Cancer Res* **64**, 7486–7490.

Oh P, Li Y, Yu J, *et al.* (2004). Subtractive proteomic mapping of the endothelial surface in lung and solid tumours for tissue-specific therapy. *Nature* **429**(6992), 629–635.

Pasqualini R, Koivunen E, Ruoslahti E (1997). Alpha v integrins as receptors for tumor targeting by circulating ligands. *Nat Biotechnol* **15**, 542–546.

Pasqualini R, Ruoslahti E (1996). Organ targeting *in vivo* using phage display peptide libraries. *Nature* **380**, 364–366.

Pepper MS (2001). Role of the matrix metalloproteinase and plasminogen activator-plasmin systems in angiogenesis. *Arterioscler Thromb Vasc Biol* **21**, 1104–1117.

Perez-Atayde AR, Sallan SE, Tedrow U, Connors S, Allred E, Folkman J (1997). Spectrum of tumor angiogenesis in the bone marrow of children with acute lymphoblastic leukemia. *Am J Pathol* **150**, 815–821.

Pfeifer A, Kessler T, Silletti S, Cherech DA, Verma IM (2000). Suppression of angiogenesis by lentiviral delivery of PEX, a noncatalytic fragment of matrix metalloproteinase 2. *Proc Natl Acad Sci U S A* **97**, 12227–12232.

Plow EF, Haas TA, Zhang L, Loftus J, Smith JW (2000). Ligand binding to integrins. *J Biol Chem* **275**, 21785–21788.

Ponnazhagan S, Mahendra G, Kumar S, *et al.* (2004). Adeno-associated virus 2-mediated antiangiogenic cancer gene therapy: long-term efficacy of a vector encoding angiostatin and endostatin over vectors encoding a single factor. *Cancer Res* **64**, 1781–1787.

Raikwar SP, Temm CJ, Raikwar NS, Kao C, Molitoris BA, Gardner TA (2005). Adenoviral vectors expressing human endostatin–angiostatin and soluble Tie2: enhanced suppression of tumor growth and antiangiogenic effects in a prostate tumor model. *Mol Ther* **12**, 61091–61100.

Rajotte D, Arap W, Hagedorn M, Koivunen E, Pasqualini R, Ruoslahti E (1998). Molecular heterogeneity of the vascular endothelium revealed by *in vivo* phage display. *J Clin Invest* **102**, 430–437.

Reynolds PN, Nicklin SA, Kaliberova L, *et al.* (2001). Combined transductional and transcriptional targeting improves the specificity of transgene expression *in vivo*. *Nat Biotechnol* **19**(9), 838–842.

Ribatti D, Scavelli C, Roccaro AM, Crivellato E, Nico B, Vacca A (2004). Hematopoietic cancer and angiogenesis. *Stem Cells Dev* **13**, 484–495.

Ribatti D, Vacca A, Nico B, Fanelli M, Roncali L, Dammacco F (1996). Angiogenesis spectrum in the stroma of B-cell non-Hodgkin's lymphomas. An immunohistochemical and ultrastructural study. *Eur J Haematol* **56**, 45–53.

Richardson TB, Kaspers J, Porter CD (2004). Retroviral hybrid LTR vector strategy: functional analysis of LTR elements and generation of endothelial cell specificity. *Gene Ther* **11**(9), 775–783.

Roberts AB, Sporn MB, Assoian RK, *et al.* (1986). Transforming growth factor type beta: rapid induction of fibrosis and angiogenesis *in vivo* and stimulation of collagen formation *in vitro*. *Proc Natl Acad Sci USA* **83**, 4167–4171.

Ruoslahti E, Pierschbacher MD (1987). New perspectives in cell adhesion: RGD and integrins. *Science* **238**, 491–497.

Ruoslahti E (2002). Specialization of tumour vasculature. *Nat Rev Cancer* **2**, 83–90.

Russell DW, Miller AD, Alexander IE (1994). Adeno-associated virus vectors preferentially transduce cells in S phase. *Proc Natl Acad Sci U S A* **91**(1989), 15–19.

Salven P, Hattori K, Heissig B, Rafii S (2002). Interleukin-1alpha promotes angiogenesis *in vivo* via VEGFR-2 pathway by inducing inflammatory cell VEGF synthesis and secretion. *FASEB J* **16**, 1471–1473.

Sappino AP, Skalli O, Jackson B, Schurch W, Gabbiani G (1988). Smooth-muscle differentiation in stromal cells of malignant and non-malignant breast tissues. *Int J Cancer* **41**, 707–712.

Sato TN, Tozawa Y, Deutsch U, *et al.* (1995). Distinct roles of the receptor tyrosine kinases Tie-1 and Tie-2 in blood vessel formation. *Nature* **376**, 70–74.

Schindler U, Baichwal VR (1994). Three NF-kappa B binding sites in the human E-selectin gene required for maximal tumor necrosis factor alpha-induced expression. *Mol Cell Biol* **14**, 5820–5831.

Schuch G, Heymach JV, Nomi M, *et al.* (2003). Endostatin inhibits the vascular endothelial growth factor-induced mobilization of endothelial progenitor cells, *Cancer Res* **63**, 8345–8350.

Sedlacek HH (2001). Pharmacological aspects of targeting cancer gene therapy to endothelial cells. *Crit Rev Oncol Hematol* **37**, 169–215.

Senzer N, Mani S, Rosemurgy A, *et al.* (2004). TNFerade biologic, an adenovector with a radiation-inducible promoter, carrying the human tumor necrosis factor alpha gene: a phase I study in patients with solid tumors. *J Clin Oncol* **22**, 592–601.

Shaked Y, Bocci G, Munoz R, *et al.* (2005). Cellular and molecular surrogate markers to monitor targeted and non-targeted antiangiogenic drug activity and determine optimal biologic dose. *Curr Cancer Drug Targets* **5**, 551–559.

Sharma S, Odunsi K (2005). Targeted therapy for epithelial ovarian cancer, *Exp Opin Ther Targets* **9**, 501–513.

Shen Y, Nemunaitis J (2006). Herpes simplex virus 1 (HSV-1) for cancer treatment. *Cancer Gene Ther* **13**, 11975–11992.

Shichinohe T, Bochner BH, Mizutani K, *et al.* (2001). Development of lentiviral vectors for antiangiogenic gene delivery. *Cancer Gene Ther* **8**, 879–889.

Shichiri M, Hirata Y (2001). Antiangiogenesis signals by endostatin. *FASEB J* **15**(6): 1044–1053.

Shih T, Lindley C (2006). Bevacizumab: an angiogenesis inhibitor for the treatment of solid malignancies. *Clin Ther* **28**(111), 779–802.

Sim BK, MacDonald NJ, Gubish ER (2000). Angiostatin and endostatin: endogenous inhibitors of tumour growth. *Cancer Metastasis Rev* **19**, 81–90.

Song W, Sun Q, Dong Z, Spencer DM, Nunez G, Nor JE (2005). Antiangiogenic gene therapy: disruption of neovascular networks mediated by inducible caspase-9 delivered with a transcriptionally targeted adenoviral vector. *Gene Ther* **12**(4), 320–329.

Srivastava A, Lusby EW, Berns KI (1983). Nucleotide sequence and organization of the adeno-associated virus 2 genome. *J Virol* **45**(2), 555–564.

Staba MJ, Mauceri HJ, Kufe DW, Hallahan DE, Weichselbaum RR (1998). Adenoviral TNF-alpha gene therapy and radiation damage tumor vasculature in a human malignant glioma xenograft. *Gene Ther* **5**, 293–300.

Stone D, Liu Y, Shayakhmetov D, Li ZY, Ni S, Lieber A (2007). Adenovirus-platelet interaction in blood causes virus sequestration to the reticuloendothelial system of the liver. *J Virol* **81**(9), 4866–4871.

Summerford C, Samulski RJ (1998). Membrane-associated heparan sulfate proteoglycan is a receptor for adeno-associated virus type 2 virions. *J Virol* **72**(2), 438–445.

Tandle A, Blazer DG 3rd, Libutti SK (2004). Antiangiogenic gene therapy of cancer: recent developments. *J Transl Med* **2**, 22.

Thomas CE, Ehrhardt A, Kay MA (2003). Progress and problems with the use of viral vectors for gene therapy. *Nat Rev Genet* **4**, 346–358.

Thorne SH, Tam BY, Kirn DH, Contag CH, Kuo CJ (2006). Selective intratumoral amplification of an antiangiogenic vector by an oncolytic virus produces enhanced antivascular and anti-tumor efficacy. *Mol Ther* **13**(5C), 938–946.

Thorpe PE, Chaplin DJ, Blakey DC (2003). The first international conference on vascular targeting: meeting overview. *Cancer Res* **63**, 1144–1147.

Tolsma SS, Volpert OV, Good DJ, Frazier WA, Polverini PJ, Bouck N (1993). Peptides derived from two separate domains of the matrix protein thrombospondin-1 have anti-angiogenic activity. *J Cell Biol* **122**, 497–511.

Trinchieri G (2003). Interleukin-12 and the regulation of innate resistance and adaptive immunity. *Nat Rev Immunol* **3**(2), 133–146.

Vacca A, Ribatti D, Presta M, *et al.* (1999). Bone marrow neovascularization, plasma cell angiogenic potential, and matrix metalloproteinase-2 secretion parallel progression of human multiple myeloma. *Blood* **93**, 3064–3073.

Vacca A, Ribatti D, Roncali L, *et al.* (1994). Bone marrow angiogenesis and progression in multiple myeloma. *Br J Haematol* **87**, 503–508.

Varda-Bloom N, Shaish A, Gonen A, *et al.* (2001). Tissue-specific gene therapy directed to tumor angiogenesis. *Gene Ther* **8**, 819–827.

Varghese S, Rabkin SD, Liu R, Nielsen PG, Ipe T, Martuza RL (2006). Enhanced therapeutic efficacy of IL-12, but not GM-CSF, expressing oncolytic herpes simplex virus for transgenic mouse derived prostate cancers. *Cancer Gene Ther* **13**, 253–265.

Verheul HM, Voest EE, Schlingemann RO (2004). Are tumours angiogenesis-dependent? *J Pathol* **202**, 5–13.

Wang X, Liu F, Li K, Li X, Xu JS (2002). Retrovirus-mediated gene transfer of human endostatin inhibits growth of human liver carcinoma cells SMMC7721 in nude mice. *World J Gastroenterol* **8**(6), 1045–1049.

White K, Nicklin SA, Baker AH (2007). Novel vectors for *in vivo* gene delivery to vascular tissue. *Exp Opin Biol Ther* **7**, 6809–6821.

Wickham T, Mathias J, Cheresh P, Nemerow DA (1993). Integrins alpha v beta 3 and alpha v beta 5 promote adenovirus internalization but not virus attachment. *Cell* **73**, 2309–2319.

Wu LH, Song GL, Diao SY, *et al.* (2003). [Inhibition of tumor angiogenesis in nude mice by adenovirus-mediated PF4 p17–70 cDNA transfection.] *Zhonghua Xue Ye Xue Za Zhi* **24**, 426–429.

Wu MY, Wu XY, Li QS, Zheng RM (2006). Expression of Egr-1 gene and its correlation with the oncogene proteins in non-irradiated and irradiated esophageal squamous cell carcinoma. *Dis Esophagus* **19**, 267–272.

Yang JC, Haworth L, Sherry RM, *et al.* (2003). A randomized trial of bevacizumab, an anti-vascular endothelial growth factor antibody, for metastatic renal cancer. *N Engl J Med* **349**, 427–434.

Yao VJ, Ozawa MG, Varner AS, *et al.* (2006). Anti-angiogenic therapy decreases integrin expression in normalized tumor blood vessels. *Cancer Res* **66**, 2639–2649.

Yoshida S, Ono M, Shono T, *et al.* (1997). Involvement of interleukin-8, vascular endothelial growth factor, and basic fibroblast growth factor in tumor necrosis factor alpha-dependent angiogenesis. *Mol Cell Biol* **17**, 4015–4023.

Young LS, Mautner V (2001). The promise and potential hazards of adenovirus gene therapy. *Gut* **48**(5), 733–736.

Yu JL, Rak JW, Coomber BL, Hicklin DJ, Kerbel RS (2002). Effect of p53 status on tumor response to antiangiogenic therapy. *Science* **295**, 1526–1528.

Ziche M, Donnini S, Morbidelli L (2004). *Curr Drug Targets* **5**, 485–493.

19

Radiosensitization in Viral Gene Therapy

Jula Veerapong, Kai A. Bickenbach and **Ralph R. Weichselbaum**

19.1 Introduction

Since the discovery of X-rays by German physicist Wilhelm Roentgen in 1995, the properties of ionizing radiation (IR) have been utilized in many applications. In particular, the DNA damaging effects of IR have been exploited in the form of radiotherapy for the treatment of cancer. However, since the DNA-damaging potential of IR can be indiscriminate with respect to irradiated tissue, efforts throughout the years have been focused on widening the therapeutic window of radiation to maximize tumour control and to spare normal tissue. In the past few decades, radiotherapy has been combined with other anti-cancer modalities such as chemotherapeutic agents to attempt to improve the therapeutic index. This strategy has met with mixed results and further novel radio-enhancing strategies are being pursued and viral vectors have been investigated to maximize therapeutic potential. Other strategies have focused on improvements in the physical delivery of radiation therapy.

Gene therapy offers a new approach to increase the therapeutic index in radiotherapy. One gene therapy strategy is the delivery of transgenes to cancer cells that increase susceptibility to radiation therapy. Another strategy is the use of conditionally replication competent viruses to increase the anti-tumour effects of radiation. Delivery of transgenes may be accomplished by viral and non-viral vectors. In this chapter, we focus on the role of viral vectors in gene therapy.

19.2 Adenovirus

The adenovirus is a non-enveloped virus that contains a linear double-stranded DNA genome. Adenoviruses are categorized by genome size, composition, homology and organization. They are grouped from A to F. Of the many serotypes, the most commonly studied are the group C serotypes 1, 2, 5, and 6. The genome of the prototypical adenoviral vector for gene therapy is based on Type 5, which consists of a linear 36 kb strand of DNA (Bent *et al.*, 1993).

Adenoviral infection is mediated by the binding of the fibre protein knob to cell surface receptors. The most common cell surface receptor is the CAR (Coxsackievirus and adenovirus receptor). Following binding, the virion is endocytosed and escapes the endosome, making its way to the nucleus. Once in the nucleus, viral transcription and replication begin. Viral transcription occurs with sequential transcription of three sets of genes. These are defined by their temporal relationship as early, delayed early, and late genes. The early gene products (E1, E2, E3, and E4) are responsible mainly for viral transcription, viral genome replication, and inhibition of host cell apoptosis. The late genes encode proteins necessary for viral assembly (Benihoud *et al.*, 1999).

Viral Therapy of Cancer Edited by Kevin J. Harrington, Richard G. Vile and Hardev S. Pandha

The products of the early genes E1A and E1B are the first viral proteins produced after viral infection. E1A is an important activator of viral gene transcription and is also involved in inducing a G_1–S transition in the host cells, enabling DNA replication. Many adenoviral vectors are based on the strategy of deleting E1A and part of E1B to make them replication deficient. This ensures viral recombination will not occur in the host and offers an additional 5–6 kb of space to package additional DNA for gene therapy (Rao *et al.*, 1992).

Adenoviral vectors have advantages for gene therapy. First, they can be linked to a promoter that can be used to drive a transgene of interest, thereby allowing selective control of transgene expression. Second, adenoviral vectors have very efficient gene transfer, and most human cell types are susceptible, allowing for cross applicability among different tumour types. Third, the adenoviral genome does not integrate itself into the host gene. This ensures that the vector will not be mutagenic and will be safe in human hosts.

The major disadvantage of adenoviral vectors is the host immune response (Gilgenkrantz *et al.*, 1995). The E1-deleted virus, though replication deficient, still expresses both early and late viral genes. This expression leads to a strong T-cell mediated response against the host cell and results in cell lysis. Another disadvantage is that many patients have been exposed to adenoviruses in the past and have already developed antibodies against viral proteins. The final disadvantage of these vectors is the limited amount of DNA that can be inserted. The genome is 36 kb and allows for a maximum of 8.5 kb for transgene insertion.

19.3 Adeno-associated viruses

Adeno-associated viruses (AAV) are small, non-enveloped viruses that contain a linear single-stranded DNA genome. These viruses are naturally replication deficient. Of the many known serotypes, the principal type used in gene therapy is AAV Type 2. Infection with AAV may result in either a latent or productive infection. Productive infection requires the co-infection of a helper virus. This helper virus may be adenovirus, herpes or vaccinia. Additionally, exposure to genotoxic stress can also activate a productive infection. In the absence of co-infection or genotoxic stress, the AAV establishes latency by integrating into the host cell genome at the long arm of chromosome 19 at the AAVS1 site (Li *et al.*, 2005).

The genome of AAV is approximately 4.7 kb and is flanked by inverted terminal repeats (ITR) of 145 bases, which serve as origins of DNA replication. There are two open reading frames (ORF) between the ITR called Rep (replication) and Cap (capsid). The ITR are required *in cis* for genome replication and integration. The Rep reading frame encodes four non-structural proteins, while the Cap reading frame encodes structural protein Vp1 through 3. AAV vectors used in gene therapy are created by deleting Rep and Cap ORF and replacing them with target transgenes (Collaco *et al.*, 1999).

The advantage of the AAV vector is that it is intrinsically replication deficient and is therefore non-pathogenic. Unlike adenoviruses, most patients have not been previously exposed to AAVs and consequently do not have pre-existing neutralizing antibodies. There is also a decreased immune response because the viral protein in the Rep and Cap reading frames are deleted. Furthermore, deletion of the Rep reading frame prevents viral genome integration, making the AAV vector non- mutagenic. AAVs are able to infect both dividing and non-dividing cells and can infect a wide variety of tissue types. The main disadvantage of the AAV vector is its size limitation of transgene DNA to 4 kb (Qing *et al.*, 1999; Summerford *et al.*, 1999).

19.4 Herpes simplex viruses

Herpes simplex viruses (HSV) are large enveloped double-stranded DNA viruses that are neurotropic and infect epithelial cells. HSV-1 infection may result in a latent state in infected sensory neurons. The virus can be reactivated from its dormant state by multiple stimuli to recommence its lytic cycle. The HSV-1 genome encodes over 90 proteins and is approximately 150 kb. The genome is represented by two stretches of DNA: unique long and unique short. These sequences of DNA are flanked by inverted repeat segments. Unique region genes are present as single copies. Those in the repeated segments are present as two copies. Viral replication is sequentially regulated by a segment of genes. The first genes to be expressed are the alpha genes which are involved in regulatory functions, followed by beta genes, which are involved in nucleic acid synthesis and DNA replication. Finally, gamma genes are expressed and they encode structural proteins.

Replication-deficient herpes simplex viruses are categorized as helper virus dependent or independent. Helper virus dependent vectors are also known as amplicons. They arise spontaneously by recombination and are amplified by serial passages at high multiplicities of infection. Their genomes are defective and have a large amount of viral DNA removed. At a minimum, the amplicons contain only a terminal alpha sequence and origin of viral DNA synthesis. Because they lack much of the genome needed for replication, this must be provided by a 'helper virus'. An example is a HSV with the alpha 4 gene deleted. In this construct, an amplicon is transfected into an alpha 4 expressing cell line and this cell line is superinfected with the alpha 4 deleted herpes simplex virus. The advantage of this system is a large gene payload since virtually all 150 kb of the HSV genome is available for use. Moreover, the transgenes are expressed for a long time, as there is no oncolysis or latency period established. Unfortunately, amplicons can become contaminated with helper virus DNA. Additionally, the yield is several orders of magnitude less than normal viral infection, and they are unstable through multiple serial passages (Kwong and Frankel, 1985; Stavropolous and Strathdee, 1998).

Helper virus independent vectors have deletions in viral genes that are essential for viral replication. A cell line that complements these defects is required to manufacture these viruses. The advantage of this system is that these vectors replicate to high titres, unlike amplicons. These viruses can also fit up to 40 kb of transgene DNA, which is less than that of amplicons but greater than that of adenoviral vectors. However, the disadvantage is that deletion of essential genes may result in altered expression of the desired transgene. HSV-1 vectors may result in an antiviral immune response, and as with all viral vectors, there is a possible of recombination and reversion of wild type phenotype.

Replication competent neuro-attenuated HSV-1 viruses have been employed for the purposes of oncolytic gene therapy. R3616 is a prototypic herpes simplex virus with a deletion of the $\gamma_1 34.5$ gene encoding for neurovirulence. The product of this gene, the infected cell protein (ICP) 34.5 functions to block the host response to viral infection. Upon infection of the host cell, large amounts of viral RNA accumulate (Jaquemont and Roizman, 1975; Kozak and Roizman, 1975). This accumulation leads to the activation of double-stranded RNA-dependent protein kinase (PKR), whose function is to phosphorylate on the α-subunit of the eukaryotic initiation factor 2 (eIF-2α). This results in the complete shutoff of protein synthesis, thereby blocking viral replication (Katze, 2995). HSV-1 overcomes this defence through ICP 34.5, which functions to recruit protein phosphatase 1-α to dephosphorylate eIF-2α (Dachs *et al.*, 1997). This allows protein synthesis and viral replication to proceed unhindered by PKR activation (Leib *et al.*, 2000; Breyer *et al.*, 2001). Mutant viruses that lack the $\gamma_1 34.5$ gene are unable to replicate in normal cells, and therefore are highly attenuated in animal models. However, in some tumours $\Delta\gamma_1 34.5$ mutant viruses replicate unhindered and cause tumour lysis (Mezhir *et al.*, 2006). The oncolytic potential of these viruses has been tested in phase I trials and they have been shown to be clinically safe in escalating doses for cancer patients (Markert *et al.*, 2000; Rampling et al., 2000). Unfortunately, the usefulness of $\Delta\gamma_1 34.5$ mutant viruses is limited by the variability in human tumour cells to support viral replication of these mutants (Chou *et al.*, 1994; Andreansky *et al.*, 1997; Bennett *et al.*, 2002; Nakamura *et al.*, 2002).

Radiation and viral therapy can be combined for an enhanced effect. It has been demonstrated that ionizing radiation enhances gene expression from late viral promoters in transduced cells in the absence of other viral gene products. In tumours infected with the attenuated virus, ionizing radiation increases 13.6-fold above baseline the gene expression from a late viral promoter as early as 2 h after virus infection. The radiation-dependent up-regulation of late viral genes is mediated by the p38 pathway with its effect being abolished by p38 inhibitors or a p38 dominant-negative construct. This suggests that ionizing radiation up-regulates late promoters active in the course of viral DNA synthesis. Hence, radiation may be used to up-regulate cytotoxic genes introduced into tumour cells by viral vectors for oncolytic therapy (Table 19.1).

19.5 Enhancing the effect of radiation by delivering tumour suppressor genes

Dysfunctional tumour suppressor genes are the most common genetic lesions found in human cancers (Roth, 2006). Delivering functional tumour suppressor transgenes by means of a viral vector is one strategy to render cancer cells susceptible to ionizing radiation. By switching on pro-apoptotic pathways that manipulate intracellular responses to radiation, enhancement of overall cell death can be achieved.

19.5.1 p53

One of the most commonly studied aberrations in human cancers is the defect in the tumour suppressor gene p53. Expression of p53 in cancer cells results in G_1 growth arrest and apoptosis. An aberration in p53 expression would expectedly result in unchecked proliferation and resistance to most forms of cancer therapy. Replication deficient adenoviruses expressing wild type p53 have been demonstrated to radiosensitize multiple cancer types with p53 mutations, including brain (Lang *et al.*, 1998), colon (Spitz *et al.*, 1996), head and neck (Chang *et al.*, 1997; Pirollo *et al.*, 1997), ovary (Gallardo *et al.*, 1996), and prostate cancers (Colletier *et al.*, 2000).

19.5.2 E2F-1

E2F-1 is a transcription factor that may play a central role in the cellular DNA damage response. Genotoxic stress triggers protein kinases, leading to an accumulation of E2F-1 and subsequent induction of apoptosis (Stevens and La Thangue, 2004). *In vivo* and *in vitro* studies in gliomas have demonstrated that overexpression of E2F-1 leads to tumour growth arrest by activation of an S_1 checkpoint and apoptosis (Fueyo *et al.*, 1998). Studies performed in p53 wild type and p53 null prostate cancer cell lines have shown that induction of apoptosis by E2F-1 enhances the tumoricidal response to radiation (Nguyen *et al.*, 2005). Experiments in fibrosarcoma cells suggest that the effect of radiosensitization by E2F-1 may be independent of p53 status (Pruschy *et al.*, 1999).

19.5.3 p21

p21 is a cyclin-dependent kinase inhibitor that has a role in modulating cell cycle progression. Expression of p21 results in an accumulation of cells in the G_0/G_1 phase and cell differentiation, preventing cell cycle progression. Introduction of p21 via adenoviral vectors have demonstrated suppression of growth in malignant tumours *in vivo*, thus checking cell proliferation. p21 has tumoricidal properties, inducing tumour suppression and apoptosis in cervical cancer.

19.6 Virus-directed enzyme prodrug therapy

One strategy of gene therapy employs the viral delivery of a gene encoding for an exogenous prodrug converting enzyme into tumour cells and the subsequent systemic administration of the non-toxic prodrug correlate. This approach of delivering genes to sensitize tumour cells to a prodrug is often termed 'suicide gene therapy'. A central concept in suicide gene therapy is an approach fostering tumour-targeted chemotherapy. The goals of this treatment are to circumvent the inherent problems associated with conventional chemotherapy and ultimately widen the therapeutic ratio in the treatment of cancer (Fischer *et al.*, 2005).

In order for virus-directed enzyme prodrug therapy to be successful, several conditions need to be met. First, the prodrug enzyme gene should be selectively transduced and expressed in the target tumour cells. There should be little or no expression in the surrounding normal tissues. Second, the non-toxic prodrug should be a selective

Table 19.1 Common viral vectors

Virus	Type of virus	Size of transgene	Advantages	Disadvantages
Adenovirus	Linear non-enveloped dsDNA	8.5 kb	Efficient transfer Does not integrate into host genome	Host immune response Pre-existing neutralizing antibodies Size limitation
AAV	Linear non-enveloped ssDNA	4 kb	Replication deficient No pre-existing neutralizing Ab Decreased immune response	Size limitation
HSV	Linear enveloped dsDNA	150 kb	Large payload Prolonged transgene expression	Possibility of recombination and reversion Integration into host genome

substrate for the exogenous enzyme, with activation occurring only within the tumour microenvironment. Third, the converted drug should be highly diffusible, resulting in killing of neighbouring tumour cells in which gene transduction of the prodrug enzyme has not taken place. This effect of inducing cytotoxicity in adjacent non-expressing tumour cells is known as the 'bystander effect'.

19.6.1 HSV-thymidine kinase

One of the most widely utilized suicide gene is the HSV-1 thymidine kinase (tk) enzyme, which confers susceptibility to nucleoside analogues such as ganciclovir (GCV), aciclovir (ACV), and bromo-vinyl-deoxyuridine (BVdUrd) (Fischer *et al.*, 2005). Viral thymidine kinase phosphorylates these analogs into monophosphates, which in turn become phosphorylated into triphosphates by endogenous host cellular kinases. The resultant triphosphates are incorporated into replicating host DNA by DNA polymerase, producing chain termination and single-strand breaks. Because DNA synthesis occurs in the S phase, the primary effect of the nucleoside analogues is cycle dependent killing.

Both adenoviral and retroviral vectors have been used extensively to transduce tumour cells to express HSV thymidine kinase. However, the relatively inefficient gene transfer by these vectors can be a potential impediment to the effectiveness of tk-directed suicide gene therapy. Nevertheless, it has been demonstrated *in vivo* that as little as 10 per cent in transduction efficiency in the tumour cell population is necessary to effect complete eradication of the tumour (Freeman *et al.*, 1993). This observation is attributed to the bystander effect, whereby only a fraction of cancer cells are essentially required to be transfected in order to achieve substantial tumoricidal activity. Several studies have been undertaken to elucidate the mechanism of HSV-tk mediated bystander effect. For example, investigations performed in gliomas and various cell lines proposed that gap junction formation between neighbouring cells may result in a high level of intracellular communication and contribute to the transfer of phosphorylated ganciclovir nucleotides (Elshami *et al.*, 1996; Dilber *et al.*, 1997; Touraine *et al.*, 1998). Cytokines are proposed to play a role in mediating the bystander effect. In a study using an HSV-tk xenogeneic model, increased interleukin (IL)-1 and IL-6 mRNA transcripts by polymerase chain reaction and IL-1 by immunohistochemistry lends support that immunostimulation may be contributory factor (Freeman *et al.*, 1995). The tk/ganciclovir complex has been shown to induce p53 accumulation and p53-mediated translocation of CD95, resulting in the formation of the death-inducing signalling complex and activation of the caspase-dependent apoptotic pathway.

Several studies have investigated the role of adding radiation treatment to thymidine kinase prodrug therapy to enhance tumoricidal activity. One study reported that combining radiation with BVdUrd in retrovirally transduced HSV-tk human glioma enhanced tumour cytotoxicity by a ratio of 1.9 (Kim *et al.*, 1994, 1995). Another report demonstrated *in vitro* and *in vivo* that an adenovirally delivered mutant HSV-tk combined with ACV radiosensitized rat glioma such that cell kill was pronounced and survival was prolonged. The mechanism of radiosensitization is not fully elucidated, but the nucleoside analogs may impair DNA so as to render it vulnerable to radiation or it may impair the cell's ability to repair radiation damage. Alternatively, thymidine kinase may cause an imbalance of the nucleoside pool (Valerie *et al.*, 2001).

19.6.2 Cytosine deaminase

Cytosine deaminase (CD) gene therapy and administration of 5-flucytosine (5-FC) has been utilized with success as well in experimental tumour models delivered by adenoviral and retroviral vectors. CD is an enzyme found in bacteria and fungi that deaminates cytosine to uracil, but it is not normally expressed in mammalian cells. More importantly and clinically relevant, CD also converts the non-toxic 5-FC into the exceptionally toxic 5-fluorauracil (5-FU). 5-FU is metabolized into 5-fluoro-2′-deoxyuridine monophosphate (FdUMP) and 5-fluorouridine triphosphate (FUTP) in both normal cells and tumour. fdUMP binds thymidylate synthase into a covalently bound ternary complex, inhibiting the synthesis of thymidylate from 2′-deoxyuridylate (dUMP). The cell essentially undergoes a 'thymidineless death' as DNA synthesis is disrupted. In addition, FUTP can be incorporated by nuclear transcriptional enzymes instead of uridine triphosphate (UTP) into RNA, thus interfering with RNA synthesis and processing.

In contrast to the HSV-tk/GCV enzyme/prodrug regimen, CD/5-FC therapy does not require cell-to-cell contact for the bystander effect. Some studies have suggested that successful exploitation of the bystander effect in the CD/5-FC gene therapy model requires an immunocompetent host, as T-cell dependent killing may augment killing of neighboring cells (Kuriyama *et al.*, 1999). It is generally agreed upon that very little gene transfer, as low as 1–5 per cent, is necessary to achieve significant regression of tumour (Huber *et al.*, 1994).

5-FU itself has been used quite extensively as an antineoplastic agent and it has long been known to be a radiosensitizer. Its enhancement of the radiation response has been attributed to the killing of cells in the S phase cycle, which are relatively radioresistant. CD/5-FC enzyme/prodrug gene therapy using an adenoviral vector has been demonstrated to improve radiation response in human colorectal xenografts (Hanna *et al.*, 1996).

19.6.3 Nitroreductase

CB1954, 5-aziridiyl-2,4-dinitrobenzamide, is a tumour inhibitory nitrophenylaziridine that was discovered to have cytotoxic activity in the Walker rat carcinosarcoma cell line. Its selective tumour inhibitory effect in this cell line was attributed to the activity of an NADPH dehydrogenase known as DT-diaphorase (DTD). This enzyme reduces CB1954 to 5-aziridinyl-4-hydroxylamino-2-nitrobenzamide in the presence of reduced nicotinamide adenine dinucleotide (NADH) or reduced nicotinamide adenine dinucleotide phosphate (NAPDH), resulting in interstrand crosslinking of DNA (Bridgewater *et al.*, 1995). The ability of human DTD to bioactivate CB1954 is greatly reduced when compared to Walker rat DTD. In contrast, a nitroreductase (NTR) purified from *Escherichia coli* has a markedly increased ability than either human or rat DTD to convert CB1954 into a cytotoxic species. Its bioactivity is due to its ability to reduce either the 2- or 4-nitro group of CB1954 and its greater substrate affinity (Knox *et al.*, 1992). Based on these findings, CB1954 has been transduced by adenoviral and retroviral vectors into multiple cell lines, including colorectal, pancreatic, and ovarian cancer (Green *et al.*, 1997; McNeish *et al.*, 1998). Studies using the CB1954/NTR enzyme/prodrug approach have demonstrated significant antitumor effects with only 5 per cent gene transfer efficiency. Apparent bystander effects are ascribed to a cell-permeable metabolite produced in transduced cells (Bridgewater *et al.*, 1997; Djeha *et al.*, 2000). CB1954 is a robust radiosensitizer with a greater than 10-fold increase in efficiency with the addition of radiation (Patterson *et al.*, 2003).

19.6.4 Cytochrome P450

The oxazaphosphorines, cyclophosphamide (CPA) and ifosphamide (IFA), are chemotherapeutic prodrugs that are activated by the cytochrome P450 (CYP) enzyme family. Cytochrome P450 induces hydroxylation of the CPA and IFA into a hydroxycyclophosphoramide that degrades into a mustard metabolite and acrolein, ultimately leading to DNA crosslinking and protein alkylation, respectively. Gene transfer of cytochrome P450 2B1 using a retrovirus vector with subsequent treatment with CPA has been demonstrated to produce decreased tumour volumes in rat glioma. Even with 10 per cent gene transfer, a bystander effect attributable to a diffusible metabolite present in the milieu was found to be present (Wei *et al.*, 1995). Whereas the cytochrome P450 alkylating agent prodrugs have not been found to sensitize cells to radiation, irradiation combining CYP 4B1 with a radioinducible promoter in an adenoviral construct has resulted in improve cytotoxic activity (Table 19.2; Hsu *et al.*, 2003).

19.6.5 Double suicide gene therapy

While the technique of single modality enzyme/prodrug gene therapy was being refined and explored, many of its limitations were becoming apparent. Though single enzyme/prodrug regimens have demonstrated varying degrees of efficacy in animal models, relative enthusiasm for its use has been tempered by incomplete tumour

Table 19.2 Virus directed enzyme prodrug therapy

Converting enzyme	Prodrug	Action of drug
HSV-thymidine kinase	Ganciclovir, aciclovir, bromovinyl-deoxyuridine	Antimetabolite
E. coli cytosine deaminase	5-Fluorocytosine	Antimetabolite
Human cytochrome P450 2B1	cyclophosphamide, ifosfamide	Alkylator
E. coli XGPRT	6-thioxanthine	Antimetabolite
E. coli DeoD	6-methylpurine-2′-deoxyribonucleoside	Antimetabolite
E. coli nitroreductase	CB1954 (5-aziridin-1-yl)-2,4-dinitrobenzamide	Alkylator

regression and recurrence upon cessation of treatment (Rogulski *et al.*, 1997). In light of this, some investigators sought to merge two different pairings of enzymes and prodrugs in an approach known as double suicide gene therapy. One of the earlier approaches examined the effect of expressing an *E. coli* CD/HSV-tk fusion gene in rat gliosarcoma cells along with concomitant radiation. These investigators found that double suicide gene therapy exhibited increased cytotoxicity over single gene therapy and potentiated the effects of radiation in a synergistic manner (Rogulski *et al.*, 1997). They further applied their findings in an orthotopic model of prostate cancer, using in a trimodal regimen that consisted of a lytic, replication-competent adenoviral vector to transduce CD/HSV-tk, 5FC/GCV treatment, and external beam radiation. Their results provided the groundwork for introducing adenovirus-mediated double suicide gene therapy into clinical trial for locally advanced prostate cancer.

19.6.6 Radiosensitization and the bystander effect

Practical application of suicide gene therapy is subject to multiple variables, including an effective delivery vector, adequate gene transfer efficiency, and efficacious prodrug/enzyme interaction. Its feasibility is enormously reliant on the bystander effect. Yet perhaps one of the more important roles of suicide gene therapy is in the augmentation of targeted killing in radiation therapy. The coupling of gene therapy with irradiation in a deliberate strategy has the potential to be a formidable regimen in the treatment of cancer.

19.6.7 Transgenes with radioresponsive promoters

The early growth response-1 (EGR-1) gene has been identified as a radiation inducible gene. Egr-1 gene product, synonymous with nerve growth factor-1-A, Krox24, TIS8, or Zif268, is a zinc finger transcription factor belonging to a group of immediate-early response genes mediating cell proliferation, differentiation, and possibly death. Egr-1 contains three tandem-repeat zinc finger motifs in its DNA-binding domain which allow it to regulate the activity of other genes (Yan *et al.*, 2000; Thiel and Cibelli, 2002). Its promoter was originally studied by Datta *et al.* (1992) who demonstrated through deletion analysis that a specific motif [CC (A + T-rich$_6$)GG] called the CArG sequence in regions of the 5′ promoter was necessary and sufficient for induction of Egr-1 by IR. Based on this result, it was hypothesized that the Egr-1 promoter could be linked to a therapeutic transgene, and IR could thus provide both spatial and temporal control to gene therapy. Tumour necrosis factor-alpha (TNF-α) was identified as a potential transgene for this experiment. TNF-α is a cytokine that has both a direct cytotoxic effect on tumour cells as well as on the endothelium of the blood vessels supplying the tumours (Mauceri *et al.*, 1996; Staba *et al.*, 1998). It has also been show to sensitize tumour cells to the effects IR (Hallahan *et al.*, 1989, 1990). This activation of Egr-1.TNF was originally

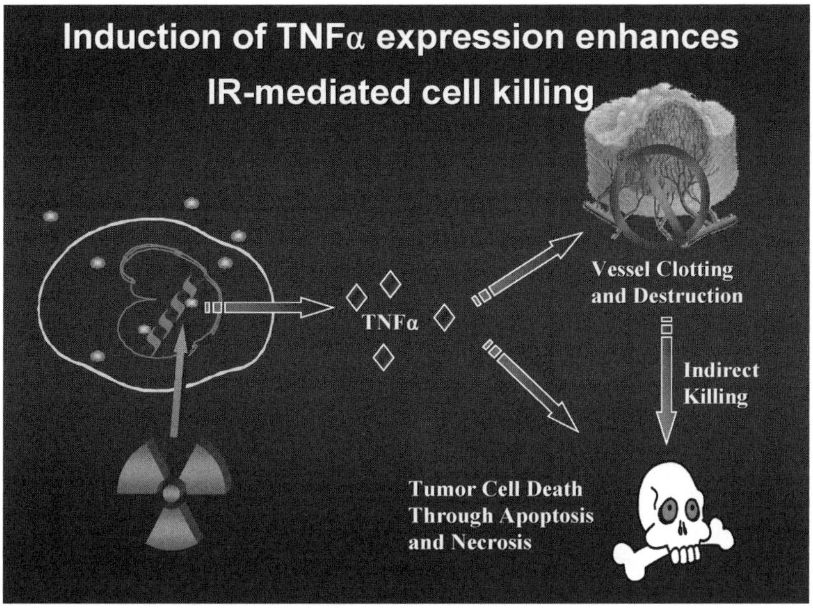

Figure 19.1 Egr-1.TNFα constructs may be used to potentiate effects of ionizing radiation

tested by transfecting Egr-1.TNF construct into human leukaemia cells (HL-525). These cells were then injected into xenografts of a human head and neck cancer SQ-20B. Animals treated with both the Egr-1.TNF construct and IR demonstrated a greater increase in tumour cure than either agent alone (Weischelbaum *et al.*, 1994). Importantly, there was no increase in systemic toxicity. A similar experiment was conducted using liposomes to transfect the Egr-1.TNF construct into xenografts of the murine fibrosarcoma line P4L. Again, the combination of IR and Egr-1.TNF demonstrated a greater reduction in tumour volumes than either agent alone (Seung *et al.*, 1995).

This paradigm was extended to a delivery strategy using a replication deficient adenoviral vector Ad.Egr.TNF. This vector contains the CArG regions of the Egr-1 promoter linked upstream to a cDNA of human TNF-α. Several studies were conducted in human xenografts employing IR and this vector. The combination of IR and Ad.Egr.TNF led to a greater tumour cell kill than either agent alone (Mauceri *et al.*, 1996; Chung *et al.*, 1998; Staba *et al.*, 1998). Again, there was no increase in systemic toxicity with the combination of Ad.Egr.TNF and IR. Histopathological analysis of the xenografts demonstrated widespread intratumoral vessel thrombosis and tumour necrosis with the combination of Ad.Egr.TNF and IR (Mauceri *et al.*, 1996; Staba *et al.*, 1998). This indicated that the tumour microvasculature was a potential target for this paradigm of gene therapy (Figure 19.1).

The activation of the Egr-1 promoter is mediated by DNA damage and reactive oxygen intermediates (ROI) through the CArG elements (Nose *et al.*, 1991; Datta *et al.*, 1992). Based on these results, Ad.Egr.TNF has also been studied with chemotherapeutic regimens that induce ROIs. Park *et al.* (2002) examined the use of cisplatin with Ad.Egr.TNF in colorectal and pancreatic xenografts. Cisplatin was chosen because it is used to treat a wide variety of tumours and is also a radiosensitizer (Mancini *et al.*, 2003; Armanios *et al.*, 2004). It was demonstrated that injection of Ad.Egr.TNF into xenografts and treatment with IR led to increased levels of TNF-α and a greater reduction in tumour volume than either agent alone (Park *et al.*, 2002). Lopez *et al.* (2004) extended these findings to several other chemotherapeutic agents including 5-flurouracil, doxorubicin, cyclophosphamide, gemcitabine, and paclitaxel. All of these agents were able to induce the production of TNF-α from the vector in both PC3, a prostate cancer cell line, and PROb, a colorectal cell line. Furthermore, the free radical scavenger *N*-acetylcysteine abolished the effect, confirming that the induction was due to chemotherapeutic induction of ROIs. This study also demonstrated that the combination of doxorubicin and Ad.Egr.TNF treatment to both PC3 and PROb xenografts had a significantly greater reduction in tumour volume than either agent alone.

19.6.8 Clinical experience

Ad.Egr.TNF has been used in several clinical trials. It was first used in a phase I trial in patients with a variety of solid tumours. Thirty-six patients received 6 weeks of intratumoral injections of Ad.Egr.TNF weekly in doses ranging from 4×10^7 to 4×10^{11} particle units (p.u.). Radiotherapy began on week 2 and consisted of a total dose range of 20–66.6 Gy. The typical regimen was 1.8–2.0 Gy daily Monday–Friday depending on the tumour type and location. There were objective responses in 14 of 30 evaluable patients. Five patients had a complete response, three of whom were patients with melanoma. A dose dependent effect was seen in the study with 12 of 22 patients (55 per cent) having an objective response with the higher doses. The response rate was only two of eight patients (25 per cent) in the group with the lower dose. The only grade 2 toxicities were constitutional symptoms of fevers, chills, and pain at the injection site. Serum TNF-α levels remained low, and no significant toxicities were seen due to TNF-α. More importantly no viral particles were recoverable from the patients' urine, blood, or sputum (Senzer *et al.*, 2004).

A second phase I trial was conducted in soft tissue sarcomas. Fourteen patients with extremity sarcomas were administered intratumoral injections of Ad.Egr.TNF at doses ranging from 4×10^9 to 4×10^{11} p.u. Injections were given twice weekly during week 1 and then once per week on weeks 2–5. Ionizing radiation was delivered during weeks 2–5 to a total dose of 50 Gy. Eleven of 13 evaluable patients (85 per cent) achieved a response, and with two patients having a complete response and nine partial responses. There was extensive tumour necrosis observed in pathological analysis of those patients that underwent resection. Only constitutional symptoms were observed in the study. One patient withdrew from the study due to grade 2 toxicity. This study demonstrated that Ad.Egr.TNF is well tolerated and effective for soft tissue extremity sarcomas (Mundt *et al.*, 2004; Table 19.3).

Several phase II trials are currently underway in both pancreatic and oesophageal cancers. In one trial, patients with unresectable pancreatic cancer are receiving intratumoral injections of Ad.Egr.TNF via endoscopic ultrasound or by percutaneous injection through computerized tomography or ultrasound guidance. Escalating doses of the vector are being administered weekly, and concomitant chemoradiotherapy is being administered with 5-FU and 50.4 Gy over 5 weeks. Response is being evaluated by spiral CT. Interim analysis has demonstrated a high response rate for the 4×10^{11} p.u. dose with six of seven patients (86 per cent) demonstrating a greater than 25 per cent tumour shrinkage. In addition, there was a median survival of greater than 15 months in the high dose of Ad.Egr.TNF. Five of 11 patients in the 4×10^{11} p.u. dose group were able to undergo resection, while four of five patients had negative margins.

A similar trial is studying TNFerade and chemotherapy was conducted in patients with esophageal cancer. Patients are receiving weekly injections of Ad.Egr.TNF by endoscopic ultrasound for a total of 5 weeks. Chemotherapy with cisplatin and 5-FU is being administered concomitantly with 45 Gy of radiotherapy. Results from the trial are still pending, but there have been several complete pathological responses in the treatment groups. An increase in thromboembolic events was noticed in this trial and the trial is currently on hold, even though no direct evidence linked to TNFerade to these thrombotic events.

19.7 Conclusions

In conclusion, data from these studies supports the use of gene therapy with chemo- and radiation-inducible promoters and DNA damaging agents

Table 19.3 Phase I clinical trials (Ad.Egr.TNF)

Investigators	Patient #	Tumour types	Regimen	Response
Senzer *et al.* (2004)	36 for toxicity 30 for tumour response	6 Pancreatic 5 Non-small cell lung 5 Breast 5 Colorectal 4 Sarcoma 3 Melanoma 3 Head and neck 5 Other	6 weeks of intratumoral with concomitant radiation (30–60 Gy) 7 dose levels between 4×10^7 to 4×10^{11} p.u.	21 of 30 responded: • 7 minimal responses • 9 partial reponses • 5 complete responses
Mundt *et al.* (2004)	14	Extremity sarcoma	Ad.Egr.TNF administered with single-daily fractionated radiation therapy to total of 50 Gy 3 escalating doses of4×10^9 to 4×10^{11} p.u. Intratumoral injections twice weekly during week 1 and once weekly during weeks 2–5	13 evaluable patients: • 2 received treatment for • 9 partial responses palliation • 2 complete responses

in the treatment of solid tumours. Further studies will delineate the true clinical role of these agents.

References

Andreansky S, *et al.* (1997). Evaluation of genetically engineered herpes simplex viruses as oncolytic agents for human malignant brain tumors. *Cancer Res* **57**(8), 1502–1509.

Armanios M, *et al.* (2004). Adjuvant chemotherapy for resected adenocarcinoma of the esophagus, gastro-esophageal junction, and cardia: phase II trial (E8296) of the Eastern Cooperative Oncology Group. *J Clin Oncol* **22**(22), 4495–4499.

Benihoud K, Yeh P, Perricaudet M (1999). Adenovirus vectors for gene delivery. *Curr Opin Biotechnol* **10**(5), 440–447.

Bennett JJ, *et al.* (2002). Comparison of safety, delivery, and efficacy of two oncolytic herpes viruses (G207 and NV1020) for peritoneal cancer. *Cancer Gene Ther* **9**(11), 935–945.

Bett AJ, Prevec L, Graham FL (1993). Packaging capacity and stability of human adenovirus type 5 vectors. *J Virol* **67**(10), 5911–5921.

Breyer B, *et al.* (2001). Adenoviral vector-mediated gene transfer for human gene therapy. *Curr Gene Ther* **1**(2), 149–162.

Bridgewater JA, *et al.* (1995). Expression of the bacterial nitroreductase enzyme in mammalian cells renders them selectively sensitive to killing by the prodrug CB1954. *Eur J Cancer* **31A**(13–14): 2362–2370.

Bridgewater JA, *et al.* (1997). The bystander effect of the nitroreductase/CB1954 enzyme/prodrug system is due to a cell-permeable metabolite. *Hum Gene Ther* **8**(6), 709–717.

Chang EH, *et al.* (1997). Restoration of the G1 checkpoint and the apoptotic pathway mediated by wild-type p53 sensitizes squamous cell carcinoma of the head and neck to radiotherapy. *Arch Otolaryngol Head Neck Surg* **123**(5), 507–512.

Chou J, *et al.* (1994). Differential response of human cells to deletions and stop codons in the gamma(1) 34.5 gene of herpes simplex virus. *J Virol* **68**(12), 8304–8311.

Chung TD, *et al.* (1998). Tumor necrosis factor-alpha-based gene therapy enhances radiation cytotoxicity in human prostate cancer. *Cancer Gene Ther* **5**(6), 344–349.

Collaco RF, Cao X, Trempe JP (1999). A helper virus-free packaging system for recombinant adeno-associated virus vectors. *Gene* **238**(2), 397–405.

Colletier PJ, *et al.* (2000). Adenoviral-mediated p53 transgene expression sensitizes both wild-type and null p53 prostate cancer cells *in vitro* to radiation. *Int J Radiat Oncol Biol Phys* **48**(5), 1507–1512.

Dachs GU, *et al.* (1997). Targeting gene therapy to cancer: a review. *Oncol Res* **9**(6–7): 313–325.

Datta R, *et al.* (1992). Ionizing radiation activates transcription of the EGR1 gene via CArG elements. *Proc Natl Acad Sci U S A* **89**(21), 10149–10153.

Dilber MS, *et al.* (1997). Gap junctions promote the bystander effect of herpes simplex virus thymidine kinase *in vivo*. *Cancer Res* **57**(8), 1523–1528.

Djeha AH, *et al.* (2000). Expression of *Escherichia coli* B nitroreductase in established human tumor xenografts in mice results in potent antitumoral and bystander effects upon systemic administration of the prodrug CB1954. *Cancer Gene Ther* **7**(5), 721–731.

Elshami AA, *et al.* (1996). Gap junctions play a role in the 'bystander effect' of the herpes simplex virus thymidine kinase/ganciclovir system *in vitro*. *Gene Ther* **3**(1), 85–92.

Fischer U, *et al.* (2005). Mechanisms of thymidine kinase/ganciclovir and cytosine deaminase/5-fluorocytosine suicide gene therapy-induced cell death in glioma cells. *Oncogene* **24**(7), 1231–1243.

Freeman SM, *et al.* (1993). The 'bystander effect': tumor regression when a fraction of the tumor mass is genetically modified. *Cancer Res* **53**(21), 5274–5283.

Freeman SM, *et al.* (1995). The role of cytokines in mediating the bystander effect using HSV-TK xenogeneic cells. *Cancer Lett* **92**(2), 167–174.

Fueyo J, *et al.* (1998). Overexpression of E2F-1 in glioma triggers apoptosis and suppresses tumor growth *in vitro* and *in vivo*. *Nat Med* **4**(6), 685–690.

Gallardo D, Drazan KE, McBride WH (1996). Adenovirus-based transfer of wild-type p53 gene increases ovarian tumor radiosensitivity. *Cancer Res* **56**(21), 4891–4893.

Gilgenkrantz H, *et al.* (1995). Transient expression of genes transferred *in vivo* into heart using first-generation adenoviral vectors: role of the immune response. *Hum Gene Ther* **6**(10), 1265–1274.

Green NK, *et al.* (1997). Sensitization of colorectal and pancreatic cancer cell lines to the prodrug 5-(aziridin-1-yl)-2,4-dinitrobenzamide (CB1954) by retroviral transduction and expression of the E. coli nitroreductase gene. *Cancer Gene Ther* **4**(4), 229–238.

Hallahan DE, *et al.* (1989). Increased tumor necrosis factor alpha mRNA after cellular exposure to ionizing radiation. *Proc Natl Acad Sci U S A* **86**(24), 10104–10107.

Hallahan DE, *et al.* (1990). The interaction between recombinant human tumor necrosis factor and radiation in 13 human tumor cell lines. *Int J Radiat Oncol Biol Phys* **19**(1), 69–74.

Hanna NN, *et al.* (1997). Virally directed cytosine deaminase/5-fluorocytosine gene therapy enhances radiation response in human cancer xenografts. *Cancer Res* **57**(19), 4205–4209.

Hsu H. *et al.* (2003). Combined radiation and cytochrome CYP4B1/4-ipomeanol gene therapy using the EGR1 promoter. *Anticancer Res* **23**(3B), 2723–2728.

Huber BE, *et al.* (1994). Metabolism of 5-fluorocytosine to 5-fluorouracil in human colorectal tumor cells transduced with the cytosine deaminase gene: significant antitumor effects when only a small percentage of tumor cells express cytosine deaminase. *Proc Natl Acad Sci U S A* **91**(17), 8302–8306.

Jacquemont B, Roizman B (1975). RNA synthesis in cells infected with herpes simplex virus. X. Properties of viral symmetric transcripts and of double-stranded RNA prepared from them. *J Virol* **15**(4), 707–713.

Katze MG (1995). Regulation of the interferon-induced PKR: can viruses cope? *Trends Microbiol* **3**(2), 75–78.

Kim JH, *et al.* (1994). Selective enhancement by an antiviral agent of the radiation-induced cell killing of human glioma cells transduced with HSV-tk gene. *Cancer Res* **54**(23), 6053–6056.

Kim JH, *et al.* (1995). Selective enhancement of radiation response of herpes simplex virus thymidine kinase transduced 9L gliosarcoma cells *in vitro* and *in vivo* by antiviral agents. *Int J Radiat Oncol Biol Phys* **33**(4), 861–868.

Knox RJ, *et al.* (1992). The bioactivation of 5-(aziridin-1-yl)-2,4-dinitrobenzamide (CB1954) – II. A comparison of an *Escherichia coli* nitroreductase and Walker DT diaphorase. *Biochem Pharmacol* **44**(12), 2297–2301.

Kozak M, Roizman B (1975). RNA synthesis in cells infected with herpes simplex virus. IX. Evidence for accumulation of abundant symmetric transcripts in nuclei. *J Virol* **15**(1), 36–40.

Kuriyama S, *et al.* (1999). Cytosine deaminase/5-fluorocytosine gene therapy can induce efficient anti-tumor effects and protective immunity in immunocompetent mice but not in athymic nude mice. *Int J Cancer* **81**(4), 592–597.

Kwong AD, Frenkel N (1985). The herpes simplex virus amplicon. IV. Efficient expression of a chimeric chicken ovalbumin gene amplified within defective virus genomes. *Virology* **142**(2), 421–425.

Lang FF, *et al.* (1998). Enhancement of radiosensitivity of wild-type p53 human glioma cells by adenovirus-mediated delivery of the p53 gene. *J Neurosurg* **89**(1), 125–132.

Leib DA, *et al.* (2000). Specific phenotypic restoration of an attenuated virus by knockout of a host resistance gene. *Proc Natl Acad Sci U S A* **97**(11), 6097–6101.

Li C, *et al.* (2005). Adeno-associated virus vectors: potential applications for cancer gene therapy. *Cancer Gene Ther* **12**(12), 913–925.

Lopez CA, *et al.* (2004). Chemoinducible gene therapy: a strategy to enhance doxorubicin antitumor activity. *Mol Cancer Ther* **3**(9), 1167–1175.

Mancini R, *et al.* (2003). Hepatic arterial infusion (HAI) of cisplatin and systemic fluorouracil in the treatment of unresectable colorectal liver metastases. *Anticancer Res* **23**(2C), 1837–1841.

Markert JM, *et al.* (2000). Conditionally replicating herpes simplex virus mutant, G207 for the treatment of malignant glioma: results of a phase I trial. *Gene Ther* **7**(10), 867–874.

Mauceri HJ, *et al.* (1996). Tumor necrosis factor alpha (TNF-alpha) gene therapy targeted by ionizing radiation selectively damages tumor vasculature. *Cancer Res* **56**(19), 4311–4314.

McNeish IA, *et al.* (1998). Virus directed enzyme prodrug therapy for ovarian and pancreatic cancer using retrovirally delivered *E. coli* nitroreductase and CB1954. *Gene Ther* **5**(8), 1061–1069.

Mezhir JJ, *et al.* (2006). Ionizing radiation: a genetic switch for cancer therapy. *Cancer Gene Ther* **13**(1), 1–6.

Mundt AJ, *et al.* (2004). A Phase I trial of TNFerade biologic in patients with soft tissue sarcoma in the extremities. *Clin Cancer Res* **10**(17), 5747–5753.

Nakamura H, *et al.* (2002). Regulation of herpes simplex virus gamma(1)34.5 expression and oncolysis of diffuse liver metastases by Myb34.5. *J Clin Invest* **109**(7), 871–882.

Nguyen KH, *et al.* (2005). Adenoviral-E2F-1 radiosensitizes p53wild-type and p53null human prostate cancer cells. *Int J Radiat Oncol Biol Phys* **63**(1), 238–246.

Nose K, *et al.* (1991). Transcriptional activation of early-response genes by hydrogen peroxide in a mouse osteoblastic cell line. *Eur J Biochem* **201** (1), 99–106.

Park JO, *et al.* (2002). Transcriptional control of viral gene therapy by cisplatin. *J Clin Invest* **110**(3), 403–410.

Patterson AV, Saunders MP, Greco O (2003). Prodrugs in genetic chemoradiotherapy. *Curr Pharm Des* **9**(26), 2131–2154.

Pirollo KF, *et al.* (1997). p53 mediated sensitization of squamous cell carcinoma of the head and neck to radiotherapy. *Oncogene* **14**(14), 1735–1746.

Pruschy M. *et al.* (1999). E2F-1 has properties of a radiosensitizer and its regulation by cyclin A kinase is required for cell survival of fibrosarcoma cells lacking p53. *Cell Growth Differ* **10**(3), 141–146.

Qing K, *et al.* (1999). Human fibroblast growth factor receptor 1 is a co-receptor for infection by adeno-associated virus 2. *Nat Med* **5**(1), 71–77.

Rampling R, *et al.* (2000). Toxicity evaluation of replication-competent herpes simplex virus (ICP 34.5 null mutant 1716) in patients with recurrent malignant glioma. *Gene Ther* **7**(10), 859–866.

Rao L, *et al.* (1992). The adenovirus E1A proteins induce apoptosis, which is inhibited by the E1B 19-kDa and Bcl-2 proteins. *Proc Natl Acad Sci U S A* **89**(16), 7742–7746.

Rogulski KR, *et al.* (1997). Pronounced antitumor effects and tumor radiosensitization of double suicide gene therapy. *Clin Cancer Res* **3**(11), 2081–2088.

Roth JA (2006). Adenovirus p53 gene therapy. *Exp Opin Biol Ther* **6**(1), 55–61.

Senzer N, *et al.* (2004). TNFerade biologic, an adenovector with a radiation-inducible promoter, carrying the human tumor necrosis factor alpha gene: a phase I study in patients with solid tumors. *J Clin Oncol* **22** (4), 592–601.

Seung LP, *et al.* (1995). Genetic radiotherapy overcomes tumor resistance to cytotoxic agents. *Cancer Res* **55** (23), 5561–5565.

Spitz FR, *et al.* (1996). Adenoviral-mediated wild-type p53 gene expression sensitizes colorectal cancer cells to ionizing radiation. *Clin Cancer Res* **2**(10), 1665–1671.

Staba MJ, *et al.* (1998). Adenoviral TNF-alpha gene therapy and radiation damage tumor vasculature in a human malignant glioma xenograft. *Gene Ther* **5**(3), 293–300.

Stavropoulos TA, Strathdee CA (1998). An enhanced packaging system for helper-dependent herpes simplex virus vectors. *J Virol* **72**(9), 7137–7143.

Stevens C, La Thangue NB (2004). The emerging role of E2F-1 in the DNA damage response and checkpoint control. *DNA Repair (Amst)* **3**(8–9): 1071–1079.

Summerford C, Bartlett JS, Samulski RJ (1999). AlphaVbeta5 integrin: a co-receptor for adeno-associated virus type 2 infection. *Nat Med* **5**(1), 78–82.

Thiel G, Cibelli G (2002). Regulation of life and death by the zinc finger transcription factor Egr-1. *J Cell Physiol* **193**(3), 287–292.

Touraine RL, *et al.* (1998). The bystander effect in the HSVtk/ganciclovir system and its relationship to gap junctional communication. *Gene Ther* **5**(12), 1705–1711.

Valerie K, *et al.* (2001). Substantially improved *in vivo* radiosensitization of rat glioma with mutant HSV-TK and acyclovir. *Cancer Gene Ther* **8**(1), 3–8.

Wei MX, *et al.* (1995). Diffusible cytotoxic metabolites contribute to the *in vitro* bystander effect associated with the cyclophosphamide/cytochrome P450 2B1 cancer gene therapy paradigm. *Clin Cancer Res* **1** (10), 1171–1177.

Weichselbaum RR, *et al.* (1994). Gene therapy targeted by radiation preferentially radiosensitizes tumor cells. *Cancer Res* **54**(16), 4266–4269.

Yan SF, *et al.* (2000). Egr-1: is it always immediate and early? *J Clin Invest* **105**(5), 553–554.

20
Radioisotope delivery

Inge D.L. Peerlinck and **Georges Vassaux**

20.1 Introduction

Iodine is an essential molecule for the normal functioning of the thyroid. In 1896 Baumann reported for the first time the ability of the thyroid cells to concentrate Iodine (Dohan *et al.*, 2003). Iodine is scarce in the environment and consequently the dietary intake is low. The capacity of the thyroid follicular cells to concentrate iodine up to 40 times the plasma level is therefore essential for the production of thyroid hormones.

Over the years radioactive iodine isotopes have been used for the diagnosis of thyroid diseases, the treatment of hyperthyroidism and benign thyroid diseases, and the diagnosis and treatment of thyroid cancer. Thyroid cancer was treated for the first time with radioiodine in 1941 (Chung, 2002) but it was not until 1996 that Dai *et al.* isolated the rat cDNA that encodes the protein involved in the uptake of iodine, the sodium iodine symporter (NIS) (Dai *et al.*, 1996). In the same year the human homolog was isolated using the cDNA sequence of the rat sodium iodine symporter (Smanik *et al.*, 1996).

Even after more than 60 years radioiodine and iodine analogues used for thyroid scintigraphy still play an important diagnostic and therapeutic role in thyroid diseases and thyroid cancer (Meller and Becker, 2002). Furthermore, the recent discovery of the NIS has permitted the investigation of new strategies extending the use of iodine therapy thanks to the ectopic expression of NIS with therapeutic vectors. The present review describes new approaches not only to the diagnosis and treatment of thyroid cancer but also to the diagnosis and treatment of a variety of cancers in other tissues.

20.2 History of iodine therapy

20.2.1 Thyroid

The discovery in 1896 by Baumann (Dohan *et al.*, 2003) that thyroid cells concentrate iodine has provided the basis for diagnostic scintigrafic imaging with radioiodine and other scintigrafic agents like pertechnetate. In contrast to normal thyroid tissue cancerous nodules exhibit a reduced or absent iodine uptake while autonomously functioning thyroid tissue shows an increase in iodine uptake, which is an important tool in the differential diagnosis of thyroid abnormalities.

The first treatment of thyroid cancer with radioiodine in 1941 (Chung, 2002) led to the development of a new clinical discipline, nuclear medicine. Radioiodine is used alone or in conjunction with surgery for the ablation of thyroid follicular cells in hyperthyroidism (Moka *et al.*, 2002) and as adjuvant therapy in thyroid carcinomas. Despite showing reduced or absent iodine uptake in imaging studies compared to normal thyroid tissue, differentiated thyroid cancer cells

Viral Therapy of Cancer Edited by Kevin J. Harrington, Richard G. Vile and Hardev S. Pandha

still let in sufficient ^{131}I to be destroyed by the radiation.

After destruction of the thyroid, complete thyroid hormone replacement is necessary and possible.

The role of NIS in the thyroid

NIS mediates iodine uptake in the thyroid and several other tissues like stomach and salivary glands. The transport of iodine across the basolateral membrane of the thyroid follicular cells is the first step in the production of thyroid hormones T_3 and T_4. Two sodium ions are translocated inward 'downhill' together with inward 'uphill' translocation of one iodine ion across the basolateral membrane which makes the sodium iodine transporter a symporter as both substrates are transported at the same time in the same direction. The driving force is the sodium gradient generated and maintained by the sodium potassium adenosine triphosphatase (Levy *et al.*, 1998a; De La Vieja *et al.*, 2000; Riedel *et al.*, 2001). NIS activity is blocked by the specific competitive inhibitors perchlorate and thiocyanate (Levy *et al.*, 1998a; Riedel *et al.*, 2001; Dohan *et al.*, 2003). It was in 1936 that Barker *et al.* (Dohan *et al.*, 2003) discovered for the first time that patients treated with thiocyanate for hypertension develop hypothyroidism and/or goitre.

In the thyroid, in contrast to other tissues, iodine is translocated from the cytoplasm across the apical membrane towards the colloid by pendrin, a chloride/iodine transporter and by the apical iodine transporter (Levy *et al.*, 1998a; Dohan *et al.*, 2003; Spitzweg and Morris, 2002). Iodine is then linked to thyrosyl residues within thyroglobulin. This step is catalysed by the thyroid-specific enzyme thyroid peroxidase and called iodine organification. Thyroid hormones T_3 and T_4 are synthesized by the pairing of two iodotyrosine residues and stored in the colloid until they are needed. Iodine accumulation in the thyroid in contrast to other tissues is stimulated by thyroid stimulating hormone (Levy *et al.*, 1998a; Riedel *et al.*, 2001; Spitzweg and Morris, 2002; Dohan *et al.*, 2003). This organification of iodine is the key to the success of radioiodine therapy.

NIS and radioiodine treatment

Radioactive iodine trapped in the cells decays by emission of β particles (electrons), which results in the release of free radicals that induce cell damage. The longer radioiodine is retained by the cells the more damage is inflicted.

In differentiated papillary and follicular thyroid carcinoma NIS mediated iodine accumulation has permitted diagnostic scintigraphic imaging and has been the key to successful treatment with radioiodine. Thyroid nodules that accumulate iodine in the same amount or more than normal surrounding thyroid tissue are usually benign, whereas most cancerous nodules contain less iodine compared to healthy thyroid tissue. In normal thyroid tissue iodine uptake is about 1 per cent per gram of the administered dose compared to 0.1 to 0.001 per cent in neoplastic tissues (Filetti *et al.*, 1999). The effective half live of iodine ranges on average from 3 to 5 days in thyroid cancers whereas it ranges from 6 to 8 days in normal thyroid tissue (Filetti *et al.*, 1999). Despite the fact that iodine uptake is reduced relative to the surrounding normal tissue it is still sufficient in most cases to enable the administered radioiodine to destroy tumour remnants after surgery and metastases, and to enable total body scans to detect recurrence. Intensive thyroid-stimulating hormone (TSH) stimulation is usually performed before administration of radioiodine to patients with thyroid cancer in order to increase the NIS expression and thus the ability of the thyroid cancer cells to take up radioiodine.

Differentiated carcinomas represent the vast majority of thyroid carcinomas (94 per cent) and survival rates for these thyroid cancers are high (98 per cent for papillary and 92 per cent for follicular carcinoma). The remaining 6 percent of thyroid carcinomas consist of medullary carcinomas (5 per cent) and anaplastic and poorly differentiated carcinomas (1 per cent). Survival rates for these cancers are 80 per cent and 13 per cent, respectively, despite multimodal treatment with surgery, radioiodine, THS suppressive thyroxine treatment, chemotherapy and radiotherapy (Spitzweg and Morris, 2004).

The molecular characterization and cloning of the NIS has been a major step forward in thyroid cancer research. It has permitted researchers to create specific NIS antibodies by which further study at a molecular level of the distribution and expression of the NIS protein in thyroid cells and the rest of the body was possible.

Several groups have attempted to establish a possible correlation between the reduced iodine uptake in thyroid cancer cells and the expression of NIS mRNA or protein by reverse transcription–polymerase chain reaction, Northern blot analysis or immunohistochemistry. As shown by several review articles that have been published on the subject (Filetti *et al.*, 1999; De La Vieja *et al.*, 2000; Heufelder *et al.*, 2001; Riedel *et al.*, 2001; Spitzweg and Morris, 2002), the results are diverse. NIS may be absent, decreased or surprisingly over expressed in differentiated thyroid carcinomas; although iodine uptake is consistently decreased. Reduced iodine uptake of differentiated thyroid cancers can therefore not be explained by lower NIS expression. Presumably defective targeting, or insufficient retention in the plasma membrane, or other factors affecting NIS regulation in cancerous cells may play a role. To date NIS gene mutations have not been found in thyroid cancer cells (Spitzweg and Morris, 2002; Dohan and Carrasco, 2003). Strategies that provide a means to re-establish the therapeutic efficacy of radioiodine in thyroid cancer cells by up-regulating iodine transport have been investigated with variable success. Retinoic acid increases radioiodine uptake in some thyroid carcinomas while down regulating iodine accumulation in surrounding normal thyroid tissue (Filetti *et al.*, 1999; Shen *et al.*, 2001; Spitzweg and Morris, 2002). Demethylation treatment with 5-azacytidine or sodium butyrate may restore NIS expression and/or iodine uptake (Spitzweg and Morris, 2002). DNA methylation may therefore be one of the posttranslational mechanisms responsible for loss of NIS expression in thyroid carcinomas. *In vitro* studies have suggested the histone deacetylase inhibitor depsipeptide may be capable of restoring NIS expression as well as iodine accumulating activity in poorly differentiated and anaplastic thyroid carcinomas (Spitzweg and Morris, 2002, Kitazono *et al.*, 2001).

20.2.2 Breast

Scintigrafic studies over the years have demonstrated that iodine is also accumulated in tissues other than the thyroid. Iodine concentrating extrathyroidal tissues include the salivary and lacrimal glands, the gastric mucosa, the lactating mammary gland, the choroid plexus, the thymus, the adrenal gland, the lung, the heart and the ciliary body of the eye (Filetti *et al.*, 1999; Tazebay *et al.*, 2000; Heufelder *et al.*, 2001; Dohan *et al.*, 2003). Iodine uptake in other tissues is similar to the thyroid in that it is also inhibited by thiocyanate and perchlorate, and that iodine concentration gradients of the same magnitude are generated.

In contrast to the thyroid, non-thyroidal tissues lack the ability to organify iodine and TSH does not influence the iodine uptake in the cells. Salivary glands and gastric mucosa concentrate thiocyanate in contrast to the thyroid where it is metabolised and eliminated after uptake (De La Vieja *et al.*, 2000).

The isolation of NIS and the production of monoclonal antibodies have permitted investigators to further study the distribution of NIS in tissues. Surprisingly the iodine transport in extrathyroidal tissues appears to be mediated by the same NIS, which however is regulated and processed differently in each tissue.

The role of NIS in the normal breast

Iodine transport in the mammary gland occurs only late in pregnancy and during lactation. Iodine is transferred to the newborn via the milk. A sufficient provision of iodine is important for thyroid hormone production in the newborn, which is essential for normal development of the nervous system, skeletal muscles and lungs. The mammary gland NIS (mgNIS) was first identified and characterized by Tazebay *et al.* (2000). mgNIS expression was non-existent in mammary glands outside pregnancy, was increasingly detectable in rat mammary glands towards the end of pregnancy, and was intensely expressed during lactation. mgNIS expression during lactation was regulated in a reversible manner by suckling. *In vivo* studies performed by Dohan and colleagues (Riedel *et al.*, 2001; Dohan *et al.*, 2003) on oophorectomized mice demonstrated that the combination of β-oestradiol, oxytocin and prolactin in the absence of progesterone induced the highest levels of mgNIS expression. This hormonal combination resembles the relative hormonal levels during lactation in mice. The other lactogenic hormones insulin and cortisol, and retinoic acid also increase iodine uptake and mgNIS protein expression (Spitzweg and Morris, 2002).

The role of NIS in breast cancer

The finding that NIS was present and active in mammary glands, and the more than 60 years' long experience with radioiodine as a unique way to detect and destroy thyroid cancer cells without harm to other healthy tissues, led researchers to believe that iodine injection and imaging could be a diagnostic and therapeutic tool in other cancers in which NIS is functionally expressed.

Tazebay *et al.* (2000) demonstrated that in contrast to normal breast tissue where mgNIS is only expressed in late pregnancy and during lactation, functional expression of mg NIS is present in human breast carcinomas and experimental mammary carcinomas in transgenic mice. The normal human breast tissue samples they studied did not express mgNIS. Furthermore functional, specific and perchlorate-inhibited mgNIS expression was demonstrated in experimental mammary carcinomas in non-gestational and non-lactating transgenic mice, either carrying an activated *ras* oncogene or overexpressing the *neu* oncogene by scintigraphic imaging and immunoblot analysis (Tazebay *et al.*, 2000). Since these studies several other groups have found mgNIS expression in human breast cancer samples (Spitzweg and Morris, 2002; Dohan *et al.*, 2003; Dadachova and Carrasco, 2004). Moon and co-workers (Moon *et al.*, 2001) have demonstrated pertechnetate accumulation in primary breast tumours in humans *in vivo*. Twenty-five breast cancer patients underwent scintigraphy and active uptake in the tumour was found in 4. This observation shows that iodine uptake is present in human patients with breast cancer *in vivo*. In this study the thyroidal uptake of radioisotope had not been suppressed, therefore it is possible that a larger proportion of breast cancers would have accumulated pertechnetate if the thyroid had been down regulated prior to the scan.

These findings strongly suggest that the NIS is up-regulated in breast cancer, which opens the path for radioiodine as a novel diagnostic and therapeutic tool in breast cancer. Additional studies and clinical trials are necessary to determine how well NIS is expressed in human breast carcinomas and how effective radioiodine is for treatment of breast cancer.

20.3 Genetic therapy

20.3.1 Principles

The characterization and cloning of the NIS gene together with the more than 60 years experience in treating thyroid disease with radioiodine have been the motor for the development of a novel gene therapy strategy. Functional NIS genes are expressed in different types of cancer cells by means of various gene delivery vectors and ectopically expressed within the cell membranes of cancer cells. This process allows iodine uptake in these cells and renders them susceptible to radioiodine therapy. Specific targeting of the hNIS gene to malignant cells can be achieved by the use of tissue specific promoters; thereby maximizing cancer specific effects and minimizing toxic side effects in non-malignant cells.

20.3.2 NIS

The human NIS gene (hNIS) is localized on chromosome 19p12–13.2 and consists of an open reading frame of 1929 nucleotides. The coding region comprises 15 exons interrupted by 14 introns and it codes for a 3.9-kb mRNA, which in its turn generates a glycoprotein of 643 amino acids with a molecular mass ranging from 70 to 90 kDa as a result of variable levels of glycosylation (Smanik *et al.*, 1997). The secondary structure of the hNIS is currently thought to consist of 13 transmembrane segments of which the amino-terminus is located extracellularly and the carboxyl terminus intracellularly. Three of its Asp residues, position 225, 485 and 497 are glycosylated but glycosylation is not essential for the function, stability or targeting of the cell membrane of the NIS (Levy *et al.*, 1998b).

20.3.3 NIS and imaging

Introduction

Molecular imaging has been broadly defined as the *in vivo* characterization and measurement of biological processes at the cellular and molecular level using remote imaging detectors (Sharma *et al.*, 2002).

Molecular imaging methods employed to detect and visualize gene expression *in vivo* in animal models, and eventually in humans, require expression of a reporter gene that alters the distribution of a radioactive tracer compound, activates a substrate, or concentrates a contrast agent. The devices used for molecular imaging need to be able to detect very low amounts of reporter probes that are designed to accumulate in cells that express the reporter ectopically (Penuelas *et al.*, 2005a). The ideal molecular imaging technique needs to be highly specific, highly sensitive, have a high temporal–spatial resolution and be non-invasive (Rudin *et al.*, 2005). A number of technologies have been developed for non-invasive molecular imaging.

Optical imaging

Optical imaging can be performed when the reporter genes that are expressed encode for bioluminescent (Lucifease) or fluorescent proteins (such as green fluorescent protein) or when fluorescent dyes are used as ligands. These can be detected externally by the use of sensitive photon detection systems. The emission of light photons from photo-proteins in cells expressing the transgene transmits through the tissues after activation by the enzyme in the absence of background noise; this is because mammalian cells do not normally express photo-proteins. Promising results have been published using these techniques (Contag *et al.*, 2000; Becker *et al.*, 2001; Contag and Ross, 2002; Skoch *et al.*, 2005), but the disadvantage is the poor spatial resolution. The anatomical definition is high at the surface but deteriorates quickly with increasing depth due to the scattering of light in tissues (Rudin *et al.*, 2005). While in small animals optical imaging techniques can provide informative data it is highly unlikely that this techniques could be used for *in vivo* gene imaging in humans (Rudin *et al.*, 2005, Penuelas *et al.*, 2005a). In addition, the reporter genes so far described are of xenotropic origin. As such they will be recognized by the immune system as 'non-self' and the cells expressing them may be destroyed by the immune system, adding an extra level of variability to the methodology.

Radionucleotide-based imaging technologies

Radionucleotide-based imaging technologies are positron emission tomography (PET) and single photon emission computed tomogaphy (SPECT). Radionucleotides or radio-labelled molecules are injected intravenously in trace quantities and are retained in tissues as a result of binding to a receptor, cell entrapment due to enzyme-catalysed conversion, or intracellular uptake through a transporter (Penuelas *et al.*, 2005a). These modalities can be performed repeatedly *in vivo* in humans, are highly sensitive, have a high spatial resolution and can give precise quantitative data. A non-invasive, clinically applicable method for quantitative imaging of transgene expression in human subjects would enable clinicians to define the location(s), magnitude and persistence of gene expression over time (Gambhir *et al.*, 2000b). We will describe these techniques below.

SPECT SPECT (Levin, 2005), like PET was originally developed for human use and subsequently adapted for the imaging of small laboratory animals. Single-photon emitting isotopes are isotopes that result in the emission of gamma ray photons or high-energy X-ray photons. The photon is emitted directly from the radioactive atom. Photon-emitting isotopes that are commonly used and transported by the NIS are ^{99m}Tc, ^{123}I, ^{131}I and ^{125}I. Each of these isotopes has different energy emission and different half-life. Single photons are emitted by the isotope in all directions and are attenuated as they travel through the body. The photons that leave the body can be detected by rotating detector panels around the object. The exact provenance of the photon can be determined by the use of collimators consisting of a well-defined configuration of holes for the photons to enter. Photons that hit the collimator at the wrong angle do not make it through the holes and are filtered out. The acquired data are analysed and reconstructed to create three-dimensional tomographic images. The spatial resolution depends on how well the photons can be collimated. High spatial resolutions (1–2 mm) have been obtained by using magnifying collimators, e.g. pinhole collimator. Marsee and co-workers have validated the use of SPECT for the *in vivo* imaging of NIS transgene expression (Marsee *et al.*, 2004). Pulmonary NIS-expressing tumours as small as 3 mm in diameter were detectable in nude mice by ^{125}I SPECT with a pinhole collimator. SPECT can provide quantitative information on the level of gene expression as accurate as PET (Meikle *et al.*, 2005).

PET (Levin, 2005) Positron emitting isotopes are required for PET scanning. A positron is a particle with the same mass as an electron but with a positive charge. Positron emitting radionuclides are generally produced in cyclotrons that have to be in or near the laboratory or hospital where the scanning takes place, since these isotopes are usually short lived. ^{124}I is a positron emitter transported by NIS. Its half-life is 4.2 days.

Emitted positrons interact with atoms they encounter in their path. They travel for some distance depending on their energy and the surrounding matter. When the velocity is low they may combine with an electron in the vicinity and annihilate. During the process of annihilation two photons with energy of 511 MeV are released simultaneously in opposite directions at an angle of 180° (if positron and electron are at rest when they combine). Detectors are arranged in a ring around the subject and register only photon pairs that arrive within a certain time span of each other at an angle of 180°. The acquired data are analysed and a three-dimensional image is reconstructed.

Levin (2005) has recently written an extensive review of all the imaging modalities including technical specifications, drawbacks and advantages. PET and SPECT can both provide accurate quantitative information on gene expression *in vivo* in small animal studies. Both imaging modalities are available for clinical use in humans; however, since human subjects have a greater mass spatial resolution is reduced and the information obtained less accurate. In order to increase the amount of information obtained from images molecular imaging modalities can be combined. PET/CT and SPECT/CT have recently become available for clinical use, combining exact high-resolution anatomical information from CT with the molecular information obtained from PET and SPECT.

Molecular imaging of gene expression

The first step towards molecular imaging is the delivery of the transgene to the tissue of interest by any currently available vector, e.g. adenovirus, retrovirus, measles virus, lentivirus, liposome, nano-particles, etc. Any promoter of choice can drive expression of the transgene. Promotors can be cell specific, restricting expression to certain cells such as cancer cells and can be constitutive, leading to permanent transcription, or inducible, leading to controlled expression (Gambhir *et al.*, 2000b). Upon expression of the transgene (reporter/marker gene) the biodistribution of a tracer molecule is altered, leading to its concentration at the site of reporter gene expression (Vassaux and Groot-Wassink, 2003). Currently three types of reporter genes are available: enzymes, receptors and transport proteins.

Enzyme The herpes simplex virus 1 thymidine kinase (HSV1-tk) gene is an example of a reporter gene encoding for an enzyme. Inside the cell expressing the transgene, the HSV1-Tk gene is transcribed into HSV1-Tk mRNA, which is then transported to the cytosol and translated to the protein HSV1 thymidine kinase enzyme (HSV1-TK) on the ribosomes. Upon intravenous injection, the radiotracer diffuses in the subject and can cross the plasma membrane of cells. However, in cells in which HSV-Tk is expressed, this radiolabelled substrate is phosphorylated and trapped within the cells. The degree of accumulation of these radiolabelled substrates reflects the level of HSV1-TK enzyme activity and thus the level of HSV1-tk gene expression (Gambhir *et al.*, 2000b).

Several thymidine analogues have been found that are preferential substrates for the HSV1-TK rather than cellular thymidine kinase. Radiolabelled derivatives can be used for imaging and therapy. 5-iodo-2′-fluoro-2′-deoxy-1-β-D-arabinofuranosyl-uracyl (FIAU) can be labelled with several different radionucleotides including ^{99m}Tc, ^{123}I, ^{124}I and ^{131}I, which makes it appropriate for imaging and therapy with PET and SPECT. Gambhir's group has investigated mutant HSV1-tk reporter genes, which present a higher specificity to the reporter probe 9-(4-[^{18}F]fluoro-3-hydroxymethylbuthyl)guanine (FHBG), providing a further increase in the sensitivity (Gambhir *et al.*, 2000a). Several investigators are in the process of creating and evaluating other reporter probes with improved sensitivity and specificity (Penuelas *et al.*, 2005a). Studies in cell cultures and small animals have shown that the [^{18}F]FDBG PET *in vivo* signal correlates with transgene expression as determined by mRNA and protein levels (Gambhir *et al.*, 1999). [For a review of animal studies see Min and Gambhir (2004).] [^{18}F]FDBG was used recently to monitor the *in vivo* expression of HSV1-sr39tk-C6 glioma (stable transfected cell lines) xenografts in nude mice and PET images were able to predict the response to therapy (Yaghoubi *et al.*, 2005). PET imaging with [^{18}F] FDBG has also been used to monitor HSV1-tk expression after intratumoural injection of a Ad-CMV-tk in patients with hepatocellular carcinoma (Penuelas *et al.*, 2005b). Radiotracer accumulation was detected in all lesions treated with an adenoviral dose above 10^{12} viral particles, sometimes as early as 50–60 min after injection of the radiotracer. Fused PET/CT images provided a precise anatomic–metabolic correlation between the tumour location and the site of HSV1-tk expression. These data demonstrate that PET could monitor transgene expression in cancer patients.

[$^{125/123}$I]-D-FIAU is a useful radiotracer for imaging HSV1-tk gene expression in nude mice by high resolution SPECT imaging (Choi *et al.*, 2005).

Receptor Intracellular or extracellular receptors can be expressed as transgenes and specific radiolabelled ligands can be developed that bind reversibly or irreversibly. An example of such a receptor is the dopamine D2 receptor (D2R) with its ligand 3-(2′-[^{18}F]fluoroethyl)spiperone (FESP), which can be imaged by PET (Gambhir *et al.*, 2000b). The D2R is endogenously expressed in the striatum.

The D2R delivered to the cell by its appropriate vector is transcribed to D2R mRNA and then translated in the cytosol to D2R protein. FESP can bind to extracellular and intracellular D2 receptors leading to accumulation of the ligand in the cells expressing the transgene. The levels of accumulation of FESP reflect the level of transgene expression (Gambhir *et al.*, 2000b). *In vivo* imaging studies in mice after intravenous injection of a replication deficient adenovirus carrying a D2R reporter gene driven by the CMV promoter with [^{18}F]FESP-PET have shown overwhelming expression of the reporter gene in the liver. The amount of radiotracer retained in the liver, as determined by region of interest measurements from the PET images, was proportional to the amount of hepatic FESP present as determined by well counting after death, D2R dependent [^{3}H] spiperone binding and the D2R mRNA levels (MacLaren *et al.*, 1999). Ectopic expression of the D2R sensitizes cells to circulation adrenergic signals and therefore a mutated receptor (D80RA) has been constructed that uncouples ligand binding from intracellular signalling (Liang *et al.*, 2001). In a direct comparison of the D2R/FESP combination and the HSV1-tk/FPCV (8-[^{18}F]fluoropenciclovir) system, similar results were obtained (Iyer *et al.*, 2001).

Transporter Transport proteins can be expressed in the cell membrane as reporter genes and they have a high specificity for certain compounds. They use active transport to concentrate the radiolabelled compound into a defined compartment such as the cytosol (Vassaux and Groot-Wassink,

2003). NIS is an example of such a transporter gene.

NIS is a unique imaging tool for cancer gene therapy. NIS is not a foreign protein since it is expressed endogenously in the thyroid and several other tissues and is thus not immunogenic. The endogenous NIS expression is limited to a small number of tissues, which results in limited background interference for exogenous NIS function. Iodine and isotopes transported by NIS itself are the tracers and therefore radiochemistry is not required. This offers significant logistic and cost-effective advantages. Decaying isotope does not produce cold tracers but disappears rapidly from the system. The radionucleotides are specific to the NIS expressing cells, which further reduces unwanted background noise.

Shimura *et al* performed the first imaging experiment with NIS (Shimura *et al.*, 1997). Undifferentiated rat thyroid cancer cells were stably transfected with the NIS gene and transplanted into rats. ^{125}I was injected intraperitoneally and radioiodine accumulation in the tumours was visualized by autoradiography. Iodine uptake in the tumour peaked 90 minutes after injection and diminished rapidly, attaining 50 per cent after 6 h. Following this first series of experiments, a number of studies have been performed using the same principles (Mandell *et al.*, 1999; Boland *et al.*, 2000; Spitzweg *et al.*, 2001; Cho, 2002; Cho *et al.*, 2002; Groot-Wassink *et al.*, 2002, 2004a, 2004b, Dingli *et al.*, 2003, 2004, 2006; Dwyer *et al.*, 2005, 2006).

In our laboratory we have performed adenoviral biodistribution studies in nude mice using hNIS as a reporter gene (Groot-Wassink *et al.*, 2002). Intravenous injection of the non-replicating Ad-CMV-NIS construct and subsequent (after 48 h) injection of ^{125}I lead to accumulation of radioactivity as measured by post mortem γ counting. Accumulation occurred mainly in the liver and the adrenal glands, and to a lesser extent in the lungs, spleen and pancreas. These data were confirmed by Northern blot analysis of total hNIS RNA. Images could be obtained *in vivo* with PET after injection of ^{124}I. In a second study we demonstrated that the combination of ^{124}I and PET imaging can provide quantitative data on gene expression *in vivo* (Groot-Wassink *et al.*, 2004b). We found a correlation between the intensity of the PET signal and iodine uptake measured by post mortem γ counting, the level of hNIS mRNA as determined by quantitative real time polymerase chain reaction, and the number of hNIS positive cells as seen by immunohistochemistry using an hNIS specific antibody. These data validate the iodine uptake induced by ectopic expression of NIS as a quantitative reporter system to evaluate gene expression. We assessed the feasibility of using the NIS reporter gene with ^{124}I/PET imaging to visualize cancer selectivity of different promoters incorporated in non-replicating adenoviral vectors (Groot-Wassink *et al.*, 2004a). After intravenous injection in tumour-free animals of a non-replicating adenovirus encoding the human telomerase promoters (Ad-hTR-NIS and Ad-hTERT-NIS) driving the NIS gene, ^{124}I/PET demonstrated a lack of signal, suggesting that the promoters used were non-functional in normal tissues. This conclusion was confirmed by post mortem gamma-counting, immunohistochemistry and quantification of hNIS mRNA levels. When injected in subcutaneous experimental human tumours in nude mice, the Ad-hTR-NIS virus-injected tumours showed increased iodine uptake compared to tumours injected with a control virus. This was similar to Ad-CMV-NIS injected tumours, suggesting that the hTR promoter is a potent cancer specific promoter. Altogether, these data suggest that PET imaging of the NIS reporter gene could be applied to measure the activity of cancer selective promoters in humans, providing unique information on the pattern of gene transfer and transgene expression in patients (Groot-Wassink *et al.*, 2004a).

20.3.4 Therapeutic gene therapy

The NIS is not only an imaging reporter gene. The extensive experience in the treatment of thyroid cancer provides sufficient evidence that NIS can also serve as a therapeutic transgene.

The vectors that are currently available for cancer gene therapy have so far failed to transduce every single tumour cell. In our study, in which all animals received intratumoral injection with a

NIS-expressing adenovirus, the tumours showed heterogeneity of expression, with areas of high-level expression adjacent to regions of low expression (Groot-Wassink *et al.*, 2004a). The combination of NIS as a transgene and the use of radioisotopes has the advantage that non-transduced cells can still be eradicated by the bystander effect. ^{131}I and other isotopes transported by NIS like ^{188}Re-perrhenate and ^{211}At (astatine) kill non-transduced cells by radiation emission.

Many *in vitro* studies have been performed on stable cell lines or cells expressing hNIS or rNIS under control of general or tissue specific promoters using different methods of gene transfer. These studies have demonstrated a cytotoxic effect when cultured cells are treated with ^{131}I (see (Chung, 2002; Cho, 2002; Spitzweg and Morris, 2002) for overviews). Mandell *et al.* (1999) demonstrated using clonogenic assays that rNIS transduced melanoma, ovarian, liver and colon cancer cell lines could be selectively killed by treatment with ^{131}I.

The therapeutic efficacy of this approach *in vivo* after treatment of tumour xenografts in mice is variable. Shimura *et al.* performed the first trial to investigate the effect of ^{131}I on tumour growth *in vivo* (Shimura *et al.*, 1997). Malignantly transformed rat thyroid cells, which do not concentrate iodine, were stably transfected with the rat NIS and injected subcutaneously into Fisher 344 rats, where they formed subcutaneous tumours. Two intraperitoneal injections of 500 μCi of ^{131}I were administered one at week 2 and one at week 3 after transplantation of the tumour. The calculated dose delivered to the tumour was approximately 4 Gy. The size of the tumour, measured at week 4 was not reduced compared to controls. Boland *et al.* injected an adenoviral vector, Ad-CMV-rNIS into subcutaneous human tumours (SiHa = cervical or MCF7 = breast) established in nude mice (Boland *et al.*, 2000). Doses of 30, 60 or 90 μCi of ^{131}I were administered by intraperitoneal injection to tumour-bearing mice. No differences could be detected in tumour volume between Ad-NIS transfected tumours and controls. The authors concluded that perhaps the dose of radioiodine was insufficient, or the clearance of radioiodine from the tumour too rapid, or perhaps in the latter case due to the low transduction rate of adenovirus to the tumour *in vivo*. Using higher radioiodine doses, Cho *et al.* (2002) reported that in rats implanted with F98 glioma cells in which the NIS was introduced by retroviral mediated transfer treatment with ^{131}I (4 mCi, three times) increased survival time by 48 per cent (30 vs. 45 days). Tumours seemed to be reduced for a period of time but eventually re-grew. The authors concluded that radioiodine treatment of NIS-bearing tumours had a cytostatic and/or cytoreductive effect.

Spitzweg *et al.* (2001) injected prostate cancer xenografts in nude mice with an adenovirus carrying the NIS gene under control of the cytomegalovirus (CMV) promoter. Two weeks before radioiodine administration mice were switched to a low iodine diet and received T_4 supplementation in their drinking water to maximize radioiodine uptake in the tumour and minimize uptake in the thyroid. Administration of 3 mCi of ^{131}I by a single intraperitoneal injection resulted in a tumour volume reduction of 84 per cent.

Dwyer and co-workers noted a 53 per cent reduction in tumour volume in ovarian tumour xenografts infected with adenoviral CMV-NIS after a single 3 mCi intraperitoneal dose of ^{131}I (Dwyer *et al.*, 2006). The same group reported tumour regression to 17 per cent of the original tumour size in breast cancer xenografts in mice that were injected with an adenoviral MUC1/NIS construct after administration of a therapeutic dose of ^{131}I (Dwyer *et al.*, 2005). These results suggest that different tumours appear to have different radio-sensitivities. The best results obtained so far were reported by Dingli *et al.* (2003). To assess the efficacy of ^{131}I therapy in conjunction with hNIS transfer they established multiple myeloma xenografts resistant to measles virus (MV) in severe combined immunodeficiency mice. They injected these mice intravenously with a single dose of recombinant MV-hNIS. Tumours regressed completely after treatment with a single therapeutic dose of ^{131}I given 9 days after viral injection, without evidence of recurrence up to 5 months after therapy.

These data demonstrate the potential of the NIS gene as a therapeutic transgene.

This variety of responses may be due the differences in sensibility to radiotherapy in tumours. Haematopoietic tumours are more sensitive than soft tissue tumours. Effective therapy with radioiodine also depends on the level of gene transfer and the retention of iodine in the cells. Thyroid cells, in contrast to other cells, have the capability to organify iodine, which increases the time radioiodine can deposit energy. However, some differentiated thyroid carcinomas can still be treated successfully with radioiodine despite having lost the ability to organify iodine. Dingli *et al.* recently demonstrated that the iodine retention in cells exogenously expressing NIS is a dynamic process consisting of uptake, slow efflux and re-uptake (Dingli *et al.*, 2004). In thyroid cells iodine is transported from the cell to the colloid by means of transporters (pendrin and apical iodine transporter). Non-thyroid cancer cells expressing NIS do not contain these transporters and they leak iodine at a slower rate than thyroid cells. Efflux of iodine from the cells and re-uptake into the cells depends on the level of NIS expression. Co expression of the thyroid peroxidase gene with NIS might prolong isotope retention (Huang *et al.*, 2001). Thyroid peroxidase increased isotope activity in the cells but did not influence the retention time. In an attempt to reduce the efflux of radioiodine from the cells lithium has been used in conjunction with radioiodine with only marginal effects (Haberkorn *et al.*, 2003). All the studies described above have used cancer cells stably transfected with the NIS to create xenografts in animals, or have used different vectors to deliver NIS to xenografts of human cancers in transgenic animals. Xenograft models in small animals are not ideal representations of potential cancer treatment in patients and cannot be extrapolated to therapeutic applications of gene therapy in humans.

Faivre *et al.* (2004) have created an animal model of liver cancer. Hepatic carcinomas were induced in immunocompetent Wistar rats by diethylnitrosamine administration. The adenoviral CMV-rNIS vector was administered directly into the portal vein. Pretreating the rats with thyroxine inhibited thyroid uptake of iodine. They found a long-term retention of iodine in normal and cancerous liver (572 ± 110 per cent hours and 294 ± 50 per cent hours) after transfer of the rNIS gene, which was mainly attributable to recirculation and reuptake. Iodine was cleared rapidly from the hepatocytes when NIS activity was blocked by perchlorate. They also showed that tumour growth was strongly inhibited and survival improved by ^{131}I therapy (40 Gy) after rNIS gene transfer in hepatocarcinoma-bearing rats. In some cases complete regression could be obtained. The authors did not observe any ^{131}I-related liver damage in 2 months follow up in Wistar rats, despite the high radiation dose, suggesting that liver carcinoma cells are more sensitive to radiation than normal hepatocytes. These results suggest that intra-arterial NIS based gene therapy for hepatocellular carcinoma is an approach worth investigating in humans.

20.3.5 Therapy and imaging

For gene therapy with hNIS to work in clinical settings a number of conditions need to be fulfilled.

hNIS is endogenously expressed in a number of tissues and is therefore non-immunogenic. However, endogenous expression also suggests that these tissues show iodine uptake when imaged and will potentially be damaged by therapeutic doses of isotopes. Thyroid uptake of iodine can be blocked by administration of thyroxine which suppresses TSH release from the pituitary gland, which in turn down-regulates thyroid function including NIS expression and function (Cho, 2002). Overloading the gland with iodine also reduces radioiodine uptake in the thyroid.

Stimulated secretion of the salivary glands is protective during radioiodine treatment and protective pharmacological agents (e.g. amifostine) have been described (Bohuslavizki *et al.*, 1999). Although there are reports of gastric damage associated with radioiodine treatment, overall complications appear to be rare. The dose-limiting tissues in radioiodine treatment of the thyroid are blood and bone marrow, restricting the maximal dose to 11.1 GBq (Menzel *et al.*, 1996). Iodine is excreted via the kidneys in to the bladder, which may cause damage to this organ. Frequent voiding is therefore advised.

The development of secondary malignancies after radioiodine treatment of thyroid cancer was initially a concern but an increased incidence has not been confirmed and this treatment is considered safe (Pauwels *et al.*, 2000).

Another limitation to gene therapy in humans is the lack of safe and efficient gene delivery systems that can be administered systemically and that have tumour specific gene expression. The use of viruses for systemic gene delivery is not ideal because of their immunogenic properties. The future will have to show whether newer vectors like liposomes and nanomolecules will be less immunogenic and better systemic gene delivery vectors. Tumour-specific gene expression can be obtained by using cancer-specific promoters. It has been demonstrated that transgenic mice promoter fragments from the human telomerase gene can provide cancer selective expression of the transgene when incorporated in a replication deficient adenoviral vector. In this system PET imaging using NIS as a reporter gene could visualize the activity of these promoters (Groot-Wassink *et al.*, 2004a). Imaging using NIS as a reporter gene could provide information on the pattern of gene transfer and transgene expression in patients.

In order for radionucleotide treatment to be efficient a high level of gene transfer is required. Penuelas *et al.* (2005b) noted that after intratumoral injection of an adenoviral vector expressing the HSV1-tk gene, detectable tumour transduction was only observed after a certain dose was injected but increasing this dose did not enhance tracer accumulation in the tumour. Furthermore, the same dose of adenoviral vector injected did not yield the same effect in every patient, which emphasizes the need for individual gene therapy. The use of NIS in combination with radioactive tracers could provide a bystander effect destroying non-transduced adjacent cells by radiation.

The β-emission of ^{131}I covers 0.2 to 2.4 mm. Other radioisotopes that are also transported by the NIS might compensate for the relative short path length of ^{131}I. ^{188}Re is a β-emitter that has been shown to deliver a radiation dose that is 4.5 times higher than ^{131}I (Dadachova *et al.*, 2002). ^{211}At is an α-emitting isotope, transported by the NIS that might also deliver higher radiation doses.

In order for radioisotope treatment to be efficient isotopes have to be retained in the cells for a sufficiently long time. In contrast to thyroid cells, cells that ectopically express NIS do not have the capability to organify iodine. However, iodine might leak out more slowly from the cells because they lack the pendrin and apical iodine transporter proteins which in the thyroid are responsible for rapid iodine transport to the colloid (Dingli *et al.*, 2004). *In vivo* experiments by Faivre and others have demonstrated that the lack of organification may not pose a problem in clinical gene therapy with NIS, since long-term retention is most likely a result of a dynamic process consisting of uptake, efflux and re-uptake (Dingli *et al.*, 2004; Faivre *et al.*, 2004).

PET and SPECT are nuclear imaging technologies available for both animal and human studies. The use of PET for repeated imaging of gene expression in humans might be difficult because of the relative long half-life (4.2 days) of ^{124}I, which increases the dose delivered to the body. The radioisotopes used in SPECT, ^{99}Tc (6 h) and ^{123}I (13.2 h) have half-lives of minutes, which may render them more amenable for repeated imaging. The currently available SPECT cameras in humans, although much cheaper then PET, have the disadvantage that they can only provide semi-quantitative information on gene expression, which could be overcome by combining different imaging modalities such as SPECT/CT. SPECT, in contrast to PET, can also be employed to image the effect of therapeutic doses of radioiodine with the isotope ^{131}I and ^{188}Re, which are both γ-emitters. The development of new radiotracers and improved imaging appliances is therefore required.

The use of hNIS for cancer gene therapy can benefit from the extensive experience gained from treating thyroid cancer with radioiodine. In future it could be possible to visualize gene expression by non-invasive imaging, to calculate the required therapeutic dose by quantitative imaging techniques and thus accomplish the concept of individualized dosimetry and cancer gene therapy.

20.4 Conclusion

Preclinical studies clearly indicate that imaging and gene therapy are feasible in animal models. How these studies translate into clinical studies will have to be established in the future.

References

Becker A, Hessenius C, Licha K, *et al.* (2001). Receptor-targeted optical imaging of tumors with near-infrared fluorescent ligands. *Nat Biotechnol* **19**, 327–31.

Bohuslavizki KH, Klutmann S, Bleckmann C, *et al.* (1999). Salivary gland protection by amifostine in high-dose radioiodine therapy of differentiated thyroid cancer. *Strahlenther Onkol* **175**, 57–61.

Boland A, Ricard M, Opolon P, *et al.* (2000). Adenovirus-mediated transfer of the thyroid sodium/iodide symporter gene into tumors for a targeted radiotherapy. *Cancer Res* **60**, 3484–3492.

Cho JY (2002). A transporter gene (sodium iodide symporter) for dual purposes in gene therapy: imaging and therapy. *Curr Gene Ther* **2**, 393–402.

Cho JY, Shen DH, Yang W, *et al.* (2002). *In vivo* imaging and radioiodine therapy following sodium iodide symporter gene transfer in animal model of intracerebral gliomas. *Gene Ther* **9**, 1139–1145.

Choi SR, Zhuang ZP, Chacko AM, *et al.* (2005). SPECT imaging of herpes simplex virus type1 thymidine kinase gene expression by [(123)I]FIAU(1). *Acad Radiol* **12**, 798–805.

Chung JK (2002). Sodium iodide symporter: its role in nuclear medicine. *J Nucl Med* **43**, 1188–1200.

Contag CH, Ross BD (2002). It's not just about anatomy: *in vivo* bioluminescence imaging as an eyepiece into biology. *J Magn Reson Imaging* **16**, 378–387.

Contag CH, Jenkins D, Contag PR, Negrin RS (2000). Use of reporter genes for optical measurements of neoplastic disease *in vivo*. *Neoplasia* **2**, 41–52.

Dadachova E, Carrasco N (2004). The Na/I symporter (NIS): imaging and therapeutic applications. *Semin Nucl Med* **34**, 23–31.

Dadachova E, Bouzahzah B, Zuckier LS, Pestell RG (2002). Rhenium-188 as an alternative to Iodine-131 for treatment of breast tumors expressing the sodium/iodide symporter (NIS). *Nucl Med Biol* **29**, 13–18.

Dai G, Levy O, Carrasco N (1996). Cloning and characterization of the thyroid iodide transporter. *Nature* **379**, 458–460.

De La Vieja A, Dohan O, Levy O, Carrasco N (2000). Molecular analysis of the sodium/iodide symporter: impact on thyroid and extrathyroid pathophysiology. *Physiol Rev* **80**, 1083–1105.

Dingli D, Diaz RM, Bergert ER, O'Connor MK, Morris JC, Russell SJ (2003). Genetically targeted radiotherapy for multiple myeloma. *Blood* **102**, 489–496.

Dingli D, Bergert ER, Bajzer Z, O'Connor MK, Russell SJ, Morris JC (2004). Dynamic iodide trapping by tumor cells expressing the thyroidal sodium iodide symporter. *Biochem Biophys Res Commun* **325**, 157–166.

Dingli D, Kemp BJ, O'Connor MK, Morris JC, Russell SJ, Lowe VJ (2006). Combined I-124 positron emission tomography/computed tomography imaging of NIS gene expression in animal models of stably transfected and intravenously transfected tumor. *Mol Imaging Biol* **8**, 16–23.

Dohan O, Carrasco N (2003). Advances in Na(+)/I(−) symporter (NIS) research in the thyroid and beyond. *Mol Cell Endocrinol* **213**, 59–70.

Dohan O, De La Vieja A, Paroder V, *et al.* (2003). The sodium/iodide Symporter (NIS): characterization, regulation, and medical significance. *Endocr Rev* **24**, 48–77.

Dwyer RM, Bergert ER, O'Connor MK, Gendler SJ, Morris JC (2005). *In vivo* radioiodide imaging and treatment of breast cancer xenografts after MUC1-driven expression of the sodium iodide symporter. *Clin Cancer Res* **11**, 1483–1489.

Dwyer RM, Bergert ER, O'Connor MK, Gendler SJ, Morris JC (2006). Sodium iodide symporter-mediated radioiodide imaging and therapy of ovarian tumor xenografts in mice. *Gene Ther* **13**, 60–66.

Faivre J, Clerc J, Gerolami R, *et al.* (2004). Long-term radioiodine retention and regression of liver cancer after sodium iodide symporter gene transfer in wistar rats. *Cancer Res* **64**, 8045–8051.

Filetti S, Bidart JM, Arturi F, Caillou B, Russo D, Schlumberger M (1999). Sodium/iodide symporter: a key transport system in thyroid cancer cell metabolism. *Eur J Endocrinol* **141**, 443–457.

Gambhir SS, Barrio JR, Phelps ME, *et al.* (1999). Imaging adenoviral-directed reporter gene expression in living animals with positron emission tomography. *Proc Natl Acad Sci U S A* **96**, 2333–2338.

Gambhir SS, Bauer E, Black ME, *et al.* (2000a). A mutant herpes simplex virus type 1 thymidine kinase reporter gene shows improved sensitivity for imaging reporter gene expression with positron emission tomography. *Proc Natl Acad Sci U S A* **97**, 2785–2790.

Gambhir SS, Herschman HR, Cherry SR, *et al.* (2000b). Imaging transgene expression with radionuclide imaging technologies. *Neoplasia* **2**, 118–138.

Groot-Wassink T, Aboagye EO, Glaser M, Lemoine NR, Vassaux G (2002). Adenovirus biodistribution and noninvasive imaging of gene expression *in vivo* by positron emission tomography using human sodium/iodide symporter as reporter gene. *Hum Gene Ther* **13**, 1723–1735.

Groot-Wassink T, Aboagye EO, Wang Y, Lemoine NR, Keith, WN, Vassaux G (2004a). Noninvasive imaging of the transcriptional activities of human telomerase promoter fragments in mice. *Cancer Res* **64**, 4906–4911.

Groot-Wassink T, Aboagye EO, Wang Y, Lemoine NR, Reader AJ, Vassaux G (2004b). Quantitative imaging of Na/I symporter transgene expression using positron emission tomography in the living animal. *Mol Ther* **9**, 436–442.

Haberkorn U, Kinscherf R, Kissel M, *et al.* (2003). Enhanced iodide transport after transfer of the human sodium iodide symporter gene is associated with lack of retention and low absorbed dose. *Gene Ther* **10**, 774–780.

Heufelder AE, Morgenthaler N, Schipper ML, Joba W (2001). Sodium iodide symporter-based strategies for diagnosis and treatment of thyroidal and nonthyroidal malignancies. *Thyroid* **11**, 839–847.

Huang M, Batra RK, Kogai T, *et al.* (2001). Ectopic expression of the thyroperoxidase gene augments radioiodide uptake and retention mediated by the sodium iodide symporter in non-small cell lung cancer. *Cancer Gene Ther* **8**, 612–618.

Iyer M, Barrio JR, Namavari M, *et al.* (2001). 8-[18F] Fluoropenciclovir: an improved reporter probe for imaging HSV1-tk reporter gene expression *in vivo* using PET. *J Nucl Med* **42**, 96–105.

Kitazono M, Robey R, Zhan Z, *et al.* (2001). Low concentrations of the histone deacetylase inhibitor, depsipeptide (FR901228), increase expression of the Na(+)/I(−) symporter and iodine accumulation in poorly differentiated thyroid carcinoma cells. *J Clin Endocrinol Metab* **86**, 3430–3435.

Levin CS (2005). Primer on molecular imaging technology. *Eur J Nucl Med Mol Imaging* **32**(Suppl 2): S325–345.

Levy O, De La Vieja A, Carrasco N (1998a). The Na+/I− symporter (NIS): recent advances. *J Bioenerg Biomembr* **30**, 195–206.

Levy O, De La Vieja A, Ginter CS, Riedel C, Dai G, Carrasco N (1998b). N-linked glycosylation of the thyroid Na+/I− symporter (NIS). Implications for its secondary structure model. *J Biol Chem* **273**, 22657–22663.

Liang Q, Satyamurthy N, Barrio JR, *et al.* (2001). Noninvasive, quantitative imaging in living animals of a mutant dopamine D2 receptor reporter gene in which ligand binding is uncoupled from signal transduction. *Gene Ther* **8**, 1490–1498.

MacLaren DC, Gambhir SS, Satyamurthy N, *et al.* (1999). Repetitive, non-invasive imaging of the dopamine D2 receptor as a reporter gene in living animals. *Gene Ther* **6**, 785–791.

Mandell RB, Mandell LZ, Link CJ Jr (1999). Radioisotope concentrator gene therapy using the sodium/iodide symporter gene. *Cancer Res* **59**, 661–668.

Marsee DK, Shen DH, Macdonald LR, *et al.* (2004). Imaging of metastatic pulmonary tumors following NIS gene transfer using single photon emission computed tomography. *Cancer Gene Ther* **11**, 121–127.

Meikle SR, Kench P, Kassiou M, Banati RB (2005). Small animal SPECT and its place in the matrix of molecular imaging technologies. *Phys Med Biol* **50**, R45–61.

Meller J, Becker W (2002). The continuing importance of thyroid scintigraphy in the era of high-resolution ultrasound. *Eur J Nucl Med Mol Imaging* **29**(Suppl 2): S425–438.

Menzel C, Grunwald F, Schomburg A, *et al.* (1996). 'High-dose' radioiodine therapy in advanced differentiated thyroid carcinoma. *J Nucl Med* **37**, 1496–1503.

Min JJ, Gambhir SS (2004). Gene therapy progress and prospects: noninvasive imaging of gene therapy in living subjects. *Gene Ther* **11**, 115–125.

Moka D, Dietlein M, Schicha H (2002). Radioiodine therapy and thyrostatic drugs and iodine. *Eur J Nucl Med Mol Imaging* **29**(Suppl 2), S486–491.

Moon DH, Lee SJ, Park KY, *et al.* (2001). Correlation between 99mTc-pertechnetate uptakes and expressions of human sodium iodide symporter gene in breast tumor tissues. *Nucl Med Biol* **28**, 829–834.

Pauwels EK, Smit JW, Slats A, Bourguignon M, Overbeek F (2000). Health effects of therapeutic use of 131I in hyperthyroidism. *Q J Nucl Med* **44**, 333–339.

Penuelas I, Haberkorn U, Yaghoubi S, Gambhir SS (2005a). Gene therapy imaging in patients for oncological applications. *Eur J Nucl Med Mol Imaging* **32** (Suppl 2), S384–403.

Penuelas I, Mazzolini G, Boan JF, *et al.* (2005b). Positron emission tomography imaging of adenoviral-mediated transgene expression in liver cancer patients. *Gastroenterology* **128**, 1787–1795.

Riedel C, Dohan O, De La Vieja A, Ginter CS, Carrasco N (2001). Journey of the iodide transporter NIS: from its molecular identification to its clinical role in cancer. *Trends Biochem Sci* **26**, 490–496.

Rudin M, Rausch M, Stoeckli M (2005). Molecular imaging in drug discovery and development: potential and limitations of nonnuclear methods. *Mol Imaging Biol* **7**, 5–13.

Sharma V, Luker GD, Piwnica-Worms D (2002). Molecular imaging of gene expression and protein function *in vivo* with PET and SPECT. *J Magn Reson Imaging* **16**, 336–351.

Shen DH, Kloos RT, Mazzaferri EL, Jhian SM (2001). Sodium iodide symporter in health and disease. *Thyroid* **11**, 415–425.

Shimura H, Haraguchi K, Miyazaki A, Endo T, Onaya T (1997). Iodide uptake and experimental 131I therapy in transplanted undifferentiated thyroid cancer cells expressing the Na+/I- symporter gene. *Endocrinology* **138**, 4493–4496.

Skoch J, Dunn A, Hyman BT, Bacskai BJ (2005). Development of an optical approach for noninvasive imaging of Alzheimer's disease pathology. *J Biomed Opt* **10**, 11007.

Smanik PA, Liu Q, Furminger TL, *et al.* (1996). Cloning of the human sodium lodide symporter. *Biochem Biophys Res Commun* **226**, 339–345.

Smanik PA, Ryu KY, Theil KS, Mazzaferri EL, Jhiang SM (1997). Expression, exon-intron organization, and chromosome mapping of the human sodium iodide symporter. *Endocrinology* **138**, 3555–3558.

Spitzweg C, Dietz AB, O'Connor MK, *et al.* (2001). *In vivo* sodium iodide symporter gene therapy of prostate cancer. *Gene Ther* **8**, 1524–1531.

Spitzweg C, Morris JC (2002). The sodium iodide symporter: its pathophysiological and therapeutic implications. *Clin Endocrinol (Oxf)* **57**, 559–574.

Spitzweg C, Morris JC (2004). Gene therapy for thyroid cancer: current status and future prospects. *Thyroid* **14**, 424–434.

Tazebay UH, Wapnir IL, Levy O, *et al.* (2000). The mammary gland iodide transporter is expressed during lactation and in breast cancer. *Nat Med* **6**, 871–878.

Vassaux G, Groot-Wassink T (2003). *In vivo* noninvasive imaging for gene therapy. *J Biomed Biotechnol* **2003**, 92–101.

Yaghoubi SS, Barrio JR, Namavari M, *et al.* (2005). Imaging progress of herpes simplex virus type 1 thymidine kinase suicide gene therapy in living subjects with positron emission tomography. *Cancer Gene Ther* **12**, 329–339.

21

Radioprotective Gene Therapy: Current Status and Future Goals

Joel S. Greenberger and **Michael W. Epperly**

21.1 Introduction

Both primary and locally recurrent cancers of the lung, oesophagus, colon, rectum, bladder, and retroperitoneum require sophisticated chemoradiotherapeutic approaches (Stickle *et al.*, 1999; Dawson *et al.*, 2001; Kies *et al.*, 2001; Martinez *et al.*, 2001; Socinski *et al.*, 2001; Rosenman *et al.*, 2002; Wolski *et al.*, 2005). In the past 30 years, improvements in combination chemotherapy (van Rongen *et al.*, 1993) and the availability of sophisticated radiotherapy techniques have stretched the limit of normal tissue tolerance (Socinscki *et al.*, 2001).

Radiation oncologists are acutely aware of the normal tissue toxicities of focused irradiation beams relative to the tumour volume being targeted. The term 'therapeutic ratio' (Rubin and Casaret, 1968; Hall, 1999) is a mainstay of the specialty of radiation oncology. While a ratio has two numerical components (tumour response/normal tissue toxicity), the majority of efforts at improving this ratio focus on strategies for tumour radiosensitization (the numerator). Sensitizer drugs delivered by local injection or intratumour placement lead to the problem of incomplete or subtotal tumour penetration. Thus, proponents of this strategy try to optimize 'bystander radiation sensitization' attempting to define the parameters by which delivery of a radiosensitizer drug or transgene product to a subpopulation of tumour cells, still results in more effective radio controllability with no increase in side effects.

Others have attempted to focus on the denominator of the 'therapeutic ratio' by attempting to provide relative radiation protection of normal tissue. Such protection, if effective and specific for normal tissue, would allow dose escalation of radiation to the tumour target volume, and thus could improve the therapeutic ratio as effectively as would specific tumour radiosensitization. Proponents of this strategy deal with the reverse problem of 'bystander normal tissue protection' realizing that effective radiation protection whether delivered by drug or transgene product, would deliver the protective agent to a subpopulation of normal tissues within the normal tissue transit volume.

Strategies of radiation protection first and foremost involve radiotherapy beam localization, and the use of blocking techniques to protect normal tissue. Templated hand-inserted blocks have now been replaced by multileaf collimator shaped beams and motion sensitivity techniques for respiratory gating (beam off during inspiration), and intensity modulated radiotherapy, moving multileaf collimator leaves in real time during dose distribution to an irregularly shaped tumour

Viral Therapy of Cancer Edited by Kevin J. Harrington, Richard G. Vile and Hardev S. Pandha

target volume (Goto *et al.*, 1987; Flanders *et al.*, 2002; Wang *et al.*, 2002; Scagliotti and Turrisi, 2003; Phillips *et al.*, 2004; Ramirez *et al.*, 2004; Metaxis *et al.*, 2005).

Pharmacologic radioprotectors have included experiments with WR2721 (amifostine) (Perry *et al.*, 2005) and other free radical scavenger agents (Damron *et al.*, 2001; Johnstone *et al.*, 2001; Sonis *et al.*, 2001; Wijers *et al.*, 2001). Organ specific delivery of such agents has been difficult and distribution of radiation protection to the tumour within a target volume has been a source of concern with respect to improving the therapeutic ratio.

We have developed a strategy of radioprotective gene therapy using the antioxidant enzyme manganese superoxide dismutase (SOD) (Epperley *et al.*, 1998, 1999a,b,c, 2000a,b,c,d, 2001a,b,c, 2002a,b, 2003a,b,c,d, 2004a; Greenberger *et al.*,1998, 2001; Zwacka *et al.*, 1998; Stickle *et al.*, 1999, Pearce *et al.*, 2001; Wang *et al.*, 1998; Guo *et al.*, 2003a,b). While both extracellular and cytoplasmic copper/zinc SOD metalloenzyme (Fernandez-Pol *et al.*, 1982; Oberley, 1982; Oberley and Buettner, 1982; Oberley and Oberley, 1986; Beckman *et al.*, 1989; Oberley *et al.*, 1989; Spitz *et al.*, 1990Bravard *et al.*, 1992; St Clair *et al.*, 1992, 1997; Xu *et al.*, 1999; Church *et al.*, 1993; Li JJ *et al.*, 1995, 1998; Liu, 1996; Yan *et al.*, 1996; Zhong *et al.*, 1996, 1997; Liu *et al.*, 1997; Gonzalez-Zuleta *et al.*, 1998; Li N *et al.*, 1998; Manna *et al.*, 1998; Li S *et al.*, 2000) as well as mitochondrial localized MnSOD contribute to the antioxidant defenses of normal tissue, it is the mitochondrial localized MnSOD which has proven to be highly radioprotective (Epperley *et al.*, 2003d). Radiation protection of cells in culture, tissues *in vivo* (including oral cavity/oropharynx, oesophagus, lung, intestine, and bladder) have been demonstrated in animal model systems using MnSOD transgene delivered by plasmid liposome, Herpes virus, adenovirus, or retrovirus (Guo *et al.*, 2003a, Epperley *et al.*, 1998; 1999a,b,c, 2000a,b,c, 2001a,b,d, 2002a,b 2003a,b, c,d, 2004a; Greenberger *et al.*, 1998, 2001; Zwacka *et al.*, 1998; Stickle *et al.*, 1999; Pearce *et al.*, 2001; Wong *et al.*, submitted).

Our laboratory has focused on the development of organ specific radiation protection, and strategies have been developed to maximize normal tissue uptake, during short duration transgene delivery. Plasmid liposomes have proven to be a safe and effective method for organ specific delivery of the MnSOD transgene. Coupled with the decreased concern for the potential infectivity or toxicity of the viral vectors, plasmid liposomes have been shown to be safe and effective in several animal model systems of organ specific ionizing irradiation damage. Furthermore, MnSOD-PL intra-oesophageal delivery to the pig has been shown to decrease photofrin-mediated photodynamic therapy-induced acute and chronic toxicity (stricture) (Perry *et al.*, 2005).

This chapter will summarize current results and plan for further studies using MnSOD-Plasmid Liposome delivery to specific organs primarily in the mouse model. These strategies, once translated to the clinic, may provide methods (by which) to increase the therapeutic ratio for the management of patients with carcinoma of the head and neck, oesophageal cancer, non-small cell lung cancer, as well as epithelial neoplasms of the abdomen and pelvis.

21.2 Organ-specific radiation protection: oral cavity/oropharynx

There is increasing evidence that mitochondrial expression of the antioxidant enzyme MnSOD if elevated acutely can provide radioprotection to normal tissues including oral cavity, oesophagus, lung, bladder and intestine. The difference in management of redox balance between normal tissues and tumour cells is associated with a paradoxical induction of radiosensitivity in squamous cell tumours by overexpression of the same transgene. Current research efforts are designed to optimize gene therapy using plasmid transfer of MnSOD transgene, but also to identify small molecules that can replace this large molecule.

21.2.1 Antioxidant capacity of solid tumours on the oral cavity/oropharynx

Squamous tumours of the head and neck are commonly hypoxic and show decreased mitochondrial cytochrome *c* oxidase activity (Oberley and Buettner, 1979; Greenberger *et al.*, 2001). Ionizing irradiation induces superoxide and two superoxide molecules are dismutated by superoxide dismutase to produce H_2O_2. H_2O_2 neutralization following irradiation is carried out by either GSH/peroxidase oxidizing to oxidized glutathione or glutathione disulfide (GSSG) and water (see Figure 21.1 below), or alternatively, haemoproteins are oxidized and produce cellular damage. These two common ways that H_2O_2 is metabolized in cells determine a non-toxic or toxic outcome. Furthermore, two groups of investigators have hypothesized that H_2O_2 generation in tumours overexpressing an introduced MnSOD is of potential therapeutic benefit (Larsson and Cerutti, 1989; Spitz *et al.*, 1990;

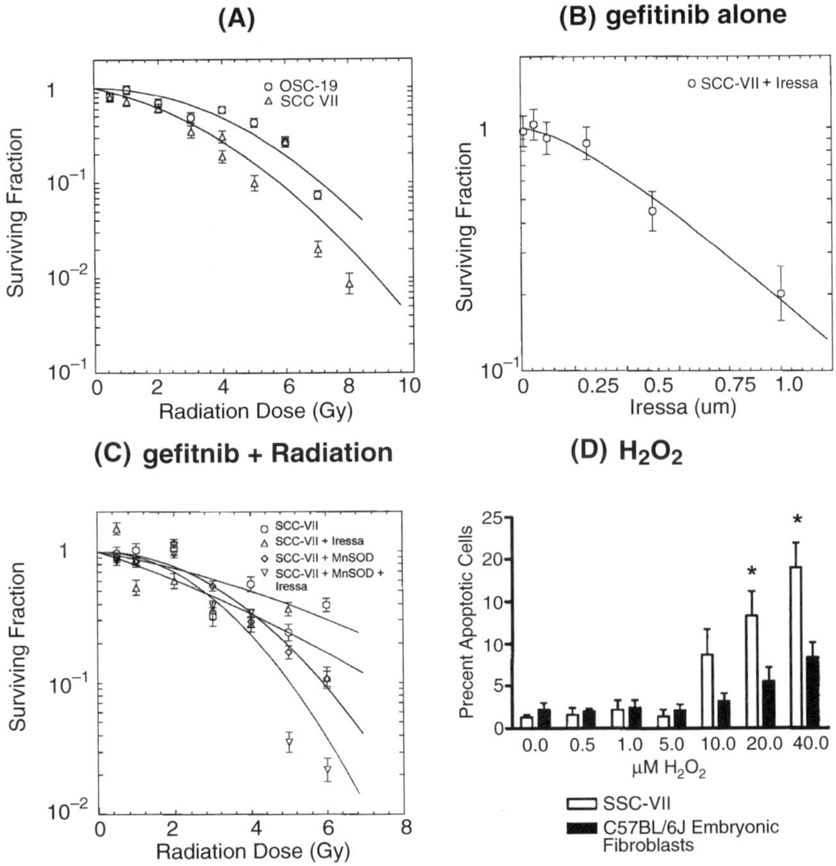

Figure 21.1 *In vitro* irradiation survival curves of SCC-VII tumour cells compared to OSC-19 show additive effects of HA-MnSOD and/or gefitinib. Cells were grown *in vitro* and irradiated to 0 to 8 Gy alone (A), treated with Iressa alone (B) or both agents (C) and plated at cell concentrations of 500, 1000 or 5000 cells/well. Seven days later, the cultures were stained with crystal violet, colonies of greater than 50 cells were counted, and the data analyzed using linear quadratic and single-hit, multi-target models. The plating efficiencies were 15.5 ± 1.8 per cent or 15.6 ± 4.5 per cent for SCC-VII or OSC-19, respectively. Both cell lines were radioresistant as shown by D_0 of 2.4 ± 0.5 Gy or 3.041 ± 1.075 Gy for SCC-VII or OSC-19 cell lines, respectively. Both were sensitive to gefitinib killing, and showed increased radiation killing in the presence of Iressa and/or HA-MnSOD (C) added 24 h prior to irradiation. HA-MnSOD pre-treated cells were more sensitive than untreated cells to Iressa or irradiation plus Iressa ($P = 0.0264$ or $P < 0.0001$, respectively). The sensitivity of SCC-VII cells to H_2O_2 compared to C57BL/6J embryonic fibroblasts is shown in (D). H_2O_2 was added to the cells over a broad range and the percent apoptotic cells determined by a Tunel assay 24 h later. There were significantly more apoptotic cells (*) in SCC-VII than C57BL/6J embryonic fibroblasts at 20 or 40 μM H_2O_2 ($P = 0.0328$ or 0.0062, respectively)

Amstad *et al.*, 1991; Zhong *et al.*, 1997; Kelner *et al.*, 1995; Schmidt *et al.*, 1995; Ceriello *et al.*, 1996; Li S *et al.*, 2000; Tatsuma *et al.*, 2000).

21.2.2 Rationale for antioxidant gene therapy approaches in radiation protection of the head and neck region

Several lines of evidence suggest that a transient increase in the levels of antioxidant gene product within normal tissues can provide protection from ionizing irradiation or chemoradiotherapy (CRT)-induced cellular damage. Experiments done in nontumour-bearing mice have demonstrated that ionizing irradiation induces a rapid increase in cytokine levels for transforming growth factor-β (TGF-β)1–β2, interleukin (IL)-1, and tumour necrosis factor-α (TNF-α), and that peak cytokine elevation decreases rapidly within 24–48 hours. A second, slow increase in levels of TNF-α within the irradiated lung occurs around day 50 and continues until day 100, when a second peak of elevation of TGF-β1 is detected. Finally, a second peak of elevation of TGF-β1–β2 is coincident with the appearance of organizing alveolitis/fibrosis and pulmonary death (Epperley *et al.*, 1999b). In the oesophagus model, cytokine elevation is also detected rapidly after irradiation, but different patterns of cytokines are detected compared to those in the lung associated with different cell types within the tissues that differentiate these organs (Epperley *et al.*, 2001a). A similar pattern of an acute peak cytokine production associated with inflammation, and a second latent period peak is associated with fibrosis in both organs. The administration of MnSOD-PL in single dose prior to single fraction irradiation or in several doses during fractionated irradiation has been shown to decrease the magnitude and duration of both the acute peak elevation in cytokine production and the late peak elevation (Epperley *et al.*, 1999b, 2001a). MnSOD-PL administration to C57BL/6J or C3H/HeNHsd mice decreases irradiation-induced organizing alveolitis and oesophageal stricture, respectively (Epperley *et al.*, 1999b, 2000a). *In vitro* experiments utilizing MnSOD overexpressing hematopoietic progenitor cell lines, embryo fibroblast lines, and freshly explanted oesophagus and lung have demonstrated a decrease in irradiation-induced apoptosis, cytokine induction and cell killing (Epperley *et al.*, 1999c, 2002a).

The mechanism of MnSOD-PL radiation protection is in part at the level of the mitochondria (Epperley *et al.*, 2002a). Substitution of cytoplasm localized Cu/ZnSOD in either plasmid/liposomes or adenovirus, introduced comparable biochemical levels of enzyme, but did not induce radiation protection (Epperley *et al.*, 1999a; Greenberger *et al.*, 2001; Li S *et al.*, 2000). A specific mitochondrial leader sequence on the MnSOD protein product has been associated with its mitochondrial concentration and radiation protective capacity (Epperley *et al.*, 2003d). Overexpression of other mitochondrial localized transgene products Bcl-xl or Bcl-2 in the same cell lines *in vitro* also confers radiation protection (Epperley *et al.*, 2002a, 2003c). The mechanism of MnSOD action in the mitochondria is not yet known, but recent experiments have demonstrated in MnSOD transgene overexpressing cells in culture that initial irradiation-induced DNA strand breaks (repaired within 15–20 min after irradiation) and translocation of stress-activated protein kinases (SAP) p38 and Jnk1, as well as Bax from nucleus to mitochondria, are unchanged by overexpression of the MnSOD transgene product (Epperley *et al.*, 2002a). In striking contrast, cells overexpressing MnSOD transgene did not demonstrate the significant distal steps associated with apoptosis that follow p38, Jnk1, and Bax translocation to the mitochondria; namely, mitochondrial membrane depolarization, cytochrome *c* leakage into the cytoplasm, and activation of caspase-3, poly-ADP-ribosyl polymerase and DNA fragmentation (Epperley *et al.*, 2002a). The mitochondrial localization of the MnSOD transgene has been demonstrated both *in vitro* and *in vivo* in the oesophagus to be associated with decreased irradiation-induced apoptosis (Epperley *et al.*, 2000b, 2002a, 2003d).

21.2.3 Development of gene therapy for radiation protection of the oral cavity/oropharynx

Gene therapy strategies for radiation protection (unique to our laboratory) (Epperley *et al.*, 1999b,

2001a) have recently been applied to the oral cavity and oropharynx (Guo *et al.*, 2003a) and other organ-specific radioprotection (Epperley *et al.*, 1998, 1999a,b,c; Greenberger *et al.*, 1998; Zwacka *et al.*, 1998; Stickle *et al.*, 1999). Intraoesophageal administration of MnSOD-PL has been demonstrated to protect the mouse oesophagus from CRT-induced oesophagitis from both single fraction and fractionated irradiation-induced oesophageal damage (Stickle *et al.*,1999; Epperley *et al.*, 2001a,d). Multiple administrations of MnSOD-PL were effective in maintaining elevated levels of messenger RNA and transgene product in the mouse model (Stickle *et al.*,1999; Epperley *et al.*, 2001a,d). Cervical oesophagus specimens removed from mice at serial time points after irradiation demonstrated a protective effect of this gene therapy approach (Stickle *et al.*,1999; Epperley *et al.*, 2001a,d). Administration of the MnSOD transgene in plasmid/liposomes or adenovirus reduced both the acute and chronic toxicity of total lung irradiation in the mouse model (Epperley *et al.*, 1998, 1999a,b, 2000c, 2001b). In contrast, orthotopic Lewis Lung carcinoma tumours in the mediastinum were not protected by intratracheal injection of MnSOD-PL (Epperley *et al.*, 2000b). This result was attributed to the ability of PL to penetrate only the local organ at the site of contact. It was also discovered that mice receiving intraoral or intratracheal MnSOD-PL gene therapy demonstrated greater radiation tumour killing and longer survival (Epperley *et al.*, 2000b; Guo *et al.*, 2003a,b). In the oral cavity and oropharynx, radiotherapy treatment complications are known to be of a more complex nature involving both salivary gland and mucosal targets for organ specific radioprotection.

A new class of biological response modifiers has recently focused on inhibition of the critical epidermal growth factor receptor (EGFR) mediation of squamous cell tumour growth (Wakeling *et al.*, 2002). Several approaches have included antibody to EGFR (Soulieres *et al.*, 2004) and inhibition of the tyrosine kinase mediated signal transduction steps of the activated EGFR (Kauschke *et al.*, 1999; Fukuoka *et al.*, 2002; Kris *et al.*, 2002; Mendelsohn and Baselga, 2003; Richardson *et al.*, 2003;). The latter category of agents includes the agent ZD1839 (gefitinib) (Tamagno *et al.*, 2003), which has recently been reported to be effective therapy (Kagan *et al.*, 1994; Kanemura *et al.*, 2003; Khalil *et al.*, 2003) in phase II trials in patients with advanced squamous cell tumours of the aerodigestive tract (Kauschke *et al.*, 1999; Fukuoka *et al.*, 2002; Kris *et al.*, 2002; Wakeling *et al.*, 2002). In recent reports, response rates of 18–20 per cent were detected in patients who had recurred after initial chemotherapy approaches (Kauschke *et al.*, 1999; Kris *et al.*, 2002). Combined modality approaches with irradiation arebeing considered; however, side effects of Iressa, including skin rash and dry skin suggest a potential for added toxicity to radiotherapy. While Iressa appears to be a potentially valuable addition to chemoradiotherapy of head and neck cancers (Kagan *et al.*, 1994; Kanemura *et al.*, 2003), concerns for possible additive toxicities to normal tissues remain (Kagan *et al.*, 1994; Schmidt-Ullric *et al.*, 1997; Carter *et al.*, 1998; Kavanagh *et al.*, 1998; Kanemura *et al.*, 2003; Khalil *et al.*, 2003; Hagan *et al.*, 2003; Eriksen *et al.*, 2004). The effect of Iressa on redox changes within tumours is not yet known.

21.2.4 Usefulness of MnSOD-PL gene therapy for ameliorating toxicity of chemoradiation of unresectable cancers of the head and neck

A major problem with combined modality therapy, principally CRT of head and neck cancer, is mucositis. Patients receiving 180 or 200 cGy per day of fractionated irradiation alone typically develop significant oral cavity and oropharyngeal mucositis by the third or fourth week of a 7-week radiotherapy (RT) treatment course (Kies *et al.*, 2001; Martinez *et al.*, 2001; Dawson *et al.*, 2001; Sonis *et al.*, 2001). In the setting of combination chemotherapy with weekly platinum-based or taxane-based chemotherapeutic agents, mucositis is usually detectable by the end of the second week. Other complications include decreased saliva production and late osteoradionecrosis in high radiation dose volumes. Improved RT delivery techniques, including conformal and intensity modulated radiotherapy (IMRT) (Dawson *et al.*,

2001; Kies *et al.*, 2001; Martinez *et al.*, 2001; Munter *et al.*, 2004; Astrenidou *et al.*, 2004; Vanhoefer *et al.*, 2004), and the use of high-dose-rate (HDR) brachytherapy (Kies *et al.*, 2001; Martinez *et al.*, 2001), have facilitated improvements in dose distribution and dose escalation. The availability of new chemotherapeutic drugs, including cisplatin, carboplatin, docetaxel, paclitaxel, gemcitibine, etoposide (Kies *et al.*, 2001; Martinez *et al.*, 2001), and others, have facilitated sophisticated programs of combined modality chemotherapy and radiotherapy to both radiosensitize tumours in the head and neck region and to prevent or decrease distant metastasis. Improved techniques of hyperalimentation and supportive care have provided increased resources in managing the toxicity of high intensity CRT of squamous cell carcinoma (SCC) of the head and neck region (Kies *et al.*, 2001; Martinez *et al.*, 2001; Dawson *et al.*, 2001). Combined modality programs have also included new techniques of neo-adjuvant CRT prior to surgical resection (Martinez *et al.*, 2001; Dawson *et al.*, 2001). Despite advances in the technical delivery of RT, radiation therapy treatment planning (RTP), and the use of new radiosensitizing drugs, local control of T_3N_1–T_3N_2–T_3N_3 carcinoma of the head and neck remains unacceptably low and recurrence of cancer both locally and distantly after CRT protocols remains at a suboptimal level (Dawson *et al.*, 2001; Kies *et al.*, 2001; Martinez *et al.*, 2001). A major problem upon which most cancer treatment centres are focusing remains the toxicity of CRT (Damron *et al.*, 2001; Johnstone *et al.*, 2001; Sonis *et al.*, 2001; Wijers *et al.*, 2001). Toxicity both limits the ability to deliver full doses of chemo RT and prevents dose escalation.

Numerous approaches have been taken to attempt to decrease oral cavity and oropharyngeal toxicity of CRT. These have included the institution of hyperfractionation (Martinez *et al.*, 2001) or hypofractionation (Dawson *et al.*, 2001) RT regimens, split-course RT techniques, improvements in treatment planning with reduced field size modification, and the recent usage of radioprotective pharmacological compounds designed to protect normal tissue. Amifostine (*s*-2 (aminopropylamino)-ethyl phosphorothioic acid, WR2721, Ethyol) is a well-characterized radio- and chemoprotective agent (Damron *et al.*, 2001; Werner-Wasik *et al.*, 2001) and has been shown to decrease RT-induced sialadenitis, although a decrease in mucositis and oropharyngeal toxicity has been less consistently observed (Damron *et al.*, 2001). Other pharmacological approaches toward decreasing oropharyngitis have included administration of pilocarpine (Roesink *et al.*, 1999; Horiot *et al.*, 2000; Johnstone *et al.*, 2001), antibiotics to reduce oral cavity flora (Wijers *et al.*, 2001; Ertekin *et al.*, 2004), pentoxyfilline (Ozturk *et al.*, 2004), adrenergic compounds, glutamine (Huang *et al.*, 2000), cytokines including granulocyte–macrophage colony-stimulating factor (GM-CSF), G-CSF, M-CSF (Makkonen *et al.*, 2000), use of a complex glucan (betafectin), kGF-palifermin (Cohen *et al.*, 2003; Emmanuouilides *et al.*, 2003 Stiff *et al.*, 2003), or erythropoietin (Martinez *et al.*, 2001). Preclinical and clinical trials using these materials have met with incomplete success at reducing treatment-related toxicity and morbidity from irradiation of tumours of the oral cavity and oropharynx (Maurici *et al.*, 1998; Regine *et al.*, 2001; Werner-Wasik *et al.*, 2001). Our studies with MnSOD-PL suggest it may be an effective radioprotector (Guo *et al.*, 2003a).

21.2.5 A potential beneficial added observation of antitumour effect of radiation protection using MnSOD-PL

A potential antitumour effect of MnSOD-PL gene therapy has been reported (Epperley *et al.*, 2000b; Li S *et al.*, 2000; Guo *et al.*, 2003b). Cerutti and co-workers (Larsson and Cerutti, 1989; Amstad *et al.*, 1991; Kelner *et al.*, 1995; Schmidt *et al.*, 1995; Ceriello *et al.*, 1996; Tatsuma *et al.*, 2000) and Oberley *et al* (Zhong *et al.*, 1999; Li S *et al.*, 2000; Liu *et al.*, 2000; Li Z *et al.*, 2001Wang *et al.*, 2001; Zhao *et al.*, 2001; Zhang *et al.*, 2002;) have shown that the toxicity of H_2O_2 generated by dismutation of SOD in tumours by MnSOD can be therapeutically advantageous (Kagan *et al.*, 1994; Kanemura *et al.*, 2003; Khalil *et al.*, 2003). SCC cell lines of the head and neck region were established from a large number of patients and each tested for intrinsic levels of production of antioxidant proteins. The majority of tumour cell lines demonstrated stably decreased levels of MnSOD production, the mechanism of which included transcriptional shut-off, mutation in the promoter region of the MnSOD gene, and other redox changes within tumour cells both *in vitro* and *in vivo* (St Clair *et al.*, 1992; Xu *et al.*, 1999). Introduction of the MnSOD transgene to these SCC head and neck tumour cell lines *in vitro* has demonstrated sensitization of the tumours to ionizing irradiation and bis-chloronitrosyl urea chemotherapy (Li S *et al.*, 2000). It is has been hypothesized that the redox balance within tumour cells has adapted to a decreased MnSOD bioavailability, rendering cells sensitive to H_2O_2 toxicity (Spitz *et al.*, 1990). Hydrogen peroxide (H_2O_2), the natural catabolic product of MnSOD biochemical action when made abundant by introduction by the MnSOD transgene, renders the cells susceptible to peroxide-induced death. The available evidence suggests that normal tissue protection by MnSOD-PL gene therapy should not adversely affect irradiation killing of SCC tumours *in vivo* and may actually be an antitumour agent (Figure 21.1).

21.2.6 HA-MnSOD effects on reducing the repopulation (cytoprotective response) induced by irradiation in tumours and the role of EGFR-TKI

Clinical doses of irradiation activate mitogen-activated protein (MAP) kinase and stress- activated protein (SAP) kinase pathways (Hagan *et al.*, 2003; Eriksen *et al.*, 2004). MAP kinase increases the irradiation induction of strand breaks and DNA damage. The MAP kinase pathway is also cytoprotective such that its inhibition may be radiosensitizing (Schmidt-Ullrich *et al.*, 1997). EGF binding to EGFR and tyrosine phosphorylation leads to MAPK activation. Blocking this pathway by Iressa could also block irradiation induction of this cytoprotective pathway and thereby increase tumour kill. What role the increased H_2O_2 production generated by HA-MnSOD expression in tumour cells has in the process is not known. Questions include the following:

1. Does MnSOD stimulated H_2O_2 enhance the EGFR-TKI effect?
2. Does MnSOD-mediated H_2O_2 production enhance or inhibit MAPK induction by irradiation?
3. Will MnSOD overexpression reverse or enhance the synergistic irradiation and gefitinib interaction?
4. Does HA-MnSOD disrupt the potentially beneficial effect of gefitinib, preventing irradiation-induction of the survival/repopulation response by tumours after irradiation?

MnSOD overexpressing cells have been shown to down regulate cytoprotective gene products vascular endothelial growth factor-1, TNF-α, and IL-β, while a stress response gene known as *GADD53* (involved in repair of DNA double strand breaks) is increased 3.3-fold (Epperley *et al.*, 2004). Increasing MnSOD levels in tumour cells can modulate other downstream effecter genes. Furthermore, free radicals have been shown to mediate upregulation of gene expression by TNF-α (Wang *et al.*, 2002). The available evidence suggests that overexpression of MnSOD in tumours could facilitate altered gene regulation through increased H_2O_2 production, which could further enhance the antitumour effects of combining irradiation with gefitinib (Takeyanna *et al.*, 2000; Kanai *et al.*, 2001;Kaimori *et al.*, 2003; Naderi *et al.*, 2003). Since MnSOD overexpression decreases TNF-α production (Epperley *et al.*, 1999b), this result may also represent an antitumour effect, which could be further enhanced by gefitinib.

There is published evidence to suggest that H_2O_2 stimulates the EGF signal transduction pathway in neuroblastoma cell lines (Kaimori *et al.*, 2003) and mucin producing epithelial cells (Lin *et al.*, 1995; Kavanagh *et al.*, 1998; Takeyanna etal., 2000; Kanai *et al.*, 2001; Guiffre *et al.*, 2002; Tampo *et al.*, 2002; Campiglio *et al.*, 2004; Tumb *et al.*, 2004; Weinmann *et al.*, 2004). Thus, MnSOD-mediated generation of H_2O_2 in tumours may beneficially modulate the EGF signaling pathway to increase its susceptibility to gefitinib. Uptake of HA-MnSOD in tumours in the mouse model has been quite successful (Figure 21.2) and effects on antioxidant pools have been encouraging (Figure 21.3).

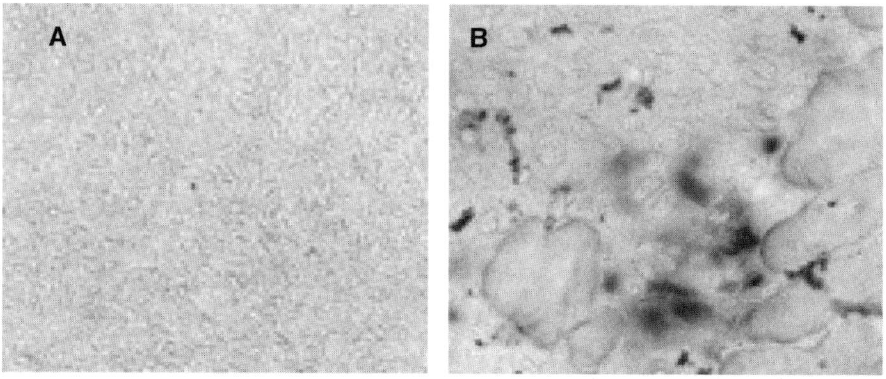

Figure 21.2 Immunohistochemical detection of HA-MnSOD in mouse orthotopic tumours. C3H/heNHsd mice were injected with 1×10^6 SCC-VII tumour cells into the cheek. Seven days later, with establishment of a 0.5 cm tumour, mice were injected i.v. with HA-MnSOD-PL and sacrificed 24 h later. The tumour was excised, frozen in OCT, and sectioned. Sections were stained with a rabbit anti-HA antibody followed by an alkaline phosphatase-conjugated anti-rabbit antibody, and stained for alkaline phosphatase activity. (A) A section of tumour from a control mouse. (B) A tumour from a mouse injected i.v. with HA-MnSOD-PL. Mice receiving intraoral HA-MnSOD-PL showed no detectable uptake similar to that in control mouse tumour

21.2.7 MnSOD-PL treatment to ameliorate the dose-limiting oesophageal toxicity of irradiation

A primary dose-limiting toxicity of chemoradiotherapy of non-small cell carcinoma of the lung is irradiation-induced oesophagitis (Bhatnagar *et al.*, 2002; Wolski *et al.*, 2005; Kahl *et al.*, 2004; Chapet *et al.*, 2005). Despite the availability of carboplatin- and taxol-based chemotherapy regimens and respiratory gated, intensity modulated radiotherapy treatment programmes, the chemoradiotherapy of non-small cell carcinoma of the lung, results in over one third to one half of the patients suffering local recurrence as a component of treatment failure (Socinski *et al.*, 2001; Bhatnagar *et al.*, 2003; Wolski *et al.*, 2005). The inability to deliver primary target volume doses in excess of 75 Gy (lung corrected dose) has been determined to be a major factor in the high frequency of tumour recurrence in the irradiated volume (Stickle *et al.*, 1999; Rosenman *et al.*, 2002; Houghton *et al.*, 2004; Kahl *et al.*, 2004; Ahn *et al.*, 2005). Doses of 90–95 Gy have been identified as necessary for effective local control of unresectable tumours; however, such doses are not achievable in the current setting of chemoradiotherapy (Stickle *et al.*, 1999; Rosenman *et al.*, 2002;). Given the significant oesophageal toxicity which is detectable in one third to one half of patients receiving current protocols, new approaches toward tumour radiosensitization and/or normal tissue radiation protection are clearly required to improve the therapeutic ratio in chemoradiotherapy of lung cancer (Socinski *et al.*, 2001; Bhatnagar *et al.*, 2003).

21.2.8 How MnSOD-PL gene therapy may interrupt the mechanisms of oesophageal irradiation damage

Data from clinical and experimental models clearly document the irradiation dose, target volume and fraction size dependence of oesophageal damage (Stickle *et al.*, 1999; Epperley *et al.*, 2001a; Socinski *et al.*, 2001; Rosenman *et al.*, 2002; Agarwal *et al.*, 2004; Houghton *et al.*, 2004; Perry *et al.*, 2005 Wolski *et al.*, 2005) measured as clinical dehydration, dysphagia weight loss, histopathologic ulceration and basal layer cell apoptosis (Epperley *et al.*, 2001a). We have demonstrated that intraoesophageal administration of MnSOD-PL decreases the incidence and severity of chemoradiotherapy induced oesophagitis in a mouse model (Epperley *et al.*, 2001a; Agarwal *et al.*, 2004) and decreases photodynamic therapy induced oesophageal stricture in a porcine model

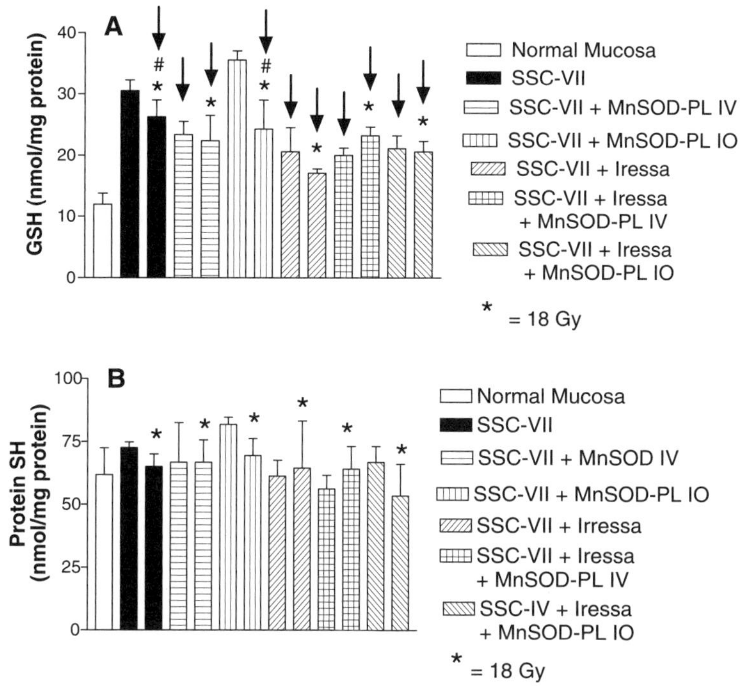

Figure 21.3 Changes in antioxidant pools 30 min after irradiation of orthotopic SCC-VII tumours. C3H/HeNHsd mice were injected with 1×10^6 SCC-VII cells into the mucosa of the oral cavity. When tumours reached 0.5 mm^3, mice were divided into the following subgroups: (1) control, (2) intravenously (i.v.) injected with MnSOD-PL (200 μg plasmid DNA) 24 h before irradiation, (3) intraorally (i.o.) given MnSOD-PL (200 μg of plasmid DNA), (4) gefitinib (250 mg/kg i.v.), (5) gefitinib plus i.v. injected MnSOD-PL, or (6) Iressa plus i.o. MnSOD-PL. MnSOD-PL was given 24 hr before irradiation while gefitinib was injected 30 min before irradiation. Half the mice in each group were irradiated to 18 Gy to the oral cavity (*). Thirty min after irradiation, tumours were removed, frozen in liquid nitrogen, homogenized, and levels of GSH (A) and total thiols (B) determined. SCC-VII tumours had higher intrinsic levels of GSH than normal mucosa. Following 18 Gy, there was a reduction of GSH detected in tumours. Intravenous MnSOD-PL reduced the GSH levels in non-irradiated tumour, and adding 18 Gy caused a further but non significant decrease. (We hypothesize that i.v. MnSOD-PL increased production of H_2O_2 and depleted GSH while i.o. MnSOD-PL, in which tumour was not exposed to uptake of the PL, had no effect on intratumour GSH levels. Irradiation decreased GSH levels more in the i.v. MnSOD-PL group. gefitinib-induced tumour cell toxicity decreased the levels of GSH when delivered alone or with irradiation or MnSOD-PL. Total thiols were not detectably altered by irradiation, MnSOD-PL pretreatment or Iressa measuring at the 30-min post-irradiation time point. We expect to see more significant changes in tumour compared to normal tissue in GSH and total thiols at later time points and in fractionation irradiation experiments). Arrows indicate significant difference from non-irradiated SCC-VII cells

(Epperley *et al.*, 2002a, 2005). Mitochondrial specific localization of MnSOD protein has been shown to be required for radiation protection *in vitro* and *in vivo*, strongly implicating the mitochondrial membrane as a subcellular target for radiation protection (Epperley *et al.*, 2003c, 2004a). In both cell lines *in vitro* (Greenberger and Epperley, 2004; Epperley *et al.*, 2003d) and tissue *in vivo* (Epperley *et al.*, 2003d), MnSOD-PL administration prior to irradiation modulates depletion of antioxidant reserves including glutathione, total thiols, and decreases irradiation-induced lipid peroxidation (Epperley *et al.*, 2004b, 2005). The effects on tissue repair including stem cell repopulation of intraoesophageal delivery of the antioxidant protein MnSOD through gene therapy are not yet known. (Note: The University of Pittsburgh Cancer Institute and Department of Radiation Oncology have begun a phase I/II clinical trial of twice weekly swallowed MnSOD-PL for oesophageal radioprotection of patients with unresectable non-small cell lung cancer as part of the Lung SPORE P50-CA090440, Jill Siegfried, PI, IRB approval date 10/7/05.)

21.2.9 MnSOD-PL may help tissue repair of the irradiated oesophagus: during tissue repair oesophageal progenitors for squamous epithelium (candidate stem cells) show evidence of donor bone marrow origin

There is an ongoing controversy in understanding stem cell self renewal and differentiation in tissue repair. Bone marrow derived cells (both hematopoietic and stromal–mesenchymal stem cell) have been demonstrated to engraft and differentiate to non-haematopoietic tissue including liver, lung, intestine, brain, skin, oesophagus, muscle and other organs (Fabisciak *et al.*, 1998; Jackson *et al.*, 1999; Jiang *et al.*, 2002; Korbling *et al.*, 2002; Theise and Krause, 2002; Herzog *et al.*, 2003; Roberts and Wakefield, 2003; Smalley *et al.*, 2003; Epperley *et al.*, 2004c; Ishikawa *et al.*, 2004; Prindull *et al.*, 2004; Spyridonidis *et al.*, 2004; Lapidos *et al.*, 2004; Dezawa *et al.*, 2005; Kucia *et al.*, 2005; Metaxis *et al.*, 2005; Sato *et al.*, 2005; Zhao *et al.*, 2005). Others contest that these findings do not demonstrate stem cell plasticity arguing instead that cell fusion may explain the results (Epperley *et al.*, 2000c; Krause *et al.*, 2001; Murray *et al.*, 2004; Nygren *et al.*, 2004; Sherwood *et al.*, 2004; Ogle *et al.*, 2005). Detection of fusion of bone marrow cells with epithelial cells led some researchers to argue for an alternative explanation of the reported bone marrow origin of liver, beta-islet cells of the pancreas, striated muscle, cardiac muscle, and other tissues (Anklesaria *et al.*, 1987; Epperley *et al.*, 2000c; Krause *et al.*, 2001; Houghton *et al.*, 2004). We have isolated a subpopulation of non-adherent oesophageal stem cell candidates (ESC) by either a side population (SP) cell sorting technique or 7-day *in vitro* preplate technique (Smalley *et al.*, 2003; Zipori, 2004). Intraoesophageal MnSOD-PL administration prior to irradiation enhances the oesophageal engraftment and survival of these intravenous (i.v.) injected ESC cells (Niu *et al.*, 2005). We have also demonstrated that bone marrow contains a subpopulation of cells which, when injected i.v., repopulates the irradiated recipient oesophagus, and differentiates to oesophageal squamous cells following explant *in vitro* (Smalley *et al.*, 2003). Our work is supported by a recent study showing that marrow stromal cells migrate and differentiate into gastric epithelial cancer (Daum, 1985). In this gastric cancer model, chronic inflammation by *Helicobacter fecalis* infection was required to recruit bone marrow origin cells (Daum, 1985). However, an alternative explanation for both our oesophageal and the gastric cancer models (Daum, 1985; Smalley *et al.*, 2003) is that synkaryons formed by initial fusion of donor marrow cells with recipient epithelial cells and then lost the epithelial cell specific chromosomes leaving cells of apparent donor marrow origin (Ogle *et al.*, 2005). Whether stem cell plasticity, fusion or both mechanisms occur in the response to irradiation oesophagogastric damage, it is important now to evaluate critically whether fusion is a marker of toxicity or of a sign of repair. Methods by which to enhance the transplantability of stem cell populations derived from non-haematopoietic organs into other organs or back into bone marrow should be enhanced by this knowledge. If fusion is indeed a harbinger of toxicity then minimizing this outcome would be of therapeutic benefit. Our current data show clear enhancement of stem cell engraftment to the oesophagus (Figure 21.4).

21.2.10 MnSOD-PL treatment may ameliorate irradiation toxicity to the oesophageal microenvironment

The bone marrow transplantation literature suggests that the irradiation damaged microenvironment limits the effectiveness of both repopulation by resident stem cells and homing, engraftment, and repopulation by circulating stem cells (Blomberg *et al.*, 1998; Prigozhina *et al.*, 1999; Chao *et al.*, 2004; Abedi *et al.*, 2005; Alyea *et al.*, 2005; Bauer *et al.*, 2005;

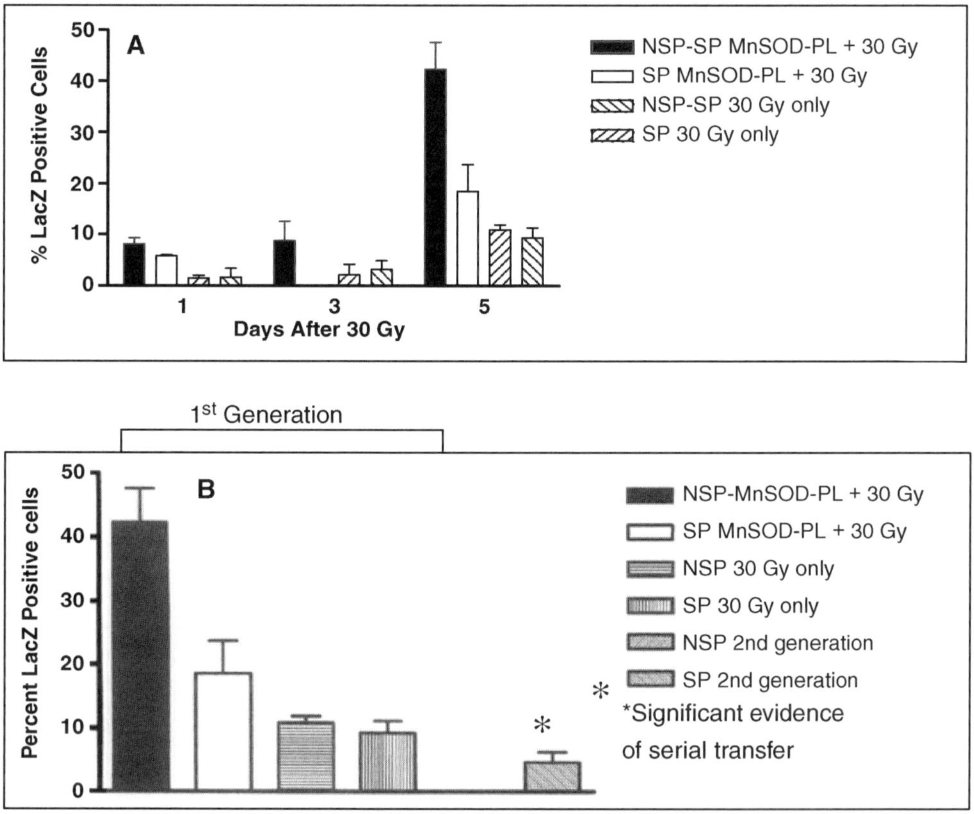

Figure 21.4 MnSOD-PL intra-oesophageal treatment improves engraftment and serial transfer of i.v. injected marrow derived oesophageal progenitor cells. C57BL/6J female mice received 30 Gy to the oesophagus. A subgroup received MnSOD-PL pre-irradiation. Male ROSA marrow cells were injected i.v. at 1, 3 or 5 days after irradiation. Oesophagus was removed at each time point at 1, 3 or 5 days and assayed for percentage donor origin cells (A). At day 5 (optimal time for transplant), SP and non-SP cells were removed and serially transferred to a second generation of 30 Gy oesophagus irradiated mice (B) (at least 10 mice per group-these second recipients were not MnSOD-PL pretreated). Only SP cells transferred the donor LacZ+ $G418^r$ and markers to the second generation recipients (*). The data show better engraftment in MnSOD-PL treated recipient oesophagus (A), and show serial transplant of SP cells to second-generation recipients (B). The data support the observation of high levels of apoptosis at day 5 in the irradiated oesophagus but not earlier. We will test the effect of MnSOD-PL treatment of the donor cells and/or second generation recipients' oesophagus as part of the grant

Martino *et al.*, 2005; Mohty *et al.*, 2005; Satwani *et al.*, 2005; Scheffold *et al.*, 2005; Zhang *et al.*, 2005). Bystander killing of stem cells by cytokines released from irradiated stromal cells may in part explain this limitation (Hendricks *et al.*, 2003; Yoon *et al.*, 2005). Haematopoietic stem cell and stromal cell seeding and repopulation of high dose compared to low dose irradiated mouse limb marrow have been demonstrated (Terry and Travis, 1989), but the stability of engraftment into high dose irradiated sites has not been critically evaluated. Enhanced survival of hemibody (abdomen) irradiated mice follows i.v. bone marrow injection (Epperley *et al.*, 2002c) attributable to engraftment of marrow-origin intestinal stem cells, but the effectiveness was limited by higher irradiation doses (Jackson *et al.*, 2001). Migration of bone marrow origin endothelial progenitor cells into irradiated tissues enhances revascularization but is also radiation dose limited (Anklesaria *et al.*, 1989; Krause *et al.*, 2001; Bauer *et al.*, 2005; Gorak *et al.*,

2005; Martino *et al.*, 2005; Satwani *et al.*, 2005). These data suggest that stem cells for non-haematopoietic tissue, while capable of migration into irradiated sites (Gorbunov *et al.*, 2000; Jackson *et al.*, 2001; Epperley *et al.*, 2002c) suffer irradiation-dose-dependent injury induced by the microenvironment (Blomberg *et al.*, 1998; Prigozhina *et al.*, 1999; Miranda *et al.*, 2000; Morgan, 2003; Chao *et al.*, 2004; Abedi *et al.*, 2005; Zhao *et al.*, 2005). Reduced total body irradiation (TBI) dose decreases the toxicity of marrow transplant; however, increased relapse of malignancy has been problematic (Gorak *et al.*, 2005). Separation of irradiation-induced cell depletion for niche clearing from delayed toxicity is very important to elucidate. Our data show that this model system (Figure 21.5) is ideal to prove the hypothesis that MnSOD-PL treatment protects themicroen vironment.

21.2.11 MnSOD-PL gene therapy may enhance stem cell repopulation of the oesophagus through effects on the microenvironment

The controversy over plasticity, or fusion, of bone marrow transplantation derived stem cells into epithelial cells may be resolved by understanding irradiation effects on the microenvironment. There is evidence that the irradiated microenvironment of the recipient induces toxic biochemical and molecular biologic changes in homing and engrafting non-irradiated stem cell populations. Irradiated bone marrow stromal cell cultures induce nitric oxide synthase (NOS), and free radicals associated with the oxidative stress response in co-cultured non-irradiated haematopoietic cells (Prigozhina *et al.*, 1999). Engrafted breast tissue into the cleared mammary fat pad of recipient mice is altered by

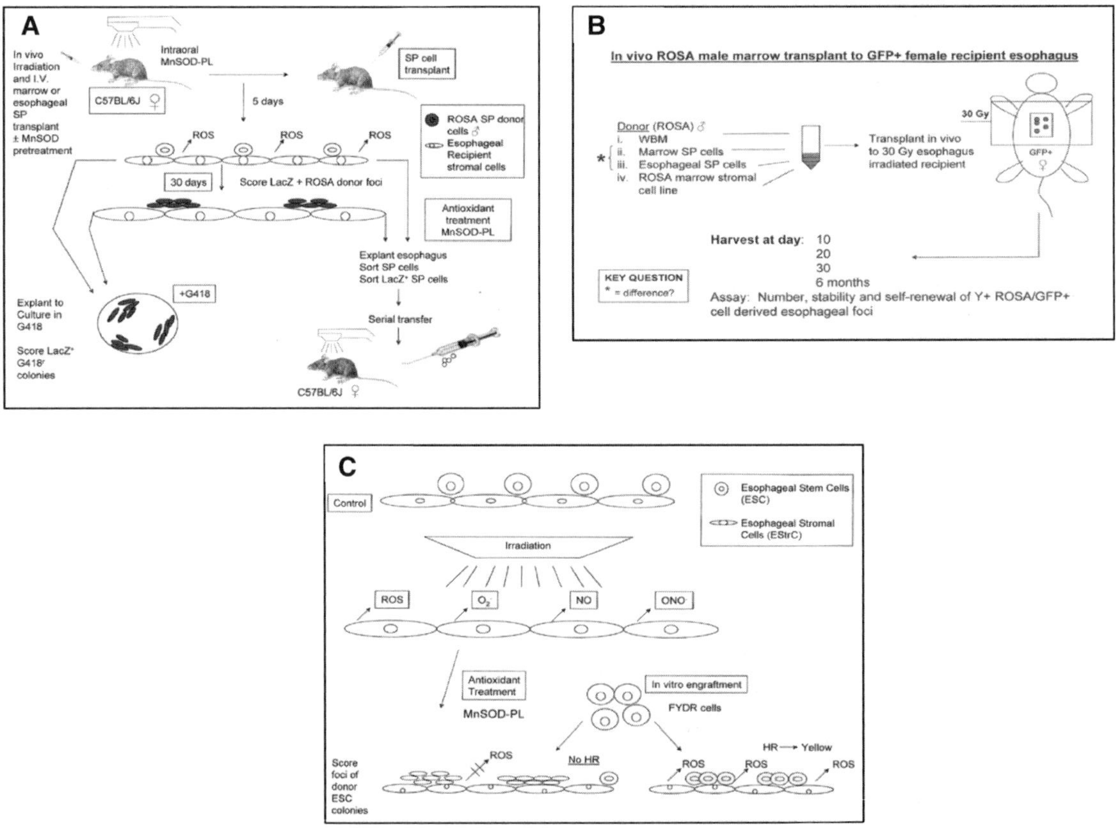

Figure 21.5 Experimental paradigm showing MnSOD-PL treatment of the oesophagus microenvironment in which enhances engraftment of ROSA stem cells (*in vivo*, A, B and *in vitro*, C)

irradiation of the mammary microenvironment (Houghton *et al.*, 2004). Bone marrow stem cell transplantation in both experimental models and clinical protocols is more effective in low dose TBI recipients, compared to those receiving conventional 'marrow clearing' or ablative doses (Chao *et al.*, 2004; Koh and Chao, 2004). Decreased self-renewal of serially transplanted haematopoietic stem cells (Mauch *et al.*, 1980; Eckner *et al.*,1982; Koh and Chao, 2004; Jang *et al.*, 2004) or stem cell exhaustion may in fact be attributable in part to toxic effects of exposure of transplanted haematopoietic stem cells to serial generations of irradiated toxic microenvironments each of which induces toxic changes in homing stem cells. If the irradiated microenvironment of recipient organs induces oxidative stress-related changes in engrafting stem cells, the antioxidant defences in those stem cells, once depleted, could lead to differentiation or death. Stem cell depletion hastens organ failure, as well as limits evidence that marrow can differentiate into muscle (Krause *et al.*, 2001) or liver (Epperley *et al.*, 2000c; Lagasse *et al.*, 2000). Oxidative stress in the irradiated organ toxic microenvironment might also induce fusion of donor with recipient cells (Epperley *et al.*, 2000c; Ogle *et al.*, 2005).

Recent data indicate that the irradiated organ microenvironment continues to induce oxidative stress in circulating or engrafting cells long after irradiation. Reactive oxygen species (ROS) including superoxide, hydroxyl radical, and reactive nitrogen species (RNS) including nitric oxide and peroxynitrite, as well as H_2O_2 have been demonstrated in irradiated mouse and rat lung, weeks to months after single fraction irradiation (Mitchell *et al.*, 1998; Wang *et al.*, 1998; Kanai *et al.*, 2002; Kang *et al.*, 2003). While the histopathology of lung may return to normal weeks after irradiation, recent evidence indicates that an oxidative stress environment continues to persist and results in not only continued free radical production (Mitchell *et al.*, 1991; Kanai *et al.*, 2002), but also accumulation of thrombomodulin in endothelial cells (Paris *et al.*, 2001), and alteration of sphingolipids in irradiated intestine (Epperley *et al.*, 1999b). Continuous maintenance of an oxidative stress inducing environment may maintain the production of inflammatory cytokines (Agarwal *et al.*, 2004) including TGF-β (Kamata *et al.*, 2005), IL-1, and TNF-α, which are associated with both inflammatory cell recruitment and fibroblast progenitor cell migration into irradiated targets to produce the late effect of fibrosis (Kagan *et al.*, 2004; Meng *et al.*, 2002). While irradiation clears space by inducing apoptosis of non-SP committed progenitor cells, facilitating niche formation for engrafted progenitors (Niu *et al.*, 2005), we hypothesize that irradiation adds the deleterious effect of continued production of ROS by the microenvironment, which limits engraftment (Niu *et al.*, 2005). We have shown that irradiated stromal cells induce toxicity including apoptosis in non-irradiated cocultivated donor cells and that the toxicity is mediated by ROS (Gorbunov *et al.*, 2000).

21.2.12 New experimental models now facilitate analysis of cell fusion and/or stem cell plasticity in marrow mediated repair of the irradiated oesophagus

Circulatory origin of stem cells in epithelial tissue reconstitution can be regulated by the injury and repair response itself (Epperley *et al.*, 2004c; Ogle *et al.*, 2005). Marrow progenitor cell fusion with epithelial organ progenitor cells could lead to synkaryon formation indicating that a fusion intermediate heterokaryon step had occurred. Intermediate fusion steps would likely be detectable both *in vitro* and *in vivo* but the timing of detection might be critical as was the case in observing early and transient post-irradiation endothelial cell apoptosis during 'target cell shifting' in the irradiated intestine (Paris *et al.*, 2001). Carefully designed experiments to determine the role of irradiation dose, and the contribution of ROS production by the irradiated microenvironment to cell fusion, organ reconstitution and delayed failure may answer these questions. Cell fusion might occur between the donor marrow stem cell progenitors and cells of the oesophageal microenvironment or with recipient oesophageal squamous epithelial progenitor cells (Ogle *et al.*, 2005). Cell fusion could occur between donor SP and endogenous radiation resistant oesophageal SP cells (Ogle *et al.*, 2005). Is cell fusion a required event that facilitates tissue repair or is it a harbinger of toxicity? We maintain that quantitating the contribution of donor stem cells to effective tissue repair requires analysis of function *in vivo* at both early and delayed time points. Clonal assays of irradiation effects on engrafting cells *in situ* are currently limited and include apotag (Epperley *et al.*, 2005a) or target protein (such as ATM) phosphorylation (Bakkenist and Kastan, 2003).

We have developed a novel system for measuring oesophageal microenvironmental irradiation effects on tissue repair by quantitation of: (a) donor oesophageal or bone marrow SP cell fusion with cells of the microenvironment and (b) ROS induced homologous recombination (HR) in donor cells from FYDR mice (Mosvas *et al.*, 2005; Figure 21.5). Utilizing our established model of oesophageal irradiation *in vivo* for quantitation of reconstitution with bone marrow chimeric SP cells (Agarwal *et al.*, 2004; Smalley and Ashworth, 2003) and utilizing MnSOD-PL administration to limit ROS production by the irradiated oesophagus we can now determine the role of bone marrow cell plasticity and/or fusion in tissue repair of the oesophagus.

21.3 MnSOD-PL treatment reduces pulmonary irradiation damage

21.3.1 Clinical problem: dose limiting normal lung tolerance

Ionizing irradiation damage to the lung is dependent upon total dose, volume treated, and fraction size (Gopal *et al.*, 2003). These basic principles of irradiation biology have dictated the parameters for thoracic tumour irradiation and continue to govern current treatment techniques, including respiratory-gated intensely modulated radiotherapy (IMRT), and image guided, or PET/CT guided radiotherapy (Yorke *et al.*, 2002; Marks, 2002; Wu *et al.*, 2004; Roof *et al.*, 2003). Normal lung tissue irradiation tolerance remains a major dose-limiting factor in clinical radiotherapy of tumour volumes in the chest (Gopal *et al.*, 2003). Because of inherent dose and volume limitations in thoracic radiotherapy, a significant number of patients with non-small cell carcinoma of the lung and oesophageal cancer suffer local recurrence after the current maximal safe doses of 65–75 cGy (lung corrected dose) (Yorke *et al.*, 2002; Marks, 2002). A recent analysis has indicated that doses of 90 Gy (lung corrected dose) are necessary to control the majority of unresectable non-small cell carcinomas of the lung (Kotton *et al.*, 2004). The data is consistent with the current 30–50 per cent local recurrence rate of non-small cell lung cancer even in the current era of effective chemoradiotherapy utilizing agents such as taxol and carboplatin (Bjhatnagar *et al.*, 2002; Flanders *et al.*, 2002).

A significant number of patients present with locally recurrent lung cancer, after previous 'maximum tolerance dose' irradiation (Roof *et al.*, 2003). Depending upon the time since previous radiotherapy, a second course of radiotherapy is usually limited to 50 Gy (lung corrected dose) and while various formulas have been generated to suggest that normal tissues 'forget' previous irradiation dose, there is no consensus among radiation oncologists or radiobiologists that a time interval is ever reached when a safe retreatment programme can approximate the dose and fractionation scheme utilized in the first treatment programme (Roof *et al.*, 2003). Therefore, most patients with local recurrence cannot be treated with a curative intent. Pulmonary irradiation damage is 'remembered' by normal tissue and limits effective retreatment (Carpenter *et al.*, 2005). Regrettably, patients who suffer local recurrence usually cannot be controlled by radiotherapy retreatment (Goto *et al.*, 1987; van Rongen *et al.*, 1993; Socinski *et al.*, 2001; Flanders *et al.*, 2002; Rosenman *et al.*, 2002; Wang *et al.*, 2002; Emami *et al.*, 2003; Gopal *et al* 2003b; Roof *et al.*, 2003; Scagliotti *et al.*, 2003; Wu *et al.*, 2003; Frazier *et al.*, 2004; Phillips *et al.*, 2004; Rakovitch *et al.*, 2004; Ramirez *et al.*, 2004). These two problems represent a currently unsolvable conundrum, which demands critical analysis and construction of new approaches to a rational solution (Wu *et al.*, 2004).

21.3.2 MnSOD-PL interrupts several steps in the mechanism of irradiation pulmonary damage

The mechanism of ionizing irradiation damage to the lung is not fully understood. The acute inflammatory reaction of swelling of endothelial cells, alveolar exudate and transudate, is dependent upon total dose, fraction size, and volume treated (Rubin and Casarett, 1968; Hall, 1999; Marks, 2002; Roof *et al.*, 2003; Wu VWC *et al.*, 2004). Those patients who can be effectively managed and recover from the acute injury by receiving non-steroidal anti-inflammatory agents and/or appropriate treatment breaks, recover lung function as well as normal histopathological evidence of pulmonary architecture (Rubin and Casarett, 1968). The mechanism of late irradiation fibrosis, which is predictable at 110–120 days in the mouse model (Franko and Sharplin, 1994; Dileto and Travis, 1996), and 6 months–2 years in humans (Rubin and Casarett, 1968; Hall, 1999), is unknown. The mechanism of the late effects of radiation damage to the lung, as in other organs, remains one of the great mysteries in radiation biology, the solution to which represents a great challenge for both clinical radiotherapy and basic radiation biology (Dileto and Travis, 1996; Franko and Sharplin, 1994; Epperley *et al.*, 1998, 1999a,b; Zwacka *et al.*, 1998; Hall, 1999).

Much information about late ionizing irradiation-induced pulmonary damage has been collected (Franko and Sharplin, 1994; Dileto and Travis, 1996). Macrophage and fibroblast migration into the lung from bone marrow origin and through the circulation, has been established as a mechanism of proliferating lesions in pulmonary fibrosis (Epperley *et al.*, 2003b). The bone marrow origin of fibroblast progenitor cells has also been demonstrated in bleomycin-induced pulmonary fibrosis (Ortiz *et al.*, 2003; Hashimoto *et al.*, 2004). Modulation or reversibility of the late lesion has been suggested by dependence of the late pulmonary lesion upon elevation of inflammatory cytokines, TGF-β, TNF-α, IL-1, and others at the time of the proliferating fibrotic response (Epperley *et al.*, 1999b, 2002b). Much evidence suggests that TGF-β elevation is critical to the initiation and/or maintenance of fibrosis (Feister *et al.*, 1988; Harrison *et al.*, 1988; Hu *et al.*, 2003; Masszi *et al.*, 2004). The molecular sequence of events in TGF-β signalling have shown a critical role of the Smad3 pathway in fibroblast responses to TGF-β as distinct from that in epithelial cells (Wilde *et al.*, 1983; Rosen and Gordon, 1987; Rosen, 1990; Bhowmick *et al.*, 2003). Administration of agents which modulate inflammatory cytokine production (Kroemer *et al.*, 1998; Epperley *et al.*, 1999b; Suzuki *et al.*, 2004) decrease the magnitude and/or slows fibrosis. The origin of the cells producing inflammatory cytokines in the lung, after a long latent period during which no such elevation is detected, remains unknown (Bhowmick *et al.*, 2003; Demicourt and Dowdy, 2003; Leask and Abraham, 2004; Ma *et al.*, 2004). The rapid proliferation of fibrotic areas in the lungs after a long latent period from irradiation first suggested an extrapulmonary origin of the proliferating fibroblasts, and led to experiments documenting the bone marrow/circulatory origin of fibroblast progenitors (Epperley *et al.*, 2003b). The cellular and molecular signals from the lung which elicit fibroblast migration from the bone marrow and how ionizing irradiation many months previously initiates these events remain unknown.

The dose and volume threshold for induction of pulmonary fibrosis in re-irradiation experiments appears to increase as the latent period between the first irradiation treatment course and second course increases (Novakova-Jiresova *et al.*, 2004). As in other tissues, lung tissue appears to 'forget' irradiation injury (Novakova-Jiresova *et al.*, 2004; Carpenter *et al.*, 2005), but not totally. Understanding the mechanism of lung recovery from irradiation might provide knowledge of both cellular and molecular targets, which could then be modulated by pharmacologic approaches or even normal stem cell repopulation of the irradiated organ (Engelhardt, 2001; Herzog *et al.*, 2003; Kotton *et al.*, 2004). New interventions are required to facilitate both dose escalation for initial radiotherapy, and if necessary, more effective re-treatment dose (Wu VWC *et al.*, 2004).

21.3.3 MnSOD-PL intrapulmonary delivery effects in irradiation-induced endothelial cell interactions with circulating marrow origin macrophages and fibroblast progenitors

Pulmonary endothelial cells have been demonstrated to upregulate vascular cell adhesion molecule-1 (VCAM-1) and intracellular adhesion molecule-1 (ICAM-1) but not E-selectin, P-selectin, or L-selectin at 120–140 days after irradiation in the C57BL/6J mouse model (Epperley *et al.*, 1999b, 2002b). Selective knockout of either VCAM-1 or ICAM-1 does not abrogate late irradiation fibrosis (Epperley *et al.*, 2004c); however, experiments with dual knockout of both adhesion molecules have not been reported. Accumulation of macrophages in the lung shortly preceding accumulation and proliferation of fibroblasts has been documented in pulmonary irradiation fibrosis/organizing alveolitis (Epperley *et al.*, 2003b). In another model of fibrosis, T-cell and B-cell infiltrates preceded macrophage and fibroblast accumulation in the lungs (Epperley *et al.*, 2005b). Macrophage L-selectin homozygous deletion recombinant negative (knockout) mice demonstrated decreased irradiation-induced pulmonary organizing alveolitis (Epperley *et al.*, 2004c) suggesting that specific macrophage surface receptors that facilitate homing to endothelial cells might be involved in initiation of the late lesion and provide further evidence for linkage of endothelial cells with macrophages (Zheng *et al.*, 2000; Bompais *et al.*, 2004; Kaminski *et al.*, 2004; Takahashi *et al.*, 2004). Recent BuDR labelling studies of lung at the time of initiation and progression of fibrosis demonstrated no significant proliferation of endothelial cells or bone-marrow derived macrophages, but significant proliferation of bone marrow origin fibroblasts (Epperley *et al.*, 2003b). Induction of irradiation fibrosis/organizing alveolitis in bone marrow chimeric mice demonstrated no significant migration of bone marrow origin endothelial cells to the lungs (Epperley *et al.*, 2003b). These data suggest that resident pulmonary endothelial cells upregulate adhesion molecules and elicit the migration of both macrophages and fibroblasts, the latter of which proliferate to form the fibrotic lesion (Haribabu *et al.*, 1999; Bompais *et al.*, 2004; Hashimoto *et al.*, 2004a; Imhof and Aurrand-Lions, 2004; Kaminski *et al.*, 2004; Takahashi *et al.*, 2004).

The mechanism by which pulmonary resident endothelial cells upregulate cell surface adhesion molecules, many months after irradiation is unknown. Thrombomodulin accumulation in endothelial cells may be an initiating factor (Folz *et al.*, 1999). The absence of detectably increased BuDR uptake in these endothelial cells (Epperley *et al.*, 2002b) contradicts one prevailing model that the mechanism of the late lesion is due to delayed response of a 'slowly proliferating' cell population within the organ (Hall, 1999). Molecular signals by which a 'threshold' is reached to initiate cell surface adhesion molecule upregulation in endothelial cells, and initiate late fibrosis, are unknown (Leask and Abraham, 2004). The role of T-cell and B-cell infiltrates potentially preceding macrophage accumulation in irradiation pulmonary fibrosis is unknown (Epperley *et al.*, 2005b).

An understanding of the mechanism of the late fibrosis/organizing alveolitis irradiation lesion of the lung might provide valuable information for therapeutic intervention (Franko and Sharplin, 1994; Dileto and Travis, 1996; Li AG *et al.*, 2004). The role of inflammatory cytokines, and inflammatory cells (macrophages) in initiation of late fibrosis, if understood with respect to irradiation pulmonary damage, might prove to be of value in understanding late changes in other irradiated organs (Rubin and Casarett, 1968). Common cell phenotypes involved in hyperbaric oxygen lung damage (Oury *et al.*, 2002), bleomycin-induced lung damage (Feister *et al.*, 1988; Ortiz *et al.*, 2003; Flanders *et al.*, 2003; Hashimoto *et al.*, 2004b), and ionizing irradiation-induced fibrosis/organizing alveolitis (Epperley *et al.*, 2003b) suggest that a common pathway may be involved. Idiopathic pulmonary fibrosis, inflammatory and fibrotic reactions in organ rejection following lung transplant (Selman and Pardo, 2002), as well as pulmonary responses to inhaled chemical toxins (Oury *et al.*, 2002) show common histopathologic and pathophysiological parameters (Kaufman *et al.*, 2001, 2004; Barbarin *et al.*, 2004; Upadhayay *et al.*, 2004; Mowery *et al.*, 2004; Perry *et al.*, 2004; Song *et al.*, 2004). The ability to control ionizing irradiation dose as well as time, volume, and fractionation make the radiation biology model of lung fibrosis an ideal way to dissect the steps involved in late pulmonary damage (Franko and Sharplin, 1994; Dileto and Travis, 1996). Identification of the appropriate targets should facilitate design of small molecule pharmacological agents which could prevent the late lesion (Springer *et al.*,

1978; Rees *et al.*, 2004). To date, no such target has been identified with respect to pulmonary ionizing irradiation-induced late damage.

21.4 MnSOD-PL gene therapy down-modulates marrow cell migration to the lungs

Administration of MnSOD-PL prior to irradiation significantly delays the onset of late pulmonary cytokine elevation (Epperley *et al.*, 1998, 1999a,b) and reduces marrow origin macrophage progenitor and stromal cell migration to form organizing alveolitis/fibrosis (Epperley *et al.*, 1998, 2003e). Administration of MnSOD-PL at delayed time points after irradiation was also effective in decreasing the damage response (Epperley *et al.*, 2003e). Persistent elevation of free radical production in the irradiated lung weeks to months after irradiation (Kroemer *et al.*, 1998) suggests that a radiation-injured cell population continues to express the damage response (although not clinically detectable as histopathological damage) long after irradiation (Gu Q *et al.*, 2004; Fakhrzadeh *et al.*, 2004; Bhowmick *et al.*, 2004; Nishioka *et al.*, 2004; Lu *et al.*, 2004).

There is a critical importance of macrophage cell surface receptors including the CD18 and VLA-4 pathway for adhesion to pulmonary endothelial cells in late pulmonary irradiation damage. Studies with a novel poly (I) poly (C) inducible cre-FAS conditional lethal mouse strain (Epperley *et al.*, 2005b), which develops pulmonary fibrosis at a known background level when subsets of T and B cells are removed, will provide a model by which to test the role of T and B cells in macrophage recruitment at the time of their known migration to the lungs (Epperley *et al.*, 2003b). These experiments should conclusively prove whether macrophages or T and B cells represent the source of the elevated TGF-β at the time of detection of organizing alveolitis (Epperley *et al.*, 1999b). We have carried out experiments with Smad3–/– (Flanders *et al.*, 2002, 2003; Suzuki *et al.*, 2004) bone marrow chimeric mice which demonstrate that pulmonary TGF-β elevation in irradiated lung less effectively induces pulmonary migration of Smad3–/– compared to Smad3+/+ fibroblasts (Bhowmick *et al.*, 2004; Du and Hannon, 2004; Lu *et al.*, 2004; Nishioka *et al.*, 2004; Oslund *et al.*, 2004). Reconstituting clonal Smad3–/– bone marrow fibroblast cell lines with the transduced Smad3 transgene has

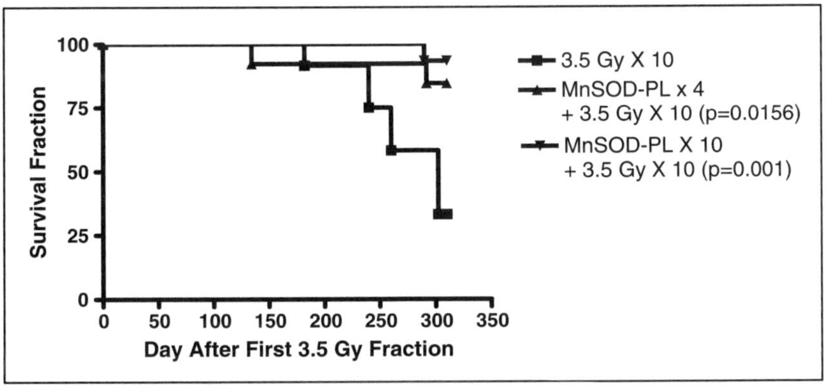

Figure 21.6 Survival after fractionated irradiation to lungs and inhalation gene therapy with MnSOD. C57BL/6NHsd mice inhaled MnSOD-PL (200 μg plasmid DNA) either Monday and Wednesday mornings for 2 weeks (four doses) or Monday through Friday mornings for 2 weeks or daily Monday through Friday for 2 weeks (10 doses). The MnSOD-PL pre-treated as well as control mice were irradiated in the afternoons to 3.5 Gy to the pulmonary cavity Monday through Friday for 2 weeks (10 fractions) and followed for the development of pulmonary organizing alveolitis/fibrosis

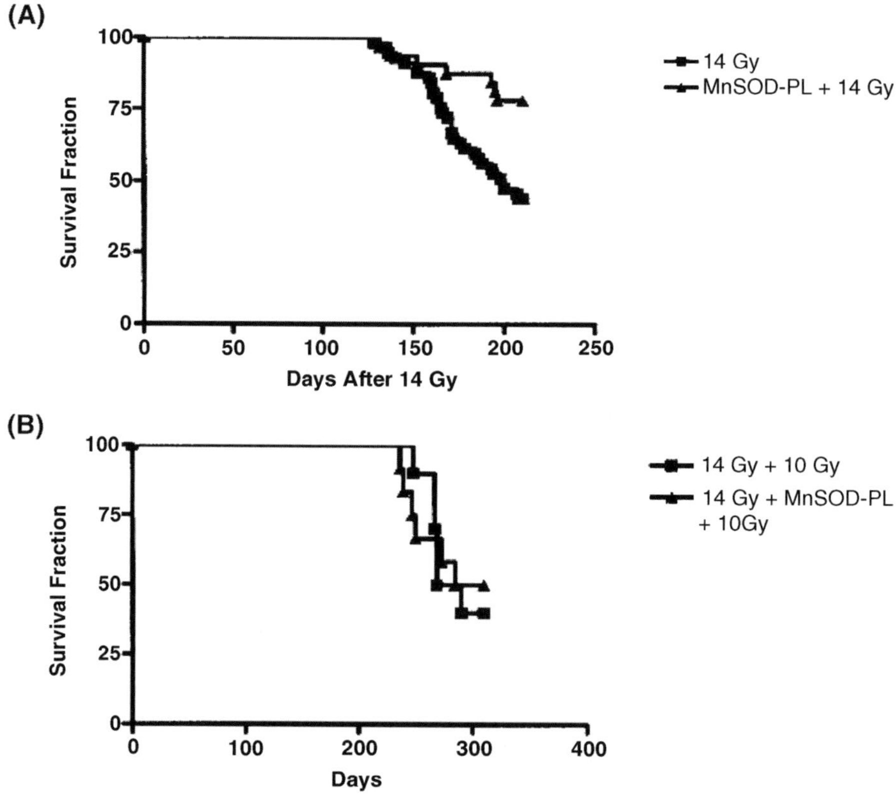

Figure 21.7 Survival of mice 'retreated' to second course of 10 Gy lung irradiation 200 days after firstcourse of 14 Gy. MnSOD-PL treatment before second course increased survival (experiment in progress). C57BL/6J mice were injected intratracheally with MnSOD-PL (250 μg plasmid DNA) and irradiated 24 h later along with control mice to 14 Gy to the pulmonary cavity. A) Mice injected with MnSOD-PL had an increased survival at day 200 compared to the control irradiated mice ($P = 0.0030$) (half of these groups are still being followed). A subgroup of each was then reirradiated. One subgroup was injected intratracheally with MnSOD-PL 250 μg plasmid DNA). Both subgroups were reirradiated at day 210 (each adjusted to 100 per cent). Mice were followed for the development of premorbid fibrosis at which time the mice were sacrificed (experiment in progress)

been achieved by us and reverses several newly discovered *in vitro* biological changes mediated by the Smad3–/– genotype (including increased saturation density growth and constitutive adipocytogenesis). These cellular resources facilitate experiments to prove the critical role of TGF-β signalling in irradiation lung fibrosis. We have developed a model of retreatment irradiation of the mouse lung and demonstrated an effective protocol of fractionated inhalation of MnSOD-PL during fractionated radiotherapy (Wandzioch *et al.*, 2004). We have quantitated the effects of MnSOD-PL on macrophages and stromal cells in a clinically relevant therapy model (Epperley *et al.*, 2003b). Inhalation gene therapy prolongs survival from lung irradiation (Figure 21.6) and decreases the toxicity of lung reirradiation treatment (Figure 21.7).

21.5 MnSOD-PL systemic administration for radiation protection from TBI

We have clear evidence of TBI protection. Intravenous administration of MnSOD-PL prior to 9.5

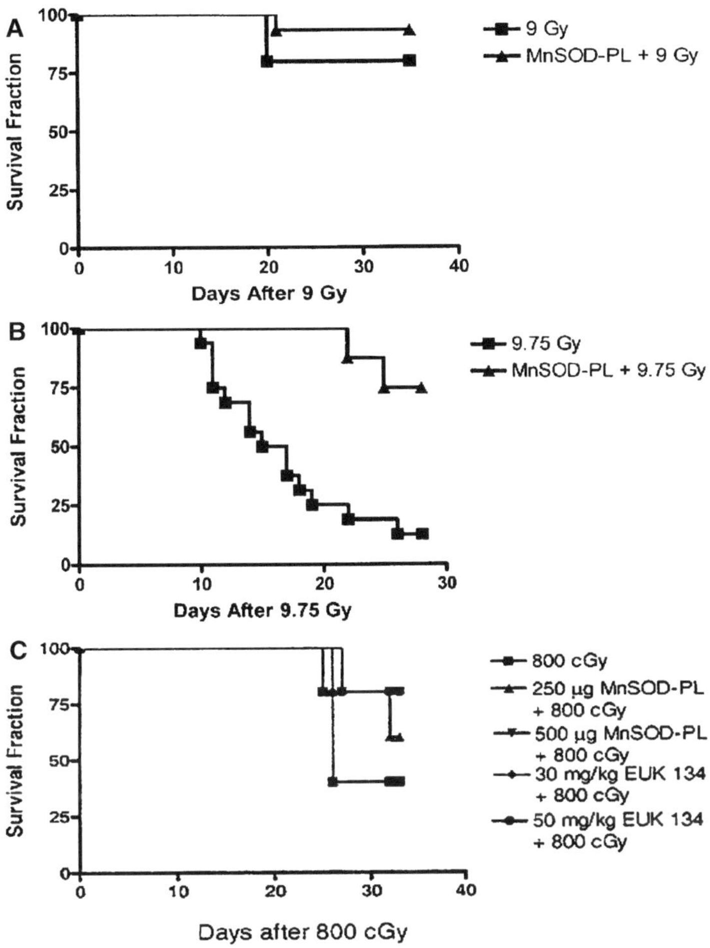

Figure 21.8 MnSOD-PL systemically I.V. (A) 9 Gy, (B) 9.75 Gy or (C) compared to EUK-134. (C) MnSOD-PL systemic i.v. injection protects C57BL/6J adult female mice against WBI LD50/30 800 cGy. Groups of 20 mice received i.v., no treatment or MnSOD-PL 24 h before irradiation. C57BL/6J female mice were injected i.v. with either 250 or 500 µg MnSOD-PL 24 h before irradiation or intraperitoneally with EUK-134 (30 or 50 mg/kg) 1 hour before irradiation. Control mice as well as groups injected with MnSOD-PL or EUK received WBI to 800 cGy. The mice were followed for survival. The mice receiving MnSOD-PL at both doses and the mice receiving EUK-134 had increased survival compared to irradiated control mice

Gy TBI increased the survival of both female and male C57BL/6J mice (Figure 21.8).

21.6 Summary and future directions

The success of MnSOD-PL radioprotective gene therapy in animal model systems is being translated to the clinic for oesophagus radioprotection in patients with non-small cell carcinoma of the lung who receive carboplatin and taxol and 7 weeks' external beam radiotherapy for the treatment of unresectable lung cancers. Twice weekly delivered oral/swallowed MnSOD-PL is being evaluated in a three tiered dose escalation protocol, phase I/II study. The development of inhalation gene therapy models for lung protection should next be translated to the clinic with nebulizer delivery of twice weekly inhaled MnSOD-PL. This approach may be of potential benefit in minimizing lung toxicity of patients with unresectable carcinoma of the oesophagus or those with lung cancer who may require significant volumes of normal lung in the transit volume. The paradoxical and beneficial radiosensitizing effect of MnSOD-PL in the SOD-poor or depleted and thus altered redox environment of tumours allows studies of intravenous administration of MnSOD-PL for potential simultaneous radiosensitization of tumour and protection of normal tissue. These experiments are in progress in orthotopic models of carcinomas of the floor of the mouth in the mouse model.

Systemic i.v. administration of MnSOD-PL has provided significant protection from total body irradiation of the mouse using the LD75/30 and LD25/30 doses. Small molecule analogues of MnSOD are being evaluated for systemic delivery as these may provide a more safe and effective as well as rapidly delivered alternative to MnSOD. However, in the event that small molecules are not as effective as is delivery of the MnSOD-transgene (perhaps attributable to the requirement for nuclear synthesis and transport to the mitochondria intracellularly) gene-gun delivery of liposome particles to the skin or patch delivery technique could be a method by which to deliver systemic doses of MnSOD-PL for total body radioprotection. Gastric juice resistant oral MnSOD-PL for intratracheal administration might be another approach to develop. Such strategies would be of value in protecting more tissues during TBI for marrow transplant and/or in the counter terrorism arena, for protecting first responders during a nuclear accident or willful terrorist event using radioactive materials. Combination of MnSOD-PL with other transgene products acting as potentially synergistic steps within the redox balance system (catalase, glutathione peroxidase, or in addition of NOS a major component of peroxynitrite) are being evaluated for potential synergy with MnSOD-PL gene therapy for radioprotection of both specific organs and total body protection.

References

Abedi M, Greer DA, Foster BM, *et al.* (2005). Critical variables in the conversion of marrow cells to skeletal muscle. *Blood* **106**(4), 1488–1495.

Aebersold DM, Burri P, Beer KT, *et al.* (2001). Expression of hypoxia-inducible factor-1α: A novel predictive and prognostic parameter in the radiotherapy of oropharyngeal cancer. *Cancer Res* **61**, 2911–2916.

Agarwal A, Perry Y, Epperly MW, Luketich JD, Greenberger JS (2004). Prevention of photodynamic therapy (PDT)-induced esophageal stricture by overexpression of manganese superoxide dismutase (MnSOD). ASTRO, Atlanta, GA, October, 2004. *Int J Radiat Oncol Biol Phys* **60**(1), Suppl. #1, #2028:5355.

Ahn SJ, Kahn D, Zhou S, *et al.* (2005). Dosimetric and clinical predictors for radiation-induced esophageal injury. *Int J Radiat Oncol Biol Phys* **61**(2), 335–347.

Alsner J, Sorensen SB, Overgaard J (2001). TP53 mutation is related to poor prognosis after radiotherapy, but not surgery, in SCC of the head and neck. *Radiother Oncol* **59**, 179–185.

Alyea EP, Kim HT, Ho V, *et al.* (2005). Comparative outcome of nonmyeloablative and myeloablative allogeneic hematopoietic cell transplantation for patients older than 50 years of age. *Blood* **105**(4), 1810–1814.

Amstad P, Peskin A, Shah G, *et al.* (1991). The balance between Cu, Zn-superoxide dismutase and catalase affects the sensitivity of mouse epidermal cells to oxidative stress. *Biochemistry* **30**, 9305–9313.

Amstad P, Moret R, Cerutti P (1994). Glutathione peroxidase compensates for the hypersensitivity of Cu,Zn-superoxide dismutase overproducers to oxidant stress. *J Biol Chem* **269**(3), 1606–1609.

Anklesaria P, FitzGerald TJ, Kase K, Ohara A, Bentley S, Greenberger JS (1989). Improved hematopoiesis in anemic $S1/S1^d$ mice by therapeutic transplantation of a hematopoietic microenvironment. *Blood* **74**, 1144–1152.

Anklesaria P, Kase KR, Glowacki J, *et al.* (1987). Engraftment of a clonal bone marrow stromal cell line in vivo stimulates hematopoietic recovery from total body irradiation. *Proc Natl Acad Sci U S A* **84**, 7681–7685.

Astreinidou E, Dehnad H, Terhaard CHJ, Raaijmakers CPJ (2004). Level II lymph nodes and radiation-induced xerostomia. *Int J Radiat Oncol Biol Phys* **58**(1), 124–131.

Bahri S, Flickinger JC, Kalend A, *et al.* (1999). Results of multifield conformal radiation therapy of non-small cell lung carcinoma using multileaf collimation beams. *Radiat Oncol Invest Clin Basic Res* **7**, 297–308.

Bakkenist CJ, Kastan MB (2003). DNA damage activates ATM through intermolecular autophosphorylation and dimer association. *Nature* **421**(30), 499–504.

Barbarin V, Arras M, Misson P, *et al.* (2004). Characterization of the effect of interleukin-10 on silica-induced lung fibrosis in mice. *Am J Respir Cell Mol Biol* **31**, 78–85.

Bauer TR, Gu Y, Tuschong LM, *et al.* (2005). Nonmyeloablative hematopoietic stem cell transplantation corrects the disease phenotype in the canine model of leukocyte adhesion deficiency. *Exp Hematol* **33**, 706–712.

Beckman BS, Balin AK, Allen RG (1989). Superoxide dismutase induces differentiation of erythroleukemia cells. *J Cell Physiol* **139**, 370–376.

Bevers EM, Comfurius P, Dekkers DWC, Swaal RFA (1999). Lipid translocation across the plasma membrane of mammalian cells. *Biochim Biophys Acta* **1439**, 317–330.

Bhatnagar A, Flickinger JC, Bahri S, *et al.* (2002). Update on results of multi-field conformal radiation therapy of nonsmall-cell lung carcinomas using multileaf collimated beams. *Clin Lung Cancer* **3**(4), 259–264.

Bhowmick NA, Ghiassi M, Aakre M, Brown K, Singh V, Moses HL (2003). TGF-beta-induced RhoA and p160ROCK activation is involved in the inhibition of Cdc25A with resultant cell-cycle arrest. *Proc Natl Acad Sci U S A* **100**(26), 15548–15553.

Bhowmick NA, Chytil A, Plieth D, *et al.* (2004). TGF-beta signaling in fibroblasts modulates the oncogenic potential of adjacent epithelia. *Science* **303**(5659), 848–851.

Blomberg M, Rao S, Reilly J, *et al.* (1998). Repetitive bone marrow transplantation in nonmyeloablated recipients. *Exp Hematol* **26**, 320–324.

Bompais H, Chagraoui J, Canron X, *et al.* (2004). Human endothelial cells derived from circulating progenitors display specific functional properties compared with mature vessel wall endothelial cells. *Blood* **103**(7), 2577–2584.

Bonniaud P, Martin G, Margetts PJ, *et al.* (2004). Connective tissue growth factor is crucial to inducing a profibrotic environment in 'fibrosis resistant' Balb/c mouse lungs. *AJRCMB* **31**, 510–516.

Bravard A, Sabatier L, Hoffschir F, Luccioni C, Dutrillaux B (1992). SOD2: a new type of tumor suppressor gene? *Int J Cancer* **51**, 475–480.

Brusa G, Benvenuti M, Mazzacurati L, *et al.* (2003). p53 loss of function enhances genomic instability and accelerates clonal evolution of murine myeloid progenitors expressing the p210BCR-ABL tyrosine kinase. *Haematol/J Hematol* **88**(6), 628–635.

Campigilio M, Lopcatelli A, Olgati C, *et al.* (2004). Inhibition of proliferation and induction of apoptosis in breast cancer cells by the epidermal growth factor receptor (EGFR) tyrosine kinase inhibitor ZD1839 ('Iressa') is independent of EGFR expression level. *J Cell Physiol* **198**(2), 259–268.

Camargo FD, Finegold M, Goodell MA (2004). Hematopoietic myelomonocytic cells are the major source of hepatocyte fusion partners. *J Clin Invest* **113**(9), 1266–1272.

Cao S, Greenberger E, Epperly MW, Kanai AJ, Greenberger JS (2005). Absence of nNOS increases longevity of long term bone marrow cultures and radiation resistance. ASH 47th Annual Meeting, Atlanta, Georgia, December 10–13.

Caron E, Self AJm Hall A (2000). The GTPase Rap1 controls functional activation of macrophage integrin alphaMbeta2 by LPS and other inflammatory mediators. *Curr Biol* **10**(16), 974–978.

Carrasco YR, Fleire SJ, Cameron T, Dustin ML, Batista FD (2004). LFA-1/ICAM-1 interaction lowers the threshold of B cell activation by facilitating B cell adhesion and synapse formation. *Immunity* **20**(5), 589–599.

Carpenter M, Epperly MW, Agarwal A, *et al.* (2005). Inhalation delivery of manganese superoxide dismutase-plasmid/liposomes (MnSOD-PL) protects the murine lung from irradiation damage. *Gene Ther* **12**, 685–693.

Carter S, Auer KL, Reardon DB, *et al.* (1998). Inhibition of the mitogen activated protein (MAP) kinase cascade potentiates cell killing by low dose ionizing irradiation in A431 human squamous carcinoma cells. *Oncogene* **16**, 2787–2796.

Ceriello A, dello Russo P, Amstad P, Cerutti P (1996). High glucose induces antioxidant enzymes in human endothelial cells in culture. Evidence linking hyperglycemia and oxidative stress. *Diabetes* **45**(4), 471–477.

Chao N F, Koh L-P, Long GD, *et al.* (2004). Adult recipients of umbilical cord blood transplants after nonmyeloablative preparative regimens. Biol Blood Marrow Transpl **10**, 569–575.

Chapet O, Thomas E, Kessler ML, Fraass BA, Ten Haken RK (2005). Esophagus sparing with IMRT in lung tumor irradiation: and EUD-based optimization technique. *Int J Radiat Oncol Biol Phys* **63**(1), 179–187.

Ch'ang H-J, Maj JG, Paris F, *et al.* (2005). ATM regulates target switching to escalating doses of radiation in the intestines. *Nat Med* **11**(5), 484–490.

Charrier S, Michaud A, Badaoui S, *et al.* (2004). Inhibition of angiotensin I-converting enzyme induces radioprotection by preserving murine hematopoietic short-term reconstituting cells. *Blood* **104**(4), 978–983.

Cheng T, Rodrigues N, Shen H, *et al.* (2000). Hematopoietic stem cell quiescence maintained by $p21^{cip1/waf1}$. *Science* **10**(287), 1804–1808.

Chuang K-P, Huang Y-F, Hsu Y-L, Liu H-S, Chen H-C, Shieh C-C (2004). Ligation of lymphocyte function-associated antigen-1 on monocytes decreases very late antigen-4-mediated adhesion through a reactive oxygen species-dependent pathway. *Blood* **104**(13), 4046–4051.

Church SL, Grant JW, Ridnour LA, *et al.* (1993). Increased manganese superoxide dismutase expression suppresses the malignant phenotype of human melanoma cells. *Proc Natl Acad Sci U S A*, **90**, 3113–3117.

Cohen EEW, Rosen F, Stadler WM, *et al.* (2003). Phase II trial of ZD1839 in recurrent or metastatic squamous cell carcinoma of the head and neck. *J Clin Oncol* **21** (10), 1980–1987.

Collett D (1994). *Modeling Survival Data in Medical Research.* Chapman & Hall, London.

Damron TA, Margulies B, Biskup D, Spadaro JA (2001). WR2721 before fractionated irradiation protects bone growth in rats better than fractionation alone. *Int J Radiat Oncol Biol Phys* **50**(1), 479–483.

Correll PH, Morrison AC, Lutz MA (2004). Receptor tyrosine kinases and the regulation of macrophage activation. *J Leukoc Biol* **75**(5), 731–737.

Dai Q, Liu J, Chen J, Durrant D, McIntyre TM, Lee RM (2004). Mitochondrial ceramide increases in UV-irradiated HeLa cells and is mainly derived from hydrolysis of sphingomyelin. *Oncogene* **23**, 3650–3658.

Daniels CE, Wilkes MC, Edens M, *et al.* (2004). Imatinib mesylate inhibits the profibrgenic activity of TGF-B and prevents bleomycin-mediated lung fibrosis. *J Clin Invest* **114**(9), 1308–1311.

Daum G (1985). Lipids of mitochondria. *Biochim Biophys Acta* **822**, 1–42.

Dawson LA, Myers LL, Bradford CR, *et al.* (2001). Conformal re-irradiation of recurrent and new primary head and neck cancer. *Int J Radiat Oncol Biol Phys* **50** (2), 377–385.

Delanian S, Baillet F, Huart J, Lefaix JL, Maulard C, Housset M (1994). Successful treatment of radiation-induced fibrosis using liposomal Cu/Zn superoxide dismutase: clinical trial. *Radiother Oncol* **32**(1), 12–20.

Dent P, Reardon DB, Park JS, *et al.* (1999). Radiation-induced release of transforming growth factor α activates the epidermal growth factor receptor and mitogen-activated protein kinase pathway in carcinoma cells, leading to increased proliferation and protection from radiation-induced cell death. *Mol Biol Cell* **10**, 2493–2506.

Denicourt C, Dowdy SF (2003). Another twist in the transforming growth factor beta-induced cell-cycle arrest chronicle. *Proc Natl Acad Sci U S A* **100**(26), 15290–12591.

Dezawa M, Ishikawa H, Itokazu Y, *et al.* (2005). Bone marrow stromal cells generated muscle cells and repair muscle degeneration. *Science* **309**, 314–318.

Dileto CL, Travis EL (1996). Fibroblast radiosensitivity *in vitro* and lung fibrosis *in vivo*: comparison between a fibrosis-prone and fibrosis-resistant mouse strain. *Radiat Res* **146**(1), 61–67.

Eckner RJ, Hettrick KL, Greenberger JS, Bennett M (1982). Extended self-renewal capacity of pluripotential hematopoietic stem cells: association with persistent Friend spleen focus forming virus. *Cell* **31**, 731–738.

Emami B, Mirkovic N, Scott C, *et al.* (2003). The impact of regional nodal radiotherapy (dose/volume) on regional progression and survival in unresectable non-small cell lung cancer: an analysis of RTOG data. *Lung Cancer* **41**(2), 207–214.

Emmanuouilides C, Spielberger R, Stiff P, *et al.* (2003). Palifermin treatment of mucositis in transplant patients reduces health resource use: Phase 3 results. *Blood* **102**(11), 251a.

Engelhardt JF (2001). Stem cell niches in the mouse airway. *Am J Respir Cell Mol Biol* **24**(6), 649–52.

Epperly, MW, Bray JA, Kraeger S, *et al.* (1998). Prevention of late effects of irradiation lung damage by manganese superoxide dismutase gene therapy. *Gene Ther* **5**, 196–208.

Epperly MW, Bray JA, Esocobar P, Bigbee WL, Watkins S, Greenberger JS (1999a). Overexpression of the human MnSOD transgene protects athymic nude mice from irradiation-induced organizing alveolitis. *Int J Radiat Oncol Phys* **43**(1), 169–181.

Epperly MW, Travis EL, Sikora C, Greenberger JS (1999b). Magnesium superoxide dismutase (MnSOD) plasmid/liposome pulmonary radioprotective gene therapy: Modulation of irradiation-induced mRNA for IL-1, TNF-α, and TGF-β correlates with delay of organizing alveolitis/fibrosis. *Biol Blood Marrow Transplant* **5**, 204–214.

Epperly MW, Bray JA, Krager S, *et al.* (1999c). Intratracheal injection of adenovirus containing the human MnSOD transgene in subclones of murine hematopoietic progenitor cell line 32D cl 3 decreases irradiation-induced apoptosis but does not alter G2/M or G1/S phase cell cycle arrest. *Radiat Oncol Invest Clin Basic Res* **7**, 331–342.

Epperly MW, Sikora C, Defilippi S, *et al.* (2000a). Plasmid/liposome transfer of the human manganese superoxide dismutase (MnSOD) transgene prevents ionizing irradiation-induced apoptosis in human esophagus organ explant culture. *Int J Cancer (Radiat Oncol Invest)* **90**(3), 128–137.

Epperly MW, Defilippi S, Sikora C, Gretton J, Kalend K, Greenberger JS (2000b). Intratracheal injection of manganese superoxide dismutase (MnSOD) plasmid/liposomes protects normal lung but not orthotopic tumors from irradiation. *Gene Ther* **7**(12),1011–1018.

Epperly MW, Epstein CJ, Travis EL, Greenberger JS (2000c). Decreased pulmonary radiation resistance of manganese superoxide dismutase (MnSOD)-deficient mice is corrected by human manganese Superoxide dismutase-plasmid/liposome (SOD2-PL) intratracheal gene therapy. *Radiat Res* **154**(4), 365–374.

Epperly MW, Gretton JA, DeFilippi SJ, *et al.* (2001a). Modulation of radiation-induced cytokine elevation associated with esophagitis and esophageal stricture by manganese superoxide dismutase-plasmid/liposome (SOD-PL) gene therapy. *Radiat Res* **155**, 2–14.

Epperly MW, Travis EL, Whitsett JA, Epstein CJ, Greenberger JS (2001b). Overexpression of manganese superoxide dismutase (MnSOD) in whole lung or alveolar type II (AT-II) cells of MnSOD transgenic mice does not provide intrinsic lung irradiation protection. *Radiat Oncol Invest* **96**, 11–21.

Epperly MW, Sikora CA, Shields D, Goff JP, Greenberger JS (2001c). Bone marrow origin of esophageal stem cells. The 30th Annual Meeting of the International Society for Experimental Hematology (ISEH), Tokyo, Japan. August 25, 28.

Epperly MW, Kagan VE, Sikora CA, *et al.* (2001d). Manganese superoxide dismutase-plasmid/liposome (MnSOD-PL) administration protects mice from esophagitis associated with fractionated irradiation. *Int J Cancer (Radiat Oncol Invest)* **6**, 221–231.

Epperly MW, Defilippi S, Sikora C, Gretton J, Greenberger JS (2002). Radioprotection of lung and esophagus by overexpression of the human manganese superoxide dismutase transgene. *Milit Med* **167**(1), 071.

Epperly MW, Sikora C, Defilippi S, *et al.* (2002a). Manganese superoxide dismutase (SOD2) inhibits radiation-induced apoptosis by stabilization of the mitochondrial membrane. *Radiat Res* **157**(5), 568–577.

Epperly MW, Sikora CA, DeFilippi SJ, *et al.* (2002b). Pulmonary irradiation-induced expression of VCAM-1 and ICAM-1 is decreased by MnSOD-PL gene therapy. *Biol Blood Bone Marrow Transplant* **8**(4), 175–187.

Epperly MW, Guo HL, Jefferson M, *et al.* (2003). Cell phenotype specific duration of expression of epitope-tagged HA-MnSOD in cells of the murine lung following intratracheal plasmid liposome gene therapy. *Gene Ther* **10**, 163–171.

Epperly MW, Guo H, Jefferson M, *et al.* (2003a). Cell phenotype specific kinetics of expression of intratracheally injected manganese superoxide dismutase-plasmid/liposomes (MnSOD-PL) during lung radioprotective gene therapy. *Gene Ther* **10**, 163–171.

Epperly MW, Guo H, Gretton JE, Greenberger JS (2003b). Bone marrow origin of myofibroblasts in irradiation pulmonary fibrosis. *Am J Respir Cell Mol Biol* **29**, 213–224.

Epperly MW, Bernarding M, Gretton J, Jefferson M, Nie S, Greenberger JS (2003c). Overexpression of the transgene for manganese superoxide dismutase (MnSOD) in 32D cl 3 cells prevents apoptosis induction by TNF-α, IL-3 withdrawal and ionizing irradiation. *Exp Hematol* **31**(6), 465–474.

Epperly MW, Gretton JE, Bernarding M, Nie S, Rasul B, Greenberger JS (2003d). Mitochondrial localization of copper/zinc superoxide dismutase (Cu/ZnSOD) confers radioprotective functions *in vitro* and *in vivo*. *Radiat Res* **160**, 568–578.

Epperly MW, Guo HL, Bernarding M, Gretton J, Jefferson M, Greenberger JS (2003e). Delayed intratracheal injection of manganese superoxide dismutase (MnSOD)-plasmid/liposomes provides suboptimal protection against irradiation-induced pulmonary injury compared to treatment before irradiation. *Gene Ther Mol Biol* **7**, 61–68.

Epperly MW, Osipov AN, Martin I, *et al.* (2004a). Ascorbate as a 'redox-sensor' and protector against irradiation-induced oxidative stress in 32D cl 3 hematopoietic cells and subclones overexpressing human manganese superoxide dismutase. *Int J Radiat Oncol Biol Phys* **58**(3), 851–861.

Epperly MW, Guo H, Shen H, *et al.* (2004b). Bone marrow origin of cells with capacity for homing and differentiation to esophageal squamous epithelium. *Radiat Res* **162**, 233–240.

Epperly MW, Shen H, Jefferson M, Greenberger JS (2004c). *In vitro* differentiation capacity of esophageal progenitro cells with capacity for homing and repopulation of the ionizing irradiation-damaged esophagus. *In Vivo* **18**, 675–686.

Epperly MW, Hongliang G, Shields D, *et al.* (2004d). Correlation of ionizing irradiation-induced late pulmonary fibrosis with long-term bone marrow culture fibroblast progenitor cell biology in mice homozygous deletion recombinant negative for endothelial cell adhesion molecules. *In Vivo* **18**, 1–14.

Epperly MW, Carpenter M, Agarwal A, Mitra P, Nie S, Greenberger JS (2004e). Intraoral manganese superoxide dismutase-plasmid/liposome (MnSOD-PL) radioprotective gene therapy decreases ionizing irradiation-induced murine mucosal cell cycling and apoptosis. *In Vivo* **18**, 401–410.

Epperly MW, Fink M, Hricisak L, Zhang X, Greenberger JS (2005). Incubation of 32D cl 3 factor-dependent hematopoietic cells in ethyl pyruvate (EP) increases radiation resistance. Radiation Research Society 2005 Annual Meeting, Abstract #PP300, October 17.

Epperly MW, Tyurina Y, Nie S, Zhang X, Kagan V, Greenberger JS (2005a). MnSOD-plasmid liposome decreases ionizing irradiation-induced lipid peroxidation of the esophagus. *In Vivo* **19**, 997–1004.

Epperly MW, Cao S, Shields D, Greenberger JS (2005b). Increased longevity of hematopoiesis in continuous bone marrow cultures derived from SMAD 3–/– mice. *Exp Hematol* **33**(3), 353–362.

Eriksen JG, Steiniche T, Askaa J, Alsner J, Overgaard J (2004). The prognostic value of epidermal growth factor receptor is related to tumor differentiation and the overall treatment time of radiotherapy in squamous cell carcinomas of the head and neck. *Int J Radiat Oncol Biol Phys* **58**(2), 561–566.

Ertekin MV, Koc M, Karslioglu I, Sezen O (2004). Zinc sulfate in the prevention of radiation-induced oropharyngeal mucositis: A prospective, placebo-controlled, randomized study. *Int J Radiat Oncol Biol Phys* **58**(1), 167–174.

Fabisciak JP, Tyurina YY, Tyurin VA, Lazo JS, Kagan VE (1998). Random versus selective membrane phospholipid oxidation in apoptosis: role of phosphatidylserine. *Biochemistry* **37**, 13781–13790.

Fakhrzadeh L, Laskin JD, Gardner CR, Laskin DL (2004). Superoxide dismutase-overexpressing mice are resistant to ozone-induced tissue injury and increases in nitric oxide and tumor necrosis factor-alpha. *Am J Respir Cell Mol Biol* **30**(3), 280–7.

Feister AJ, Browder B, Willis HE, Mohanakumar T, Ruddy S (1988). Pertussis toxin inhibits human neutrophils responses mediated by the 42-kilodalton IgG Fc receptor. *J Immunol* **141**, 228–233.

Fernandez-Pol JA, Hamilton PD, Klos DJ (1982). Correlation between the loss of the transformed phenotype and an increase in superoxide dismutase activity in a revertant subclone of sarcoma virus-infected mammalian cells. *Cancer Res* **42**, 609–617.

Flanders KC, Sullivan CD, Fujii M, *et al.* (2002). Mice lacking Smad3 are protected against cutaneous injury induced by ionizing irradiation. *Am J Pathol* **160**(3), 1057–1068.

Flanders KC, Major CD, Arabshahi A, *et al.* (2003). Interference with transforming growth factor-beta/Smad 3 signaling results in accelerated healing of wounds in previously irradiated skin. *Am J Pathol* **163** (6), 2247–2257.

Folz RJ, Abushamaa AM, Suliman HB (1999). Extracellular superoxide dismutase in the airways of transgenic mice reduces inflammation and attenuates lung toxicity following hyperoxia. *J Clin Invest* **103**, 1055–1066.

Franko AJ, Sharplin J (1994). Development of fibrosis after lung irradiation in relation to inflammation and lung function in a mouse strain prone to fibrosis. *Radiat Res* **140**(3), 347–355.

Frazier RC, Vicini FA, Sharpe MB, *et al.* (2004). Impact of breathing motion on whole breast radiotherapy: a dosimetric analysis using active breathing control. *Int J Radiat Oncol Biol Phys* **58**(4), 1041–1047.

Friedrich G, Soriano P (1991). Promoter traps in embryonic stem cells: a genetic screen to identify and mutate developmental genes in mice. *Genes Dev* **5**, 1513–1523.

Fukuoka M, Yano S, Giaccone G, *et al.* Final results from a phase II trial of ZD1839 ('Iressa') for patients with advanced non-small-cell lung cancer (IDEAL-1). Proceedings of ASCO, Abstract #2629, 2002.

Gansauge S, Gansauge F, Gause H, Poch B, Schoenberg MH, Beger HG (1997). The induction of apoptosis in proliferating human fibroblasts by oxygen radicals is associated with a p53- and p21 induction. *FEBS Lett* **404**, 6–10.

Garcia-Barros M, Paris F, Cordon-Cardo C, *et al.* (2003). Tumor response to radiotherapy regulated by endothelial cell apoptosis. *Science* **300**, 1155–1157.

Gonzalez-Zulueta M, Ensz LM, Mukhina G, *et al.* (1998). Manganese superoxide dismutase protects nNOS neurons from NMDA and nitric oxide-mediated neurotoxicity. *J Neurosci* **18**, 2040–2055.

Gopal R, Starkschall G, Tucker SL, *et al.* (2003a). Effects of radiotherapy and chemotherapy on lung function in patients with non–small-cell lung cancer. *Int J Radiat Oncol Biol Phys* **56**(1), 114–120.

Gopal R, Tucker SL, Komaki R, Liao Z, Forster KM, Stevens C, Kelly JF, Starkschall G (2003b). The relationship between local dose and loss of function for irradiated lung. *Int J Radiat Oncol Biol Phys* **56**(1), 106–113.

Gorak E, Geller N, Srinivasan R, *et al.* (2005). Engraftment syndrome after nonmyeloblative allogeneic hematopoietic stem cell transplantation: incidence and effects on survival. *Biol Blood Marrow Transplant* **11**, 542–550.

Gorbunov NV, Pogue-Geile KL, Epperly MW, *et al.* (2000). Activation of the nitric oxide synthase 2 pathway in the response of bone marrow stromal cells to high doses of ionizing radiation. *Radiat Res* **154**, 73–86.

Goto T, Kino T, Hatanaka H, *et al.* (1987). Discovery of FK506, a novel immunosuppressant isolated from streptomyces. *J Am Chem Soc* **109**, 5031–5033.

Goto Y, Hogg JC, Whalen B, Shih CH, Ishii H, Van Eeden SF (2004). Monocyte recruitment into the lungs in pneumococcal pneumonia. *AJRCMB* **30**(5), 620–626.

Greenberger JS, Bahri S, Jett J, Belani C, Kalend A, Epperly MW (1998). Considerations in optimizing radiation therapy for non-small cell lung cancer. (International Symposium on Thoracic Malignancies). *Chest* **113**(1), 465–525.

Greenberger JS, Epperly MW. Pleiotrophic stem cell and tissue effects of ionizing irradiation protection by MnSOD-plasmid liposome gene therapy. In: Columbus F (ed.), *Progress in Gene Therapy.* Nova Science Publications.

Greenberger JS, Epperly MW, Gretton J, *et al.* (2003). Radioprotective gene therapy. *Curr Gene Ther* **3**, 183–195.

Greenberger JS, Epperly MW (2004). Radioprotective antioxidant gene therapy: potential mechanisms of action. *Gene Ther Mol Biol* **8**, 31–44.

Greenberger JS, Guo HL, Epperly MW (2003). Title: Role of adhesion molecules in irradiation-induced organizing alveolitis/fibrosis in C57BL/6J mice. ASTRO, September, 2003. *Int J Radiat Oncol Biol Phys* **57**(2), 5157.

Greenberger JS, Kagan VE, Pearce L, Boriseniao G, Epperly MW (2001). Modulation of redox signal transduction pathways in the treatment of cancer. *Antioxidants Redox Signaling* **3**(3), 347–359.

Gu Q, Wang D, Wang X, *et al.* (2004). Basic fibroblast growth factor inhibits radiation-induced apoptosis of HUVECs. II. The RAS/MAPK pathway and phosphorylation of BAD at serine 112. *Radiat Res* **161**(6), 703–711.

Guiffre A, Barone MC, Brunori M, *et al.* (2002). Nitric oxide reacts with the single-electron reduced active site of cytochrome c oxidase, *J Biol Chem* **277**, 22402–22406.

Guo H, Seixas-Silva J, Epperly MW, *et al.* (2003). Prevention of irradiation-induced oral cavity mucositis by plasmid/liposome delivery of the human manganese superoxide dismutase (MnSOD) transgene. *Radiat Res* **159**, 361–370.

Guo HL, Seixas-Silva JA, Epperly MW, Gretton JE, Shin DM, Greenberger JS (2003a). Prevention of irradiation-induced oral cavity mucositis by plasmid/liposome delivery of the human manganese superoxide dismutase (MnSOD) transgene. *Radiat Res* **159**, 361–370.

Guo HL, Epperly MW, Bernarding M, *et al.* (2003b). Manganese superoxide dismutase-plasmid/liposome (MnSOD-PL) intratracheal gene therapy reduction of irradiation-induced inflammatory cytokines does not protect orthotopic Lewis Lung carcinomas. *In Vivo* **17**, 13–22.

Hagan MP, Yacoub A, Dent P (2003). The switch from DNA repair to apoptosis: Discovery of a refractory period for radiation-induced EGFR-MAPK signaling following irradiation. *Int J Radiat Oncol Biol Phys* **57** (2), Suppl 1050.

Hall EJ (1999). *Radiobiology for the Radiologist*, 4th edition. JB Lippincott Inc., Philadelphia, PA.

Hanna E, Shrieve DC, Ratanatharathorn V, *et al.* (2001). A novel alternative approach for prediction of radiation response of SCC of the head and neck. *Cancer Res* **61**, 2376–2380.

Hao Z, Hampel B, Yagita H, Rajewsky K (2004). T cell-specific ablation of Fas leads to Fas ligand-mediated lymphocyte depletion and inflammatory pulmonary fibrosis. *J Exp Med* **199**, 1355–1365.

Hardman RA, Afshari CA, Barrett JC (2001). Involvement of mammalian MLH1 in the apoptotic response to peroxide-induced oxidative stress. *Cancer Res* **61**, 1392–1397.

Haribabu B, Zhelev DV, Pridgen BC, Richardson RM, Ali H, Snyderman R (1999). Chemoattractant receptors activate distinct pathways for chemotaxis and secretion. *J Biol Chem* **274**(52), 37087–37092.

Harrison DE, Astle CM, Lerner C (1988). Number and continuous proliferation pattern of transplanted primitive immunohematopoietic stem cells. *Proc Natl Acad Sci U S A* **85**, 822–826.

Hashimoto N, Jin H, Liu T, Chensue SW, Phan SH (2004a). Bone marrow-derived progenitor cells in pulmonary fibrosis. *J Clin Invest* **113**(2), 243–251.

Hashimoto N, Kawabe T, Imaizumi K, *et al.* K. Hasegawa Y (2004b). CD40 plays a crucial role in lipopolysaccharide-induced acute lung injury. *Am J Respir Cell Mol Biol* **30**(6), 808–815.

Henderson RB, Hobbs JA, Mathies M, Hogg N (2003). Rapid recruitment of inflammatory monocytes is independent of neutrophil migration. *Blood* **102**(1), 328–35.

Hendricks CA, Almeida KH, Stitt MS, *et al.* (2003). Spontaneous mitotic homologous recombination at an enhanced yellow fluorescent protein (EYFP) cDNA direct repeat in transgenic mice. *Proc Natl Acad Sci U S A* **100**(11), 6325–30.

Herzog EL, Chai L, Krause DS (2003). Plasticity of marrow-derived stem cells. *Blood* **102**(10):3483–3493.

Hochrein H, Shortman K, Vremec D, Scott B, Hertzog P, O'Keeffe M (2001). Differential production of IL-12, IFN-alpha, and IFN-gamma by mouse dendritic cell subsets. *J Immunol* **166**(9), 5448–5455.

Holt PG, Haining S, Nelson DJ, Sedgwick JD (1994). Origin and steady-state turnover of class II MHC-bearing dendritic cells in the epithelium of the conducting airways. *J Immunol* **153**(1), 256–61.

Hong KU, Reynolds SD, Watkins S, Fuchs E, Stripp BR (2004). Basal cells are a multipotent progenitor capable of renewing the bronchial epithelium. *Am J Pathol* **164**(2).

Horiot J-C, Lipinski F, Schraub S, *et al.* (2000). Post-radiation severe xerostomia relieved by pilocarpine: a prospective French cooperative study. *Radiother Oncol* **55**, 233–239.

Houghton J, Stoicov C, Nomura S, *et al.* (2004). Gastric cancer originating from bone marrow-derived cells. *Science* **306**(5701), 1568–71.

Hu B, Wu Z, Phan SH (2003). Smad3 mediates transforming growth factor-beta-induced alpha-smooth muscle actin expression. *Am J Respir Cell Mol Biol* **29**(3 Pt 1), 397–404.

Huang E-Y, Leung SW, Wang C-J, *et al.* (2000). Oral glutamine to alleviate radiation-induced oral mucositis: a pilot randomized trial. *Int J Radiat Oncol Biol Phys* **46**(3), 535–539.

Imhof BA, Dunon D (1995). Leukocyte migration and adhesion. *Adv Immunol* **58**, 345–416.

Imhof BA, Aurrand-Lions M (2004). Adhesion mechanisms regulating the migration of monocytes. *Nat Rev Immunol* **4**(6), 432–444.

Ishikawa F, Yasukawa M, Yoshida S, *et al.* (2004). Human cord blood and bone marrow derived CD34+ cells regenerate gastrointestinal epithelial cells. *FASEB J* **18**, 1958–1967.

Ito K, Hirao A, Arai F, *et al.* (2004). Regulation of oxidative stress by ATM is required for self-renewal of haematopoietic stem cells. *Nature* **431**, 947–1000.

Jackson KA, Mi T, Goodell MA (1999). Hematopoietic potential of stem cells isolated from murine skeletal muscle. *Proc Natl Acad Sci U S A* **96**(25), 14482–14486.

Jackson KA, Majka SM, Wang H, *et al.* (2001). Regeneration of ischemic cardiac muscle and vascular endothelium by adult stem cells. *J Clin Invest* **107**, 1395–1402.

Jang Y-Y, Collector MI, Baylin SB, Diehl AM, Sharkis SJ (2004). Hematopoietic stem cells convert into liver cells within days without fusion. *Nat Cell Biol* **6**(6), 532–537.

Jahnukainen K, Jahnukainen T, Salmi TT, Svechnikov K, Eksborg S, Soder O (2001). Amifostine protects against early but not late toxic effects of doxorubicin in infant rats. *Cancer Res* **61**, 6423–6427.

Jia J, Wang B, Wu A, Cheng G, Li Z, Dong S (2002). A method to construct a third-generation horseradish peroxidase biosensor: self-assembling gold nanoparticles to three-dimensional sol-gel network. *Anal Chem* **74**, 2217–2223.

Jian D, Hannon GJ (2004). Suppression of p160ROCK bypasses cell cycle arrest after Aurora-A/STK15 depletion. *Proc Natl Acad Sci U S A* **101**, 8975–8980.

Jiang Y, Vaessen B, Lenvik T, Blackstad M, Reyes M, Verfaillie CM (2002). Multipotent progenitor cells can be isolated from postnatal murine bone marrow, muscle, and brain. *Exp Hematol* **30**(8), 896–904.

Johnstone PA, Peng YP, May BC, Inouye WS, Niemtzow RC (2001). Acupuncture for pilocarpine-resistant xerostomia following radiotherapy for head and neck malignancies. *Int J Radiat Oncol Biol Phys* **50** (2), 353–357.

Kagan VE, Epperly MW, Jefferson M, Bernarding M, Greenberger JS. *In vitro* and *in vivo* detection of irradiation-induced free radicals. ASTRO, *IJROBP* #82, p. 50, Vol. 54, No. 2, Suppl., 2002.

Kagan VE, Borisenko GG, Tyurina YY, *et al.* (2004). Oxidative lipidomics of apoptosis: redox catalytic interactions of cytochrome c with cardiolipin and phosphatidylserine. *Free Radic Biol Med* **37**(12), 1963–1985.

Kagan VE, Yalowich JC, Day BW, Goldman R, Gantchev TG, Stoyanovsky DA (1994). Ascorbate is the primary reductant of the peroxyl radical of etoposide in the presence of thiols both in cell homogenates and in model systems. *Biochemistry* **33**(32), 9651–9660.

Kahl C, Mielcarek M, Iwata M, Harkey MA, Storer B, Torok-Storb B (2004). Radiation dose determines the degree of myeloid engraftment after nonmyeloablative stem cell transplantation. *Biol Blood Marrow Transplant* **10**, 826–833.

Kaimori J, Takenaka M, Nakajima H, *et al.* (2003). Induction of glia maturation factor-B in proximal tubular cells leads to vulnerability to oxidative injury through the p38 pathway and changes in antioxidant enzyme activities. *J Biol Chem* **278**(35), 33519–33527.

Kaji H, Canaff L, Lebrun JJ, Goltzman D, Hendy GN (2001). Inactivation of menin, a Smad3-interacting protein, blocks transforming growth factor type beta signaling. *Proc Natl Acad Sci U S A* **98**(7), 3837–3842.

Kamata H, Honda S, Maeda S, Chang L, Hirata H, Karin M (2005). Reactive oxygen species promote TNFa-induced death and sustained JNK activation by inhibiting MAP kinase phosphatases. *Cell* **120**, 649–661.

Kaminski A, Pohl CB, Sponholz C, *et al.* (2004). Up-regulation of endothelial nitric oxide synthase inhibits pulmonary leukocyte migration following lung ischemia-reperfusion in mice. *Am J Pathol* **164**(6), 2241–2249.

Kang SK, Rabbani ZN, Folz RJ, *et al.* (2003). Overexpression of extracellular superoxide dismutase protects mice from radiation-induced lung injury. *Int J Radiat Oncol Biol Phys* **57**(4), 1056–1066.

Kang SK, Rabbani ZN, Folz, RJ, *et al.* (2003). Overexpression of extracellular superoxide dismutase protects mice from radiation-induced ling injury. *Int J Radiat Oncol Biol Phys* **57**, 1056–1066.

Kanai AJ, Pearce LL, Clemens PR, *et al.* (2001). Identification of a neuronal nitric oxide synthase in isolated cardiac mitochondria using electrochemical detection. *Proc Natl Acad Sci U S A* **98**, 14126–14131.

Kanai AJ, Zeidel ML, Lavelle JP, *et al.* (2002). Manganese superoxide dismutase gene therapy protects against irradiation-induced cystitis. *Am J Physiol (Renal Physiol)* **44**, 1152–1160.

Kanemura T, Suzuki H, Aoshiba K, Yokohori N, Nagai A (2003). Acute lung injury in lung cancer patients treated with ZD1839 and bleomycin-induced pulmonary fibrosis in mice. *Proc Am Soc Clin Oncol* **22**, 652.

Karawajew L, *et al.* (1990). Flow sorting of hybrid hybridomas using the DNA stain Hoechst 33342. *J Immunol Methods* **129**(2), 277–282.

Kaufman J, Graf BA, Leung EC, *et al.* (2001). Fibroblasts as sentinel cells: role of the CDcd40-CDcd40 ligand system in fibroblast activation and lung inflammation and fibrosis. *Chest* **120**(1 Suppl), 53S–55S.

Kaufman J, Sime PJ, Phipps RP (2004). Expression of CD154 (CD40 ligand) by human lung fibroblasts: differential regulation by IFN-gamma and IL-13, and implications for fibrosis. *J Immunol* **172**(3), 1862–1871.

Kauschke SG, Knorr A, Heke M, *et al.* (1999). Two assays for measuring fibrosis: reverse transcriptase-polymerase chain reaction of collagen alpha(1) (III) mRNA is an early predictor of subsequent collagen deposition while a novel serum N-terminal procollagen (III) propeptide assay reflects manifest fibrosis in carbon tetrachloride-treated rats. *Anal Biochem* **275** (2), 131–140.

Kavanagh BD, Dent P, Schmidt-Ullrich RK, Chen P, Mikkelsen RB (1998). Calcium-dependent stimulation of mitogen-activated protein kinase activity in A431 cells by low doses of ionizing radiation. *Radiat Res* **149**(6), 579–587.

Kawada H, Ogawa M (2001). Bone marrow origin of hematopoietic progenitors and stem cells in murine muscle. *Blood* **98**(7), 2008–2013.

Kelner MJ, Bagnell R, Montoya M, Estes L, Uglik SF, Cerutti P (1995). Transfection with human copper zinc superoxide dismutase induces bi-directional alterations in other antioxidant enzymes, proteins, growth factor response and paraquat resistance. *Free Radic Biol Med* **18**(3), 497–506.

Kelley TJ, Elmer HL, Corey DA (2001). Reduced Smad3 protein expression and altered transforming growth factor-beta1-mediated signaling in cystic fibrosis epithelial cells. *AJRCMB* **25**(6), 732–738.

Khalil MY, Grandis JR, Shin DM (2003). Targeting epidermal growth factor receptor: novel therapeutics in the management of cancer. *Exp Rev Anticancer Ther* **3**(3), 367–380.

Kies MS, Haraf DJ, Rosen F, *et al.* (2001). Concomitant infusional docitaxol and fluorouracil, oral hydroxyurea, and hyperfractionated radiation for locally advanced squamous head and neck cancer. *J Clin Oncol* **19**(7), 1961–1969.

Kiziltepe T, Yan A, Dong M, Jonnalagadda VS, Dedon PC, Engelward BP (2005). Delineation of the chemical pathways underlying nitric oxide-induced homologous recombination in mammalian cells. *Chem Biol* **12**, 357–369.

Koh L-P, Chao NF (2004). Umbilical cord blood transplantation in adults using myeloablative and nonmyeloablative preparative regimens. *Biol Blood Marrow Transplant* **10**, 1–22.

Kopp H-G, Avecilla ST, Hooper AT, *et al.* (2005). Tie2 activation contributes to hemangiogenic regeneration after myelosuppression. *Blood* **106**(2), 505–515.

Korbling M, Katz RL, Khanna A, *et al.* (2002). Hepatocytes and epithelial cells of donor origin in recipients of peripheral-blood stem cells. *N Engl J Med* **346**(10), 738–746.

Kotton DN, Summer R, Fine A (2004). Lung stem cells: new paradigms. *Exp Hematol* **32**(4), 340–343.

Krause DS, Theise ND, Collector MI, *et al.* (2001). Multi-organ, multi-lineage engraftment by a single bone marrow-derived stem cell. *Cell* **105**, 369–377.

Kretschmer A, Moepert K, Dames S, Sternberger M, Kaufmann J, Klippel A (2003). Differential regulation of TGF-beta signaling through Smad2, Smad3 and Smad4. *Oncogene* **22**(43):6748–63.

Kris MG, Natale RB, Herbst RS, *et al.* (2002). A phase II trial of ZD1839 ('Iressa') in advanced non-small cell lung cancer (NSCLC) patients who had failed platinum- and docetaxel-based regimens (IDEAL 2). Proceedings of ASCO, Abstract #1166.

Kroemer G, Dallaporta B, Resche-Rigon M (1998). The mitochondrial death/life regulator in apoptosis and necrosis. *Annu Rev Physiol* **60**, 619–642.

Kucia M, Reca R, Dawn B, Ratajczak J, Ratajczak MZ (2005). Bone marrow as a home of heterogenous populations of non-hematopoietic stem cells. *Leukemia* **19**, 1118–1127.

Lagasse E, Connors H, Al-Dhalimy M, *et al.* (2000). Purified hematopoietic stem cells can differentiate into heptocytes *in vivo*. *Nat Med* **6**(11):1229–1232.

Lapidos KA, Chen YE, Earley JU, *et al.* (2004). Transplanted hematopoietic stem cells demonstrate impaired sarcoglycan expression after engraftment into cardiac and skeletal muscle. *J Clin Invest* **114** (11):1577–1585.

Larsson R, Cerutti P (1989). Translocation and enhancement of phosphotransferase activity of protein kinase C following exposure in mouse epidermal cells to oxidants. *Cancer Res* **49**(20), 5627–5632.

Leask A, Abraham DJ (2004). TGF-B signaling and the fibrotic response. *FASEB J* **18**, 816–827.

Lechpammer S, Epperly MW, Zhou S, Nie S, Glowacki J, Greenberger JS (2005). Antioxidant pool regulated adipocyte differentiation Sod2-/- bone marrow stromal cells. *Exp Hematol* **33**, 1201–1208.

Leong SS, Tan EH, Fong KW, *et al.* (2003). Randomized double-blind trial of combined modality treatment with or without amifostine in unresectable stage III non-small-cell lung cancer. *J Clin Oncol* **21**(9), 1767–1774.

Ley K, Kansas GS (2004). Selectins in T-cell recruitment to non-lymphoid tissues and sites of inflammation. *Nat Rev Immunol* **4**(5), 325–335.

Li AG, Lu S-L, Zhang M-X, Deng C, Wang X-J (2004). Smad3 knockout mice exhibit a resistance to skin chemical carcinogenesis. *Cancer Res* **64**, 7836–7845.

Li J-J, Oberley LW, Fan M, Colburn NH (1998). Inhibition of AP-1 and NF-$\kappa\beta$ by manganese-containing superoxide dismutase in human breast cancer cells. *FASEB J* **12**, 1713–1723.

Li JJ, Oberley LW, St. Clair DK, Ridnour LA, Oberley TD (1995). Phenotypic changes induced in human breast cancer cells by overexpression of manganese-containing superoxide dismutase. *Oncogene* **10**, 1989–2000.

Li N, Oberley TD, Oberley LW, Zhong W (1998). Inhibition of cell growth in NIH/3T3 fibroblasts by overexpression of manganese Superoxide dismutase: mechanistic studies. *J Cell Physiol* **175**, 359–369.

Li S, Yan T, Yan J-Q, Oberley TD, Oberley LW (2000). The role of cellular glutathione peroxidase redox regulation in the suppression of tumor cell growth by manganese superoxide dismutase. *Cancer Res* **60**, 3927–3939.

Li Z, Khaletskiy A, Wang J, Wong JYC, Oberley LW, Li J-J (2001). Genes regulated in human breast cancer cells overexpressing manganese-containing superoxide dismutase. *Free Radic Biol Med* **30**(3), 260–267.

Lin F, Jackson VE, Girotti AW (1995). Amplification and hyperexpression of the catalase gene in selenoperoxidase-deficient leukemia cells. *Arch Biochem Biophys* **317**(1), 7–18.

Lin AL, Johnson DA, Wu Y, Wong G, Ebersole JL, Yeh C-K (2001). Measuring short-term γ-irradiation effects on mouse salivary gland function using a new saliva collection device. *Arch Oral Biol* **46**, 1085–1089.

Liu R. Effects of overexpression of manganese superoxide dismutase and endothelial nitric oxide synthase on tumor biology of human oral carcinoma SCC-25 cells. PhD Thesis, University of Iowa, May 1996.

Liu R, Oberley TD, Oberley LW (1997). Transfection and expression of MnSOD cDNA decreases tumor malignancy of human oral squamous cell carcinoma SCC-25 cells. *Hum Gene Ther* **8**, 585–595.

Liu R, Buettner GR, Oberley LW (2000). Oxygen free radicals mediate the induction of manganese superoxide dismutase gene expression by TNF-α. *Free Radic Biol Med* **28**(8), 1197–1205.

Lu T, Burdelya LG, Swiatkowski SM, *et al.* (2004). Secreted transforming growth factor β2 activates NF-B, blocks apoptosis, and is essential for the survival of some tumor cells. *Proc Natl Acad Sci U S A* **101**, 7112–7117.

Ma Y, Liu H, Tu-Rapp H, *et al.* (2004). Fas ligation on macrophages enhances IL-1R1-Toll-like receptor 4 signaling and promotes chronic inflammation. *Nat Immunol* **5**(4), 380–387.

Makkonen TA, Minn H, Jekunen A, Vilja P, Tuominen J, Joensuu H (2000). Granuloctye macrophage-colony stimulating factor (GM-CSF) and sucralfate in prevention of radiation-induced mucositis: a prospective randomized study. *Int J Radiat Oncol Biol Phys* **46**(3), 525–534.

Manna SK, Zhang HJ, Yan T, Oberley LW, Aggarwal BB (1998). Overexpression of manganese superoxide dismutase suppresses tumor necrosis factor-induced apoptosis and activation of nuclear transcription factor-$\kappa\beta$ and activated protein-1. *J Biol Chem* **273**, 13245–13254.

Mantell LL, Horowitz S, Davis JM, Kazzaz JA (1999). Hyperoxia-induced cell death in the lung–the correlation of apoptosis, necrosis, and inflammation. *Ann N Y Acad Sci* **887**, 171–80.

Marks LB (2002). Dosimetric predictors of radiation-induced lung injury. *Int J Radiat Oncol Biol Phys* **54** (2), 313–6.

Martinez JC, Villar A, Cabezon MA, *et al.* (2001). Hyperfractionated chemoradiation with carbogen breathing, with or without erythropoietin: A stepwise developed treatment schedule for advanced head and neck cancer. *Int J Radiat Oncol Biol Phys* **50**(1), 47–53.

Martino R, Perez-Simon JA, Moreno E, *et al.* (2005). Reduced-intensity conditioning allogeneic blood stem cell transplantation with fludarabine and oral busulfan with or without pharmacokinetically targeted busulfan dosing in patients with myeloid leukemia ineligible for conventional conditioning. *Biol Blood Marrow Transplant* **11**, 437–447.

Masszi A, Fan L, Rosivall L, *et al.* (2004). Integrity of cell-cell contacts is a critical regulator of TGF-B1-induced epithelial-to-myofibroblast transition. *Am J Pathol* **165**, 1955–1967.

Matsuura I, Denissova NG, Wang G, He D, Long J, Liu F (2004). Cyclin-dependent kinases regulate the antiproliferative function of SMADs. *Nature* **430**, 226–300.

Mauch P, Greenberger JS, Botnick LE, Hannon EC, Hellman S (1980). Evidence for structured variation in self-renewal capacity within long-term bone marrow cultures. *Proc Natl Acad Sci U S A*, **77**, 2927–2930.

Maurici D, Monti P, Campomenosi P, *et al.* (2001). WR2721 (WR2721) restores transcriptional activity of specific p53 mutant proteins in a yeast functional assay. *Oncogene* **20**, 3533–3540.

McKinney-Freeman SL, Jackson KA, Camargo FD, Ferrari G, Mavilio F, Goodell MA (2002). Muscle-derived hematopoietic stem cells are hematopoietic in origin. *Proc Natl Acad Sci U S A* **99**(3), 1341–1346.

McLeod SJ, Shum AJ, Lee RL, Takei F, Gold MR (2004). The Rap GTPases regulate integrin-mediated adhesion, cell spreading, actin polymerization, and Pyk2 tyrosine phosphorylation in B lymphocytes. *J Biol Chem* **279**(13), 12009–12019.

Mechalakos J, Yorke E, Mageras GS, *et al.* (2004). Dosimetric effect of respiratory motion in external beam radiotherapy of the lung. *IJROBP* **71**, 191–200.

Mendelsohn J, Baselga J (2003). Status of epidermal growth factor receptor antagonists in the biology and treatment of cancer. *J Clin Oncol* **21**(14):2787–2799.

Meng TC, Fukada T, Tonks NK (2002). Reversible oxidation and inactivation of protein tyrosinase phosphorylation *in vivo*. *Mol Cell* **8**, 387–399.

Metaxis Y, Zeiser R, Schaitt-Greff, *et al.* (2005). Human hematopoietic cell transplantation results in generation of donor-derived epithelial cells. *Leukemia* **19**, 1287–1289.

Mi Z, Mai J, Lu X, Robbins PD (2000). Characterization of a class of cationic peptides able to facilitate efficient protein transduction in vitro and in vivo. *Mol Ther* **2** (4), 339–352.

Miranda KM, Espey MG, Jourd'heuil D, *et al.* (2000). The chemical biology of nitric oxide. In: Ignarro LJ (ed.), *Nitric Oxide Biology and Pathobiology*. Academic Press, San Diego, pp. 41–56.

Mitchell JB, DeGraff W, Kaufman D, *et al.* (1991). Inhibition of oxygen-dependent radiation-induced damage by the nitroxide superoxide dismutase mimic, tempol. *Arch Biochem Biophys* **289**, 62–70.

Mohty M, de Lavallade H, Ladaique P, *et al.* (2005). The role of reduced intensity conditioning allogeneic stem cell transplantation in patients with acute myeloid leukemia: a donor vs no donor comparison. *Leukemia* **19**, 916–920.

Morgan WF (2003). Non-targeted and delayed effects of exposure to ionizing radiation: I. radiation-induced genomic instability and bystander effects *in vitro*. *Radiat Res* **159**, 567–580.

Mork J, Lie AK, Glattre E, *et al.* (2001). Human papillomavirus infection as a risk factor for squamous cel carcinoma of the head and neck. *N Engl J Med* **344**, 1125–1131.

Mothersill C, Seymour CB (2004). Radiation-induced bystander effects – implications for cancer. *Nat Rev Cancer* **4**, 158–164.

Movsas B, Scott C, Langer C, *et al.* (2005). Randomized trial of Amifostine in locally advanced non-small cell lung cancer patients receiving chemotherapy and hyperfractionated radiation: radiation therapy oncology group trial 98–01. *J Clin Oncol* **23**, 2145–2154.

Mowery P, Yang Z-Q, Gordon EJ, *et al.* (2004). Synthetic glycoprotein mimics inhibit L-selectin-mediated rolling and promote L-selectin shedding. *Chem Biol* 725–732.

Munter MW, Karger CP, Hoffner SG, *et al.* (2004). Evaluation of salivary gland function after treatment of head-and-neck tumors with intensity-modulated radiotherapy by quantitative pertechnetate scintigraphy. *Int J Radiat Oncol Biol Phys* **58**(1), 175–184.

Murley JS, Kataoka Y, Weydert CJ, Oberley LW, Grdina DJ (2002). Delayed cytoprotection after enhancement of Sod2 (MnSOD) gene expression in SA-NH mouse sarcoma cells exposed to WR-1065, the active metabolite of amifostine. *Radiat Res* **158**, 101–109.

Murray CE, Soonpaa MH, Reinecke H, *et al.* (2004). Haematopoietic stem cells do not transdifferentiate into cardiac myocytes in myocardial infarcts. *Nature* **428**, 564–568.

Naderi J, Hung M, Pandey S (2003). Oxidative stress-induced apoptosis in dividing fibroblasts involves activation of p38 MAP kinase and overexpression of Bax: Resistance of quiescent cells to oxidative stress. *Apoptosis* **8**, 91–100.

Naparstek E, Pierce J, Metcalf D, *et al.* (1986). Induction of growth alterations in factor-dependent hematpoietic progenitor cell lines by co-cultivation with irradiated bone marrow stromal cell lines. *Blood* **67**, 1395–1403.

Neben S, Anklesaria P, Greenberger JS, Mauch P (1993). Quantitation of murine hematopoietic stem cells in vitro by limiting dilution analysis of cobblestone area formation on a clonal stromal cell line. *Exp Hematol* **21**, 438–444.

Nishikawa T, Ramesh R, Munshi A, Chada S, Meyn RE (2004). Adenovirus-mediated mda-7 (IL24) gene therapy suppresses angiogenesis and sensitizes NSCLC xenograft tumors to radiation. *Mol Ther* **9** (6), 818–828.

Nishioka A, Ogawa Y, Mima T, *et al.* (2004). Histopathologic amelioration of fibroproliferative change in rat irradiated lung using soluble transforming growth factor-beta (TGFβ) receptor mediated by adenoviral vector *Int J Radiat Oncol Biol Phys* **58**(40), 1235–1241.

Niu Y, Shen H, Epperly MW, *et al.* (2005). Protection of esophageal multi-lineage progenitors of squamous epithelium (stem cells) from ionizing irradiation by manganese superoxide dismutase-plasmid/liposome (MnSOD-PL) gene therapy. *In Vivo* **19**, 965–974.

Novakova-Jiresova A, van Gameren MM, Coppes RP, Kampinga HH, Groen HJM (2004). Transforming growth factor-B plasma dynamics and post-irradiation lung injury in lung cancer patients. *Radiother Oncol* **71**, 183–189.

Nygren JM, Jovinge S, Breitbach M, *et al.* (2004). Bone marrow-derived hematopoietic cells generate cardiomyocytes at a low frequency through cell fusion, but not transdifferentiation. *Nat Med* **10**(5), 494–501.

Oberley LW (1982). Superoxide dismutase and cancer. In: Oberley LW (ed.), *Superoxide Dismutase.* Vol. II. Boca Raton, FL, CRC Press, Chapter 6.

Oberley LW, Buettner GR (1979). Role of superoxide dismutase in cancer: a review. *Cancer Res* **39**, 1141–1149.

Oberley LW, Oberley TD (1986). Free radicals, cancer, and aging. In: Johnson Jr JE, Walford R, Harmon D, Miquel J (eds), *Free Radicals, Aging, and Degenerative Diseases.* Alan R. Liss Inc., New York, pp. 325–381.

Oberley LW, McCormick ML, Sierra-Rivera E, Kasemset St. Clair D (1989). Manganese superoxide dismutase in normal and transformed human lung fibroblasts. *Free Radic Biol Med* **6**, 379–384.

Ogle BM, Cascalho M, Platt JL (2005). Biological implications of cell fusion. *Mol Cell Biol* **6**, 567.

Ortiz LA, Gambelli F, McBride C, *et al.* (2003). Mesenchymal stem cell engraftment in lung is enhanced in response to bleomycin exposure and ameliorates its fibrotic effects. *Proc Natl Acad Sci U S A* **100**(14), 8407–8411.

Oslund KL, Miller LA, Usachenko JL, Tyler NK, Wu R, Hyde DM (2004). Oxidant-injured airway epithelial cells upregulate thioredoxin but do not produce interleukin-8. *Am J Respir Cell Mol Biol* **30**, 597–604.

Oury TD, Schaefer LM, Fattman CL, Choi A, Weck KE, Watkins SC (2002). Depletion of pulmonary EC-SOD after exposure to hyperoxia. *Am J Physiol Lung Cell Mol Physiol* **283**(4), L777–84.

Ozturk B, Egehan I, Atavci S, Kitapci M (2004). Pentoxifylline in prevention of radiation-induced lung toxicity in patients with breast and lung cancer: A double-blind randomized trial. *Int J Radiat Oncol Biol Phys* **58**(1), 213–219.

Pajonk F, Schlessmann S, Guttenberger, Henke M (2001). Epithelial cells in the peripheral blood of patients with cancer of the head and neck: incidence, detection and possible clinical significance. *Radiother Oncol* **59**, 213–217.

Paris F, Fuks Z, Kang A, *et al.* (2001). Endothelial apoptosis as the primary lesion initiating intestinal radiation damage in mice. *Science* **293**, 293–296.

Patel AB, Edelman MJ, Kwok Y, Krasna MJ, Suntharalingam M (2004). Predictors of acute esophagitis in patients with non-small-cell lung carcinoma treated with concurrent chemotherapy and hyperfractionated radiotherapy followed by surgery. *IJROBP* **60**(4), 1106–12.

Pearce LL, Epperly MW, Greenberger JS, Pitt B, Peterson J (2001). Identification of respiratory complexes I and III as mitochondrial sites of damage following exposure to ionizing radiation and nitric oxide. *Nitric Oxide: Biol Chem* **5**(2), 128–136.

Peng H, Usas A, Gearhart B, Young B, Olshanski A, Huard J (2004). Development of a self-inactivating tet-on retroviral vector expressing bone morphogenetic protein 4 to achieve regulated bone formation. *Mol Ther* **9**(6), 885–894.

Perry SS, Wang H, Pierce LJ, Yang AM, Tsai S (2004). Spangrude GJ. L-selectin defines a bone marrow analog to the thymic early T-lineage progenitor. *Blood* **103**(8), 2990–6.

Perry Y, Epperly MW, Fernando H, *et al.* (2005). Photodynamic therapy induced esophageal stricture-an animal model: from mouse to pig. *J Surg Res* **123**, 67–74.

Phillips RJ, Burdick MD, Hong K, *et al.* (2004). Circulating fibrocytes traffic to the lungs in response to CXCL12 and mediate fibrosis. *J Clin Invest* **114**(3), 438–444.

Phillips GL, Meisenberg B, Reece DE, *et al.* (2004). Amifostine and autologous hematopoietic stem cell support of escalating-dose melphalan: A phase I study. Biol. *Blood Bone Marrow Transplant* **10**, 473–483.

Prigozhina TB, Gurevitch O, Slavin S (1999). Nonmyeloablative conditioning to induce bilateral tolerance after allogeneic bone marrow transplantation in mice. *Exp Hematol* **27**, 1503–1510.

Prindull G, Zipori D (2004). Environmental guidance of normal and tumor cell plasticity: epithelial mesenchymal transitions as a paradigm. *Blood* **103**(8), 2892–2899.

Raja S, Luketich JD, Kelly L, Gooding WE, Finkelstein SD, Godfrey TE (2002). Rapid, quantitative reverse transcriptase-polymerase chain reaction: Application to intraoperative molecular detection of occult metastases in esophageal cancer. *J Thorac Cardiovasc Surg* **123**(3), 475–483.

Rakovitch E, Tsao M, Ung Y, Pignol J-P, Cheung P, Chow E (2004). Comparison of the efficacy and acute toxicity of weekly versus daily chemoradiotherapy for non-small-cell lung cancer: a meta-analysis. *Int J Radiat Biol Phys* **58**(1), 196–203.

Ramirez AM, Takagawa S, Sekosan M, Jaffe HA, Varga J, Roman J (2004). Smad3 deficiency ameliorates experimental obliterative bronchiolitis in a heterotopic tracheal transplantation model. *Am J Pathol* **165**(4), 128.

Rees DC, Congreve M, Murray CW, Carr R (2004). Fragment-based lead discovery. *Nature* **3**, 660–672.

Regine WF, Valentino J, Arnold SM, *et al.* (2001). High-dose intra-arterial cisplatin boost with hyperfractionated radiation therapy for advanced squamous cell carcinoma of the head and neck. *J Clin Oncol* **19**(14), 333–339.

Reisdorf P, Lawrence DA, Sivan V, Klising E, Martin MT (2001). Alteration of transforming growth factor-beta1 response involves down-regulation of Smad3 signaling in myofibroblasts from skin fibrosis. *Am J Pathol* **159**(1), 263–72.

Remy I, Montmarquette A, Michnick SW (2004). PKB/Akt modulates TGF-beta signalling through a direct interaction with Smad3. *Nat Cell Biol* **6**(4), 358–65.

Richardson CM, Swinson DE, Richardson D, Swain WA, O'Byrne KJ (2003). Epidermal growth factor receptor (EGFR) and cyclooxygenase (COX)-2 in non-small cell lung cancer (NSCLC):Growth factors and cytokines enhance the *in vitro* anti-tumor activity of selective EGFR and/or COX-2 inhibitors. *Proc Am Soc Clin Oncol* **22**, 654.

Robert F, Ezekiel MP, Spencer SA, *et al.* (2001). Phase I study of anti-epidermal growth factor receptor antibody Cetuximab in combination with radiation therapy in patients with advanced head and neck cancer. *J Clin Oncol* **19**(13), 3234–3240.

Roberts AB, Wakefield LM (2003). The two faces of transforming growth factor beta in carcinogenesis. *Proc Natl Acad Sci U S A* **100**(15), 8621–8623.

Roesink JM, Konings AWT, Terhaard CHJ, Battermann JJ, Kampinga HH, Coppes RP (1999). Preservation of the rat parotid gland function after radiation by prophylactic pilocarpine treatment: radiation dose dependency and compensatory mechanisms. *Int J Radiat Oncol Biol Phys* **45**(2), 483–489.

Roof KS, Fidias P, Lynch TJ, Ancukiewicz M, Choi NC (2003). Radiation dose escalation in limited-stage small-cell lung cancer. *Int J Radiat Oncol Biol Phys* **57**(3), 701–8.

Rosen H, Gordon S (1987). Monoclonal antibody to the murine type 3 complement receptor inhibits adhesion of myelomonocytic cells *in vitro* and inflammatory cell recruitment *in vivo*. *J Exp Med* **166**, 1685–1701.

Rosen H (1990). Role of CR3 in induced myelomonocytic recruitment: insights from *in vivo* monoclonal antibody studies in the mouse. *J Leukoc Biol* **48**, 465–469.

Rosenman JG, Halle JS, Socinski MA, *et al.* (2002). High-dose conformal radiotherapy for treatment of stage IIIA/IIIB non-small-cell lung cancer: technical issues and results of a phase I/II trial. *Int J Radiat Oncol Biol Phys* **54**(2), 348–56.

RTOG toxicity scoring criteria. Available at: http://www.nog.org/members/main.html.

Rubin P, Casarett GW (1968). *Clinical Radiation Pathology*. WB Saunders, Philadelphia, PA.

Rugo RE, Almeida KH, Hendricks CA, Jonnalagadda VS, Engelward BP (2005). A single acute exposure to a chemotherapeutic agent induces hyper-recombination in distantly descendant cells and in their neighbors. *Oncogene* **00**, 1–10.

Sakakeeny MA, Greenberger JS (1982). Granulopoiesis longevity in continuous bone marrow cultures and factor dependent cell line generation: Significant variation among 28 inbred mouse strains and outbred stocks. *J Natl Cancer Inst* **68**, 305–317.

Sato M, Muragaki Y, Saika S, Roberts AB, Ooshima A (2003). Targeted disruption of TGF-beta1/Smad3 signaling protects against renal tubulointerstitial fibrosis induced by unilateral ureteral obstruction. *J Clin Invest* **112**(10), 1486–94.

Sato Y, Araki H, Kato J, *et al.* (2005). Human mesenchymal stem cells xenografted directly to rat liver are differentiated into human hepatocytes without fusion. *Blood* **106**(2), 756–764.

SatwaniP,HarrisonL,MorrisE,DelToroG,CairoMS(2005). Reduced intensity allogeneic stem cell transplantation in adults and children with malignant and non-malignant diseases: end of the beginning and future challenges. *Biol Blood Marrow Transplant* **11**, 403–422.

Scagliotti GV, Turrisi AT 3rd (2003). Docetaxel-based combined-modality chemoradiotherapy for locally advanced non-small cell lung cancer. *Oncologist* **8** (4), 361–374.

Scheffold C, Scheffold YC, Cao TM, Gworek J, Shizura JA (2005). Cytokines and cytotoxic pathways in engraftment resistance to purified allogeneic hematopoietic stem cells. *Biol Blood Marrow Transplant* **11**, 1–12.

Schmidt KN, Amstad P, Cerutti P (1995). The roles of hydrogen peroxide and superoxide as messengers in the activation of transcription factor NF-kB. *Chem Biol* **2**, 13–22.

Schmidt-Ullrich RK, Mikkelsen RB, Dent P, *et al.* (1997). Radiation-induced proliferation of the human A431 squamous carcinoma cells is dependent of EGFR tyrosine phosphorylation. *Oncogene* **15**, 1191–1197.

Schneeberger EE, Vu Q, LeBlanc BW, Doerschuk CM (2000). The accumulation of dendritic cells in the lung is impaired in CD18-/- but not in ICAM-1-/- mutant mice. *J Immunol* **164**(5), 2472–2478.

Selman M, Pardo A (2002). Idiopathic pulmonary fibrosis: an epithelial/fibroblastic cross-talk disorder. *Respir Res* **3**, 3–12.

Sherwood RI, Christensen JL, Conboy IM, *et al.* (2004). Isolation of adult mouse myogenic progenitors: functional heterogeneity of cells within and engrafting skeletal muscle. *Cell* **119**, 543–554.

Shimonaka M, Katagiri K, Nakayama T, *et al.* (2003). Rap1 translates chemokine signals to integrin activation, cell polarization, and motility across vascular endothelium under flow. *J Cell Biol* **161**(2), 417–27.

Shin DM, Khuri FR, Murphy B, *et al.* (2001). Combined interferon-alfa, 13-cis-retinoic acid, and alpha tocopherol in locally advanced head and neck squamous cell carcinoma: novel bioadjuvant Phase II trial. *J Clin Oncol* **19**(12), 3010–3016.

Sipkins DA, Wei X, Wu JW, *et al.* (2005). In vivo imaging of specialized bone marrow endothelial microdomains for tumor engraftment. *Nature* **435**, 969–974.

Smalley M, Ashworth A (2003). Stem cells and breast cancer: a field in transit. *Nat Rev Cancer* **3**, 832–842.

Socinski MA, Rosenman JG, Halle J, *et al.* (2001a). Dose-escalating conformal thoracic radiation therapy with induction and concurrent carboplatin/paclitaxel in unresectable stage IIIA/B nonsmall cell lung carcinoma. A modified Phase I/II trial. *Cancer* **92**, 1213–1223.

Song SU, Hong YJ, Oh IS, *et al.* (2004). Regeneration of hyaline articular cartilage with irradiated transforming growth factor B1-producing fibroblasts. *Tissue Eng* **10** (5/6),665.

Sonis ST, Oster G, Fuchs H, *et al.* (2001). Oral mucositis and the clinical and economic outcomes of hematopoietic stem cell transplantation. *J Clin Oncol* **19**(8), 2201–2205.

Soulieres D, Senzer N, Vokes EE, Hidalgo M, Agarwala SS, Siu LL (2004). Multicenter phase II study of Erlotinib, an oral epidermal growth factor receptor tyrosine kinase inhibitor, in patients with recurrent or metastatic squamous cell cancer of the head and neck. *J Clin Oncol* **22**, 77–85.

Spector JA, Mehrara BJ, Luchs JS, *et al.* (2000). Expression of adenovirally delivered gene products in healing osseous tissues. *Ann Plast Surg* **44**, 522–528.

Spitz DR, Elwell JH, Sun Y, *et al.* (1990). Oxygen toxicity in control and H_2O_2-resistant Chinese hamster fibroblasts. *Archiv Biochem Biophys* **279**, 249–260.

Spotl L, *et al.* (1995). Cell membrane labeling with fluorescent dyes for the demonstration of cytokine-induced fusion between monocytes and tumor cells. *Cytometry* **21**(2), 160–9.

Springer T, GalfrÒ G, Secher DS, Milstein C (1978). Monoclonal xenogeneic antibodies to murine cell surface antigens: identification of novel leukocyte differentiation antigens. *Eur J Immunol* **8**, 539–551.

Spyridonidis A, Schmitt-Graff A, Tomann T, *et al.* (2004). Epithelial tissue chimerism after human hematopoietic cell transplantation is a real phenomenon. *Am J Pathol* **164**(4), 1147–1155.

St Clair DK, Wan XS, Oberley TD, Muse KE, St Clair WH (1992). Suppression of radiation-induced neoplastic transformation by overexpression of mitochondrial superoxide dismutase. *Molec Carcinogenesis* **6**, 238–242.

St Clair DK, Wan XS, Kuroda M, Vichitbandha S, Tsuchida E, Urano M (1997). Suppression of tumor metastasis by manganese Superoxide dismutase is associated with reduced tumorigenicity and elevated fibronectin. *Oncol Rep* **4**, 753–757.

Stickle RL, Epperly MW, Klein E, Bray JA, Greenberger JS (1999). Prevention of irradiation-induced esophagitis by plasmid/liposome delivery of the human manganese superoxide dismutase (MnSOD) transgene. *Radiat Oncol Invest Clin Basic Res* **7**(6), 204–217.

Stiff P, Bensinger W, Emmanouilides C, *et al.* (2003). Treatment of mucositis with palifermin improves patient function and results in a clinically meaningful reduction in mouth and throat soreness (MTS): Phase 3 results. *Blood* **102**(11),194a.

Sudbo J, Kildal W, Risberg B, Koppang HS, Danielsen HE, Reith A (2001). DNA content as a prognostic marker in patients with oral leukoplakia. *N Engl J Med* **344**, 1270–1280.

Sumiyoshi H, Mor N, Lee SY, *et al.* (2004). Esophageal muscle physiology and morphogenesis require assembly of a collagen XIX-rich basement membrane zone. *J Cell Biol* **166**(4), 591–600.

Suzuki E, Kapoor V, Cheung H-K, *et al.* (2004). Soluble type II transforming growth factor-B receptor inhibits established murine malignant mesothelioma tumor growth by augmenting host antitumor immunity. *Clin Cancer Res* **10**, 5907–5918.

Takahasi M, Nakamura T, Toba T, Kajiwara N, Kato H, Shimizu Y (2004). Transplantation of endothelial progenitor cells into the lung to alleviate pulmonary hypertension in dogs. *Tissue Eng* **10**(5/6),771.

Takeyanna K, Dabbagh K, Shim JJ, Dao-Pick T, Ueki IF, Nadel JA (2000). Oxidative stress causes mucin synthesis via transactivation of epidermal growth factor recepetor: role of neutrophils. *J Immunol* **164**, 1546–1552.

Tamagno E, Robino G, Obbili A, *et al.* (2003). H_2O_2 and 4-hydroxynonenal mediate amyloid B-induced neuronal apoptosis by activating JNKs and $p38^{MAPK}$. *Exp Neurol* **180**, 144–155.

Tampo Y, Kotamraju S, Chitambar CR, *et al.* (2002). Oxidative stress-induced iron signaling is responsible for peroxide-dependent oxidation of dichlorodihydrofluorescein in endothelial cells. *Circ Res* **92**(1), 56–63.

Tatsuma T, Mori H, Fujishima A (2000). Electron transfer from diamond electrodes to heme peptide and peroxidase. *Chemistry* **72**, 2919–2924.

Terry NH, Travis EL (1989). The influence of bone marrow depletion on intestinal radiation damage. *Int J Radiat Oncol Biol Phys* **17**, 569–573.

Tertov VV, *et al.* (1989). Use of lipophilic fluorescent probes for the isolation of hybrid cells in flow cytometry. *J Immunol Methods* **118**(1), 139–43.

Thaller SR, Salzhauer MA, Rubinstein AJ, Thion A, Tesluk H (1998). Effect of insulin-like growth factor type I on critical size calvarial bone defects in irradiated rats. *J Cranialfacial Surg* **9**(2), 138–141.

Theise ND, Krause DS (2002). Toward a new paradigm of cell plasticity. *Leukemia* **16**(4), 542–548.

Thirion P, Holmberg O, Collins CD, *et al.* (2004). Escalated dose for non-small-cell lung cancer with accelerated hypofractionated three-dimensional conformal radiation therapy. *IJROBP* **71**, 163–166.

Tohyama Y, Katagiri K, Pardi R, Lu C, Springer TA, Kinashi T (2003). The critical cytoplasmic regions of the alphaL/beta2 integrin in Rap1-induced adhesion and migration. *Mol Biol Cell* **14**(6), 2570–82.

Tumbar T, Guasch G, Greco V, *et al.* (2004). Defining the epithelial stem cell niche in skin. *Science* **303**, 359–363.

Upadhyay D, Bundesmann M, Panduri V, Correa-Meyer E (2004). Kamp DW. Fibroblast growth factor-10 attenuates H_2O_2-induced alveolar epithelial cell DNA damage: role of MAPK activation and DNA repair. *Am J Respir Cell Mol Biol* **31**, 107–113.

van Rongen E, Thames HD, Travis EL (1993). Recovery from radiation damage in mouse lung: interpretations in terms of two rates of repair. *Radiat Res* **133**, 225–233.

Vanhoefer U, Tewes M, Rojo F, *et al.* (2004). Phase I study of the humanized antiepidermal growth factor receptor monoclonal antibody EMD72000 in patients with advanced solid tumors that express the epidermal growth factor receptor. *J Clin Oncol* **22**, 175–184.

Vermaelen KY, Carro-Muino I, Lambrecht BN, Pauwels RA (2001). Specific migratory dendritic cells rapidly transport antigen from the airways to the thoracic lymph nodes. *J Exp Med* **193**(1), 51–60.

Verrecchia F, Tacheau C, Schorpp-Kistner M, Angel P, Mauviel A (2001). Induction of the AP-1 members c-Jun and JunB by TGF-beta/Smad suppresses early Smad-driven gene activation. *Oncogene* **20**(18), 2205–11.

Wakeling AE, Guy SP, Woodburn JR, *et al.* (2002). ZD1839 (Iressa): An orally active inhibitor of epidermal growth factor signaling with potential for cancer therapy. *Cancer Res* **62**, 5749–5754.

Wandzioch E, Kolterud A, Jacobsson M, Friedman SL, Carlsson L (2004). Lhx2-/- mice develop liver fibrosis. *Proc Natl Acad Sci U S A* **101**(47),16549–16554.

Wanda PF, *et al.* (1982). A general method for heterokaryon detection using resonance energy transfer and a fluorescence-activated cell sorter. *J Histochem Cytochem* **30**(12), 1297–300.

Wang HP, Qian SY, Schafer FQ, Domann FE, Oberley LW, Buettner GR (2001). Phospholipid hydroperoxide glutathione peroxidase protects against singlet oxygen-induced cell damage of photodynamic therapy. *Free Radic Biol Med* **30**(8), 825–835.

Wang J, Zheng H, Sung C-C, Richter KK, Hauer-Jensen M (1998). Cellular sources of transforming growth factor-b isoforms in early and chronic enteropathy. *Am J Pathol* **153**, 1531–1540.

Wang J, Zheng H, Ou X, Fink LM, Hauer-Jensen M (2002). Deficiency of microvascular thrombomodulin and up-regulation of protease-activated receptor-1 in irradiated rat intestine. Possible link between endothelial dysfunction and chronic radiation fibrosis. *Am J Pathol* **160**(6), 2063–2070.

Wardley AM, Booth D, Roberts D, Roberts SA, Scarffe JH, Potten CS (1998). A quantitative histometric murine *in vivo* model of radiation-induced oral mucositis. *Arch Oral Biol* **43**, 567–577.

Wasabaugh CH, Ontell MP, Ontell M (2004). Nonmuscle stem cells fail to significantly contribute to regeneration of normal muscle. *Gene Ther* **11**, 1724–1728.

Weinmann M, Marini P, Jendrossek V, *et al.* (2004). Influence of hypoxia on TRAIL-induced apoptosis in tumor cells. *Int J Radiat Oncol Biol Phys* **58**, 386–396.

Wen FQ, Liu X, Kobayashi T, *et al.* (2004). Interferon-gamma inhibits transforming growth factor-beta production in human airway epithelial cells by targeting Smads. *AJRCMB* **30**(6), 816–822.

Werner-Wasik M, Axelrod RS, Friedland DP, *et al.* (2001). Preliminary report on reduction of esophagitis by amifostine in patients with non-small-cell lung cancer treated with chemoradiotherapy. *Clin Lung Cancer* **2**(4), 284–289.

Weydert CJ, Smith BB, Xu L, *et al.* (2003). Inhibition of oral cancer cell growth by adenovirus MnSOD plus BCNU treatment. *Free Radic Biol Med* **34**(3), 316–329.

Wijers OB, Levendag PC, Harms ER, *et al.* (2001). Mucositis reduction by selective elimination of oral flora in irradiated cancers of the head and neck: A placebo-controlled double-blinded randomized study. *Int J Radiat Oncol Biol Phys* **50**(1), 343–352.

Wilde DB, Marrack P, Kappler J, Dialynas DP, Fitch FW (1983). Evidence implicating L3T4 class II MHC antigen reactivity; monoclonal antibody GK1.5 (anti-L3T4a) blocks class II MHC antigen-specific proliferation, release of lymphokines, and binding by cloned murine helper T lymphocyte lines. *J Immunol* **131**, 2178–2183.

Williams JP, Hernady E, Johnston CJ, *et al.* (2004). Effect of administration of lovastatin on the development of late pulmonary effects after whole-lung irradiation in a murine model. *Radiat Res* **161**(5), 560–7.

Wolfraim LA, Fernandez TM, Mamura M, *et al.* (2004). Loss of SMAD 3 in acute t-cell lymphoblastic leukemia. *N Engl J Med* **351**, 552–559.

Wolski MJ, Bhatnagar A, Flickinger JC, Belani CP, Ramalingam S, Greenberger JS (2005). Multivariate analysis of survival, local control, and time to distant metastases in patients with unresectable non-small cell lung carcinoma treated with three-dimensional conformal radiotherapy with or without concurrent chemotherapy. *Clin Lung Cancer* **7**(2), 100–107.

Wong HY, Epperly MW, Godfrey T, Greenberger JS, Luketich JD. Manganese superoxide dismutase (MnSOD) gene insertion protects normal cells during photodynamic therapy. *J Cardiothorac Surg* (submitted).

Wu K-L, Jiang G-L, Liao Y, *et al.* (2003). Three-dimensional conformal radiation therapy for non–small-cell lung cancer: A Phase I/II dose escalation clinical trial. *Int J Radiat Oncol Biol Phys* **57**(5), 1336–1344.

Wu VWC, Kwong DLW, Sham JST (2004). Target dose conformity in 3-dimensional conformal radiotherapy and intensity modulated radiotherapy. *Radiother Oncol* **71**, 201–206.

Wurzler KK, DeWeese TL, Sebald W, Reddi AH (1999). Radiation-induced impairment of bone healing can be overcome by recombinant human bone morphogenetic protein-2. *J Cranialfacial Surg* **9**(2), 131–137.

Wynn RF, Hart CA, Corradi-Perini C, *et al.* (2004). A small proportion of mesenchymal stem cells strongly expresses functionally active CXCR4 receptor capable of promoting migration to bone marrow. *Blood* **104**(9), 2643–2650.

Xu Y, Krishnan A, Wan SX, *et al.* (1999). Mutations in the promoter reveal a cause for the reduced expression of the human manganese superoxide dismutase gene in cancer cells. *Oncogene* **18**, 93–102.

Yan T, Oberley LW, Zhong W, St. Clair DK (1996). Manganese-containing superoxide dismutase overexpression causes phenotypic reversion in SV40-transformed human lung fibroblasts. *Cancer Res* **56**, 2864–2871.

Yoon YS, Lee JH, Hwang SC, Choi KS, Yoon G (2005). TGF-Beta1 induces prolonged mitochondrial ROS generation through decreased complex i.v. activity with senescent arrest in Mv1Lu cells. *Oncogene* **24**, 1895–1903.

Yorke ED, Jackson A, Rosenzweig KE, *et al.* (2002). Dose-volume factors contributing to the incidence of radiation pneumonitis in non-small-cell lung cancer patients treated with three-dimensional conformal radiation therapy. *Int J Radiat Oncol Biol Phys* **54** (2), 329–39.

Zhang Y, Cao HJ, Graf B, Meekins H, Smith TJ, Phipps RP (1998). CD40 engagement up-regulates cyclooxygenase-2 expression and prostaglandin E2 production in human lung fibroblasts. *J Immunol* **160**(3), 1053–1957.

Zhang Y, Zhao W, Zhang HJ, Domann FE, Oberley LW (2002). Overexpression of copper zinc superoxide dismutase suppresses human glioma cell growth. *Cancer Res* **62**, 1205–1212.

Zhang Z, Liao Z, Jin J, *et al.* (2005). Dose-response relationship in locoregional control for patients with stage II-III esophageal cancer treated with concurrent chemotherapy and radiotherapy. *Int J Radiat Oncol Biol Phys* **61**(3), 656–64.

Zhao W, Spitz DR, Oberley LW, Robbins MEC (2001). Redox modulation of the pro-fibrogenic mediator plasminogen activator inhibitor-1 following ionizing irradiation. *Cancer Res* **61**, 5537–5543.

Zhao Y, Zhan Y, Burke KA, Anderson WF (2005). Soluble factor(s) from bone marrow cells can rescue lethally irradiated mice by protecting endogenous hematopoietic stem cells. *Exp Hematol* **33**, 428–434.

Zheng H, Wang J, Koteliansky VE, Gotwals PJ, Hauer-Jensen M (2000). Recombinant soluble transforming growth factor B type II receptor ameliorates radiation enteropathy in mice. *Gastroenterology* **119**, 1286–1296.

Zhong W, Oberley LW, Oberley TD, Yan T, Domann FE, St Clair DK (1996). Inhibition of cell growth and sensitization to oxidative damage by overexpression of manganese superoxide dismutase in rat glioma cells. *Cell Growth Differ* **7**, 1175–1186.

Zhong W, Oberley LW, Oberley TD, St Clair DK (1997). Suppression of the malignant phenotype of human glioma cells by overexpression of manganese superoxide dismutase. *Oncogene* **14**, 481–490.

Zhong W, Yan T, Lim R, Oberley LW (1999). Expression of superoxide dismutases, catalase, and glutathione peroxidase in glioma cells. *Free Radic Biol Med* **27** (11/12), 1334–1345.

Zipori D (2004). The nature of stem cells: state rather than entity. *Nat Rev* **5**, 873–883.

Zwacka RM, Dudus L, Epperly MW, Greenberger JS, Engelhardt JF (1998). Redox gene therapy protects human IB-3 lung epithelial cells against ionizing radiation-induced apoptosis. *Hum Gene Ther* **9**, 1381–1386.

22

Chemoprotective Gene Delivery

Michael Milsom, Axel Schambach, David Williams and **Christopher Baum**

22.1 Introduction

Chemoprotection mediated by transfer and expression of genes that confer resistance to cytotoxic drugs is a promising approach in the treatment of cancer and for conditional enrichment of gene-modified cells in the treatment of various other inherited and acquired disorders. The approach is best explored in the haematopoietic system. Proof of principle has been obtained in preclinical studies and large animal models. Case reports of clinical studies have also been encouraging. The major limitations to be addressed in future research are potential side effects of protein over-expression, residual DNA damage caused by cytotoxic drugs in unprotected and partially protected cells, leukaemia-promoting co-factors such as replication stress and forced clonal amplification, and insertional mutagenesis. The wide field of viral and non-viral vectors offers many solutions to these problems. In combination with refined drug regimens promoting engraftment and selection of gene-modified cells, progress in vector technology may thus pave the way for successful clinical trials.

22.2 The promise of chemoselection strategies

Gene transfer to achieve chemoprotection is a fundamental approach in gene therapy, with two major aims: to protect healthy cells and tissues from cytotoxic chemotherapy, and to establish a conditional selective advantage in gene-modified cells. If chemoprotection is mediated at the level of a self-renewing stem cell and reliably inherited to progeny cells, major parts of a given organ or tissue might express the desired phenotype, especially if the selective pressure is applied repeatedly. In theory, the protected cell pool may then gradually outcompete all unmodified cells [reviewed in (Sorrentino, 2002; Milsom and Fairbairn, 2004; Neff *et al.*, 2006) This is why stably integrating vectors such as those derived from retroviruses are considered particularly promising for chemoprotection.

If gene-modified cells have no spontaneous selective advantage, the only alternative to chemoprotection is the introduction of genes that enhance the fitness of self-renewing cells by modulating apoptosis, cell cycle control and/or differentiation (Sauvageau *et al.*, 1995; Jin *et al.*, 2000); however, these mechanisms may transform cells if the phenotype is not rendered fully reversible.

Chemoprotection is most advanced in the context of the haematopoietic system. On top of the haematopoietic hierarchy resides the rare population of haematopoietic stem cells (HSC). These cells have an enormous regenerative capacity and give rise to all mature cell lineages of the lymphohaematopoietic system (Figure 22.1), such that their genetic modification should be sufficient to

Viral Therapy of Cancer Edited by Kevin J. Harrington, Richard G. Vile and Hardev S. Pandha

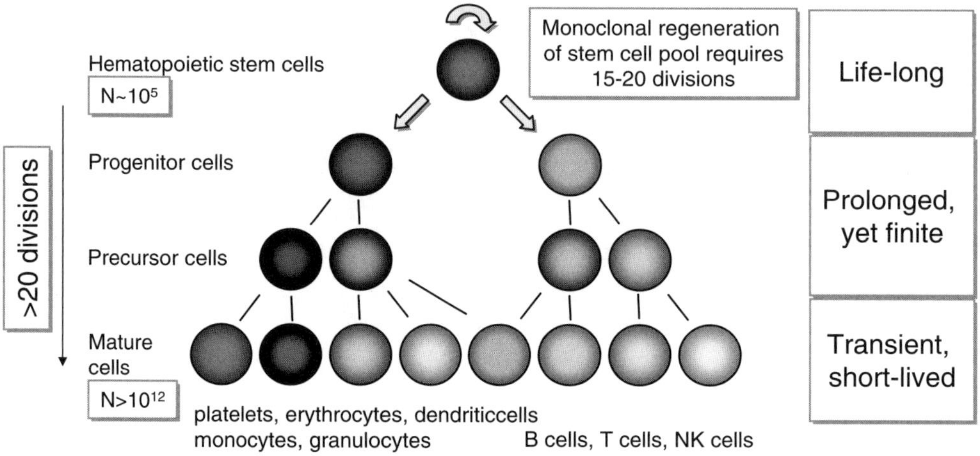

Figure 22.1 Schematic representation of the haematopoietic system. At the pinnacle of the organ hierarchy are the haematopoietic stem cells (HSC). It can be estimated that a single HSC would need to undergo up to 20 self-renewal divisions to regenerate the physiological stem cell pool comprising $\sim 10^5$ cells. Genetic material that is stably introduced into these cells may persist for the lifetime of the organism. With gradual differentiation into highly proliferative progenitor and precursor cells, the formation of the various haematopoietic cell types is initiated and at the same time, the cell pool is greatly amplified. An adult human organism has more than 10^{12} mature haematopoietic cells (in the major lineages of erythrocytes, platelets, neutrophils, monocytes and other professional antigen-presenting cells, B lymphocytes, T lymphocytes and natural killer (NK) cells). Many of these cell types are relatively short-lived (days to a few months), with the exception of memory T and B cells. Genetic chemoprotection established in these various compartments will last according to the life-span and self-renewal potential of the respective cell type. The replicative stress required to establish a fully functional haematopoietic system out of a single precursor cell is indicated

protect the large population of progeny cells from cytotoxic damage. If efficient and safe *in vivo* selection by chemoprotection can be achieved in humans, then this will increase the contribution of gene-modified cells to haematopoiesis, thus overcoming modest transduction rates of human HSC that have been observed to date in clinical trials and large animal models (Neff *et al.*, 2006). Hence, a selective advantage will make it more likely that a therapeutically relevant level of gene-modified cells will be attained. The interest in chemoprotection strategies therefore extends into treatment strategies for non-malignant disorders of haematopoiesis such as inborn errors of metabolism, or severe acquired infections for which no alternative treatment exists (Sorrentino, 2002; Persons *et al.*, 2003a; Davis *et al.*, 2004; Milsom and Fairbairn, 2004; Richard *et al.*, 2004; Neff *et al.*, 2006, Schambach *et al.*, 2006d). As the number of long-term repopulating haematopoietic cells amenable to genetic modification is limited (in current approaches typically less than 10 per cent of the endogenous cell pool is harvested for ex vivo gene transfer), chemoprotection may be crucial to increase the contribution of gene-modified cells to haematopoiesis. This is even more important if no or dose-reduced conditioning of endogenous haematopoiesis is used prior to transplantation of gene-modified cells. If the chemoprotective principle is stably established at the top of the haematopoietic differentiation hierarchy, i. e. in HSC, chemoprotection may even allow for a conditional enrichment of the gene-modified population if the level of chimerism declines to subtherapeutic levels months or even years after infusion of gene-modified cells.

Following engraftment of chemoprotected cells, their steady enrichment can thus be expected with each cycle of chemotherapy administered. The level of chimerism to be achieved varies depending on the nature of the underlying disorder (Neff *et al.*, 2006) (Figure 22.2). For gene therapy of

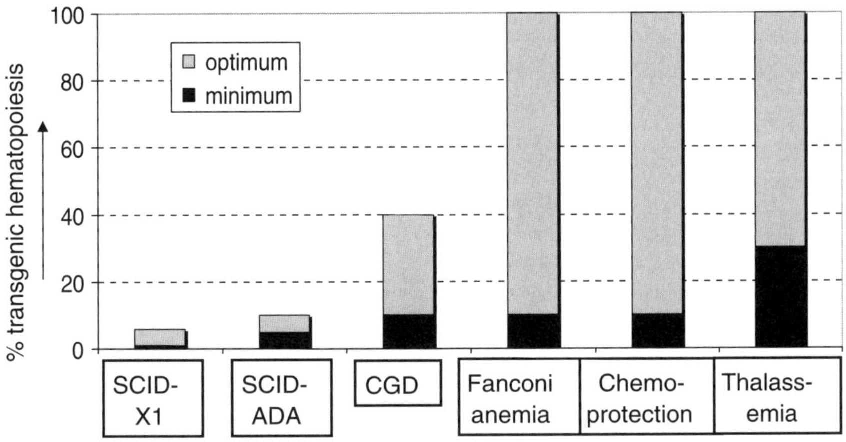

Figure 22.2 Schematic view of the degree of gene-correction (minimum and optimum) required to cure paradigmatic haematopoietic diseases. For SCID-X1 the optimum may be even higher than indicated given that side effects related to insertional mutagenesis can be reduced. Further details can be found in the text and Neff *et al.* (2006)

severe combined immunodeficiency (SCID)-X1 where the therapeutic cDNA is sufficient to confer a selective advantage, far less than 5 per cent of the stem cell pool needs to contain the corrective gene, which can be achieved without any additional selection of conditioning. For gene therapy of SCID-ADA, a higher contribution of gene-modified HSCs is required to obtain optimal results, as suggested from an ongoing clinical trial; non-myelobalative conditioning prior to infusion of gene-modified cells appears to be sufficient to reach this goal (Aiuti *et al.*, 2002). Fortunately, in these two disorders the therapeutic gene is sufficient to confer a selective advantage such that the additional expression of a chemoprotective principle does not seem to be required. For gene therapy of chronic granulomatous disease (CGD), gene-corrected myeloid cells probably need to exceed 10 per cent (Ott *et al.*, 2006). In this case, the therapeutic gene does not confer a selective advantage at the level of self-renewing stem and progenitor cells. As high levels of chimerism may be critical for the long-term success of gene therapy of CGD, chemoselection continues to be an important consideration in this disorder. In gene therapy of Fanconi anaemia (caused by errors in one of the various DNA repair proteins contributing to the Fanconi multiprotein complex), improvement of the severe anaemia can be expected if gene-modified cells contribute more than 10 per cent of the haematopoietic output; however, to reduce the risk of leukaemia development originating from the uncorrected cells, a complete chimerism might have to be achieved. In thalassaemia, clinical benefit is expected when gene-corrected erythrocytes constitute more than 30 per cent of the circulating population; however, a full prevention of the various complications associated with this disease might require substantially higher levels of chimerism. Chemoprotection of bone marrow cells has similar requirements as gene therapy of Fanconi anaemia. First clinical benefits are expected when at least 10 per cent of the haematopoietic cell pool is gene-modified, but again, the prevention of long-term side effects might require up to 100 per cent chimerism [Figure 22.2, adapted from (Neff *et al.*, 2006)]. Similar considerations apply to many other genetic disorders of haematopoiesis.

In the light of the demand for genetic principles that allow a conditional selection of gene-modified cells, it is encouraging to note that convincing evidence for the feasibility of chemoprotection has been obtained not only in mice but also in large animal models (Sorrentino, 2002; Milsom and Fairbairn, 2004; Neff *et al.*, 2006). Especially in large animals and humans, polyclonal chemoprotection might be achieved, given that a large

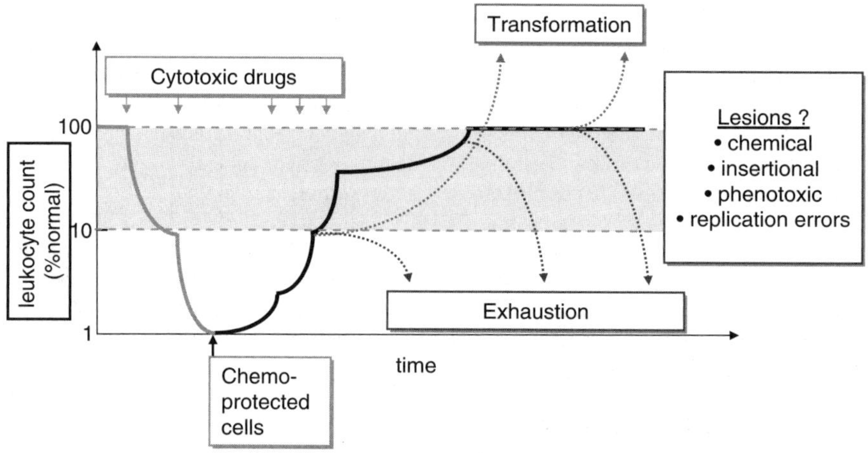

Figure 22.3 Hypothetical outcome of bone marrow chemoprotection with the two major limitations, transformation and exhaustion. Cytotoxic drugs administered before delivery of chemoprotected cells will result in severe haematopoietic damage with a strong reduction of leukocyte count. After infusion of chemoprotected (drug-resistant) cells, haematopoiesis recovers well even if cytotoxic drugs are administered more frequently and with shorter intervals. The major lesions that might give rise to transformation or exhaustion are listed

number of HSCs can be gene-modified and engrafted (Neff *et al.*, 2005). Ideally, this will result in stable, unstressed haematopoiesis even in the presence of increasing cytotoxic dose levels and treatment intensity (Figure 22.3).

While the goal for the patient suffering from non-malignant disorders of haematopoiesis is restricted to the conditional enrichment of gene-modified therapeutic cells, the advantages of the chemoprotection approach for cancer patients may include the following:

- increased therapy intensity resulting in an increased rate of tumour cell kill;
- reduced risk of infectious complications and potentially also reduced severity of mucosal damage due to preservation of neutrophil and lymphocyte counts and function;
- increased cancer immunosurveillance as chemotherapy no longer suppresses immunity;
- reduced risk of bleeding due to preservation of thrombocyte counts;
- increased tissue oxygenation because of maintenance of erythrocyte counts which not only improves systemic fitness but might even contribute to increased tumour cell kill.

All these anticipated benefits are highly appealing and are difficult to achieve by conventional pharmaceutical approaches.

The genes (and their encoded proteins) eligible for chemoprotection should fulfil the following criteria:

- non-immunogenic;
- non-toxic when being over-expressed;
- robust detoxification or prevention of cytotoxic cell damage induced by clinically approved chemotherapeutic agents.

The three major players investigated to date are MDR1, DHFR and MGMT, although many other chemoprotective principles may also be useful.

- Multidrug resistance 1 (MDR1) is a paradigmatic member of the larger family of ABC transporters (Ambudkar *et al.*, 2003). These efflux pumps reduce the cytoplasmic concentration of a relatively wide range of cancer drugs such as anthracyclines, vinca alkaloids and taxanes. Different ABC transporters are encoded in the human genome, each recognizing a distinct, partially overlapping spectrum of drugs. Most of the cytotoxic agents recognized by ABC transporters act in a cell-cycle dependent manner. This may contribute to the observed difficulty to select HSCs by chemoprotection based on MDR1 (Sorrentino *et al.*, 1992; Podda *et al.*, 1992; Blau *et al.*, 1997; Turner *et al.*, 2000; Carpinteiro *et al.*, 2002).

- Dihydrofolate reductase (DHFR) is an enzyme involved in nucleotide metabolism. Point mutants of this enzyme are resistant to antifolates such as methotrexate. This is the first chemoselection principle that was explored in mammals (Williams *et al.*, 1987, Corey *et al.*, 1990). The major drawback of this selection principle, especially in primates, is its low efficiency in non-proliferating cells, which might be overcome by modifying the cytotoxic drug regimen (May *et al.*, 1995; Allay *et al.*, 1998; Persons *et al.*, 2004).
- O^6-methylguanine-DNA-methyltransferase (MGMT) is a protein that repairs certain DNA lesions induced by alkylating agents such as BCNU or temozolomide (Gerson, 2004). The repair process inactivates MGMT, such that the level of expression is directly correlated with the repair capacity. However, MGMT cannot repair DNA methylation or alkylation other than those occurring at the O^6 position of guanine. As clinically used alkylating agents are not specific for this lesion, MGMT mediated chemoprotection cannot be expected to leave the cells unperturbed. Various MGMT mutants such as MGMTP140K have been described that are insensitive to the pharmaceutical inhibitor named O^6-benzylguanine, unlike the wild type protein. These mutants appear to be particularly promising for *in vivo* selection of HSCs when combining the alkylating agents with the inhibitor treatment (Davis *et al.*, 1999; Ragg *et al.*, 2000; Lee *et al.*, 2001; Zielske and Gerson, 2002; Zielske *et al.*, 2003).

Human clinical trials have been performed with both MDR and MGMT (Hesdorffer *et al.*, 1998; Moscow *et al.*, 1999; Abonour *et al.*, 2000; Cornetta *et al.*, 2006) [and additional references in (Neff *et al.*, 2006)]. However, evidence of successful chemoprotection and *in vivo* selection is still limited. This may be explained by low gene transfer rates, insufficient levels of expression, and/or inappropriate stringency of the chemotherapeutic regimen applied. Some of these studies have dampened the enthusiasm for future clinical trials. To revive the approach, its efficiency must be improved. At the same time, it is important to foresee and counteract potential biological limitations, such as the induction of malignant transformation in gene-modified cells, and the exhaustion of haematopoiesis (Figure 22.3).

22.3 The limitations of chemoselection strategies

In the example of phase I clinical trials of gene therapy for X-linked SCID, the survival advantage conferred to lymphatic cells and their progenitors by retroviral delivery of the common gamma chain (the subunit of the interleukin receptors that is defective in these patients) has been central to the therapeutic efficacy (Cavazzana-Calvo *et al.*, 2000; Hacein-Bey-Abina *et al.*, 2002; Gaspar *et al.*, 2004). While this represented the first clear success of gene therapy in the haematopoietic system, to date three patients of the largest clinical trial have subsequently developed a clonal lymphoproliferative disorder driven by retrovirally transduced cells (Hacein-Bey-Abina *et al.*, 2003). Analysis of the proviral integration sites in the leukaemic clones for each of the three patients revealed activation of the same proto-oncogene (LMO2) by the powerful retroviral enhancer element. Whilst it is not clear whether this activation of LMO2 acts coordinately with the transgene and/or the immunocompromised phenotype to promote cellular transformation (Davé *et al.*, 2004; Pike-Overzet *et al.*, 2006; Shou *et al.*, 2006; Thrasher *et al.*, 2006; Woods *et al.*, 2006), this trial as well as related observations in mouse models (Li *et al.*, 2002; Modlich *et al.*, 2005; Montini *et al.*, 2006) and non-human primates (Seggewiss *et al.*, 2006) demonstrate that insertional mutagenesis is a reality and potentially dose-limiting side effect of random transgene insertion (Baum *et al.*, 2003, 2006a; Kohn *et al.*,

2003; Nienhuis *et al.*, 2006). A clonal survival advantage triggered by insertional mutagenesis may thus contribute towards leukaemogenesis, especially if key pathways regulating apoptosis, differentiation and self-renewal are being affected. Hence, it is an important exercise to consider which factors may increase or decrease the risk of therapy related transformation in the context of a retroviral mediated chemoprotection strategy.

As indicated above, there are only a few patients treated in gene therapy clinical trials employing a chemoprotective strategy in which appreciable numbers of chemoprotected haematopoietic cells have been engrafted (Abonour *et al.*, 2000). Therefore, we must consider animal models in order to elucidate the potential risk associated with this approach. As discussed previously, a large number of studies have used the murine model to investigate the feasibility of chemoprotective/selective gene therapy strategies. Despite this abundance of articles [reviewed in Sorrentino (2002); Milsom and Fairbairn (2004), Neff *et al.* (2006)], leukaemic complications have only been documented under experimental conditions that involved an unusually high vector copy number in transduced cells. In the first report, a myeloproliferative syndrome appeared to result from the transduction of murine bone marrow cells with a retrovirus encoding MDR1 (Bunting *et al.*, 1998, 2000). However, in a number of model systems that employed lower gene transfer rates (Sellers *et al.*, 2001; Heyworth *et al.*, 2002; Licht *et al.*, 2002) no evidence was obtained that the sole overexpression of the efflux pump encoded by MDR1 may transform cells.

More recent studies support the conclusion that a high vector copy number is an important trigger of cellular transformation (Modlich *et al.*, 2005, 2006): Combinatorial insertional mutagenesis of proto-oncogenes was discovered in single leukaemogenic cell clones arising after transduction of murine haematopoietic cells with MDR1 or fluorescent proteins such as red fluorescent protein (DsRed2) (Modlich *et al.*, 2005, 2006). These studies thus revealed that retroviral vector mediated transgenesis may have a rather narrow therapeutic index: multiple vector integrations into self-renewing cells should be avoided, especially if the vector contains enhancer elements that can transactivate cellular genes over relatively large distances (>100 kb). Importantly, increasing evidence reveals that so-called self-inactivating lentiviral or gammaretroviral vectors, which are devoid of strong viral enhancers in the long terminal repeats, reduce the risk of insertional mutagenesis (Montini *et al.*, 2006; Modlich *et al.*, 2006).

Also of major interest is the data from a small number of gene therapy chemoprotective regimens that have been performed in larger animal systems. Of particular note is a recent publication which describes the incidence of acute myeloid leukaemia in a rhesus macaque, driven by cells transduced with a vector encoding DHFR (Seggewiss *et al.*, 2006). In this study, granulocyte colony-stimulating factor/stem cell factor mobilized autologous $CD34^+$ cells were transduced with a gammaretroviral vector which co-expressed DHFR and EGFP via an internal ribosomal entry site (IRES). Unusually, within the first year post-engraftment, transduced peripheral blood cells (PBC) comprised up to 80 per cent of the myeloid compartment, resulting from the emergence of a dominant transduced clone which contained two distinct pro-viral insertion sites. The prevalence of gene modified cells then decreased to a stable level of 1–4 per cent of PBC for the next 4 years, with cell counts remaining in the normal range throughout. Two years post transplant, the animal received one round of chemotherapy and displayed transient *in vivo* selection, likely at the progenitor level. At 5 years post-transplant, the animal developed acute myeloid leukaemia derived from the same transduced clone that had promoted the initial unusual enhanced myeloid engraftment. Significantly, one of the two proviral integration sites within this clone was identified as proximal to the anti-apoptotic gene *BCL-2A1*, raising the possibility that this insertion comprised a pro-leukaemic hit. Nonetheless, because of the prolonged latency period we must assume that a series of other genetic lesions was required to fully transform this clone (Seggewiss *et al.*, 2006). However, it is not possible to discern whether the chemotherapeutic regimen was instrumental in provoking the emergence of this leukaemic clone or whether the kinetics of leukaemogenesis would have been identical or even accelerated without this of treatment.

While similar DHFR containing retroviral vectors were used to transduce HSC in other large animal studies, this remains the only adverse event reported to date. Therefore it would seem unlikely that ectopic delivery of the DHFR transgene as such had any significant role to play in promoting clonal proliferation. Hence, thus far, there appears to be little evidence that points to any of the commonly used chemoprotective genes themselves contributing towards transformation/leukaemogenesis.

Despite the fact that none of the proposed chemoprotective genes appears to be acutely transforming, in the wider field of gene therapy, there has been an increasing focus upon the propensity of an integrated proviral backbone to drive leukaemogenesis through activation of cellular proto-oncogenes. As discussed above, studies in mice indicate that gammaretroviral vectors containing a range of transgene cassettes are able to drive the emergence of preleukaemic and leukaemic clones through insertional mutagenesis (Li *et al.*, 2002, Du *et al.*, 2005b; Kustikova *et al.*, 2005; Modlich *et al.*, 2005, 2006; Montini *et al.*, 2006). In the same studies, it has also become apparent that applying a proliferative stress, such as prolonged cell culture before cell infusion or serial transplantation, can lead to the development of a leukaemic clone (Li *et al.*, 2002; Modlich *et al.*, 2006). Indeed, it would seem that simply transplanting gene-modified cells selects clones harboring retroviral integrants in the vicinity of genes which promote engraftment and/or long-term self-renewal. With this in mind, it is worth contemplating the additional risk caused by the magnitude of the replicative stress that will be placed on transduced HSC during the cytotoxic regime (Figures 22.1 and 22.3).

In addition to any selective events which may occur during *ex vivo* culture/transplant (Modlich *et al.*, 2005), gene protected HSC and progenitors will be required to expand many fold post-chemotherapeutic insult in order to support haematopoiesis in the patient (Figure 22.1). This expansion may constitute a similar proliferative stress to that enforced upon cells during serial transplantation, a procedure used in animal models to test the robustness of HSC self-renewal. To what extent chemically induced *in vivo* selection results in the establishment of dominant stem cell clones that are prone to leukaemogenesis or exhaustion remains to be determined in experimental models which span an appropriate length of time. Therefore, it is important to allow for long-term follow-up of ongoing canine and primate studies.

A related issue is the potential to selectively expand clones which contain more than one vector insertion (Kustikova *et al.*, 2003; Modlich *et al.*, 2005). This may be considered more likely to occur in the MGMT chemoprotection system where absolute levels of transgene expression determine the number of genotoxic lesions which may be repaired. Zielske and Gerson analysed the copy number in K562 cells transduced with lentiviral vectors expressing MGMT(P140K) grafts, both before and after chemoselection (Zielske and Gerson, 2004). They showed that drug selection primarily operates at the levels of MGMT protein rather than transgene copy number. In this study, cells with high transgene copy numbers (>4) were not enriched. Possible reasons include high copy number being toxic to the cell, or high levels of MGMT protein resulting in a deleterious effect at the level of the self-renewal. Whatever the mechanism, enrichment of transduced cells containing more than one proviral integration site is certainly undesirable due to the increased likelihood of vector mediated insertional mutagenesis providing pro-leukaemic hits (Modlich *et al.*, 2005). Thus, it is particularly important to limit copy number in systems where a significant *in vivo* expansion is expected in order to guarantee that cells with a single integrant form the vast majority of input cells.

A further consideration in the evaluation of the risk of a chemoprotective strategy is the effect that such a regimen would have on the occurrence of secondary therapy-related leukaemias due to the genotoxic nature of chemotherapeutic compounds. For example, the newly introduced alkylating agent temozolomide has recently been shown to cause profound mutagenesis in the haematopoietic system (Geiger *et al.*, 2006). First, when taking into account transduced protected cells, resistance genes will increase the proportion of cells which are able to survive a given dose of drug. This is likely the product of cells not having accumulated enough genetic lesions to engage the apoptotic pathway. However, it is conceivable that the result of chemoprotection is an accumulation of cells which harbour sublethal frequencies of mutagenic lesions, particularly after multiple rounds of treatment. Therefore there is a theoretical risk that a dose intensified regimen may result in the mutagenic transformation of transduced HSC or HSC that express the chemoprotective genes at suboptimal levels.

A more pronounced concern involves the fate of non-transduced cells which will be far more susceptible to chemotherapeutic insult. If chemoprotection does indeed facilitate cytotoxic dose intensification in cancer patients then it is not unreasonable to assume that, as well as increased extra-haematopoietic toxicity, the rate of therapy related transformation may increase in non-protected cells. On the other hand, there is some evidence to suggest that protecting the haematopoietic system may reduce collateral toxicity in other tissues such as the gut (Zhao *et al.*, 1997). If this phenomenon is mediated by a mechanism that does not reduce the number of genetic lesions in the non-protected tissue (e.g. by reduced production of pro-inflammatory cytokines and maintained cellular defence against invading pathogens), then an increased transformation rate is the likely outcome.

Acute toxicity (rather than mutagenicity) in unprotected tissue represents another limitation. In murine studies of MGMT-mediated chemoprotection, relatively high death rates can be observed if the chemotherapeutic regimen is too stringent. In large animals and humans, dose adjustment may be more appropriate because monitoring for side effects is technically easier and more informative (due to the availability of daily patient reports, frequent blood counts, screening for gastrointestinal bleeding, etc.). However, severe and potentially delayed organ toxicity (such as induction of lung fibrosis by BCNU, nerve toxicity by taxanes or myocardial damage by anthracyclines) remains a serious concern. As in any medical intervention, these treatment-related risks must be balanced against the prognosis of the underlying disease.

22.4 Which expression level of chemoprotective genes is appropriate?

Another point for consideration in the chemoselection strategy is at what level is it necessary to express the drug resistance gene in order to confer adequate clinical benefit? In some cases, the drug resistance gene in question encodes an enzyme; therefore above a certain threshold level of expression there will not be a directly proportional relationship between cellular concentration of drug resistance protein and concentration of drug that is toxic to the cell. However, in the case of MGMT, since the number of O^6 lesions repaired will, at best, be equimolar to the frequency of MGMT molecules within a given cell, a high level of expression will be critical in conferring resistance to an aggressive chemotherapeutic regime. Therefore, we must consider the appropriate expression level depending on the type of gene expressed.

Part of this consideration is the biological limit to overexpression; toxicity may result from an increased metabolic load, potentially inducing a growth disadvantage in the absence of selection. Furthermore, there might be more protein-specific adverse side effects of (non-physiological) high levels of protein: MDR1, an ABC-transporter protein that recognizes a wide range of endogenous and exogenous substrates, may serve as an example. In our murine models, we were not able to achieve stable engraftment levels of MDR1-vector modified cells *in vivo*, in stark contrast to cells transduced by a great many other vectors encoding different transgenes (unpublished data). Indeed, a reliable enrichment of HSCs was difficult to obtain, even following chemoselection with relatively high levels of drug that lead to severe systemic toxicity.

While this might be related to a relative lack of HSC toxicity of the drug regimen used, a relevant alternative explanation remains a selective disadvantage conferred by over-expression of this efflux pump. The only condition that allowed researchers to demonstrate the persistence of MDR1-expressing cells *in vivo* involved the transduction of cells with a relatively high gene copy number. However, these are conditions that promote preleukaemic or leukaemic transformation of haematopoietic cells, such that it cannot be excluded that persistent expression of MDR1 is primarily related to co-existing insertional mutagenesis. Whether or not other resistance proteins induce a selective disadvantage upon high levels of ectopic expression remains to be explored.

For future studies exploring the feasibility and safety of vector-mediated chemoprotection, we would therefore propose the following three steps to potential success:

1. Define the vector backbone providing optimal levels of constitutive expression
2. Explore side effects of both constitutive transgene expression and insufficient chemoprotection in competitive repopulation assays
3. Develop systems for regulated transgene expression.

22.5 Vector design to achieve optimal expression levels

The development of suitable constructs for constitutive expression of chemoresistance genes follows established principles of designing transgene expression cassettes. These include the use of a suitable enhancer-promoter to initiate transcription on the cell type of choice, the optimization of RNA processing to provide maximal mRNA integrity, export, stability and translation, and the insulation of the expression cassette against disturbing influences of the cellular genome (Ellis, 2005; Baum *et al.*, 2006b). Additionally, the cellular genome also needs to be protected against side effects that are potentially triggered by the insertion of the transgene cassette (Montini *et al.*, 2006; Modlich *et al.*, 2006).

Traditionally, the strongest enhancer-promoter elements active in mammalian cells were derived from viruses. For HSC and their progeny, various enhancer-promoter sequences of murine leukaemia viruses, the avian retrovirus Rous sarcoma virus, the cytomegalovirus, or hybrids of these elements with cellular promoters have been proposed to achieve maximal gene expression levels (Baum *et al.*, 1995; Challita *et al.*, 1995; Ramezani *et al.*, 2000, 2003; Logan *et al.*, 2002). While this search for powerful enhancer-promoter elements showed increasing success, safety concerns also accumulated that such unphysiological sequences might be particularly dangerous in activating neighboring cellular genes. As a cell is not expected to be able to distinguish viral from cell-derived sequences, it appears likely that the risk of insertional mutagenesis primarily reflects the strength of the enhancer and not its developmental origin. It may not only reflect coincidence that all cases of malignant transformation observed after gammaretroviral or lentiviral gene delivery into haematopoietic cells involved constructs containing strong viral enhancer- promoter sequences (Li *et al.*, 2002; Hacein-Bey-Abina *et al.*, 2003; Du *et al.*, 2005a; Modlich *et al.*, 2005; Lagresle-Peyrou *et al.*, 2006; Modlich *et al.*, 2006; Seggewiss *et al.*, 2006; Shou *et al.*, 2006; Woods *et al.*, 2006).

Widely used cellular constitutive promoters include the elongation factor 1 alpha (EF1a) promoter and the phosphoglycerate kinase (PGK) promoter (Woods *et al.*, 2000; Ramezani *et al.*, 2000, 2003, Salmon *et al.*, 2000; Schambach *et al.*, 2006b). The former exists in two forms, the longer version of which contains an untranslated first exon followed by an intron. Compared to a powerful MLV-derived enhancer-promoter such as that from the polycythemic strain of the spleen-focus-forming virus (SFFVp), the expression levels of EF1a and PGK promoters are 3–10-fold lower (per transgene copy) (Schambach *et al.*, 2006b). The SFFVp enhancer-promoter is strong enough to mediate relatively high levels of chemoprotection from a single vector copy, as demonstrated for MDR1 and MGMT P140K (Baum *et al.*, 1995; Schambach *et al.*, 2006b). Although constructs driven by EF1a and PGK may also confer chemoprotection, there may already be an increased risk of selecting cells with multiple insertions when using these promoters. A systematic search for promoters associated with genes that are relatively highly expressed in HSC and their progeny may lead to the discovery of even stronger cellular enhancer–promoters.

At least as important as the discovery of suitable promoters is the improvement of mRNA processing (Krall *et al.*, 1996; Kim *et al.*, 1998; Hildinger *et al.*, 1999; Zufferey *et al.*, 1999; Schambach *et al.*, 2000; 2006a; Hope, 2002; Kraunus *et al.*, 2004; Lee *et al.*, 2004). Post-transcriptional mRNA processing is particularly suboptimal in traditional retroviral vectors which place the promoter in the long terminal repeats (LTRs). This configuration results in a relatively large untranslated leader region that precedes the cDNA of choice and contains highly folded structures such as the retroviral packaging signal and upstream sequences comprising R, U5 and the retroviral primer binding site (PBS). An intron generated by splice sites in the 5′ untranslated region (UTR) promotes mRNA export and translation, both leading to higher transgene expression (Krall *et al.*, 1996; Kim *et al.*, 1998; Hildinger *et al.*, 1999; Kraunus *et al.*, 2004). If the packaging signal is contained within the intron, it will be maintained in all retroviral transgenes incorporated into target cells. An intron including the packaging signal may also be formed in the SIN vector context, by placing the promoter in between the PBS and the splice donor. Surprisingly, this configuration enhances splicing of the gammaretroviral intron (Kraunus *et al.*, 2004). Similar constructs may be generated in the context of lentiviral (Dull *et al.*, 1998) or spumaviral (Trobridge *et al.*, 2002) backbones. However, if the intron does not contain the packaging signal, it will be lost by a splice event in the packaging cell. Gammaretroviral LTR vectors are typically designed to undergo balanced splicing of the leader in the packaging cell, such that enough genomic RNA is formed to generate high vector titers (Krall *et al.*, 1996; Hildinger *et al.*, 1999; Lee *et al.*, 2004). In the transduced target cells, the retroviral intron may be more efficiently processed than in the packaging cell, thus improving post-transcriptional mRNA processing.

Human immunodeficiency virus-based vectors may maintain introns even if these do not contain the packaging signal. This 'stabilization' of the genomic mRNA in packaging cells may depend on the presence of the rev-responsive element (RRE) and its interaction with Rev expressed in the packaging cell. Rev then guides the lentiviral transcript into CRM-1 dependent nuclear export pathway, potentially preventing splicing. In target cells that do not express Rev, the intron may work with full efficiency. However, it has been demonstrated that strong splice sites are only partially 'protected' by the Rev-RRE interaction (Chang and Sharp, 1989; Zaiss *et al.*, 2002). Therefore, a considerable percentage of the integrated proviruses may have lost the intron, potentially increasing intercellular variability of transgene expression.

Improved 3′ end processing may also significantly improve vector performance. In the context of LTR-driven retroviral vectors, combining the posttranscriptional regulatory element (PRE) of the woodchuck hepatitis virus (wPRE) in the 3′ UTR with an alternatively spliced intron in the 5′ UTR currently represents the optimal design to achieve high titers and protein expression (Schambach *et al.*, 2000). However, increasing evidence suggests that the LTR configuration with strong enhancers is particularly prone to induce insertional upregulation of cellular proto-oncogenes (Montini *et al.*, 2006; Modlich *et al.*, 2006). The advent of lentiviral vectors has revived the so-called self-inactivating (SIN) configuration of retroviral vectors (Dull *et al.*, 1998). As originally proposed by Gilboa *et al.* for gammaretroviral constructs (Yu *et al.*, 1986), SIN constructs promise increased vector biosafety by eliminating active enhancer-promoters from the LTR and rather moving these elements into the region 3′ of the packaging signal, upstream of the cDNA(s) to be expressed. While SIN vectors initially suffered from low titers, improving the plasmid's 5′ promoters to efficiently drive the full-length retroviral mRNA has overcome this drawback (Dull *et al.*, 1998; Schambach *et al.*, 2006c). This 5′ promoter will be lost after transduction of target cells, as it is not part of the retroviral genomic RNA.

The internal expression cassette of SIN vectors may also be placed in antisense orientation to the genomic RNA of the vector. Although this architecture might reduce viral titers by formation of antisense RNA in packaging cells, some transgene cassettes appear to be more stable when cloned in this orientation. The best studied example is the complex erythroid cell-specific globin cassette which destabilizes retroviral and lentiviral vectors when cloned in the sense direction of the vector genome, probably due to the presence of cryptic termination motifs. Lentiviral SIN vectors containing the same cassette in the antisense orientation can be produced at reasonable titers (Leboulch *et al.*, 1994; May *et al.*, 2000; Pawliuk *et al.*, 2001; Persons *et al.*, 2003b; Hanawa *et al.*, 2004; Imren *et al.*, 2004; Puthenveetil *et al.*, 2004). However, in the majority of cases, the sense orientation of the internal expression cassette is preferred.

Internal bidirectional promoters may even allow the simultaneous and co-ordinated expression of two cDNAs from a single integrated vector genome (Amendola *et al.*, 2005). Bidirectional promoters also tend to form antisense mRNA in packaging cells, and are therefore often suboptimal when aiming for high vector titres. Again, using a powerful 5' promoter to drive the genomic RNA in packaging cells may partially overcome this limitation. Alternative ways of co-expression include the use of IRES or the self-cleaving 2A proteinase sequence derived from the foot-and-mouth disease virus. IRES elements typically act in a very context-dependent manner, depending on the nature of the flanking sequences and the cell type investigated. Although an increasing number of IRESes (both cellular and viral) has been defined over the recent years, to date none of these sequences has emerged as a standard for the design of co-expression constructs. The 2A sequences have a different mode of action in that they mediate the co-translational separation of a nascent polypeptide chain *in cis* (Klump *et al.*, 2001; de Felipe, 1999, 2004; Milsom *et al.*, 2004; Chinnasamy *et al.*, 2006). This leaves a stretch of $\sim$20 amino acids at the C-terminus of the N-terminal protein chain and a single proline at the N-terminus of the C-terminal portion of the protein. Fusion proteins in which the 2A sequence failed to trigger cleavage may be formed with a frequency of about 10 per cent of the N-terminal protein part (Schiedlmeier *et al.*, 2003). The frequency of the uncleaved fusion protein may be cell-type dependent. The C-terminal protein usually represents not more than 15 per cent (1/7) of the N-terminal part, unless it has a relatively longer half-life. Using degenerated 2A sequences several proteins can be expressed from a single precursor polyprotein. Of course, all major co-expression strategies can be combined to potentially express four or more different proteins from a single integrated vector copy. However, as a rule of thumb, more complex and larger expression cassettes typically compromise vector titres and increase the risk of forming rearranged vector genomes (de Felipe and Izquierdo, 2003).

Rearranged vector genomes can be triggered by cryptic splice sites present within cDNAs or regulatory sequences such as promoters, insulators, matrix attachment regions, cis-elements introduced to affect other aspects of mRNA processing, or larger polylinkers. In the context of chemoprotection, the best studied case of a cryptic intron is that present in the relatively large human MDR1 cDNA (Sorrentino *et al.*, 1995). Although the major cryptic intron of MDR1 is readily identified by sequencing of shortened forms of mRNA and reverse transcribed vector genomes, alternative splicing of this relatively large cDNA is difficult to suppress since cDNAs with mutations in cryptic splice sites tend to undergo alternative splicing mediated by neighboring cryptic splice sites that become activated once a more dominant splice site has been removed (Cmejlova *et al.*, 2003). Interestingly, the extent of cryptic splicing in MDR1 depends upon the background of the cell in which it is expressed, reminiscent of differentiation-specific alternative splicing of cellular genes (Cmejlova *et al.*, 2003). Codon-optimization of open reading frames may contribute to improved mRNA stability, and promote both nuclear export and translation (Wagner *et al.*, 2000).

22.6 Exploring side effects of continued transgene expression and insufficient chemoprotection

Having defined a vector configuration that mediates sufficient levels of transgene expression from a single vector copy, extended biosafety studies need to be conducted in preclinical models to test the safety and efficiency of genetic chemoprotection. To date, no standards have been appointed for such studies. We prefer the C57BL6 mouse model of bone marrow transplantation since it allows to distinguish host and donor cells on the basis of a polymorphism on the common leukocyte antigen (CD45) and has a relatively low incidence of endogenous tumor formation if recipient mice are irradiated with a dose that eliminates the majority of the endogenous HSC and observed for up to 7 months (Modlich *et al.*, 2005). Moreover, activation of endogenous retroviruses seems to be an infrequent event in this mouse strain. This model thus allows studies of competitive repopulation, testing the hypothesis that the cell population expressing the chemoprotective transgene(s) confers a selective advantage or disadvantage. The integrity of haematopoiesis can be monitored by conventional cytology, flow cytometry and histopathology, following recommended standards (Kogan *et al.*, 2002; Morse *et al.*, 2002). In the case of leukaemic complications, the model has been shown to allow the detection of insertional leukaemias (Li *et al.*, 2002; Modlich *et al.*, 2005). Serial transplantation of engineered haematopoietic cells is required to demonstrate that chemoprotection has been achieved at the level of the long-term repopulating HSC.

In the murine model, the most convincing evidence of chemoprotection has been obtained with MGMT and DHFR (Allay *et al.*, 1998; Ragg *et al.*, 2000; Sawai *et al.*, 2001). Studies with retrovirally expressed MDR1 have not convincingly shown that HSC can be protected from dose-limiting side effects of chemotherapy, although transient and in some cases more prolonged selection could be demonstrated (Carpinteiro *et al.*, 2002; Podda *et al.*, 1992; Sorrentino *et al.*, 1992). This may be a consequence of the cytotoxic agents used. If these are not toxic at the level of a poorly replicating stem cell, selection can hardly be demonstrated (Blau *et al.*, 1997). However, as discussed above, it is also possible that constitutive over-expression of MDR1 or other drug resistance genes impairs the competitive fitness of HSC. Further studies exploring this hypothesis are required.

Side effects of insufficient chemoprotection may be overlooked unless assays are designed that are sensitive enough to demonstrate the presence of chromosomal aberrations and other mutations in transduced and untransduced cells. This issue also deserved greater attention in future studies (Seggewiss *et al.*, 2006).

22.7 The future: inducible expression of drug resistance genes

A perfect solution to benefit from the promises of chemoprotection and to avoid complications related to over-expression of drug-resistance genes would be the inducible expression of the drug resistance phenotype. Regulation can be achieved at the level of the protein, e.g. by fusing MGMT with the oestrogen-receptor domain that mediates conditional nuclear translocation in the presence of tamoxifen, as exemplified with the Cre recombinase (Littlewood *et al.*, 1995; Metzger *et al.*, 1995). More elegant would be inducible transcription, e.g. using lentiviral or gammaretroviral vectors that employ the tetracycline-dependent transactivator and corresponding expression cassette (Kafri *et al.*, 2000; Vigna *et al.*, 2002; Barde *et al.*, 2005). If the expression unit is potent enough, sufficient levels of protein expression may be achieved in the induced state. Cells may be allowed to convert to the repressed state when the serum levels of the cytotoxic agents have declined (Figure 22.4). The remaining challenge is the potential immunogenicity and toxicity of the artificial transactivator whose expression is required to regulate the expression cassette of interest. In the context of chemoprotection which typically involves hardly immunogenic proteins that are also not toxic unless being expressed at very high levels, residual background activity of the uninduced promoter appears to be a negligible limitation of the inducible expression strategy. Alternatively, episomal vectors might be also be sufficient (Rund *et al.*, 1998), if the following three conditions are fulfilled: chemotherapy cycles are not given over prolonged periods of time, the initial rate of episome establishment in haematopoietic cells can be improved, and the duration of episome maintenance in cycling cells can be prolonged. Episomal plasmids containing a scaffold

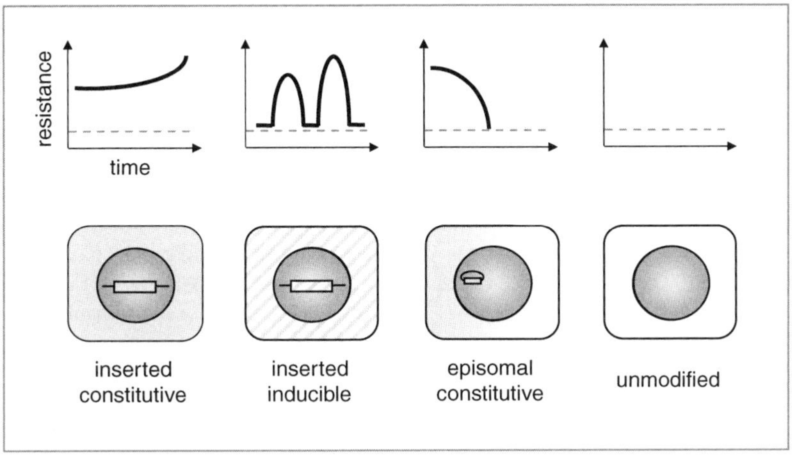

Figure 22.4 Three major principles to establish and express drug-resistance genes in somatic cells. Further details in the text

attachment region have been shown to be maintained in several cell types (Bode and Maass, 1988; Baiker *et al.*, 2000; Jenke *et al.*, 2004), but initial reports from haematopoietic cells have been less convincing (Papapetrou *et al.*, 2006). Future studies need to address these issues in great detail before human clinical trials can be started.

Acknowledgements

We are grateful for financial support by the Bundesministerium für Bildung und Forschung (BMBF), the Deutsche Forschungsgemeinschaft (DFG), the European Union (EU), and the National Institutes of Health (NIH).

References

Abonour R, Williams DA, Einhorn L, *et al.* (2000). *Nat Med* **6**, 652–658.

Aiuti A, Slavin S, Aker M, *et al.* (2002). *Science* **296**, 2410–2413.

Allay JA, Persons DA, Galipeau J, *et al.* (1998). *Nat Med* **4**, 1136–1143.

Ambudkar SV, Kimchi-Sarfaty C, Sauna ZE, Gottesman MM (2003). *Oncogene* **22**, 7468–7485.

Amendola M, Venneri MA, Biffi A, Vigna E, Naldini L (2005). *Nat Biotechnol* **23**, 108–116.

Baiker A, Maercker C, Piechaczek C, *et al.* (2000). *Nat Cell Biol* **2**, 182–184.

Barde I, Zanta-Boussif MA, Paisant, *et al.* (2006). *Mol Ther*, **13**, 382–390.

Baum C, Hegewisch-Becker S, Eckert HG, Stocking C, Ostertag W (1995). *J Virol* **69**, 7541–7547.

Baum C, Dullmann J, Li Z, *et al.* (2003). *Blood* **101**, 2099–2114.

Baum C, Kustikova O, Modlich U, Li Z, Fehse B (2006a). *Hum Gene Ther* **17**, 253–263.

Baum C, Schambach A, Bohne J, Galla M (2006b). *Mol Ther* **13**, 1050–1062.

Blau CA, Neff T, Papayannopoulou T (1997). *Blood* **89**, 146–154.

Bode J, Maass K (1988). *Biochemistry* **27**, 4706–4711.

Bunting KD, Galipeau J, Topham D, Benaim E, Sorrentino BP (1998). *Blood* **92**, 2269–2279.

Bunting KD, Zhou S, Lu T, Sorrentino BP (2000). *Blood* **96**, 902–909.

Carpinteiro A, Peinert S, Ostertag W, *et al.* (2002). *Int J Cancer* **98**, 785–792.

Cavazzana-Calvo M, Hacein-Bey S, de Saint Basile G, *et al.* (2000). *Science* **288**, 669–672.

Challita PM, Skelton D, el-Khoueiry A, Yu XJ, Weinberg K, Kohn DB (1995). *J Virol* **69**, 748–755.

Chang DD, Sharp PA (1989). *Cell* **59**, 789–795.

Chinnasamy D, Milsom MD, Shaffer J, *et al.* (2006). *Virol J* **3**, 14.

Cmejlova J, Hildinger M, Cmejla R, *et al.* (2003). *Gene Ther* **10**, 1061–1065.

Corey CA, DeSilva AD, Holland CA, Williams DA (1990). *Blood* **75**, 337–343.

Cornetta K, Croop J, Dropcho E, *et al.* (2006). *Cancer Gene Ther* **13**, 886–895.

Davé U, Jenkins NA, Copeland NG (2004). *Science* **303**, 333.

Davis BM, Humeau L, Dropulic B (2004). *Mol Ther* **9**, 160–172.

Davis BM, Roth JC, Liu L, Xu-Welliver M, Pegg AE, Gerson SL (1999). *Hum Gene Ther* **10**, 2769–2778.

de Felipe P (2004). *Genet Vaccines Ther* **2**, 13.

de Felipe P, Izquierdo M (2003). *J Gen Virol* **84**, 1281–1285.

de Felipe P, Martin V, Cortes ML, Ryan M, Izquierdo M (1999). *Gene Ther* **6**, 198–208.

Du Y, Jenkins NA, Copeland NG (2005a). *Blood* **106**, 3932–3939.

Du Y, Spence SE, Jenkins NA, Copeland NG (2005b). *Blood* **106**, 2498–2505.

Dull T, Zufferey R, Kelly M, *et al.* (1998). *J Virol* **72**, 8463–8471.

Ellis J (2005). *Hum Gene Ther* **16**, 1241–1246.

Gaspar HB, Parsley KL, Howe S, *et al.* (2004). *Lancet* **364**, 2181–2187.

Geiger H, Schleimer D, Nattamai KJ, Dannenmann SR, Davies SM, Weiss BD (2006). *Blood* **107**, 3010–3011.

Gerson SL (2004). *Nat Rev Cancer* **4**, 296–307.

Hacein-Bey-Abina S, Le Deist F, Carlier F, *et al.* (2002). *N Engl J Med* **346**, 1185–1193.

Hacein-Bey-Abina S, Von Kalle C, Schmidt M, *et al.* (2003). *Science* **302**, 415–419.

Hanawa H, Hargrove PW, Kepes S, Srivastava DK, Nienhuis AW, Persons DA (2004). *Blood* **104**, 2281–2290.

Hesdorffer C, Ayello J, Ward M, *et al.* (1998). *J Clin Oncol* **16**, 165–172.

Heyworth CM, Gagen D, Edington KG, Fairbairn LJ (2002). *Leukemia* **16**, 106–111.

Hildinger M, Abel KL, Ostertag W, Baum C (1999). *J Virol* **73**, 4083–4089.

Hope T (2002). *Curr Top Microbiol Immunol* **261**, 179–189.

Imren S, Fabry ME, Westerman KA, *et al.* (2004). *J Clin Invest* **114**, 953–962.

Jenke AW, Stehle IM, Eisenberger T, *et al.* (2004). *Proc Natl Acad Sci USA* **101**, 11322–11327.

Jin L, Zeng H, Chien S, *et al.* (2000). *Nat Genet* **26**, 64–66.

Kafri T, van Praag H, Gage FH, Verma IM (2000). *Mol Ther* **1**, 516–521.

Kim SH, Yu SS, Park JS, Robbins PD, An CS, Kim S (1998). *J Virol* **72**, 994–1004.

Klump H, Schiedlmeier B, Vogt B, Ryan M, Ostertag W, Baum C (2001). *Gene Ther* **8**, 811–817.

Kogan SC, Ward JM, Anver MR, *et al.* (2002). *Blood* **100**, 238–245.

Kohn DB, Sadelain M, Glorioso JC (2003). *Nat Rev Cancer* **3**, 477–488.

Krall WJ, Skelton DC, Yu XJ, *et al.* (1996). *Gene Ther* **3**, 37–48.

Kraunus J, Schaumann DHS, Meyer J, *et al.* (2004). *Gene Ther* **11**, 1568–1578.

Kustikova OS, Wahlers A, Kuehlcke K, *et al.* (2003). *Blood* **102**, 3934–3937.

Kustikova OS, Fehse B, Dllmann J, *et al.* (2005). *Science* **308**, 1171–1174.

Lagresle-Peyrou C, Yates F, Malassis-Seris M, *et al.* (2006). *Blood* **107**, 63–72.

Leboulch P, Huang GM, Humphries RK, *et al.* (1994). *EMBO J* **13**, 3065–3076.

Lee JT, Yu SS, Han E, Kim S (2004). *Gene Ther* **11**, 94–99.

Lee K, Gerson SL, Maitra B, Koc ON (2001). *J Hematother Stem Cell Res* **10**, 691–701.

Li Z, Dullmann J, Schiedlmeier B, *et al.* (2002). *Science* **296**, 497.

Licht T, Haskins M, Henthorn P, *et al.* (2002). *Proc Natl Acad Sci U S A* **99**, 3123–3128.

Littlewood TD, Hancock DC, Danielian PS, Parker MG, Evan GI (1995). *Nucl Acids Res* **23**, 1686–1690.

Logan AC, Lutzko C, Kohn DB (2002). *Curr Opin Biotechnol* **13**, 429–436.

May C, Gunther R, McIvor RS (1995). *Blood* **86**, 2439–2448.

May C, Rivella S, Callegari J, *et al.* (2000). *Nature* **406**, 82–86.

Metzger D, Clifford J, Chiba H, Chambon P (1995). *Proc Natl Acad Sci U S A* **92**, 6991–6995.

Milsom MD, Fairbairn LJ (2004). *J Gene Med* **6**, 133–146.

Milsom MD, Woolford LB, Margison GP, Humphries RK, Fairbairn LJ (2004). *Mol Ther* **10**, 862–873.

Modlich U, Bohne J, Schmidt M, *et al.* (2006). Cell-culture assays reveal the importance of retroviral vector design for insertional genotoxicity. *Blood* **108**, 2545–2553.

Modlich U, Kustikova O, Schmidt M, *et al.* (2005). *Blood* **105**, 4235–4246.

Montini E, Cesana D, Schmidt M, *et al.* (2006). *Nat Biotechnol* **24**, 687–696.

Morse HC, Anver MR, Fredrickson TN, *et al.* (2002). *Blood* **100**, 246–258.

Moscow J A, Huang H, Carter C, *et al.* (1999). *Blood* **94**, 52–61.

Neff T, Beard BC, Peterson LJ, Anandakumar P, Thompson J, Kiem HP (2005). *Blood* **105**, 997–1002.

Neff T, Beard BC, Kiem HP (2006). *Blood* **107**, 1751–1760.
Nienhuis AW, Dunbar CE, Sorrentino BP (2006). *Mol Ther* **13**, 1031–1049.
Ott MG, Schmidt M, Schwarzwaelder K, *et al.* (2006). *Nat Med* **12**, 401–409.
Papapetrou EP, Ziros PG, Micheva ID, Zoumbos NC, Athanassiadou A (2006). *Gene Ther* **13**, 40–51.
Pawliuk R, Westerman KA, Fabry ME, *et al.* (2001). *Science* **294**, 2368–2371.
Persons DA, Allay ER, Sawai N, *et al.* (2003a). *Blood* **102**, 506–513.
Persons DA, Hargrove PW, Allay ER, Hanawa H, Nienhuis AW (2003b). *Blood* **101**, 2175–2183.
Persons DA, Allay JA, Bonifacino A, *et al.* (2004). *Blood* **103**, 796–803.
Pike-Overzet K, de Ridder D, Weerkamp F, *et al.* (2006). *Nature* **443**, E5; discussion E6–7.
Podda S, Ward M, Himelstein A, *et al.* (1992). *Proc Natl Acad Sci U S A* **89**, 9676–9680.
Puthenveetil G, Scholes J, Carbonell D, *et al.* (2004). *Blood* **104**, 3445–3453.
Ragg S, Xu-Welliver M, Bailey J, *et al.* (2000). *Cancer Res* **60**, 5187–5195.
Ramezani A, Hawley TS, Hawley RG (2000). *Mol Ther* **2**, 458–469.
Ramezani A, Hawley TS, Hawley RG (2003). *Blood* **101**, 4717–4724.
Richard E, Robert E, Cario-Andre M, *et al.* (2004). *Gene Ther* **11**, 1638–1647.
Rund D, Dagan M, Dalyot-Herman N, *et al.* (1998). *Hum Gene Ther* **9**, 649–657.
Salmon P, Kindler V, Ducrey O, Chapuis B, Zubler RH, Trono D (2000). *Blood* **96**, 3392–3398.
Sauvageau G, Thorsteinsdottir U, Eaves CJ, *et al.* (1995). *Genes Dev* **9**, 1753–1765.
Sawai N, Zhou S, Vanin EF, Houghton P, Brent TP, Sorrentino BP (2001). *Mol Ther* **3**, 78–87.
Schambach A, Wodrich H, Hildinger M, Bohne J, Kraussl ich HG, Baum C (2000). *Mol Ther* **2**, 435–445.
Schambach A, Bohne J, Baum C, *et al.* (2006a). *Gene Ther* **13**, 641–645.
Schambach A, Bohne J, Chandra S, *et al.* (2006b). *Mol Ther* **13**, 391–400.
Schambach A, Zychlinsk D, Galla M, *et al.* (2006c). *Gene Ther.* **Jun 8**, (epub ahead of print)
Schambach A, Schiedlmeier B, Kuhlcke K, *et al.* (2006d). *Gene Ther* **13**, 1037–1047.
Schiedlmeier B, Klump H, Will E, *et al.* (2003). *Blood* **101**, 1759–1768.
Seggewiss R, Pittaluga S, Adler RL, *et al.* (2006). *Blood* **107**, 3865–3867.
Sellers SE, Tisdale JF, Agricola BA, *et al.* (2001). *Blood* **97**, 1888–1891.
Shou Y, Ma Z, Lu T, Sorrentino BP (2006). *Proc Natl Acad Sci U S A* **103**, 11730–11735.
Sorrentino BP (2002). *Nat Rev Cancer* **2**, 431–441.
Sorrentino BP, Brandt SJ, Bodine D, *et al.* (1992). *Science* **257**, 99–103.
Sorrentino BP, McDonagh KT, Woods D, Orlic D (1995). *Blood* **86**, 491–501.
Thrasher AJ, Gaspar HB, Baum C, *et al.* (2006). *Nature* **443**, E5–6; discussion E6–7.
Trobridge G, Josephson N, Vassilopoulos G, Mac J, Russell DW (2002). Improved foamy virus vectors with minimal viral sequences. *Mol Ther* **6**, 321–328.
Turner SD, Rafferty JA, Fairbairn LJ, *et al.* (2000). *Leukemia* **14**, 1796–1802.
Vigna E, Cavalieri S, Ailles L, *et al.* (2002). *Mol Ther* **5**, 252–261.
Wagner R, Graf M, Bieler K, *et al.* (2000). *Hum Gene Ther* **11**, 2403–2413.
Williams DA, Hsieh K, DeSilva A, Mulligan RC (1987). *J Exp Med* **166**, 210–218.
Woods NB, Fahlman C, Mikkola H, *et al.* (2000). *Blood* **96**, 3725–3733.
Woods NB, Bottero V, Schmidt M, von Kalle C, Verma IM (2006). *Nature* **440**, 1123.
Yu SF, von Ruden T, Kantoff PW, *et al.* (1986). *Proc Natl Acad Sci U S A* **83**, 3194–3198.
Zaiss AK, Son S, Chang LJ (2002). *J Virol* **76**, 7209–7219.
Zhao SC, Banerjee D, Mineishi S, Bertino JR (1997). *Hum Gene Ther* **8**, 903–909.
Zielske SP, Gerson SL (2002). *Mol Ther* **5**, 381–387.
Zielske SP, Gerson SL (2004). *Mol Ther* **9**, 923–931.
Zielske SP, Reese JS, Lingas KT, Donze JR, Gerson SL (2003). *J Clin Invest* **112**, 1561–1170.
Zufferey R, Donello JE, Trono D, Hope ZJ (1999). *J Virol* **73**, 2886–2892.

Index

5T4 107–8, 287

AAV *see* adeno-associated virus
ABC transporters 380–1, 384
aciclovir (ACV) 26
  herpes simplex virus 118, 120, 126
  radiosensitization 318
  retroviruses 76
adeno-associated virus (AAV) 55–68
  amplicon vectors 31
  angiogenesis 60, 300, 301, 302–3
  anti-oncogenic vectors 58–9
  antisense oligonucleotides 251
  bacteriophages 302–3
  clinical trials 62–3
  combination therapies 60
  cytotoxic gene therapies 58
  gene correction 58
  gene-directed enzyme prodrug therapy 257, 267–8
  immune responses 61–2
  lifecycle 55–6
  limitations 217, 218
  molecular chemotherapy 59
  radiosensitization 314, 317
  recombinant vectors 55–7, 59–63
  retargeting 62
  RNA interference pathway 246, 248, 250
  serotypes 56–7
  T helper 1/cytotoxic T lymphocytes 60–1
  transductional targeting 62
  transgene expression 59–60, 61, 62
  tropism modifications 62
  tumour-specific delivery 62
adenosine deaminase 88
adenovirus death protein 4, 265
adenoviruses 1–17
  angiogenesis 299–301
  antisense oligonucleotides 251
  clinical trials 7–10, 13–14
  first generation 3, 5, 13
  fusogenic membrane glycoproteins 125–6
  gene replacement therapy 11–12
  gene-directed enzyme prodrug therapy 257–60, 264–6
  immunomodulation 7, 10, 282, 286
  limitations 217
  oncolytic vectors 12–13, 257–60, 264–6
  quantitation/toxicity 13–14
  radiosensitization 313–14, 317
  replication-defective vectors 257–60, 264, 299–300
  retargeting 6–7
  RNA interference pathway 248, 250
  second generation 3, 5
  serotypes 1, 2
  structure/lifecycle 1–5
  suicide gene therapy 7, 10–11
  third generation 3, 5–6
  transcription/replication 2–5
  transcriptional targeting 7
  transductional targeting 6–7, 10
adoptive transfer 278
alphaviruses 217–28
  angiogenesis 221–2
  apoptosis 221
  delivery systems 217
  development of vectors 223–4
  gene expression vectors 218, 220–1
  genome 218–19
  immunomodulation 222, 282
  lifecycle 219–20
  oncolytic vectors 222–3
  replication-defective vectors 218, 221

Viral Therapy of Cancer Edited by Kevin J. Harrington, Richard G. Vile and Hardev S. Pandha

alphaviruses (*Continued*)
  Semliki Forest virus 217–19, 221–5, 283
  Sindbis virus 218–19, 221–5
  vaccines 222
  Venezuelan equine encephalitis virus 218, 221–2
ALV *see* avian leukosis virus
ALVAC vectors 108–9, 284, 286
amifostine 342
amplicon vectors 24, 25, 30–2
anaplastic astrocytomas 25
angiogenesis 295–312
  adeno-associated virus 60, 300, 301, 302–3
  adenoviruses 299–301
  alphaviruses 221–2
  bacteriophages 302–3
  DF3 promoters 304
  endogenous negative regulators 295–6, 298
  growth factors 295–7
  herpes simplex virus 300, 302
  lentiviruses 301–2
  measles virus 300
  preproendothelin-1 promoters 304
  retroviruses 300, 301
  RGD targeting 298, 305–6
  targeting tumour vasculature 297–9
  TNFerade 305
  transcriptional targeting 303–5
  transductional targeting 305–6
  treatment responsive promoters 304–5
  tumour development/metastasis 295–7
angiostatin 60, 295–6, 298, 300
antiangiogenic gene delivery *see* angiogenesis
anticancer immunogens 101–2
antigen-presenting cells (APC)
  immunomodulation 277, 278
  lentiviral vectors 86
  Newcastle disease virus 174–5, 177
  poxviruses 102
antigen-specific therapies 277–8
antioncogenic vectors 58–9
antioxidant gene therapy (MnSOD-PL) 342–54
antisense oligonucleotides 241, 243–4
  delivery methods 248, 251–2
  innate immunity 247
  off-target effects 244–6
  RNA interference pathway 243–4, 248
APC *see* antigen-presenting cells
apoptosis
  alphaviruses 221
  chemoprotective gene delivery 377, 382, 384
  herpes simplex virus 31, 35–6
  reovirus 145
  tumour-suppressor genes 230–5
  vaccinia virus 157, 158
attenuation 192, 197, 211, 212
ATV-NDV antitumour vaccine 176–80
autologous tumour cells 176–80
avian leukosis virus (ALV) 72
avipoxviruses 95, 99, 108–10, 281–2

β-galactosidase 162–3
β-globin 88–9, 91
β-thalassaemia 88–9, 91
B-cell malignancies 208
B7.1 103–4
bacille Calmette–Guérin (BCG) 277
bacterial artificial chromosome 117
bacteriophages 302–3
BCG *see* bacille Calmette–Guérin
Bin1 gene 34–5
bispecific antibodies/fusion proteins 179–80
bone marrow (BM)
  chemoprotective gene delivery 370–80
  lentiviral vectors 87–8, 90
  Newcastle disease virus 177
  progenitor cells 90, 353–4, 355–7
  radioprotective gene therapy 350–7
  retroviruses 72
  vaccinia virus 154
breast cancers 329
bystander effects
  gene-directed enzyme prodrug therapy 255–6, 320
  herpes simplex virus 31
  radioprotective gene therapy 341, 351
  radiosensitization 320
  retroviruses 76
  vaccinia virus 160–1, 166

C-type retroviral vectors 72
CAM *see* cellular adhesion molecules
cancer-associated fibroblasts 297
candidate stem cells 350
canine distemper virus (CDV) 206
canstatin 295–6
Cap gene 55–6, 57
CAR *see* coxsackie and adenovirus receptor
carboplatin 234, 348, 354
carboxylesterase 259, 262
carboxypeptidases 258, 261
carcinoembryonic antigen (CEA) 104–5, 280, 282, 284–5
CB1954 258, 261
CDV *see* canine distemper virus
CEA *see* carcinoembryonic antigen

CED *see* convection-enhanced delivery
cell cycle disruption 145
cellular adhesion molecules (CAM) 109
cervical intraepithelial neoplasia (CIN) 289
CGD *see* chronic granulomatous disease
chemoprotective gene delivery 377–91
  acute toxicity 384, 387–8
  advantages 377–81
  apoptosis 377, 382, 384
  chemoselection strategies 377–84
  clinical trials 381, 389
  dihydrofolate reductase 381–2
  disadvantages 381–4, 387–8
  expression levels 384–7
  haematopoietic stem cells 377–81, 383–6, 388
  inducible drug resistance genes 388–9
  methylguanine-DNA-methyltransferase 381, 383–4, 385, 388
  multidrug resistance 380–1, 382, 384–5, 387–8
  transgene expression 381–2, 384–6, 387–8
chemoradiotherapy (CRT) 344, 345–6, 348, 354
chemoselection strategies 377–84
chronic granulomatous disease (CGD) 379
chronic lymphocytic leukaemia (CLL) 252
CIN *see* cervical intraepithelial neoplasia
cisplatin
  adenoviruses 11, 13
  antisense oligonucleotides 251
  herpes simplex virus 39, 128–9
  tumour-suppressor genes 233, 234
CLL *see* chronic lymphocytic leukaemia
CMV *see* cytomegalovirus
Colorado tick fever virus (CTFV) 139
colorectal cancer 287
combination therapies
  adeno-associated virus 60
  herpes simplex virus 39–42, 127–9
  measles virus 211
  reovirus 147–8
  tumour-suppressor genes 233–5
  vesicular stomatitis virus 194
convection-enhanced delivery (CED) 36–7
coxsackie and adenovirus receptor (CAR) 1–2, 6–7, 13, 313
CPT-11 259, 262
Cre recombinase 388
CRT *see* chemoradiotherapy
CTFV *see* Colorado tick fever virus
CTL *see* cytotoxic T lymphocytes
cyclophosphamide 37
cystic fibrosis 5, 63
cytochrome P450 259, 261–2, 319
cytomegalovirus (CMV) 7, 250, 333–6
cytopathic effect 208–9
cytoprotective responses 347–8, 352–3
cytosine deaminase 31, 256, 261, 318–19
cytotoxic gene therapies 58
cytotoxic T lymphocytes (CTL)
  adeno-associated virus 60–1
  immunomodulation 279, 281–2, 286, 288
  Newcastle disease virus 175, 177
  poxviruses 102, 108
  vesicular stomatitis virus 189, 193, 195, 197

danger model of immunity 174, 175–7, 197
dendritic cells (DC)
  adeno-associated virus 62
  immunomodulation 277, 278
  lentiviral vectors 90–1
  Newcastle disease virus 174–5
deoxyribozymes 241–2
DF3 promoters 304
dihydrofolate reductase 381–2
direct tumour killing *see* angiogenesis
docetaxel 106, 233–4
dominant negative approaches 241, 252
dopamine D2 receptors 333
double suicide gene therapy 319–20
doxorubicin 233
drug resistance genes 388–9

E2F-1 316
early growth response-1 (EGR-1) gene 320–2
EBV *see* Epstein–Barr virus
EC *see* endothelial cells
ECM *see* extracellular matrix
EEV *see* extracellular enveloped virion
EF1a 385
EGFR *see* epidermal growth factor receptors
EGP-2 6
EGR-1 320–2
EHBA *see* extrahepatic biliary atresia
elongation factor 1 alpha (EF1a) 385
endostatin 60, 295–6, 298, 300
endothelial cells (EC) 295–9, 301, 355–7
env gene 70, 71
epidermal growth factor receptors (EGFR)
  herpes simplex virus 31, 38
  measles virus 209–10
  radioprotective gene therapy 345, 347–8
  reovirus 142–3
  vaccinia virus 155
episomal plasmids 388–9
Epstein–Barr virus (EBV) 31

etoposide 233
evasion strategies 161–4
extracellular enveloped virion (EEV) 96–7, 152–3, 158–60, 161
extracellular matrix (ECM) 22, 33, 36
extrahepatic biliary atresia (EHBA) 142

factor VIII 88
Fanconi anaemia 89, 379
FBP *see* folate-binding protein
FGF *see* fibroblast growth factor
Fhit gene 230–1
FIAU 118, 126
fibroblast growth factors (FGF) 56, 295–7, 302
fibroblast progenitor cells 355–7
fludarabine 259, 262
fluorodeoxyuridine (FUDR) 40
5-fluorouracil/5-fluorocytosine
  adenoviruses 13
  gene-directed enzyme prodrug therapy 256, 258, 260–1
  herpes simplex virus 31, 40, 126, 127
  radiosensitization 318–19
  retroviruses 76
  tumour-suppressor genes 233
  vaccinia virus 160–1
FMG *see* fusogenic membrane glycoproteins
folate-binding protein (FBP) 160
fowlpox 95, 110
FUDR *see* fluorodeoxyuridine
FUS1 nanoparticles 235
fusogenic membrane glycoproteins (FMG) 125–6

G207 27, 32, 122–3, 129
GAG *see* glycosaminoglycans
gag gene 70–1
GALV *env* R- *see* Gibbon Ape leukaemia virus
gamma knife radiotherapy (GKR) 41
gammac gene 73–4, 75
ganciclovir (GCV)
  adenoviruses 7, 10–11
  gene-directed enzyme prodrug therapy 255–8, 260
  herpes simplex virus 26, 31, 39–40, 118, 120, 126–7
  radiosensitization 318
  retroviruses 75–7
gatekeepers 230
GBM *see* glioblastoma multiforme
GCV *see* ganciclovir
GDEPT *see* gene-directed enzyme prodrug therapy
GDNF *see* glial cell-line derived neurotrophic factor
gefitinib 343, 347, 349
Genasense 252
gene correction 58
gene-directed enzyme prodrug therapy (GDEPT) 10, 255–76
  adeno-associated virus 257, 267–8
  adenoviruses 257–60, 264–6
  bystander effects 255–6, 320
  CB1954/nitroreductase 258, 261
  clinical trials 260, 322–3
  CPT-11/carboxylesterase 259, 262
  cytochrome P450 319
  cytosine deaminase 318–19
  double suicide gene therapy 319–20
  enzyme-prodrug systems 255–62
  fludarabine/purine nucleoside phosphorylase 259, 262
  5-fluorocytosine/cytosine deaminase 256, 258, 260–1
  ganciclovir/thymidine kinase 255–8, 260
  gene delivery vectors 262–8
  herpes simplex virus 257–9, 266–7, 318
  lentivirus 257
  nitrogen mustard prodrugs/carboxypeptidase G2 258, 261
  nitroreductase 319
  oxazaphosphorines/cytochrome P450 259, 261–2
  preclinical studies 257–9
  radiosensitization 316–22
  retroviruses 257–60, 263–4
  selenomethionine/methioninase 259, 262
  Sindbis virus 257, 267
  transgene expression 320–2
  vaccinia virus 258–9, 266
  vesicular stomatitis virus 257, 267
gene expression vectors 218, 220–1
gene replacement therapy 11–12
gene silencing 84–5, 242–7, 253
genetic prodrug activation therapy (GPAT) 10
Gibbon Ape leukaemia virus (GALV *env* R-) 125–6, 127
GKR *see* gamma knife radiotherapy
glial cell-line derived neurotrophic factor (GDNF) 30
glioblastoma multiforme (GBM)
  herpes simplex virus 24–5, 27, 37
  immunomodulation 283
  Newcastle disease virus 178
  retroviruses 76
glycoproteins
  adenoviruses 125–6
  alphaviruses 219
  herpes simplex virus 21–3, 37–8, 125–6
  vesicular stomatitis virus 193
glycosaminoglycans (GAG) 21–2

GM-CSF *see* granulocyte–macrophage colony-stimulating factor
gp100 110, 286
GPAT *see* genetic prodrug activation therapy
granulocyte–macrophage colony-stimulating factor (GM-CSF)
 adeno-associated virus 58, 62
 adenoviruses 10
 herpes simplex virus 28, 32, 125, 127–8, 129
 immunomodulation 284–5
 poxviruses 102–3, 104–6
 radioprotective gene therapy 346
 retroviruses 78–9
 vaccinia virus 166
 vesicular stomatitis virus 196
gutless vectors 3, 5–6

haematopoietic stem cells (HSC) 87–9, 377–81, 383–6, 388
hepatocellular carcinoma (HCC) 193–4
herpes simplex virus (HSV) 19–53, 115–37
 amplicon vectors 24, 25, 30–2
 angiogenesis 300, 302
 apoptosis 31, 35–6
 clinical trials 119, 129–30
 combination therapies 39–42, 127–9
 convection-enhanced delivery 36–7
 distribution 33, 36
 efficacy 32–42
 extracellular matrix 22, 33, 36
 fusogenic membrane glycoproteins 125–6
 gene regulation/particle assembly 23–4
 gene-directed enzyme prodrug therapy 257–9, 266–7
 growth mechanisms 34–5
 HSV-1/HSV-2 intertypic recombinants 123
 ICP6$^-$ mutations 122–3
 ICP34.5$^-$ mutations 120–5, 128–9
 ICP47 mutants 124–5, 129
 immunomodulation 127–8, 283
 indoleamine-2,3-dioxygenase 34–5
 infection mechanism 21–3
 innate immunity 33–6
 interferons 26, 33–4, 37
 ionizing radiation therapy 41–2
 local inflammation 37
 multiple mutations 122–5
 neurovirulence gene 120–2
 nomenclature 21
 novel mutant combinations 39
 nuclear factors 26, 35
 nucleotide metabolism gene mutants 118, 120
 oncolytic vectors 24–8, 32–3, 36, 41–2, 115–37, 257–9, 266–7
 prodrug activation therapy 126–7
 promoter-mediated tumour cells 38–9
 properties 117–18
 radiosensitization 314–18
 recombinant vectors 32
 replication-defective vectors 25, 28–30, 257–8, 266–7, 315
 retargeting 37–8
 serotypes 19
 single mutations 118–22
 specificity 36–42
 structure/lifecycle 20–1, 115–17
 suicide gene therapy 40–1
 transgene expression 29–30, 31, 35, 40
 tumour-selective replication 117, 118–25
 US2 mutants 123–4
 US11 mutants 124–5, 129
 vaccination 32
herpes simplex virus thymidine kinase (HSV-tk) 30, 39–40
 adeno-associated virus 59
 angiogenesis 302
 gene-directed enzyme prodrug therapy 255–6
 oncolytic vectors 25–7, 118, 120, 126
 radioisotope delivery 332–4
 radiosensitization 318
 retroviruses 75–7
 suicide gene therapy 10–11
herpes virus entry mediators (HVEM) 22–3
high capacity vectors 3, 5–6
HIV *see* human immunodeficiency virus
HPV *see* human papillomavirus
HSC *see* haematopoietic stem cells
HSV *see* herpes simplex virus
HSV-tk *see* herpes simplex virus thymidine kinase
HTLV-1 *see* human T lymphotropic virus-1
human immunodeficiency virus (HIV)
 adeno-associated virus 61
 lentiviral vectors 83–4, 89, 91
 measles virus 206
 retroviruses 74
 vaccinia virus 164, 165
 vesicular stomatitis virus 196
human papillomavirus (HPV) 107, 251, 289
human T lymphotropic virus-1 (HTLV-1) 190
HVEM *see* herpes virus entry mediators
hyperthermia 153–4

ICAM-1 *see* intracellular adhesion molecule-1
IFN *see* interferons

IGF *see* insulin-like growth factor
IL *see* interleukins
immunogenicity 101–2, 146
immunomodulation 277–94
 5T4 287
 adeno-associated virus 61–2
 adenoviruses 7, 10, 282, 286
 adoptive transfer 278
 alphaviruses 222, 282
 antigen-presenting cells 277, 278
 carcinoembryonic antigen 280, 282, 284–5
 clinical trials 283–9
 colorectal cancer 287
 delivery methods 104–10, 278–80
 evasion strategies 161–4
 future prospects 289–90
 gp100 286
 herpes simplex virus 127–8, 283
 human papillomaviruses 289
 lentiviral vectors 86–7
 MAGE antigens 285
 measles virus 205–6, 210–11
 melanoma 285–6
 MUC-1 278, 286–7
 Newcastle disease virus 173, 174–5, 179–80
 NY-ESO-1 protein 289
 poxviruses 95, 102–10, 280–2, 284–6
 prostate cancer 287–9
 PSA 287–9
 retroviruses 78–9
 Semliki Forest virus 283
 Sindbis virus 283
 solid tumours 286–7
 suicide gene therapy 280
 tumour-associated antigens 277–8, 280–9
 tyrosinase 285–6
 vaccinia virus 102–7, 161–4, 281, 284–5, 288
 vesicular stomatitis virus 194–5, 198–9
 virotherapy 280
immunosuppression 163, 211
IMRT *see* intensely modulated radiotherapy
IMV *see* intracellular mature virion
indoleamine-2,3-dioxygenase 34–5
inducible drug resistance genes 388–9
infectious subvirion particles (ISVP) 140–1
infiltrating T lymphocytes (TIL) 78, 277
inflammation 37
inherent targeting 72, 75
innate immunity 33–6, 246–7
insertional mutagenesis 72, 73–5
insulin-like growth factor (IGF) 209, 232, 252
intensely modulated radiotherapy (IMRT) 354
interferons (IFN)
 adeno-associated virus 58, 60
 angiogenesis 296
 herpes simplex virus 26, 33–4, 37
 immunomodulation 279–80, 282
 innate immunity 246–7
 lentiviral vectors 86–7
 Newcastle disease virus 171, 175–6, 178
 poxviruses 109
 reovirus 143, 145
 retroviruses 78–9
 vaccinia virus 161–2
 vesicular stomatitis virus 189–99
interleukins (IL)
 adeno-associated virus 58, 62
 adenoviruses 10
 alphaviruses 221, 222
 angiogenesis 296
 herpes simplex virus 32, 38, 127–8
 immunomodulation 279–80, 282, 286–7
 measles virus 206
 Newcastle disease virus 180
 poxviruses 97, 103, 104, 108–9
 radioprotective gene therapy 344, 353, 355
 radiosensitization 318
 retroviruses 72, 78–9
 vaccinia virus 161–3, 164
intracellular adhesion molecule-1 (ICAM-1) 355–6
intracellular mature virion (IMV) 96–7, 152–3, 158–9
iodine therapy *see* sodium iodide symporter
ionizing radiation 41–2, 129, 233–5
 see also radioprotective gene therapy
irinotecan 40
ISVP *see* infectious subvirion particles

JAM receptors 145

Kruppel associated box (KRAB) domain 250
KSA 109

lacZ gene 26, 27
latency active promoter 2 (LAP2) 20, 30
latency-associated transcripts 20
lentiviral vectors (LV) 72, 83–94
 angiogenesis 301–2
 cancer treatments 89–91
 clinical trials 91
 development 83–5
 endogenous gene silencing 84–5
 gene-directed enzyme prodrug therapy 257
 haematopoietic stem cells 87–9
 immune responses 86–7

limitations/safety 83–4
RNA interference pathway 248, 250
transgene expression 84, 85–7
leukaemogenesis 382–3
LMO2 gene 73, 84
local inflammation 37
luciferase 155, 164
lung irradiation damage 354–60
LV *see* lentiviral vectors

macrophage colony-stimulating factor (MCSF) 58
macrophages 355–7
MAGE antigens 285
major histocompatibility complex (MHC)
adenoviruses 1–2
alphaviruses 219, 221
herpes simplex virus 124
immunomodulation 278–9, 281, 286
lentiviral vectors 86
measles virus 209
retroviruses 78
vesicular stomatitis virus 190
malignant gliosarcomas 26
manganese superoxide dismutase (MnSOD) 342–54, 357–60
MAPK *see* mitogen-activated kinases
MART-1 *see* melanoma target antigen
matrix metalloproteinases (MMP) 36, 90, 298, 302
MC38 tumours 157
MCSF *see* macrophage colony-stimulating factor
MDR *see* multidrug resistance
measles virus (MV) 205–15
angiogenesis 300
animal models 207–8
attenuation 211, 212
clinical trials 211–12
combination therapies 211
genome 206–7
imaging 210
immune responses 205–6, 210–11
oncolytic vectors 207–8, 210
pathogenesis 205–6, 207–8
radioisotope delivery 335
receptors 207
retargeting 209–10
safety 211–12
specificity 208–9
structure 206–7
systemic therapy 210–11
vaccine strain 206, 211–12
melanoma 27, 285–6
melanoma target antigen (MART-1) 78–9, 286
memory T cells 175, 177–9
metachromatic leukodystrophy (MLD) 88
metastatic melanoma 27
methioninase 259, 262
methotrexate 233, 381
1-methyl-tryptophan 35
O^6-methylguanine-DNA-methyltransferase (MGMT) 381, 383–4, 385, 388
MHC *see* major histocompatibility complex
micro RNA (miRNA) 242–3, 245–7, 252
mitogen-activated kinases (MAPK) 347
mitomycin C 128–9
MLD *see* metachromatic leukodystrophy
MMP *see* matrix metalloproteases
MnSOD *see* manganese superoxide dismutase
modified vaccinia Ankara (MVA)
clinical trials 161, 285, 289
development 152
immunomodulation 98–9, 107–8
viral vectors 281
molecular imaging of gene expression 332–4
MUC-1 278, 286–7, 304
mucins 104
multidrug resistance (MDR) 380–1, 382, 384–5, 387–8
MVA *see* modified vaccinia Ankara
Myc gene 34–5
myeloid leukaemia 382–3

natural killer (NK) cells 72–3, 285, 287
NC-RNA *see* nucleocapsid protein complexes
NDV *see* Newcastle disease virus
nerve growth factors (NGF) 22
neurovirulence gene 120–2
Newcastle disease virus (NDV) 171–86
ATV-NDV antitumour vaccine 176–80
autologous tumour cells 176–80
bispecific antibodies/fusion proteins 179–80
clinical trials 172–4, 177–9
combination therapies 194
danger model of immunity 174, 175–7
development 171
immunomodulation 173, 174–5, 179–80
memory T cells 175, 177–9
oncolytic vectors 171–2, 181
pathogenicity 171–2
recombinant vectors 179, 180–1
retargeting 180
safety 174, 181
selectivity/specificity 174, 180, 181
structure/lifecycle 171–2
systemic application 173

NF *see* nuclear factors
NIS *see* sodium iodide symporter
nitric oxide synthase (NOS) 352
nitrogen mustard prodrugs 258, 261
nitroreductase 261, 319
NK *see* natural killer
non-Hodgkin lymphoma 208
non-small cell lung cancer 232, 234
NOS *see* nitric oxide synthase
nuclear factors (NF)
  herpes simplex virus 26, 35, 120
  reovirus 142, 144, 145
nucleocapsid protein complexes (NC-RNA) 69
nucleotide metabolism gene mutants 118, 120
NY-ESO-1 107, 289

oesophageal irradiation damage 348–54
off-target gene silencing 244–6
oligonucleotide agents 241, 243–51
oncogenesis 73–5
oncolytic vectors
  adenoviruses 5, 12–13, 257–60, 264–6
  alphaviruses 222–3
  gene-directed enzyme prodrug therapy 257–60, 263–7
  herpes simplex virus 24–8, 32–3, 36, 41–2, 115–37, 257–9, 266–7
  measles virus 207–8, 210
  Newcastle disease virus 171–2, 181
  poxviruses 95, 99–101, 102–4
  reovirus 139, 142–6
  retroviruses 83, 257–8, 263–4
  vaccinia virus 100, 102–4, 151–69, 258–9, 266
  see also vesicular stomatitis virus
OncoVEX 28
ONXY-015 13
open reading frames (ORF) 218–19
optical imaging 331
oral cavity cancers 342–54
orbiviruses 139
ORF *see* open reading frames
oropharynxeal cancers 342–54
orphan viruses 139
ovalbumin 61
oxazaphosphorines 259, 261–2

p21 3–4, 316
p38 344
p53
  adenoviruses 3–4, 11–12
  alphaviruses 221
  herpes simplex virus 34
  poxviruses 109
  radiosensitization 316, 318
  retroviruses 77–8
  tumour-suppressor genes 230–5
  vaccinia virus 156–7
p450 127
PCNA *see* proliferating cell nuclear antigen
PDGF *see* platelet-derived growth factor
peripheral blood cells 382
PET *see* positron emission tomography
phosphoglycerate kinase (PGK) 385
platelet-derived growth factor (PDGF) 142–3, 146, 295, 300, 303
PNP *see* purine nucleoside phosphorylase
pol gene 70–1
polio virus 120–1
polyepitopes 108
positron emission tomography (PET) 331–4, 337, 354
poxviruses 95–114
  anti-cancer immunogens 101–2
  avipoxviruses 95, 99, 108–10, 281–2
  classification 95, 96
  clinical trials 102–10
  development 151–2
  gene delivery 95, 99–100, 104–10
  immunomodulation 95, 102–10, 280–2, 284–6
  modified vaccinia Ankara 98–9, 107–8, 152, 156, 281, 285, 289
  oncolytic vectors 95, 99–101, 102–4
  recombinant vectors 102–3, 105–7, 280, 284–5
  selectivity 156
  structure/lifecycle 95–7
  Yatapoxviruses 99
  see also vaccinia virus
preproendothelin-1 (PPE-1) promoters 304
pro gene 70
prodrug activation therapy 126–7
progenitor cells 90, 353–4, 355–7
proliferating cell nuclear antigen (PCNA) 121
promoter-mediated tumour cells 38–9
prostate cancer 287–9
prostate specific antigen (PSA) 105–7, 287–9
protein kinase R (PKR)
  herpes simplex virus 26, 34, 39, 118, 120–2, 124
  innate immunity 246–7
  lentiviral vectors 89
  radiosensitization 315
  reovirus 143–5
  vesicular stomatitis virus 189, 190, 192, 196
proto-oncogenes 74, 84
PSA *see* prostate specific antigen
pseudoreceptors 210

pseudovirions 251
pulmonary irradiation damage 354–60
purine nucleoside phosphorylase (PNP) 259, 262

R3616 27
radioisotope delivery 327–40
 breast cancers 329
 genetic therapy 330–7
 imaging 331–4, 336–7
 molecular imaging of gene expression 332–4
 optical imaging 331
 PET/SPECT imaging 331–4, 337
 sodium iodide symporter 327–40
 therapeutic gene therapy 334–7
 thyroid cancers 327–30, 336
radiolabeling 164
radioprotective gene therapy 341–75
 antitumour effects 346–7
 bystander effects 341, 351
 chemoradiotherapy 344, 345–6, 348, 354
 cytoprotective responses 347–8, 352–3
 dose-limiting toxicity 348, 354
 epidermal growth factor receptors 345, 347–8
 head/neck cancers 342–7
 mechanistic effects 348–50, 354–5
 microenvironment effects 350–3
 MnSOD-PL delivery 342–54, 357–60
 oesophageal irradiation damage 348–54
 oral cavity/oropharynxeal 342–54
 pulmonary irradiation damage 354–60
 stem cell plasticity/fusion 350, 353–4
 therapeutic ratio 341–2, 348
 tissue repair 350
 total body irradiation 352, 358–9, 360
radiosensitization 313–26, 341
 adeno-associated viruses 314, 317
 adenoviruses 313–14, 317
 bystander effects 320
 clinical trials 322–3
 double suicide gene therapy 319–20
 gene-directed enzyme prodrug therapy 316–22
 herpes simplex virus 314–18
 transgene expression 320–2
 tumour suppressor genes 316
radiovirotherapy 210
Raf kinases 142
Ras signalling 142–4, 147–8
RB gene 229, 230–1
reactive nitrogen species (RNS) 353
reactive oxygen species (ROS) 353–4
recombinant vectors
 adeno-associated virus 55–7, 59–63
 herpes simplex virus 32
 Newcastle disease virus 179, 180–1
 poxviruses 102–3, 105–7, 280, 284–5
 vaccinia virus 288
 vesicular stomatitis virus 192–3
reovirus 139–50
 apoptosis 145
 cell cycle disruption 145
 clinical trials 142, 146–8
 combination therapies 147–8
 human infection 141–2
 immunogenicity 146
 lifecycle 141
 oncolytic vectors 139, 142–6
 preclinical experience 145–6
 replication 140–1
 selective tumour cell cytotoxicity 139, 145–7
 structure 139–40
Rep gene 55–9
replication-competent *see* oncolytic vectors
replication-defective vectors
 adenoviruses 257–60, 264, 299–300
 alphaviruses 218, 221
 gene-directed enzyme prodrug therapy 257–60, 263–4, 266–8
 herpes simplex virus 25, 28–30, 257–8, 266–7, 315
 retroviruses 73, 76, 257–60, 263–4
 see also adeno-associated virus; Sindbis virus
retargeting
 adeno-associated virus 62
 adenoviruses 6–7
 herpes simplex virus 37–8
 measles virus 209–10
 Newcastle disease virus 180
retinoblastoma 229
retroviruses 69–81
 angiogenesis 300, 301
 clinical trials 78–9
 gene-directed enzyme prodrug therapy 257–60, 263–4
 genome 69–70
 immunomodulation 78–9
 inherent targeting 72, 75
 insertional mutagenesis 72, 73–5
 integration 71–2, 74
 lifecycle 70–1
 limitations/safety 72, 75, 217, 218
 oncogenesis 73–5
 oncolytic vectors 257–8, 263–4
 replication-defective vectors 73, 76, 257–60, 263–4
 structure 69
 suicide gene vectors 75–7

retroviruses (*Continued*)
transductional targeting 74–5, 79
transgene expression 73, 74
translational targeting 70–1
tumour suppressor genes 77–8
vectors 71–2
X-linked severe combined immune deficiency 72–5, 80
see also lentiviruses
rev-responsive element (RRE) 386
RGD targeting 298, 305–6
ribonucleotide reductase 26, 118
ribozymes 241–2
RNA interference pathway 241
antisense oligonucleotides 243–4, 248
delivery methods 247–51
gene silencing 242–7, 253
innate immunity 246–7
mechanism 242–3
off-target effects 244–6, 253
research applications 252
therapeutic applications 252–3
viral vectors 246, 248–51
virosomes 251
RNS *see* reactive nitrogen species
ROS *see* reactive oxygen species
rotaviruses 139
RRE *see* rev-responsive element

SAGE *see* serial analysis of gene expression
SAP *see* stress-activated protein kinases
SCC *see* squamous cell carcinoma
selectins 355–6
selective tumour cell cytotoxicity 139, 145–7
selenomethionine 259, 262
Semliki Forest virus (SFV) 217–19, 221–5, 283
serial analysis of gene expression (SAGE) 298
serine protease inhibitors (serpins) 156–7, 159
SFV *see* Semliki Forest virus
SGT *see* suicide gene therapy
short hairpin RNA (shRNA) 242, 246, 248–51
short interfering RNA (siRNA) 242, 244–9, 251–3
sickle cell disease 88, 91
signal transducers and activators of transcription (STAT) 34
simian immunodeficiency virus (SIV) 61
SIN vectors 84, 88, 91, 386–7
Sindbis virus 218–19, 221–5
gene-directed enzyme prodrug therapy 257, 267
glycoproteins 193
immunomodulation 283
single photon emission computed tomography (SPECT) 331–2, 337
siRNA *see* short interfering RNA; small inhibitory RNAs
SIV *see* simian immunodeficiency virus
Skp-2 250
SLAM-induced fusion 209–10, 211
Smad3 fibroblasts 357–8
small inhibitory RNAs (siRNA) 31
smallpox 95, 98, 151, 280–1
SNV *see* spleen necrosis virus
sodium iodide symporter (NIS) 210, 327–40
breast cancers 329
genetic therapy 330–7
imaging 331–4, 336–7
molecular imaging of gene expression 332–4
optical imaging 331
PET/SPECT imaging 331–4
radioiodine treatment 328–9
therapeutic gene therapy 334–7
thyroid cancers 327–30, 336
solid tumours 286–7, 342–3
somatostatin receptors 164–5
SPECT *see* single photon emission computed tomography
spleen necrosis virus (SNV) 72
squamous cell carcinoma (SCC) 232–3, 235, 346–7
STAT *see* signal transducers and activators of transcription
stem cell plasticity/fusion 350, 353–4
stress-activated protein kinases (SAP) 344, 347
suicide gene therapy (SGT)
adenoviruses 7, 10–11
double 319–20
herpes simplex virus 40–1
immunomodulation 280
retroviruses 75–7
see also gene-directed enzyme prodrug therapy
systemic therapy 210–11

T cells
lentiviral vectors 85, 91
leukaemias 72, 73–5
Newcastle disease virus 174–7
poxviruses 103–4, 109
retroviruses 78–9
T helper 1 (Th1) 60–1
TAA *see* tumour-associated antigens
taxol 348, 354
TBI *see* total body irradiation
telomerase reverse transcriptase promoters 7
TEM cells 90

temozolomide 25
tetracyclines 250
TGF *see* transforming growth factor
TGS *see* transcriptional gene silencing
Th1 *see* T helper 1
thalassaemia 88–9, 91, 379
therapeutic ratio 341–2, 348
thrombospondins 295–6, 299–300
thymidine kinase (TK)
  gene-directed enzyme prodrug therapy 255–8, 260
  immunomodulation 281
  poxviruses 100–1
  vaccinia virus 154–6
  see also herpes simplex virus thymidine kinase
thyroid cancers 327–30, 336
thyroid stimulating hormone (TSH) 328, 336
Tie2 299–300
TIL *see* tumour infiltrating lymphocytes
tissue inhibitor of metalloproteinase-2 (TIMP-2) 30, 31
TK *see* herpes simplex virus thymidine kinase; thymidine kinase
TLR *see* toll-like receptors
TNF *see* tumour necrosis factor
TNFerade 305, 322
toll-like receptors (TLR) 189, 196, 198, 246–7
total body irradiation (TBI) 352, 358–9, 360
TRA *see* tumour rejection antigens
TRAIL 31, 35–6
transcriptional gene silencing (TGS) 243, 245–6
transcriptional targeting 7, 303–5
transductional targeting
  adeno-associated virus 62
  adenoviruses 6–7, 10
  angiogenesis 305–6
  retroviruses 74–5, 79
transforming growth factor (TGF) 344, 353, 355, 357–8
transgene expression
  adeno-associated virus 59–60, 61, 62
  chemoprotective gene delivery 381–2, 384–6, 387–8
  gene-directed enzyme prodrug therapy 320–2
  herpes simplex virus 29–30, 31, 35, 40
  lentiviral vectors 84, 85–7
  radioprotective gene therapy 342–54
  radiosensitization 320–2
  retroviruses 73, 74
  tumour-suppressor genes 233
transgenesis 87
translational targeting 70–1
treatment responsive promoters 304–5
TRICOM 279–80, 284–5, 288–9
TSH *see* thyroid stimulating hormone
tumour-associated antigens (TAA) 90
  immunomodulation 277–8, 280–9
  Newcastle disease virus 174–5, 177–9
  poxviruses 101–2, 107–8
tumour infiltrating lymphocytes (TIL) 78, 277
tumour necrosis factor (TNF)
  adeno-associated virus 58
  adenoviruses 12
  angiogenesis 296, 300, 305
  herpes simplex virus 22, 31, 35–6, 37, 40–1
  radioprotective gene therapy 344, 347, 353, 355
  radiosensitization 320–2
  reovirus 145
  retroviruses 78–9
tumour rejection antigens (TRA) 176
tumour-selective replication simplex virus 117, 118–25
tumour-specific delivery 62
tumour-suppressor genes 229–39
  apoptosis 230–5
  clinical trials 232–3, 234–5
  combination therapies 233–5
  conceptual development 231
  discovery 229–30
  future prospects 235
  gatekeepers 230
  lifecycle 230–1
  multimodality anticancer treatment 233–5
  p53 230–5
  preclinical studies 233–4
  radiosensitization 316
  retroviruses 77–8
  technology development 231–2, 235
  transgene expression 233
tumstatin 295–6
tyrosinase 107, 108, 285–6

uracil N-glycosylase (UNG) 118, 123

vaccinia growth factor (VGF) 100–1, 154, 155, 164
vaccinia virus (VV) 95, 97–8, 151–69
  antitumour effects 153–61, 163–4
  apoptosis 157, 158
  bystander effects 160–1, 166
  clinical trials 165–6
  gene delivery 104–7
  gene-directed enzyme prodrug therapy 258–9, 266
  imaging 164–5
  immunomodulation 102–7, 161–4, 281, 284–5, 288
  limitations 164

vaccinia virus (VV) (*Continued*)
  oncolytic vectors 100, 102–4, 151–69, 258–9, 266
  recombinant vectors 288
  selectivity 153–60, 166
  structure/lifecycle 151–3
variola virus 95, 98
vascular cell adhesion molecule-1 (VCAM-1) 355–6
vascular endothelial growth factor (VEGF)
  adeno-associated virus 60
  angiogenesis 295–8, 300–1, 303–4
  herpes simplex virus 30
VCAM-1 *see* vascular cell adhesion molecule-1
VDEPT *see* gene-directed enzyme prodrug therapy
VEGF *see* vascular endothelial growth factor
Venezuelan equine encephalitis (VEE) virus 218, 221–2
vesicular stomatitis virus (VSV) 84, 86, 187–203
  antitumour responses 195–7
  attenuation 192, 197
  combination therapies 194
  danger model of immunity 197
  delivery strategies 198–9
  gene-directed enzyme prodrug therapy 257, 267
  host range 188–9
  immune responses 194–5, 198–9
  in vitro replication 190–1, 193–4
  lysis 190–1
  pathogenesis 188–9
  recombinant vectors 192–3
  safety 198, 199
  selectivity 192, 194, 197
  structure/lifecycle 187–8, 197
  therapeutic index 192, 194, 195
  type I interferon response 189–99
VGF *see* vaccinia growth factor
vincristine 39
virosomes 251
virotherapy 280
VSV *see* vesicular stomatitis virus
VV *see* vaccinia virus

Wiskott-Aldrich syndrome (WAS) 85

X-linked severe combined immune deficiency 72–5, 80

Yaba-like disease (YLD) 99, 162–3
Yatapoxviruses 99, 162